1

ANNUAL REVIEW
OF IMMUNOLOGY

ANNUAL REVIEW OF IMMUNOLOGY

VOLUME 5, 1987

WILLIAM E. PAUL, *Editor*
National Institutes of Health, Bethesda, Maryland

C. GARRISON FATHMAN, *Associate Editor*
Stanford University, Stanford, California

HENRY METZGER, *Associate Editor*
National Institutes of Health, Bethesda, Maryland

ANNUAL REVIEWS INC   4139 EL CAMINO WAY   P.O. BOX 10139   PALO ALTO, CALIFORNIA 94303-0897

ANNUAL REVIEWS INC.
Palo Alto, California, USA

International Standard Serial Number : 0732-0582
International Standard Book Number : 0-8243-3005-6

Annual Review and publication titles are registered trademarks of Annual Reviews
Inc.

Annual Reviews Inc. and the Editors of its publications assume no responsibility
for the statements expressed by contributors to this *Review*.

TYPESETTING AND COMPOSITION BY AUP TYPESETTERS LTD., GLASGOW, SCOTLAND
TYPESETTING PRODUCTION COORDINATOR, FRED PICKHARD
PRINTED AND BOUND IN THE UNITED STATES OF AMERICA

Annual Review of Immunology
Volume 5, 1987

CONTENTS

(*continued*)   v

vi   CONTENTS (continued)

SOME RELATED ARTICLES IN OTHER *ANNUAL REVIEWS*

From the *Annual Review of Biochemistry*, Volume 55 (1986)

Lectins as Molecules and as Tools, Halina Lis and Nathan Sharon

From the *Annual Review of Cell Biology*, Volume 2 (1986)

The Role of Protein Kinase C in Transmembrane Signalling, Ushio Kikkawa and Yasutomi Nishizuka

T-Cell Activation, H. Robson MacDonald and Markus Nabholz

The Directed Migration of Eukaryotic Cells, S. J. Singer and Abraham Kupfer

G Proteins: A Family of Signal Transducers, Lubert Stryer and Henry R. Bourne

From the *Annual Review of Genetics*, Volume 20 (1986)

The myc *Oncogene: Its Role in Transformation and Differentiation*, Michael D. Cole

Germ-Line Transformation of Mice, Richard D. Palmiter and Ralph L. Brinster

From the *Annual Review of Pharmacology and Toxicology*, Volume 27 (1987)

Molecular and Cellular Basis of Chemically Induced Immunotoxicity, Michael I. Luster, James A. Blank, and Jack H. Dean

Prostaglandins, Leukotrienes, and Platelet-Activating Factor in Shock, Giora Feuerstein and John M. Hallenbeck

From the *Annual Review of Medicine*, Volume 38 (1987)

Live Attenuated Varicella Vaccine, Anne A. Gershon

Immune Disregulation of Hematopoiesis, Kenneth F. Mangan

IgA Nephropathy, A. R. Clarkson, A. Woodroffe, I. Aarons, Y. Hiki, and G. Hale

Leukocyte Adhesion Deficiency: An Inherited Defect in the Mac-1, LFA-1, and p150,95 Glycoproteins, Donald C. Anderson and Timothy A. Springer

Chronic Epstein-Barr Virus Infection, James F. Jones and Stephen E. Straus

IgG Subclass Deficiencies, H. D. Ochs and R. J. Wedgwood

Endotoxins and Disease Mechanisms, David C. Morrison and John L. Ryan

Human Lymphocyte and Lymphoma Homing Receptors, Sirpa Jalkanen, Nora Wu, Robert F. Bargatze, and Eugene C. Butcher

ANNUAL REVIEWS INC. is a nonprofit scientific publisher established to promote the advancement of the sciences. Beginning in 1932 with the *Annual Review of Biochemistry*, the Company has pursued as its principal function the publication of high quality, reasonably priced *Annual Review* volumes. The volumes are organized by Editors and Editorial Committees who invite qualified authors to contribute critical articles reviewing significant developments within each major discipline. The Editor-in-Chief invites those interested in serving aś future Editorial Committee members to communicate directly with him. Annual Reviews Inc. is administered by a Board of Directors, whose members serve without compensation.

ANNUAL REVIEWS OF
Anthropology
Astronomy and Astrophysics
Biochemistry
Biophysics and Biophysical Chemistry
Cell Biology
Computer Science
Earth and Planetary Sciences
Ecology and Systematics
Energy
Entomology
Fluid Mechanics
Genetics
Immunology

Materials Science
Medicine
Microbiology
Neuroscience
Nuclear and Particle Science
Nutrition
Pharmacology and Toxicology
Physical Chemistry
Physiology
Phytopathology
Plant Physiology
Psychology
Public Health
Sociology

SPECIAL PUBLICATIONS

Annual Reviews Reprints:
  Cell Membranes, 1975–1977
  Immunology, 1977–1979

Excitement and Fascination
  of Science, Vols. 1 and 2

Intelligence and Affectivity,
  by Jean Piaget

Telescopes for the 1980s

A detachable order form/envelope is bound into the back of this volume.

Michael Sela

Ann. Rev. Immunol. 1987. 5 : 1–19

A PERIPATETIC AND PERSONAL VIEW OF MOLECULAR IMMUNOLOGY FOR ONE THIRD OF THE CENTURY

Michael Sela

Department of Chemical Immunology, The Weizmann Institute of Science, Rehovot, Israel 76100

INTRODUCTION

I always felt as inclined toward literature, history, and Latin as toward science, but as far back as I can remember I wanted to be a scientist. Most probably this was because of my loving esteem for an uncle, an inorganic chemist who, after working at a Kaiser Wilhelm Institute in Berlin, spent ten years in Moscow as a foreign guest. He visited us often in Poland when he was going on holiday—or for science—to the West. He returned to Poland to live just before World War II broke out. The last I heard of him was that both he and my aunt had committed suicide by taking cyanide in the Baranowicze ghetto, to avoid deportation.

After my first 11 years in Poland and another 5 in Rumania, I reached Palestine in February 1941, together with my parents. That year I started studying chemistry and physics at the Hebrew University on Jerusalem's Mount Scopus. With Andor Fodor, the first professor appointed to the University when it was founded in 1925, I did my master's thesis on the synthesis of some derivatives of glutamine. I worked under the supervision of Noah Lichenstein, who taught me order and precision in the laboratory, something which I also found helpful—when properly applied—in thought processes.

Interested in biological macromolecules, I then began graduate studies

1

0732–0582/87/0410–0001$02.00

with Kurt H. Mayer at the School of Chemistry of the University of Geneva. Within less than a year, however, I went to work in Italy to aid the survivors of Hitler's camps and to participate in efforts leading to the declaration of the independence of Israel. It was only after a couple of additional years as secretary of the Commercial Section of the Legation of Israel in Prague that I returned home in the autumn of 1950 to work with Ephraim Katchalski in the Department of Biophysics at the Weizmann Institute of Science. The Institute was officially opened in November, 1949, but was still at that stage only an extension of the original Daniel Sieff Research Institute built by Dr. Chaim Weizmann himself in 1934.

PROTEINS AND PROTEIN MODELS

The Katchalski brothers, Ephraim and his older brother Aharon, were the heads, respectively, of the Departments of Biophysics and of Polymer Research at the Weizmann Institute. In 1973 Ephraim became President of the State of Israel, changing his name to Katzir in sad memory of his brother, murdered by terrorists in the Tel Aviv airport in 1972, who had earlier adopted the Hebrew name of Katzir.

Ephraim first synthesized polylysine in the early 1940s and was successfully exploring the use of polyamino acids as protein models. With his collaborators he prepared and studied many physical, chemical and biological properties of several polymers of trifunctional amino acids such as polyarginine, polyaspartic acid, polyhistidine, and polyserine, as well as the polyimino acids polyproline and polyhydroxyproline (1, 2). Among the early syntheses was the one of poly-L-tyrosine, which was part of the subject of my PhD thesis. My thesis research included also the synthesis of poly-p-amino-L-phenylalanine. Later on I was directly involved with the synthesis of polytryptophan and polycyclohexylalanine. Of special interest was the spectrophotometric titration of polytyrosine and of copolymers of tyrosine with positively or negatively charged amino acids; this showed the influence of the vicinal electrostatic field on the ease of ionization of the phenolic hydroxyl group. We took part in intense seminars with both the Katchalski-Katzir brothers, while Cohn & Edsall's book on *Proteins, Peptides and Amino Acids as Dipolar Ions* served as our Bible. The accent was on physicochemical properties of proteins, polyamino acids, and polyelectrolytes generally. Ephraim was a remarkable teacher—stimulating, inspiring, patient, and always friendly.

I shall describe later how the work with polyamino acids brought me into immunology. What I would like to mention here is that we could use as initiators of polymerization of N-carboxyamino acid anhydrides (the monomers from which polyamino acids were built) not only mono-functional small molecules but also macromolecules possessing several amino groups. If proteins were used as such polyvalent initiators, we ended up with polypeptidyl proteins, whereas when polylysine or polyornithine were used, we had for the first time multichain polyamino acids. Incidentally, the idea occurred to me following a lecture delivered by Herman Mark (Mr. Polymer) in a movie house in Rehovot in 1954 (we did not yet have an auditorium). Mark Stahmann also prepared such polypeptidyl proteins in Wisconsin.

In 1955, I was back in Europe—this time as a young scientist presenting the work on multichain polyamino acids at the International Congress of Chemistry in Zurich and, a week later, the work on spectrophotometric titration of polymers and copolymers of amino acids at the International Congress of Biochemistry in Bruxelles. This was a wonderful occasion to meet scientists whom until then I had known only by reputation and through their papers.

In addition, I felt ready to go abroad for a postdoctoral period to work with proteins rather than with protein models. Thus, I arrived in the laboratory of Chris Anfinsen in Bethesda in 1956. The friendship between us—which I cherish until today—resulted in prolonged stays in Bethesda in 1956–1957, 1960–1961, and 1973–1974. For his part, Chris came on several occasions on sabbaticals to Rehovot and has been an active, extremely valuable member of the Board of Governors of the Weizmann Institute of Science. From the National Institutes of Health we sent out our first joint paper, which Chris was in a hurry to prepare for a *Festchrift* honoring Linderstrom-Lang. We finished it before I got my US driver's license or rented an apartment. The topic was the selective splitting of protein chains by trypsin at arginine residues, after lysine residues were reversibly blocked by carbobenzoxy groups.

The main excitement of that period was the series of initial experiments on the total reduction of RNase and the recovery of full enzymatic activity after total reoxidation (3). This was accomplished even though statistical considerations (worked out later with Shneior Lifson when I returned home) pointed to more than 100 various ways in which the four disulfide bridges could reform. These studies, to which I was happy to make a further contribution during my second visit to Bethesda, led to the richly deserved award to Chris of the 1972 Nobel prize in chemistry. I learned a lot from Chris, a dedicated scientist who has an incredible flair for discovering the right problem and the most elegant experiment to solve it, a flair matched only by his literary talent. The "elder statesman cum Nobel laureate" status never agreed with him, and he felt always more comfortable when he could return to the bench—a luxury he has enjoyed less in recent years.

During that period, 1956–1957, I also enjoyed working on the physical chemistry of both RNase and polyproline with another lifelong friend, Bill Harrington, who soon thereafter moved to Johns Hopkins University in Baltimore.

On a lighter level, when I arrived in Bethesda everybody was writing the sequences of amino acids in straight lines, the disulfide bridges—at straight angles to the chain—occupied as much space as ten residues. I found a huge hole-puncher with which I made round pieces of paper, wrote the name of one residue on each piece of paper, and played with them until the half-cystines of a bridge were touching each other. This gave rise to the well-known "swan" shape of ribonuclease, and it delights me to see that proteins still are often schematically presented like this.

Another recollection concerns something that happened shortly after my second arrival at NIH in 1960. One day Marshall Nirenberg came and asked me whether I had some poly-L-phenylalanine and whether I knew its solubility properties. I did not have the polymer in Bethesda, but I did ask Nirenberg why he was interested. Through these conversations I became one of the first to know about the breaking of the genetic code, UUU leading to Phe. While I was somewhat skeptical of the story, I immediately looked for and found, hidden somewhere in an experimental section of a paper in the *Journal of the American Chemical Society*, that poly-L-phenylalanine was insoluble in all the solvents we had tested, with the exception of a saturated solution of anhydrous hydrogen bromide in glacial acetic acid. Since on that very day I was preparing just such a solution (used to remove carbobenzoxy groups) in the lab, I gladly gave the reagent to Nirenberg and was touched and surprised when he acknowledged this in the classical paper that resulted in his receiving the Nobel Prize. But the real point of the story lies elsewhere: Why did we try to use such a peculiar solvent? The truth of the matter is that years earlier, together with the late Arieh Berger in Rehovot, we were investigating the mechanism of polymerization leading to linear and multichain polyamino acids. One day I had two test tubes—one with polyphenylalanine and one with polycarbobenzoxylysine—stuck in an ashtray on my desk. Arieh came to decarbobenzoxylate the lysine polymer, a reaction with hydrogen bromide in glacial acetic acid during which carbon dioxide is released. He took the wrong test tube away with him and then returned, puzzled because the material had dissolved and he could not see any carbon dioxide evolution. At once we realized the mistake, and I noted in my lab book that, at long last, we had found a solvent for poly-L-phenylalanine.

SYNTHETIC ANTIGENS

Once polymers and copolymers of amino acids were available, there was great interest in comparing their properties with those of proteins. Of special interest was the possibility that they possessed distinct biological properties. A great effort was put into investigating whether they possessed any enzymatic (catalytic) properties or whether they could serve as efficient enzyme inhibitors. From the beginning I was interested in examining whether these synthetic macromolecules were antigenic, and I now relate here how this story developed.

My PhD thesis was concerned with polytyrosine, which is a polymeric chain of phenols and poly-p-aminophenylalanine, a polymeric chain of anilines. It was natural to produce polypeptidic azo dyes from them. I reasoned that these should serve as synthetic models for azoproteins, of which one rare example, provided by Landsteiner, was the attachment of haptens including peptides, via an azo bond, to proteins. Reading Landsteiner's book *The Specificity of Serological Reactions*, I came across the statement that gelatin is probably not antigenic because it contains no tyrosine. This led me to study the increase of antigenicity of gelatin upon attachment of tyrosine peptides. To do these studies, the amino groups of the protein were used to initiate the polymerization of the tyrosine mono-mer, as mentioned before. The continuation of this study was the PhD thesis of Ruth Arnon, and we showed that limited tyrosylation enhanced immunogenicity without significantly changing specificity, whereas more extensive tyrosylation converted gelatin into a potent immunogen pro-voking mainly antibodies to tyrosyl peptides. It was at this time that we clearly defined the notion of immunogenicity and distinguished it from antigenic specificity.

While visiting Europe in 1954, I went to meet Sir Charles Harington in London. It was in Harington's lab just before the outbreak of World War II that John Humphrey attached carbobenzoxytyrosine to gelatin, trying to convert it into a better immunogen. Harington was the man who elucidated the structure of thyroxine and predicted that it is formed from two diiodotyrosine residues by oxidation rather than simple dismutation. I wanted to tell him that the availability of polytyrosine had allowed Sara Sarid and myself to find out that, after alkaline incubation of iodinated polytyrosine, the hydrolysate contained 2% of thyroxine and 2% of serine. We had thus proved the validity of his prediction.

When I first came to the United States in 1956 with my late wife Margalit and my daughters Irit and Orlee, we arrived by ship and the late David and Sara Rittenberg were the first to welcome us to New York. The next day David took me for lunch to the cafeteria at the Columbia Medical School where we were joined by a scientist whose name I could not catch. He was interested in my initial immunochemical results and was quite encouraging. What a surprise it was to learn, after he left, that I had just spent an hour with Elvin Kabat; he has remained a close friend since that day.

As a result of the immunological studies of tyrosylated gelatin, we assumed—with Ruth Arnon—that gelatin is not necessary for immunogenicity; we therefore replaced it with multichain poly-DL-alanine as the carrier for peptides of tyrosine and glutamic acid and showed that the resulting copolymer, denoted (T,G)-A--L, led to specific precipitable antibodies in experimental animals. At that stage Sara Fuchs joined us, and we synthesized numerous linear and multichain polyamino acids and tested them for immunogenicity and antigenic specificity (4, 5). Let me mention also among the pioneers in immunological studies of polyamino acids Mark Stahmann and Paul Maurer as well as Paul Doty and Tom Gill III.

Obviously, those interested in polyamino acids were delighted with our results, mentioned above. I have often been asked how our first reports were accepted by the immunological community. My answer is that they were very positively received and caused a great interest in the whole synthetic approach and its new possibilities.

The availability of synthetic antigens permitted a systematic elucidation of the molecular basis of antigenicity. We could learn a lot about the role of size, composition, and shape, as well as about the accessibility of those parts of the molecule crucial for immunogenicity in the nature of the immune response. As a matter of fact, we learned that it was possible, provided one wanted it enough to invest the necessary effort, to prepare synthetic immunogens leading to antibodies of essentially any specificity.

Thus, synthetic antigens were prepared with specificity directed to sugars, nucleosides, glycolipids, lipids, pyridoxal, various haptens used in immunochemical studies, and an increasingly large array of small and larger peptides. While in most cases a good immunogen has a molecular weight of at least several thousand daltons, dinitrophenyl-hexalysine and arsanil-trityrosine are by themselves capable of triggering an efficient immune response. Thus, the minimal size for a molecule to be immunogenic depends largely on its chemical nature.

Although the electrical charge may be important in defining the antigenic specificity of an epitope, it is not a minimum necessary cause for immunogenicity: we could prepare water-soluble amino acid copolymers devoid of charge that were immunogenic. Polymers of D-amino acids were immunogenic only when they were administered in minute amounts and led to no secondary response. They also resembled some polysaccharides in inducing immune paralysis. Their immunogenicity was thymus-independent, as was that of several other polymers such as linear $(ProGlyPro)_n$ and multichain polymers of L-proline locked in with terminal side chains of D-Tyr and D-Glu. The common denominator of all these "thymus-independent" antigens was that they not only possessed repeating antigenic determinants, they also were metabolized slowly, if at all. With Michael Taussig, we investigated the phenomenon of competition between antigenic determinants, using synthetic antigens. Earlier Israel Schechter had shown that peptides of D-amino acids compete successfully with similar sequences of L-amino acids.

One of the most fascinating aspects of our studies with synthetic antigens had to do with the steric conformation of the immunogen and of its epitopes. In studies with Israel and Bilha Schechter, we distinguished between conformational (conformation-dependent) and sequential determinants (6) and showed how the same peptide—TyrAlaGlu—may lead to antibodies recognizing the sequence (when attached to multichain poly-DL-alanine) or recognizing an epitope defined by conformation (when the tripeptide was polymerized to give an α-helical structure). In addition, we could demonstrate for the first time, by circular dichroism measurements, how antibodies to the α-helical polymer could help transconform into a helical shape a small polymer that was not yet helical (7). These studies led us directly to use of proteins and to synthesis of a macromolecule in which a synthetic "loop" peptide derived from hen egg white lysozyme was attached to branched polyalanine (8). The antibodies formed reacted with intact lysozyme through the "loop" region, but the reaction was totally abolished when the disulfide bond within the "loop" was opened and thus the three-dimensional structure was collapsed. In this connection it should be remembered that partial degradation products of proteins may still be immunogenic. Furthermore, the sera we were investigating might have contained a myriad of antibodies against degradation products derived from the original immunogen.

In later years we showed that a peptide of 20 amino acid residues derived from the coat protein of MS2 bacteriophage provided, after conjugation with an appropriate carrier, the formation of antibodies that cross-reacted with the intact virus. More recently, tens of peptides analogous to segments of proteins have been prepared that may lead to antibodies cross-reacting with the intact protein. Nevertheless, one should remember that many similar peptides have been prepared that were not capable of provoking antiprotein responses. The extent of cross-reactivity will depend entirely on the probability that the free peptide will be able to attain the conformation that it possesses in the native protein. This capability may be prevented either when the segment is too short, not yet able to possess a stabilized correct conformation, or when it is too long and possesses a preferred stabilized conformation different from the one capable of cross-reaction. If the protein segment is more flexible, the chance of cross-reaction is higher, even though there will be cases when a small peptide is capable of having a relatively rigid conformation similar to the one it possesses within a native protein. Thus, antibodies to the rigid helix (Pro-Gly-Pro)$_n$ cross-reacted with collagen. The availability to the outside (hydrophilicity) is also very important to immunopotency: "Whatever sticks out most, is most immunopotent."

Studies making use of synthetic antigens, of antibody specificity, of immunological tolerance, of the role of net electrical charge in defining the nature of antibodies, and of delayed hypersensitivity led us to the inevitable conclusion that an immunogen is much more than an antigenic determinant attached to an inert carrier. Unfortunately, we did not know that separate recognition of antigen by T and B cells was the explanation for our results. However, we did clearly state that the "carrier" had a crucial role in defining the nature of the immune response towards an epitope. In a way similar to the cooperation between T and B cells for antibody formation, delayed hypersensitivity might be the result, we suggested, of cooperation between two distinct sets of T cells.

GENETIC CONTROL OF IMMUNE RESPONSE

In 1961, on my way back from my second visit to Bethesda, I stopped in London to plan experiments with John Humphrey to determine whether there were any molecules of antigen inside a cell-making antibody. The story of this collaboration, which also included Israel Schechter in Rehovot and Hugh McDevitt and Ite Askonas in London, is interesting. Our conclusion that the antigen (T, G)-A--L, does not have to be present in antibody-producing cells, is told in Humphrey's prefatory chapter to Volume 2 of the Annual Review of Immunology (9). Gus Nossal and his colleagues had reported results pointing in the same direction, but these could be criticized because the antigen was tagged with radioactive iodine, an external marker. The latter could possibly move to another molecule, in which case positive results would not necessarily mean that the antigen is present within the cell; however, negative results did not have to mean that it was absent. In our case, the antigen was internally labeled with tritium, and the above criticism would not hold. Our results confirmed those of Nossal and thus were compatible with the clonal selection theory. Later on, John Humphrey extended these results to a synthetic antigen with tritium as an internal marker and iodine as an external one, as well as to an antigen, TIGAL, in which iodine became an internal marker because it was an integral part of the epitope.

I refer to the above experiments here because, before working with tritiated (T, G)-A--L, Humphrey and McDevitt injected rabbits with non-radioactive antigen. A few weeks later I met John Humphrey at a WHO meeting in Geneva, and he informed me that the sandylop rabbits they immunized produced no antibodies. During that discussion we first considered the genetic make-up of the animal as one possible explanation of this result. Within a short time it was clear that New Zealand rabbits produced as many antibodies as our rabbits in Rehovot and that Himalayan rabbits were almost an order of magnitude better. Working with Hugh McDevitt, we switched to inbred strains of mice and showed for the first time the determinant-specific genetic control of immune response in mice (10). In several papers that followed, the study was extended to additional polypeptides, and the specificity was further analyzed. This led to the unequivocal proof of the genetic control of the immune response. Some time later, using our multichain synthetic polypeptides that possessed tyrosine, histidine or phenylalanine, Hugh McDevitt was able to show for the first time the link between immune response and the major histocompatibility locus of the mouse (11), which in turn led to our present day understanding of immune response genes and their products.

Baruj Benacerraf was the first, with his colleagues, to describe genetic control of immune response to synthetic antigens in guinea pigs. They used haptenated (DNP, penicillin) polylysine antigens in strains of guinea pigs and a copolymer of L-glutamic acid and L-lysine; it was apparent that strain-13 guinea pigs could not make antibodies to any haptens attached to the polylysine carrier. Thus, both carrier and determinant may affect genetic control. This observation was much more easily understood later, when the knowledge of various functions of different immunocytes was elucidated. Chemical determinants stimulating helper and suppressor responses are distinct and can be present simultaneously within the same molecule. Thus, in a collaborative study with Benacerraf and colleagues, we showed that the attachment of a tyrosine peptide to the carboxyl terminus of a (Glu, Ala) polypeptide converted this immunogenic molecule to an immunosuppressive molecule in mice bearing the appropriate H-2 type.

My first meeting with Baruj took place at an Antibody Workshop in La Jolla in January, 1961, and it too led to a lifelong friendship. He played a paramount role in producing order in the entire area of immunogenetics. In those days the Antibody Workshops were a most exciting forum for discussing new discoveries in molecular and cellular immunology. In addition to those workshops taking place in La Jolla, I remember especially the workshop held in August, 1961, and organized by Rodney Porter in London, as well as the one I organized in Rehovot in 1966. (At the end of the Rehovot workshop, the participants planted a mango tree. I pass near it on my way to the lab each day, and it has grown to a considerable size.)

By 1973, in collaboration with Edna Mozes, we had shown that genetic differences existed in the response of inbred strains of mice to unique epitope clusters of a protein; I refer to the loop region of lysozyme. Later, Gene Shearer joined us in the study of the differences in frequency of specific precursor cells. He continued his interest in the genetic control of the immune response after his return to the United States and discovered that haptenated cells will provoke a cellular immune response to the hapten only if the cells share the major histocompatibility complex determinants with the animals into which they are administered—a finding that has contributed enormously to current understanding of the nature of antigen recognition.

One corollary to the story of (T, G)-A--L was our worry that so much research was being carried out on a polypeptide prepared by polymeric techniques. So, with Edna Mozes, Michal Schwartz, and later on Behnaz Parhami, we built a series of defined tetrapeptides composed of two tyrosine and two glutamic acid residues. We attached them to multichain poly-DL-alanine, and after an initial survey, we settled on two: TyrTyrGluGlu and TyrGluTyrGlu. We chose these because we were amazed that two molecules so similar in their chemistry on paper would lead to such vastly different biological results: They do not cross-react immunologically with each other, and only one (TyrTyrGluGlu) cross-reacts strongly with the randomly prepared (T, G)-A--L. In addition, they are under different genetic controls, and only the TyrTyrGluGlu response in mice is linked to H-2. The TyrTyrGluGlu polymer is thymus-dependent, whereas the TyrGluTyrGlu polymer is thymus-independent. The first one penetrates easily through the macrophage membrane; the other sticks to it. While normally antibodies to (T, G)-A--L cross-react strongly with Tyr-TyrGluGlu and hardly at all with TyrGluTyrGlu, in animals tolerant to TyrTyrGluGlu the response to (T, G)-A--L was different: A different antibody population had been triggered, obviously one incapable of cross-reacting with TyrTyrGluGlu. However, some antibodies definitely cross-reacted with TyrGluTyrGlu (12). We are now investigating their physicochemical properties, since they seem particularly relevant to our understanding of the molecular basis of immunological phenomena.

The ordered tetrapeptides provided the proof that low responsiveness to (T, G)-A--L was not at the level of a structural gene in the B cell. After injection with the antigen (in this case TyrTyrGluGlu-A--L, complexed with methylated bovine serum albumin), low responders produced antibodies indistinguishable from those produced in the high responders. These similarities included affinity, isoelectric focusing of IgG antibody populations, and idiotypes.

It has been an enormous and pleasurable surprise to observe the rapid development and great expansion of the field of study of the genetic control of immune response and its link to the immune response genes (part of the whole major histocompatibility complex in each species) and equally so, to see the short time span between initial basic observations and studies directly relevant to human health.

The close linkage of a disease such as ankylosing spondylitis to one specific histocompatibility locus is remarkable. In this connection, together with Carmi Geller-Bernstein we investigated genetic aspects of hypersensitivity diseases in humans. Within large atopic families we found a linkage between HLA haplotype and hay fever, as well as a marked heritability of IgE (high) and IgA (low) levels (13).

ANTIBODIES

David Givol started his PhD with me around 1960, and we tried to find out whether one could learn more about antibodies to simple peptides by making use of the techniques then available for studying proteins. It was an exciting period, what with Rodney Porter and Gerald Edelman elucidating the chemical structure of the immunoglobulin G molecule, which was the scaffold on which much of our chemical, biological, and genetic understanding of immunology has been built. Givol's interest in the field led him to many relevant observations, including—after periods spent with Anfinsen and Porter—the chemical mapping by affinity labeling of the antibody combining site and the isolation and characterization of the Fv fragment (the piece of immunoglobulin composed of the variable portion of the heavy chain and the variable portion of the light chain). With Hans Cahnmann, Ruth Arnon, and later Mira Lahav, we succeeded in preparing an analog of the (Fab) dimer by splitting the IgG molecule with cyanogen bromide rather than with enzymes. This was the first use of a selective chemical cleavage of peptide chains to prepare a biologically active protein fragment.

Polymerization of alanine on macromolecules was of interest not only in the preparation of synthetic antigens but also in studies (with Israel Schechter) on the size and nature of the combining sites of antibodies produced against poly-L-, poly-D-, and poly-DL-alanyl proteins. In studies similar to those of Kabat on antibodies to polysaccharides, we determined the size of the cavity, in terms of amino acid residues of the peptide determinants, the properties of each subsite, and the immunodominant role of the subsite interacting with the most exposed portion of the antigen. Moreover, we found out that the predominance of anti-D-alanyl antibodies in antisera against poly-DL-alanyl proteins is apparently due to competition between determinants composed of L-, D-, and DL-amino acids, with the D sequence being the most efficient.

In a collaborative effort with Otto Kratky and Ingrid Pilz in Graz, we investigated by small angle X-ray scattering the properties of anti-poly-D-alanyl antibodies in the absence and presence of the peptide epitope. These studies followed their earlier work with Gerald Edelman, in which they utilized the same techniques to probe into an immunoglobulin-G molecule. Reaction with the D-alanine tetrapeptide resulted in a significant volume contraction and decrease of the radius of gyration of the antibodies, due to a change in antibody conformation upon interaction with the ligand. This volume contraction was lost after reduction and carboxamido-methylation of the single inter-heavy-chain disulfide bridge. Moreover, no changes were observed when the tetrapeptide reacted with Fab and Fab′ dimer, both of which lack both the hinge region and the inter-heavy-chain disulfide bridge (14).

Earlier we had found out that poly-DL-alanylation increased protein solubility and that proteins insoluble in distilled water, e.g. gliadin and myosin, became water-soluble, with myosin keeping all its ATPase activity. (This last was learned with Isidore Edelman, now at Columbia University Medical School.) Later, Fred Karush, who was visiting the Weizmann Institute on a sabbatical, showed that extensive alanylation of equine antibodies did not greatly alter their affinity for homologous hapten. The poly-DL-alanylation of immunoglobulin permitted the total reduction of all the disulfide bridges within the molecule without insolubilizing the light- and heavy-chain products. This allowed for controlled reoxidation leading to the correct association of light and heavy chains and the recovery of both antigenic and antibody activity.

When Israel Pecht returned to Rehovot from Mandred Eigen's laboratory, he extended the interest in the antibody combining site to its kinetic aspects. Using temperature-jump methodologies Pecht was able to show that the reaction of haptens with the first DNP-specific IgA myeloma (MOPC 315; identified by Michael Potter and Herman Eisen) was not accompanied by conformational changes. Some years later, using methods of higher resolution, Doron Lancet examined several other systems kinetically and showed that conformational adjustments are the rule rather than the exception.

Besides the above studies I have been involved over the years with the detection, isolation, insolubilization, and biosynthesis of antibodies. I would like first to mention viroimmunoassay with chemically modified bacteriophage used to detect minute amounts of antigens and antibodies (work done with Joseph Haimovich). To T4 bacteriophage haptens such as dinitrophenyl and arsanil, small molecules such as penicillin, prostaglandin, angiotensin, and nucleosides, and various peptides, as well as macromolecules such as proteins and nucleic acids were covalently attached. The bacteriophages thus modified were still viable and could be neutralized by antibodies to the molecules attached. The technique was extremely sensitive for the detection of antibodies. Moreover, the inactivation could be inhibited by the free molecules and thus would serve as a most sensitive technique for their detection and quantitation. Olavi Makela in Finland independently developed similar techniques. While we successfully used the modified bacteriophage in many studies, the technique is more cumbersome in standardization than enzymoimmunoassay and has not been accepted for routine use. We used it with Michael Feldman and Harriet Gershon to show that one cell makes only one kind of antibody.

Of great interest to me was the observation we made with Edna Mozes on the inverse electrical net charge relationship between an antigen and the antibodies it provoked (15). The more positively charged the antigen, the more negatively charged was the antibody. We showed that this depended on the net charge of the intact antigen, not on local clustering of charges around the epitope. Thus we proved that the epitope is recognized while the antigen is still intact. The phenomenon holds for different classes of antibodies and is valid also at the cellular level, but not for thymus-independent antigens. More recently, with Lawrence Steinman we showed that in delayed type hypersensitivity the inverse charge relationship prevails also for the primary immunological recognition of antigen bound to macrophages (16). Obviously, the decision as to what type of antibodies or sensitized cells are selected for a certain epitope is taken while the antigenic molecule is still intact—a fact that must be taken into consideration when we discuss the "processed antigenic fragments" recognized by the T-cell receptor.

Even at a time when both immunologists and protein chemists regarded antibodies, because of their enormous heterogeneity, as a special category—totally different from other proteins—I was convinced that, to the contrary, it was through the study of the structure and genetics of immunoglobulins that we would learn more about proteins. Similarly, the reaction of the epitope with the antibody receptor on the cell membrane is a well-studied model relevant for the interaction of any ligand with a membrane receptor. We made an early effort, reported in 1971, to detect antibody-like structures on the cell surface with bacteriophage modified with dinitrophenyl and penicilloyl groups (17).

SYNTHETIC VACCINES

The studies mentioned earlier focused on conformational and sequential epitopes and on the successful attachment of the synthetic loop peptide of lysozyme to a multichain poly-DL-alanine carrier, which resulted in a conjugate provoking antibodies that react with intact lysozyme (8). These led to the inevitable conclusion that a new approach to vaccination was possible. We reasoned that synthetic vaccines might be a reality in the future (18), for the simple reason that if these conclusions hold for one protein, they may hold for others, including e.g. viral coat proteins and bacterial toxins. Of course, it is not sufficient to have just a synthetic epitope that will provoke antibodies to the protein. I shall not repeat here the arguments I have made before as to why there is place for improvement of vaccines as used today, but for a synthetic vaccine to be successful, it should contain at least five ingredients: (a) specificity; (b) built-in adjuvanticity; (c) the correct genetic background; (d) the capacity to cope with antigenic competition; and (e) the correct "texture," i.e. form that will give persistent and long-lasting immune protection. Much of the experimental work was done in collaboration with Ruth Arnon and various other colleagues.

We first synthesized a peptide from the amino-terminus of the carcinoembryonic antigen of the colon; this showed a weak cross-reaction with the intact antigen. The first study related to viruses was the PhD thesis of Harry Langbeheim, who synthesized a peptide from the envelope of the MS2 bacteriophage. He showed that the synthetic peptide inhibited phage neutralization by antiphage antibodies and that the same peptide, after attachment to multichain poly-DL-alanine, elicited antibodies capable of neutralizing the virus (19). Similarly, Ruth Arnon succeeded in preparing a conjugate of a synthetic peptide derived from influenza hemagglutinin, and it provoked antibodies and protected mice against influenza challenge (20). With Chaim Jacob we showed that tetanus toxoid coupled with synthetic peptides of the B subunit of cholera toxin led to the formation of antibodies neutralizing the toxic activity of native cholera toxin (21).

In all the above studies, we used Freund's adjuvant or water soluble peptidoglycan as an adjuvant. A short while later, in collaboration with Louis Chedid and Francoise Audibert, we used their synthetic muramyl dipeptides to prepare totally synthetic conjugates in which a synthetic antigenic determinant and a synthetic adjuvant were linked covalently to a synthetic carrier. The resulting conjugate, when administered in aqueous solution into experimental animals, provoked the formation of protective antibodies. When the muramyl dipeptide was bound covalently, it was much more efficient than when it was first mixed with the antigen. We prepared, with Ruth Arnon, such totally synthetic antigens, and these led to neutralization of a virus, MS2 (22), as well as to protection against diphtheria and cholera. Ruth Arnon also did it for influenza.

In contrast to the easy acceptance of synthetic antigens, the notion of synthetic vaccines met a certain reluctance period, followed by much activity in both industrial and academic laboratories. Possibly due to the practical interest that the results may have, developments in this field have been hurt by secrecy and unwarranted priority claims. However, in the past few years the whole area of synthetic vaccines has expanded enormously, and there is a fair chance that in the near future some agents, such as hepatitis B, malaria, or foot-and-mouth disease, will reach the stage of clinical application. It gives me an enormous satisfaction that I played some role in the inception of the concept (23). It seems that competition between antigenic determinants may be avoided and thus that one molecule possessing specific epitopes for several different diseases may be envisaged. In view of the linkage between the immune response and the main histocompatibility locus of each species, it is possible that the choice of the correct macromolecular carrier will depend on the genetic background of the individual to be immunized. The major challenge now is how to make synthetic vaccines of the future in such a way that they will lead to long-lasting immunity. Whether the solution will come by choosing the right carrier, and/or epitope, wrapping the antigen in the right envelope, or creating micelles or iscoms, I do not know, but I have no doubt the challenge will be successfully met.

MULTIPLE SCLEROSIS

Another study directly related to polyamino acids as protein models has been in the area of experimental allergic encephalomyelitis and, later on, in multiple sclerosis. The basic protein of the myelin sheath of the brain is capable, when administered in complete Freund's adjuvant, of provoking the disease experimentally in many animal species. Knowing that, I wondered whether a simple synthetic basic copolymer could do the same thing. The rationale for such a hypothesis was that demyelination is really delipidation. Therefore, the basic protein might be simply a carrier (a "schlepper") for some acidic lipids, and the autoimmune response might be directed against such lipid epitopes. Thus, we hoped to be able to induce the disease with some basic copolymers, in a way similar to the successful onset of the experimental disease with the natural basic encephalitogen. In experiments with Ruth Arnon and other colleagues, we failed to induce the disease with any of the copolymers we prepared, but some of them, when administered in aqueous solution, efficiently suppressed the onset of the disease induced by the natural basic encephalitogen in guinea pigs and rabbits. The copolymer we primarily used, denoted Cop 1, was composed of a small amount of glutamic acid, a much larger amount of lysine, some tyrosine, and a major share of alanine (24).

In later studies, with Ruth Arnon and Dvora Teitelbaum, we showed that Cop 1 is capable of suppressing the experimental disease in baboons and rhesus monkeys and that there is a weak but significant immunological cross-reaction between Cop 1 and the basic encephalitogen of the myelin sheath of the brain. When an analog of Cop 1 made from D-amino acids was tested, it had no suppressing capacity nor did it cross-react immunologically with the basic protein. We then demonstrated that Cop 1 was not generally immunosuppressive, and Ruth Arnon showed that the specific action occurs in experimental animals via suppressor T cells.

At that stage we were joined by Oded Abramsky from the Hebrew University-Hadassah Medical School in Jerusalem. We first did toxicological studies with animals and then, in Jerusalem, treated several cases of advanced multiple sclerosis, which showed that there were essentially no side effects. We then embarked on a collaborative project with Murray Bornstein and his colleagues at the Albert Einstein College of Medicine in the Bronx, with our devoted technician Israel Jacobson diligently preparing the Cop 1 for the clinical trials. Murray Bornstein organized a two-year double blind trial with 50 patients suffering from the exacerbating-remitting form of the disease. The results have only recently become available (25). Six of 23 placebo and 14 of 25 Cop 1–treated patients were exacerbation-free. Average number of exacerbations over 2 years were 2.7 for placebo and 0.6 for Cop-1 patients. Disability changes also favored Cop 1. Differences were even more pronounced in less involved patients. Finally, except for local irritation at injection sites, there were essentially no undesirable side reactions.

Two double-blind trials involving the chronic progressive form of multiple sclerosis are going on at the Baylor University in Houston and at the Albert Einstein College of Medicine in the Bronx. The encouraging results of the double-blind trial with the exacerbating remitting form of the disease will certainly lead to additional clinical trials, in the hope that Cop 1 may eventually become a useful drug for certain stages of the disease.

IMMUNOTARGETING OF DRUGS

In the last dozen years I have become more and more involved in immunological research with possible clinical application. Besides the studies on multiple sclerosis and on synthetic vaccines, I found myself very interested in the immunotargeting of drugs, especially in cancer. I distinctly remember discussions with Henri Isliker at WHO in Geneva in 1962. He was very enthusiastic about the subject and had Jean-Claude Jaton work on it for his PhD. Even though Jean-Claude spent several years afterwards in my laboratory, he synthesized and investigated antigens based on multichain polyproline. However, we did not get to immunotargeting until many years later, when with Esther Hurwitz, Ronald Levy, Meir Wilchek and Ruth Arnon, we bound daunomycin and adriamycin via a dextran bridge to antibodies against antigens of leukemia, lymphoma, and plasmacytoma. We showed that these are effective as "guided missiles" both in vitro and in vivo (26). An especially successful example was the result of our collaboration with our Japanese colleagues Hidematsu Hirai and Yutaka Tsukada in which we could show a chemotherapeutic effect against hepatoma in rats. In these experiments, intravenous injection was made of daunomycin attached covalently via a dextran bridge to monoclonal anti-rat α-fetoprotein antibodies. This was successful in preventing the death of more than half the animals even when administration started several days after transfer of hepatoma cells (27).

In the last decade, studies on immunotoxins have been started. In these investigations the toxin (or a part of it), attached covalently to antibodies, was shown to be responsible for killing the target cell. It seems a sound possibility that immunotargeting either toxins or drugs may become an important additional tool in the armamentarium against cancer, one whose possible use can be extended to other diseases.

CONCLUDING REMARKS

My introduction to immunology took place a third of a century ago, from the molecular side, as a protein chemist. I have been fortunate enough to follow the development of a most interesting and exciting branch of science through elucidation of some of its essential basic concepts. While my mentors were mainly Ephraim Katchalski-Katzir and Christian Anfinsen, I lacked the benefit of a great teacher in immunology, and so to some extent, I was self-taught. Thus, I had to "figure out" many things for myself. Immunology expanded so much during this period that on several occasions I predicted that it would have to reach a plateau. Every time I mentioned this, however, I was wrong. It is still an area into which enter many, both experienced scientists from other disciplines and young ones who choose it as their first field of specialization.

At an early stage I accepted the correctness, in a broad sense, of the clonal selection theory, but I was never deeply involved in arguments about immunological theories. I enjoyed them or criticized them, but I always thought that their main value was in helping to plan experiments and to examine results more critically, thus, to reach a better understanding of immune phenomena. Taken this way, theories are fruitful whether they are right or wrong; when one is too dogmatic, however, it interferes with the progress of science.

I have spent the last 36 years at the Weizmann Institute of Science, as a student, a senior scientist, a professor, a departmental chairman, a dean, and the president. My commitment to this Institute and to the development of immunology—everywhere, but especially in Israel—is a matter of record. I always deeply believed in the importance of international cooperation for the progress of science. I have tried hard to cement the close links between the Weizmann and institutions such as the Max-Planck, Pasteur, and Rockefeller, and to further the development of international organizations such as EMBO and IUIS.

My credo in science has always been: If something is not worth doing *at all*, it will not help if it is done very well. On the other hand, if it is worth doing, why should *I* do it? Is it not sufficiently obvious for somebody else to do it? Only if I found it not "sufficiently obvious" would I embark wholeheartedly on a project. In short, it has been a tremendous joy and privilege to do research, and it still is.

When I received the invitation to write this prefatory chapter I was flattered, awed, ready to accept the challenge, but also surprised. I do not feel that my active interest in immunology has abated as yet. So, perhaps, a more correct title for this essay would have been "So Far!".

Literature Cited

1. Katchalski, E., Sela, M. 1958. Synthesis and chemical properties of poly-α-amino acids. *Adv. Protein Chem.* 13 : 243
2. Sela, M., Katchalski, E. 1959. Biological properties of poly-α-amino acids. *Adv. Protein Chem.* 14 : 391
3. Sela, M., White, F. H. Jr., Anfinsen, C. B. 1957. Reductive cleavage of disulfide bridges in ribonuclease. *Science* 125 : 691
4. Sela, M. 1966. Immunological studies with synthetic polypeptides. *Adv. Immunol.* 5 : 29
5. Sela, M. 1969. Antigenicity: Some molecular aspects. *Science* 166 : 1365
6. Sela, M., Schechter, B., Schechter, I., Borek, F. 1967. Antibodies to sequential and conformational determinants. *Cold Spring Harbor Symp. Quant. Biol.* 32 : 537
7. Schechter, B., Conway-Jacobs, A., Sela M. 1971. Conformational changes in a synthetic antigen induced by specific antibodies. *Eur. J. Biochem.* 20 : 321
8. Arnon, R., Maron, E., Sela, M., Anfinsen, C. B. 1971. Antibodies reactive with native lysozyme elicited by a completely synthetic antigen. *Proc. Natl. Acad. Sci. USA* 68 : 1450
9. Humphrey, J. H. 1984. Serendipity in immunology. *Ann. Rev. Immunol.* 2 : 1
10. McDevitt, H. O., Sela, M. 1965. Genetic control of the antibody response. I. Demonstration of determinant-specific differences in response to synthetic polypeptide antigens in two strains of inbred mice. *J. Exp. Med.* 122 : 517
11. McDevitt, H. O., Chinitz, A. 1969. Genetic control of the antibody response: Relationship between immune response and histocompatibility (H-2) type. *Science* 163 : 1207
12. Schwartz, M., Parhami, B., Mozes, E.,

Sela, M. 1979. Changes in the specificity of antibodies to a random synthetic branched polypeptide in mice tolerant to its ordered analogs. *Proc. Natl. Acad. Sci. USA* 76: 5286

13. Geller-Bernstein, C., Kenett, R., Keret, N., Bejerano, A., Reichman, B., Tsur, S., Lahav, M., Sela, M. 1984. Survey of immunoglobulin levels in atopic families. *Asian Pac. J. Allergy Immunol.* 2: 181

14. Pilz, I., Schwarz, E., Durchschein, W., Licht, A., Sela, M. 1980. Effect of cleaving interchain dissulfide bridges on the radius of gyration and maximum length of anti-poly (D-alanyl) antibodies before and after reaction with tetraalanine hapten. *Proc. Natl. Acad. Sci. USA* 77: 117

15. Sela, M., Mozes, E. 1966. Dependence of the chemical nature of antibodies on the net electrical charge of antigens. *Proc. Natl. Acad. Sci. USA* 55: 445

16. Teitelbaum, D., Steinman, L., Sela, M. 1977. Unprimed spleen cell populations recognize macrophage bound antigen with opposite net electric charge. *Proc. Natl. Acad. Sci. USA* 74: 1693

17. Sulica, A., Haimovich, J., Sela, M. 1971. Detection of antibody-like structures on cell surfaces with chemically modified bacteriophages. *J. Immunol.* 106: 721

18. Sela, M. 1974. Vaccins Synthétiques—Un Rêve ou Une Réalité, in La Microbiologie Cent Ans Après Pasteur Symposium, Paris, 1972. *Bull. Inst. Pasteur* 72: 73

19. Langbeheim, H., Arnon, R., Sela, M. 1976. Antiviral effect on MS-2 coliphage obtained with a synthetic antigen. *Proc. Natl. Acad. Sci. USA* 73: 4636

20. Muller, G. M., Shapira, M., Arnon, R. 1982. Anti-influenza response achieved by immunization with a synthetic conjugate. *Proc. Natl. Acad. Sci. USA* 79: 569

21. Jacob, C. O., Sela, M., Arnon, R. 1983. Antibodies against synthetic peptides of the B subunit of cholera toxin: Cross reaction and neutralization of the toxin. *Proc. Natl. Acad. Sci. USA* 80: 7611

22. Arnon, R., Sela, M., Parant, M., Chedid, L. 1980. Anti-viral response elicited by a completely synthetic antigen with built-in adjuvanticity. *Proc. Natl. Acad. Sci. USA* 77: 6769

23. Sela, M., Arnon, R., Jacob, C. O. 1986. Synthetic peptides with antigenic specificity for bacterial toxins. In *Synthetic Peptides as Antigens*, ed. R. Porter, J. Whelan, p. 194. Chichester: Wiley

24. Teitelbaum, D., Meshorer, A., Hirshfeld, T., Arnon, R., Sela, M. 1971. Suppression of experimental allergic encephalomyelitis by a synthetic polypeptide. *Eur. J. Immunol.* 1: 242

25. Bornstein, M. B., Crystal, H., Drexler, E., Kielson, M., Miller, A., Slagle, S., Spada, V., Wassertheir-Smoller, S., Weiss, W., Weitzman, M., Arnon, R., Jacobson, I., Teitelbaum, D., Sela, M. 1986. A double-blind, randomized, placebo-controlled clinical trial of a synthetic polypeptide (Copolymer 1) in the exacerbating-remitting multiple sclerosis patient. Submitted

26. Levy, R., Hurwitz, E., Maron, R., Arnon, R., Sela, M. 1975. The specific cytotoxic effects of daunomycin conjugated to anti-tumor antibodies. *Cancer Research* 35: 1182

27. Tsukada, Y., Hurwitz, E., Kashi, R., Sela, M., Hibi, N., Hara, A., Hirai, H. 1982. Chemotherapy by intravenous administration of conjugates of daunomycin with monoclonal and conventional anti-rat α-fetoprotein antibodies. *Proc. Natl. Acad. Sci. USA* 79: 7896

Ann. Rev. Immunol. 1987. 5 : 21–42

ACTIVATION OF THE FIRST COMPONENT OF COMPLEMENT

Verne N. Schumaker

Department of Chemistry and Biochemistry and the Molecular Biology Institute, University of California, Los Angeles, California 90024

Peter Zavodszky

Institute of Enzymology, Biological Research Center of the Hungarian Academy of Sciences, Budapest H-1502, Hungary

Pak H. Poon

Department of Chemistry and Biochemistry and the Molecular Biology Institute, University of California, Los Angeles, California 90024

INTRODUCTION

Recent physical and chemical studies have contributed substantially to our knowledge of the structures of subcomponents C1q, C1r, and C1s and their assembly into C1. Kinetic studies have provided new insight into the activation process, particularly the central role played by C1-inhibitor: This protein appears to be loosely associated with the C1 complex and functions in preventing spontaneous activation and in regulating activation by immune complexes. New probes of C1 activation—monoclonal antibodies directed against C1q and their (Fab)$_2$ and Fab fragments—have been used to show that activation requires multivalent binding to C1q, but not the presence of the antibody Fc piece. These structural and kinetic data may be assembled to provide a plausible picture of the C1 activation mechanism; moreover, this raises new questions and poses interesting problems for future studies.

Models for the activation of C1 have been suggested (1–4). For a comprehensive review of C1, see (5). Other recent reviews include (6–10).

21

0732–0582/87/0410–0021$02.00

THE STRUCTURE AND FLEXIBILITY OF THE C1 SUBCOMPONENTS

Structure of C1q

From amino acid sequence and electron microscope observations, Reid & Porter intuited a detailed model for C1q: six A chains, six B chains, and six C chains were assembled to form six collagen triple helixes beginning close to the N-terminus and continuing to about residue 88–90; the remaining 125–135 residues were folded to form the globular C1q head. A disulfide bond connected each pair of A and B chains within a single triple helix, and a second disulfide connected C chains in adjacent triple helixes to form a pair of helixes and heads. Three such pairs then assembled to form the familiar C1q bouquet. The deviation in collagen sequence occurred at the kink where the arms radiated outward from the central stem (12–15; Figure 1B).

Reid & Porter's model was supported by a variety of physical studies, including sedimentation analyses (16–18), neutron diffraction patterns (19, 20), and additional electron microscope investigations (21). Combining the sedimentation coefficient of 10.2S with the mol wt of 459,300 (10) and a partial specific volume of 0.756 ml g^{-1} (20), we can calculate a frictional ratio of 1.87, consistent with the extended structure presented by the C1q bouquet. The radius of gyration of 14 nm (19) obtained by neutron diffraction also supported an open structure with the arms forming an angle of about 60° with the central stem; a small secondary maximum suggested that this configuration was unique (19). Another neutron diffraction study (20) found a radius of gyration at infinite contrast of 12.8 nm; the stem-arm angle was estimated as 45°($\pm$5°). These angles bracket the average angle of 50° reported in a detailed electron microscope study of C1q conformation; in addition, the electron microscope study suggested that the angle was not constant but that each arm could rotate about a semi-flexible joint located at the kink. A broad distribution of arm angles, ranging from 20° to 90°, was seen from these microscope images (21).

Atomic Resolution Model for C1q

That the C1q molecule as described by Porter & Reid is compatible with steric and energetic restraints at the atomic level is shown by a detailed molecular model recently published by Kilchherr et al (22). This model provides new insights into the symmetry and flexibility of C1q.

Space-filling and Kendrew wire models were constructed that retained the planarity and *trans* configuration of the peptide amide bonds and the torsional angles of residues in the triple helix for the regions of collagen-like sequence (22). The kink region was modelled using an interactive computer display. Reasonable but arbitrary side-chain positions were selected. The set of atomic coordinates resulting from this procedure was subject to energy minimization refinement, using a computer program that "changes bond lengths, bond angles, torsion angles and Van der Waals distances stepwise in order to minimize the conformational energy of the whole molecule" (22). Another computer program located the best interaction edges for assembling the six triple helixes to form the stem of the C1q molecule. Using these construction techniques, a single model was developed that met these requirements: the disulfide bond between the A and B chains is sterically possible; the sequence irregularities match in the kink region, and a maximum number of gly-X-Y units may be assembled into triple helixes. The torsional angles all fell within allowed regions on the Ramachandran plot, and only for two residues in the A chain (Arg 38 and Thr 39) did the pair of dihedral angles deviate from the triple helical dihedral angles.

In the model proposed by Kilchherr et al (22), all of the tripeptide units may be placed in a triple helix; in the kink region, an "almost perfect but bended triple helix is formed in this way. A bad Van der Waals contact which would arise between the methyl group of Ala C36 and the peptide carbonyl group of Ile A37 in a straight triple helix is relaxed by the kink and by the extra residue in the long segment...." It was also inferred from the model that in the kink "only limited flexibility is expected, which should not be much different from that of normal straight collagen helices." These authors question the interpretation of a semi-flexible joint at the kink, proposing instead that the triple helix is not rigid and that flexibility along the length of the collagenous arm may account for most of the variation in cone angle observed by Schumaker et al (21).

When the N-terminal halves of the six triple helixes were assembled, good hydrophobic and electrostatic interactions were found between residues located along two edges of each helix. These permitted construction of a model with almost perfect six-fold symmetry, and with the kink oriented radially so that the arms and heads pointed away from the central axis. Almost the only exception to the rule of six-fold symmetry were the three disulfide bonds connecting the six C chains and located at the N-terminal end of the stem. Even here, the six cysteines were centrally located so that the alternate disulfide pairings could be formed without disturbing the interactions between the helixes.

About 8% of the weight of C1q is carbohydrate. About two thirds of the carbohydrate is present as glucosylgalactosyl residues linked N-terminal to hydroxylysine (69%); the rest is linked to asparaginyl residues in the globular C-terminal domains (15, 23, 24). Since 11 out of 14 hydroxylysine residues in the stem may be glycosylated, the 2 residues lying within the interaction edges were examined. "There is no problem in orienting the side-chain of HylB35 in such a way that the disaccharide unit is placed outside the interaction edge. This is possible also for HylB32 but with more difficulty" (22).

Structure of Clr_2

Isolated Clr_2 is a dimer in the presence or absence of Ca^{++}; it undergoes pH-dependent dissociation to monomers at pH 5 (25). Each monomeric unit is composed of a single polypeptide chain containing 688 amino acids. Electron microscopy shows the dimer to be composed of two dumbbell-like monomers; each monomer is composed of two prominent globular domains joined by an interconnecting strand. The dimer has the shape of an asymmetric X (27; Figure 1A); dimerization apparently occurs through contact sites located near the junction between interconnecting strand and the γ-B domain (see below; 26, 27, 28). Upon activation the polypeptide chain is cleaved between an arg-ile bond (29) to yield the N-terminal A chain (446 residues), disulfide bonded to the C-terminal B chain (242 residues). Further autolytic cleavages of the A chain yield the N-terminal α fragment (211 residues), coincident with the N-terminal "interaction" domain; the β fragment (68 residues) which forms interconnecting strand; and the γ fragment (167 residues) which, together with the B chain, forms the C-terminal "catalytic" γ-B domain (35a, number of residues from M. G. Colomb, G. T. Alaud, personal communication).

Activated $\overline{Clr}_2$ is a serine protease, with the characteristic active-site seryl, histidyl, and aspartyl residues located on the B chain at positions 39, 94, and 191 (counting from the newly formed N-terminus) (30). Activation is accompanied by both an increase in intrinsic fluorescence and changes in the far UV circular dichroism spectrum, indicative of substantial conformational rearrangement which probably generates the $\overline{Clr}$ active site (31).

Upon prolonged incubation at 37°C, $\overline{Clr}_2$ undergoes additional autocatalytic cleavage (32). The fragment containing the C-terminal catalytic domains, called $(\overline{Clr} \text{ II})_2$, which still retains catalytic activity, has been visualized as two globular domains, connected by a short, curved rod (27, 28). The site through which $\overline{Clr}$ associates with $\overline{Cls}$ was not present on these catalytic domains; association is the function of the smaller, N-terminal domains (27, 28).

Clr_2 can self-activate; thus, at neutral pH especially in the absence of Ca^{++}, the dimeric proenzyme is an enzyme capable of cleaving itself to become the fully activated enzyme (33). The reversible dissociation that occurs when the pH is lowered coincides with the loss of ability to autoactivate (35a,b); this supports the suggestion (26, 34, 35a,b) that activation involves one monomeric subcomponent of the dimer cleaving the other. Upon the addition of Ca^{++}, $C1r_2$ tends to associate, and spontaneous activation no longer occurs (34, 35a,b).

Structure of C1s

In the absence of Ca^{++}, C1s is visualized as an asymmetric dumbbell, with a large, spherical globular domain about 8 nm in diameter and a smaller domain of about 5×6 nm (28), probably connected by flexible chain. Addition of Ca^{++} causes dimerization, with the two dumbbells joining to form a linear chain through contacts between the smaller "interaction" domains (27). When activated C1s is further digested with plasmin, cleavage yields several polypeptides: $\alpha1$, $\alpha2$, β, and γ-B; the γ and B chains are joined by a disulfide bond. From electron microscopy, N-terminal sequencing, and analogy with C1r, Villiers et al (28) tentatively conclude that the N-terminal $\alpha1$ and $\alpha2$ peptides of C1s form the smaller domain, joined through an interconnecting strand, β, to the larger C-terminal domain, γ-B. According to their model, the smaller N-terminal domain is the Ca^{++}-binding interaction domain through which dimerization of $C1s_2$ occurs, or through which C1s joins to $C1r_2$. The C-terminal γ-B domain contains the catalytic site, with the characteristic seryl, histidyl, and aspartyl residues being located on the B chain. The B chain has been completely sequenced (36).

Although very sensitive to proteolytic activation, C1s does not autoactivate (37–39); in the C1 complex it is cleaved by activated $\overline{C1r}_2$ (40), becoming a serine esterase specific for cleaving C2 and C4.

Structure of $C1r_2C1s_2$

When equivalent amounts of $C1r_2$ and C1s are mixed in the presence of Ca^{++}, they associate to form a flexible, linear chain of four subcomponents, called tetramer, with the sequence C1s-C1r-C1r-C1s (41). The linear structure visualized by electron microscopy is compatible with a translational frictional coefficient of 2.0, derived from hydrodynamic studies, and with a radius of gyration (at infinite contrast) of 17 nm, obtained from neutron diffraction studies (9, 41, 43). The ordering of the subcomponents has been established by using a ferritin-avidin complex to mark the position of biotin-labelled C1s at the two ends of the tetramer. In addition, the characteristic shapes of the subcomponents allow the order to be directly visualized by high-resolution electron microscopy (27). The tetramer appears to be joined through C1r-C1s contacts between the smaller, or N-terminal, interaction domains. In contrast, the C-terminal catalytic domains of C1s appear located at the extreme ends of the tetramer, and the catalytic domains of C1r appear to be joined by flexible chain to the center of the tetramer (Figure 1A; 27). This identification of the interaction domains is compatible with the results of proteolytic fragmentation studies, which show that the isolated catalytic domains of $\overline{C1s}$ or $\overline{C1r_2}$ will not associate with the other subcomponent (35a).

The electron microscope studies of the tetramer have led to the conclusion (9, 41, 65) that $C1r_2C1s_2$ possessed an axis of two-fold rotational symmetry, since many of the molecules showed an inverted S-shape on the carbon support. This result is readily explained by hypothesis of a molecule

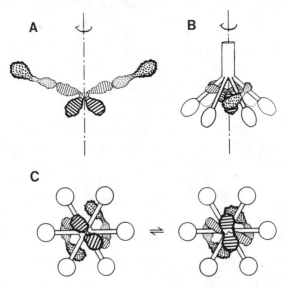

Figure 1   Structure and mechanism of activation of C1.

A. The structure of the $C1r_2C1s_2$ tetramer is illustrated, with the centrally located, X-shaped $C1r_2$, crosshatched; and the two terminally located, dumbbell-shaped $C1s$ subcomponents, stippled. Catalytic domains are emphasized by the thick outlines, and the twofold rotational axis indicated by the vertical line.

B. A side view of the symmetrical model for C1 shows C1q (unshaded) and the tetramer symmetrically located among the collagenous arms of C1q. According to the model, most of the $C1r_2$ is located inside the cone, but the interaction domains protrude to connect with the interaction domains of $C1s$. The terminal $C1s$ subcomponents are wrapped around the outside of the cone. The twofold axis of the tetramer coincides with the sixfold axis of C1q and is indicated by the vertical line.

C. Two views of C1 through the open end of the cone indicate the proposed activation mechanism. The complex is shown in two conformations: To the left is the open conformation in which the catalytic domains of $C1s$ are separated from those of $C1r$ by an intervening arm of C1q. On the right, arm movement, either bending or rotation, brings the $C1s$ catalytic domains into contact with the catalytic domains of $C1r$, inducing a conformational rearrangement to generate the catalytic sites. One $C1r$ then cleaves the other to initiate activation. All illustrations are drawn to the same scale to facilitate comparison. (Redrawn from illustrations in references 3 and 27.)

possessing a two-fold axis perpendicular to the support, one side of which displays an enhanced reactivity with the hydrophobic carbon film. Two-fold symmetry is also compatible with dimeric subcomponent composition, with the labelling experiments showing that the two C1s subcomponents lie at the ends of the linear chain, and with the locations of the interaction and catalytic domains symmetrical along the linear chain.

Spontaneous activation of the $C1r_2C1s_2$ tetramer in the presence of Ca^{++} is slow, even upon incubation at 37°C (45, 74). Addition of EDTA causes dissociation into $C1r_2$ and $C1s$ subcomponents. Spontaneous activation of the $C1r_2$ then occurs, as previously described, and in turn, the activated $\overline{C1r_2}$ cleaves and activates C1s. Thus, Ca^{++} binding appears to play at least two roles in the tetrameric complex: It is responsible for the association between the interaction domains of C1s and C1r, and it is involved in the stabilization of the catalytic domains of $C1r_2$ that prevents spontaneous activation. The results of Ca^{++} binding studies are interesting: The unactivated dimeric $C1r_2$ appears to possess a single Ca^{++} binding site, while each C1s monomer has a single site. With activation, the number of Ca^{++} binding sites on $\overline{C1r_2}$ increases to 3, the number on $\overline{C1s}$ remains the same, and a total of 5.0 ± 0.4 sites are found on the activated tetramer (46). The odd numbers 1, 3, and 5 are only consistent with twofold symmetry if a Ca^{++} binding site coincides with the rotational axis, located at the junction between the two C1r subcomponents in the dimeric and tetrameric structures. Since the $C1r_2$ dimeric structure is retained in the presence of EDTA, that single Ca^{++} cannot be essential for association. It may function, rather, to stabilize the proenzymic form (34). A mechanism is suggested by Villiers et al (44): "In a proposed autocatalytic mechanism for C1r in which the pro-site of one monomer activates the pro-site of the other, there is need for rotational mobility of the subunits to bring the reacting group of first the C1r and subsequently the C1s subunits into apposition. A role which calcium might play in this mechanism would be to prevent such rotational freedom, thereby preventing the autoactivation" (46). The single Ca^{++} bound by each C1s is probably involved in the association between subcomponents, either through induction of conformational changes forming the interaction sites, or through direct participation as Ca^{++} bridges (46). This association, however, may also involve Ca^{++}-independent forces (25).

STRUCTURE OF C1-INHIBITOR

C1-inhibitor is a 104,000-dalton glycoprotein, of which approximately 35% is carbohydrate; the carbohydrate is 17% sialic acid and 12% hexoses (47). The molecule is highly asymmetric, with a translational frictional coefficient of about 2.74. Electron micrographs of the rotatory shadowed material show a molecule shaped rather like a match, with a bulbous head and a long, rod-like tail. The length has been estimated from the shadowed micrographs to be 33 nm (48). Direct determination of amino acid sequence (49, 50) and the nucleotide sequence derived from cloning (51a,b) suggest that C1-inhibitor belongs to a superfamily of inhibitors which includes anti-thrombin III and α-1-antitrypsin.

Activated $\overline{C1}$, $\overline{C1r_2C1s_2}$, $\overline{C1s}$, and $\overline{C1r}$ are all rapidly inactivated by formation of tight complexes with C1-inhibitor; upon complexation, a small peptide is released from the inhibitor (50, 52). The complexes resist boiling in sodium dodecylsulfate, but they can be dissociated with hydroxylamine; these facts suggest that a covalent ester bond has been formed between the components (53).

The facts that enzymic activity is lost and that C1-inhibitor will not interact with diisopropyl-$\overline{C1s}$ (54) suggest that C1-inhibitor interacts with the catalytic domains of activated $\overline{C1r}$ or $\overline{C1s}$. Sedimentation data are compatible with hypothesis of a complex formed of one C1-inhibitor and one activated-C1s in the absence of Ca^{++} and with the possibility of a dimer of this complex in the presence of Ca^{++} (53). With $\overline{C1r_2}$, the complex composed of one inhibitor and one activated $\overline{C1r}$ is formed in the absence of Ca^{++} (53). With $\overline{C1r_2C1s_2}$, the complex probably has the form Inh-$\overline{C1s}$-$\overline{C1r}$-Inh (55).

The kinetics of the interaction between C1-inhibitor and activated $\overline{C1s}$ has been investigated (56, 57), and in the most recent study (58), the rate of the bimolecular reaction, as measured by fluorescence changes and loss of enzymatic activity, was shown to be the same as the rate of formation of the stable complex, as measured by separation by high pressure exclusion chromatography in SDS. The authors conclude, "This argues strongly against the accumulation of significant quantities of reversible intermediate under the conditions employed."

In serum, the concentration of C1-inhibitor is 137 ± 11 μg ml^{-1} (59), and the molar ratio of C1 to C1-inhibitor, approximately 1 : 7.

Recently, another important function for C1-inhibitor has been discovered: It prevents spontaneous activation of unactivated C1 (60). It exerts an important regulatory role under physiological conditions, to prevent activation before C1 complex encounters immune complexes.

STRUCTURE OF THE C1 COMPLEX

Physical and Ultrastructural Data

When C1q and the $C1r_2C1s_2$ tetramer are mixed in equimolar proportions, one molecule of C1q and one of tetramer spontaneously reassemble to form the C1 complex (61). This interaction is reversible with a dissociation constant (K_d) measured at 5°C, of about 15 nm for the unactivated tetramer (62, 63) and about 140 nm for the activated tetramer (62). Thus, activation apparently weakens the interactions between the tetramer and C1q by about one order of magnitude. These values for K_d, determined at 5°C, may underestimate the strength of the interaction at 37°C. The interaction at the higher temperature appeared to be stronger, although no evaluation of K_d at the higher temperature was provided (63).

Initial attempts to visualize the C1 complex by electron microscopy failed; most of the complexes dissociated on the carbon support and were seen as individual C1q and $C1r_2C1s_2$. This problem was solved by cross-linking the C1 with a water-soluble carbodiimide (64). This reagent forms peptide bonds between exposed amino and carboxyl groups that are already close together; sometimes it is called a "zero-length" cross-linking reagent. Cross-linking did not affect the sedimentation rate of C1 (64), and structural rearrangements were probably minimal; to date, only electron micrographs of cross-linked complexes have been examined. In a gallery of cross-linked C1 (65), both side- and top-view images showed the tetramer folded around or among the C1q arms in the region between the C1q heads and the stem. A portion of the collagenous arms free of the folded tetramer was visible in many images. Top-view images showed that the C1q arms had a significantly reduced range of flexibility in the C1 complex; apparently the tetramer made contacts with most of the C1q arms, tending to hold them at a constant angle (65).

Neutron scattering from solution and sedimentation measurements also were compatible with a model for the C1 complex in which the elongated tetramer was folded extensively and centered on the C1q. Thus, the radii of gyration of C1q and $C1r_2C1s_2$ were 12.8 nm and 17 nm, respectively (20). If the centers of these molecules were brought into coincidence without a change in shape, the calculated radius of gyration would be approximately 14.7 nm. The measured value for the radius of gyration of C1 was considerably smaller—only 12.6 nm (20); clearly, the tetramer must be assembled with C1q in a folded configuration.

Stokes radii may be determined from sedimentation coefficients of 10.2S for C1q and 15.9S for C1 (61, 66), partial specific volumes of 0.756 ml g^{-1} and 0.754 ml g^{-1} (20), and molecular weights of 460,000 and 784,000, respectively. From these values, Stokes radii of 9.7 nm for C1q and 10.7 nm for C1 are calculated. Thus, the hydrodynamic friction generated by translational motion through the aqueous solvent is only 10% greater for C1 than for C1q. This is compatible with a compact structure for the tetramer, which may be partially located inside the cone of the spreading C1q arms but may not be entirely protected from the solvent.

Symmetrical Models

In order to understand how the subcomponents interact to form a functional C1 complex, a plausible model for the folding of the tetramer around or among the C1q arms is required. Symmetrical models for C1 proposed by three laboratories (1, 3, 27, 65) are described first.

Examples of macromolecular complexes formed of symmetrical but nonidentical subcomponents include aspartate transcarbamylase—composed of two symmetrical trimers of catalytic subunits and three symmetrical dimers of regulatory subcomponents (67, 68)—and Eco R1 restriction endonuclease—a symmetrical protein dimer that binds to a symmetrical DNA palindrome (69, 70). In each case, the complex is formed by bringing the subcomponents together along their axes of symmetry, automatically replicating the contact sites, and thus maximizing the energy of interaction.

C1q shows sixfold symmetry (22) in the central region where the tetramer was observed to bind. Moreover, the tetramer is flexible and can be arranged in space to possess a twofold axis. In order to form a symmetrical structure from two objects that possess sixfold and twofold symmetries, it is necessary that the axes of symmetry be made to coincide; thus, the only way that symmetry can be maintained and the tetramer be positioned between the stem and heads of C1q is if the $C1r_2$ portion is aligned symmetrically *within* the cone formed by the spreading C1q arms. Then the protruding ends of the tetramer must be folded around or within the cone to form an S-shaped structure. In the symmetrical model proposed by Poon et al (65), Schumaker et al (3), and Weiss et al (27), the protruding ends are wrapped around the outside of the cone (Figures 1B,C). Other variants are possible for this wrapping process; in the model proposed by Colomb et al (1, 2), the tetramer loops around a single C1q arm and then is brought directly inside to form a figure 8.

The symmetrical models for C1 possess an additional appealing feature: The S-shaped or figure-8 fold brings the catalytic domains of C1s, located at the ends of the tetramer, into the vicinity of the catalytic domains of $C1r_2$, located at the center of the tetramer (Figures 1B,C). This feature, first emphasized by Colomb et al (1, 2), is employed below to suggest a molecular mechanism for C1 activation.

Asymmetrical Models

An asymmetrical model is one for which the sixfold axis of C1q is not aligned with the twofold axis of the tetramer. Although measurements of radii of gyration measurements provide no information about the location of the folded tetramer on C1, the shape of the neutron-scattering curve obtained at higher angles may theoretically be used to test models developed for C1. A detailed analysis is provided by Perkins (71), who considers W-shaped, Y-shaped, S-shaped, P-shaped and O-shaped models for the tetramer folded on C1q. All of these, with the exception of the S-shaped model, are asymmetrical. Moreover, for the W, Y, and P models, tetramer extends into the region of the C1q heads, and this disagrees with the electron microscope images. From Perkins's analysis, it seems clear that a variety of configurations for the folded tetramer are compatible with the neutron-scattering curves; apparently, the amount of information available is insufficient to distinguish between different structures as long as the tetramer is folded.

Among the asymmetrical models, only the O-shaped one seems a serious possibility; this model is discussed in detail by Cooper (5). In Cooper's version, the tetramer is first folded around its own twofold axis, aligning one half with the other. Then the folded tetramer is wrapped once around the outside of the cone of C1q, so that the catalytic domains of $C1r_2$ at one end of the tetramer are brought into contact with the catalytic domains of $C1s_2$ at the other. Among the advantages of this model is the apparent freedom of access of tetramer to its binding site on C1q, and to C2, C4, and to C1-inhibitor. It also seems compatible with hydrodynamic (20) and electron microscope observations (65).

The authors of this review favor the symmetrical models because in them the interaction energies between C1q and tetramer are maximized. Entropic and kinetic factors may compensate, however, and the asymmetrical models cannot be eliminated on the basis of energetic considerations alone.

SPONTANEOUS ACTIVATION OF C1

Kinetic Data

Reassembled in solution from purified subcomponents, C1 became activated rapidly, even in the absence of immune complexes or other activators: At 37°C, a half-life of 4 min for spontaneous activation was reported by Ziccardi (45). In Ziccardi's experiments, the reaction was first order; the half-life was independent of the initial *macromolecular* C1 concentration and was not affected by p-methylsulfonylfluoride. "These data indicate that C1 spontaneously activates by an intramolecular autocatalytic mechanism . . ." (45).

The rate of spontaneous activation depended strongly upon temperature. Over the range 20°–37°C, the activation energy was 19.1 kcal mol^{-1}; thus, at 20°C, the half-life rose to 22 min (45). Compared with 0.693 k_{cat}^{-1} for most enzymes-substrate complexes, the 4-min half-life of the initial cleavage was slower by orders of magnitude. The subsequent activation steps, presumably involving the cleavage of the second C1r by the first $\overline{C1r}$, and then the cleavages of the two C1s, were much more efficient; the rates of cleavage of C1r and C1s were virtually indistinguishable during spontaneous activation (45).

Ziccardi's results showed that only associated C1 was activatable; its concentration could be calculated from the K_d for the equilibration between free C1q, free tetramer, and C1. Therefore, activation of the remaining free tetramer was much slower; reassociation of the free sub-components to form new C1 presumably became rate-limiting (45). Not all laboratories have reported first-order kinetics for spontaneous activation of C1; indeed, both a lag phase and second-order kinetics have been observed (2, 72–74). Time and temperature of preincubation of the mixed subcomponents and the ratios of tetramer to C1q may be important considerations.

The rate of spontaneous activation of C1 increased as the ionic strength was lowered; thus, dropping the ionic strength by about a factor of two, from 0.15 M to 0.08 M, increased the rate by a factor of six (75). Ziccardi suggested that the increase in rate was related to the enhanced binding between C1q and $C1r_2C1s_2$ at the lower ionic strength; however, 75% of these subcomponents were already in the C1 form at physiological concentrations and in 0.15 M salt. An alternative explanation is that tighter association induced a conformational rearrangement of the tetramer on C1q, leading to the sixfold rate increase.

In developing a model for C1 activation, an important corollary of Ziccardi's studies is that immune complexes or other activators are not required for rapid activation of C1 in vitro.

Inhibition of Spontaneous Activation

Rapid, spontaneous activation at 37°C raises an interesting question: Why is C1 unactivated in the serum of humans and animals until it is triggered by immune complexes? When labelled C1 is added to normal human serum, the rate of spontaneous activation decreases by orders of magnitude (45, 60; also Bianchino et al—unpublished observations from our laboratory). Preincubation of the serum at 56°C destroys this inhibitory effect (45). Similarly, addition of C1-inhibitor to reconstituted C1 strongly inhibits spontaneous activation; pretreatment of C1-inhibitor at 56°C destroys the inhibition (60).

The amount of C1-inhibitor required to exert its in vitro effect is 25–35% of the physiological concentration, when C1 is also present at physiological concentrations (60, 76). The same amount of inhibitor is required in vivo: Hereditary angioedema is a disorder in which C1-inhibitor is either present in low concentration or in a nonfunctional form and in which complement pathway activation is detected. In this disorder, "down to a C1-INH level of approximately 0.075 g/l (38% of normal) apparent C1-INH functions were found within the normal range, while below this level functional adequacy of C1-INH could no longer be ascertained" (77).

In vivo, there are normally seven molecules of C1-inhibitor for each molecule of C1, and inhibition appears to require a two- to three-fold excess of inhibitor. The precise amount is difficult to determine experimentally, for when activation begins, inhibitor is consumed by forming irreversible complexes with activated C1rC1s, dropping the inhibitor concentration further and accelerating the rate of activation. The (2–3):1 ratio would be most simply explained if it were demonstrated that C1-inhibitor formed tight complexes with C1, composed of two molecules of inhibitor and one of C1 (3). Preliminary sedimentation experiments by Ziccardi (60) and current experiments in our own laboratory (Poon et al, unpublished) failed to provide physical evidence for complex formation. However, Ziccardi has been able to demonstrate substantial interaction between C1 and C1-inhibitor by zone centrifugation in the presence of a uniform concentration of nitrophenylguanidinobenzoate (NPGB), a reversible, synthetic inhibitor of C1 activation. The NPGB concentration was 100 μM, sufficient to prevent activation of C1, but not to prevent the C1-inhibitor from binding to C1. Increasing the NPGB concentration to 500 μM did displace the C1-inhibitor (78). From Ziccardi's data, a minimum value of 10^{-6} M can be estimated for the K_d between C1-inhibitor and C1 in the presence of 100 μM NPGB at 20°C.

IMMUNE COMPLEX–INDUCED ACTIVATION OF C1

The Role of Antibody

Signal transmission from the IgG antigen-binding site to the Fc region via changes in flexibility or the distribution of conformational states (79, 80) has been discussed as a critical feature for C1 binding and/or activation. For IgG antibody, the binding of antigen is not essential; chemically cross-linked IgG binds and activates C1 in the absence of antigen (81). The IgG antibody does not have to be bivalent, because immune complexes formed from univalent antibody are efficient activators (82, 83). It does not appear that the hinge angle between the Fab arms of IgG becomes locked upon binding to C1q, because the Fab arms of IgG bound to C1q in solution retain segmental flexibility (84). The critical factor appears to be the extent of clustering of the Fc on the surface of the immune complex (85, 86). Thus, the C1q subcomponent of C1 is a multivalent ligand that binds in a cooperative fashion to a cluster of antibody Fc exposed on the surface of an immune complex (87). Evidently, multivalent binding through the C1q heads is necessary for C1 activation (88); whether it is also sufficient may still be debated, "but the evidence continues to favor multiple binding as the only condition required . . . in the case of IgG" (89).

For the activation of C1 by pentameric IgM antibody, a conformational change in the structure of the antibody appears to be essential. C1q binds weakly to IgM in solution at physiological ionic strength (90); moreover, when C1 binds to IgM in solution, it does not activate (78). However, when C1 binds to single molecules of IgM attached to antigen on the cell surface, activation occurs readily (91). Activation is related to hapten density; simultaneous binding to antigen by several Fabs from the same IgM is probably required to induce a conformational change in IgM (91, 92), perhaps exposing multiple C1q binding sites. Allosteric and distortive mechanisms for signal transmission from the antigen binding site have been suggested (93, 94).

Activation Kinetics in the Absence of C1-Inhibitor

In the absence of C1-inhibitor, spontaneous activation of C1 proceeded rapidly at 37°C. Addition of immune complexes increased the activation rate by a factor of seven to yield a half-life of less than a minute; no lag phase was observed by Ziccardi (75). In contrast, other workers have reported pronounced lag phases during activation by immune complexes (95) or chemically cross-linked oligomers (96).

In general, the appearance of a lag phase involves a multiple-step reaction sequence in which the concentration of an intermediate increases until a steady state is reached:

1. $C1q + C1r_2 C1s_2 \rightleftharpoons C1$

2. $C1 + IC \rightleftharpoons C1.IC$

3. $C1.IC \rightleftharpoons C1^\bullet.IC$

4. $C1^\bullet.IC \rightleftharpoons C\bar{1}.IC.$

In this scheme, C1 is reformed in step 1 and bound to immune complexes (IC) in step 2. Depending upon experimental conditions, either of these steps might account for a lag phase. Even though step 2 is written as a second-order reaction, it might also involve a slow rearrangement of bound C1q heads after a rapid initial binding to the immune complex. Dodds et al (95) and Tschopp et al (96) attributed the lag phase to step 3 and introduced an intermediary form of C1, here written $C1^\bullet$, which represents the first component with a conformationally altered $C1r^\bullet$, "an unsplit form of subcomponent C1r which has a catalytic site able to . . . cause the hydrolysis of C1r" (95). An intermediary form of C1r with a complete active site has been detected by fluorescence (97) and by use of an inhibitor (98).

Activation Kinetics in the Presence of C1-Inhibitor

At 37°C, the addition of C1-inhibitor did not prevent activation of C1 by antibody-coated erythrocytes (EA) or tetanus-antitetanus immune complexes (IC), although it did inhibit activation by DNA and heparin (60). Thus, binding to EA or IC at 37°C released the C1 from the inhibitory influence of C1-inhibitor. At 20°C, however, activation by EA or IC was completely suppressed by C1-inhibitor (60).

Activation in the presence of C1-inhibitor was investigated with a monoclonal antibody (1H11) directed against the C1q heads (76). Activation was stoichiometric and cooperative—that is, maximal activation was achieved at a ratio of one antibody binding site to one C1q head; and activation varied with the 3.5 power of antibody concentration. Generated from 1H11 by proteolytic cleavage, Fab$_2$ also activated efficiently and cooperatively, with a Hill coefficient of 2.0. Monovalent Fab did not show significant activation.

Ultracentrifuge studies showed that cross-linked complexes composed of two or more C1q were formed when either 1H11 or its Fab$_2$ were added. Electron micrographs showed some complexes composed of two C1q brought together along a mutual symmetry axis, with heads touching and stems pointing in opposite directions, presumably held together by multiple Fab$_2$ cross-linking the C1q heads. The structure is roughly analogous to the open cage formed by joining the fingertips of two hands. Activation of C1 occurred at the Fab$_2$ concentrations where complex formation began; complex formation probably involved the C1 being cross-linked by multiple antibody molecules, compatible with the observed cooperativity. The authors suggested that the cross-linking of C1 sterically prevented access of C1-inhibitor to C1r located inside the cone of C1q arms; this exclusion of inhibitor then resulted in spontaneous activation.

Study of another monoclonal antibody, mAB 130, which is believed to bind to the collagenous arms in the region immediately adjacent to the heads of C1q, has suggested that, "bivalency of mAb 130 was a requirement for C1 activation. . . . The sedimentation behavior . . . indicated a predominance of complexes of 1–3 mAbs bound to a C1q molecule. . . . Our data support a recently proposed mechanism for C1 activation, in which distortion of the six flexible C1q arms provides the activating signal for C1r$_2$C1s$_2$," (R. Hoekzema, M. Martins, M. C. Brouwer, and C. E. Hack, personal communication).

SPECULATION: A MODEL FOR THE ACTIVATION OF C1

An activation model for C1 must explain: (*a*) how the addition of C1q to tetramer induces spontaneous activation; (*b*) how binding to immune complexes at 37°C releases C1 from the inhibitory action of C1-inhibitor, and why this does not occur at 20°C; (*c*) how binding of C1 to immune complexes generates an additional activation signal; and (*d*) how activated C1s and C1r can interact with C2, C4, and C1-inhibitor. The eclectic model developed below is principally taken from References 1, 3, 4, 22, and 27.

A Model for Spontaneous Activation

A model for spontaneous activation of C1 is suggested by the symmetrical structure of the C1 complex (65), illustrated in Figure 1C, in which the tetramer is folded to bring the catalytic domains of C1s into the vicinity of the catalytic domains of C1r (1, 2). Thus, all four of the proenzyme domains are brought together, permitting access of the $\overline{C1r}$ enzyme, located in the middle of the tetramer, to the C1s located terminally. In the spontaneous activation model, C1q provides a template around which the tetramer is bent, and it also provides a barrier that separates the proenzyme domains of C1s (located outside the cone) from the proenzyme domains of C1r (located inside the cone) (3, 4). The model assumes that thermal energy moves the C1q arms sufficiently to allow contact between the catalytic domains of C1r and C1s, as indicated by the reversed arrows of Figure 1C (3). It has been suggested that this contact may induce a conformational change, temporarily converting the C1r proenzyme into an enzyme (27). An analogy exists for the induction of proteolytic activity in a proenzyme through conformational change arising from macromolecular interaction. Activation of plasminogen is triggered by the binding of the dimeric streptokinase, which induces a conformational change to generate an enzymatically active form of plasminogen. This then proceeds to activate other molecules of plasminogen (99). The analogy seems particularly apt, for both C1s and streptokinase are proenzymes yet play no direct catalytic role in the activation process. Thus, activated C1s, subsequently inactivated with diisopropylfluorophosphate, will serve in the C1 complex to permit activation of $C1r_2$ (95). In the model, this contact between C1s and C1r induces a conformational change in the C1r proenzyme such that it develops enzymatic activity and cleaves and activates the other C1r (27). Then the activated C1r cleaves the proenzyme C1r (1). These initial proteolytic events may also create new swivels allowing rotation of the C1r catalytic domains to face and activate C1s.

A Model for Immune Complex–Induced Activation

RELEASE FROM INHIBITION BY C1-INHIBITOR   The binding of C1 to an immune complex releases it from the inhibitory influence of the macromolecular C1-inhibitor. Small molecule inhibitors such as NPGB are still effective inhibitors (95), a fact clearly illustrated by their use during C1 isolation on IgG-Sepharose affinity columns. In the model, the release from inhibition is due to steric factors that exclude the highly asymmetric C1-inhibitor glycoprotein from the binding sites located on the catalytic domains of C1r (3). When C1 is bound by multiple heads to a cluster of IgG-Fc on an immune complex, the catalytic domains of the C1r are protected from interaction with the macromolecular inhibitor by the cage formed by the spreading C1q arms and by those portions of the tetramer wrapped around the outside of the cage. Small molecule inhibitors still diffuse through the cage to reach their binding sites on C1r, to prevent activation (3).

Ziccardi (60) found C1-inhibitor more effective in preventing activation by immune complexes at 20°C than at 37°C. Assuming that the binding sites are located on the outsides of the C1q heads (that is, facing away from the axis) as proposed by Kilchherr et al (22), we suggest the C1, with inhibitor still attached, initially binds with its axis parallel to the surface of the immune complex. Initial binding would be followed by the reversible release of inhibitor and reorientation of the cone axis perpendicular to the surface to develop a maximal binding configuration between the C1q heads and the clustered Fc. Thus, the rate of attachment of C1 to immune complexes would not be impeded by inhibitor, and C1 could be bound with inhibitor attached. Rearrangement would require the release of inhibitor; according to the model, this is reversible and temperature-dependent.

Ziccardi also found (60) inhibitor effective in preventing activation by nonimmune activators such as DNA and heparin. DNA and heparin are elongated polyanions that bind electrostatically to C1q. Simultaneous binding by the inhibitor probably prevented activation.

GENERATION OF THE ACTIVATION SIGNAL   The activation rate increased sevenfold when immune complexes were added in the absence of C1-inhibitor (75); this observation implies that binding generates an additional activation signal. C1q is a flexible molecule and, according to the model, binding to an asymmetric cluster of Fc distorts the cone formed by the spreading C1q arms (100). This distortion is the activation signal; arm movement allows the catalytic domains of the C1s access to the interior of the cone, which permits contact with the catalytic domains of the C1r. This induces the same conformational change described above for the spontaneous activation of C1 (3, 27, 100). An alternate mechanism involves rotation of C1q arms, also induced by binding to a cluster of Fc. "A transmission of a signal by a rotational movement appears to be possible in view of the low energies needed to change the pitch of the collagen supercoil from 30 to 45 residues per turn...the kink region... may act as a 'bearing' for the arm rotations" (22). The C1s catalytic domain, bound to the rotating arm, would be brought inside the cone where it would be available for interaction with C1r.

A Model for Interaction with C4, C2, and C1-Inhibitor

The binding of activated tetramer to C1q is weaker, by about an order of magnitude, than the binding of unactivated tetramer (62). The model suggests that the contact sites between the C1q arms and the $\overline{C1s}$ catalytic domains are altered by activation of the tetramer, and that the ends of the tetramer may become free to extend away from the C1q cone into the surrounding medium. There, they are available to complex with and to catalyze the cleavage of the peptide chain of the soluble C4 (3). The $\overline{C4b}$ component, with its newly exposed thioester, attaches covalently to nearby amino and hydroxyl groups, some of which are located on adjacent Fab arms of the clustered antibody molecules (101). C2 then binds to the $\overline{C4b}$, and it, in turn, is cleaved by the activated $\overline{C1s}$. This second cleavage would also be facilitated by the freedom of the tetramer, tethered on C1q, to extend sufficiently to permit the catalytic domain of activated $\overline{C1s}$ to reach the susceptible bond on the C2 polypeptide. Finally, the catalytic domains of activated $\overline{C1s}$, no longer tightly bound to the C1q cone, are readily accessible to inactivation by C1-inhibitor (102, 103).

Literature Cited

1. Colomb, M. G., Arlaud, G. J., Villiers, C. L. 1984. Activation of C1. *Philos. Trans. R. Soc. London B* 306: 283–92
2. Colomb, M. G., Arlaud, G. J., Villiers, C. L. 1984. Structure and activation of C1: Current concepts. *Complement* 1: 69–80
3. Schumaker, V. N., Hanson, D. C., Kilchherr, E., Phillips, M. L., Poon, P. H. 1986. A molecular mechanism for the activation of the first component of complement by immune complexes. *Mol. Immunol.* 23: 557–65
4. Schumaker, V. N. 1986. C1 structure and antibody recognition. In *Effectors of the Immune System*, ed. E. Podack, p. 1. Boca Raton, Fla: Chem. Rubber Co. In press
5. Cooper, N. R. 1985. The classical complement pathway: Activation and regulation of the first complement component. *Adv. Immunol.* 37: 151–216
6. Lachmann, P. J., Hughes-Jones, N. C. 1984. Initiation of complement activation. *Springer Semin. Immunopathol.* 7: 143–62
7. Ziccardi, R. J. 1983. The first component of human complement (C1): Activation and control. *Springer Semin. Immunopathol.* 6: 213–30
8. Loos, M. 1982. The classical complement pathway: Mechanism of activation of the first component by antigen-antibody complexes. *Prog. Allergy* 30: 135–92
9. Schumaker, V. N., Strang, C. H., Siegel, R. L., Phillips, M. L., Poon, P. H. 1982. Electron microscopy of the

first component of human complement. *Surv. Immunol. Res.* 1 : 305–15

10. Reid, K. B. M. 1983. Proteins involved in the activation and control of the two pathways of human complement. *Biochem. Soc. Trans.* 11 : 1–12

11. Deleted in proof

12. Reid, K. B. M., Porter, R. R. 1975. The structure and mechanism of activation of the first component of complement. *Contemp. Top. Mol. Immunol.* 4 : 1–22

13. Porter, R. R., Reid, K. B. M. 1979. Activation of the complement system by antibody-antigen complexes : The classical pathway. *Adv. Protein Chem.* 33 : 1–71

14. Reid, K. B. M., Gagnon, J., Frampton, J. 1982. Completion of the amino acid sequences of the A and B chains of subcomponent C1q of the first component of human complement. *Biochem. J.* 203 : 559–69

15. Reid, K. B. M. 1979. Complete amino acid sequences of the three collagen-like regions present in subcomponent C1q of the first component of human complement. *Biochem. J.* 179 : 367–71

16. Reid, K. B. M., Lowe, D. M., Porter, R. R. 1972. Isolation and characterization of C1q, a subcomponent of the first component of complement, from human and rabbit sera. *Biochem. J.* 130 : 749–63

17. Schumaker, V. N., Calcott, M. A., Spiegelberg, H. L., Muller-Eberhard, H. J. 1976. Ultracentrifuge studies of the binding of IgG of different subclasses to the C1q subunit of the first component of complement. *Biochemistry* 15 : 5175–81

18. Liberti, P. A., Paul, S. M. 1978. Gross conformation of C1q: A subcomponent of the first component of complement. *Biochemistry* 17 : 1952–58

19. Gilmour, S., Randall, J. T., Willan, K. Js., Dwek, R. A., Torbet, J. 1980. The conformation of subcomponent C1q of the first component of human complement. *Nature* 285 : 512–14

20. Perkins, S. J., Villiers, C. L., Arlaud, G. J., Boyd, J., Burton, D. R., Colomb, M. G., Dwek, R. A. 1984. Neutron scattering studies of subcomponent C1q of first component C1 of human complement and its association with subunit $C1r_2$-$C1s_2$ within C1. *J. Mol. Biol.* 179 : 547–57

21. Schumaker, V. N., Poon, P. H., Seegan, G. W., Smith, C. A. 1981. A semi-flexible joint in the C1q subunit of the first component of human complement. *J. Mol. Biol.* 148 : 191–97

22. Kilchherr, E., Hofmann, H., Steigemann, W., Engel, J. 1985. A structural model of the collagen-like region of C1q comprising the kink region and the fibre-like packing of the six triple helices. *J. Mol. Biol.* 186 : 403–15

23. Shinkai, H., Yonemasu, K. 1979. Hydroxylysine-linked glycosides of human complement subcomponent C1q and of various collagens. *Biochem. J.* 177 : 847–52

24. Mizuochi, T., Yonemasu, K., Yamashita, K., Kobata, A. 1979. The asparagine-linked sugar chains of subcomponent C1q of the first component of human complement. *J. Biol. Chem.* 253 : 7404–9

25. Arlaud, G. J., Chesne, S., Villiers, C. L., Colomb, M. G. 1980. A study on the structure and interactions of the C1 subcomponents C1̄r and C1̄s in the fluid phase. *Biochem. Biophys. Acta* 616 : 105–15

26. Arlaud, G. J., Gagnon, J., Villiers, C. L., Colomb, M. G. 1986. Molecular characterization of the catalytic domains of human complement protease C1r. *Biochemistry* 25 : 5177–82

27. Weiss, V., Fauser, C., Engel, J. 1986. Functional model of subcomponent C1 of human complement. *J. Mol. Biol.* Submitted

28. Villiers, C. L., Arlaud, G. J., Colomb, M. G. 1985. Domain structure and associated functions of subcomponents C1r and C1s of the first component of human complement. *Proc. Natl. Acad. Sci. USA* 82 : 4477–81

29. Arlaud, G. J., Gagnon, J. 1985. Identification of the peptide bond cleaved during activation of human C1r. *FEBS Lett.* 180 : 234–38

30. Arlaud, G. J., Gagnon, J. 1983. Complete amino acid sequence of the catalytic chain of human complement subcomponent C1̄r. *Biochemistry* 22 : 1758–64

31. Villiers, C. L., Arlaud, G. J., Colomb, M. G. 1984. Autoactivation of human subcomponent C1r involves structural changes reflected in modifications of intrinsic fluorescence, circular dichroism and reactivity with monoclonal antibodies. *Biochem. J.* 215 : 369–75

32. Gagnon, J., Arlaud, G. J. 1985. Primary structure of the A chain of human complement-classical-pathway enzyme C1̄r. *Biochem. J.* 225 : 135–42

33. Ziccardi, R. J., Cooper, N. R. 1976. Physiochemical and functional characterization of the C1r subunit of the first

complement component. *J. Immunol.* 116: 496–503

34. Ziccardi, R. J., Cooper, N. R. 1976. Activation of C1r by proteolytic cleavage. *J. Immunol.* 116: 504–9

35. Arlaud, G. J., Villiers, C. L., Chesne, S., Colomb, M. G. 1980. Purified proenzyme C1r. Some characteristics of its activation and subsequent proteolytic cleavage. *Biochim. Biophys. Acta* 616: 116–29

35b. Villiers, C. L., Duplaa, A.-N., Arlaud, G. J., Colomb, M. G. 1982. Fluid phase activation of proenzymic C1r purified by affinity chromatography. *Biochim. Biophys. Acta* 700: 118–26

36. Carter, P. E., Dunbar, B., Fothergill, J. E. 1983. The serine proteinase chain of human complement component $C\overline{1}s$. *Biochem. J.* 215: 565–71

37. Gigli, I., Porter, R. R., Sim, R. B. 1976. The unactivated form of the first component of human complement, C1. *Biochem. J.* 157: 541–48

38. Arlaud, G. J., Reboul, A., Colomb, M. G. 1977. Proenzymic C1s associated with catalytic amounts of C1r. Study of the activation kinetics. *Biochim. Biophys. Acta* 487: 227–35

39. Sakai, K., Stroud, R. M. 1974. The activation of C1s with purified C1r. *Immunochemistry* 11: 191–96

40. Valet, G., Cooper, N. R. 1974. Isolation and characterization of the proenzyme form of subunit C1s of the first complement component. *J. Immunol.* 112: 339–50

41. Tschopp, J., Villiger, W., Fuchs, H., Kilchherr, E., Engel, J. 1980. Assembly of subcomponents C1r and C1s of the first component of complement: Electron microscope and ultracentrifugal studies. *Proc. Natl. Acad. Sci. USA* 77: 7014–18

42. Deleted in proof

43. Boyd, J., Burton, D. R., Perkins, S. J., Villiers, C. L., Dwek, R. A., Arlaud, G. J. 1983. Neutron scattering studies of the isolated $C1r_2C1s_2$ subunit of first component of human complement in solution. *Proc. Natl. Acad. Sci. USA* 80: L3769–73

44. Deleted in proof

45. Ziccardi, R. J. 1982. Spontaneous activation of the first component of human complement (C1) by an intramolecular catalytic activity. *J. Immunol.* 128: 2500–4

46. Villiers, C. L., Arlaud, G. J., Painter, R. H., Colomb, M. G. 1980. Calcium binding properties of the C1 subcomponents C1q, C1r, and C1s. *FEBS Lett.* 117: 289–94

47. Haupt, H., Heimburger, N., Kranz, T., Schwick, H. G. 1970. Ein Betrag zur Isolierung und Charakterisierung des $C\overline{1}$-Inaktivators aus Humanplasma. *Eur. J. Biochem.* 17: 254–61

48. Odermatt, E., Berger, H., Sano, Y. 1981. Size and shape of human C1-inhibitor. *FEBS Lett.* 131: 283–85

49. Harrison, R. A. 1983. Human $C\overline{1}$ inhibitor: Improved isolation and preliminary structural characterization. *Biochemistry* 22: 5001–7

50. Salvesen, G. S., Catanese, J. J., Kress, L. F., Travis, J. 1985. Primary structure of the reactive site of human $C\overline{1}$-Inhibitor. *J. Biol. Chem.* 260: 2432–36

51a. Davis, A. E., Whitehead, A. S., Harrison, R. A., Dauphinais, A., Bruns, G. A. P., Cicardi, M., Rosen, F. S. 1986. Human inhibitor of the first component of complement, C1: Characterization of cDNA clones and localization of the gene to chromosome 11. *Proc. Natl. Acad. Sci. USA* 83: 3161–65

51b. Tosi, M., Duponchel, C., Bourgarel, P., Colomb, M., Meo, T. 1986. Molecular cloning of human C1 inhibitor: Sequence homologies with α_1-antitrypsin and other members of the serpins superfamily. *Gene* 42: 265–72

52. Weiss, V., Engel, J. 1983. Heparin-stimulated modification of $C\overline{1}$-inhibitor by subcomponent $C\overline{1}s$ of human complement. *Hoppe-Seyler's Z. Physiol. Chem.* 364: 295–301

53. Chesne, S., Villiers, C. L., Arlaud, G. J., LaCroix, M. G., Colomb, M. G. 1982. Fluid phase interaction of $C\overline{1}$ inhibitor (C1 Inh) and the subcomponents $C\overline{1}r$ and $C\overline{1}s$ of the first component of complement, C1. *Biochem. J.* 201: 61–70

54. Bing, D. H. 1969. Nature of the active site of a subunit of the first component of human complement. *Biochemistry* 8: 4503–10

55. Ziccardi, R. J., Cooper, N. R. 1979. Active disassembly of the first component of complement, $C\overline{1}$, by $C\overline{1}$-inactivator. *J. Immunol.* 123: 788–92

56. Sim, R. B., Arlaud, G. J., Colomb, M. G. 1980. Kinetics of reaction of human $C\overline{1}$-inhibitor with the human complement system proteases $C\overline{1}r$ and $C\overline{1}s$. *Biochim. Biophys. Acta* 612: 433–49

57. Nilsson, T., Wiman, B. 1983. Kinetics of the reaction between human C1-esterase inhibitor and $C\overline{1}r$ or $C\overline{1}s$. *Eur. J. Biochem.* 129: 663–67

58. Lennick, M., Brew, S. A., Ingham, K.

C. 1986. Kinetics of interaction of C̄1-inhibitor with C̄1s. *Biochemistry* 25: 3890–98

59. Ziccardi, R. J., Cooper, N. R. 1980. Development of an immunochemical test to assess C̄1 inactivator function in human serum and its use for the diagnosis of hereditary angioedema. *Clin. Immunol. Immunopathol.* 15: 465–71

60. Ziccardi, R. J. 1982. A new role for C1-inhibitor in homeostasis: Control of activation of the first component of human complement. *J. Immunol.* 128: 2505–8

61. Ziccardi, R. J., Cooper, N. R. 1977. The subunit composition and sedimentation properties of human C1. *J. Immunol.* 118: 2047–52

62. Siegel, R. C., Schumaker, V. N. 1983. Measurement of the association constants of the complexes formed between intact C1q or pepsin-treated C1q stalks and the unactivated or activated C1r$_2$C1s$_2$ tetramers. *Mol. Immunol.* 20: 53–66

63. Ziccardi, R. J. 1985. Nature of the interaction between the C1q and C1r$_2$C1s$_2$ subunits of the first component of human complement. *Mol. Immunol.* 22: 489–94

64. Strang, C. J., Siegel, R. C., Phillips, M. L., Poon, P. H., Schumaker, V. N. 1982. Ultrastructure of the first component of human complement: Electron microscopy of the crosslinked complex. *Proc. Natl. Acad. Sci. USA* 79: 586–90

65. Poon, P. H., Schumaker, V. N., Phillips, M. L., Strang, C. J. 1983. Conformation and restricted segmental flexibility of C1, the first component of human complement. *J. Mol. Biol.* 168: 563–77

66. Siegel, R. C., Schumaker, V. N., Poon P. H. 1981. Stoichiometry and sedimentation properties of the complex formed between the C1q and C1r$_2$C1s$_2$ subcomponents of the first component of complement. *J. Immunol.* 127: 2447–52

67. Monaco, H. L., Crawford, J. L., Lipscomb, W. N. 1978. Three-dimensional structures of aspartate carbamoyltransferase from *Escherichia coli* and of its complex with cytidine triphosphate. *Proc. Natl. Acad. Sci. USA* 75: 5276–80

68. Gerhart, J. C., Schachman, H. K. 1965. Distinct subunits for the regulation and catalytic activity of aspartate transcarbamylase. *Biochemistry* 4: 1054–62

69. Modrich, P. 1979. Structures and mechanisms of DNA restriction and modification enzymes. *Q. Rev. Biophys.* 12: 315–69

70. Frederick, C. A., Grable, F., Melia, M., Samudzi, C. L., Jen-Jacobson, L., Bi-Cheng, W., Greene, P., Boyer, H. W., Rosenberg, J. M. 1984. Kinked DNA in crystalline complex with EcoR1 endonuclease. *Nature* 309: 327–31

71. Perkins, S. J. 1985. Molecular modelling of human complement subcomponent C1q and its complex with C1r$_2$C1s$_2$ derived from neutron-scattering curves and hydrodynamic properties. *Biochem. J.* 228: 13–26

72. Lepow, I. H., Naff, G. B., Pensky, J. 1965. Mechanisms of activation of C1 and inhibition of C1 esterase. In *CIBA Foundation Symposium: Complement*, ed. G. E. W. Wolstenholme, J. Knight, pp. 74–98. London: Church Hill. 388 pp.

73. Medicus, R. G., Chapuis, R. M. 1981. The physiologic mechanism of C1 activation. *Hoppe-Seyler's Z. Physiol. Chem.* 362: 17 (Abstr.)

74. Lin, T.-Y., Fletcher, D. S. 1980. Activation of a complex of C1r and C1s subcomponents of human complement C1 by the third subcomponent C1q. *J. Biol. Chem.* 255: 7756–62

75. Ziccardi, R. J. 1984. The role of immune complexes in the activation of the first component of human complement. *J. Immunol.* 132: 283–88

76. Kilchherr, E., Schumaker, V. N., Phillips, M. L., Curtiss, L. K. 1986. Activation of the first component of human complement, C1, by monoclonal antibodies directed against different domains of subcomponent C1q. *J. Immunol.* 137: 255–62

77. Spath, P. J., Wuthrich, B., Butler, R. 1984. Quantification of C1-inhibitor functional activities by immunodiffusion assay in plasma of patients with hereditary angioedema—evidence of a functionally critical level of C1-inhibitor concentration. *Complement* 1: 147–59

78. Ziccardi, R. J. 1985. Demonstration of the interaction of native C1 with monomeric immunoglobulins and C1-inhibitor. *J. Immunol.* 134: 2559–63

79. Zavodszky, P., Jaton, J. C., Venyaminov, S. Y., Medgyesi, G. A. 1981. Increase of conformational stability of homogeneous rabbit immunoglobulin G after hapten binding. *Mol. Immunol.* 18: 39–46

80. Pecht, I. 1982. Dynamic aspects of anti-

body function. In *The Antigens*, ed. M. Sela, 6: 1–68. New York: Academic Press. 453 pp.

81. Tschopp, J., Schulthess, R., Engel, J., Jaton, J. C. 1980. Antigen-independent activation of the first component of complement C1 by chemically cross-linked rabbit IgG oligomers. *FEBS Lett.* 112: 152–54

82. Couderc, J., Kazatchkine, M. D., Ventura, M., Duc, H. T., Maillet, F., Thobie, N., Liacopoulos, P. 1985. Activation of the human classical complement pathway by a mouse monoclonal hybrid IgG1-2a monovalent anti-TNP antibody bound to TNP-conjugated cells. *J. Immunol.* 134: 486–91

83. Watts, H. F., Anderson, V. A., Cole, V. M., Stevenson, G. T. 1985. Activation of complement pathways by univalent antibody derivatives with intact Fc zones. *Mol. Immunol.* 22: 803–10

84. Hanson, D. C., Shumaker, V. N. 1985. Immunoglobulin G antibody bound to the C1q subcomponent of human complement exhibits segmental flexibility. *J. Mol. Biol.* 183: 377–83

85. Circolo, A., Battista, P., Borsos, T. 1985. Efficiency of activation by anti-hapten antibodies at the red cell surface: Effect of patchy vs random distribution of hapten. *Mol. Immunol.* 22: 207–14

86. Hughes-Jones, N. C., Gorick, B. D., Howard, J. C. 1983. The mechanism of synergistic complement-mediated lysis of rat red blood cells by monoclonal IgG antibodies. *Eur. J. Immunol.* 13: 635–41

87. Dower, S. K., Segal, D. M. 1981. C1q binding to antibody coated cells—predictions from a single multivalent binding model. *Mol. Immunol.* 18: 823–29

88. Metzger, H. 1978. The effects of antigen on antibodies: Recent studies. *Contemp. Top. Mol. Immunol.* 7: 119–52

89. Painter, R. H. 1984. C1q receptor site on human immunoglobulin G. *Can. J. Biochem. Cell Biol.* 62: 418–25

90. Poon, P. H., Phillips, M. L., Schumaker, V. N. 1985. Immunoglobulin M possesses two binding sites for complement subcomponent C1q, and soluble 1:1 and 2:1 complexes are formed in solution at reduced ionic strength. *J. Biol. Chem.* 260: 9357–65

91. Borsos, T., Chapuis, R. M., Langone, J. J. 1981. Distinction between fixation of C1 and the activation of complement by natural IgM antihapten antibody: Effect of cell surface hapten density.

Mol. Immunol. 18: 18: 863–68

92. Karush, G., Chua, M.-M., Rodwell, J. D. 1979. Interaction of a bivalent ligand with IgM anti-lactose antibody. *Biochemistry* 18: 2226–32

93. Crossland, K. D., Koshland, M. E. 1983. Expression of Fc effector function in homogeneous murine anti-ars IgM. In *Protein Conformation as an Immunological Signal*, ed. F. Celada, V. N. Schumaker, E. E. Sercarz, pp. 59–72. New York: Plenum

94. Feinstein, A., Richardson, N. E., Gorick, B. D., Hughes-Jones, N. C. 1983. Immunoglobulin M conformational change is a signal for complement activation. See Ref. 93, pp. 47–57

95. Dodds, A. W., Sim, R. B., Porter, R. R., Kerr, M. A. 1978. Activation of the first component of human complement (C1) by antibody-antigen aggregates. *Biochem. J.* 175: 383–90

96. Tschopp, J. 1982. Kinetics of activation of the first component of complement (C1) by IgG oligomers. *Mol. Immunol.* 19: 651–57

97. Kasahara, Y., Takahashi, K., Nagasawa, S., Koyama, J. 1982. Formation of a conformationally changed C1r, a subcomponent of the first component of human complement as an intermediate of its autocatalytic reaction. *FEBS Lett.* 141: 128–31

98. Niinobe, M., Ueno, Y., Hitomi, Y., Fujii, S. 1984. Detection of intermediary C1r with complete active site, using a synthetic proteinase inhibitor. *FEBS Lett.* 172: 159–62

99. Castellino, F. J. 1979. A unique enzyme-protein substrate modifier reaction: Plasmin/streptokinase interaction. *TIBS* 4: 1–5

100. Hanson, D. C., Siegel, R. C., Schumaker, V. N. 1985. Segmental flexibility of the C1q subcomponent of human complement and its possible role in the immune response. *J. Biol. Chem.* 260: 3576–83

101. Campbell, R. D., Dodds, A. W., Porter, R. R. 1980. The binding of human complement component C4 to antibody-antigen aggregates. *Biochem. J.* 189: 67–80

102. Ziccardi, R. J. 1981. Activation of the early components of the classical complement pathway under physiologic conditions. *J. Immunol.* 126: 1769–73

103. Tenner, A. J., Frank, M. M. 1986. Activator-bound C1 is less susceptible to inactivation by C1 inhibition than is fluid-phase C1. *J. Immunol.* 137: 625–30

Ann. Rev. Immunol. 1987. 5 : 43–64

HISTOINCOMPATIBLE BONE MARROW TRANSPLANTS IN HUMANS

Reginald A. Clift and Rainer Storb

Fred Hutchinson Cancer Research Center, 1124 Columbia, Seattle, Washington 98104

INTRODUCTION

Bone marrow transplantation constitutes one of the strongest possible challenges to the mechanisms of transplantation immunity. Because the transplanted marrow can restore both hematopoietic and immunologic function, extremely powerful immunosuppressive techniques can be used to facilitate engraftment. However, despite the effectiveness with which the host immune response is ablated, the development of graft-versus-host disease by the newly implanted donor immune system exposes every host tissue to the mechanisms of immunologic rejection. In this setting antigenic disparities expressed on any tissue can influence the outcome of the transplantation procedure. All attempts to transplant human marrow except between identical twins involve such antigenic disparities and therefore could be described as histoincompatible.

This review surveys the results of human marrow transplantation where donor and recipient are not genotypically identical for all detectable antigens of the major human histocompatibility complex (MHC)—the human leukocyte antigen (HLA) system.

HISTORICAL PERSPECTIVE

The first marrow transplants were performed in mice in the early 1950s (1, 2). Infusions of syngeneic, allogeneic, or xenogeneic marrow were shown to protect irradiated mice against early mortality from total body irradiation. As a consequence it was hoped that marrow transplantation

43

0732–0582/87/0410–0043$02.00

could escape the histocompatibility constraints that restrict transplants of other tissues. The survival rates of mice receiving syngeneic marrow were soon shown to be far superior to those of recipients of allogeneic or xenogeneic marrow (3). Uphoff demonstrated that this was a consequence of graft-versus-host disease (GVHD), which did not develop when donor and recipient were identical for antigens of the H-2 system—the MHC of the mouse (4, 5).

The human MHC was not identified with any clarity until the mid-1960s, and the lack of any method for selecting human donor–recipient combinations that were identical for MHC was a major impediment to the development of clinical marrow transplantation. Despite this incapacity many attempts at human marrow transplantation were made during the decade ending in the early 1960s (6). In retrospect, most of the donor-recipient pairs clearly must have been HLA incompatible. The results were almost uniformly unsuccessful, with the notable exception of transplants between identical twins. Before the mid-1960s clinical practice had other pertinent limitations: The lack of a platelet and granulocyte transfusion technology and the inability to maintain platelet levels in irradiated patients were major obstacles to a successful outcome. Very few effective antibiotics were available, and attempts to prevent and treat gram-negative sepsis were particularly ineffective. Further progress in clinical marrow transplantation had to await advances in all these fields. The problems of histocompatibility needed to be addressed in a model that could accommodate the lack of syngeneic donors and provide a reasonable parallel to the human outbred situation. Studies in the dog suggested that, even in the absence of complete isogeneic identity, genotypic identity for the MHC could be beneficial (7).

The development of alloantisera for HLA typing (8, 9) and later of the mixed leukocyte culture test (MLC) (10, 11) pointed the way by the mid-1960s to selection of human MHC-"compatible" donor-recipient pairs. However, the technology was relatively primitive; the development of the concepts of matching by selection of inherited haplotypes (12) was most influential in opening up clinical marrow transplantation. Family typing to select HLA genotypically identical sibling donors permitted the use of multispecific antisera. The first successful human marrow transplants in which these concepts were used to select donors were reported in 1968. One patient with Wiskott Aldrich Syndrome (WAS) received a transplant from an HLA genotypically identical sibling (13), and a patient with severe combined immunodeficiency disease (SCID) received a transplant from a sibling with an HLA genotype that was haploidentical with one antigen mismatched (14). Santos and his group in Baltimore first reported the treatment of leukemia by use of marrow transplantation between HLA-identical siblings (15). The first report of such a transplant for severe aplastic anemia came from Thomas and his colleagues in Seattle (16). Since then the experience with human marrow transplantation has grown steadily; by the end of 1985 many thousands of transplants between siblings with identical HLA genotypes had been reported (17–19).

This burgeoning use of allogeneic marrow transplantation provides eloquent testimony to its increasing success. Several factors in addition to the selection of histocompatible donors have had an important influence on this improved record. The supportive care techniques before 1968 were very inadequate by the standards of 1985.

THE HLA SYSTEM

This review does not describe in detail the latest state of knowledge of the structure of the HLA system or techniques for identifying its components. For a complete treatment of current knowledge of the HLA system the reader is referred elsewhere (20). However, when marrow transplants are made from donors whose HLA genotypes are not identical with those of the recipients, understanding the results requires some consideration of the HLA system. Moreover, knowledge of this complex system has developed slowly over the 15 years covered by this review, and a perspective that accommodates this increased knowledge is especially important in considering phenotypic identity. The judgment that two individuals have phenotypically identical antigens is very sensitive to improvements in the acuity with which the antigens can be perceived. Thus, if the donor and recipient each had the antigen A19 in 1970, their HLA phenotypes were identical at the relevant locus. However, by 1981 one of them could be distinguished as A30 and the other as A31; thus, they would now be considered disparate at the A locus.

For most of the period under review the contemporary assumptions were that the HLA system was composed of antigens determined by alleles at four closely linked loci—HLA A, B, C, and D. The HLA A, B, and C antigens were identified by the use of alloantisera and were referred to as class-I determinants, whereas HLA D antigens were described as class-II determinants and were detected by the one-way MLC. In 1967 it was frequently impossible to detect two antigens at each class-I locus in circumstances where homozygosity could not be expected. Such "amorphs" were presumed to be a consequence of missing specificities in the available alloantisera. During the next decade the repertoire of antisera was enlarged and refined, and by 1980 amorphs were very unusual at the A and B loci. Increased use of unrelated donors has introduced some problems in identifying racially associated variants both of antigenic structure and of linkage disequilibrium; however, present capabilities for identifying the class-I products are both comprehensive and highly discriminating. The development of antisera that recognize antigens determined at the HLA-C locus has been disappointingly slow, and our understanding of the influence of this locus on marrow transplantation is inadequate. Failure to identify HLA-C antigens has been relatively unimportant in the recognition of genotypic identity because the locus lies between the A and B loci. However, there is an unmet need to evaluate the influence of the C locus when "unrelated" haplotypes are involved.

Our understanding of the HLA system has been revised most extensively in relation to the class-II products. In 1967 we only dimly perceived the existence of a locus, separate from the HLA-B locus, that influenced the MLC reaction. The development of techniques using homozygous typing cells (21) had little effect on the recognition of HLA genotypic identity, but their use introduced some difficult decisions in categorizing compatibility when unrelated haplotypes were being considered. The demonstration of the DR loci by B-cell typing (22) produced only a small increase in complexity because of the substantial genetic disequilibrium at the D and DR loci. The recent major expansion of knowledge of the D region with the delineation of the DP, DQ, and Dw loci cannot be evaluated yet, but it will presumably soon be extended by molecular biologists. Structural knowledge of the region of chromosome 6 that contains the HLA information as well as of the use of non-HLA determinants such as C4a, Bf, and GLO has been vital to understanding the inheritance of structural components of the MHC.

In selecting donors other than HLA genotypically identical siblings researchers have emphasized the concept of haplotype, which provides a useful basis for categorizing the nature of the "mismatch." Almost every patient has a relative with a haplotype presumably identical to one of the patient's own. Thus, the description in the literature of disparities for HLA antigens is usually couched in terms of the disparity at each of the detectable loci on one haplotype. Each haplotype is usually considered to be determined by alleles segregating at three loci—HLA A, B, and D. Little consideration is usually given to HLA C because of the location of the C locus within the haplotype; more importantly, the full complement of HLA-C antigens can only rarely be identified. For most of the period under review, disparity at the D region of the HLA complex has been evaluated in a yes/no fashion, with the recipient and donor regarded as either matched or mismatched (usually on the basis of the MLC reaction).

The use of the terms "haploidentical" and "haplomismatched" in reporting the use of histoincompatible donors has created much confusion. Usually the term haploidentical refers to genotypic identity for one haplotype with the range of possibilities for the other haplotype extending from phenotypic identity to disparity at each of the three major loci.

TREATMENT STRATEGIES FOR MISMATCHED TRANSPLANTS

To make marrow transplantation available to a larger population of patients, two main approaches have been employed. One involves searching for donors with close antigenic similarity to the patient and then using the same regimens employed for transplantation between HLA genotypically identical siblings. The other involves accepting a larger degree of antigenic disparity and devising special conditioning regimens to make such incompatibility tolerable. These approaches are to a large extent complementary.

The search for acceptable donors must start in the patient's family because this is where antigenic similarities are most likely to occur. Thus, fixing identity for one haplotype by using inheritance provides a very powerful tool for simplifying both the search and the analysis of results. Most transplant teams have adopted the policy of using two categories of donor. One category contains donors who are genotypically identical with the recipients for one HLA haplotype and are further classified on the basis of similarities on the nonidentical haplotype. The other category consists of donors unrelated to recipients with whom, however, they fortuitously share all definable HLA antigens. These categories differ in several important ways. The extent to which phenotypic identity parallels genotypic identity depends on the sophistication of the typing technique. However, genotypic identity for a haplotype has the potential for contributing more to the successful selection of donors than phenotypic identity even when the typing is extremely perceptive. Important but hitherto undetected histocompatibility information may be lodged within the linear information of the MHC. The influence of such information on the outcome of marrow transplantation has been shown in canine studies that demonstrate the superiority of DLA region identity by inheritance over phenotypic identity for detectable elements of the region (23). Of course, the use of related donors also has the advantage of using the shared inheritance of numerous other antigenic polymorphisms distributed throughout the genome. Interestingly, reports from Seattle (24–26) and elsewhere (27, 28) suggest that phenotypic identity between donor and recipient for one haplotype when there is genotypic identity for the other is superior to any other degree of mismatch. Yet phenotypic identity for both haplotypes where donor and recipient are unrelated has produced disappointing results (29).

Many attempts to design conditioning regimens have aimed at overcoming the two major disadvantages of histoincompatibility in marrow transplantation. One of these is the propensity for inadequate graft function and for graft rejection in mismatched transplants, and the other is the likelihood of increased incidence and severity of GVHD. These problems are linked because some of the approaches aimed at reducing the problem of GVHD have exacerbated the problem of achieving stable engraftment. Moreover, the phenomena of rejection and of severe GVHD are both influenced by the same factor—incompatibility for tissue antigens. Success in achieving engraftment in the patient with a propensity to reject is likely to involve a particularly high probability of severe acute GVHD. A basic conceptual problem exists in designing "new and improved" regimens for the prevention of GVHD in patients receiving mismatched transplants. GVHD is the major determinant of success in allogeneic marrow transplantation even when donor and recipient are identical in both HLA genotypes; a great need exists for regimens more effective in preventing and treating this condition in the recipients of histocompatible transplants. The use of regimens specifically designed to prevent GVHD in mismatched transplants implies that the inherent toxicity of these regimens prohibits their use in the matched situation. Protocols using such regimens are not favorable to a comparative evaluation of the risks of histoincompatibility.

MISMATCHED TRANSPLANTS FOR IMMUNODEFICIENCY

When an HLA genotypically identical donor is available, marrow transplantation is the treatment of choice for most forms of severe immunodeficiency disease (30) and the cure rate is better than 70% (31). Indeed, when the transplant is performed soon after birth, before the onset of infections, some groups have reported long-term disease-free survival rates of greater than 90% (32). Patients with most types of genetic immunodeficiency can undergo transplants without any pretransplant immunosuppressive conditioning; posttransplant immunosuppression for prophylaxis against GVHD is not usually given. Despite this, the incidence of GVHD is low. The reason for this is unclear but may reflect the very young age of most of these patients. Most patients achieve complete immune competence which is sometimes a result of cooperation between transplanted T cells and host type-B cells (33). Patients with certain types of severe immunodeficiency such as adenosine deaminase deficiency (ADA −) and WAS have the ability to reject tissue transplants; they therefore require immunosuppression in order to achieve successful marrow engraftment, even from donors with identical HLA genotypes. As mentioned above, the first report of the use of HLA genotypically identical marrow involved a patient with WAS who was conditioned with cyclophosphamide (CY) (13).

The gratifying degree of success using HLA-identical sibling donors was accompanied by frustration resulting from the fact that less than 40% of patients with SCID had fully matched donors; together these experiences encouraged early attempts at histoincompatible marrow transplantation. Successful marrow transplantation between siblings who were selected as well-matched on the basis of HLA typing was first reported in 1968 by a Minneapolis team. They treated a patient with severe combined immunodeficiency disease with marrow transplantation from his sister (14). These siblings were HLA genotypically identical at the B and D loci but mismatched at one of the A loci as a result of recombination. It was anticipated that disparity for class-I antigens would be more easily tolerated than class-II differences. By 1976 a total of 17 transplant teams in Europe and North America had submitted data to the International Bone Marrow Transplant Registry on 48 SCID patients treated with allogeneic marrow transplantation, and this experience was reported in 1977 (34). The 6-month survival rates were: 63% when donor and recipient were HLA genotypically identical, 38% when they were phenotypically identical or differed only for class-I antigens, and 5% for 19 recipients of marrow from donors differing for class-II determinants. The lesser degree of success using donors not identical with the recipient in HLA genotype was in part due to failure to achieve immune reconstitution but also was due to an increased incidence of GVHD. In a recent review, O'Reilly examined the results of transplantation from 21 related, partially matched donors (30 and 35, reviewed in 32). All donor-recipient pairs were HLA-D identical, and 10 were phenotypically identical for HLA A, B, and D. The long-term disease-free survival was 30% compared with 56% for a similar population with transplants from HLA genotypically identical siblings. The most common transplant-related problem was failure to engraft.

Animal studies have suggested that mature T lymphocytes in the donor marrow are responsible for the development of GVHD (36–38). Attempts to improve the results of marrow transplantation from donors other than those that are identical in both HLA genotypes have therefore concentrated on the removal of T cells from the marrow inoculum. These attempts have included treatment of the donor marrow with monoclonal antibodies directed at mature T lymphocytes in the presence of rabbit (39, 40) or human complement (41). Another widely practiced technique for T-cell depletion has involved processing the marrow with soybean lectin (42, 43). Depletion after E-rosette formation is often used to supplement these procedures. Failure to achieve engraftment as well as both early and late graft rejection have been major problems when similar techniques are used to facilitate transplantation from donors with identical HLA genotypes in the treatment of patients with leukemia (39). Such complications are less common when T cell–depleted, HLA-identical marrow is used for the immune reconstitution of children with SCID (35), presumably because the immunodeficiency state reduces the probability of immunological rejection. However when T-cell depletion is used with mismatched donors, difficulties have arisen in engraftment even in patients with SCID (31, 32, 41). A review of the European experience reported failure of engraftment in 17 of 63 marrow transplants for the treatment of immunodeficiency (44). The incidence of acute GVHD was low, however, and the success rate overall was high (61%). O'Reilly reviewed the outcome in 35 patients with SCID at 5 major centers who were transplanted with lectin-treated parental marrow mismatched for HLA A, B, and D; 29 were reported well, at home without GVHD (32). T-cell reconstitution occurred in 28, but only 10 exhibited B-cell reconstitution and humoral immune function. The circumstances of SCID lend themselves to repeated attempts at transplantation because failure of engraftment is not necessarily associated with myeloid aplasia. With histoincompatible transplants (especially when the marrow is T-cell depleted), some interesting features are the very slow response by many patients and the occasional success, despite repeated earlier failures, when fourth and fifth attempts are made at transplantation (32, 45).

All reports of the use of T cell–depleted marrow for treatment of SCID claim convincingly that incidence and severity of GVHD is markedly reduced. This reduction may be due to the increased frequency of graft rejection, which removes from analysis those transplants most likely to result in GVHD. It is frequently difficult to determine from published reports the degree of mismatch in these transplants. The donors are most commonly (and redundantly) described as haploidentical parents, which implies a substantial degree of antigenic disparity. However, antigen sharing is a common genetic curiosity in families containing children with congenital immunodeficiencies, and probably many of the donor/recipient combinations are not disparate at all major loci of the nonshared haplotype. The published results are certainly such that a randomized controlled trial of the effect of T-cell depletion on the incidence and severity of GVHD in this setting cannot be contemplated. These impressive results have stimulated studies of ways to overcome the problems of engraftment associated with T-cell depletion.

The circumstances of SCID are favorable to the availability of a parent as a marrow donor. Experience with the use of histocompatible unrelated donors for the treatment of this condition is rare and inadequate for useful analysis (32).

MISMATCHED TRANSPLANTS FOR APLASTIC ANEMIA

Marrow transplantation from an HLA genotypically identical donor is effective treatment for patients with severe aplastic anemia (46–48). When this type of therapy was first studied, graft rejection was a major cause of failure; approximately 40% of previously transfused patients rejected their grafts (49). This phenomenon was almost entirely limited to patients who had received blood-product transfusions before transplantation (50). Patients receiving HLA-identical marrow transplants for the treatment of leukemia very rarely rejected their grafts. These patients were prepared for transplantation with regimens containing both cyclophosphamide (CY) and total body irradiation (TBI), whereas patients with aplastic anemia were conditioned with high-dose CY only. Graft rejection mechanisms thus appeared more likely to be expressed when the conditioning regimen did not contain TBI. Experience with the use of conditioning regimens containing TBI for patients with aplastic anemia was very disappointing. Although the rejection rate for patients receiving TBI fell, the overall survival rate was low due to patients developing GVHD and interstitial pneumonia (51). Studies in the dog suggested that the infusion of suspensions of peripheral blood leukocytes from the donor after the marrow infusion might enhance engraftment (52). Following use of a regimen including the infusion of donor peripheral blood leukocytes after the marrow, the Seattle team lately reports that the 3-year disease-free survival rate is about 70% for previously transfused patients who receive marrow transplants from siblings who were identical in HLA genotype (51).

The use of high doses of antithymocyte globulin (ATG) provides an alternative treatment for patients with severe aplastic anemia (53–55). In Seattle this form of therapy is associated with a 3-year survival of approximately 50% (56). Full recovery of marrow function is slow, and the Seattle report is that if a response is not under way within 2 months, recovery is rare (56). The reason for the effectiveness of ATG therapy is not known but is perhaps a consequence of an immune etiology for the aplastic anemia. Methods of forecasting the response to ATG are being studied (57).

Experience with the use of histoincompatible marrow for the treatment of severe aplastic anemia has been disappointing. The Seattle experience is as follows: By November 1982, 17 patients with severe aplastic anemia had received transplants from donors with whom they were geno-typically identical for one haplotype and with whom they had similarities on the other (58, 59). Details of these transplants and the outcome are presented in Table 1. All except 1 of these patients were conditioned for transplant with the contemporary regimen for treatment of aplastic anemia (i.e. without TBI). Only 2 of these patients became long-term survivors, and these were the only recipients phenotypically identical with their donors for all detectable HLA antigens. Of the 15 patients who died, none survived more than 9 months from transplant, and only 3 survived more than 4 months. The principal causes of death were rejection of the marrow trans-plant and failure to engraft. This high incidence of graft failure is not surprising. In a setting where rejection occurs in transplants, with identical HLA genotypes, it is likely to be more common when disparity for antigens determined by the HLA system invites additional sensitization. For many years these early failures discouraged transplantation from mismatched

Table 1 Mismatched related transplants for aplastic anemia

Mismatched[a] antigens	UPN[b]	Age	Donor relationship	"Take"[c]	AGVH	Survival[d]	Comments
Phenotypically identical							
Nil	671	13	Mother	Yes	0	>3299	Alive & well
Nil	863	19	Sibling	Yes	3	>827	Alive with chronic GVHD
One antigen mismatched							
A	715	3	Mother	Yes	4	27	GVHD
A	755	32	Father	No	0	34	Rejection
A	813	14	Mother	Yes	4	227	GVHD
A	948	44	Mother	NE	NE	9	Infection
B	861	18	Father	No	0	20	Rejection
D	327	9	Sibling	No	0	68	Rejection
D	640	12	Sibling	Yes	4	18	GVHD
D	855	14	Mother	No	0	20	Rejection
D	996	14	Sibling	No	0	120	Rejection
D	1587	17	Sibling	Yes	2	96	Viral infection
D	1674	27	Sibling	Yes	3	130	GVHD
D	1859	11	Father	No	0	34	Rejection
Two antigens mismatched							
A & B	355	27	Father	No	0	35	Rejection
A & B	1478	23	Father	No	0		Rejection
Second graft from same donor				Yes	3	88	Infection
A & D	1212	20	Father	No	0	59	Rejection

[a] All patients were genotypically haploidentical with donor.
[b] Unique Patient Number.
[c] Evidence of successful engraftment.
[d] Days post-transplant as of 7/86.
[e] Present status or cause of death.

donors for patients with aplastic anemia. However, early favorable experience with a regimen including TBI for transplantation of patients with preleukemia has encouraged the use of such a regimen to study marrow transplantation for severe aplastic anemia in patients who failed to show a response within 2 months of their completing ATG treatment. Only 3 patients had been entered into this protocol by the end of 1985, and 2 of these survive beyond 1 year from transplant (R. Storb, unpublished observation).

Although the successful use of HLA phenotypically identical related (59) and unrelated donors has been reported (60–63), other transplant teams have encountered major problems in using other than HLA genotypically identical donors for marrow transplants to patients with aplastic anemia. Thus, in 1983 Filipovich reported the Minnesota experience in which 6 children with aplastic anemia were given transplants from mismatched related donors using conditioning regimens containing either TBI or a combination of CY and total lymphoid irradiation (TLI) (27). Of these patients 3 died without achieving engraftment and only one survived long-term. Hows has reported experience from Hammersmith, England (64, 65), in doing transplants for 13 patients with aplastic anemia from 8 matched unrelated donors and 5 mismatched related donors. One patient achieved autologous marrow reconstitution ; 6 patients had primary failure of marrow engraftment ; and 5 patients developed severe GVHD. There were 3 engrafted survivors. Surprising success can occasionally be obtained. Thus, in 1978 O'Reilly and his colleagues reported experience in transplanting marrow from an HLA-A, -B, and -D mismatched sibling into an infant with asplastic anemia ; they used a regimen of cytosine arabinoside, thioguanine, and CY (66). In this case the only similarity on the mismatched haplotype was in the GVHD direction, and this was due to HLA-D homozygosity in the patient. After moderately severe GVHD, the patient was alive with donor-type hemopoiesis at the time of the report, 8 months after transplant. This patient was only 18 months old at the time of transplant, and it may be that this was an important factor favoring success.

MISMATCHED TRANSPLANTS FOR LEUKEMIA

The design of treatment for patients with hematologic and other malignancies has been transformed by techniques to rescue patients from the marrow ablative effects of high-dose chemotherapy or radiotherapy. Marrow transplantation from an identical twin or a donor with an identical HLA genotype is an extremely important option for those patients with leukemia who have such a family member. It is unfortunate that less than

40% of patients in the United States and an even smaller proportion of patients in most other countries have such donors. This has engendered a major interest in seeking other sources of marrow for these patients. Autologous marrow can be used in some situations, particularly in the case of solid tumors that do not involve the marrow. However, for most hematologic malignancies, allogeneic transplantation is clearly preferable if a suitable donor can be found; there has thus been intense interest in studying histoincompatible marrow transplants. Several hundred such transplants have now been reported (26, 43, 67–69). Such extensive experience should permit conclusions about the increased hazard associated with histoincompatibility. Studies of this question require that patients with varying patterns of incompatibility be given transplants and that they undergo regimens of predictable efficacy. Most experience with this aspect of the problem has been accumulated by the Seattle group.

The prevailing theme of the Seattle studies has been that patients receiving HLA mismatched marrow should be treated with the same regimens used for conditioning patients with the same disease who are receiving genotypically identical marrow. Donor marrow was not T-cell depleted, and posttransplant immunosuppression consisted of methotrexate administered in a standard protocol (70). Reports have described two groups of patients. One consisted of patients who were HLA genotypically identical with their donors for one haplotype and who had a definable similarity on the other. The other group consisted of patients with no HLA genotypic identity who were phenotypically identical with their (usually unrelated) donors. The first report of these studies was published in 1979 (58), and the most recent summary of this experience appeared in 1986 (71). With regard to the patients receiving genotypically haploidentical marrow, several clearcut conclusions may be drawn from these studies. A surprising degree of success is possible when the donor and recipient are either phenotypically identical or mismatched for only one antigen on the unshared haplotype (71). For such patients the probability of survival is indistinguishable from that of patients with the same disease situation who are receiving marrow from HLA genotypically identical siblings. However, there was a significant difference in the incidence of acute GVHD (Figure 1). For patients with acute nonlymphoblastic leukemia (ANL) in first remission who were receiving marrow from genotypically identical sibling donors, the probabilities of developing grade 2 or worse GVHD were (a) 0.58 for patients over the age of 20 and (b) 0.38 for patients under this age (72). The probability of developing grade 2 or worse GVHD for mismatched patients receiving phenotypically identical marrow was 0.37. For patients transplanted while in remission, from donors with one antigen mismatched, this probability was 0.62. There was no detectable correlation between rates of survival or incidence of GVHD and the effect of antigens determined by different loci.

It is more difficult to evaluate whether, for patients receiving marrow from donors who are mismatched for 2 antigens, survival rates are worse than those of patients mismatched with their donors for one antigen. This is so because only patients with advanced disease were eligible to be transplanted in the face of larger antigenic disparity. When the survival was compared of patients with differing degrees of mismatching transplanted either in remission or in the chronic phase of chronic myelogenous leukemia (CML), it was shown that a significant difference existed between the results for patients incompatible at 1 locus and for those incompatible at 2 loci (71) (Figure 2). It is unfortunate that in this comparison nearly all the patients transplanted in first remission or chronic phase were mismatched for only one antigen. The probability of developing grade 2 or worse GVHD was greater (0.72) for patients mismatched for two antigens than for those with one antigen mismatched (71). Although this difference is relatively small, the probability of developing grade 4 GVHD was three times greater in patients with a two-antigen mismatch than with one (0.3 versus 0.1; $p < 0.005$) (72). A small number of transplants from donors with a three-antigen mismatch have been performed in Seattle, usually in young patients with advanced disease (R. Clift, personal observation). The number and the circumstances of such transplants do not permit statistical evaluation.

Graft rejection or failure to engraft are both extremely rare occurrences in patients receiving unmodified marrow from donors with identical HLA genotypes. In the Seattle series, the probability of delayed myeloid engraft-

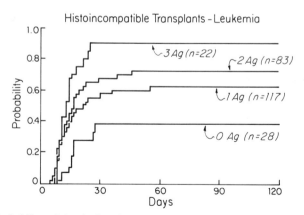

Figure 1 Probability of developing Grade 2 or worse acute GVHD for patients with leukemia receiving marrow transplants with differing degrees of histoincompatibility. All patients were genotypically haploidentical with their donors.

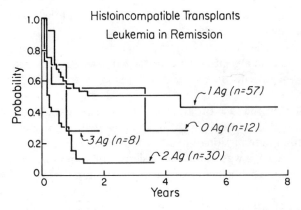

Figure 2 Probability of survival for patients with leukemia, transplanted in remission from donors with differing degrees of histoincompatibility. All patients were genotypically haploidentical with their donors.

ment or of graft rejection was significantly increased in patients receiving mismatched transplants compared with that of transplant patients with identical HLA genotypes (24, 71). Moreover, as Table 2 shows, the likelihood of such complications increased with the degree of mismatch (71). There has not yet been an analysis of the effect of vectors of compatibility in the Seattle cases. It would be interesting to know whether the incidence of GVHD was reduced in cases where the recipient was homozygous for HLA antigens and whether the incidence of rejection was reduced where the donor had such homozygosity.

The Seattle experience with the use of HLA phenotypically identical unrelated donors for the treatment of leukemia has been less extensive (71). The first report of such a case was a child with poor-risk acute

Table 2 Histoincompatible transplants for leukemia—quality of engraftment

Number of disparate antigens[a]	Number of transplants	Poor graft function[b] (%)	Graft failure[c]
0	26	7	1
1	117	20	3
2	83	22	13
3	22	10	0

[a] All patients were genotypically haploidentical with donor.
[b] Actuarial risk of granulocytes not reaching $1000/mm^3$ by day 40.
[c] Failure to engraft or rejection after engraftment.

lymphoblastic leukemia in first remission. This patient possessed two very common haplotypes and was given a transplant of marrow from a staff member with whom she was phenotypically identical at all detectable HLA loci (73). She had an uneventful course with no GVHD but died of recurrent disease on day 711. Together with the excellent results of transplantation from haploidentical donors phenotypically HLA identical for all detectable HLA antigens, this experience encouraged a study of the results of transplantation from phenotypically identical, unrelated donors. This is a cooperative study which involves several blood centers in the search for donors. Because the search for matched unrelated donors was expected to take a long time, it was decided to concentrate on patients with CML in chronic phase or ANL in first remission—two relatively stable disease situations. A preliminary report of the first 8 patients reveals a very high incidence of GVHD (71).

The only other reports of large numbers of patients transplanted from mismatched donors without the use of special regimens are from Powles and his colleagues (74). The study reported 34 patients treated for acute leukemia with marrow transplants from mismatched related donors with whom they were one haplotype HLA genotypically identical. These patients were conditioned with the same regimen employed for HLA genotypically identical transplants [CY and TBI with postgraft immunosuppression using Cyclosporine (CSA)]. In a follow-up (43) they reported that 5 of the 16 patients mismatched for all antigens, and 6 of 18 patients with some shared antigens on the nonidentical haplotype were alive beyond 2.5 years from transplant. They also reported a high incidence of massive pulmonary edema with increased vascular permeability and associated cerebral edema. This syndrome had not been reported from Seattle. The Seattle regimen used methotrexate instead of CSA and it was considered that this "leaky capillary syndrome" may have been a complication of CSA exacerbated by the presence of HLA incompatibility.

The Wisconsin group has reported the use of marrow T cell depleted with the monoclonal antibody anti–CT-2 and a very intensive conditioning regimen to treat 29 patients with hematologic malignancy (75). Of these, 8 had HLA genotypically identical donors; all engrafted promptly; none received posttransplant immunosuppression; none had significant acute GVHD, and 2 became long-term survivors. Of these 8 patients 2 developed lymphoproliferative disorders. Twenty-one patients received marrow from mismatched donors with varying degrees of incompatibility. Engraftment was a major problem in this group. Of the 17 patients who lived for more than 30 days, 4 failed to achieve engraftment, and several of those with successful grafts developed severe GVHD.

Engraftment failure is clearly a problem in patients with leukemia receiving marrow, depleted of T cells by a variety of techniques, from HLA identical donors (76–79). The concept that at least part of this failure is immunologically mediated and may respond to more intensive pre-transplant immunosuppression is reasonable, supported by the results of animal experiments, and consistent with but not proven by the results of clinical trials (25, 80–82). Attempts to overcome this problem have there-fore focussed on using conditioning regimens with a greater immuno-suppressive capability. Such regimens have employed more chemotherapy, and/or higher doses of TBI or TLI (82–85). Some of these approaches have produced initially encouraging results, but there is inadequate ex-perience to justify conclusions as to their effectiveness.

Studies of the use of HLA phenotypically identical unrelated donors for the treatment of leukemia have also used T-cell depletion with results not dissimilar to those with related mismatched donors (65).

DISCUSSION

Substantial progress has been made in unravelling the complexities associ-ated with the use of donors other than HLA-identical siblings. Ten years ago it was generally accepted that transplants in which the donor and recipient were mismatched for any HLA determinants were likely to be extremely hazardous. The only type of incompatibility believed to be even marginally acceptable was for other than a single class-II determinant, and much importance was attached to the level of reactivity in the MLC. Today it is clear that incompatibility for the products of a single HLA locus in a haploidentical combination has no impact on the survival of patients transplanted for the treatment of leukemia, although the incidence of GVHD is increased in such transplants. It is also clear that there is no difference between disparity for class-I and for class-II products in this regard. This apparent lack of impact on survival from doubling of the incidence of acute GVHD is testimony to the effectiveness of current regimens for the treatment of grade 2 GVHD. The difference in results for mismatched transplants between patients transplanted for leukemia and those transplanted for the treatment of aplastic anemia is real and appears to be a consequence of the increased rejection rate for aplastic anemia patients. The reason for this is probably the difference in conditioning regimens, but this has not been clearly demonstrated to be the cause. The success of high-dose ATG therapy for aplastic anemia has obstructed the study of mismatched transplants in this disease. However, the current studies in Seattle involve aplastic patients who have failed ATG therapy and who are prepared with the same regimens used for transplanting patients with leukemia; these studies may shed light on the rejection problem.

The objective of all these efforts is to extend the option of marrow transplantation to the 60–70% of patients who lack HLA identical siblings. This objective will not be much advanced by demonstrating that the use of minimally matched relatives can be made successful because the number of patients with such donors will be extremely small. However, the strategy that should be pursued in this field is plain. Regimens are needed that will permit successful transplantation between donors and recipients without regard to the degree of disparity for HLA antigens. In this endeavor it is important to isolate the causes of failure. An evaluation of regimens for the prevention and treatment of GVHD is best made in circumstances where the problems of rejection have been overcome. Selection of the best setting for such a study is crucial. Disparity for only one antigen probably provides too small a challenge, and the most profitable setting is likely to be one where donor and recipient are HLA genotypically identical for one haplotype and mismatched for two antigens on the other. It is important to develop conditioning regimens that will ensure consistently successful engraftment in this category of transplant. When this result has been achieved, regimens for the prevention of GVHD in this setting can be studied. The objective is that these regimens should produce results in mismatched related transplants no worse than those between HLA genotypically identical pairs. Even when this has been accomplished, it is quite likely that the results of transplants from unrelated donors will be worse than those when donor and recipient are related with the same degree of HLA disparity. This would reflect the greater likelihood of mismatching for presently undetectable antigenic systems in the unrelated setting. Mismatching in these systems produces the GVHD and rejection responsible for so much morbidity and mortality in matched related transplants. Progress in reducing the immunological hazards of HLA genotypically identical transplants should have a major impact in reducing these hazards in the unrelated setting.

When the mechanisms of antigen-mediated immunologic rejection are understood and tamed, it is likely that new toxic consequences of histoincompatibility will be uncovered. In particular it is likely that the cellular cooperation so necessary for the effective functioning of one of the most complex organs in the body will be impaired when the cooperating cells are genetically different with a resulting increase in infection and malignant transformation.

Literature Cited

1. Lorenz, E., Uphoff, D., Reid, T. R., Shelton, E. 1951. Modification of irradiation injury in mice and guinea pigs by bone marrow injections. *J. Natl. Cancer Inst.* 12: 197

2. Congdon, C. C., Lorenz, E. 1954. Humoral factor in irradiation protection: Modification of lethal irra-

diation injury in mice by injection of rat bone marrow. *Am. J. Physiol.* 176: 298

3. Barnes, D. W. H., Loutit, J. F. 1955. Spleen protection. The cellular hypothesis. In *Radiology Symposium 1954*, ed. Z. M. Bacq, P. Alexander, p. 134. London: Butterworth

4. Uphoff, D. E. 1957. Genetic factors influencing irradiation protection by bone marrow. I. The F_1 hybrid effect. *J. Natl. Cancer Inst.* 19: 123

5. Uphoff, D. E., Law, L. W. 1958. Genetic factors influencing irradiation protection by bone marrow. II. The histocompatibility-2 (H-2) locus. *J. Natl. Cancer Inst.* 20: 617

6. Bortin, M. M. 1970. A compendium of reported human bone marrow transplants. *Transplantation* 9: 571

7. Epstein, R. B., Storb, R., Ragde, H., Thomas, E. D. 1968. Cytotoxic typing antisera for marrow grafting in littermate dogs. *Transplantation* 6: 45

8. Dausset, J. 1958. Iso-leuco-anticorps. *Acta Haematol.* (Basel) 20: 156

9. Amos, D. B., Bashir, H., Boyle, W., MacQueen, M., Tiilikainen, A. 1969. A simple micro cytotoxicity test. *Transplantation* 7: 220

10. Bach, F. H., Voynow, N. K. 1966. One-way stimulation in mixed leukocyte cultures. *Science* 153: 545

11. Dupont, B. 1980. HLA factors and bone marrow grafting. In *Cancer: Achievements, Challenges and Prospects for the 1980s*, ed. J. H. Burchenal, H. F. Oettgen, p. 683. New York: Grune & Stratton

12. Ceppellini, R., van Rood, J. J. 1974. The HL-A system. I. Genetics and molecular biology. *Semin. Hematol.* 11: 233

13. Bach, F. H., Albertini, R. J., Anderson, J. L., Joo, P., Bortin, M. M. 1968. Bone-marrow transplantation in a patient with the Wiskott-Aldrich syndrome. *Lancet* ii: 1364

14. Gatti, R. A., Allen, H. D., Meuwissen, H. J., Hong, R., Good, R. A. 1968. Immunological reconstitution of sex-linked lymphopenic immunological deficiency. *Lancet* ii: 1366

15. Santos, G. W., Burke, P. J., Sensenbrenner, L. L., Owens, A. H. 1969. Marrow transplantation and graft-versus-host disease in acute monocytic leukemia. *Exp. Hematol.* 18: 20

16. Thomas, E. D., Buckner, C. D., Storb, R., Neiman, P. E., Fefer, A., Clift, R. A., Slichter, S. J., Funk, D. D., Bryant, T. I., Lerner, K. G. 1972. Aplastic anaemia treated by marrow transplantation. *Lancet* i: 284

17. Bacigalupo, A., Van Lint, M. T., Congiu, M., Pittaluga, P. A., Occhini,

D., Marmont, A. M. 1986. Treatment of SAA in Europe 1970–1985: A report of the SAA Working Party. In *Proceedings of XII Annual Meeting of the European Cooperative Group for Bone Marrow Transplantation, Courmayeur, Italy, 1986*, p. 19. London: MacMillan

18. Gratwohl, A., Zwaan, F. G., Hermans, J., Lyklema, A. 1986. Bone marrow transplantation for leukaemia in Europe. Report from the Leukaemia Working Party. See Ref. 17, p. 177

19. Gale, R. P., Champlin, R. E. 1983. Bone marrow transplantation in leukemia. Critical analysis and controlled clinical trials. In *Recent Advances in Bone Marrow Transplantation, UCLA Symposia on Molecular and Cellular Biology, New Series*, Vol. 7, ed. R. P. Gale, p. 71. New York: Liss

20. Bodmer, W. F. 1984. The HLA system, 1984. In *Histocompatibility Testing 1984*, ed. E. D. Albert, M. P. Baur, W. R. Mayr, p. 11. Berlin: Springer Verlag

21. Dupont, B., Hansen, J. A., Yunis, E. J. 1976. Human mixed-lymphocyte culture reaction: Genetics, specificity, and biological implications. *Adv. Immunol.* 23: 107

22. Danilovs, J. A., Ayoub, G., Terasaki, P. I. 1980. B lymphocyte isolation by thrombin-nylon wool. In *Histocompatibility Testing 1980*, ed. P. I. Terasaki, p. 287. Los Angeles: Univ. Calif. Los Angeles Press

23. Storb, R., Weiden, P. L., Graham, T. C., Lerner, K. G., Thomas, E. D. 1977. Marrow grafts between DLA-identical and homozygous unrelated dogs. Evidence for an additional locus involved in graft-versus-host disease. *Transplantation* 24: 165

24. Clift, R. A., Hansen, J. A., Thomas, E. D., Buckner, C. D., Sanders, J. E., Mickelson, E. M., Storb, R., Singer, J. W., Goodell, B. W. 1979. Marrow transplantation from donors other than HLA-identical siblings. *Transplantation* 28: 235

25. Hansen, J. A., Clift, R. A., Beatty, P. G., Mickelson, E. M., Nisperos, B., Martin, P. J., Thomas, E. D. 1983. Marrow transplantation from donors other than HLA genotypically identical siblings. See Ref. 19, p. 739

26. Beatty, P. G., Clift, R. A., Mickelson, E. M., Nisperos, B., Flournoy, N., Martin, P. J., Sanders, J. E., Stewart, P., Buckner, C. D., Storb, R., Thomas, E. D., Hansen, J. A. 1985. Marrow transplantation from related donors other than HLA-identical siblings. *N. Engl. J. Med.* 313: 765

27. Filipovich, A. H., Ramsay, N. K. C.,

McGlave, P., Quinones, R., Winslow, C., Heintz, K. J., Arthur, D., Kersey, J. H. 1983. Mismatched bone marrow transplantation at the University of Minnesota : Use of related donors other than HLA MLC identical siblings and T cell depletion. See Ref. 19, p. 769

28. Delfini, C., Nicolini, G., Lucarelli, G., Galimberti, M., Paradisi, O., Donati, M., Valentini, M. 1986. Bone marrow donors other than HLA-genotypically identical siblings for patients with thalassaemia. See Ref. 17, p. 126

29. Hansen, J. A., Thomas, E. D. 1983. The immunogenetics of clinical bone marrow transplantation. In Bone Marrow Transplantation, A Technical Workshop, ed. R. S. Weiner, E. Hackel, C. A. Schiffer, p. 23. Arlington, Va : Com. Tech. Workshops, Am. Assoc. Blood Banks

30. Good, R. A., Kapoor, N., Pahwa, R. N., West, A., O'Reilly, R. J. 1981. Current approaches to the primary immunodeficiencies. In Progress in Immunology IV, ed. M. Fougereau, J. Dausset, p. 907. New York : Academic Press

31. Parkman, R., Rappeport, J. M., Geha, R., Rosen, F. S., Nathan, D. G. 1983. Bone marrow transplantation. Lymphoid and hematopoietic immunodeficiencies. See Ref. 19, p. 173

32. O'Reilly, R. J., Brochstein, J., Dinsmore, R., Kirkpatrick, D. 1984. Marrow transplantation for congenital disorders. Semin. Hematol. 21 : 188

33. Griscelli, C., Durondy, A., Virelizyer, J. L., Ballet, J. J., Daguillard, F. Selective depletion of precursor T cells associated with apparently normal B lymphocytes in severe combined immunodeficiency. J. Pediatr. 93 : 404

34. Bortin, M. M., Rim, A. A. 1977. Severe combined immunodeficiency disease. J. Am. Med. Assoc. 238 : 591

35. O'Reilly, R. J., Kapoor, N., Kirkpatrick, D., Flomenberg, N., Pollack, M. S., Dupont, B., Good, R. A., Reisner, Y. 1983. Transplantation of hematopoietic cells for lethal congenital immunodeficiencies. Birth Defects 19 : 129

36. Korngold, R., Sprent, J. 1978. Lethal graft-versus-host disease after bone marrow transplantation across minor histocompatibility barriers in mice : Prevention by removing mature T cells from marrow. J. Exp. Med. 148 : 1687

37. Kolb, H. J., Rieder, I., Netzel, B., Grosse-Wilde, H., Scholtz, S., Schaffer, E., Kolb, H., Thierfelder, S. 1979. Antilymphocytic antibodies and marrow transplantation : VI. Graft-versus-host tolerance in DLA-incompatible dogs after in vitro treatment of bone marrow with absorbed antithymocyte globulin. Transplantation 27 : 242

38. Wagemaker, G., Vriesendorp, H. M., van Bekkum, D. W. 1981. Successful bone marrow transplantation across major histocompatibility barriers in rhesus monkeys. Transplant. Proc. 13 : 875

39. Martin, P. J., Hansen, J. A., Buckner, C. D., Sanders, J. E., Deeg, H. J., Stewart, P., Appelbaum, F. R., Clift, R., Fefer, A., Witherspoon, R. P., Kennedy, M. S., Sullivan, K. M., Flournoy, N., Storb, R., Thomas, E. D. 1985. Effects of in vitro depletion of T cells in HLA-identical allogeneic marrow grafts. Blood 66 : 664

40. Martin, P. J., Hansen, J. A., Storb, R., Thomas, E. D. 1985. A clinical trial of in vitro depletion of acute graft-versus-host disease (GVHD). Transplant. Proc. 17 : 486

41. Morgan, G., Linch, D. C., Knott, L. T., Davies, E. G., Sieff, C., Chessells, J. M., Hale, A., Waldman, H., Levinsky, R. J. 1986. Successful haploidentical mismatched bone marrow transplantation in severe combined immunodeficiency : T cell removal using CAMPATH-I monoclonal antibody and E-rosetting. Br. J. Haematol. 62 : 421

42. Reisner, Y., Kapoor, N., Kirkpatrick, D., Pollack, M. S., Cunningham-Rundles, S., Dupont, B., Hodes, M. Z., Good, R. A., O'Reilly, R. J. 1983. Transplantation for severe combined immunodeficiency with HLA-A,B,D,DR incompatible parental marrow cells fractionated by soybean agglutinin and sheep red blood cells. Blood 6 : 341

43. Powles, R., Pedrazzini, A., Crofts, M., Clink, H., Millar, J., Bhattia, G., Perez, D. 1984. Mismatched family bone marrow transplantation. Semin. Hematol. 21 : 182

44. Fischer, A. (For the Working Party of the EGBMT) 1986. Bone marrow transplantation for immunodeficiencies in Europe 1969–1985. See Ref. 17, p. 133

45. O'Reilly, R. J., Dupont, B., Pahwa, S., Grimes, E., Smithwick, E. M. 1977. Reconstitution in severe combined immunodeficiency by transplantation of marrow from an unrelated donor. N. Engl. J. Med. 297 : 182

46. Storb, R., Thomas, E. D., Buckner, C. D., Appelbaum, F. R., Clift, R. A., Deeg, H. J., Doney, K., Hansen, J. A., Prentice, R. L., Sanders, J. E., Stewart, P., Sullivan, K. M., Witherspoon, R. P. 1984. Marrow transplantation for aplastic anemia. Semin. Hematol. 21 : 27

47. Storb, R., Thomas, E. D., Weiden, P. L., Buckner, C. D., Clift, R. A., Fefer, A., Fernando, L. P., Giblett, E. R.,

Goodell, B. W., Johnson, F. L., Lerner, K.G., Neiman, P. E., Sanders, J. E. 1976. Aplastic anemia treated by allogeneic bone marrow transplantation: A report on 49 new cases from Seattle. *Blood* 48: 817

48. UCLA Bone Marrow Transplant Team. 1976. Bone marrow transplantation in severe aplastic anemia. *Lancet* ii: 921

49. Storb, R., Thomas, E. D., Buckner, C. D., Clift, R. A., Fefer, A., Fernando, L. P., Giblett, E. R., Johnson, F. L., Neiman, P. E. 1976. Allogeneic marrow grafting for treatment of aplastic anemia: A follow-up on long-term survivors. *Blood* 48: 485

50. Storb, R., Thomas, E. D., Buckner, C. D., Clift, R. A., Deeg, H. J., Fefer, A., Goodell, B. W., Sale, G. E., Sanders, J. E., Singer, J., Stewart, P., Weiden, P. L. 1980. Marrow transplantation in thirty "untransfused" patients with severe aplastic anemia. *Ann. Intern. Med.* 92: 30

51. Storb, R. 1979. Decrease in the graft rejection rate and improvement in survival after marrow transplantation for severe aplastic anemia. *Transplant. Proc.* 11: 196

52. Storb, R., Epstein, R. B., Bryant, J., Ragde, H., Thomas, E. D. 1968. Marrow grafts by combined marrow and leukocyte infusions in unrelated dogs selected by histocompatibility typing. *Transplantation* 6: 587

53. Speck, B., Gratwohl, A., Nissen, C., Leibundgut, R., Ruggero, D., Osterwalder, B., Burri, H., Cornu, P., Jeannet, M. 1981. Treatment of severe aplastic anemia with antilymphocyte globulin or bone-marrow transplantation. *Br. Med. J.* 282: 860

54. Camitta, B., O'Reilly, R. J., Sensenbrenner, L., Rappeport, J., Champlin, R., Doney, K., August, C. 1982. Severe aplastic anemia: A controlled trial of antilymphocyte globulin therapy. *Blood* 60(Suppl. A): 165 (Abstr. 588)

55. Champlin, R., Ho, W., Gale, R. P. 1983. Antithymocyte globulin treatment in patients with aplastic anemia. A prospective randomized trial. *N. Engl. J. Med.* 308: 113

56. Doney, K., Dahlberg, S. J., Monroe, D., Storb, R., Buckner, C. D., Thomas, E. D. 1984. Therapy of severe aplastic anemia with anti-human thymocyte globulin and androgens: The effect of HLA-haploidentical marrow infusion. *Blood* 63: 342

57. Doney, K. C., Torok-Storb, B., Dahlberg, S., Buckner, C. D., Martin, P., Hansen, J. A., Thomas, E. D., Storb, R. 1984. Immunosuppressive therapy of severe aplastic anemia. In *Aplastic Anemia: Stem Cell Biology and Advances in Treatment*, ed. N. S. Young, A. S. Levine, R. K. Humphries, p. 259. New York: Liss.

58. Clift, R. A., Hansen, J. A., Thomas, E. D. (For the Seattle Marrow Transplant Team) 1981. The role of HLA in marrow transplantation. *Transplant. Proc.* 13: 234

59. Storb, R., Thomas, E. D., Appelbaum, F. R., Clift, R. A., Deeg, H., Doney, K., Hansen, J. A., Prentice, R. L., Sanders, J. E., Singer, J. W., Shulman, H., Stewart, P. S., Sullivan, K. M., Dahlberg, S. J., Buckner, C. D., Witherspoon, R. P. 1984. Marrow transplantation for severe aplastic anemia: The Seattle experience. See Ref. 57, p. 297

60. Tricot, G. J. K., Jansen, J., Zwaan, F. E., Eernisse, J. G., Sabbe, L., Van Rood, J. J. 1981. Successful bone marrow transplantation for aplastic anemia using an HLA phenotypically identical parent. *Transplantation* 31: 86

61. Duquesnoy, R. J., Zeevi, A., Marrari, M., Hackbarth, S., Camitta, B. 1983. Bone marrow transplantation for severe aplastic anemia using a phenotypically HLA-identical SB-compatible unrelated donor. *Transplantation* 35: 566

62. Gordon-Smith, E. C., Fairhead, S. M., Chipping, P. M., Hows, J., James, D. C. O., Dodi, A., Batchelor, J. R. 1982. Bone-marrow transplantation for severe aplastic anemia using histocompatible unrelated volunteer donors. *Br. Med. J.* 285: 835

63. Speck, B., Zwaan, F. E., van Rood, J. J., Eernisse, J. G. 1973. Allogeneic bone marrow transplantation in a patient with aplastic anemia using a phenotypically HL-A-identical unrelated donor. *Transplantation* 16: 24

64. Hows, J., Yin, J., Marsh, J., Fairhead, S., Batchelor, J. R., Gordon-Smith, E. C. 1986. Treatment of acquired aplastic anaemia at Hammersmith, 1980–1985. See Ref. 17, p. 22

65. Hows, J. M., Yin, J., Jones, L., Apperley, J., Econimou, K., James, D. C. O., Batchelor, J., Goldman, J. M., Gordon-Smith, E. C. 1986. Allogeneic bone marrow transplantation with volunteer unrelated donors. See Ref. 17, p. 125

66. O'Reilly, R. J., Pahwa, D., Kirkpatrick, M., Sorell, A., Kapadia, N., Kapoon, N., Hansen, J. A., Pollack, M., Schutzer, S. E., Good, R. A., Dupont, B. 1978. Successful transplantation of marrow from an HLA-A, -B, -D mismatched

heterozygous sibling donor into an HLA-D-homozygous patient with aplastic anemia. *Transplant. Proc.* 10: 957

67. Powles, R. L., Morgenstern, G. R., Crofts, M., Evans, B., Lawler, S. D. 1983. Mismatched family bone marrow transplantation. See Ref. 19, p. 757

68. Gingrich, R. D., Howe, C. W. S., Goeken, N. E., Ginder, G. D., Kugler, J. W., Tewfik, H. H., Klaasen, L. W., Armitage, J. D., Fyfe, M. A. 1985. The use of partially matched, unrelated donors in clinical bone marrow transplantation. *Transplantation* 39: 526

69. Barrett, A. J., Beard, J., McCarthy, D., Shaw, P. J., Hugh-Jones, K., Hobbs, J. R., James, D. C. O. 1986. Bone marrow transplant for leukaemia with matched unrelated donors. See Ref. 17, p. 127

70. Thomas, E. D., Storb, R., Clift, R. A., Fefer, A., Johnson, F. L., Neiman, P. E., Lerner, K. G., Glucksberg, H., Buckner, C. D. 1975. Bone-marrow transplantation. *N. Engl. J. Med.* 292: 832, 895

71. Hansen, J. A., Beatty, P. G., Anasetti, C., Clift, R. A., Martin, P. J., Sanders, J., Sullivan, K., Buckner, C. D., Storb, R., Thomas, E. D. 1986. Treatment of leukemia by marrow transplantation from donors other than HLA genotypically identical siblings. In *Recent Advances in Bone Marrow Transplantation, UCLA Symposia on Molecular and Cellular Biology* New Series, ed. R. P. Gale, R. Champlin, Vol. 53. Alan R. Liss, Inc. In press

72. Clift, R. A., Beatty, P. G., Thomas, E. D., Buckner, C. D., Weiden, P., McGuffin, R. (For the Seattle Marrow Transplant Team) 1985. Marrow transplantation from mismatched donors for the treatment of malignancy. *Transplant. Proc.* 17: 445

73. Hansen, J. A., Clift, R. A., Thomas, E. D., Buckner, C. D., Storb, R., Giblett, E. R. 1980. Transplantation of marrow from an unrelated donor to a patient with acute leukemia. *N. Engl. J. Med.* 303: 565

74. Powles, R. L., Kay, H. E. M., Clink, H. M., Barrett, A., Depledge, M. H., Sloane, J., Lumley, H., Lawler, S. D., Morgenstern, G. R., McIlwain, T. J., Dady, P. J., Jamieson, B., Watson, J. C., Leigh, M., Hedley, D., Filshie, J. 1983. Mismatched family donors for bone-marrow transplantation as treatment for acute leukaemia. *Lancet* i: 612

75. Trigg, M. E., Billing, R., Sondel, P. M., Exten, R., Hong, R., Bozdech, M. J.,

Horowitz, S. D., Finlay, J. L., Moen, R., Longo, W., Erickson, C., Peterson, A. 1985. Clinical trial depleting T lymphocytes from donor marrow for matched and mismatched allogeneic bone marrow transplants. *Cancer Treat. Rep.* 69: 377

76. Martin, P. J., Hansen, J. A., Buckner, C. D., Sanders, J. E., Deeg, H. J., Stewart, P., Appelbaum, F. R., Clift, R., Fefer, A., Witherspoon, R. P., Kennedy, M. S., Sullivan, K. M., Flournoy, N., Storb, R., Thomas, E. R. 1985. Effects of in vitro depletion of T cells in HLA-identical allogeneic marrow grafts. *Blood* 66: 664

77. Martin, P. J., Hansen, J. A., Storb, R., Thomas, E. D. 1985. A clinical trial of in vitro depletion of T cells in donor marrow for prevention of acute graft-versus-host disease. *Transplant. Proc.* 17: 486

78. Patterson, J., Prentice, H. G., Brenner, M. K., Gilmore, M., Blacklock, H., Hoffbrand, A. V., Janossy, G., Skeggs, D., Ivory, K., Apperley, J., Goldman, J., Burnett, A., Gribben, J., Alcorn, M., Pearson, C., McVickers, I., Hann, I., Reid, C., Wendle, D., Bacagalupo, A., Robertson, A. G. 1985. Graft rejection following T cell depleted BMT. *Br. J. Haematol.* 61: 562

79. Hale, G., Waldmann, H. 1986. Depletion of T-cells with Campath-1 and human complement. Analysis of GvHD and graft failure in a multi-centre study. See Ref. 17, p. 93

80. Barrett, A. J., Poynton, C. H., Flanagan, P., Shaw, P. J., McCarty, D. 1986. Factors associated with rejection in bone marrow depleted of T lymphocytes by Campath-1 monoclonal antibody. See Ref. 17, p. 95

81. Reisner, Y., Kapoor, N., Pollack, S., Freidrich, W., Kirkpatrick, D., Shank, B., Csurny, R., Pollack, M. S., Dupont, B., Good, R. A., O'Reilly, R. J. 1986. See Ref. 71, p. 355

82. Bozdech, M. J., Sondel, P. M., Trigg, M. E., Longo, W., Kohler, P. C., Flynn, B., Billing, R., Anderson, S. A., Hank, J. A., Hong, R. 1985. Transplantation of HLA-identical T-cell-depleted marrow for leukemia: Addition of cytosine arabinoside to the pretransplant conditioning prevents rejection. *Exp. Hematol.* 13: 1201

83. Sondel, P. M., Bozdech, M. J., Trigg, M. E., Hong, R., Finlay, J. L., Kohler, P.C., Longo, W., Hank, J. A., Billing, R., Steeves, R., Flynn, B. 1985. Additional immunosuppression allows engraftment following HLA-mismatched T cell-

depleted bone marrow for leukemia. *Transplant. Proc.* 17 : 460

84. Martin, P. J., Hansen, J., Storb, R., Thomas, E. D. 1986. Human marrow transplantation : An immunological perspective. In *Advances in Immunology.* In press

85. Or, R., Weshler, Z., Lugassy, G., Steiner-Salz, D., Galun, E., Weiss, L., Sanuel, S., Pollack, A., Rachmilewitz, E. A., Waldmann, H., Slavin, H. 1985. Total lymphoid irradiation (TLI) as adjunct immunosuppressor for preventing late graft failure (LGF) associated with T-cell depleted marrow allograft. *Exp. Hematol.* 13 : 409 (Abstr. 209)

Ann. Rev. Immunol. 1987. 5 : 65–84

LIPID MEDIATORS PRODUCED THROUGH THE LIPOXYGENASE PATHWAY

Charles W. Parker

Howard Hughes Medical Institute at Washington University School of Medicine, 4939 Audubon, St. Louis, Missouri 63110

INTRODUCTION

The most important known lipid mediators of inflammation are either metabolites of long-chain fatty acids, particularly arachidonic acid (AA), or 1-alkyl-2-acetyl analogues of phosphatidyl choline (platelet activating factor—PAF) (1–8). Because of the scope of the subject, this chapter reviews only the mediators derived from AA and emphasizes lipoxygenase products.

Arachidonate Metabolites

Arachidonic acid (eicosatetraenoic acid) is a C20 fatty acid with double bonds at the 5–6, 8–9. 11–12. and 14–15 positions. The multiplicity of double bonds provides a number of potential sites for oxidation which, together with the various double bond rearrangements that are possible, permits a number of bioreactive lipids to be formed. Since AA itself and most of its initial metabolites have 20 carbons, its metabolites have been termed eicosanoids. The two major routes of AA metabolism in mammalian cells are the lipoxygenase (LO) and the cyclooxygenase (CO) pathways (Figure 1). Both enzymes act largely or entirely on unesterified AA. Lipoxygenases differ in the location of the double bond on the AA molecule where enzymatic attack is initiated (9, 10). They include the 12-lipoxygenases (12-LO, Figure 1—top left) of which the enzyme in platelets is the best characterized; the 5-lipoxygenase (5-LO, Figure 1—top right) which is prominent in leukocytes and mast cells; and a 15-LO which is also found in leukocytes, tracheal epithelial cells and other cell types (not shown).

65

0732–0582/87/0410–0065$02.00

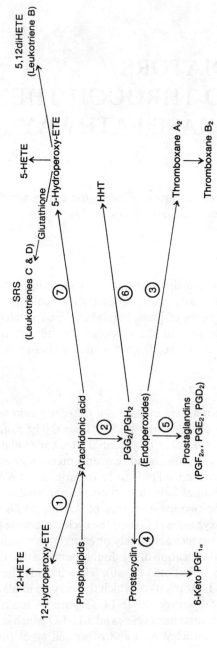

Figure 1 Pathways of arachidonate metabolism. (From 5, as modified in 6.)

Recent studies have indicated that products of the 5-lipoxygenase (5-LO) pathway are of special significance in allergic and inflammatory responses but that other lipoxygenases and their products may be also important. The lipoxygenases are primarily soluble enzymes that differ from one another not only in the site of oxidation but also in the mechanism for their activation. Lipoxygenases produce monohydroperoxy fatty acids (HPETEs) as their initial products; these in turn are converted to monohydroxy, dihydroxy, and trihydroxy fatty acids (HETEs, diHETES and triHETEs, respectively). Keto and epoxy derivatives may also be formed. The term *leukotriene* has been applied to lipoxygenase derivatives with at least three conjugated double bonds; these include leukotriene B and the sulfidoleukotrienes (leukotrienes C, D, and E). The 5-HPETE readily rearranges enzymatically and nonenzymatically to the epoxide leukotriene A_4 (LTA_4), which is then converted to the other 5-LO -products. Whereas a single lipoxygenase can produce di- and tri- as well as mono-HETEs, LTB_4 formation involves a widely distributed third enzyme, (LTA_4 hydrolase, LTB_4 synthetase), in addition to 5-LO and LTA_4 synthetase (11). Sequential action of 15 and 5, or 12 and 5, lipoxygenases in different cell types can also occur, giving rise to still more products with distinctive biological activities. Trihydroxy products formed by the successive action of 15-LO and 5-LOs have been termed lipoxins and are described in more detail below. An eicosanoid frequently may exist in several structural forms depending on the number, location, and configuration of the double bonds. By convention the number of double bonds is indicated by a subscript (for example, when AA is converted to leukotriene C, the original four double bonds are retained and LTC_4 is formed). This is true of leukotrienes in general: Double bond rearrangement occurs, but the original number of double bonds is ordinarily maintained. C20 fatty acids with three (eicosatrienoic acid) or five (eicosapentaenoic acid) double bonds may also serve as eicosanoid precursors. The metabolites either lack the missing double bond or have the extra one. C20 fatty acids with three or five double bonds are normally much less prevalent in mammalian tissues than AA, but special diets can be used to increase their levels. The metabolites differ from the usual metabolites in rates of formation, biologic activities, and catabolism. For example, LTB_5 is much less active than LTB_4 in its effects on leukocytes; of perhaps greater significance is a fall in prostaglandin formation (12). Dietary manipulations, such as the use of eicosapentaenoic acid are being evaluated for their effects on leukotriene metabolism and possible value in the treatment of allergic and autoimmune diseases. Products related to the eicosanoids can also be produced from long-chain fatty acids with numbers other than 20, provided the appropriate double bonds are present. However, AA is normally by far the predominant eicosanoid precursor, and in the ensuing discussion the emphasis is on AA itself and its metabolites.

The CO pathway (lower part of Figure 1) gives the endoperoxides, PGG and PGH, which in turn are converted to prostacyclin; it also gives the various prostaglandins (PGs) or thromboxane A_2 (TxA_2). The precise pattern depends on the cell. The cyclooxygenases are associated with microsomes and show marked structural and enzymatic similarities in different cells and tissues. The CO products are not discussed in detail in this review.

5-Lipoxygenase Products

CELLULAR SOURCES   Known or strongly suspected sources of 5-LO products (LTB_4, LTC_4, LTD_4, 5-HETE) include monocytes, neutrophils, eosinophils, basophils, and mast cells, various organs such as the lung, lymph nodes, spleen and skin, and indeed any organ that is undergoing an inflammatory response (13–20). Various immunological stimuli may stimulate lipoxygenase product formation; these stimuli include microorganisms containing bound immunoglobulins, and complement, other phagocytic stimuli, aggregated immunoglobulins, and multivalent anti-receptor antibodies. Nonimmunologic stimuli that stimulate this pathway include A-23187, FMLP, cobra venom, serum-activated zymosan, purified activated complement components, lectins, and 48/80 (21). Often the strongest stimulus is the divalent cation ionophore A-23187, but if a response is to occur, Ca^{2+} must be present in the medium. In the lung, reactive oxygen metabolites may contribute to lipoxygenase activation (22). Product formation usually begins within a few minutes after the stimulus, but the process also declines with equal rapidity. Possible factors in the decline include exhaustion of substrate, negative feedback inhibition of the enzyme by 5-HETE or products of other lipoxygenases, compensatory adjustments of intracellular Ca^{2+} concentrations, and inactivation or degradation of the enzyme.

Leukotriene B_4 (LTB_4)

Known and probable sources for LTB_4 in vivo are mast cells, basophils, neutrophils, monocytes, and possibly lymphocytes. Recent reports suggest the possibility that LTB_4 may be produced by T lymphocytes after mitogenic activation (23, 24), but a role for contaminating cells, particularly monocytes, is often difficult to exclude. Our own studies indicate that lymphocytic cell lines produce an AA metabolite that chromatographs in a way similar but not identical to that of LTB_4 and which is presumably either a different dihydroxyeicosanoic acid or perhaps not a lipoxygenase product at all. The maximal production of LTB_4 by neutrophils occurs within the first 5–10 min after cell activation (6, 25). Early in the response much of the LTB_4 produced initially remains associated with the cells, but ultimately most of the LTB_4 is released into the medium either as LTB_4 itself or one of its metabolites. At 37°C, human neutrophils rapidly metabolize LTB_4 to its 20-OH and 20-COOH derivatives (26), both of which are considerably less active in aggregation and granule enzyme release than is LTB_4 itself.

LTB_4 appears to be a major mediator of leukocyte activation. Its effects include stimulation of cell aggregation, lysosomal enzyme release, and nondirected migration and chemotaxis. This activation spectrum is very similar to that of other activators of leukocyte function, such as C5a, F-met-leu-phe (F-MLP), LTB_4, PAF, and ATP, all of which also stimulate these responses. LTB_4 is particularly potent in inducing leukocyte aggregation; it produces responses at concentrations in the nanomolar or high picomolar range. It is almost as active as F-MLP in this system (8, 27, 28). LTB_4 and F-MLP show other similarities; both rapidly enhance Ca^{2+} and Na^+ influx in neutrophils (29), although F-MLP produces a more marked response than LTB_4. LTB_4 stimulates chemokinesis in human neutrophils at micromolar concentrations or below (30, 31, 32), but substantially higher concentrations are required if albumin is present. Cell movement is also enhanced in eosinophils, lymphocytes, and monocytes, indicating a general effect on cell migration. LTB_4 is also chemotactic in neutrophils and presumably other cell types, although considerably higher concentrations are required. Fixed tissues cells such as keratinocytes produce LTB_4 molecules with chemotactic activity (33) which may contribute to the inflammatory response during tissue injury. LTB_4 also promotes lysosomal enzyme release. However, the significance of the response is uncertain since LTB_4 is only effective in the presence of cytochalasins, and these themselves produce complex effects on leukocyte function. LTB_4 has also been reported to increase the number of surface receptors for C3b in polymorphonuclear leukocytes and eosinophils (6). Because mature leukocytes are very limited in their capacity for synthesizing proteins, this is likely to be an effect on receptor availability that results from granule translocation or altered organization of the plasma membrane.

The aggregating or chemotactic activity of LTB_4 for leukocytes from various species may be compared with that of its disasteroisomers and metabolites which remain the original double-bond configurations but are hydroxylated or carboxylated at the C20 position. LTB_4 is more potent than the other agents by a factor of 10 or greater. This suggests that LTB_4 is acting on cells through specific receptors. Direct binding studies with radiolabeled LTB_4 with human leukocytes have confirmed the existence of saturable and reversible binding sites inhibited by excess unlabeled ligand (27, 28, 34, 35). Because LTB_4 is rapidly metabolized at 37°C, the results of the binding studies are more easily interpreted at lower temperatures, particularly at 4°C, but this in turn makes correlations with chemotaxis or aggregation rather difficult. Several laboratories made detailed analysis of the binding curves with human polymorphonuclear leukocytes; this study indicates the existence of at least two classes of binding sites of differing affinities. The higher-affinity receptor is active in the concentration range where LTB_4 produces cell aggregation. Chemotactic responses occur at considerably higher LTB_4 concentrations and may involve the lower-affinity receptor. It is interesting that rat peripheral blood and peritoneal leukocytes possess high-affinity binding sites comparable to those on human cells, but the rat cells lack the lower-affinity binding sites (28). Rat cells aggregate readily in response to LTB_4 but fail to show a chemotactic response, even though they are highly responsive to other agents such as F-MLP. Since rats are widely used as models of inflammation, these observations have potential relevance for pharmacologic screening studies.

Very little is known about the factors controlling the number of LTB_4 receptors on cells or their possible variation in various disease states (8). LTB_4 receptors appear apparently de novo when the immature leukocyte cell lines are induced to differentiate. These changes are associated with other evidence for cell maturation and therefore cannot be considered to represent a specific induction process. Because their protein synthesizing activity is very limited, there is considerable doubt that mature polymorphonuclear leukocytes can regulate LTB_4 receptor levels by synthesizing new LTB_4 receptors. However, regulation through internalization and reexpression on the surface, shedding, or granule redistribution is probably occurring. Since polymorphonuclear leukocytes turn over rapidly, the production of new cells may be the most effective mechanism for restoring receptor numbers on cells in the periphery.

Like histamine and adenosine, LTB_4 must be considered as a possible mediator or modulator of T-lymphocyte function. Nonspecific or partially specific LO inhibitors such as eicosatetraynoic acid (ETYA) and nordihydroguaretic acid markedly decrease the DNA synthetic response to lectin and nonlectin mitogens; it also decreases much earlier responses such as AIB transport of amino isobutyric acid that usually increases within the first several hours of stimulation (36, 37). DNA synthesis in human and murine lymphocytes is inhibited by 15-HETE, 5-HETE, and their hydroperoxides, although relatively high concentrations are required (38). Because the HETEs are also lipoxygenase inhibitors, these observations provide indirect evidence that lipoxygenase products may be necessary for the response; other interpretations, however, are obviously possible. On the other hand LTB_4 is a potent inhibitor of transformation.

The inhibition of mitogenesis may be exerted through suppressor T cells. LTB_4 can increase DNA synthesis in T 8–bearing (suppressor phenotype), human-peripheral-blood T lymphocytes, whereas the T 4–bearing (helper) T cells are suppressed (39, 40, 41). Studies with fluorescent LTB_4-albumin conjugates indicate that both helper and suppressor T cells bind to LTB_4, but only a minority of each population is involved in the interaction (40). The significance of the changes in DNA synthesis is not altogether clear. The enhancement in T-8 cells is usually less than 20% and the suppression in other T cells less than 50%, even when a range of LTB_4 concentrations is evaluated. Nonetheless, in some experiments changes can be seen at picomolar concentrations of LTB_4; thus, a response could be occurring in vivo. LTB_4 could play a role in regulating the balance between helper and suppressor T lymphocytes by promoting the suppressor arm of the response (42). As a possible mechanism for this effect, LTB_4 reportedly can convert the T4 to the T8 T-cell phenotype; such a process has been described for T cells stimulated by histamine and adenosine (41). The possible role of 5-LO products including LTB_4 in responses of lymphocytes to interleukins is considered in more detail below.

Slow Reacting Substances (SRSs): Leukotrienes C, D, and E

The slow reacting substances (SRSs) are 6-sulfido-leukotrienes with glutathione, cysteinyl glycyl, and cysteinyl side chains (leukotrienes C, D, and E, respectively). They are known especially for their activities in producing slowly evolving but sustained contractile responses in airway and gastrointestinal smooth muscle preparations. LTC_4 is produced first, and under most circumstances, it is rapidly converted to LTD_4 and less rapidly to LTE_4 (6). Enzymes present in human plasma produce these conversions. The formation of SRSs is stimulated by antigen in guinea pig lung fragments sensitized either with IgE or IgG_1 antibodies. In humans, release appears to occur primarily through IgE-dependent mechanisms; IgG_4 immunoglobulins may also have this capability. It has been difficult or impossible to demonstrate IgE-dependent SRS release following stimulation with antigen in isolated rat peritoneal mast cells (16), even though these cells release histamine under the same circumstances and produce SRS in response to the nonimmunologic stimulus A-23187. Intestinal mast cells are more active in producing SRS, and SRS formation is also demonstrable in mast cells cultured from murine bone marrow preparations (43). Whether human mast cells show the same type of diversity is uncertain. Evidence suggests that highly purified human lung mast cells make enough SRS to account for most or all of the response in chopped lung preparations (L. Lichtenstein, personal communication). The possible role of macrophages (6, 44) as a source of SRS in IgE-mediated allergic reactions in human lungs has also been considered. Macrophages contain considerable amounts of 5-LO activity and, depending on the circumstances, may or may not produce LTC_4 in substantial quantity. Macrophages can be quite active in the formation of both cyclooxygenase [especially thromboxane A_2 (PGE_2)] and 5-LO products (45, 46, 47), but the proportion of the various metabolites varies markedly depending on the stimulus. The mechanism of this variation is incompletely understood. Among the factors that regulate AA metabolism in macrophages are nonlipid mediators of immediate hypersensitivity such as histamine and serotonin and lymphokines (48). There is a current interest in an apparently selective suppressive effect of smoking on 5-LO activity in alveolar macrophages, although the significance of this suppression in emphysema induced by smoking is unknown (49).

Since glutathione is the predominant intracellular thiol and is directly utilized for LTC_4 synthesis, it is obviously important in SRS biosynthesis. A number of 5-LO inhibitors have been synthesized that are possibly useful in inhibiting leukotriene formation, but their eventual applicability is uncertain at present. One factor involved in the control of SRS synthesis is the intracellular level of reduced glutathione. When cells from the rat mastocytoma line RBL-1 are depleted of reduced glutathione by use of pharmacologic inhibitors, then SRS biosynthesis is inhibited almost completely (50). Decreases of as little as 30–40% in intracellular concentrations of reduced glutathione produce easily detectable reductions in SRS production. LTC_4 production in human lung cells is also apparently controlled by reduced glutathione (51). Since intracellular levels of reduced glutathione are subject to metabolic control, such fluctuations may be involved in the control of SRS biosynthesis (51). The mechanism for the conjugation with glutathione is still not completely clear. The 5-HPETE can directly couple with thiols through leukotriene A (26), but the reaction probably involves an S-glutathione transferase (LTC_4 synthetase) with a particularly high affinity for the oxidized AA (52, 53, 54). The enzymes involved in converting LTC_4 to LTE_4 are likely to be some of the usual peptidases present in plasma and tissue. The available evidence indicates that LTE_4 or its N-acetylated derivative is a significant metabolite of the sulfido leukotrienes in the urine and elsewhere (55).

The biologic activities of SRS have recently been reviewed (6, 16, 56). The activity that received most attention is the spasmogenic action of SRS on smooth muscle. Slow reacting substance is considerably more selective in the spectrum of smooth muscle preparations that it affects than are the primary prostaglandins. Concentrations of LTD_4 as low as 100 pM ml^{-1} produce easily detectable contractile responses on smooth muscle strips from guinea pig ilea and trachea. On a molar basis, LTD_4 is about 60–100 times more active than histamine on the guinea pig ileum. LTC_4 and LTE_4 are somewhat less active than LTD_4 (by factors of 1.5 and 10, respectively, in the ileal system) but are still much more potent than histamine. In addition to its high potency, SRS produces an exceptionally long sustained contractile response without tachyphylaxis, and this raises the possibility of an unusually prolonged action in vivo (57). SRS appears to act at least in part by stimulating Ca^{2+} channels in some muscle cells (56). Because of the suspected importance of SRS in human bronchial asthma, its effect on lung smooth muscle is of particular interest. The SRS leukotrienes act both on central and peripheral airway smooth muscle (58). LTC_4 and LTD_4 also have been reported to act on cutaneous blood vessels, producing increases in vascular permeability and potentiating the action of other vasoactive agents (58). These facts indicate that they could act as mediators in urticarial reactions as well. LTC_4 and LTD_4 also act on blood vessels elsewhere, and although high concentrations are required, they show many of the effects on leukocytes seen with LTB_4. The in vitro effects of SRS suggest that it could be a major mediator in the allergic reactions in the lung that produces bronchospasm. Its role is supported by in vivo studies both in humans receiving leukotrienes by aerosol (59) and in experimental animals. As one example, in rhesus monkeys (60) with ascaris allergy, results indicate that alterations in pulmonary resistance, respiratory rate, and compliance in response to aerosolized antigen are inhibited both by selective lipoxygenase inhibitors and by the SRS end-organ antagonist, FPL-55712. However, the results of treatment with SRS antagonists in humans are needed for a more definitive evaluation of their role in this species. LTC_4 effects on mucin production (61, 62) and ciliary action (63) in the trachea and bronchi have also been reported, and this suggests actions of possible importance in asthma other than those on airway muscle contraction per se. These leukotrienes may also act as mediators during hypopoxic pulmonary vasoconstriction (64).

In addition, some evidence indicates that lipoxygenase products may serve as amplifiers of other anaphylactic stimuli. C5a, a biologically active fragment of C5, induces prolonged smooth muscle contraction in the isolated guinea pig trachea. As little as 1 μM ETYA markedly reduces the contractile response to C5a (65). A cyclooxygenase inhibitor, acetyl salicylate, has no effect on 5a-induced contractions; in fact, it enhanced acetylcholine-induced contractions. FPL-55712, a selective antagonist of SRS-A, almost completely inhibits the response to C5a. These data indicate that the C5a-induced tracheal contraction is mediated by lipoxygenase products, most likely LTC_4 or LTD_4.

The evidence for other potentially important biologically actions of LTC_4 and LTD_4 continues to mount (8):

1. Myeloid colony formation reportedly is stimulated by sulfido-leu-kotrienes (66).

2. Promotion of proliferation by LTC_4 in glomerular epithelial cells has been described (66, 67).

Both of these observations call to mind earlier indirect evidence in lymphocytes (see below) that lipoxygenase products may be involved in mitogenesis (34, 68). In this connection, PGD_2 is reported to have the potentially antagonistic effect of lowering nuclear DNA polymerase activity in mastocytoma cells (69).

3. Slow reacting substance exerts a sustained change in Purkinjie nerve cell activity, and this raises the possibility of important effects in the central or peripheral nervous system (70). An action of this kind might help explain the increased reflex irritability of the tracheal-bronchial tree and the lung in asthma.
4. SRS also affects chemotaxis and granule enzyme release in leukocytes. Although it is much less potent than LTB_4 in this regard, LTC_4 in anterior pituitary cells is a highly effective and selective secretagogue for at least two pituitary hormones, LH and LHRH (71, 72).
5. The 5-LO products such as LTC_4 also have a possible role in cholesterol ester accumulation in macrophages; this may be relevant to the development of atherosclerosis (73).
6. A possible role has been reported for 5-LO products, including LTC_4, in preventing polyspermic fertilization in sea urchin eggs (74).

Evidence suggests that specific cellular receptors exist for LTC_4, LTD_4, and LTE_4 (67, 75, 76, 77, 78). Aside from a limited amount of work in defined smooth-muscle cell lines, experiments have been conducted with heterogeneous tissues such as the lung, intestinal, or cardiac muscle, making it difficult or impossible to correlate binding with the biological response of the tissue. Nevertheless, the evidence for specific receptors is strongly suggestive. There appear to be two major types of receptors, one binds LTC_4 especially well and another binds LTD_4 and LTE_4 primarily (76). The existence of two types of receptors presents the same potential problems for the development of end-organ antagonists as does the existence of multiple receptors for histamine.

The Possible Role of 5-LO Products in Intracellular Signalling and Responses to Interleukins

The potency of leukotrienes and other 5-LO products as inflammatory mediators acting after release into the medium does not preclude the possibility they may act as mediators in the cells in which they originate. The possibility that secretion of granules may be in part controlled by 5-HETE, its hydroperoxide, and by dihydroxylated eicosanoids such as

LTB$_4$, has been investigated in some detail by a number of laboratories. Final answers, however, are not available. There is no doubt that exogenous 5-HETE, 5-HPETE, and LTB$_4$ can initiate or potentiate secretion from neutrophils and mast cells or that several 5-LO antagonists can markedly decrease secretion (79, 80). Moreover, as already noted, evidence is accumulating that LTC$_4$ may be important in the release of several pituitary hormones. However, because other studies indicate that secretion can occur in neutrophils or mast cells under conditions in which little or no 5-LO product formation is demonstrable (81), alternative mechanisms for secretion are possible.

A role for lipoxygenase products has also been suggested in the production of interleukin-1 (IL-1) by human monocytes stimulated with *Staphylococcus albus* (82). A comparison was made of the effect on IL-1 production of drugs inhibiting either the cyclooxygenase or the lipoxygenase pathway. Ibuprofen inhibited the production of PGE$_2$ but had no effect on IL-1 release. ETYA, which is more effective at inhibiting LO and CO, prevented IL-1 production from human monocytes when ETYA was added to the incubation fluid prior to cell activation. After cell activation, however, ETYA was ineffective. The IL-1 response to staphylococci was also inhibited by BW755C (3-amino-1-3-trifluoromethylphenyl-2-pyrazoline), which inhibits both pathways. BW755C also inhibited the IL response to endotoxin and muramyl dipeptide.

Activation of LOs may be involved in the action of IL-1 as well as in its formation, and this may also be true for interleukin 2 (IL-2). Farrar & Humes (83) have studied the role of AA metabolites in the induction of IL-2 synthesis by IL-1 and also in IL-2 stimulation of proliferation and interferon-γ (IFN-γ) production in T cells. In cell lines that are specifically responsive to IL-1 or IL-2, each interleukin stimulates lipoxygenation of arachidonic acid in its respective target cell. Lymphocytes apparently make small amounts of lipoxygenase products shortly after stimulation by lectin (84). The ability of the two interleukins to induce HETE formation is apparently correlated with the induction of secondary lymphokine secretion. For selective and partially selective pharmacologic inhibitors of AA metabolism, the data suggest that the participation of lipoxygenase activity is required both for induction of IL-2 production by IL-1 and for regulation of the proliferation and secretion of IFN-γ by IL-2. The same requirement for lipoxygenase activity was seen when phorbol myristate acetate (PMA) was used as a secretory stimulant, and this suggests a similar mode of action for stimulation-secretory activity between PMA and interleukins. These results call to mind earlier studies which show that ETYA and other lipoxygenase inhibitors markedly inhibited lymphocyte mitogenesis (85). Studies performed with an endogenous inhibitor of 5-

lipoxygenase (15-HETE) further demonstrated that this enzyme system apparently requires IL 2–dependent proliferation and IFN-γ production for induction (83). Although leukotrienes could replace IL-2 for secretion of IFN-γ, they had no effect on promotion of growth IL-2. The results suggest both that IL-1, IL-2, and PMA may each stimulate the lipoxygenase pathway and that the stimulation is involved in the signal transduction process that regulates secretory activity and/or cellular proliferation. The possibility that lipoxygenase products may function in lymphocyte-mediated cytotoxicity has also been suggested, but the effect appears to be on the inductive rather than the effector phase of the response (86). IL-1 also regulates AA metabolism in chrondocytes and endothelial cells that activate PGE_2 or prostacyclin production, depending on the cell (87, 88). Thus, eicosanoids appear to be involved in the action of IL-1 as well as in its formation.

While confirmation is needed, the observations with IL-2 suggest that LOs in lymphocytes are subject to long-term regulatory influences that affect enzyme biosynthesis. This may also be true for other leukocytes. The human granulocyte cell line H-60 shows marked increases in 5-LO product formation after it has been induced to differentiate in vitro (89). Of course, many other enzymes show increases in activity under these conditions; there is no evidence that the effect on 5-LO is selective. The production of lipoxygenase metabolites by mature monocytes may be markedly increased by sustained exposure to naturally occurring stimuli including pathogenic microorganisms; again, this raises the possibility of induction of increased enzyme levels. For example, a recent study of macrophages from animals infected with *Leishmania Donovani* (90) indicates that exposure to the organism for 24 hr in vitro is associated with a marked increase in the ability of the cells to produce LTC_4 when they are additionally stimulated by lipopolysaccharide or zymogen. This might suggest that more 5-LO is available in infected cells (as might occur due to increased enzyme biosynthesis), but the production of metabolites through the CO pathway is also increased and activation of phospholipase may be occurring. Again, this is presumably not a selective effect on the 5-LO alone, because activation of other enzyme systems and increased expression of receptors is almost certainly occurring. It is interesting that parasites may themselves produce leukotrienes; this raises interesting questions as to the possible role of the parasite products on the initial response of the host (8).

Interactions Between 5-Lipoxygenase Products and Other Lipid Mediators

Lipoxygenase may also affect the production of other biologically active lipids including platelet activating factor (PAF) and thromboxane A_2. In human neutrophils stimulated with A-23187, the production of PAF was greatly augmented by 5-HETE, 5-HPETE, and LTB_4 (91). In endothelial cells LTC_4 and LTD_4 generally stimulate AA metabolism as well as increase PAF (92). In the endothelial cell system, the 5-LO products may be acting by augmenting phospholipase A_2 activation. The 5-LO products also stimulate PAF production in macrophages (93). In perfused lungs, leukotrienes considerably augment thromboxane formation (94). PAF and thromboxin A_2 are highly active autocoids; thus, the potentiation of their formation by lipoxygenase products could markedly amplify local tissue responses. In the next section other evidence is discussed which indicates that 12-LO and 15-LO products can regulate PG and leukotriene production (95).

15-LIPOXYGENASE (15-LO) PRODUCTS   15-LOs have been reported present in polymorphonuclear leukocytes and monocytes and in high concentration in tracheal epithelial cells (96). In pharmacologic concentrations, 15-HETE and 15-HPETE can act as inhibitors of 5-LOs, whereas the 5-LO pathway is stimulated (97). The physiologic significance of this inhibition in the control of 5-LO product formation is uncertain. Lipoxins are a mixture of six different isomers of 5, 6, 15 or 5, 14, 15 trihydroxy eicosatetraenoic acids (lipoxins A and B, respectively). They can be prepared in quantity by incubating 15-HPETE with A-23187 stimulated leukocytes (98). Once the 15-HPETE is available, much of the subsequent processing appears to be nonenzymatic (99, 100). Whether the lipoxins are produced in significant amounts under physiological conditions remains to be determined. Thus far, the amounts obtained directly from biologic sources have been minute. They are nonetheless of considerable interest because of their biologic activities. One of their effects is to stimulate phospholipase C (101), an important regulatory enzyme in a variety of cellular systems. An apparently selective effect on natural killer cell activity has also been described (102).

CONTROL OF ARACHIDONATE RELEASE FROM CELLULAR STORES   The AA present in mammalian cells is primarily derived either from shorter-chain essential fatty acids or from AA itself in the diet and is carried to and from cells through the blood or lymphatic fluid. Almost all of the AA present in tissues is esterified in phospholipids and triglycerides. With the exception of the 5-LO, which requires activation at the time cells are stimulated, the rate-limiting step in AA metabolism is thought to be the availability of free AA released enzymatically by phospholipases or lipases when cells are activated by antigens and other stimuli. Ordinarily a stimulated cell is assumed to supply its own unesterified AA, but it should be kept in mind that phospholipids, triglycerides, and unesterified fatty acids are present in the blood in chylomicrons or are bound reversibly to the lipoproteins and albumin present in plasma. In inflamed tissues and exudates and especially within the blood stream itself, this plasma-bound lipid may conceivably serve as another source of rapidly available AA. Moreover, where cell densities are high, a cell may utilize AA derived from nearby cells. For example, in human–peripheral blood leukocytes stimulated with lectins, some of the AA released from lymphocytes is metabolized by monocytes, which appear to be considerably more active in this regard than are lymphocytes (102a).

Despite the interest in AA metabolism, the mechanism of release of esterified AA is still incompletely understood. In the older literature almost all of the emphasis was on phospholipase A_2. Subsequent studies in mast cells and platelets raised the strong possibility that when phospholipase C is activated the diacylglycerol that is formed is metabolized in part by lipases releasing free AA (7). Ferber and his colleagues (103) have suggested another possible mechanism for AA release in which the enzymatic reaction catalyzing the exchange of fatty acids between phospholipids (acyl CoA : lysophosphatide acyltransferase) undergoes reversal. Further elucidation of the relative importance of these pathways is needed.

Because of the importance of lipid mediators in inflammation and the potent anti-inflammatory properties of glucocorticoids, the possibility that glucocorticoids act at least in part by inhibiting lipid mediator formation has been considered repeatedly (104). The anti-inflammatory action of glucocorticoids has been attributed to the induction of a group of phospholipase A_2-inhibitory proteins, collectively called lipocortin. These proteins are thought to control biosynthesis of the lipid mediators of inflammation by inhibiting the phospholipase A_2-mediated hydrolysis of phospholipids. Lipocortin-like proteins have been isolated from various cell types, including monocytes, neutrophils, and renal medullary cell preparations. The predominant active form is a protein with an apparent molecular weight of 40,000. Partially purified preparations of lipocortin show anti-inflammatory activity in various in vivo model systems. With the use of amino acid sequence information obtained from purified rat lipocortin, DNA complementary to human lipocortin has been cloned and expressed in *Eschericia coli*. Their studies confirm that lipocortin is a potent inhibitor of phospholipase A_2.

In addition to the evidence for involvement of lipocortin in the anti-inflammatory action of steroids, a possible action of cyclic AMP agonists at this level needs to be kept in mind. A number of the pharmacologic agents inhibiting AA release in human lymphocytes in response to lectin are also cAMP agonists and inhibitors of transformation; the three activities appear to correlate rather closely (105). The inhibition of AA release is demonstrable within seconds to minutes; this indicates a direct action at the level of the plasma membrane rather than remote effects on intracellular metabolism.

Regulation of lymphocyte metabolism is also possible at another level, involving the participation of low-density lipoprotein (LDL) receptors (reviewed in Gurr—106). Lymphocytes have receptors for a subspecies of LDL that inhibits their response to mitogens. Unlike cholesterol biosynthesis, which involves a regulatory pathway, LDL does not require internalization for inhibition. The relationship of the inhibition by LDL to possible changes in AA metabolism and to overall effects on lipid metabolism is poorly understood at present.

It is interesting that lipid metabolism in macrophages and perhaps in atheromatous plaques in blood vessels may be regulated in part by lipoxygenase products (107). Accumulation of cholesterol esters was induced by incubating cells with acetylated low-density lipoprotein. The stable prostacyclin analogue ZK 36 374, prostaglandin E_2, and cyclooxygenase inhibitor indomethacin showed no effect on cellular storage of the esthers. By contrast, however, two inhibitors of lipoxygenase activity, nordihydroguaretic acid and BW 755 C, markedly suppressed the accumulation of cholesterol esters in monocyte-derived macrophages. These observations suggest still another possible role for lipoxygenase products in helping control the intracellular storage of lipids. Increased cholesterol deposition may in turn promote lipoxygenase activity (108).

Literature Cited

1. Samuelsson, B. 1983. Leukotrienes: Mediators of immediate hypersensitivity reactions and inflammation. *Science* 220: 568–75
2. Samuelsson, B. 1984. Lipoxins: A novel series of compounds formed from arachidonic acid. In *Advances in Prostaglandin, Thromboxane, and Leukotriene Research*, ed. J. E. Pike, D. R. Morton, 14: 45. New York: Raven
3. Goetzl, E. J. 1980. Oxygenation products of arachidonic acid as mediators of hypersensitivity and inflammation. *Med. Clin. North Am.* 65: 809–28
4. Lewis, R. A., Austen, K. F. 1981. Mediation of local homeostasis and inflammation by leukotrienes and other mast cell-dependent compounds. *Nature* 293: 103–8
5. Parker, C. W. 1983. Immunopharmacology of slow reacting substances. In *Immunopharmacology of the Lung*, ed. H. H. Newball, p. 25. New York: Marcel Dekker
6. Parker, C. W. 1984. Mediators: Release and function. In *Fundamental Immunology*, ed. W. E. Paul, p. 697. New York: Raven
7. Parker, C. W. 1985. Leukotriene and prostaglandin in the immune system. In *Advances in Prostaglandin, Thromboxane and Leukotriene Research*, ed. O. Hayaishi, S. Yamamoto, 15: 167–68. New York: Raven
8. Parker, C. W. 1985. Lipid mediators and inflammation. In *New Trends in Allergy II*, ed. J. Ring, pp. 132–57. Heidelberg: Springer Verlag
9. Vliegenthart, J. F. G., Veldink, G. A., Verhagen, J., Slappendel, S. 1983. Lipoxygenases from plant and animal origin. In *Biological Oxidations*, ed. H. Sund, V. Ullrick, p. 203. Berlin: Springer-Verlag
10. Stenson, W. F., Parker, C. W. 1983. Metabolites of arachidonic acid. *Clin. Rev. Allergy* 1: 369–84
11. Izumi, T., Shimizu, T., Seyama, Y., Ohishi, N., Takaku, F. 1986. Tissue distribution of leukotriene A_4 hydrolase activity in guinea pig. *Biochem. Biophys. Res. Commun.* 135: 139–45

12. Terano, T., Salmon, J. A., Higgs, G. A., Moncada, S. 1986. Eicosapentaenoic acid as a modulator of inflammation. *Biochem. Pharmacol.* 35: 779–85

13. Czarnetzki, B. M., Konig, W., Lichtenstein, L. 1976. Eosinophil chemotactic factor (ECF). I. Release from polymorphonuclear leukocytes by the calcium ionophore A23187. *J. Immunol.* 117: 229–34

14. Scott, W. A., Pawlowski, N. A., Murray, H. W., Reach, M., Zrike, J., Cohn, Z. A. 1982. Regulation of arachidonic acid metabolism by macrophage activation. *J. Exp. Med.* 155: 1148–60

15. Jorg, A., Henderson, W. R., Murphy, R. C., Klebanoff, S. J. 1982. Leukotriene generation by eosinophils. *J. Exp. Med.* 155: 390–402

16. Parker, C. W. 1982. The chemical nature of slow-reacting substances. In *Advances in Inflammation Research*, ed. I. Weissmann, 4: 1–24. New York: Raven

17. Murphy, R. C., Hammarstrom, S., Samuelsson, B. 1979. Leukotriene C: A slow-reacting substance from murine mastocytoma cells. *Proc. Natl. Acad. Sci. USA* 76: 4275–79

18. Morris, H. R., Taylor, G. W., Piper, P. J., Samhoun, M. N., Tippins, J. R. 1980. Slow reacting substances (SRSs): The structure identification of SRSs from rat basophil leukemia (RBL-1) cells. *Prostaglandins* 19: 185–201

19. Orning, L., Hammarstrom, S., Samuelsson, B. 1980. Leukotriene D: A slow reacting substance from rat basophilic leukemia cells. *Proc. Natl. Acad. Sci. USA* 77: 2014–17

20. Watanabe-Kohno, S., Parker, C. W. 1980. Role of arachidonic acid in the biosynthesis of slow reacting substance of anaphylaxis (SRS-A) from sensitized guinea pig lung fragments: Evidence that SRS-A is very similar or identical structurally to nonimmunologically induced forms of SRS. *J. Immunol.* 125: 946–55

21. Razin, E. 1985. Activation of the 5-lipoxygenase pathway in E-mast cells by peanut agglutinin. *J. Immunol.* 134: 1142–45

22. Burghuber, O. C., Strife, R. J., Zirrolli, J., Henson, J. E., Mathias, M. M., Reeves, J. T., Murphy, R. C., Voelkel, N. F. 1985. Leukotriene inhibitors attenuate rat lung injury induced by hydrogen peroxide. *Am. Rev. Respir. Dis.* 131: 778–85

23. Goodwin, J. S., Webb, D. R. 1980. Regulation of the immune response by prostaglandins. *Clin. Immunol. Immunopathol.* 15: 106–22

24. Humes, J. L., Bonney, R. J., Pelus, L., Dahlgren, M. E., Sadowski, S. J., Kuehl, F. A. Jr., Davis, P. 1977. Macrophage synthesis and release of prostaglandins in response to inflammatory stimuli. *Nature* 269: 149–51

25. Kimura, Y., Okuda, H., Arichi, S., Baba, K., Kozawa, M. 1985. Inhibition of the formation of 5-hydroxy-6,8,11,14-eicosatetraenoic acid from arachidonic acid in polymorphonuclear leukocytes by various coumarins. *Biochim. Biophys. Acta* 834: 224–29

26. Samuelsson, B. 1982. Leukotrienes: An introduction. In *Leukotrienes and Other Lipoxygenase Products*, ed. B. Samuelsson, R. Paoletti, pp. 1–18. New York: Raven

27. Goldman, D. W., Goetzl, E. J. 1984. Heterogeneity of human polymorphonuclear leukocyte receptors for leukotriene B_4. Identification of a subset of high affinity receptors that transduce the chemotactic response. *J. Exp. Med.* 159: 1027–41

28. Kreisle, R. A., Parker, C. W., Griffin, G. L., Senior, R. M., Stenson, W. F. 1985. Studies of leukotriene B_4-specific binding and function in rat polymorphonuclear leukocytes: Absence of a chemotactic response. *J. Immunol.* 134: 3356–62

29. Shaafi, R. I., Naccache, P. H., Molski, T. F. P., Borgeat, P., Goetzl, E. J. 1981. Cellular regulatory role of leukotriene B_4: Its effects on cation homeostasis in rabbit neutrophils. *J. Cell. Phys.* 108: 401–8

30. Ford-Hutchinson, A. W., Bray, M. A., Doig, M. V., Shipley, M. E., Smith, M. J. H. 1980. Leukotriene B, a potent chemokinetic and aggregating substance released from polymorphonuclear leukocytes. *Nature* 286: 264–65

31. Goetzl, E. J. 1980. Mediators of immediate hypersensitivity derived from arachidonic acid. *New Engl. J. Med.* 303: 822–25

32. Palmer, R. M. J., Stepney, R. J., Higgs, G. A., Eakins, K. E. 1980. Chemokinetic activity of arachidonic and lipoxygenase products on leukocytes of different species. *Prostaglandins* 20: 411–18

33. Grabbe, J., Rosenbach, T., Czarnetzki, B. M. 1985. Production of LTB_4-like chemotactic arachidonate metabolites

from human keratinocytes. *J. Invest. Dermatol.* 85: 527–30

34. Parker, C. W. 1981. Arachidonic acid metabolism in activated lymphocytes. In *Mechanisms of Lymphocyte Activation*, ed. K. Resch, H. Kirchner, pp. 47–57. New York: Elsevier/North-Holland Biomedical

35. Lin, A. H., Ruppel, P. L., Gorman, R. R. 1984. Leukotriene B_4 binding to human neutrophils. *Prostaglandins* 28: 837–49

36. Kelly, J. P., Johnson, M. C., Parker, C. W. 1979. Effect of inhibitors of arachidonic acid metabolism of mitogenesis in human lymphocytes. Possible role of thromboxanes and products of the lipoxygenase pathway. *J. Immunol.* 122: 1563–71

37. Udey, M. D., Parker, C. W. 1982. Effect of inhibitors of arachidonic acid metabolism on α-aminoisobutyric acid transport in human lymphocytes. *J. Biochem. Pharm.* 31: 337–46

38. Bailey, J. M., Bryant, R. W., Low, C. E., Pupillo, M. B., Vanderhoek, J. Y. 1982. Role of lipoxygenase in regulation of PHA and Phorbol Ester induced mitogenesis. See Ref. 26, p. 341

39. Gualde, N., Atluru, D., Goodwin, J. S. 1985. Effect of lipoxygenase metabolites of arachidonic acid on proliferation of human T cells and T cell subsets. *J. Immunol.* 134: 1125–29

40. Payan, D. G., Missirian-Bastian, A., Goetzl, E. J. 1984. Human T-lymphocyte subset specificity of the regulatory effects of leukotriene B_4. *Proc. Natl. Acad. Sci. USA* 81: 3501–5

41. Rola-Pleszczynski, M. 1985. Differential effects of leukotriene B_4 and $T4^+$ and $T8^+$ lymphocyte phenotype and immunoregulatory functions. *J. Immunol.* 135: 1357–60

42. Rocklin, R. E., Bendtzen, K., Greineder, D. 1980. Mediators of immunity: Lymphokines and monokines. *Adv. Immunol.* 29: 55–136

43. Robin, J.-L., Seldin, D. C., Austen, K. F., Lewis, R. A. 1985. Regulation of mediator release from mouse bone marrow-derived mast cells by glucocorticoids. *J. Immunol.* 135: 2719–26

43a. Newball, H. H., Lichtenstein, L. 1981. Mast cells and basophils: Effector cells of inflammatory disorders in the lung. *Thorax* 36: 721–25

44. Razin, J., Romeo, L. C., Krilis, S., Liu, F.-T., Lewis, R. A., Corey, E. J., Austen, F. 1984. An analysis of the relationship between 5-lipoxygenase product generation and the secretion of preformed mediators from mouse bone

marrow-derived mast cells. *J. Immunol.* 133: 938–45

45. Fels, A. O. S., Pawlowski, N. A., Abraham, E. L., Cohn, Z. A. 1986. Compartmentalized regulation of macrophage arachidonic acid metabolism. *J. Exp. Med.* 163: 752–57

46. Fels, A. O. S., Cohn, Z. A. 1986. The alveolar macrophage. *J. Appl. Physiol.* 60: 353–69

47. Bonney, R. J., Opas, E. E., Humes, J. L. 1985. Lipoxygenase pathways of macrophages. *Fed. Proc.* 44: 2933–36

48. Hsueh, W., Jordan, R. L., Gonzalez-Crussi, F. 1985. Rabbit lymphocyte factors modulate prostaglandin biosynthesis in alveolar macrophages. *Immunology* 54: 155–61

49. Laviolette, M., Coulombe, R., Picard, S., Braquet, P., Borgeat, P. 1986. Decreased leukotriene B_4 synthesis in smokers' alveolar macrophages *in vitro*. *J. Clin. Invest.* 77: 54–60

50. Parker, C. W., Fischman, C. M., Wedner, H. J. 1980. Relationship of slow reacting substance biosynthesis to intracellular glutathione levels. *Proc. Natl. Acad. Sci. USA* 77: 6870–73

51. Sautebin, U., Vigano, T., Grassi, E., Crivellari, M. T., Galli, G., Berti, F., Mezzetti, M., Folco, G. 1985. Release of leukotrienes, induced by Ca^{++} ionophore A23187, from human lung parenchyma *in vitro*. *J. Pharmacol.* 234: 217–21

52. Abe, M., Kawazoe, Y., Tsunematsu, H., Shigematsu, N. 1985. Peritoneal macrophages of guinea pig possibly lack LTC_4 synthetase. *Biochem. Biophys. Res. Comm.* 127: 15–23

53. Bach, M. K., Brasher, J. R., Morton, D. R. Jr. 1984. Solubilization and characterization of the leukotriene C4 synthetase of rat basophil leukemia cells: A novel, particulate glutathione S-transferase. *Arch. Biochem. Biophys.* 230: 455–65

54. Kuo, C. G., Lewis, M. T., Jakschik, A. 1984. Leukotriene D_4 and E_4 formation by plasma membrane bound enzymes. *Prostaglandins* 28: 929–38

55. Orning, L., Kaijser, L., Hammarstrom, S. 1985. *In vivo* metabolism of leukotriene C_4 in man: Urinary excretion of leukotriene E_4. *Biochem. Biophys. Res. Comm.* 130: 214–20

56. Findlay, S. R., Leary, T. M., Parker, C. W., Bloomquist, E. I., Scheid, C. R. 1982. Contractile events induced by LTD_4 and histamine in smooth muscle—a function of calcium utilization. *Am. Physiol. Soc.* 243: C133–C139

57. Kellaway, C. H., Trethewie, E. R. 1940. The liberation of a slow-reacting smooth muscle-stimulating substance in anaphylaxis. *Q. J. Exp. Physiol.* 30: 121–45

58. Drazen, J. M., Austen, K. F., Lewis, R. A., Clark, D. A., Goto, G., Marfat, A., Corey, E. J. 1980. Comparative airway and vascular activities of leukotrienes C-1 and D *in vivo* and *in vitro. Proc. Natl. Acad. Sci. USA* 77: 4354–58

59. Smith, L. J., Greenberger, P. A., Patterson, R., Krell, R. D., Bernstein, P. R. 1984. The effect of inhaled leukotriene D_4 in humans. *Am. Rev. Respir. Dis.* 131: 368–72

60. Patterson, R., Harris, K. E. 1981. Inhibition of immunoglobulin E-mediated, antigen-induced monkey asthma and skin reactions by 5,8,11,14-eicosatetraenoic acid. *J. Allergy Clin. Immunol.* 67: 146–52

61. Marom, Z., Shelhamer, J. H., Kaliner, M. 1981. Effects of arachidonic acid, monohydroxyeicosatetraenoic acid and prostaglandins on the release of mucous glycoproteins from human airways *in vitro. J. Clin. Invest.* 67: 1695–1702

62. Richardson, P. S., Peatfield, A. C., Jackson, D. M., Piper, P. J. 1982. Effect of leukotriene on the output of mucins from the cat trachea. See Ref. 26, p. 178

63. Tashiro, N., Kamishima, K., Kishi, F., Kawakami, Y., Furudate, M. 1986. Action of inhaled leukotriene D_4 on lung mucociliary transport in humans: Differential effects on central and peripheral airways. Personal communication

64. Morganroth, M. L., Stenmark, K. R., Zirrolli, J. A., Mauldin, R., Mathias, M., Reeves, J. T., Murphy, R. C., Voelkel, N. F. 1984. Leukotriene C_4 production curing hypoxic pulmonary vasoconstriction in isolated rat lungs. *Prostaglandins* 28: 867–75

65. Regal, J. F., Pickering, R. J. 1981. C5a-induced tracheal contraction: Effect of an SRS-A antagonist and inhibitors of arachidonate metabolism. *J. Immunol.* 126: 313–16

66. Ziboh, V. A., Wong, T., Wu, M.-C., Yunis, A. A. 1986. Modulation of colony stimulating factor-induced murine myeloid colony formation by S-peptido-lipoxygenase products. *Cancer Res.* 46: 600–3

67. Baud, L., Sraer, J., Perez, J., Nivez, M.-P., Ardailou, R. 1985. Leukotriene C_4 binds to human glomerular epithelial cells and promotes their proliferation

in vitro. J. Clin. Invest. 76: 374–77

68. Kelly, J. P., Johnson, M. C., Parker, C. W. 1979. Effect of inhibitors of arachidonic acid metabolism of mitogenesis in human lymphocytes. Possible role of thromboxanes and products of the lipoxygenase pathway. *J. Immunol.* 122: 1563–71

69. Kawamura, M., Koshihara, Y. 1984. Prostaglandin D_2 lowers nuclear DNA polymerase activity in cultured mastocytoma cells. *Prostaglandins* 27: 517–25

70. Palmer, M. P., Mathews, R., Murphy, R. C., Hoffer, B. 1980. Leukotriene C elicits a prolonged excitation of cerebellar purkinje neurons. *Neurosci. Lett.* 18: 173–80

71. Hulting, A.-L., Lindgren, J. A., Hokfelt, T., Eneroth, P., Werner, S., Patrono, C., Samuelsson, B. 1985. Leukotriene C_4 as a mediator of luteinizing hormone release from rat anterior pituitary cells. *Proc. Natl. Acad. Sci. USA* 82: 3834–38

72. Gerozissis, K., Rougeot, C., Dray, F. 1986. Leukotriene C_4 is a potent stimulator of LHRH section. *Eur. J. Pharmacol.* 121: 159–60

73. Van der Schroeff, J. G., Havekes, L., Weenheim, A. M., Emeis, J. J., Vermeis, B. J. 1985. Suppression of cholesteryl ester accumulation in cultured human monocyte-derived macrophages by lipoxygenase inhibitors. *Biochem. Biophys. Res. Comm.* 127: 366–72

74. Schuel, H., Moss, R., Schuel, R. 1985. Induction of polyspermic fertilization in sea urchins by the leukotriene antagonist FPL-55712 and the 5-lipoxygenase inhibitor BW755C. *Gamete Res.* 11: 41–50

75. Mong, S., Scott, M. O., Lewis, M. A., Wu, H.-L., Hogaboom, G. K., Clark, M. A., Crooke, S. T. 1985. Leukotriene B_4 binds specifically to LTD_4 receptors in guinea pig lung membranes. *Eur. J. Pharmacol.* 109: 183–92

76. Hogaboom, G. K., Mong, S., Stadel, J. M., Crooke, S. T. 1985. Characterization of guinea pig myocardial leukotriene C_4 binding sites. *Molec. Pharmacol.* 27: 236–45

77. Krilis, S., Lewis, R. A., Corey, E. J., Austen, K. F. 1984. Specific binding of leukotriene C_4 to iliac segments and subcellular fractions of ileal smooth muscle cells. *Proc. Natl. Acad. Sci. USA* 82: 4529–33

78. Borgeat, P., Sirois, P. 1981. Leukotrienes: A major step in the understanding of immediate hypersensitivity reactions. *J. Med. Chem.* 24: 121–26

79. Marone, G., Hammarstrom, S., Lichtenstein, L. M. 1980. An inhibitor of lipoxygenase inhibits histamine release

from human basophils. *Clin. Immunol. Immunopath.* 17: 117–22

80. Stenson, W. F., Parker, C. W. 1980. Monohydroxyeicosatetraenoic acids (HETEs) induce degranulation of human neutrophils. *J. Immunol.* 124: 2100–4

81. Shoam, H., Razin, E. 1985. BW755C inhibits the 5-lipoxygenase in E-mast cells without affecting degranulation. *Biochim. Biophys. Acta* 837: 1–5

82. Dinarello, C. A., Bishai, I., Rosenwasser, L. J., Coceani, F. 1984. The influence of lipoxygenase inhibitors on the *in vitro* production of human leukocytic pyrogen and lymphocyte activating factor (interleukin-1). *Int. J. Immunopharmacol.* 6: 43–50

83. Farrar, W. L., Humes, J. L. 1985. The role of arachidonic acid metabolism in the activities of interleukin 1 and 2. *J. Immunol.* 135: 1153–59

84. Parker, C. W., Stenson, W. F., Huber, M. G., Kelly, J. P. 1979. Formation of thromboxane B$_2$ and hydroxyarachidonic acids in purified human lymphocytes in the presence and absence of PHA. *J. Immunol.* 122: 1572–77

85. Kelly, J. P., Parker, C. W. 1979. Effects of arachidonic acid and other unsaturated fatty acids on mitogenesis in human lymphocytes. *J. Immunol.* 122: 1556–62

86. Leung, K. H., Ehrke, M. J., Mihich, E. 1982. Modulation of the development of cell-mediated immunity: Possible role of the products of the cyclo-oxygenase and the lipoxygenase pathways of arachidonic acid metabolism. *Int. J. Immunopharmacol.* 4: 195–204

87. Chang, J., Gilman, S. C., Lewis, A. J. 1986. Interleukin 1 activates phospholipase A$_2$ in rabbit chondrocytes: A possible signal for IL 1 action. *J. Immunol.* 136: 1283–87

88. Rossi, V., Breviario, F., Ghezzi, P., Dejana, E., Mantovani, A. 1985. Prostacyclin synthesis induced in vascular cells by interleukin-1. *Science* 229: 174–76

89. Agins, A. P., Hollmann, A. B., Agarwal, K. C., Wiemann, M. C. 1985. Detection of a novel cyclooxygenase metabolite produced by human promyelocytic leukemia (HL-60) cells. *Biochem. Biophys. Res. Comm.* 126: 143–49

90. Reiner, N. E., Malemud, C. J. 1985. Arachidonic acid metabolism by murine peritoneal macrophages infected with Leishmania donovani: *In vitro* evidence for parasite-induced alterations in cyclooxygenase and lipoxygenase pathways. *J. Immunol.* 134: 556–63

91. Billah, M. M., Bryant, R. W., Siegel, M. I. 1985. Lipoxygenase products of arachidonic acid modulate biosynthesis of platelet-activating factor (1-0-alkyl-2 - acetyl - sn - glycero - 3 - phosphocholine) by human neutrophils via phospholipase A2. *J. Biol. Chem.* 260: 6899–6906

92. McIntyre, T. M., Zimmerman, G. A., Prescott, S. M. 1986. Leukotrienes C$_4$ and D$_4$ stimulate human endothelial cells to synthesize platelet-activating factor and bind neutrophils. *Proc. Natl. Acad. Sci. USA* 83: 2204–8

93. Saito, H., Hirai, A., Tamura, Y., Yoshida, S. 1985. The 5-lipoxygenase products can modulate the synthesis of platelet-activating factor (alkyl-acetyl GPC) in Ca-ionophore A23187-stimulated rat peritoneal macrophages. *Prostaglandin Leuk. Med.* 18: 271–86

94. Engineer, D. M., Morris, H. R., Piper, P. J., Sirois, P. 1978. The release of prostaglandins and thromboxanes from guinea-pig lung by slow reacting substance of anaphylaxis and its inhibition. *Br. J. Pharmacol.* 64: 211–18

95. Humes, J. L., Opas, E. E., Galavage, M., Soderman, D., Bonney, R. J. 1986. Regulation of macrophages eicosanoid production by hydroperoxy- and hydroxy-eicosatetraenoic acids. *Biochem. J.* 233: 199–206

96. Hunter, J. A., Finkbeiner, W. E., Nadel, J. A., Goetzl, E. J., Holtzman, M. J. 1985. Predominant generation of 15-lipoxygenase metabolites of arachidonic acid by epithelial cells from human trachea. *Proc. Natl. Acad. Sci. USA* 82: 4633–37

97. Vanderhoek, J. Y., Karmin, M. T., Ekborg, S. L. 1985. Endogenous hydroxy-eicosatetraenoic acids stimulate the human polymorphonuclear leukocyte 15-lipoxygenase pathway. *J. Biol. Chem.* 260: 15482–87

98. Serhan, C. N., Hamberg, M., Samuelsson, B., Morris, J., Wishka, D. G. 1986. On the stereochemistry and biosynthesis of lipoxin B. *Proc. Natl. Acad. Sci. USA* 83: 1983–87

99. Fitzsimmons, B. J., Adams, J., Evans, J. F. Leblanc, Y., Rokach, J. 1985. The lipoxins. Stereochemical identification and determination of their biosynthesis. *J. Biol. Chem.* 260: 13008–12

100. Rokach, J., Adams, J., Fitzsimmons, B. J., Girard, Y., LeBlanc, Y., Evans, J. F. 1985. Chemical and enzymatic syntheses of lipoxin A: Stereochemical assignment of natural lipoxin A and its possible biosynthesis. *Adv. Prosta-*

glandin Thromb. Leuk. Res. 15: 309–12
101. Hansson, A., Serhan, C. N., Haeggstrom, J., Ingelman-Sundberg, M., Samuelsson, B. 1986. Activation of protein kinase C by lipoxin A and other eicosanoids. Intracellular action of oxygenation products of arachidonic acid. *Biochem. Biophys. Res. Comm.* 134: 1215–22
102. Ramstedt, U., Ng, J., Wigzell, H., Serhan, C. N., Samuelsson, B. 1985. Action of novel eicosanoids lipoxin A and B on human natural killer cell cytotoxicity: Effects on intracellular cAMP and target cell binding. *J. Immunol.* 135: 3434–38
102a. Parker, C. W., Kelly, J. P., Falkenheim, S. F., Huber, M. G. 1979. Release of arachidonic acid from human lymphocytes in response to mitogenic lectins. *J. Exp. Med.* 149: 1487–1503
103. Flesch, I., Ecker, B., Ferber, E. 1984. Acyltransferase-catalyzed cleavage of arachidonic acid from phospholipids and transfer to lysophosphatides in macrophages derived from bone marrow. Comparison of different donor- and acceptor substrate combinations. *Eur. J. Biochem.* 139: 431–37
104. Wallner, B. P., Mattaliano, Hession, C., Cate, R. L., Tizard, R., Sinclair, L. K., Foeller, C., Chow, E. P., Browning, J. L., Ramachandran, K. L., Pepinsky, R. B. 1986. Cloning and expression of human lipocortin, a phospholipase A_2 inhibitor with potential anti-inflammatory activity. *Nature* 320: 77–84
105. Parker, C. W. 1982. Pharmacologic modulation of release of arachidonic acid from human mononuclear cells and lymphocytes by mitogenic lectins. *J. Immunol.* 128: 393–96
106. Gurr, M. I. 1983. The role of lipids in the regulation of the immune system. *Prog. Lipid Res.* 22: 257–87
107. Van der Schroeff, J. G., Havekes, L., Weerheim, A. M., Emeis, J. J., Vermeen, B. J. 1985. Suppression of cholesteryl ester accumulation in cultured human monocyte-derived macrophages by lipoxygenase inhibitors. *Biochem. Biophys. Res. Comm.* 127: 366–72
108. Mathur, S. N., Field, F. J., Spector, A. A., Armstrong, M. L. 1985. Increased production of lipoxygenase products by cholesterol-rich mouse macrophages. *Biochim. Biophys. Acta* 837: 13–19

Ann. Rev. Immunol. 1987. 5:85–108

THE ROLE OF SOMATIC MUTATION OF IMMUNOGLOBULIN GENES IN AUTOIMMUNITY

Anne Davidson, Rachel Shefner, Avi Livneh and Betty Diamond

Departments of Medicine and Microbiology and Immunology, Albert Einstein College of Medicine, Bronx, New York 10461

INTRODUCTION

The humoral arm of the immune system consists of cells producing immunoglobulin molecules capable of recognizing the myriad of foreign or nonself molecules to which an individual is exposed during his or her lifetime. How does the immune system produce enough unique immunoglobulin molecules to recognize the vast number of antigens it may encounter without at the same time producing harmful antibodies directed toward self antigens? The first part of this puzzle has been solved by molecular genetic technology, and several mechanisms responsible for generating the diversity of immunoglobulin molecules have been identified. The relationship of autoantibodies to antibodies directed to foreign antigens is not yet understood, however, and studies of the molecular genetic basis for autoantibody production are just now beginning. In this chapter we examine the evidence that autoantibodies are derived from genes that encode protective antibodies rather than from genes uniquely committed to autoantibody production. We discuss two possible mechanisms for their production: (*a*) that they are encoded by germline immunoglobulin genes or (*b*) that they represent a somatic diversification of those genes.

The diversity of the antibody repertoire is generated by several mechanisms. Immunoglobulin heavy-chain and light-chain genes are composed of different gene segments, VDJC and VJC respectively, that undergo

85

0732–0582/87/0410–0085$02.00

rearrangement during B-cell ontogeny. Combinatorial diversity—the random joining of different VD and J heavy-chain gene segments or V and J light-chain gene segments—leads to the production of a large number of immunoglobulin heavy and light chains from a relatively restricted number of V_H, D_H, J_H and V_L, J_L gene segments. Imprecise joining of V_H with D_H, D_H with J_H, and V_L with J_L generates further diversity of heavy- and light-chain genes. Finally, there is random association of heavy and light chains. (Reviewed 1–3.)

In addition, over the past several years it has become clear that somatic mutation of already productively rearranged immunoglobulin genes is an important mechanism in expanding antibody diversity in the normal immune response. In this paper by *somatic mutation* we mean either point mutation, gene conversion, or recruitment of a new V_H or V_L gene (4–7). Although somatic mutation of IgM antibodies has been reported, (8, 9) in general, it is a characteristic of a secondary immune response and is found preferentially in IgG antibodies (10–18, and reviewed in 4, 19). That somatic mutation occurs is indisputable. The frequency with which it occurs is still unclear, and determinations of its frequency depend on the experimental approach of the investigators, on rough estimates of cell generation time, and on other unproven assumptions.

Many investigators have studied somatic mutation occurring in vivo by using hybridoma technology to capture B cells at various times during the immune response (10, 11, 13, 14, 16). These analyses are hampered by uncertainty about the genealogy of the immortalized B cells. Furthermore, recent increased estimates of the number of murine V_H genes—from 200 (20–22) to more than 1000 (23)—and the observation that two distinct germline genes can be more than 95% homologous (23a) together make it difficult to interpret some studies of hybridoma cell lines. It is therefore not always clear when two idiotypically similar monoclonal immunoglobulins differ from each other because of somatic mutation or when they represent different but highly homologous germline genes. It is easier to be certain that somatic mutation is indeed occurring in studies of cloned tissue culture lines (8, 24, 25) or in systems where there is a restricted number of germline genes (26, 27).

Presumably the somatic mutation that occurs in vivo is a random process and can affect the affinity (24. 25) and the antigenic (28, 28a) or idiotypic (29, 30) specificity of the antibody molecule. Although it is an important process in the maturation of a B-cell response, (4, 19, 31) somatic mutation may also lead to production of degenerate antibodies (24, 25) and perhaps to antibodies directed to self antigens (28). There must, therefore, be some regulation of the products of somatic mutation. In the normal immune response antigen, the idiotypic network, and regulatory T cells all function in modulating expression of B-cell clones and in determining which antibodies from among the innumerable molecules that can be made are actually present in serum. Usually an entire antibody repertoire can be formed without causing the clonal expansion of cells that make pathogenic autoantibodies. In some individuals, however, such autoantibodies are produced. It is not yet clear whether the genes that encode antibodies in the normal immune response are also used to encode these autoantibodies or whether there are separate germline genes that encode autoreactivity. Further, it is not clear whether pathogenic autoantibodies represent germline gene sequences or somatic mutations of germline gene sequences. Molecular studies directed to these issues are still in their infancy. An understanding of autoantibody production can help resolve many questions about autoimmunity: (a) What antigens and antigenic determinants elicit these antibodies; (b) What regulates their expression; and (c) What determines their pathogenicity? Answers to these questions should lead to better therapeutic interventions for autoimmune disease.

ARE THERE SPECIFIC AUTOIMMUNE GENES?

Investigators who study autoimmunity analyze either "natural" autoantibodies which can be found in all individuals or the potentially pathologic autoantibodies of autoimmune disease. Distinguishing between these two kinds of autoantibody is important because what similarities and differences exist between these sets of autoantibodies is not clear, nor is the relation of each to normal protective antibody molecules.

Furthermore, even in autoimmune disease, not all autoantibodies are alike, and antibodies directed to the same antigen may differ in their pathogenicity (32, 33). Various studies have implicated isotype (34), affinity (35, 36), and charge (37) in the pathogenicity of autoantibodies, but no immunochemical definition of a noxious autoantibody has yet emerged. Indeed, it appears that even such a hope may be simplistic, since autoreactive antibodies may be present in one individual with no clinical symptoms, while these same antibodies may cause disease when present in another individual. For example, anti-Ro antibodies present in the serum of a mother but of no known clinical consequence can cause disease when transmitted across the placenta to the fetus (37a).

Natural autoantibodies can be made by B cells present in all individuals (38–40). These IgM antibodies are widely cross-reactive with a number of autoantigens (41, reviewed in 42); often, although not always, they have a low affinity for antigen (40). In contrast, the IgG autoantibodies that are found in individuals with autoimmune disease are frequently highly specific and often have high affinity for their autoantigen (32, 33). It is not known whether the same or different genetic mechanisms give rise to both these classes of autoantibodies. Most investigators assume, however, that the genes that encode autoantibodies—both "natural" autoantibodies and the pathogenic autoantibodies of autoimmune disease—are present in all individuals. This assumption implies that there are no immunoglobulin genes giving rise to pathogenic autoantibody formation that are present only in autoimmune individuals or autoimmune strains of mice.

A genetic relationship between autoantibodies and normal antibodies has been assumed from the structural homology between these two sets of antibodies. Anti-DNA antibodies provide an example. Anti-DNA antibodies produced by hybridomas from autoimmune mice are similar in protein sequence to antibodies elicited by foreign antigen. Eilat et al sequenced a hybridoma anti-DNA antibody from an NZB/NZW mouse and found that the heavy-chain variable-region amino acid sequence is highly homologous to a BALB/c T15 germline gene that encodes anti-phosphorylcholine antibodies (43). Similarly Kofler et al found that the heavy-chain variable region of an anti-DNA hybridoma from an MRL/*lpr* mouse is homologous to the V_H of an antibody binding to the hapten nitrophenyl (44). An idiotypic analysis of anti-DNA autoantibodies in mice developing an SLE-like syndrome has revealed that although many anti-DNA antibodies share cross-reactive idiotypes (45, 46), there are idiotype-bearing antibodies present in serum that do not have specificity for DNA. Furthermore, these SLE idiotypes can be found in the sera of unimmunized nonautoimmune mice (47–49). Thus, anti-DNA antibodies are part of a larger population of idiotype-bearing antibodies that includes molecules presumably with specificity for other antigens. While antibodies bearing a cross-reactive idiotype are, in general, thought to be encoded by the same germline gene or gene family (50), a genetic analysis of idiotypically related anti-DNA antibodies shows that they can be derived from more than one V_H gene family (51). In this study, idiotypic cross-reactivity may result when the anti-DNA antibody shares either a heavy chain or light chain with nonautoreactive antibodies but possesses a pathogenic complementary chain. Such data have, nevertheless, led to the assumption that there are no separate genes to encode autoantibodies, but that autoantibodies derive from genes used to encode antibodies in the normal repertoire. Thus, these studies do not determine whether anti-DNA antibodies and the structurally related non–DNA binding antibodies arise from the same small number of germline genes or from a larger number of highly homologous genes.

It has similarly been demonstrated in humans that there are cross-reactive idiotypes on pathogenic autoantibodies in unrelated patients with autoimmune disease (52–55) and that these idiotypes are also present on nonautoreactive antibodies (52, 56–58). Lefvert et al, for example, have demonstrated that anti–acetylcholine receptor antibodies found in patients with myasthenia gravis share a cross-reactive idiotype and that idiotype-bearing antibodies without specificity for the acetylcholine receptor can be found in healthy relatives of these patients (56). Similar studies examining idiotypes on anti-DNA antibodies in patients with SLE also demonstrate the existence of idiotypic antibodies with no autoreactivity (57). In fact, some human monoclonal anti-DNA antibodies have been found to bind to a variety of bacterial antigens (59) as have some anti–acetylcholine receptor antibodies of patients with myasthenia gravis (59a)—possible evidence that there are no unique autoantibody genes. This interpretation must be viewed with caution since (a) idiotypically cross-reactive molecules can be encoded by different genes and even by genes from different gene families, and (b) where autoantibodies cross-react with bacterial antigens, it is important to determine whether the autoantibody is pathogenic and whether the antibacterial activity is protective. Although the genes encoding some autoantibodies have been sequenced, to date, germline V_H and V_L DNA sequences have not been obtained from autoimmune strains of mice or from humans with an autoimmune diathesis. It is, therefore, not yet known if particular germline immunoglobulin gene sequences may predispose to autoimmunity or if regulatory defects and antigenic exposure are the critical determinants of autoimmunity.

More convincing evidence for the absence of particular autoimmune immunoglobulin genes has been derived from genetic experiments with autoimmune mice. Autoimmune mice have been shown to have a variety of alterations in regulation of the immune response. They demonstrate an early maturation of the B-cell response and abnormalities in regulatory T cells (reviewed 60). These defects alone may account for autoimmunity. The *lpr* gene has been shown to accelerate the onset of autoimmune disease in MRL mice. When the *lpr* gene is bred into nonautoimmune strains of mice, these mice, with no change in their germline immunoglobulin genes, now produce a variety of autoantibodies (61). This proves that the genetic material for autoantibody production is present in nonautoimmune animals. It is interesting that these mice produce fewer IgG autoantibodies and have less pathology than do MRL/*lpr* mice. The *lpr* gene can induce autoreactivity but cannot alone induce autoimmune disease.

The most convincing data that genes for autoreactivity are present in all animals was shown by Datta and Schwartz. When an autoimmune NZB mouse was mated with a nonautoimmune SWR mouse, the progeny produced autoantibodies bearing the immunoglobulin allotype of the non-autoimmune parental strain and developed an immune complex glomerulonephritis (62). This suggests that a gene in SWR mice normally either not expressed or not used to make autoantibodies can encode autoantibodies when in the right genetic environment. Whether the autoantibodies encoded by SWR genes are deposited in the kidneys and cause disease is not yet determined. Yet such studies represent the most conclusive evidence that there is not a separate class of genes designated for autoantibody production.

ARE AUTOANTIBODIES DERIVED FROM UNMUTATED GERMLINE GENES?

Jerne has proposed that autoantibodies are among the first antibodies to be produced by the individual (63). These bind to self-antigens and are encoded by germline genes. In this scheme, self recognition is the force that maintains the germline gene repertoire through evolution. Also encoded by germline genes are anti-idiotypic antibodies that suppress the expression of these autoantibodies. Foreign antigens in this model might elicit anti-idiotypes that cross-react with antigen or might elicit anti–self antibodies that have somatically mutated to avoid suppression by anti-idiotype.

Supportive evidence for this theory is mainly circumstantial. Ample data indicate the presence of B cells capable of making autoantibodies in normal human beings and normal animals. Such data have been obtained by direct measurement of serum immunoglobulin autoreactivity (64), by using polyclonal B-cell activators to induce autoantibody formation in vitro and in vivo (65), or by capturing particular autoreactive B cells with the use of hybridoma technology (38–40, 65, 66). While autoreactive antibodies can be generated from B cells of all individuals, it is not clear how these antibodies relate to the autoantibodies of autoimmune disease. In general, the autoantibodies produced by the above methods are the IgM antibodies with extensive antigenic cross-reactivity and no known pathogenicity (reviewed 42).

Evidence that these "natural" autoantibodies are germline gene encoded has been obtained in a number of experimental systems. Using hybridoma technology, a higher percent of lines with autoreactivity are generated from neonatal mice than from adult mice (67). Hybridomas from germ-free mice, assumed to express the immunoglobulin gene repertoire prior to selection by antigen, likewise show more autoreactivity than hybridomas from mice whose B-cell repertoire has been formed by antigen (68). The presence of autoantibodies both in neonatal mice and in their presumed adult analogue, germ-free mice, supports the hypothesis of germline gene origin. In mice most of these antibodies may be derived from V_H genes from the two most 3′ gene families, Q52 and 7183 (69), genes preferentially used to encode the immunoglobulin repertoire early in the development of an animal (70).

Zanetti and colleagues derived a hybridoma from a neonatal mouse making an antithyroglobulin molecule; the hybridoma is idiotypically similar to a monoclonal antithyroglobulin antibody captured in a hybridoma from a mouse with experimental autoimmune thyroiditis. While these antibodies are not entirely identical, the 30 N-terminal amino acids of the heavy chains are identical as are the mRNA transcripts encoding these residues (71). This observation suggests that if natural autoantibodies are produced by unmutated immunoglobulin genes, autoantibodies present in autoimmune disease may also reflect unmutated gene sequences. In addition, an analysis of monoclonal antibodies derived from patients with myeloma, macroglobulinemia, and other B-cell tumors reveals that a high percentage of these display autoantibody activity (72, 73). Since there is some evidence that the malignant transformation that causes these diseases may occur early in the maturation of B cells, even in the pre–B cell stage (74), the high frequency of autoreactivity among these monoclonal antibodies may suggest that autoreactivity is an intrinsic property of some germline immunoglobulin gene sequences.

That "natural" antibodies may bear some structural homology to pathogenic autoantibodies is suggested in some instances by their sharing a cross-reactive idiotype (75). Schwartz and colleagues, for example, have generated human monoclonal IgM anti-DNA antibodies by fusing peripheral-blood B cells from a patient with systemic lupus erythematosus to a drug-marked lymphoblastoid cell line. These anti-DNA antibodies are thought to be encoded by a germline immunoglobulin gene because they are IgM antibodies (a class that, in general, shows few somatic mutations) and because they are broadly cross-reactive with nucleotide and phospholipid antigens (53; 76, reviewed in 77). Some of these monoclonal anti-DNA antibodies share a cross-reactive idiotype, 16/6, with an IgM myeloma protein WEA that binds a *Klebsiella* cell wall antigen. Amino acid sequencing of four monoclonal anti-DNA antibodies reveals that while the heavy-chain variable regions are not homologous with the WEA heavy chain, the light chains are highly homologous to the WEA light chain in the N-terminal 40 amino acid residues. The first framework of all five antibodies is identical, and there is only one amino acid difference between the anti-DNA antibodies and WEA in the first hypervariable region. These data are interpreted to show that the anti-DNA antibodies are encoded by a highly conserved germline gene (78). The 16/6 anti-idiotype also binds serum antibodies in a high proportion of patients with SLE and binds pathogenic (presumably IgG) antibodies present in SLE skin (79) and kidney (80) biopsies. This suggests some structural homology between the presumably germline–encoded monoclonal IgM anti-DNA antibody and the pathogenic anti-DNA antibodies of SLE. It does not, however, define the molecular genetic basis for that homology. It may be

(a) that the switch to IgG heavy chains alone renders these IgM antibodies pathogenic, (b) that concurrent with the switch to an IgG constant region, these antibodies acquire somatic mutations that increase their affinity for autoantigen without altering their idiotypic specificity, or (c) that they are encoded by genes homologous but not identical to those that encode pathogenic IgG autoantibodies.

Molecular evidence for the hypothesis that natural autoantibodies may be derived from germline genes early in the ontogeny of the immune system has been provided by Naparstek et al (81). They showed that hybridoma antibodies obtained from unimmunized mice and expressing the V_H Id^{cr} of antiarsonate antibodies are encoded by germline genes and possess a high incidence of low-affinity anti–single stranded DNA activity. None of these antibodies bind arsonate. The proportion of Id^{cr} positive antibodies that react with single-stranded DNA is markedly reduced if one analyzes hybridomas generated following immunization with arsonate. Those that now bind arsonate have somatically mutated gene sequences, and most of these no longer bind DNA. The authors interpret these data to show that germline genes encode autoreactivity and that somatic mutation leads to a loss of autoreactivity. However, since only those cells making an antibody that binds arsonate are expanded in vivo following immunization with arsonate, the entire repertoire of somatically mutated genes was not surveyed. Immunization with another antigen might have elicited somatic mutants of the Id^{cr} gene that bind DNA with a higher affinity than the germline gene–encoded antibodies. While this experiment is consistent with the Jerne hypothesis that genes encoding "natural" autoantibodies mutate to form the antibodies to exogenous antigen, it does not explain the molecular genetic origin of autoaggressive antibodies.

It is clear from the above experiments that there are B cells present in normal humans and mice which can produce autoantibodies and that these autoantibodies are probably encoded by germline DNA sequences. It remains to be determined what the relationship is between these antibodies and autoimmunity. Although these cells are not clonally deleted, it is unclear whether all of the B cells that form hybridomas in vitro or that can be activated by B-cell mitogens to secrete autoantibodies actually produce antibody during the course of the individual's lifetime. Coutinho and colleagues have recently shown that some autoreactive B cells are activated in vivo. They found that there was a higher frequency of B cells making antithyroglobulin antibodies and anti–mouse red blood cell antibodies in the fraction of large activated B cells than in the small resting B-cell population from mouse spleens. Hybridomas derived from these activated B cells revealed that some of the antithyroglobulin antibody being produced has an affinity of 10^{-6} to 5×10^{-7} m^{-1} (69). It is clear that these autoantibodies while of relatively high affinity are nonpathogenic, perhaps due in part to the antigenic determinant they bind (82). While some B cells making "natural" autoantibodies may be activated in vivo, others, as Jerne suggests, may be constantly suppressed (63). Furthermore, because many "natural" autoantibodies have low affinity for self antigens, it is not clear whether they can be activated to secrete antibody by auto-antigen or whether some other antigenic trigger is required.

The evolution of thinking about the molecular genetic origins of auto-antibodies is exemplified in studies of human rheumatoid factors, perhaps the most studied group of autoantibodies. Monoclonal human rheumatoid factors from patients with Waldenstrom's macroglobulinemia are mostly of the K111b light-chain isotype, and many share a cross-reactive idiotype Wa related to the primary structure of the light chain of the antibody molecule (83, 84). In order to examine the structural and genetic basis for this idiotype, peptides corresponding to the second and third hypervariable regions of the light chain of a Wa-positive immunoglobulin were synthesized and antipeptide antibodies prepared (85). One of these antibodies, directed against a peptide corresponding to the second hypervariable region of the light chain of a Wa-positive monoclonal protein, was found to bind to all Wa-positive monoclonal antibodies. Reactivity with the antibody to the third hypervariable was found in a significant percentage but not all of Wa-positive antibodies (86). Amino acid sequences of the light chains expressing the Wa idiotype display extensive amino acid homology (87). These data suggest that many monoclonal rheumatoid factors arise from the same immunoglobulin gene or from closely related germline genes; perhaps they may reflect germline gene sequences.

To investigate the hypothesis at the molecular level that rheumatoid factors are encoded by germline gene sequences, Fong et al cloned and sequenced the genes encoding rheumatoid factor light chains. Cells taken from a patient with chronic lymphocytic leukemia were producing an IgM cryoglobulin exhibiting rheumatoid factor activity and were reactive with the antiserum to the second hypervariable region. These were used to clone the rheumatoid factor light-chain gene. The amino acid sequence of this cloned light-chain gene, however, revealed 14 amino acid differences from the proposed Wa-prototype sequence; this (87a) demonstrates that there may be more than one germline gene sequence giving rise to idiotypically cross-reactive light chains.

Further cloning experiments in which a human K111 probe was used to screen a genomic fetal liver library have led recently to the isolation of a clone HumKv 321, from which residues 42–95 encode a protein sequence identical to that of the Wa-prototype rheumatoid factor (88). This result proves at a molecular level that monoclonal rheumatoid factor light chains derive from highly conserved germline genes. It is still unclear, however, whether the differences in rheumatoid factor light-chain sequences are due to use of different but highly homologous germline genes or whether they are due to somatic events.

It is also unclear whether these monoclonal rheumatoid factors are a mirror of the rheumatoid factors found in patients with rheumatoid arthritis and therefore whether rheumatoid factors can be used to draw conclusions about events occurring in autoimmune disease. Two independent experiments suggest that this may not be the case. A murine monoclonal antibody recognizing Wa-positive rheumatoid factor binds kappa light chains of approximately 50% of monoclonal IgM rheumatoid factors and is a serological marker for the protein product of the gene(s) hybridizing to the HumKv 321 clone. Antibodies recognized by this anti-idiotype can be found in human serum (89). However, when rheumatoid factors from patients with rheumatoid arthritis were isolated, it was found that although the light chains express the idiotype defined by polyclonal antiserum directed to the second hypervariable region of the prototypic Wa light chain, they do not react with either the antiserum to the third hypervariable region or with the monoclonal anti-idiotype (90). These data could be interpreted to suggest either that rheumatoid factors in patients with rheumatoid arthritis arise from germline genes different from those giving rise to monoclonal rheumatoid factors or that extensive somatic diversification of the germline rheumatoid factor gene has occurred especially within the third hypervariable region resulting in loss of particular idiotypes. Another study has revealed that many rheumatoid factor light chains isolated from patients with rheumatoid arthritis belong to different V_K subgroups; this indicates that the monoclonal rheumatoid factors that express predominantly the VK111b light chain do not reflect the population of rheumatoid factors found in patients with rheumatoid arthritis (91). These studies of rheumatoid factors exemplify the difficulty of assuming that data derived from monoclonal antibodies arising from tumors or hybridomas explain the autoimmune response in vivo. Further studies are clearly needed to elucidate the genetic basis of rheumatoid factor formation. Finally, rheumatoid factors are a peculiar autoantibody: The role they play in the pathogenesis of autoimmune disease is not understood, and they are routinely produced by nonautoimmune individuals.

DOES SOMATIC MUTATION CONTRIBUTE TO AUTOANTIBODY FORMATION?

While much attention has been paid to the proposition that autoantibodies are encoded by immunoglobulin germline gene DNA sequences, evidence is accumulating that some autoantibodies in autoimmune disease may result from somatic mutation of germline DNA sequences.

Bursectomized embryonic chickens were studied for their ability to form autoantibodies. Such chickens can still produce immunoglobulins in normal amounts; but as demonstrated by isoelectric focussing, the population of antibodies produced is less diverse than that seen in normal chicken serum, and most of the light chains focus at a basic pH. The bursectomy appears to cause a defect in the ability to generate diversity. When the bursectomized chickens were immunized with a panel of foreign antigens, no specific antibody response was elicited. Moreover, no "natural" autoantibodies were detectable among the population of antibodies made by these chickens (92). These data were interpreted to suggest that even the production of nonpathogenic "natural" autoantibodies requires the ability to diversify antibody molecules somatically. What is actually required for the production of autoantibodies is not clear. Studies in mice have shown that the *xid* gene will retard autoantibody formation in autoimmune mouse strains (93). This gene interferes with the normal B-cell response to polysaccharide antigens (94). A recent study shows that B cells from NZB mice making anti-DNA antibodies will continue to proliferate when transferred into a nonautoimmune histocompatible *xid* mouse (95). This suggests that some autoimmune B cells may proliferate in the absence of T-cell help. A number of studies, however, suggest that T cells are needed both for somatic mutation of antibody molecules (96) and for an autoimmune response (60, 97–99). Pathogenic autoantibodies that are almost exclusively IgG may also require T-cell help. While the autoreactive T cells involved in autoantibody production have not been studied in a number of diseases, antigen-specific T helper cells have been cloned from patients with myasthenia gravis (100) and thyroiditis (101). Whether these T cells permit the expression of generally unexpressed germline gene sequences or whether they abet somatic diversification of immunoglobulin genes or both is not yet clear.

Anti-Sm antibodies in MRL/*lpr* mice are perhaps the best example of somatic mutation generating autoreactivity in vivo (reviewed 102). MRL/*lpr* mice spontaneously develop autoantibodies and a lupus-like syndrome (60). Like many patients with SLE, these mice produce antibodies to the nuclear antigen Sm ; (103) however, levels of these antibodies are highly variable among individual mice, and 50% of MRL/*lpr* mice never produce anti-Sm antibodies. This 50% is not genetically determined, and one cannot breed mice which reliably produce or do not produce anti-Sm antibodies (104–106). Anti-Sm antibodies are all of the IgG isotype. There are no anti-Sm antibodies of the IgM isotype, and no IgM anti-Sm precursor cells are detectable in the spleens of MRL/*lpr* mice (107). Anti-idiotypic antibodies to anti-Sm have revealed that anti-Sm antibodies belong to a large idiotypic family, only a small percent of which are Sm binding (101, 108). In sum, these data suggest that anti-Sm antibodies are not germline gene encoded but are derived by a mechanism such as somatic mutation that: (a) distinguishes IgM from IgG antibodies, (b) can alter the antigenic specificity of idiotypically related molecules, and (c) can lead to their appearance in only a subset of genetically identical animals. Until sequencing studies are done, it remains a formal possibility that an IgG constant region is necessary for Sm binding; this, however, would not explain the observation that only 50% of an inbred strain of mouse expresses these antibodies and that this phenotype does not breed true.

In our laboratory we have been studying the anti-double-stranded DNA antibodies of the human disease SLE. We generated a monoclonal anti-idiotype to anti-DNA antibodies purified from a single patient with SLE (55, 109, 110). This anti-idiotype, 3I, recognizes a determinant present on κ light chains of anti-DNA antibodies. Because 3I reactivity with isolated light chains is essentially the same for SLE patients and normals, while 3I reactivity with intact Ig is much higher for SLE patients, we believe that most heavy chains associate with 3I positive light chains in a way that obscures the 3I determinant. Some heavy chains preferentially associated with 3I positive light chains in SLE patients reveal the 3I determinant (111). The 3I anti-idiotype does not react with autoantibodies in general because it does not react with serum from patients with rheumatoid arthritis or with drug-induced SLE in which antihistone antibodies are present (109). High-titered 3I reactivity is found on anti-DNA antibodies in over 85% of more than 200 unrelated patients with SLE. In these patients 3I recognizes a variable but substantial portion of anti-DNA antibodies. In 4 patients studied, 3I reactivity was present on 40–90% of the anti-DNA antibodies in serum (109). In addition, 3I reactive antibodies have been found in skin and kidneys of patients with SLE, which suggests that among the antibodies recognized by 3I are pathogenic anti-DNA antibodies (112). However, not all the 3I reactive antibodies in SLE patients have specificity for DNA. Furthermore low-titered 3I reactivity is found in normal individuals, and none of these 3I reactive antibodies bind DNA (111). The 3I reactive anti-DNA antibodies are therefore part of a larger family of idiotypic antibodies, and this idiotype is used to make antibodies in the normal immune response.

To determine whether the preferential use of the 3I idiotype might contribute to the genetic predisposition to SLE, we examined sera from 27 members of three human kindreds with familial SLE. There were 8 patients with SLE and 19 family members who were not affected. Sera were examined for idiotype expression and DNA binding. Six of 8 lupus patients and 15 of 19 family members had high-titered 3I reactivity. When sera were examined under conditions that dissociate immune complexes, 6 patients and 4 family members were found to have anti-DNA antibodies. In all these individuals, 3I reactivity was present on anti-DNA antibodies, and anti-DNA activity was preferentially expressed by cationic 3I reactive antibodies. The 11 family members not affected showed elevated titers of 3I reactive antibodies that did not bind DNA (57). The presence of antibodies bearing SLE idiotypic reactivity in relatives of patients with SLE has also been reported by Isenberg et al (113). These findings suggest that expression of these cross-reactive idiotypes may be genetically regulated.

The finding of anti-DNA antibodies in family members raises an unsolved problem. Four individuals with no evidence of clinical disease had anti-DNA antibodies indistinguishable from those present in SLE patients. These antibodies are cationic, IgG, idiotypically cross-reactive antibodies which, like lupus antibodies, can precipitate double-stranded DNA (57). It is not clear, therefore, what accounts for the pathogenicity of autoantibodies—whether there was subclinical disease present in the four individuals with these antibodies, whether the chronicity with which autoantibodies are present is crucial, whether there are as yet undefined differences between these and pathogenic autoantibodies, or whether other immunologic or nonimmunologic variables are critical in determining pathology.

Our finding of individuals with high-titered 3I reactivity but no anti-DNA activity has led us to ask whether DNA is the antigen that elicits high-titered expression of the 3I idiotype. It is notoriously difficult to elicit anti-DNA antibodies by immunization with DNA (114). Some anti-DNA antibodies, however, cross-react with phospholipids and phosphorylated proteins, and immunization with cardiolipin leads to production of anti-DNA antibodies (115). Also, as mentioned, in two instances it has been shown that murine monoclonal anti-DNA antibodies use variable region genes highly homologous to those used in the response to the exogenous antigens nitrophenyl and phosphorylcholine (43, 44). These observations have led to the suggestion that anti-DNA antibodies may represent a pathogenic cross-reactivity of antibodies elicited by a microbial antigen. Our finding (described below) that somatic mutation can convert an anti-bacterial antibody to an anti-DNA antibody (28) has led us to suggest that anti-DNA antibodies may represent somatic mutants of antibacterial antibodies. The genetic predisposition to SLE may result, in part, from preferential use in the normal immune response of variable region genes that require only a few mutations to acquire autoreactivity.

To examine the structural differences between idiotype-bearing antibodies that bind DNA in the human disease SLE and those that do not, we have analyzed sera from 706 patients with multiple myeloma for expression of the 3I idiotype (116). Myeloma proteins are a source of monoclonal immunoglobulin molecules, some of which will reflect germline gene sequences and some of which will reflect somatic mutations. We wanted to know how many myeloma proteins express the 3I idiotype and whether they bind to double-stranded DNA. We then wanted to examine the differences between those 3I positive myelomas that bind DNA and those that do not. The distribution of the heavy- and light-chain classes of the myelomas studied reflected that of normal serum immunoglobulin. Of the 706 myelomas 82 showed high-titered 3I reactivity. All of these myelomas bear κ light chains, and this is consistent with previous data in patients with SLE which demonstrated that 3I recognizes an epitope on κ light chains of anti-DNA antibodies.

When displayed on isoelectric focussing gels, then transferred to nitrocellulose and examined for binding of radiolabelled DNA, 29 of the 82 3I reactive myelomas were found to bind to double-stranded DNA. All the DNA binding myelomas focus at a pI > 7. Not all the 3I positive cationic antibodies are DNA binding, however, and this indicates that antibody charge is not the only determinant of antigenic specificity. When DNA binding was correlated with immunoglobulin isotype we found that 26/49 IgG myelomas, 2/25 IgM myelomas, and 1/8 IgA myelomas bind double-stranded DNA. The difference between the proportion of IgG and the proportion of IgM myelomas that bind DNA was significant (p < 0.001). The high frequency of IgG compared with IgM DNA-binding myelomas suggests to us that in this system DNA binding does not typically represent germline-encoded DNA sequences as reflected in IgM antibodies but more likely is acquired by somatic mutation of germline genes. To examine this further, we analyzed sera from 23 IgM non-DNA binding myelomas, 22 IgG non-DNA binding myelomas, and 25 IgG DNA binding myelomas by 2D gel electrophoresis. This enabled us to determine the charge heterogeneity of the 3I reactive light chains in IgM and IgG myelomas. We found that there was markedly increased charge heterogeneity of the IgG light chains of both DNA and non-DNA binding myelomas compared with light chains of the IgM myelomas. We hypothesize that this charge heterogeneity reflects somatic mutations in the IgG antibodies, some of which have resulted in the acquisition of autoreactivity.

Although we feel that somatic mutation is the most likely mechanism for generation of DNA-binding 3I-positive myelomas, we are not able to exclude two additional possibilities: (a) the use of an IgG constant region may in some way change the conformation of the variable region and lead to a change in the antigenic specificity of the antibody (116a), and (b) further rearrangements of different variable region genes are occurring in the IgG myelomas. Such secondary rearrangements have been reported in an in vitro cell culture system (117), but in this instance the secondarily rearranged variable regions would have to be variable regions not used in forming IgM antibodies. The antigenic specificity of the non–DNA binding 3I positive myelomas is not known; however, three of the myeloma proteins, one of which also binds DNA, were found to bind to phosphorylcholine (A. Davidson, B. Diamond, unpublished). Interestingly, the heavy-chain variable regions of some human IgM anti-DNA monoclonal antibodies possess a high degree of homology to a germline human V_H gene, which is similar to the murine V_H that encodes anti-phosphorylcholine antibodies (118).

When the sequences of sequenced IgM anti-DNA antibodies (78) and of WEA are compared to the first 30 amino acids of 3I reactive high-affinity IgG anti-DNA antibodies isolated from a patient with SLE, one sees extensive homology in the first framework region of the light chain but many differences in the first hypervariable region (A. Davidson, B. Diamond, unpublished). It is still unclear, therefore, whether the 3I-positive IgG antibodies arise from different light-chain germline genes from the IgM monoclonal antibodies or whether they are derived from the same gene but diversify by somatic mutation.

We have generated another anti-idiotype 8.12 to anti-DNA antibodies found in SLE serum. The 8.12 recognizes a determinant on λ light chains of a subset of anti-DNA antibodies. Like 3I, it recognizes a cross-reactive idiotype on pathogenic anti-DNA antibodies but also recognizes antibodies in SLE patients and in normal individuals with no specificity for DNA (119). Analysis of 24 myeloma proteins bearing the 8.12 idiotype revealed that 12 bind DNA; 0 of 8, IgM; 1 of 2, IgA; and 11 of 14, IgG proteins (A. Livneh, B. Diamond, unpublished). The 8.12 idiotype is therefore on antibodies with and without autoreactivity, and DNA binding correlates with IgG isotype in this idiotype system also, suggesting again that specificity for DNA may be somatically generated and that the germline gene encodes antibodies with a different specificity.

The analysis that has been undertaken for both the murine anti-Sm system and the human anti-DNA system suggests that in some instances autoreactivity is not germline encoded but is acquired by a random mechanism such as somatic mutation. The genetic predisposition to autoimmune disease suggests some germline genes such as those of the 3I family may need to acquire only a few somatic mutations to generate autoreactivity. Alternatively, increased mutation rates may occur in autoimmune individuals so that the likelihood of randomly generating an autoantibody is increased. Other as yet only partially analyzed regulatory defects in the T-cell arm of the immune system in autoimmune mice and humans may permit the expansion of these ubiquitously arising but usually suppressed B-cell clones.

Recently we have been examining the role of somatic mutation in autoantibody production in a murine system. The murine BALB/c myeloma cell line S107 produces an antibody against phosphorylcholine (PC), the major antigen of the pneumococcal cell wall (120). The S107 protein bears the T15 idiotype which is the dominant idiotype present in the in vivo anti-PC response in many strains of mice and which reflects a germline gene sequence (31). We have analyzed a tissue culture–derived somatic mutant of S107, U4, that has a single amino acid substitution in the first hypervariable region of the heavy chain. The substitution of alanine for glutamic acid at residue 35 in the first hypervariable region alters the antigenic specificity of the antibody without altering expression of the T15 idiotype. The U4 protein has a markedly decreased affinity for PC but has acquired reactivity with double-stranded DNA and some other phosphorylated molecules. This change in antigenic specificity is probably related to conformational changes in the antigen binding site. The glutamic acid at residue 35 usually forms a hydrogen bond with a tyrosine at residue 94 of the light chain which is thought to be important in forming the antigen binding site. The substitution of an alanine for the glutamic acid prevents the formation of this bond and may distort the antigen binding pocket (28).

The S107-U4 system provides a model for understanding the relationship between an antibody to a bacterial antigen and an autoantibody and shows how an autoantibody can arise from an antibacterial antibody. It remains however to be determined whether the U4 protein is pathogenic and whether it arises in vivo. We have demonstrated that some anti-DNA antibodies in the serum of MRL mice in vivo bear the T15 idiotype. In addition, we have found anti-DNA antibodies in the serum of NZB/W mice bearing the T15 idiotype (112). These data support the studies of Rudikoff and colleagues, who sequenced heavy and light chains from anti-DNA hybridomas from an NZB/W mouse. They found one heavy chain highly homologous to BALB/c T15 germline sequences (43). Similarly, they support findings of Barrett & Trepicchio who sequenced an anti-DNA hybridoma from an MRL/lpr mouse and found one derived from a T15 germline gene (51). Because the T15 germline gene sequences in autoimmune strains are only now being determined, it is not possible to know from the sequences of anti-DNA antibodies derived from the T15 gene family whether they represent germline sequences or somatic mutants. Such information should soon be available.

We have recently undertaken studies in the BALB/c mouse to examine the relationship between antibacterial antibodies and autoantibodies within the T15 gene family. We have developed two protocols for eliciting T15 anti-DNA antibodies in this nonautoimmune strain. One method involves immunization with monoclonal anti-IJ^d antibody. I-J is a marker on T suppressor cells; anti-I-J antibodies and anti-idiotypes to anti-I-J antibodies can block T-suppressor-cell function in vitro (121). Mice immunized with anti-I-J^d develop anti-idiotypes to anti-I-J^d and significant titers of anti-DNA antibodies (112). It may be that the in vivo consequence of this immunization is to generate anti-idiotypic antibodies that block T-suppressor-cell function and that the presence of anti-DNA antibodies in these mice implicates T suppressor cells in the normal regulation of autoantibody formation. This is consistent with studies by Cooke and colleagues, who show that T suppressor cells can abort production of antierythrocyte antibodies and hemolytic anemia in mice (122), and also with studies by Gibson et al, who show that anti-I-J antibody given to pregnant mice causes autoantibody production in the offspring (123).

We have generated hybridoma cell lines making anti-DNA antibodies from spleen cells of anti-I-J^d immunized mice. Some of these antibodies possess T15 serologic determinants and express the T15 heavy-chain gene. Some of the hybridomas produce antibodies that bind PC and DNA; others bind only DNA. By sequencing these antibodies, we will determine whether they arose by unusual combinatorial or junctional rearrangements, or whether they reflect somatic mutations. A knowledge of the genetic mechanism involved in generating an antibody molecule and of the antigenic specificity of that antibody may provide clues about the triggers for antibody production and whether they are antigen or anti-idiotype.

The other protocol that has led to the production of T15 anti-DNA antibodies is to immunize BALB/c mice with PC-KLH. It is apparent from the S107-U4 paradigm that DNA binding can be generated from PC-binding cell lines in a very few mutational steps. Therefore, it is quite probable that during the clonal expansion of cells stimulated by immunization with the bacterial antigen PC, antibodies to DNA will be generated. Whether these are antigenically cross-reactive, or are somatic mutants of anti-PC antibodies that no longer bind PC, or are antibodies with a cross-reactive idiotype but different antigenic specificity needs elucidation but may be determined in part by analysis of antigenic specificity. It is also not clear whether these antibodies will be pathogenic or how they are regulated. We have therefore selected several T15-bearing anti-DNA antibody–producing hybridoma cell lines that were generated 7 days after immunization with PC-KLH; we will use these for sequence analysis and detailed determination of specificity. We can study the kinds of anti-DNA antibodies produced by these two immunization protocols and their relationship to the germline–encoded antibacterial antibodies in a system where the germline sequences are known. This will permit a greater understanding of the role of somatic mutation in generating DNA-binding specificity; perhaps we will also understand the regulatory mechanisms that govern expression of autoantibodies.

To understand the molecular genetic basis for autoantibody production requires a better definition of the term autoantibody. We must learn to distinguish "natural" nonpathogenic autoantibodies from noxious autoantibodies and to determine the role of antibody isotype, charge, affinity, and fine specificity in pathogenicity. Only when we can identify pathogenic autoantibodies can we determine whether they reflect unmutated gene sequences or whether the genes encoding them have undergone somatic diversification, point mutation, gene conversion, or secondary rearrangements. Most immunologists assume that there are no unique genes present only in rare individuals for the production of pathogenic autoantibodies. If so, then delineation of the genealogy of B cells making such antibodies is a first step in understanding the regulation of their production. Such information may help identify the antigenic trigger to autoimmune disease. In animal models, a number of triggers have been identified. Denatured antigen or xenogenic antigen is often used to induce autoimmunity (124, 125). Immunization with cardiolipin induces anti-DNA antibodies; this shows that a cross-reacting antigen can elicit autoantibodies (114). The studies of Onodera et al provide the first model of a viral antigen that elicits long-term autoreactivity (126, 127). A more detailed analysis of the molecular genetic origin of pathogenic autoantibodies and their antigenic and idiotypic specificity may lead to an understanding of whether they are elicited by autoantigens, native or altered in some way; whether they are cross-reactive with microbial antigens; or whether they are selected via an idiotypic network; and, finally, how these antibody molecules escape normal regulatory networks.

ACKNOWLEDGMENTS

This work was supported by NIH grants AM32771, AI10702, and CA13330. Anne Davidson is a Damon-Runyon fellow; Avi Livneh is a Camp David Fellow and a recipient of a Fulbright award. Betty Diamond is an Established Investigator of the American Heart Association and a recipient of an Irma T. Hirschl Award.

We would like to thank B. Birshtein, S. Behar, and M. Scharff for critical reading of this manuscript and Donna Jackson for her patient secretarial assistance.

Literature Cited

1. Tonegawa, S. 1983. Somatic generation of antibody diversity. *Nature* 302: 575–81
2. Honjo, T., Habu, S. 1985. Origin of immune diversity: Genetic variation and selection. *Ann. Rev. Biochem.* 54: 803–30
3. Leder, P. 1983. Genetic control of immunoglobulin production. *Hosp. Practice* 18: 73–82
4. Milstein, C. 1986. From antibody structure to immunological diversification of immune response. *Science* 231: 1261–68
5. Clearly, M. L., Meeker, T. C., Levy, S., Lee, E., Treia, M., Sklar, S., Levy, R. 1986. Clustering of extensive somatic mutations in the variable region of an immunoglobulin heavy chain gene from a human B cell lymphoma. *Cell* 44: 97–106
6. Clarke, S. H., Rudikoff, S. 1984. Evidence for gene conversion among immunoglobulin heavy chain variable region genes. *J. Exp. Med.* 159: 773–82
7. Krawinkel, V., Zoebelein, G., Bruggemann, M., Radbruch, A., Rajewsky, K. 1983. Recombination between antibody heavy chain variable region genes: Evidence for gene conversion. *Proc. Natl. Acad. Sci. USA* 80: 4997
8. Wabl, M., Burrows, P. D., von Gabain, H., Steinberg, C. 1985. Hypermutation at the immunoglobulin heavy chain locus in a pre-B cell line. *Proc. Natl. Acad. Sci. USA* 82: 479–82
9. Hartman, A. B., Rudikoff, S. 1984. V_H genes encoding the immune response to B-(1,6)-galactan: Somatic mutation of IgM molecules. *EMBO J.* 3: 3023–30
10. Rudikoff, S., Pawlita, M., Pumphrey, J., Heller, M. 1984. Somatic diversification of immunoglobulins. *Proc. Natl. Acad. Sci. USA* 81: 2162–66
11. Sablitzky, F., Wildner, G., Rajewsky, K. 1985. Somatic mutation and clonal expansion of B cells in an antigen-driven immune response. *EMBO J.* 4: 345–50
12. Gearhart, P., Johnson, N., Douglas, R., Hood, L. 1981. IgG antibodies to phosphorylcholine exhibit more diversity than their IgM counterparts. *Nature* 291: 29–34
13. Griffiths, G. M., Berek, C., Kaartinen, M., Milstein, C. 1985. Somatic mutation and the maturation of immune response to 2-phenyl oxazolone. *Nature* 312: 271–75
14. Berek, C., Griffiths, G. M., Milstein, C. 1985. Molecular events during maturation of the immune response to oxazolone. *Nature* 316: 412–18
15. Crews, S., Griffin, J., Huang, H., Calame, K., Hood, L. 1981. A single V_H gene segment encodes the immune response to phosphorylcholine: Somatic mutation is correlated with the class of the antibody. *Cell* 25: 59–66
16. McKean, D., Huppi, K., Bell, M., Standt, L., Gerhard, W., Weigert, M. 1984. Generation of antibody diversity in the immune response of Balb/c mice to influenza virus hemagglutinin. *Proc. Natl. Acad. Sci. USA* 81: 3180–84
17. Clarke, S. H., Huppi, K., Ruezinsky, D., Staudt, L., Gerhard, W., Weigert, M. 1985. Inter- and intraclonal diversity in the antibody response to influenza hemagglutinin. *J. Exp. Med.* 161: 687–704
18. Wysocki, L., Manser, T., Gefter, M. L. 1986. Somatic evolution of variable region structures during an immune response. *Proc. Natl. Acad. Sci. USA* 83: 1847–51
19. Manser, T., Wysocki, L. S., Gridley, T., Near, R. I., Gefter, M. L. 1985. The molecular evolution of the immune response. *Immunol. Today* 6: 94–101
20. Dildrop, R. 1984. A new classification of mouse V_H sequences. *Immunol. Today* 5: 85–86
21. Brodeur, P. H., Riblet, R. 1984. The immunoglobulin heavy chain variable region (IgG-V) locus in the mouse. I. One hundred IgG-V genes comprise seven families of homologous genes. *Eur. J. Immunol.* 14: 922–30
22. Cory, S., Tyler, B. M., Adams, S. M. 1981. Sets of immunoglobulin V kappa genes homologous to the cloned V kappa sequences: Implication for the number of germ line V kappa genes. *J. Mol. Appl. Genet.* 1: 103–16
23. Zeelon, E. P., Bothwell, A. L. M., Kantor, F., Schechter, I. 1981. An experimental approach to enumerate the genes coding for immunoglobulin variable regions. *Nucleic Acids Res.* 9: 3809–20
23a. Hood, L. 1986. Presented at the UCLA Symposia: B and T Cell Repertoire. Lake Tahoe, Nevada, January 1986
24. Cook, W. P., Scharff, M. D. 1977. Antigen binding mutants of mouse mye-

loma cells. *Proc. Natl. Acad. Sci. USA* 74: 5687–91

25. Cook, W. D., Rudikoff, S., Giusti, A. M., Scharff, M. D. 1982. *Proc. Natl. Acad. Sci. USA* 79: 1240–44

26. Eisen, H. N., Reilly, E. B. 1985. Lambda chains and genes in inbred mice. *Ann. Rev. Immunol.* 3: 337–65

27. Reynaud, C. A., Anguez, V., Dahan, A., Weill, J. C. 1985. A single rearrangement event generates most of the chicken immunoglobulin light chain diversity. *Cell* 40: 283–91

28. Diamond, B., Scharff, M. D. 1984. Somatic mutation of the T15 heavy chain gives rise to an antibody with autoantibody specificity. *Proc. Natl. Acad. Sci. USA* 81: 5841–44

28a. Schiff, C., Milili, M., Hue, I., Rudikoff, S., Fougereau, M. 1986. Genetic basis for expression of the idiotypic network. One unique Ig V_H germ line accounts for the major family of Ab1 and Ab3 (Ab1′) antibodies of the GAT system. *J. Exp. Med.* 163: 573–87

29. Radbruch, A., Zaiss, S., Kappen, C., Bruggemann, M., Beyreuther, K., Rajewsky, K. 1985. Drastic change in idiotypic but not antigen-binding specificity of an antibody by a single amino acid substitution. *Nature* 315: 506–8

30. Pollock, B. A., Kearney, J. F. 1984. Identification and characterization of an apparent germ line set of auto-anti-idiotypic regulatory B lymphocytes. *J. Immunol.* 132: 114–21

31. Perlmutter, R. M., Crews, S. T., Douglas, R., Sorensen, G., Johnson, N., Nivera, N., Gearhart, P., Hood, L. 1984. The generation of diversity in phosphorylcholine binding antibodies. *Adv. Immunol.* 35: 1–37

32. Hahn, B. H. 1982. Characteristics of pathogenic subpopulations of antibodies to DNA. *Arthritis Rheum.* 25: 747–52

33. Chubick, A. 1980. DNA antibodies in systemic lupus erythematosus and pseudolupus syndrome. *Adv. Intern. Med.* 26: 467–87

34. Schur, P. H., Monroe, M., Rothfield, N. 1972. The γG subclass of antinuclear and antinucleic acid antibodies. *Arthritis Rheum.* 15: 174–82

35. Leon, S. A., Green, A., Ehrlich, G. E., Poland, M., Shapiro, B. 1977. Avidity of antibodies in SLE. Relations to severity of renal involvement. *Arthritis Rheum.* 20: 23–29

36. Winfield, J. B., Faiferman, I., Koffler, D. 1977. Avidity of anti-DNA antibodies in serum and IgG glomerular eluates in patients with systemic lupus erythematosus. Association of high avidity anti-native DNA antibody with glomerulonephritis. *J. Clin. Invest.* 59: 90–96

37. Ebling, F., Hahn, B. 1980. Restricted subpopulations of DNA antibodies in kidneys of mice with systemic lupus. *Arthritis Rheum.* 23: 392–403

37a. Scott, J. S., Maddison, P. J., Taylor, P. V., Esscher, E., Scott, O., Skinner, R. P. 1983. Connective-tissue disease, antibodies to ribonucleoprotein and congenital heart block. *New Engl. J. Med.* 309: 209–12

38. Prabhakar, B. G., Segusa, J., Onodera, T., Notkins, A. L. 1984. Lymphocytes capable of making monoclonal autoantibodies that react with multiple organs are a common feature of the normal B cell repertoire. *J. Immunol.* 133: 2815–17

39. Cairns, E., Block, J., Bell, D. A. 1984. Anti-DNA autoantibody producing hybridomas of normal human lymphoid cell origin. *J. Clin. Invest.* 74: 880–87

40. Satoh, J., Prabhakar, B. S., Haspel, M. V., Ginsberg-Fellner, F., Notkins, A. L. 1983. Human monoclonal antibodies that react with multiple endocrine organs. *New Engl. J. Med.* 309: 217–20

41. Guilbert, B., Dighiero, G., Avrameas, S. 1982. Naturally occurring antibodies against nine common antigens in human sera. I. Detection, isolation and characterization. *J. Immunol.* 128: 2779–87

42. Avrameas, S. 1986. Autoreactive B cells and autoantibodies: The "know thyself" of the immune system. *Ann. Inst. Pasteur* 137D: 150–56

43. Eilat, D., Hochberg, M., Pumphrey, S., Rudikoff, S. 1984. Monoclonal antibodies to DNA and RNA from NZB/NZWF1 mice: Antigenic specificities and NH_2 terminal amino acid sequences. *J. Immunol.* 133: 489–94

44. Kofler, R., Noonan, D. S., Levy, D. E., Wilson, M. C., Moller, N. P. H., Dixon, F. J., Theofilopoulos, A. N. 1985. Genetic elements used for a murine lupus anti-DNA autoantibody are closely related to those for antibodies to exogenous antigens. *J. Exp. Med.* 161: 805–15

45. Rauch, J., Murphy, E., Roths, B., Stollar, B. D., Schwartz, R. S. 1982. A high frequency idiotypic marker of anti-DNA autoantibodies in MRL *lpr/lpr* mice. *J. Immunol.* 129: 236–41

46. Marion, T. N., Lawton, A. R., Kearney, J. F., Briles, D. 1982. Anti-

DNA autoantibodies in NZB × NZW F₁ mice are clonally heterogeneous but the majority share a common idiotype. *J. Immunol.* 128 : 668–74

47. Monier, J. C., Brochier, J., Moreira, A., Sault, C., Roux, B. 1984. Generation of hybridoma antibodies to double stranded DNA from non-autoimmune Balb/c strain : Studies on anti-idiotype. *Immunol. Lett.* 8 : 61–68

48. Datta, S. K. 1984. Anti-DNA antibody idiotypes in normal and lupus mice. In *Regulation of the Immune System*, ed. H. Cantor, L. Chess, E. E. Sercarz, pp. 877–86. New York : Liss

49. Datta, S. K., Stollar, B. D., Schwartz, R. S. 1983. Normal mice express idiotypes related to autoantibody idiotypes inherited by lupus mice. *Proc. Natl. Acad. Sci. USA* 80 : 2723–27

50. Rajewski, K., Takemori, T. 1983. Genetics expression and function of idiotypes. *Ann. Rev. Immunol.* 1 : 569–607

51. Barrett, K. S., Trepicchio, W. J. 1986. *The Vₕ genes of anti-DNA autoantibodies.* Presented at 6th International Congress of Immunology Toronto (Abstr. 3.45.25), Ontario

52. Qian, A., Fu, S. M., Reichlin, M. 1984. Cross-reactive idiotype of anti RO/ SSA antibodies : Identification of a V region marker preferentially expressed in SLE. *Arthritis Rheum.* 27 : S16 (Abstr.)

53. Shoenfeld, Y., Isenberg, D. A., Rauch, J., Madaio, M. P., Stollar, B. D., Schwartz, R. S. 1983. Idiotypic cross-reactions of monoclonal lupus autoantibodies. *J. Exp. Med.* 158 : 718–30

54. Zouali, M., Eyquem, A. 1984. Idiotype restriction in human autoantibodies to DNA in systemic lupus erythematosus. *Immunol. Lett.* 7 : 187–90

55. Solomon, G., Schiffenbauer, J., Keiser, H. D., Diamond, B. 1983. Use of monoclonal antibodies to identify shared idiotypes on human antibodies to native DNA from patients with systemic lupus erythematosus. *Proc. Natl. Acad. Sci. USA* 80 : 850–54

56. Lefvert, A. K., Pirskanen, R., Svonborg, E. 1985. Anti-idiotypic antibodies, acetylcholine receptor antibodies and disturbed neuromuscular function in healthy relatives to patients with myasthenia gravis. *J. Neuroimmunol.* In press

57. Pasquali, S. L., Fong, S., Tsoukas, C., Vaughan, J. H., Carson, D. A. 1980. Inheritance of immunoglobulin in rheumatoid factor idiotypes. *J. Clin. Invest.* 66 : 863–66

58. Halpern, R., Davidson, A., Lazo, A.,

Solomon, G., Lahita, R., Diamond, B. 1985. Familial systemic lupus erythematosus : Presence of a cross reactive idiotype in healthy family members. *J. Clin. Invest.* 76 : 731–36

59. Carroll, P., Stafford, D., Schwartz, R. S., Stollar, B. D. 1985. Murine monoclonal anti-DNA autoantibodies bind to endogenous bacteria. *J. Immunol.* 135 : 1086–90

59a. Stefansson, K., Dieperink, M. E., Richman, D. P., Gomez, C. M., Marton, L. S. 1985. Sharing of antigenic determinants between the nicotinic acetyl choline receptor and proteins in *Escherichia coli*, *Proteus vulgaris* and *Klebsiella pneumoniae*. *New Engl. J. Med.* 312 : 221–25

60. Theofilopoulos, A. N., Dixon, F. J. 1985. Murine models of systemic lupus erythematosus. *Adv. Immunol.* 37 : 269–390

61. Izui, S., Kelley, V. E., Masuda, K., Yoshida, H., Roths, J., Murphy, E. 1984. Induction of various autoantibodies by mutant gene *lpr* in several strains of mice. *J. Immunol.* 133 : 227–33

62. Gavalchin, J., Nicklas, J. A., Eastcott, J. W., Madaio, M. P., Stollar, B. D., Schwartz, R. S., Datta, S. K. 1985. Lupus prone (SWR × NZB)F₁ mice produce potentially nephritogenic autoantibodies inherited from the normal SWR parent. *J. Immunol.* 134 : 885–93

63. Jerne, N. K. 1984. Idiotypic networks and other preconceived ideas. *Immunol. Rev.* 79 : 5–25

64. Hooper, B., Whittingham, S., Mathews, J. D., Mackay, I. R., Curnow, D. H. 1972. Autoimmunity in a rural community. *Clin. Exp. Immunol.* 12 : 79–87

65. Carson, D. A., Freimak, B. D. 1986. Human lymphocyte hybridomas and monoclonal antibodies. *Adv. Immunol.* 38 : 275–311

66. Steele, E. J., Cunningham, A. J. 1978. Most IgM producing cells in the mouse secrete autoantibodies (rheumatoid factor). *Nature* 274 : 480–84

67. Dighiero, G., Lymberi, P., Holmberg, D., Lundquist, F., Coutinho, A., Avrameas, S. 1985. High frequency of natural autoantibodies in normal newborn mice. *J. Immunol.* 134 : 765–71

68. Underwood, J. R., Pederson, J. S., Chalmers, P. J., Toh, B. H. 1985. Hybrids from normal, germfree, nude and neonatal mice produce monoclonal autoantibodies to eight different intracellular structures. *J. Clin. Exp. Immunol.* 60 : 417–26

69. Portnoi, D., Freitas, A., Holberg, D., Bandeira, A., Coutinho, A. 1986. Immunocompetent auto-reactive B lymphocytes are activated cycling cells in normal mice. *J. Exp. Med.* 164: 25–35

70. Blackwell, T. K., DePinho, R. A., Reth, M. G., Yancopoulos, G. D., Alt, F. W. 1986. Regulation of genome rearrangement events during lymphocyte differentiation. *Immunol. Rev.* 89: 5–30

71. Glotz, D., Zanetti, M. 1986. Detection of a regulatory idiotype on a spontaneous neonatal self-reactive hybridoma antibody. *J. Immunol.* 137: 223–27

72. Seligmann, M., Brouet, J. C. 1973. Antibody activity of human myeloma globulins. *Sem. Hematol.* 10: 163–77

73. Dighiero, G., Guilbert, B., Fernand, J. B., Lymberi, P., Danon, F., Avrameas, S. 1983. Thirty six human monoclonal immunoglobulins with antibody activity against cytoskeletal proteins, thyroglobulin and native DNA: Immunologic studies and clinical correlations. *Blood* 62: 264–70

74. Kubagawa, H., Volger, L. B., Capra, J. D., Conrad, M. E., Lawton, A. R., Cooper, M. 1979. Studies on the clonal origin of multiple myeloma use of individually specific (idiotype) antibodies to trace the oncogenic event to its earliest point of expression in B cell differentiation. *J. Exp. Med.* 150: 792–807

75. Lymberi, P., Dighiero, G., Ternynck, T., Avrameas, S. 1985. A high frequency of cross-reactive idiotypes among murine natural autoantibodies. *Eur. J. Immunol.* 15: 702–7

76. Lafer, M., Rauch, J., Andrzejewski, C. J., Mudd, D., Furie, B., Schwartz, R. S., Stollar, B. D. 1981. Polyspecific monoclonal lupus autoantibodies reactive with both polynucleotides and phospholipids. *J. Exp. Med.* 153: 897–909

77. Schwartz, R. S., Stollar, B. D. 1985. Origins of anti-DNA antibodies. *J. Clin. Invest.* 75: 321–27

78. Atkinson, P. M., Lampman, G. W., Furie, B. C., Naparstek, Y., Schwartz, R. S., Stollar, B. D., Furie, B. 1985. Homology of the NH₂ terminal amino acid sequences of the heavy and light chains of human monoclonal lupus autoantibodies containing the dominant 16/6 idiotype. *J. Clin. Invest.* 75: 1138–43

79. Isenberg, D., Dudeney, C., Wojnaruska, F., Bhogal, B. S., Rauch, J., Schattner, A., Naparstek, Y., Duggan, D. 1985. Detection of cross reactive anti-DNA antibody idiotypes on tissue-bound immunoglobulins from skin biopsies of lupus patients. *J. Immunol.* 135: 261–64

80. Isenberg, D. A., Collins, C. 1985. Detection of cross-reactive anti-DNA antibody idiotypes on renal tissue-bound immunoglobulins from lupus patients. *J. Clin. Invest.* 76: 287–94

81. Naparstek, Y., Andre-Schwartz, J., Manser, T., Wysocki, L., Breitman, L., Stollar, B. D., Gefter, M. L., Schwartz, R. S. 1986. A single germ line V_H gene segment of normal A/J mice encodes autoantibodies characteristic of systemic lupus erythematosus. *J. Exp. Med.* 164: 614–26

82. Lu, B. Z., Coward, P. O., Schwartz, A., Strosberg, A. D. 1983. The internal image of catecholamines: Expression and regulation of a functional network. *Ann. N.Y. Acad. Sci.* 418: 240–47

83. Kunkel, H. G., Winchester, R. J., Jeslin, F. G., Capra, J. D. 1974. Similarities in the light chains of antigammaglobulins showing cross-idiotypic specificities. *J. Exp. Med.* 139: 128–36

84. Pons-Estel, B., Goni, F., Solomon, A., Frangione, B. 1984. Sequence similarities among K111b chains of monoclonal human IgM_K autoantibodies. *J. Exp. Med.* 160: 893–904

85. Chen, P. P., Goni, F., Houghten, R. A., Fong, S., Goldfien, R., Vaughan, J. H., Frangione, B., Carson, D. A. 1985. Characterization of human rheumatoid factors with seven anti-idiotypes induced by synthetic hypervariable region peptides. *J. Exp. Med.* 162: 487–500

86. Chen, P. P., Goni, F., Fong, S., Jirik, F., Vaughan, J. H., Frangione, B., Carson, D. A. 1985. The majority of human monoclonal IgM rheumatoid factors express a "primary structure-dependent" cross-reactive idiotype. *J. Immunol.* 134: 3281–85

87. Goni, F., Chen, P. P., Pons-Estel, B., Carson, D. A., Frangione, B. 1985. Sequence similarities and cross idiotypic specificity of L chains among human monoclonal IgM_K with anti-γ globulin activity. *J. Immunol.* 135: 4073–79

87a. Jirik, F. R., Sorge, J., Fong, S., Heitzmann, J. G., Curd, J. G., Chen, P. P., Goldfien, R., Carson, D. A. 1986. Cloning and sequence determination of a human rheumatoid factor light-chain gene. *Proc. Natl. Acad. Sci. USA* 83: 2195–99

88. Chen, P. P., Schwartz, R., Carson, D.

A. 1986. Genetic basis of three major cross reactive idiotypes (CRI) of human rheumatoid factor (RF). *Arthritis Rheum.* 29: S39 (Abstr.)

89. Fong, S., Chen, P. P., Gilbertson, T. A., Fox, R. I., Vaughan, J. H., Carson, D. A. 1985. Structural similarities in the κ light chains of human rheumatoid factor paraproteins and serum immunoglobulins bearing a cross-reactive idiotype. *J. Immunol.* 135: 1955–60

90. Fong, S., Chen, P. P., Gilbertson, T. A., Weber, J. R., Fox, R. I., Carson, D. A. 1986. Expression of three cross-reactive idiotypes on rheumatoid factor autoantibodies from patients with autoimmune diseases and sero- positive adults. *J. Immunol.* 137: 122–28

91. Williams, J. M., Gorevic, P. D., Looney, R. S., Abraham, G. N. 1986. Association of Vk111b immunoglobulin light chains with IgM anti-IgG autoantibodies. *Arthritis Rheum.* 29: S33 (Abstr.)

92. Jalkanen, S., Jalkanen, M., Grenfors, K., Toivanen, P. 1984. Defect in the generation of light chain diversity in bursectomized chickens. *Nature* 311: 69–71

93. Steinberg, B. S., Smothers, P. A., Frederiksen, K., Steinberg, A. D. 1982. Ability of the *xid* gene to prevent autoimmunity in (NZB × NZW)$_{F1}$ mice during the course of their natural history after polyclonal stimulation or following immunization with DNA. *J. Clin. Invest.* 70:587–97

94. Amsbaugh, D. F., Hansen, C. T., Prescott, B., Stashak, P. W., Barthold, D. R., Baker, P. 1972. Genetic control of the antibody response to type III pneumococcal polysaccharide in mice. I. Evidence that an x linked gene plays a decisive role in determining responsiveness. *J. Exp. Med.* 136: 931–47

95. Klinman, D. M., Steinberg, A. D. 1986. Proliferation of anti-DNA producing NZB B cells in a non autoimmune environment. *J. Immunol.* 137: 69–75

96. Maizels, N., Bothwell, A. 1985. The T cell independent immune response to the hapten NP uses a large repertoire of heavy chain genes. *Cell* 43: 715–20

97. Woodland, R., Cantor, H. 1978. Idiotype-specific T helper cells are required to induce idiotype-positive B memory cells to secrete antibody. *Eur. J. Immunol.* 8: 600–606

98. Smith, H. R., Steinberg, A. D. 1983. Autoimmunity—A perspective. *Ann. Rev. Immunol.* 1: 175–210

99. Corley, R. B. 1985. Somatic diversification of B cells: A role for autoreactive T lymphocytes. *Immunol. Today* 6: 196–98

100. Hohlfeld, R., Toyka, K. V., Heininger, K., Grosse-Wilde, A., Kalies, I. 1984. Autoimmune human T lymphocytes specific for acetylcholine receptor. *Nature* 310: 244–51

101. Londei, M., Bottazzo, G. F., Feldmann, M. 1985. Human T cell clones from autoimmune thyroid glands: Specific recognition of autologous thyroid cells. *Science* 228: 85–89

102. Pisetsky, D. S. 1984. Hybridomas SLE autoantibodies: Insights for the pathogenesis of autoimmune disease. *Clin. Immunol. Rev.* 3: 169–234

103. Lerner, E. A., Lerner, M. R., Janeway, C. A., Steitz, J. A. 1981. Monoclonal antibodies to nucleic acid-containing constituents: Probes for molecular biology and autoimmune disease. *Proc. Natl. Acad. Sci. USA* 78: 2737–41

104. Pisetsky, D. S., Lerner, E. A. 1982. Idiotypic analysis of a monoclonal anti-Sm antibody. *J. Immunol.* 129: 1489–92

105. Pisetsky, D. S., Semper, K. F., Eisenberg, R. 1984. Idiotypic analysis of a monoclonal anti-Sm antibody. II. Chain distribution of a common idiotypic determinant and in relationship to anti-Sm expression. *J. Immunol.* 133: 2085–89

106. Eisenberg, R. A., Craven, S. Y., Cohen, P. L. 1986. Stochastic process in the anti-Sm response in SLE mice. *Arthritis Rheum.* 29: S34 (Abstr.)

107. Eisenberg, R. A., Winfield, J. B., Cohen, P. C. 1982. Subclass reconstruction of anti-Sm antibodies in MRL mice. *J. Immunol.* 129: 2146–49

108. Pisetsky, D. S., Lerner, E. A., Carler, S. A. 1983. Idiotypic cross-reaction between MRL autoantibodies and a Balb/c myeloma. *Mol. Immunol.* 20: 615–21

109. Diamond, B., Solomon, G. 1983. A monoclonal antibody that recognizes anti-DNA antibodies in patients with systemic lupus. *Ann. N.Y. Acad. Sci.* 418: 379–85

110. Halpern, R., Schiffenbauer, J., Solomon, G., Diamond, B. 1984. Detection of masked anti-DNA antibodies in lupus sera by a monoclonal anti-idiotype. *J. Immunol.* 133: 1852–56

111. Halpern, R. 1985. *Analysis of anti-DNA antibodies in SLE by a monoclonal anti-idiotype.* PhD thesis. Albert Einstein Col. Med. 131 pp. Bronx, NY

112. Davidson, A., Chien, N., Frank, L., Halpern, R., Snapper, S., Diamond, B.

1986. *An idiotypic analysis of anti-DNA antibodies yields new insights into their genetic origins.* Presented Princess Lillian Found. Cardiol. Symp., Brussels, October 1985. In press

113. Isenberg, D. A., Shoenfeld, Y., Walport, M., Mackworth-Young, C., Dudeney, C., Todd-Pokropek, A., Brill, S., Weinberger, A. 1985. Detection of cross-reactive anti-DNA antibody idiotypes in the serum of systemic lupus erythematosus patients and of their relatives. *Arthritis Rheum.* 28: 999–1007

114. Madaio, M. P., Hodder, S., Schwartz, R. S., Stollar, B. D. 1984. Responsiveness of autoimmune and normal mice to nucleic acid antigens. *J. Immunol.* 132: 872–76

115. Rauch, J., Murphy, E., Roths, J. B., Stollar, B. D., Schwartz, R. S. 1984. Monoclonal anticardiolipin antibodies bind to DNA. *Eur. J. Immunol.* 14: 529–39

116. Davidson, A., Preud'homme, J. L., Solomon, A., Chang, M. Y., Beede, S., Diamond, B. 1986. Idiotypic analysis of myeloma proteins: Anti-DNA activity of monoclonal immunoglobulins bearing an SLE idiotype is more common in IgG than IgM antibodies. Submitted

116a. Nishinarita, S., Claflin, J. L., Lieberman, R. 1985. IgA isotype-restricted idiotypes associated with T15 id⁺ PC antibodies. *J. Immunol.* 134: 2544–49

117. Hardy, R. R., Daryl, J. L., Hayakawa, K., Jorge, A., Herzenberg, L. A., Herzenberg, L. A. 1986. Frequent λ light chain gene rearrangement and expression in Ly-1 B lymphoma with a productive κ chain allele. *Proc. Natl. Acad. Sci. USA* 83: 1438–42

118. Matthyssens, G., Rabbitts, T. H. 1980. Structure and multiplicity of genes for the human immunoglobulin heavy chain variable region. *Proc. Natl. Acad. Sci. USA* 77: 6561–65

119. Livneh, A., Halpern, A., Perkins, D., Lazo, A., Halpern, R., Diamond, B.

1986. A monoclonal antibody to a cross-reactive idiotype on cationic human anti-DNA antibodies expressing λ light chains: A new reagent to identify a potentially differential pathogenic subset. Submitted

120. Briles, D. E., Forman, C., Hudak, S., Claflin, J. L. 1982. Antiphosphorylcholine antibodies of the T15 idiotype are optimally protective against streptococcus pneumoniae. *J. Exp. Med.* 156: 1177–85

121. Murphy, D. B., Horowitz, M. C., Homer, R. J., Flood, P. M. 1985. Genetic, serological and functional analysis of I-J molecules. *Immunol. Rev.* 83: 79–104

122. Cooke, A., Hutchings, P. R., Playfair, J. H. L. 1978. Suppressor T cells in experimental autoimmune hemolytic anemia. *Nature* 273: 154–55

123. Gibson, J., Basten, A., Walker, K. Z., Loblay, R. H. 1985. A role of suppressor T cells in induction of self-tolerance. *Proc. Natl. Acad. Sci. USA* 82: 5150–54

124. Zamvil, S., Nelson, P., Trotter, J., Mitchell, D., Knobler, X., Fritz, R., Steinman, L. 1985. T cell clones specific for myelin basic protein induce chronic relapsing paralysis and demyelination. *Nature* 317: 355–58

125. Weigle, W. O. 1980. Analysis of autoimmunity through experimental models of thyroiditis and allergic encephalomyelitis. *Adv. Immunol.* 30: 159–273

126. Onodera, T., Toniolo, A., Ray, U. R., Jensen, A. B., Knazek, R. A., Notkins, A. L. 1981. Virus induced diabetes mellitus. XX. Polyendocrinopathy and autoimmunity. *J. Exp. Med.* 153: 1457–73

127. Onodera, T., Ray, U. R., Melez, K. A., Suzuki, H., Toniolo, A., Notkins, A. L. 1982. Virus induced diabetes mellitus: Autoimmunity and polyendocrine disease prevented by immunosuppression. *Nature* 297: 68

Ann. Rev. Immunol. 1987. 5: 109–26

RHEUMATOID FACTOR AND IMMUNE NETWORKS

D. A. Carson, P. P. Chen, R. I. Fox, T. J. Kipps,
F. Jirik, R. D. Goldfien, G. Silverman, V. Radoux,
S. Fong

Department of Basic and Clinical Research, Scripps Clinic and Research
Foundation, La Jolla, California 92037

INTRODUCTION

Rheumatoid factors are autoantibodies against IgG that were defined
originally by the ability to react with antigenic determinants in the Fc
region. However, some rheumatoid factors also recognize antigens in the
Fab region. High titers of these factors are characteristic of rheumatoid
arthritis, a common and debilitating chronic inflammatory disease (1).
Recent experiments have demonstrated that normal humans and animals
also synthesize rheumatoid factors during secondary immune responses
(2–4). The rheumatoid factors probably are a fundamental component of
the immune network (5). Their interaction with the Fc region of IgG may
represent a prototype of the immunoglobulin domain interactions that
have been postulated to regulate both humoral and cellular immunity
(6, 7).

In this review we summarize experiments from this and other laboratories concerning: (*a*) the genetic regulation of rheumatoid factor
structure and synthesis; (*b*) the environmental stimuli that induce rheumatoid factor production and the physiologic role of the autoantibody;
(*c*) the relation between rheumatoid factor and B-cell malignancies; and (*d*)
the significance of persistent rheumatoid factor production in autoimmune
diseases.

GENETIC CONTROL OF RHEUMATOID FACTOR SYNTHESIS IN HUMANS

Human IgM paraproteins with rheumatoid factor activity are relatively
common. Kunkel and coworkers first prepared a rabbit anti-idiotypic

109

0732–0582/87/0410–0109$02.00

antibody (anti-Wa) that recognized 60% of monoclonal human rheu-matoid factors from unrelated individuals (8). The idiotypically cross-reactive rheumatoid factors all had light chains that belonged to the minor κ IIIb variable region sub-subgroup (9). Andrews & Capra determined the complete light- and heavy-chain variable region sequences of two rheumatoid factors that reacted with the anti-Wa anti-idiotypic antibody (10). The light-chain variable regions were remarkably homologous.

The occurrence of nearly identical variable region sequences on immu-noglobulin light chains from unrelated human subjects is distinctly unusual. Therefore, we postulated that the rheumatoid factor light chains were encoded by a conserved κ variable region gene (5). To prove this hypothesis required (a) the idiotypic analysis of a much larger series of proteins and (b) the cloning of the putative rheumatoid factor light-chain gene.

In mice, anti-idiotypic antibodies have been used to trace the inheritance and expression of immunoglobulin variable region genes (11). However, typical anti-idiotypic antibodies recognize antigens that depend upon the quarternary interaction of immunoglobulin light and heavy chains. The two polypeptide chains are encoded by genes on separate chromosomes, which segregate independently during meiosis. For this reason, most anti-idiotypic antibodies are only useful for the analysis of idiotypic inheritance among related strains of mice that share a relevant light- or heavy-chain variable region gene. In contrast, anti-idiotypic antibodies that are suitable for the immunogenetic analysis of diverse human populations should preferably recognize antigenic determinants that are dependent upon pri-mary sequences of isolated light or heavy chains.

We generated a monoclonal antibody, designated 17.109, that reacted with nearly 50% of monoclonal human IgM rheumatoid factors (12–14), but with only 1–2% of pooled normal human IgM or IgG. It is important that the monoclonal anti-idiotype did bind to the isolated κ light chains of the rheumatoid factors when the anti-idiotype was tested both by western blotting and by enzyme-linked immunoassay (ELISA). It did not react with isolated heavy chains. The anti-idiotype inhibited the binding of rheumatoid factor to IgG. These data provided evidence for the exis-tence of structurally related κ light-chain variable regions on several different human monoclonal rheumatoid factors.

To define more precisely the role of the individual hypervariable regions in the formation of the cross-reactive idiotype on human rheumatoid factor κ chains, we utilized a battery of antipeptide antibodies (15–19). These reagents recognize primary sequence–dependent determinants on isolated heavy or light chains (18).

Synthetic peptides were prepared that corresponded to two light-chain hypervariable region sequences of the human monoclonal IgM rheumatoid factor Sie. The peptides were coupled to keyhole limpet hemocyanin, and the conjugates were emulsified in complete Freund's adjuvant and injected into rabbits. As tested by western blotting, the resultant antisera reacted specifically with light chains that contained the cognate hypervariable region sequence. In each case, the binding of the anti-idiotypic antibody was inhibitable by the appropriate free peptide in solution but not by an irrelevant peptide.

A total of 17 monoclonal rheumatoid factors were probed with these two antipeptide antibodies. Of these, 11 were recognized by antibodies against the second and third κ chain hypervariable regions (Table 1), and 7 also reacted with monoclonal antibody 17.109.

Table 1   Idiotype expression of 17 human monoclonal IgM-rheumatoid factors (RF)[a]

IgM RF	Idiotype						
	PSL2	PSL3	PSH3	PWH2	PWH3	PPH2	PPH3
Cur	+ +	+ +	− −	− −	− −	− −	− −
Gar	+ +	+ +	− −	− −	− −	− −	− −
Glo	+ +	+ +	− −	− −	− −	− −	− −
Got	+ +	+ +	− −	− −	− −	− −	− −
Neu	+ +	+	− −	− −	− −	− −	− −
Pal	+ +	+ +	− −	ND	ND	ND	ND
Pay	+ +	+ +	− −	− −	− −	− −	− −
Pom	− −	− −	− −	− −	− −	+ +	+ +
Sie	+ +	+ +	+ +	− −	− −	− −	− −
Wol	+ +	− −	− −	+ +	+ +	− −	− −
Boc	+ +	+ +	− −	− −	− −	− −	− −
Flo	+ +	+ +	− −	− −	− −	− −	− −
Gal	− −	+ +	+	− −	− −	− −	− −
Lew	+ +	− −	− −	− −	− −	− −	− −
She	− −	− −	− −	− −	− −	− −	− −
Lay	− −	− −	− −	ND	ND	− −	− −
Teh	+ +	ND	ND	ND	ND	ND	ND
Total positive	13	11	2	1	1	1	1
Total assayed	17	16	17	14	14	15	15
Percent positive	76	69	12	7	7	7	7

[a] Seventeen purified monoclonal IgM-RF paraproteins were fractionated by SDS-polyacrylamide gel electrophoresis under reducing conditions. The isolated heavy and light chains were transferred to nitrocellulose and probed with seven different anti-idiotypic antibodies against synthetic hypervariable region peptides. The peptides are designated by a four letter code: P, peptide; S, W, P, name of protein; L, H, light or heavy chain; 2, 3, second or third hypervariable region. Note that 13/17 proteins have the PSL2 idiotype, identified by an anti-peptide antibody against the second complementarity determining region in the light chain of the monoclonal IgM rheumatoid factor Sie (18). ND, not done.

The same rheumatoid factors were also probed with antibodies against synthetic peptides representing the complementarity determining regions in the μ chains of the Sie, Wol or Pom proteins. Antibodies against the third hypervariable regions of the heavy chains reacted strongly with intact rheumatoid factors that contained the same amino acid sequences (18, 19). In contrast, antibodies against synthetic peptides corresponding to the first and second heavy-chain hypervariable regions bound weakly or not at all to the intact rheumatoid factors and to denatured heavy chains. None of the antibodies against the heavy-chain peptides recognized more than one or two rheumatoid factors (Table 1). These results suggested that the heavy-chain hypervariable regions in human rheumatoid factors are heterogeneous.

In sum, our experiments showed that the light chains from multiple monoclonal IgM rheumatoid factors from unrelated human subjects shared three different cross-reactive idiotypic determinants. The idiotypic antigens were defined by antibodies against peptides PSL2 and PSL3 and by the monoclonal antibody 17.109. The most reasonable interpretation of these data is that the light chains derive from a conserved κ variable region germline gene or a small gene family. Frangione and colleagues reached a similar conclusion, based upon amino acid sequence analysis of the light chains from several rheumatoid factors (20, 21).

Recently, we have cloned and sequenced the germline κ variable region gene that encodes the idiotypically cross-reactive rheumatoid factor light chains (22a,b; V. Radoux, manuscript in prep.). Fortunately, the haploid human genome contains only about 25–50 κ variable region genes (23, 24). Approximately 20–40% of the genes probably belong to the κ III variable region subgroup. DNA libraries from fetal liver and placenta were probed with a rearranged κ III variable region gene or with a germline κ III variable region gene (22, 24). By restriction mapping of isolated κ III clones and the subsequent sequencing of selected subclones, it was eventually possible to obtain the entire rheumatoid-factor κ variable region gene. The translated sequence of the cloned gene includes hypervariable regions identical to those identified by the peptide-induced anti-idiotypic antibodies. From position 1 to 95 the germline sequence is identical to the amino acid sequence of four separate rheumatoid factors from unrelated individuals (Figure 1).

Kunkel and coworkers also described a second rabbit anti-idiotypic antiserum against human rheumatoid factors that recognized approximately 20% of the monoclonal anti-IgG autoantibodies (8). A limited amino acid sequence analysis of two idiotypically related rheumatoid factors (Lay and Pom) suggested that this second idiotype might be the product of a heavy-chain variable region gene (25). However, we recently

```
                                    2          3    3
                1                   4          0    4
                                               A
Vκ(RF)          EIVLTQSPGT LSLSPGERAT LSCRASQSVS SSYLAWYQQK PGQAPRLLIY
                                          ____CDR1____

IgM RF k chains
1.  CUR         ---------- ---------- ---------- ---------- ----------
2.  FLO         ---------- ---------- ---------- ---------- ----------
3.  GAR         ---------- ---------- ---------- ---------- ----------
4.  GLO         ---------- ---------- ---------- ---------- ----------
          mutations
5.  GOT   1     ---------- ---------- --------R- ---------- ----------
6.  PAY   1     ---------- ---------- ---------- --------R- ----------
7.  SIE   2     ---------- ---------- ---------- N--------- ----------
8.  NEU   4     ---------- ---------- ---------- -R-------- ----------
9.  GOL   7     ---------- ---------- -----ALLS- RG-------- --------M-


                5          5                      8          9
                0          6                      9          5
Vκ(RF)          GASSRATGIP DRFSGSGSGT DFTLTISRLE PEDFAVYYCQ QYGSSP
                __CDR2_                                     ____CDR3

IgM RF k chains
1.  CUR         ---------- ---------- ---------- ---------- ------
2.  FLO         ---------- ---------- ---------- ---------- ------
3.  GAR         ---------- ---------- ---------- ---------- ------
4.  GLO         ---------- ---------- ---------- ---------- ------
          mutations
5.  GOT   1     ---------- ---------- ---------- ---------- ------
6.  PAY   1     ---------- ---------- ---------- ---------- ------
7.  SIE   2     ---------- ---------- ---------- -D-------- ------
8.  NEU   4     ---------- ---T------ -----V---- ---------- ---A--
9.  GOL   7     ---------- ---------- ---------- ---------- ------
```

Figure 1 Amino acid sequences of a human germline κ variable region gene and nine idiotype-positive, human-rheumatoid-factor light chains. The germline κ variable region gene sequence is shown on the top. The dashed lines represent amino acids that are identical with the germline sequence. All nine light chains share two idiotypic antigens in the second and third complementarity-determining regions (22).

isolated a rearranged κ variable region gene from the rheumatoid factor–secreting B cells of a patient with chronic lymphatic leukemia (26). This gene encodes a light chain that is very similar to that of the monoclonal rheumatoid factor Pom. Again, the presence of very homologous light-chain variable region sequences on rheumatoid factors from unrelated individuals is presumptive evidence of a germline gene. It appears likely that two different variable region genes can encode rheumatoid-factor κ chains.

Shlomchik and coworkers have employed the mRNA sequencing technique to determine the primary structure of rheumatoid factors secreted by several murine hybridomas (27). The sequences of the light chains were homologous, particularly in the framework regions. Because the mouse genome may contain several hundred κ variable region genes (28), the structural similarities among the rheumatoid-factor κ chains were, therefore, considered to be highly significant.

INDUCTION OF RHEUMATOID FACTOR SYNTHESIS

Strong selective pressures must have operated to maintain the identical rheumatoid-factor κ variable region sequences among highly divergent human populations. It is possible that the rheumatoid-factor κ light chain, in association with a particular heavy chain, produces an antibody against an important environmental pathogen. However, the evidence suggests that at least two different conserved human germline κ variable region genes can be utilized for synthesizing anti-IgG autoantibody. Apparently, each can combine with a number of different heavy chains to yield a rheumatoid factor. The most plausible interpretation of these results is that the rheumatoid factors are encoded by germline genes because the autoantibodies themselves perform a beneficial role.

Experiments in humans and mice have shown that two types of stimuli regularly induce IgM rheumatoid factor: polyclonal B-cell activation (29, 30) and exposure to antigen-antibody complexes (2–4). Many different species of gram$^+$ and gram$^-$ bacteria can stimulate B lymphocytes to secrete immunoglobulin, in the relative absence of T-cell help (31). However, polyclonally induced IgG antibodies usually have low affinities for antigen. By cross-linking the Fc regions of IgG antibodies aligned on the surface of an invading microorganism, IgM rheumatoid factors effectively render them both polyvalent and polyspecific. The net result is a substantial increase in functional antibody avidity and specificity.

Many pathogenic microorganisms are first encountered at mucosal surfaces. Preliminary experiments in humans have shown that plasma cells with the major rheumatoid factor–associated cross-reactive idiotype are relatively numerous in tonsil and salivary glands (32). Van Snick & Mason showed that rheumatoid factor–precursor B cells migrated from the bone marrow to the gut-associated lymphoid tissue in a 129/Sv mouse colony that was infected with an intestinal parasite (33). Clarkson & Mellow reported that the colostrum of nursing rats contained IgM rheumatoid factor that helped to protect the suckling offspring from trypanosomal infection (34).

In vitro infection of human B lymphocytes by Epstein-Barr virus induces abundant rheumatoid factor synthesis (30). The autoantibodies are produced predominantly by the minor B-lymphocyte subset that forms rosettes with mouse erythrocytes (35). Some of the rosetting B cells also express the Leu-1 surface antigen. The Leu-1 molecule is present on nearly all mature T cells but on only a small fraction of B lymphocytes (36).

Leu-1$^+$ B cells in humans and Ly-1$^+$ B lymphocytes in mice are probably homologues (36). Murine Ly-1$^+$ B cells are commonly found in the peritoneal cavity. The cells secrete IgM antibodies against bacterial polysaccharides, as well as certain autoantibodies (37). They are thought to be resistant to T cell–dependent triggering and do not readily switch to IgG synthesis (38). Perhaps the Leu-1$^+$–Ly-1$^+$ B-lymphocyte population is the cellular source of the rheumatoid factors that regularly accompany polyclonal B-cell activation at mucosal sites of microbial invasion. This hypothesis has not been formally tested.

The frequency of rheumatoid-factor precursor B lymphocytes in the peripheral blood of a human increases severalfold between birth and young adulthood (39). The rise may occur as a result of repeated infections and/or deliberate immunization. In both humans and mice, transient synthesis of IgM rheumatoid factor regularly accompanies secondary immune responses (2–4). The rheumatoid factors may not be measurable in the serum because they combine avidly with antigen-antibody complexes. However, the cells secreting rheumatoid factor can be detected by in vitro culture under limiting dilution conditions or by somatic cell hybridization. Indeed, the frequency of rheumatoid factor–precursor cells in mouse spleen is remarkably high and may approach the frequency of cells producing specific antibody (40).

Adoptive transfer experiments have elucidated the cellular requirements for the induction of rheumatoid factor synthesis during murine secondary immune responses. Optimal production of rheumatoid factor requires the presence of T cells sensitized to the specific antigen being administered, as well as antibodies against the antigen (41, 42). The rheumatoid-factor precursor B cells can come from unimmunized or T cell–deficient mice. It is important that the rheumatoid factor elicited during secondary immune responses is directed against the IgG isotype that is dominant in the antigen-antibody complex (40–42).

These results are readily explainable if one remembers that activated B lymphocytes, as well as monocytes, can present antigen to helper T lymphocytes (Figure 2 ; 43). During secondary immune responses, antigen-antibody complexes can be taken up and processed by rheumatoid-factor precursor B lymphocytes, as well as by antigen-specific B cells, and then presented on the cell surface in association with class-II molecules. T lymphocytes that recognize the antigen would trigger the B cells to synthesize both rheumatoid factor and specific antibody.

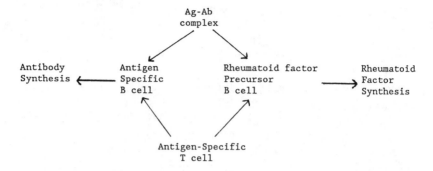

Figure 2 Induction of rheumatoid factor synthesis by antigen (Ag)-antibody (Ab) complexes. The Ag-Ab complexes bind to rheumatoid-factor precursor B cells. Subsequently, antigen is presented on the cell surface in conjunction with class-II (Ia) antigens. T cells sensitized to the antigen then trigger rheumatoid factor synthesis.

In mice, the IgGl subclass predominates during secondary immune responses to several different soluble protein antigens (2). Most IgGl antibodies do not fix complement. However, in the presence of small amounts of serum IgM rheumatoid factor, IgGl antibodies efficiently activate the complement system (44). Complement receptors are abundant in the liver and probably are more numerous than receptors for IgG. Thus, the regular appearance of IgM rheumatoid factor during secondary immune responses may promote both complement-mediated lysis and the clearance of antigen-antibody complexes.

IgM rheumatoid factor apparently represents the predominant anti-antibody produced during secondary immune responses, a finding consistent with the relative abundance of rheumatoid-factor precursor B cells in the lymphoid organs of normal animals.

RHEUMATOID FACTOR AND LYMPHOID MALIGNANCIES

Macroglobulinemia of Waldenstrom is an uncommon B-cell tumor associated with the presence of a monoclonal IgM paraprotein. Many Waldenstrom's macroglobulins are cryoproteins, and nearly 10% have anti-IgG autoantibody activity. As noted earlier, approximately 70% of the monoclonal rheumatoid factors have idiotypically cross-reactive light chains that are the product of a conserved κ variable region gene.

Chronic lymphatic leukemia is the most frequently occurring hematologic neoplasm in adults. The abnormal lymphocytes have surface IgM but usually secrete only small amounts of antibody. Preliminary experiments from this laboratory suggest that more than 10% of chronic lymphatic leukemias express the same κ chain idiotype that is prevalent among the monoclonal rheumatoid factors (T. J. Kipps, in preparation). In contrast, the same idiotype is found on no more than 1–2% of IgG myeloma proteins.

Malignant transformation of lymphocytes is presumed to be a random process. The repeated occurrence of particular idiotypes among B-lymphocyte malignancies is a remarkable phenomenon. In the patients with chronic lymphatic leukemia, the utilization of the rheumatoid factor–associated κ variable region gene does not usually lead to the production of detectable serum anti-IgG autoantibodies.

Several possibilities may explain the selective expression of the rheumatoid factor–associated κ variable region gene in human B-cell malignancies. As was already mentioned, the haploid human genome probably contains on the order of 25–50 κ variable region genes (24). Through somatic mutation and selection by antigen, this limited number of genes gives rise to the diverse array of κ chain sequences characteristic of serum IgG (23). Immature B lymphocytes may express a restricted κ chain repertoire because somatic mutations have not yet accumulated, whereas mature B lymphocytes display the diversified κ chains that are probably the product of the unique immunologic history of each individual. Therefore, random neoplastic transformation of immature B lymphocytes should yield a restricted set of idiotypic antigens, whereas the transformation of mature B lymphocytes or plasma cells should lead to the expression of mainly private idiotypic antigens.

The surface phenotype of the malignant lymphocytes in patients with Waldenstrom's macroglobulinemia has not been analyzed thoroughly, because the cells are usually confined to the bone marrow and lymph nodes. However, the circulating malignant B cells in patients with chronic lymphatic leukemia have been studied extensively. The neoplastic B cells frequently form rosettes with mouse erythrocytes and display the Leu-1 antigen (45). It is possible that chronic lymphatic leukemia (and even Waldenstrom's macroglobulinemia) represents tumors derived from a Leu-1$^+$ B-cell subpopulation that expresses germline-encoded κ variable region genes and that contains abundant rheumatoid factor precursors.

As noted earlier, some circumstantial evidence suggests that both rheumatoid factor precursors and lymphocytes positive for the Leu 1–Ly 1 antigen tend to localize in the peritoneum or at mucosal surfaces. Perhaps the transforming events that induce chronic lymphatic leukemia and Waldenstrom's macroglobulinemia occur at these sites. In this regard, we have studied a patient who developed a rapidly progressive B-cell lymphoma following an allogeneic bone marrow transplantation. The malignant B cells expressed the rheumatoid factor–associated cross-reactive idiotype; they arose in the pharyngeal lymphoid tissue. HLA typing indicated that the abnormal B cells actually came from the donor bone marrow. The cells were positive for the Epstein-Barr nuclear antigen. It is possible that in this patient, a transplanted rheumatoid-factor precursor B lymphocyte migrated to the pharyngeal lymphoid tissue and was infected by the Epstein-Barr virus at that site.

An additional explanation for the recurrent expression of the rheumatoid-factor κ variable region gene in B-cell malignancies may be suggested by the location and/or structure of the gene itself. The most Jh-proximal Vh gene segments are preferentially utilized in murine pre–B cell lines transformed by Abelson virus (46). A similar phenomenon could occur among human κ variable region genes. To resolve this issue will require more knowledge concerning the organization of the various human κ variable region genes on chromosome 2 (47).

The upstream region of the germline rheumatoid-factor κ variable region gene contains the nucleotide sequence TGGGGCGGGT on the antisense strand (22b, V. Radoux, manuscript in prep.). This sequence is located just between the conserved pentadecanucleotide and decanucleotide sequences that are thought to be involved in transcriptional regulation (48). The same sequence has not been found in the upstream regulatory regions of other germline human κ variable region genes. The TGGGGCGGGGT sequence represents a potential recognition site for the *Sp1* protein (49). This promoter-specific factor has been shown to increase the transcription of several eukaryotic genes (49). It is conceivable that the presence of an *Sp1* binding site sequence in the transcriptional regulatory region of the human rheumatoid-factor κ variable region gene facilitates its rearrangement and/or expression in malignant cells.

In any event, the collected results indicate that the idiotypes expressed by human B-cell malignancies are recurrent and nonrandom. In part, they may be predicted by the sequences of the germline-encoded light-chain genes. If this hypothesis is correct, it may be possible to create a battery of anti-idiotypic reagents that recognize the protein products of the conserved human κ variable region genes. Such anti-idiotypic antibodies would be expected to recognize a significant number of malignant B-cell clones but few normal memory B lymphocytes.

RHEUMATOID FACTOR AND AUTOIMMUNE DISEASE

As summarized earlier, synthesis of rheumatoid factor probably accompanies most secondary immune responses. In diseases associated with persistent antigen-antibody complex formation, such as subacute bacterial endocarditis, rheumatoid factors may achieve very high titers. However, the continued production of the autoantibodies depends upon antigenic stimulation (50, 51).

In patients with rheumatoid arthritis, Sjogren's syndrome, and mixed cryoglobulinemia, rheumatoid factors persist in the circulation in the absence of any known exogenous antigenic stimulus. Understanding the regulation of rheumatoid factor production may therefore yield clues concerning the immune pathogenesis of these diseases.

We have used our collection of well-characterized anti-idiotypic reagents, as well as antibodies against κ variable region subgroups, to analyze the rheumatoid factor autoantibodies in patients with autoimmune disease (13, 14, 32). The rheumatoid factors from some patients with Sjogren's syndrome or mixed cryoglobulinemia were enriched in light chains belonging to the κ-III variable region subgroup. Furthermore, a proportion of the anti-IgG autoantibodies displayed the same idiotypic antigens as the monoclonal rheumatoid factor paraproteins.

The restricted nature of the rheumatoid factors in some patients with Sjogren's syndrome or mixed cryoglobulinemia is uncommon for a typical antibody response. Usually, antigen-driven immune responses undergo time-dependent diversification. The accumulation of somatic mutations, immunoglobulin class switching, and the recruitment of new antibody-secreting clones, all contribute to antibody heterogeneity. Conceivably, synthesis of rheumatoid factor in the pathologic states referred to above is not driven by antigen but is the consequence of an autonomous proliferation of B lymphocytes. In this regard, clinical studies have shown that a significant fraction of patients presenting with mixed cryoglobulinemia harbor a lymphoma (52, 53). Neoplastic transformation of B lymphocytes is also an established complication of Sjogren's syndrome (54).

Two patients with Sjogren's syndrome who developed B-cell lymphomas had been studied prospectively at our institution. In both cases, the malignant B lymphocytes expressed the dominant rheumatoid-factor cross-reactive idiotype. An analysis of salivary gland biopsies taken from the patients prior to the onset of lymphoma revealed the presence of numerous lymphocytes and plasma cells that displayed the same idiotypic antigen (32).

To reiterate, it is possible that mixed cryoglobulinemia or Sjogren's syndrome (as well as the ill-defined syndrome of hypergammaglobulinemic purpura) are related diseases associated with abnormal growth of immature B lymphocytes. In a certain percentage of the patients, the continued proliferation of the lymphocyte clones can lead to lymphomatous degeneration.

The rheumatoid factors in the serum of patients with rheumatoid arthritis are heterogeneous. The autoantibodies contain light chains distributed among the κ-I, -II, and -III variable region subgroups (G. J. Silverman, presently in preparation). Compared to rheumatoid factor–depleted immunoglobulin from the same patients, the isolated autoantibodies react preferentially with anti-idiotypic antibody against the conserved second hypervariable region (PSL2) in the κ light chains of Waldenstrom's macroglobulins (14). However, neither anti-idiotypic antibody against the third κ light-chain hypervariable region (PSL3) nor monoclonal antibody 17.109 recognize a significant proportion of the purified anti-IgG autoantibodies from rheumatoid sera.

In patients with rheumatoid arthritis, the rheumatoid factor response is distributed among the IgM, IgG, and IgA classes (1). Compared to age-matched controls, these patients do not have a particularly increased frequency of B-cell lymphoma. Together, the results suggest that the autoantibody response is antigen driven and T-cell dependent.

The antigen(s) that induce anti-IgG autoantibody synthesis in rheumatoid arthritis have not been identified. Several years ago, Winchester et al isolated and analyzed the immune complexes that are abundant in the joint fluids of affected patients (55). The complexes predominantly consisted of IgG and IgM rheumatoid factors. No exogenous antigens were detected.

The inflamed synovial membranes of rheumatoid arthritis patients are infiltrated with T lymphocytes (56). The T cells are polyclonal, but nothing is known about their antigenic specificities. There is no reproducible evidence to indicate that the T cells react with antigens in the Fc region of IgG. Compared to the Fab region, the Fc part of the IgG molecule is exquisitely sensitive to proteolytic digestion. After ingestion of antigen-antibody complexes by rheumatoid-factor precursor B lymphocytes, or by monocyte-macrophages, the Fc portion of the IgG molecule probably is degraded rapidly and therefore is not presented on the cell surface. Thus, even if rare Fc-reactive T cells existed, they probably would not be triggered efficiently.

At this time, one can only speculate about the mechanism of T-cell sensitization in rheumatoid arthritis. Perhaps the T cells are directed against an unknown environmental pathogen. It is also conceivable that the T cells recognize an autoantigen that by chance interacts with a rheumatoid factor. Rheumatoid factors have been described that bind bacterial peptidoglycans and several different nuclear autoantigens (57–60). The double reactivity is not unexpected, considering the high frequency of rheumatoid factors in the restricted germline-encoded human antibody repertoire. Rheumatoid factors that cross-react even weakly with a second antigen, in conjunction with low-affinity IgG antibodies against the same antigen, might form stable immune complexes that bind to low-affinity antigen-presenting cells. The sensitization of T cells to self components might be promoted by repeated exposure to autoantigens presented on the surface of B cells and macrophages, adjacent to class-II determinants.

Either idiotypes or Fc receptors could be the cross-reactive antigens responsible for sustained synthesis of rheumatoid factor in patients with rheumatoid arthritis. Anti-Fab antibodies are detectable in the sera of most patients with rheumatoid arthritis and characteristically occur in association with rheumatoid factors that are specific for antigens in the Fc region (61, 62). Immune complexes have been isolated that contain both anti-Fc and anti-Fab antibody activity (63). Goldman et al have described monoclonal rheumatoid factors that react with both the Fab and Fc region of autologous IgG (64).

IgG anti-idiotypic antibodies against human rheumatoid factors are functionally trivalent. Their Fab and Fc regions both have the potential to interact with the anti-Fc autoantibodies. The anti-idiotypes may tend to form stable immune complexes despite an intrinsically low affinity for the rheumatoid factor molecule.

Immunoglobulins are constructed of globular domains that have homologous amino acid sequences. For this reason, an antibody against one immunoglobulin domain may occasionally react weakly with another domain. For instance, we generated an antibody against a synthetic peptide corresponding to an amino acid sequence in the first complementarity-determining region of the κ light chain of the human monoclonal rheumatoid factor Sie. As detected by western blotting, the anti-idiotypic antibody reacted not only with the rheumatoid factor light chain but also with γ heavy chain (65). In each case, the binding was inhibited by the synthetic peptide. The cross-reactivity of the antipeptide antibody could be attributed to the presence of a conserved Ser-Ser-Ser sequence in both the rheumatoid-factor light chain and in the first constant region domain of the γ heavy chain. Anti-idiotypic antibodies of this type have been termed "epibodies" by Bona et al (66).

Antibodies against bacterial or viral Fc-binding proteins might also express idiotypic antigens that could interact with rheumatoid factors. *Staphylococci* and *streptococci*, and possibly Herpes simplex, have surface proteins that bind Fc fragments (67–69). Furthermore, it has recently been suggested that endogenous retroviral gene segments may direct the synthesis of novel Fc-binding proteins (70). Perhaps a small proportion of the antibodies against Fc-binding proteins mimic the structure of the Fc fragment and therefore are recognized by rheumatoid factors. This hypothesis has not been tested directly. However, we have isolated anti-idiotypic antibodies against human rheumatoid factors that bear the "internal image" of the antigen, i.e. the Fc fragment of IgG (71 ; S. Fong, manuscript in preparation). The "internal image" anti-idiotypic antibodies were recognized by human, rabbit, and mouse antibodies against purified human Fc fragments. The injection of syngeneic mice with an "internal image" anti-idiotype elicited the formation of IgM antibodies against the human Fc region (S. Fong, manuscript in preparation).

IgG rheumatoid factors represent another unusual anti-antibody that has been implicated in the pathogenesis of rheumatoid arthritis (1, 72). The autoantibodies are particularly abundant in rheumatoid joint fluids (55). IgG rheumatoid factors self-associate in a concentration-dependent manner. Subsequent binding of IgM rheumatoid factors can anchor the self-associating IgG-IgG aggregates to yield large, complement-fixing, immune complexes.

In the steady state, idiotype–anti-idiotype responses presumably are kept in check by regulatory T cells. It is possible that either anti-idiotypes that by chance interact with immunoglobulin constant regions and/or rheumatoid factors that are also anti-idiotypes can upset this delicate balance by stabilizing low-affinity interactions. Under such conditions, immune complex formation would be promoted, and idiotype-reactive T lymphocytes might be triggered inappropriately.

SUMMARY

1. Rheumatoid factors represent a normal component of the immune network. The autoantibodies promote complement fixation and clearance of immune complexes. They amplify the avidity of polyclonally induced IgG.
2. Genes related to the primary structure of rheumatoid-factor light chains are widely distributed in the human population and have been conserved during the evolution and dispersion of the species. Products of these genes may be detected with anti-idiotypic antibodies against synthetic peptides corresponding to individual hypervariable regions on rheumatoid-factor light chains. Such anti-peptide antibodies provide unique reagents for analyzing the genetics of immunoglobulins in outbred populations.
3. Precursors of rheumatoid factor are abundant among immature B lymphocytes. Some of these cells may tend to localize to mucosal surfaces, where they are stimulated directly by pathogenic microorganisms with polyclonal B cell–activating properties.
4. Synthesis of rheumatoid factor regularly accompanies all secondary immune responses but is usually transient. Production of the autoantibody is T-cell dependent. The T cells may recognize antigen in an IgG-

antigen immune complex that is processed and presented by B-cell precursors of rheumatoid factor.

5. Rheumatoid factor–associated light-chain idiotypes are rare in serum IgG and on IgG myeloma proteins. They are common among monoclonal IgM proteins and on the surface of the malignant B cells from patients with chronic lymphatic leukemia.

6. The rheumatoid factors that are produced by patients with mixed cryoglobulinemia, or primary Sjogren's syndrome can share idiotypic antigens with monoclonal rheumatoid factors. Rheumatoid factor synthesis in the diseases may reflect an abnormal proliferation of B-cells that is not antigen-driven and that can degenerate into malignancy.

7. The rheumatoid factors in patients with rheumatoid arthritis are diverse and almost certainly represent the outcome of antigen-induced, T cell–dependent mechanisms. The antigens that drive the T cells have not been identified but could represent exogenous microorganisms, self components, or idiotypic antigens that fortuitously interact with rheumatoid factors.

ACKNOWLEDGMENTS

This research was supported in part by grants AM25443, AM35218, AG04100, AM33294, and AM07144. P. Chen and T. Kipps are recipients of Arthritis Investigator Awards, and F. Jirik is the recipient of a postdoctoral fellowship award from the Arthritis Foundation. S. Fong is the recipient of a Research Career Development Award from the National Institute of Aging. V. Radoux is supported by the Arthritis Society of Canada. This is publication number 4498BCR from the Research Institute of Scripps Clinic, La Jolla, California.

Literature Cited

1. Carson, D. A. 1985. Rheumatoid factor. In *Textbook of Rheumatology*, ed. W. J. Kelley, E. D. Harris, Jr., S. Ruddy, C. B. Sledge, pp. 664–76. Philadelphia: Saunders
2. Coulie, P., Van Snick, J. 1983. Rheumatoid factors and secondary immune responses in the mouse. II. Incidence, kinetics and induction mechanisms. *Eur. J. Immunol.* 13: 895–99
3. Nemazee, D. A., Sato, V. L. 1983. Induction of rheumatoid antibodies in the mouse: Regulated production of autoantibody in the secondary humoral response. *J. Exp. Med.* 158: 529–45
4. Welch, M. J., Fong, S., Vaughan, J. H., Carson, D. A. 1983. Increased frequency of rheumatoid factor precursor B lymphocytes after immunization of normal adults with tetanus toxoid. *Clin. Exp. Immunol.* 51: 299–305
5. Carson, D. A., Pasquali, J.-L., Tsoukas, C. D., Fong, S., Slovin, S. F., Lawrance, S. K., Slaughter, L., Vaughan, J. H. 1981. Physiology and pathology of rheumatoid factors. *Springer Semin. Immunopathol.* 4: 161–79
6. Jerne, N. K., Roland, J., Cazenave, P.-A. 1982. Recurrent idiotypes and internal images. *EMBO J.* 1: 243–47
7. Bona, C., Hiernaux, J. 1981. Immune response: Idiotype anti-idiotype network. *Crit. Rev. Immunol.* 2: 33–81
8. Kunkel, H. G., Agnello, V., Joslin, F.

124   CARSON ET AL

G., Winchester, R. J., Capra, J. D. 1973. Cross-idiotypic specificity among monoclonal IgM proteins with anti-gammaglobulin activity. *J. Exp. Med.* 137: 331–42

9. Kunkel, H. G., Winchester, R. J., Joslin, F. G., Capra, J. D. 1974. Similarities in the light chains of anti-gamma-globulins sharing cross-idiotypic specificities. *J. Exp. Med.* 139: 128–36

10. Andrews, D. W., Capra, J. D. 1981. Complete amino acid sequence of variable domains from two monoclonal human anti-gamma globulins of the Wa cross-idiotypic group: Suggestion that the J segments are involved in the structural correlate of the idiotype. *Proc. Natl. Acad. Sci. USA* 78: 3799–3803

11. Weigert, M., Riblet, R. 1978. The genetic control of antibody variable regions in the mouse. *Springer Semin. Immunopathol.* 1: 133–69

12. Carson, D. A., Fong, S. 1983. A common idiotype on human rheumatoid factors identified by a hybridoma antibody. *Mol. Immunol.* 20: 1081–87

13. Fong, S., Chen, P. P., Gilbertson, T. A., Fox, R. I., Vaughan, J. H., Carson, D. A. 1985. Structural similarities in the kappa light chains of human rheumatoid factor paraproteins and serum immunoglobulins bearing a cross-reactive idiotype. *J. Immunol.* 135: 1955–60

14. Fong, S., Chen, P. P., Gilbertson, T. A., Weber, J. R., Fox, R. I., Carson, D. A. 1986. Expression of three cross reactive idiotypes on rheumatoid factor autoantibodies from patients with autoimmune diseases and seropositive adults. *J. Immunol.* 137: 122–28

15. Chen, P. P., Houghten, R. A., Fong, S., Rhodes, G. H., Gilbertson, T. A., Vaughan, J. H., Lerner, R. A., Carson, D. A. 1984. Anti-hypervariable region antibody induced by a defined peptide. A new approach for studying the structural correlates of idiotypes. *Proc. Natl. Acad. Sci. USA* 81: 1784–88

16. Chen, P. P., Fong, S., Normansell, D., Houghten, R. A., Karras, J. G., Vaughan, J. H., Carson, D. A. 1984. Delineation of a cross-reactive idiotype on human autoantibodies with antibody against a synthetic peptide. *J. Exp. Med.* 159: 1502–11

17. Chen, P. P., Goni, F., Fong, S., Jirik, F., Vaughan, J. H., Frangione, B., Carson, D. A. 1985. The majority of human monoclonal IgM rheumatoid factors express a primary structure-dependent cross-reactive idiotype. *J. Immunol.* 134: 3281–85

18. Chen, P. P., Goni, F., Houghten, R. A., Fong, S., Goldfien, R. D., Vaughan, J. H., Frangione, B., Carson, D. A. 1985. Characterization of human rheumatoid factors with seven antiidiotypes induced by synthetic hypervariable-region peptides. *J. Exp. Med.* 162: 487–500

19. Goldfien, R. D., Chen, P. P., Fong, S., Carson, D. A. 1985. Synthetic peptides corresponding to third hypervariable region of human monoclonal IgM rheumatoid factor heavy chains define an immunodominant idiotype. *J. Exp. Med.* 162: 756–61

20. Pons-Estel, B., Goni, F., Solomon, A., Frangione, B. 1984. Sequence similarities among kIIIb chains of monoclonal human IgMk autoantibodies. *J. Exp. Med.* 160: 893–904

21. Goni, F., Chen, P. P., Pons-Estel, B., Carson, D. A., Frangione, B. 1985. Sequence similarities and cross-idiotypic specificity of L chains among human monoclonal IgM-K with anti-gammaglobulin activity. *J. Immunol.* 135: 4073

22a. Chen, P. P., Albrandt, K., Orida, N. K., Radoux, V., Chen, E. W., Schrantz, R., Liu, F.-T., Carson, D. A. 1986. Genetic basis for the cross reactive idiotypes on the light chains of human IgM anti-IgG autoantibodies. *Proc. Natl. Acad. Sci. USA.* In press

22b. Radoux, V., Chen, P. P., Sorge, J. A., Carson, D. A. 1986. *J. Exp. Med.* In press

23. Bentley, D. L., Rabbitts, T. H. 1981. Human Vk immunoglobulin gene number: Implications for the origin of antibody diversity. *Cell* 24: 613–23

24. Bentley, D. L. 1984. Most k immunoglobulin mRNA in human lymphocytes is homologous to a small family of germ-like V genes. *Nature* 307: 77

25. Capra, J. D., Kehoe, J. M. 1974. Structure of antibodies with shared idiotype: The complete sequence of the heavy chain variable regions of two immunoglobulin M anti-gamma globulins. *Proc. Natl. Acad. Sci. USA* 71: 4032–36

26. Jirik, F. R., Sorge, J., Fong, S., Heitzmann, J. G., Chen, P. P., Curd, J. G., Goldfien, R., Carson, D. A. 1986. Cloning and sequence determination of a human rheumatoid factor light-chain gene. *Proc. Natl. Acad. Sci. USA* 83: 2195–99

27. Schlomchik, M. J., Nemazee, D. A., Sato, V. L., Van Snick, J., Carson, D. A., Weigert, M. G. 1986. Variable region sequences of murine IgM anti-IgG monoclonal antibodies (rheumatoid factors): A structural explanation for the

high frequency of IgM anti-IgG B cells. *J. Exp. Med.* In press

28. Cory, S., Tyler, B., Adams, J. 1981. Sets of immunoglobulin V kappa genes homologous to ten cloned V kappa sequences: Implications for the number of germline V kappa genes. *J. Mol. Appl. Genet.* 1 : 103

29. Izui, S., Eisenberg, R. A., Dixon, F. J. 1979. IgM rheumatoid factors in mice injected with bacterial lipopolysaccharides. *J. Immunol.* 122 : 2096–2102

30. Slaughter, L., Carson, D. A., Jensen, F. C., Holbrook, T. L., Vaughan, J. H. 1978. In vitro effects of Epstein-Barr virus on peripheral blood mononuclear cells from patients with rheumatoid arthritis and normal subjects. *J. Exp. Med.* 148 : 1429–34

31. Rasanen, L., Mustikkamaki, V. P., Arvilommi, H. 1982. Polyclonal responses of human lymphocytes to bacterial cell walls, peptidoglycans and teichoic acids. *Immunology* 46 : 481–86

32. Fox, R. I., Chen, P. P., Carson, D. A., Fong, S. 1986. Expression of a cross reactive idiotype on rheumatoid factor in patients with Sjogren's syndrome. *J. Immunol.* 136 : 477–83

33. Van Snick, J. L., Mason, P. L. 1979. Age-dependent production of IgA and IgM autoantibodies against IgG2A in a colony of 129/Sv mice. *J. Exp. Med.* 149 : 1519–29

34. Clarkson, A. B. Jr., Mellow, G. M. 1981. Rheumatoid factor-like immunoglobulin M protects previously uninfected rat pups and dams from *Trypanosoma lewisi. Science* 14 : 186–88

35. Fong, S., Vaughan, J. H., Carson, D. A. 1983. Two different rheumatoid factor-producing cell populations distinguished by the mouse erythrocyte receptor and responsiveness to polyclonal B cell activators. *J. Immunol.* 130 : 162–64

36. Caligans-Cappio, F., Gobbi, M., Bofill, M., Janossy, G. 1982. Infrequent normal B lymphocytes express features of B chronic lymphocytic leukemia. *J. Exp. Med.* 155 : 623–26

37. Hayakawa, K., Hardy, R. R., Honda, M., Herzenberg, L. A., Steinberg, A. D., Herzenberg, L. A. 1984. Ly-1 B cells: Functionally distinct lymphocytes that secrete IgM autoantibodies. *Proc. Natl. Acad. Sci. USA* 81 : 2494–97

38. Hayakawa, K., Hardy, R. R., Herzenberg, L. A., Steinberg, A. D., Herzenberg, L. A. 1984. Ly-1 B : A functionally distinct B cell subpopulation. *Prog. Immunol.* 1554–68

39. Fong, S., Chen, P. P., Vaughan, J. H., Carson, D. A. 1985. Origin and age-associated changes in the expression of a physiologic autoantibody. *Gerontology* 31 : 236–50

40. Van Snick, J., Coulie, P. 1983. Rheumatoid factors and secondary immune responses in the mouse. I. Frequent occurrence of hybridomas secreting IgM anti-IgG autoantibodies after immunization with protein antigens. *Eur. J. Immunol.* 13 : 890–95

41. Nemazee, D. A. 1985. Immune complexes can trigger specific, T cell-dependent, autoanti-IgG antibody production in mice. *J. Exp. Med.* 161 : 242–56

42. Coulie, P. G., Van Snick, J. 1985. Rheumatoid factor (RF) production during anamnestic immune responses in the mouse. III. Activation of RF precursor cells is induced by their interaction with immune complexes and carrier-specific helper T cells. *J. Exp. Med.* 161 : 88–97

43. Lanzavecchia, A. 1985. Antigen specific interaction between T and B cells. *Nature* 314 : 537–39

44. Carson, D. A. 1984. Increase in the complement-fixing ability of murine IgG anti-lymphocyte antibodies by addition of monoclonal IgM rheumatoid factors. *J. Immunol. Methods* 68 : 103–8

45. Wang, C. Y., Good, R. A., Ammirati, P., Dymbort, G., Evans, R. L. 1980. Identification of a p69/70 complex expressed on human T cells sharing determinants with B type chronic lymphocytic leukemia cells. *J. Exp. Med.* 151 : 1539–44

46. Yancopoulos, G., Desiderio, S., Paskind, M., Kearney, J., Baltimore, D., Alt, F. 1984. Preferential utilization of the most Jh-proximal Vh gene segments in pre-B cell lines. *Nature* 311 : 727–31

47. Lotscher, E., Grzeschik, K.-H., Bauer, H. G., Pohlenz, H.-D., Straubinger, B., Zachau, H. B. 1986. Dispersed human immunoglobulin K light chain genes. *Nature* 320 : 456–58

48. Falkner, F. G., Zachau, H. G. 1984. Correct transcription of an immunoglobulin K gene requires an upstream fragment containing conserved sequence elements. *Nature* 310 : 71–74

49. Dynan, W. S., Tjian, R. 1985. Control of eukaryotic messenger RNA synthesis by sequence specific DNA binding proteins. *Nature* 316 : 774–78

50. Williams, R. C., Kunkel, H. G. 1961. Rheumatoid factor, complement and conglutinin aberrations in patients with subacute bacterial endocarditis. *J. Clin. Invest.* 41 : 666–75

51. Carson, D. A., Bayer, A. S., Eisenberg,

R. A., Lawrance, S., Theofilopoulos, A. N. 1978. IgG rheumatoid factor in subacute bacterial endocarditis: Relationship to IgM rheumatoid factor and circulating immune complexes. *Clin. Exp. Immunol.* 31: 100–3

52 Preud'homme, J. L., Seligmann, M. 1972. Anti-human immunoglobulin G activity of membrane-bound monoclonal immunoglobulin M in lymphoproliferative disorders. *Proc. Natl. Acad. Sci. USA* 69: 2132–35

53. Brouet, J. C., Clauvel, J. P., Danon, F., Klein, M., Seligmann, M. 1974. Biological and clinical significance of cryoglobulins. A report of 86 cases. *Am. J. Med.* 57: 775–88

54. Talal, N., Bunim, J. 1964. The development of malignant lymphoma in Sjogren's syndrome. *Am. J. Med.* 36: 529–40

55. Winchester, R. J., Agnello, V., Kunkel, H. G. 1970. Gamma globulin complexes in synovial fluids of patients with rheumatoid arthritis: Partial characterization and relationship to lowered complement levels. *Clin. Exp. Immunol.* 6: 689–706

56. Forre, O., Thoen, J., Lea, T., Dobloug, J. H., Mellbye, O. J., Natvig, J. B., Pahle, J., Solheim, B. G. 1982. In situ characterization of mononuclear cells in rheumatoid tissues, using monoclonal antibodies. No reduction of T8-positive cells or augmentation in T4-positive cells. *Scand. J. Immunol.* 16: 315–19

57. Bokisch, V. A., Chiao, J. W., Bernstein, D., Krause, R. M. 1973. Homogeneous rabbit 7s anti-Ig with antibody specificity for peptidoglycans. *J. Exp. Med.* 138: 1184–93

58. Hannestad, K., Johannessen, A. 1976. Polyclonal human antibodies to IgG (rheumatoid factors) which cross-react with cell nuclei. *Scand. J. Immunol.* 5: ˙541–47

59. Agnello, V., Arbetter, A., de Kasep, G. I., Powell, R., Tan, E. M., Joslin, F. 1980. Evidence for a subset of rheumatoid factors that cross-react with DNA-histone and have a distinct cross-reactive idiotype. *J. Exp. Med.* 151: 1514

60. Rubin, R. L., Balderas, R., Tan, E. M., Dixon, F. J., Theofilopoulos, A. 1984. Multiple autoantigen binding capabilities of mouse monoclonal antibodies selected for rheumatoid factor activity. *J. Exp. Med.* 159: 1429

61. Nasu, H., Chia, D., Iwaki, Y., Barnett, E. V., Terasaki, P. I. 1981. Cytotoxicity of anti-FAB antibodies against B lymphocytes in rheumatoid arthritis. *Arthritis Rheum.* 24: 1278–84

62. Birdsall, H. H., Lidsky, M. D., Rossen, R. D. 1983. Anti-Fab' antibodies in rheumatoid arthritis. Measurement of the relative quantities incorporated in soluble immune complexes in sera and supernatants from cultured peripheral blood lymphocytes. *Arthritis Rheum.* 26: 1481–92

63. Geltner, D., Franklin, E. C., Frangione, B. 1980. Antiidiotypic activity in the IgM fractions of mixed cryoglobulins. *J. Immunol.* 125: 1530–34

64. Goldman, M., Renversez, J. C., Lambert, P. H. 1983. Pathological expression of idiotypic interactions: Immune complexes and cryoglobulins. *Springer Semin. Immunopathol.* 6: 33–47

65. Chen, P. P., Fong, S., Houghten, R. A., Carson, D. A. 1985. Characterization of an epibody: An antiidiotype which reacts with both the idiotype of rheumatoid factors (RF) and the antigen recognized by RF. *J. Exp. Med.* 161: 323–31

66. Bona, C. A., Finley, S., Waters, S., Kunkel, H. G. 1982. Anti-immunoglobulin antibodies. III. Properties of sequential anti-idiotypic antibodies to heterologous anti-gamma globulins. Detection of reactivity of anti-idiotype antibodies with epitopes of Fc fragments. *J. Exp. Med.* 156: 986–99

67. Forsgren, A., Sjoquist, J. 1966. Protein A from Staphylococcus aureus. I. Pseudo immune reaction with human gamma-globulin. *J. Immunol.* 97: 822–26

68. Reis, K. J., Ayoub, E. M., Boyle, M. D. 1984. Streptococcal Fc receptors— Isolation and partial characterization of the receptor from a group C streptococcus. *J. Immunol.* 132: 3091–97

69. Johansson, P. J. H., Schroder, A. K., Nardella, F. A., Mannik, M., Christensen, P. 1986. Interaction between herpes simplex type I and induced Fc receptor and rabbit immunoglobulin G (IgG) domain. *Immunology* 58: 251–55

70. Moore, K. W., Jardieu, P., Mietz, J. A., Trounstine, M. L., Kuff, E. L., Ishizaka, K., Martens, C. L. 1986. Rodent IgGE-binding factor genes are members of an endogenous retrovirus-like gene family. *J. Immunol.* 136: 4283–90

71. Fong, S., Gilbertson, T. A., Chen, P. P., Vaughan, J. H., Carson, D. A. 1984. Modulation of human rheumatoid factor-specific lymphocyte responses with a cross-reactive anti-idiotype bearing the internal image of antigen. *J. Immunol.* 132: 1183–89

72. Pope, R. M., Teller, D. C., Mannik, M. 1975. The molecular basis of self-association of IgG rheumatoid factors. *J. Immunol.* 115: 365–73

Ann. Rev. Immunol. 1987. 5 : 127–50

DISORDERS OF PHAGOCYTE FUNCTION[1]

Daniel Rotrosen and John I. Gallin

Bacterial Diseases Section, Laboratory of Clinical Investigation, National Institute of Allergy and Infectious Diseases, National Institutes of Health, Bethesda, Maryland 20892

Introduction and Historical Background

Over a century ago Metchnikoff formulated the phagocytic theory of host defenses. In experiments with marine invertebrates, Metchnikoff noted that certain strains of the invading fungus were ingested and destroyed by phagocytes. Other strains were not attacked, and a disseminated fatal disease then developed (1). Metchnikoff predicted that abnormalities of phagocytic cells would compromise host defenses. A century later this concept was commonly demonstrated by the development of over-whelming bacterial and fungal infection in conditions resulting in severe quantitative or functional deficiencies of circulating neutrophils. Beginning with Janeway's description in 1954, of a "fatal granulomatous disease of childhood" (2), several intrinsic disorders of phagocytic cells have been characterized. In the past 15 years, application of new techniques and intense interest in the biology of phagocytic cells have greatly promoted knowledge of factors regulating leukocyte function. The present discussion is not intended to be an exhaustive review of phagocyte defects but is focused on well-characterized and historically important disorders and emphasizes recent advances in our knowledge of the underlying defects. For a detailed discussion of the evaluation and management of patients with phagocyte defects, the reader is referred to several recent reviews (3–6).

Adherence Related Functions: Margination, Chemotaxis, and Phagocytosis

The concept of margination and exudation of neutrophils dates to the early nineteenth century when Dutrochet first described egress of "vesicular globules" (i.e. leukocytes) from the vascular lumen to the extravascular

[1] The US Government has the right to retain a nonexclusive royalty-free license in and to any copyright covering this paper.

127

space in the transilluminated tadpole tail (7). Cohnheim emphasized the apparent adherence of white blood cells to the vessel wall and suggested that endothelial cells participate actively in this process (8). Development of the rabbit ear chamber early in the twentieth century fostered a renewed interest in rheological behavior and enabled long-term visualization of the microvasculature. It soundly established the importance of neutrophil aggregation and adherence as early events in the induction of the inflammatory response.

Eukaryotic cells have a negative surface charge, yielding net electrostatic forces that oppose cell-to-cell adherence. The paradoxical adherence of cells of like charge results, in part, from van der Waals attractions. Chemotactic factors cause a small but significant reduction in neurophil surface charge (9, 10) and augment adherence to endothelial cells in vitro (11).

During diapedesis, neutrophils maintain intimate contact with endothelial cells of the vessel wall. Relatively little attention has been paid to the factors that allow for the subsequent disengagement of the two cell types; functionally important similarities may exist between these events and the disaggregation of neutrophils observed in vitro. Investigation of patients with iC3b receptor (CR3) deficiency (see below) has highlighted the importance of a family of structurally and functionally related glycoproteins that play a critical role in leukocyte adherence (12–21).

Neutrophils and mononuclear phagocytes have plasma membrane receptors for synthetic formyl peptide chemoattractants (reviewed in 22, 23), analogous products of bacterial metabolism (24), complement-derived chemoattractants (25), and leukotrienes (26). Of these, the formyl peptide receptor has been the most extensively studied. Work of numerous investigators supports the concept that chemoattractant-receptor-mediated responses in the neutrophil result from increases in cytosolic free calcium alone or in concert with activation of protein kinase C (27–31). Binding of ligand to chemotactic receptors initiates phospholipase C–dependent hydrolysis of membrane phospholipids yielding inositol 1,4,5-trisphosphate, an endogenous calcium ionophore, and diacylglycerol, an activator of protein kinase C. Neutrophil chemotactic receptors utilize a guanine nucleotide regulatory protein sensitive to *Bordetella pertussis* toxin in transduction of postligation responses (27, 32) and modulation of chemotactic receptor affinity (33).

Zigmond has shown that polarized neutrophils can sense a chemoattractant gradient of 1% from the front to the back of the cell (34). On exposure to a chemotactic gradient, the "resting" neutrophil is transformed from a spherical cell with a smooth plasma membrane to a polarized, elongated cell with extensive membrane ruffling, a broad anterior lamellipodium, and a knoblike uropod in the rear. These events are associated with actin polymerization in the advancing granule-free lamellipodium, posterior displacement of the nucleus, and anterior disposition of the microtubule organizing center and cytoplasmic granules. Polymerization of actin occurs within seconds of exposure to formyl peptide and at extremely low rates of receptor occupancy (27). These facts suggest that such changes are important in the alignment and sustained movement of cells in a chemotactic gradient. Despite substantial evidence that calcium is required for optimal chemotaxis (reviewed in 23), and preliminary evidence that cytosolic calcium is elevated at the front end of the advancing neutrophil (35), a rise in cytosolic calcium is neither a required nor a sufficient stimulus for actin polymerization (30, 36). Localized changes in calcium concentration may have other effects on the motile apparatus since myosin and gelsolin are calcium dependent. Ligand-activated phospholipid metabolism may be more proximally linked to cytoskeletal changes since diacylglycerol is important (in platelets) for the α actinin–dependent association between cytoskeleton and plasma membrane (37). In addition, in a cell-free system, phosphatidylinositol 4,5-bisphosphate promotes dissociation of the actin/profilin complex, favoring actin polymerization (38).

Modulation of chemoattractant receptors contributes to directed migration; in the polarized cell, differential receptor density appears to result from translocation of an intracellular pool of spare receptors to the leading edge (39, 40). Clearance of ligand from occupied receptors may also occur by membrane-associated (41, 42) and secreted proteases (43), and it appears to be important in directed migration.

Optimal phagocytosis of opsonin-coated microorganisms and particles is mediated by receptors for either the immunoglobulin Fc-domain or for the low-molecular-weight complement-cleavage products C3b and possibly for iC3b (44). When phagocytes contact an opsonized particle, the response is highly specific and localized; advancing pseudopods are directed at each step of phagosome formation and fusion by sequential engagement of opsonin and receptor, a process termed the "zipper hypothesis" of phagocytosis (44). Under most conditions the energy required for phagocytosis derives from anaerobic glycolysis. Fc receptors are relatively more efficient in initiating ingestion, whereas C3 receptors more efficiently mediate particle attachment. When present together, immunoglobulin and complement opsonins facilitate ingestion at concentrations of either that would be ineffective if present alone. A rise in periphagosomal cytosolic calcium may accompany ingestion (35), but does not appear to be a requirement for it (45). In the macrophage, the Fc receptor acts as a ligand-dependent channel for monovalent cations (46), suggesting a possible role for sodium or potassium in intracellular signaling during Fc receptor–mediated phagocytosis.

Disorders of Adherence, Chemotaxis, and Phagocytosis

iC3b RECEPTOR DEFICIENCY   Recognition in the late 1970s of a group of patients with iC3b-receptor deficiency has stimulated intense interest in the process of leukocyte glycoproteins mediating adherence-related functions (12–21). A frequent history of consanguinity and the demonstration of subclinical defects in unaffected family members are most consistent with an autosomal recessive mode of inheritance. The clinical profile includes a history of delayed separation of the umbilical stump, poor wound healing, and absence of pus formation in a setting of recurrent bacterial and fungal infection. Focal or spreading skin and subcutaneous infections, otitis, mucositis, gingivitis, and periodontitis are common. Systemic and deep-seated infections are occasional complications. Routine laboratory studies show a moderate to marked neutrophilic leukocytosis (presumably due to deficient margination) with peripheral white blood cell counts generally in the range of 15,000 mm^{-3} to 150,000 mm^{-3}.

Elucidation of the underlying defect has greatly advanced knowledge of the molecular basis of normal leukocyte adhesion. Beginning in the early 1980s, several laboratories developed monoclonal antibodies specific for leukocyte adherence-related antigens. Mo1 antigen (found on neutrophils and monocytes) and LFA-1 antigen [found on neutrophils, monocytes, T-cells, natural killer (NK) cells, and Epstein Barr virus (EBV)–transformed B cells] consist of distinct α-chains of apparent mol wt 150,000 to 177,000, bound noncovalently to a common β-chain of mol wt 95,000. A third antigen, designated p150,95, whose function is not so well understood, consists of an identical β-subunit noncovalently linked to an α-chain that differs from those of the Mo1 and LFA-1 antigens. Different patients with similar clinical manifestations probably share the same glycoprotein deficiency, and variations in reported molecular weights most likely reflect differences among laboratories in gel electrophoresis techniques.

Leukocytes from patients severely deficient in iC3b receptor lack both the α- and β-subunits of the receptor. The deficiency of different α- and common β-chains and synthetic studies in EBV-transformed lymphocytes from affected patients suggest that the fundamental defect involves β-chain synthesis. The inability to detect α-chain must arise secondarily, since production of β-chain is thought to be necessary for normal synthesis and/or membrane insertion of α-chains; uncomplexed free α-chain is most likely degraded (15).

The severity of clinical infection correlates with the degree of glyco-protein deficiency. In addition, more severe abnormalities in aggregation (Figure 1), adherence to glass, leukocyte migration into Rebuck skin

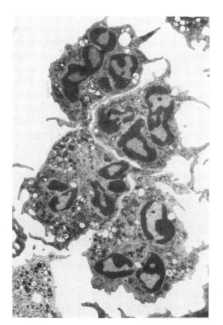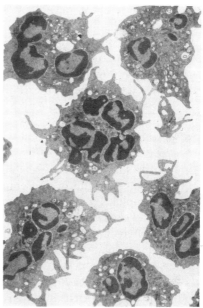

Figure 1 Formyl peptide stimulated shape and aggregation in normal (left) and iC3b receptor deficient (right) polymorphonuclear neutrophils (× 5000). Note failure to aggregate despite shape change in iC3b receptor deficient cells. (Electron micrographs courtesy of M. M. Friedman, Georgetown University.)

windows, chemotaxis, and phagocytosis of iC3b-opsonized particles are seen in patients with severe, as compared to moderate, deficiency. Defective antibody-dependent cytotoxicity and phagocytosis of IgG and C3b-coated particles have both been noted in some, but not all, iC3b receptor–deficient patients. This is consistent with the concept that adhesive glycoproteins may be required for maximal intercellular contact and optimal engagement of Fc or C3b receptors. Clinically normal family members demonstrate moderate deficiencies (40–60% of normal expression) of the adherence related antigens and mild or no in vitro functional deficits (15, 17, 18).

Diminished susceptibility to infection and improved neutrophil function follow bone marrow transplantation in two patients with iC3b receptor deficiency (19). As noted by Anderson et al (15), classification of the deficiency as severe or moderate may be an important variable when the potential risks and benefits of transplantation are weighed.

Harlan and coworkers have studied interactions between cultured endothelial cells and iC3b-deficient leukocytes or normal leukocytes incubated with monoclonal antibodies to iC3b-receptor epitopes (20, 21). Expression of the iC3b receptor appears to be necessary for neutrophil adherence to endothelial cells and for disruption of endothelial monolayers by activated neutrophils. Stimulation of endothelial cells with interleukin-1, tumor necrosis factor, or endotoxin results in the synthesis of endothelial surface proteins that augment neutrophil adherence by an iC3b receptor-dependent interaction. These observations justify speculation that modulation of iC3b-receptor function may be of therapeutic use. For example, inhibition of leukocyte adherence should prevent the disastrous consequences of unrestricted neutrophil activation, as in the adult respiratory distress syndrome.

Further studies of iC3b receptor–deficient patients will likely provide unanticipated insights into receptor processing and signal transduction in the neutrophil. In two iC3b receptor–deficient patients (17, 47), formyl peptides elicited a prolonged respiratory burst, whereas the response to phorbol myristate acetate was impaired; these facts suggest that the iC3b receptor modulates additional interactions related to organization of diverse membrane receptors.

In a widely cited report, Boxer et al (48) described a seven-month-old male with recurrent infection and disorders of adherence, spreading, chemotaxis, and phagocytosis. Defective actin polymerization was noted in vitro. The patient died following bone marrow transplantation. An unrelated patient with similar defects was subsequently identified. The disease was initially attributed to a defect in actin polymerization; in retrospect, both patients are thought to have had severe deficiency of the iC3b receptor (49). This furthers speculation that the latter may play a role in organization of the cytoskeleton and membrane receptors.

NEUTROPHIL SPECIFIC GRANULE DEFICIENCY   The first granules to appear during neutrophil maturation are the primary or azurophil granules which comprise about one third of all granules in the mature cell. Azurophil granules function predominantly in the intracellular milieu where they are involved in killing and digestion of microorganisms (50). Exocytosis of primary granule components may play a role in deactivation of chemoattractants and localization of the inflammatory process.

The secondary granules appear at approximately the metamyelocyte stage of neutrophil maturation and are called "specific" by virtue of their unique contents, including lactoferrin, cytochrome b, and vitamin B-12 binding protein (50). In addition, specific granules contain lysozyme and collagenase, also found in azurophil granules. Neutrophil specific granules are preferentially released during diapedesis and migration into tissues, and with minor membrane perturbation in vivo or in vitro (51, 52). Specific granule products activate the complement cascade to generate the chemoattractant C5a and the opsonin C3b (53) and are selectively chemotactic for monocytes, thereby initiating the "second wave" of inflammation associated with monocyte accumulation and specific immune responses (50).

Release of specific granule products is accompanied by translocation of formyl peptide and iC3b receptors to the plasma membrane (22, 23, 40, 52), facilitating proper orientation of cells during chemotaxis. Cytochrome *b* is likewise translocated from an intracellular pool that cosediments with specific granules to the plasma membrane where its coupling with components of NADPH oxidase may be important in activation of the respiratory burst (54–56). Release of specific granule lactoferrin promotes the formation of hydroxyl radical and increases the adhesiveness of the neutrophil (9).

In the mid-1970s, patients with a congenital deficiency of neutrophil specific granules were recognized (reviewed in 50). Most have been offspring of nonconsanguineous marriages, and both sexes have been affected about equally. If the disease is inherited, it is probably transmitted in an autosomal recessive fashion. All patients have had recurrent bacterial infections without increased susceptibility to a particular pathogen. The peripheral white blood count is usually normal, but on Wright's stain the neutrophils appear to lack granules (Wright's stain demonstrates only specific granules). Azurophil granules, however, are evident on peroxidase stains (Figure 2). Nuclei are bilobed, and the nuclear membrane may be distorted by blebs, clefts, and pockets. The specific granule products lactoferrin, vitamin B-12 binding protein, and cytochrome *b* are either absent or markedly reduced, and cells are deficient in the plasma membrane marker, alkaline phosphatase.

Parmley et al (57) studied neutrophils and bone marrow of a specific granule-deficient patient using a periodic acid-thiocarbohydrazide staining method to identify vicinal glycol-containing complex carbohydrates. It was inferred that the patient's cells contained "empty" specific granules since small, abnormal, elongated organelles appeared late in neutrophil maturation and were "secreted" in response to phorbol myristate acetate. The underlying defects in granule genesis probably differ among patients, because ultrastructural studies have not demonstrated "empty" granules in all patients studied (50; Figure 2).

Both accumulation of neutrophils and monocytes in Rebuck skin windows and in vitro chemotaxis of neutrophils are impaired in congenital specific granule deficiency. In one patient studied, monocyte chemotaxis in vitro was normal (58), which suggests that the impaired monocyte recruitment seen in vivo reflects deficient release of monocyte chemoattractants or of factors capable of generating chemoattractants from serum.

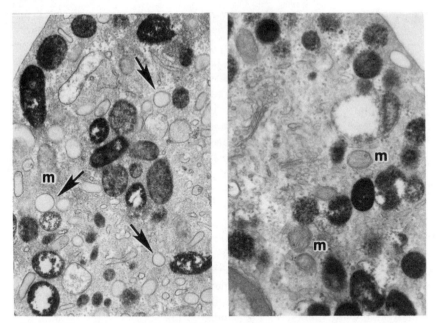

Figure 2 Peroxidase stain of normal (left) and specific granule deficient (right) poly-morphonuclear neutrophils; note darkly stained azurophil granules ($\times 27,000$). Arrows indicate specific granules; m, mitochochondrion. (Electron micrographs courtesy of M. M. Friedman, Georgetown University.)

In support of this, neutrophil secretory products from the patient failed to generate the chemoattractant C5a from serum. Neutrophils of patients with congenital deficiency of specific granules also exhibit impaired up-regulation of formyl peptide receptors, an impaired respiratory burst to certain stimuli, and deficient bacterial killing (50). Thus, patients with neutrophil specific granule deficiency constitute an important model illustrating the critical role that specific granules normally play in the evolution of the inflammatory response.

Other acquired or induced forms of neutrophil specific granule deficiency emphasize these points. For example, neutrophils of patients with thermal injury appear to be activated in the circulation; patients are deficient in specific granule constituents and demonstrate increased expression of C3b and iC3b receptors (50, 59, 60). The evolution of these abnormalities correlates temporally with the appearance of a chemotactic defect. Impaired oxidative metabolism and elevations of serum lysozyme and lactoferrin also occur following burns. The abnormalities in surface-marker expression and enzyme content correlate with the degree of neutrophil functional impairment (59, 60), suggesting that the acquired deficiency of specific granules in thermal injury may indeed play a role in the increased susceptibility of burn patients to infection.

Other models of neutrophil specific granule deficiency include neonatal cells, normal cells experimentally depleted of organelles (termed "neutrophil cytoplasts"), and HL-60 promyelocytic leukemia cells induced in vitro toward a neutrophil-like phenotype. Functional studies in these cells also support an important role for specific granules in amplification of early phases of inflammation (50).

CHEDIAK-HIGASHI SYNDROME   The Chediak-Higashi syndrome (CHS) is a rare disease arising from an autosomal-recessive inheritance and characterized by partial oculo-cutaneous albinism, photophobia, nystagmus, peripheral neuropathy, neutropenia, recurrent pyogenic infections, and characteristic giant lysosomes (Figure 3) in many tissues (reviewed in 61, 62). In CHS, neutrophil giant lysosomes contain enzymes and other constituents normally segregated in azurophil and specific granules. A similar syndrome has been described in Aleutian mink, partial albino Hereford cattle, albino whales, and beige mice. Neutropenia in CHS is moderate; white blood cell counts generally range from 1,000 mm^{-3} to 2,500 mm^{-3}. The neutropenia has been attributed to intramedullary destruction; during sepsis patients usually mount a leukocytosis, but the degree of neutrophilia may be inappropriately low. In most patients, the disease eventually enters an accelerated phase characterized by extensive nonneoplastic lymphoid infiltration, pancytopenia, serious infection, and progressive peripheral neuropathy. Central nervous system involvement, though rare, has been clearly documented; clinical manifestations include cranial nerve palsies, increased intracranial pressure, cerebellar dysfunction, and seizures (61, 62).

Factors underlying the increased susceptibility to infection in man may include abnormalities of cyclic nucleotide metabolism, disorders of microtubule assembly, impaired neutrophil and monocyte chemotaxis, and delayed phagolysosomal fusion (62). Henson has emphasized the importance of elastase in neutrophil-mediated disruption of endothelial monolayers (63); a recent report describes a selective deficiency of elastase and cathepsin G in beige mice (64). Because the defects in production and packaging of granule constituents differ in human and animal forms of CHS (62), it will be important to confirm these observations in patients with CHS.

Boxer et al (65) noted markedly elevated levels of cAMP in an eleven-month-old girl who was in the accelerated lymphoma-like stage of CHS. Treatment with ascorbate, a stimulant of leukocyte cGMP, was associated with a decrease in cAMP, and improved chemotaxis, degranulation, and bacterial killing. The authors suggested that the improvement in neutrophil function resulted from enhanced microtubule assembly modulated by the changes in cyclic nucleotide metabolism. Gallin et al (66) were unable to confirm these observations in two adult patients treated with ascorbic acid. However, in the latter study neutrophil function in CHS mice was improved by ascorbic acid without demonstrable changes in cyclic nucleotide metabolism. Therefore, it remains possible that in CHS the stage of the underlying disease determines responsiveness to therapy with ascorbate or that ascorbic acid modulates neutrophil function through noncyclic nucleotide-dependent mechanisms. Of interest, Nath et al (67) showed that increased tubulin tyrosinolation in neutrophils of CHS patients was normalized by the administration of ascorbate in vivo or in vitro.

HYPERIMMUNOGLOBULIN E (JOB'S) SYNDROME   In 1966, Davis et al (68) described two girls with persistent weeping eczematoid lesions and large subcutaneous staphylococcal abscesses termed "cold" because they lacked typical signs of inflammation. Otitis, sinusitis, staphylococcal pneumonia, furunculosis, and cellulitis were prominent features. Patients described in subsequent reports have had clinical features similar to those already noted, in addition to characteristic coarse facies (Figure 4) and striking elevations of serum IgE, but with normal levels of IgG, IgA, IgM, and IgD. Despite high levels of IgE, clinical manifestations of severe atopy are uncommon. Aside from pneumonia, deep-seated infection is infrequent and when seen, usually arises by extension from a contiguous soft-tissue site (68, 69).

The factors that account for the increased susceptibility to infection in Job's syndrome are poorly understood. Patients have a moderate defect in neutrophil chemotaxis (69–72), but the contribution of this abnormality to the increased risk of infection has been questioned. Donabedian & Gallin (70, 71) identified a chemotatic inhibitory factor produced in vitro by mononuclear cells of patients with Job's syndrome. The factors were partially purified, with molecular weights of ∼ 61,000 and 30,000–45,000; they are sensitive to proteolytic digestion and stable to 56°C.

Lymphocyte function in Job's syndrome is also abnormal (72). Delayed hypersensitivity responses to a variety of skin-test antigens are impaired in some but not all patients with Job's syndrome. Mitogen-induced and antigen-induced in vitro lymphocyte transformation is also impaired.

Patients with Job's syndrome are not merely a subset of those with severe eczema complicated by repeated superficial infections; the finding of markedly elevated anti-*Staphylococcus aureus* IgE in Job's syndrome, but not in eczema, may aid in this distinction (73). Even when extremely elevated, specific anti-*S. aureus* IgE constitutes less than 1% of total serum IgE (74; 75; S. C. Dreskin & J. I. Gallin, unpublished observations). Job's syndrome is also characterized by high levels of anti-*Candida albicans* IgE, and deficiencies of both total salivary IgA and anti-*S. aureus* IgA in serum and saliva. The patients also have elevated anti-staphylococcal IgM but not the expected excess of anti-*S. aureus* IgG (73).

In a recent study of IgE turnover in these patients Dreskin et al (76) showed that the fractional catabolic rate of IgE is inversely related to the concentration of circulating IgE. Thus, relative estimates of IgE synthetic rate based on the serum IgE level alone can over estimate the true synthetic rate as much as five fold. In addition, IgE from patients with Job's syndrome was catabolized normally in healthy volunteers, indicating that the defect in IgE clearance in Job's syndrome is not due to a structurally abnormal IgE. Modeling analysis supports previous evidence that IgE is catabolized by both intravascular and extravascular pathways and suggests that an IgE-binding site may be important in the function of the latter pathway. No substantial differences appeared in the manner in which patients with Job's syndrome metabolized IgE compared to that of patients with severe atopy and elevated IgE; this weighs against there being a unique defect in IgE metabolism in Job's syndrome.

It has been proposed that the extreme elevations of anti-*S. aureus* and anti-*C. albicans* IgE may play a role in the increased susceptibility to infection due to these organisms. The possibility that *S. aureus*– and *C. albicans*–specific IgE antibodies are elevated simply as a consequence of frequent infection seems unlikely because similar abnormalities were not seen in carefully selected control groups with recurrent infections due to the same organisms (73). Furthermore, in one child monitored from birth, elevation of anti-*S. aureus* IgE preceded clinical manifestations of the syndrome (S. C. Dreskin & J. I. Gallin, in preparation).

LOCALIZED JUVENILE PERIODONTITIS   In patients with neutropenia or functional leukocyte impairment, mucosal ulcerations, gingivitis, and periodontitis occur frequently and are often severe. In most cases, intraoral infection is associated with recurrent infection at other sites. In contrast, localized juvenile periodontitis (LJP) is an adolescent disease of the supporting structures of the permanent dentition, characterized by severe alveolar bone loss limited primarily to the first molars and incisors. Patients are not predisposed to extraoral infection (77, 78).

Genco and coworkers have demonstrated in patients with LJP a moderate but reproducible impairment of chemotaxis to formyl peptides and C5a; neutrophil adherence is normal (77, 78). Defective chemotaxis persists following aggressive local therapy and has been demonstrated in siblings prior to the development of LJP. Limited data suggest that formyl peptide receptor numbers and affinities are normal; however, receptor up-regulation or internalization is impaired. It remains unclear whether the suspected abnormalities in receptor processing in LJP are causally related or are only secondary to the chemotactic defect. If confirmed, studies indicating that LJP neutrophils are deficient in a 108-kd surface protein would suggest a role for the latter in the organizing and processing of chemotactic receptors (78).

Microbicidal Mechanisms

With the initiation of phagocytosis and following closure of the phagocytic vacuole, granules of the phagocytic cell fuse with the invaginated plasma membrane. As a result, granule contents are delivered to the phagosome, and granule membrane constituents are translocated to the limiting membrane of the newly formed phagolysosome. Activation of the respiratory burst occurs after a brief delay relative to the onset of degranulation. Ingested organisms are entrapped in a milieu of toxic oxygen species and granule components which function in killing and digestion. Thus, the microbicidal activities of phagocytic cells arise from two fundamentally distinct mechanisms: those that depend on oxidative metabolism and those that do not.

Oxygen-Independent Antimicrobial Systems

Knowledge of factors controlling degranulation in neutrophils has advanced considerably in recent years (79, 80). As is the case for other secretory cells, considerable evidence supports hypothesis of a role for calcium in the regulation of exocytosis. Other signals are clearly important: Phorbol esters are potent secretagogues and potentiate the response to formyl peptides, yet elicit no rise in cytosolic calcium. Activation of gelsolin may promote granule access to the forming phagolysosome by dissolution of the submembranous periphagosomal actin mesh, a process analogous to the proposed facilitation of exocytosis by cytochalasins. Apposition of granule and plasma membranes does not appear to be sufficient in itself to elicit membrane fusion, and additional signals or changes in membrane configuration or composition are probably required. Recent evidence indicates that neutrophils contain a synexin-like protein that promotes liposome fusion in vitro and may be important in membrane fusion in vivo (81).

In the 1950s and 1960s, Hirsch and Spitznagel isolated antimicrobial cationic proteins from rabbit neutrophils (reviewed in 82). Weiss and coworkers (83) have characterized a 59-kd protein (designated bactericidal/permeability increasing factor) from azurophil granules. Its cationic charge and hydrophobicity are thought to be important in binding to the outer membrane of gram negative organisms, which causes increased membrane permeability and failure of bacterial transport processes. Lehrer and associates have recently described a family of homologous, low molecular weight (< 3500-dalton) peptides, termed "defensins" because of their broad cytotoxic properties against bacteria, fungi, viruses, and tumor cells (84).

Aside from patients with congenital or acquired deficiency of specific granules, and patients with Chediak-Higashi syndrome, isolated defects in degranulation or deficiencies of components of the nonoxidative antimicrobial systems have not been described. It is of great interest, however, that a number of obligate intracellular parasites persist and proliferate within macrophages, possibly by virtue of their ability to prevent normal phagosomal acidification and/or phagolysosomal fusion (85). Viable *Toxoplasma gondii*, *Mycobacterium tuberculosis*, and *Chlamydia* species inhibit phagolysosomal fusion, whereas nonviable or opsonized organisms fail to do so. The virulent trypomastigote of *Trypanosoma cruzi* is able to lyse the newly formed phagosome and proliferate within the macrophage cytoplasm, whereas the avirulent epimastigote remains within the phagosome and is eventually destroyed. *Legionella pneumophila* is ingested by means of a unique form of coiling phagocytosis in which the bacterium is initially wrapped within an elongated thin pseudopod. Failure of phagolysosomal fusion is accompanied by defective phagosomal acidification and a peculiar "beading" of the phagosomal membrane by host-cell ribosomes and mitochondria. Similarly, Lehrer has suggested that inefficient killing of *Candida albicans* by normal neutrophils is related to impaired vacuolar sealing during phagocytosis (86). Elucidation of mechanisms by which microorganisms modify host defenses will likely provide new insights into factors controlling normal phagolysosomal fusion.

Oxygen-Dependent Antimicrobial Systems

Phagocytic cells including neutrophils, eosinophils, monocytes, and macrophages have in common a mechanism by which membrane perturbation and exposure to soluble stimuli initiate a burst of oxygen consumption, glucose consumption via the hexose monophosphate shunt, and production of superoxide anion (O_2^-), hydrogen peroxide (H_2O_2), hydroxyl radical ($\cdot OH$), singlet oxygen ($^1O_2$) (87, 88). The striking kinetics of the respiratory burst suggested to early observers that the responsible enzyme system, dormant in resting cells, could be rapidly activated by appropriate stimuli. Although the marked instability of the oxidase in response to temperature and salts has hampered efforts toward its purification, recent studies suggest that the oxidase is a nonmitochondrial electron transport chain comprised of an FAD-requiring flavoprotein, an unusual b-cytochrome unique to phagocytes (cytochrome b_{558}, also designated cytochrome b_{-245}), and possibly a quinone. The oxidase utilizes the reduced pyridine nucleotide NADPH, generated by the hexose monophosphate shunt, as the preferred electron donor:

$$2O_2 + NADPH \longrightarrow 2O_2^- + NADP^+ + H^+.$$

The most widely accepted view is that the NADPH binding site of the oxidase is disposed at the cytoplasmic face of the plasma membrane while the terminal component (probably the b-cytochrome) faces the extracellular or intraphagosomal space. Recent attempts to purify the oxidase suggest that the NADPH binding component is a ~ 66 kd protein (89). There is evidence that the cytochrome is translocated from an intracellular membrane compartment, with the density of specific granules, to the plasma membrane (54–56), thereby completing the respiratory chain and facilitating full expression of the oxidative burst. However, translocation does not appear to be a strict requirement for activation of the burst, and the extent of cytochrome translocation in response to various stimuli correlates poorly with superoxide generation (56). Babior has questioned a functional role for the cytochrome, primarily on the basis of kinetic data indicating that the rate of O_2^- production greatly exceeds that of cytochrome reduction (88, 90). However, as Segal (91) and others (92) noted, O_2^- production is measured aerobically, while reduction of the cytochrome is determined only under strict anaerobic conditions. Oxygen may affect the affinity of the cytochrome for its electron donor or may independently alter the bioenergetics of electron flow through the chain.

Other factors controlling activation of the respiratory burst are not well understood. It appears clear that several independent transduction pathways exist and that changes in cytosolic calcium, intracellular pH, protein phosphorylation, transmethylation reactions, and cell attachment may play a role (93).

Superoxide generated by the respiratory burst is thought to have little direct toxicity and probably exerts its microbicidal effects through the products it forms, namely H_2O_2, $\cdot OH$, and $^1O_2$. Superoxide can cross cell membranes via anion channels or by penetration of the lipid bilayer, and thus it gains access to sites where its highly reactive products exert their toxic effects (94); H_2O_2 may also diffuse across cell membranes and gain access to intracellular organelles (95). The antimicrobial activity of H_2O_2 is greatly enhanced by conversion to hypochlorous acid (HOCl) in the presence of myeloperoxidase and chloride:

$$H_2O_2 + Cl^- + H^+ \longrightarrow H_2O + HOCl.$$

Other halides can substitute, but chloride appears to be the physiologic substrate (87). The myeloperoxidase-H_2O_2-halide system can damage bacteria by incorporation of halide into the bacterial cell wall, but other mechanisms probably account for the microbicidal potency of the system. In this regard, hypochlorous acid reacts with chloride to form chlorine and potent long-lived N-chloro oxidants (117); the myeloperoxidase system decarboxylates amino acids, converting them to aldehydes, carbon dioxide, and ammonia.

In addition to its microbicidal activities, the myeloperoxidase system may modulate the inflammatory response by oxidative inactivation of soluble mediators and chemoattractants including formyl peptides, C5a, leukotrienes, and prostaglandins (96, 97).

Disorders of Oxygen-Dependent Antimicrobial Systems

CHRONIC GRANULOMATOUS DISEASE (CGD)   Since the initial descriptions of CGD in the mid-1950s, well over 300 cases have been reported (reviewed in detail in 5, 92, 98). Despite considerable variation in clinical features, all share a profound impairment in oxidative metabolism and an increased susceptibility to serious infection. The phenotypic expression of CGD is rare, affecting about 1 in 1,000,000 persons; nonetheless, the disease has engendered tremendous interest as a prototype for abnormalities of phagocyte oxidative metabolism. Elucidation of the underlying defects has been critical to a developing understanding of the structure, activation, and function of the NADPH oxidase.

Patients with the "classic" form of CGD develop serious infections early in life, usually within the first year. Common infectious syndromes include pneumonia and lung abscesses, skin and soft tissue infections, lymphadenopathy and suppurative lymphadenitis, osteomyelitis, usually involving the small bones of the hands and feet, and hepatic abscesses. Septicemia, meningitis, brain abscesses, and infection of the gastrointestinal and genitourinary tracts are not uncommon. Though severe, infection in CGD may follow a rather indolent course, characterized initially only by malaise, low-grade fever, and a mild leukocytosis or elevation in the erythrocyte sedimentation rate. The diagnosis of CGD is readily established by a defect in the ability to reduce the dye nitroblue tetrazolium (Figure 5) or by other tests of neutrophil oxidative metabolism. Diagnosis is often delayed with potentially catastrophic consequences.

S. aureus and gram negative bacilli account for the majority of serious infections, but fungal infection due to *Aspergillus* and *Candida* is not uncommon. Organisms which produce H_2O_2 but are catalase negative (e.g. streptococci, pneumococci, and lactobacilli) are not major pathogens in CGD; the persistence of H_2O_2 within the phagosome, in concert with host cell MPO, may result in bactericidal activity against these organisms. Alternatively, oxygen-independent microbicidal mechanisms may be sufficient in CGD to kill certain pathogens.

Variants of the disease have been recognized, including presentation in adolescence or young adulthood; such patients usually have a history of infection since childhood, though typically less severe than in the "classic" form of the disease. As the name suggests, patients with CGD are predisposed to granuloma formation, occasionally resulting in esophageal, antral, or genitourinary tract obstruction. Normal neutrophils inactivate chemoattractants via the MPO-H_2O_2-halide system; failure to do so, in conjuction with diminished neutrophil autooxidation, may account for prolonged leukocyte recruitment in CGD (99). Alternatively, inefficient degradation of antigen may result in chronic release of mediators, such as γ-interferon, which may play an important role in the granulomatous process (85).

Considerable evidence indicates that the CGD phenotype can arise from distinct genetic and biochemical lesions related to the structure and activation of the NADPH oxidase. While activation mechanisms and structure-function relationships for the oxidase are still poorly understood, significant progress has been made. The presence of a nonmitochondrial b-cytochrome in rabbit neutrophils was first noted in the early 1960s by Hattori: (100). The significance of that observation was not generally appreciated until 1978 when Segal et al (91) reported the presence of the same b-cytochrome in normal human neutrophils and its absence in neutrophils from four male patients with CGD. Since then several laboratories have confirmed this finding, and a current understanding of the relationship between cytochrome b_{558} and CGD transmission can be stated as follows. Nearly all males with CGD and X-linked genetics have absent or grossly abnormal cytochrome b_{558}; in female carriers the concentration of cytochrome b_{558} in the neutrophil population as a whole correlates directly with the proportion of cells demonstrating inactivation of the abnormal X-chromosome. Ohno et al (101) described a single male patient with normal levels of the cytochrome in association with apparent X-linked transmission; these facts suggest that the X-chromosome may code for additional factors important to the function of the oxidase. In about 15% of reported cases of CGD, an autosomal recessive pattern of transmission seems most likely; the vast majority of such patients have normal levels of the cytochrome, although two patients with apparent autosomal recessive transmission and cytochrome deficiency are known (101, 102). In response to these observations, Ohno et al (101) speculated that a structural gene for the cytochrome might be found on an autosomal chromosome, whereas the X-chromosome may encode a putative b-cytochrome gene activator, an "enhancer element" required for appropriate transcription, or a protein (possibly a component of the oxidase) necesssary for normal insertion of the cytochrome in the oxidase chain. Of particular interest is a gene that Orkin and coworkers (103) have recently cloned; the transcript of the gene is expressed in phagocytic myeloid cells but was absent or structurally abnormal in four patients with X-linked CGD. The nucleotide sequence of complementary DNA clones predicts a polypeptide of at least 468 amino acids with no homology to known cytochromes.

Several laboratories have noted reduced FAD in CGD neutrophils. In general, the results support a close relationship between cytochrome b_{558} deficiency and the reduced FAD content. While the FAD deficiency probably occurs only in the setting of b-cytochrome deficiency, a normal content of FAD in the setting of b-cytochrome deficiency is probably at least as common (101).

Segal et al reported a defect in phosphorylation of a 44-kd protein in stimulated neutrophils from patients with autosomal recessive but not X-linked CGD (104). This result, since confirmed in other laboratories, is consistent with the concept that the 44-kd protein is an important link in activation of the oxidase or may serve as an electron transporting molecule in the oxidase chain. Rossi et al (105) have presented data suggesting that the b-cytochrome is itself phosphorylated in stimulated guinea-pig neutrophils.

The heterogeneity of the defect in CGD is further illustrated by patients with selective defects in activation of the respiratory burst by particulate but not soluble stimuli (106) or by soluble but not particulate stimuli (107). In three patients with a greatly diminished respiratory burst, the defect has been attributed to abnormal NADPH oxidase kinetics arising from a decreased affinity of the oxidase for its substrate, NADPH (108–110). These patients were deficient in the b-cytochrome and exhibited normal or sluggish transmembrane depolarization in response to activators of the respiratory burst. In contrast, all those patients with CGD described to date exhibit greatly impaired depolarization to formyl peptides or PMA; this suggests that depolarization is closely related to activation of the respiratory burst (5). However, the nature of that relationship remains obscure since depolarization of CGD cells by calcium ionophores or extracellular K^+ does not activate the burst.

Severe deficiency of the enzyme glucose-6-phosphate dehydrogenase (G-6-PD) may result in the phenotypic expression of CGD (92). However, NADPH oxidase activity is normal in G-6-PD deficiency. The clinical features can be accounted for solely on the basis of impaired activity of the hexose monophosphate shunt, which furnishes reducing equivalents (NADPH) insufficient for superoxide production.

MYELOPEROXIDASE DEFICIENCY   The development in the early 1980s of an automated flow cytochemical system for leukocyte differential counts enabled screening of large populations for MPO deficiency. It is now apparent that MPO deficiency is a relatively common disorder, occurring in ~1 in 2000 individuals (111). Most MPO deficient patients are not at increased risk of serious infection. Disseminated *Candida* infection has been noted infrequently; in these patients, underlying immunosuppressive conditions such as poorly controlled diabetes mellitus probably contribute to the risk of infection.

Phagocytosis in MPO deficiency is normal or increased, probably due to a failure to down-regulate receptor-mediated recognition mechanisms (112). Deactivation of the respiratory burst is likewise delayed. Bactericidal activity is usually normal, whereas candidacidal activity may be moderately to severely impaired (113).

Close to 50% of patients are entirely lacking in MPO; the remainder have a partial deficiency associated with decreased amounts of structurally and functionally normal MPO (111, 114). Native MPO consists of two heavy-light protomers each with a heavy subunit of approximately 59 kd and a light subunit of approximately 14 kd. The interrelationships between subunits and the location of heme groups have not been fully established (115, 116). Recent studies indicate that MPO synthesis ceases with neutrophil maturation, and that MPO deficiency is due not to absence of the gene encoding for MPO but to a defect in cotranslational or post-translational processing of an MPO precursor that results in a failure to package the enzyme correctly into the azurophil granule (115). While current concepts of the defect would not exclude a single gene mutation, transmission patterns in MPO deficiency are not consistent with simple autosomal recessive inheritance (111).

Conclusions

Since the early descriptions of chronic granulomatous disease three decades ago, exciting advances have been made in our understanding of disorders of phagocytic cell function. Due to their striking clinical and laboratory features these syndromes remain of great importance to clinicians and basic scientists alike. The remarkable advances of the past decade will, we hope, set the stage for molecular and genetic approaches to therapy in the coming years.

Literature Cited

1. Metchnikoff, E. 1893. *Lectures on the Comparative Pathology of Inflammation.* Transl. F. A. Starling, E. H. Starling. London: Kegan Paul Trench. (From French)

2. Janeway, C. A., Gitlin, D. 1954. Hypergammaglobulinemia associated with severe, recurrent and chronic non-specific infection. *Am. J. Dis. Child.* 88: 388 (Abstr.)

3. Hill, H. R. 1984. Clinical disorders of leukocyte function. In *Contemporary Topics in Immunobiology,* ed. R. Snyderman, 14: 345–93. New York: Plenum

4. Roberts, R., Gallin, J. I. 1983. The phagocytic cell and its disorders. *Ann. Allergy* 50: 330–45

5. Gallin, J. I., Buescher, E. S., Seligmann, B. E., Nath, J., Gaither, T. E., Katz, P. 1983. Recent advances in chronic granulomatous disease. *Ann. Intern. Med.* 99: 657–74

6. Gallin, J. I. 1981. Abnormal phagocyte chemotaxis: Pathophysiology, clinical manifestations, and management of patients. *Rev. Infect. Dis.* 3: 1196–1220

7. Dutrochet, M. H. 1824. *Recherches, anatomiques et physiologiques sur la structure intime des animaux et des vegetaux, et sur leur motilite.* Paris: Bailliere

8. Cohnheim, J. 1889. *Lectures in General Pathology,* Vol. 1. London: New Sydenham Soc.

9. Gallin, J. I. 1980. Degranulating stimuli decrease the negative surface charge and increase the adhesiveness of human neutrophils. *J. Clin. Invest.* 65: 298–306

10. Hoover, R. L., Briggs, R. T., Karnovsky, M. J. 1978. The adhesive interaction between polymorphonuclear leukocytes and endothelial cells in vitro. *Cell* 14: 423–28

11. Harlan, J. M. 1985. Leukocyte-endothelial interactions. *Blood* 65: 513–25

12. Springer, T. A., Anderson, D. C. 1986. The importance of the Mac-1, LFA-1 glycoprotein family in monocyte and granulocyte adherence, chemotaxis, and migration into inflammatory sites: insights from an experiment of nature. In *Biochemistry of Macrophages, Ciba Foundation Symposium 118,* ed. D. Evered, J. Nugent, M. O'Connor, pp. 102–26. London: Pittman

13. Gallin, J. I. 1985. Leukocyte adherence-related glycoproteins LFA-1, Mo1, and p150,95: A new group of monoclonal antibodies, a new disease, and a possible opportunity to understand the molecular basis of leukocyte adherence. *J. Infect. Dis.* 152: 661–64

14. Anderson, D. C., Schmalstieg, F. C., Arnaout, M. A., Kohl, S., Tosi, M. F., Dana, N., Buffone, G. J., Hughes, B. J., Brinkley, B. R., Dickey, W. D., Abramson, J. S., Springer, T., Boxer, L. A., Hollers, J. M., Smith, C. W. 1984. Abnormalities of polymorphonuclear leukocyte function associated with a heritable deficiency of high molecular weight surface glycoproteins (gp138): Common relationship to diminished cell adherence. *J. Clin. Invest.* 74: 536–51

15. Anderson, D. C., Schmalsteilg, F. C., Finegold, M. J., Hughes, B. J., Rothlein, R., Miller, L. J., Kohl, S., Tosi, M. F., Jacobs, R. L., Waldrop, T. C., Goldman, A. S., Shearer, W. T., Springer, T. A. 1985. The severe and moderate phenotypes of heritable Mac-1, LFA-1 deficiency: Their quantitative definition and relation to leukocyte dysfunction and clinical features. *J. Infect. Dis.* 152: 668–89

16. Arnaout, M. A., Todd, R. F., Dana, N., Melamed, J., Schlossman, S. F., Colten, H. R. 1983. Inhibition of phagocytosis of complement C3- or immunoglobulin G-coated particles and of C3bi binding by monoclonal antibodies to a monocyte-granulocyte membrane glycoprotein (Mo1). *J. Clin. Invest.* 72: 171–79

17. Weisman, S. J., Berkow, R. L., Plautz, G., Torres, M., McGuire, W. A., Coates, T. D., Haak, R. A., Floyd, A., Jersild, R., Boehner, R. L. 1985. Glycoprotein-180 deficiency: Genetics and abnormal neutrophil activation. *Blood* 65: 696–704

18. Buescher, E. S., Gaither, T., Nath, J., Gallin, J. I. 1985. Abnormal adherence-related functions of neutrophils, monocytes, and Epstein-Barr virus-transformed B cells in a patient with C3bi receptor deficiency. *Blood* 65: 1382–90.

19. Fisher, A., Trung, P. H., Descamps-Latsdra, B., Lisowska-Grospierre, B., Gerota, I., Perez, N., Scheinmetzler, C., Durandy, A., Virelizier, J. L., Griscelli, C. 1983. Bone marrow-transplantation for inborn error of phagocytic cells associated with defective adherence, chemotaxis, and oxidative response during opsonized particle phagocytosis. *Lancet* 2: 473–76

20. Diener, A. M., Beatty, P. G., Ochs, H. D., Harlan, J. M. 1985. The role of

neutrophil membrane glycoprotein 150 (gp-150) in neutrophil-mediated endothelial cell injury in vitro. *J. Immunol.* 135: 537–43

21. Pohlman, T. H., Stanness, K. A., Beatty, P. G., Ochs, H. D., Harlan, J. M. 1986. An endothelial cell surface factor(s) induced in vitro by lipopolysaccharide, interleukin 1, and tumor necrosis factor-α increases neutrophil adherence by a CDw18-dependent mechanism. *J. Immunol.* 136: 4548–53

22. Sklar, L. A., Jesaitis, A. J., Painter, R. G. 1984. The neutrophil N-formyl peptide receptor: Dynamics of ligand-receptor interactions and their relationship to cellular responses. In *Regulation of Leukocyte Function*, ed. R. Synderman, 2: 29–82. New York: Plenum

23. Gallin, J. I., Seligmann, B. E. 1984. Neutrophil chemoattractant fmet-leu-phe receptor expression and ionic events following activation. See Ref. 22, pp. 83–108

24. Marasco, W. A., Phan, S. H., Krutzsch, H., Showell, H. J., Feltner, D. E., Nairn, R., Becker, E. L., Ward, P. A. 1984. Purification and identification of formyl - methionyl - leucyl - phenylalanine as the major peptide neutrophil chemotactic factor produced by *Escherichia coli. J. Biol. Chem.* 259: 5430–39

25. Chenoweth, D. E., Hugli, T. E. 1978. Demonstration of specific C5a receptor on intact human polymorphonuclear leukocytes. *Immunol.* 75: 3943–47

26. Goldman, D. W., Gifford, L. A., Olson, D. M., Goetzl, E. J. 1985. Transduction by Leukotriene B_4 receptors of increases in cytosolic calcium in human polymorphonuclear leukocytes. *J. Immunol.* 135: 525–30

27. Becker, E. L. 1985. Leukocyte stimulation: Receptor, membrane, metabolic events: Introduction and summary. *Fed. Proc.* 45: 2148–59

28. Sklar, L. A., Hyslop, P. A., Oades, Z. G., Omann, G. M., Jesaitis, A. J. 1985. Signal transduction and ligand-receptor dynamics in the human neutrophil. *J. Biol. Chem.* 26: 11461–67

29. Lew, P. D., Monod, A., Waldvogel, F. A., Dewald, B., Baggiolini, M. 1986. Quantitative analysis of the cytosolic free calcium dependency of exocytosis from three subcellular compartments in intact human neutrophils. *J. Cell Biol.* 102: 2197–2204

30. Sklar, L. A., Oades, Z. G. 1985. Signal transduction and ligand-receptor dynamics in the neutrophil. *J. Biol. Chem.*

260: 11468–75

31. Di Virgilio, F., Lew, D. P., Pozzan, T. 1984. Protein kinase C activation of physiological processes in human neutrophils at vanishingly small cytosolic Ca^{2+} levels. *Nature* 310: 691–93

32. Verghese, M. W., Smith, C. D., Charles, L. A., Jakoi, L., Snyderman, R. 1986. A guanine nucleotide regulatory protein controls polyphosphoinositide metabolism, Ca^{2+} mobilization, and cellular responses to chemoattractants in human monocytes. *J. Immunol.* 137: 271–75

33. Koo, C., Lefkowitz, R. J., Snyderman, R. 1983. Guanine nucleotides modulate the binding affinity of the oligopeptide chemoattractant receptor on human polymorphonuclear leukocytes. *J. Clin. Invest.* 72: 748–53

34. Zigmond, S. H. 1977. The ability of polymorphonuclear leukocytes to orient in gradients of chemotactic factors. *J. Cell Biol.* 75: 606–16

35. Sawyer, D. W., Sullivan, J. A., Mandell, G. L. 1985. Intracellular free calcium localized in neutrophils during phagocytosis. *Science* 230: 663–66

36. Sha'afi, R. I., Shefcyk, J., Yassin, R., Molski, T. F. P., Volpi, M., Naccache, P. H., White, J. R., Feinstein, M. B., Becker, E. L. 1986. Is a rise in intracellular concentration of free calcium necessary or sufficient for stimulated cytoskeletal-associated actin? *J. Cell Biol.* 102: 1459–63

37. Burn, P., Rotman, A., Meyer, R. K., Burger, M. M. 1985. Diacylglycerol in large alpha-actinin/actin complexes and in the cytoskeleton of activated platelets. *Nature* 314: 469–72

38. Lassing, I., Lindberg, U. 1985. Specific interaction between phosphatidylinositol 4,5-bisphosphate and profilactin. *Nature* 314: 472–74

39. Sullivan, S. J., Daukas, G., Zigmond, S. H. 1984. Asymmetric distribution of the chemotactic peptide receptor on polymorphonuclear leukocytes. *J. Cell. Biol.* 99: 1461–67

40. Fletcher, M. P., Seligmann, B. E., Gallin, J. I. 1982. Correlation of human neutrophil secretion, chemoattractant receptor mobilization, and enhanced functional capacity. *J. Immunol.* 128: 941–48

41. Aswanikumar, S., Schiffmann, E., Corcoran, B. A., Wahl, S. M. 1976. Role of a peptidase in phagocyte chemotaxis. *Proc. Natl. Acad. Sci. USA* 73: 2439–42

42. Yuli, I., Snyderman, R. 1986. Extensive

hydrolysis of n - formyl - 1 - methionyl - 1 - leucyl - [^{3}H] phenylalanine by human polymorphonuclear leukocytes. *J. Biol. Chem.* 261 : 4903–8

43. Gallin, J. I., Wright, D. G., Schiffmann, E. 1978. Role of secretory events in modulating human neutrophil chemotaxis. *J. Clin. Invest.* 62 : 1364–74

44. Wright, S. D. 1985. Cellular strategies in receptor-mediated phagocytosis. *Rev. Infect. Dis.* 7 : 395–97

45. McNeil, P. L., Wright, S. D., Silverstein, S. C., Taylor, D. L. 1986. Fc-receptor-mediated phagocytosis occurs in macrophages without an increase in average $[Ca^{++}]_i$. *J. Cell Biol.* 102 : 1586–92

46. Young, J. D. E., Unkeless, J. C., Kaback, H. R., Cohn, Z. A. 1983. Mouse macrophage Fc receptor for IgG 2b/1 in artificial and plasma membrane vesicles functions as a ligand-dependent ionophore. *Proc. Natl. Acad. Sci. USA* 80 : 1636–40

47. Nauseef, W. M., Clark, R. A. 1986. Activation and regulation of neutrophil oxidative metabolism in Mol deficiency. *Clin. Res.* 34 : 678A (Abstr.)

48. Boxer, L. A., Hedley-White, E. T., Stossel, T. P. 1974. Neutrophil actin dysfunction and abnormal neutrophil behavior. *New. Engl. J. Med.* 291 : 1093–99

49. Southwick, F. S., Holbrook, T., Howard, T., Springer, T., Stossel, T. P., Arnaout, M. A. 1986. Neutrophil actin dysfunction is associated with a deficiency of Mol. *Clin. Res.* 34 : 533A (Abstr.)

50. Gallin, J. I. 1985. Neutrophil specific granule deficiency. *Ann. Rev. Med.* 36 : 263–74

51. Wright, D. G., Gallin, J. I. 1979. Secretory responses of human neutrophils: Exocytosis of specific (secondary) granules by human neutrophils during adherence *in vitro* and during exudation *in vivo*. *J. Immunol.* 123 : 285–94

52. Zimmerli, W., Seligmann, B., Gallin, J. I. 1986. Exudation primes human and guinea pig neutrophils for subsequent responsiveness to the chemotactic peptide N-formylmethionylleucylphenylalanine and increases complement component C3bi receptor expression. *J. Clin. Invest.* 77 : 925–33

53. Wright, D. G., Gallin, J. I. 1977. A functional differentiation of human neutrophil granules: Generation of C5a by a specific (secondary) granule product and inactivation of C5a by azurophil (primary) granule products.

J. Immunol. 119 : 1068–76

54. Borregaard, N., Heiple, J. M., Simons, E. R., Clark, R. A. 1983. Subcellular localization of the *b*-cytochrome component of the human neutrophil microbicidal oxidase: Translocation during activation. *J. Cell Biol.* 97 : 52–61

55. Higson, F. K., Durbin, L., Pavlotsky, N., Tauber, A. I. 1985. Studies of cytochrome b_{-245} translocation in the PMA stimulation of the human neutrophil NADPH-oxidase. *J. Immunol.* 135 : 519–24

56. Ohno, Y., Seligmann, B. E., Gallin, J. I. 1985. Cytochrome *b* translocation to human neutrophil plasma membranes and superoxide release. *J. Biol. Chem.* 260 : 2409–14

57. Parmley, R. T., Tzeng, D. Y., Baehner, R. L., Boxer, L. A. 1983. Abnormal distribution of complex carbohydrates in neutrophils of a patient with lactoferrin deficiency. *Blood* 62 : 538–48

58. Gallin, J. I., Fletcher, M. P., Seligmann, B. E., Hoffstein, S., Cehrs, K., Mounessa, N. 1982. Human neutrophil-specific granule deficiency: A model to assess the role of neutrophil-specific granules in the evolution of the inflammatory response. *Blood* 59 : 1317–29

59. Davis, J. M., Dineen, P., Gallin, J. I. 1980. Neutrophil degranulation and abnormal chemotaxis after thermal injury. *J. Immunol.* 124 : 1467–71

60. Moore, F. D., Davis, C., Rodrick, M., Mannick, J. A., Fearon, D. T. 1986. Neutrophil activation in thermal injury as assessed by increased expression of complement receptors. *New Engl. J. Med.* 314 : 948–53

61. Blume, R. S., Wolff, S. M. 1972. The Chediak-Higashi syndrome: Studies in four patients and a review of the literature. *Medicine* 51 : 247–80

62. Klebanoff, S. J., Clark, R. A. 1978. Chediak-Higashi syndrome. In *The Neutrophil: Function and Clinical Disorders.* Amsterdam: Elsevier/North Holland Biomedical

63. Smedly, L. A., Tonnesen, M. G., Sandhaus, R. A., Haslett, C., Guthrie, L. A., Johnston, R. B. Jr., Henson, P. M., Worthen, G. S. 1986. Neutrophil mediated injury to endothelial cells. Enhancement by endotoxin and essential role of neutrophil elastase. *J. Clin. Invest.* 77 : 1233–43

64. Takeughi, K., Wood, H., Swank, R. T. 1986. Lysosomal elastase and cathepsin G in beige mice. *J. Exp. Med.* 163 : 665–77

65. Boxer, L. A., Watanabe, A. M., Rister,

M., Besch, H. R. Jr., Allen, J., Baehner, R. L. 1976. Correction of leukocyte function in Chediak-Higashi syndrome by ascorbate. *New Engl. J. Med.* 295: 1041–45

66. Gallin, J. I., Elin, R. J., Hubert, R. T., Fauci, A. S., Kaliner, M. A., Wolff, S. M. 1979. Efficacy of ascorbic acid in Chediak-Higashi Syndrome (CHS): Studies in humans and mice. *Blood* 53: 226–34

67. Nath, J., Flavin, M., Gallin, J. I. 1982. Tubulin tyrosinolation in human polymorphonuclear leukocytes: Studies in normal subjects and in patients with the Chediak-Higashi syndrome. *J. Cell Biol.* 95: 519–26

68. Davis, S. D., Schaller, J., Wedgwood, R. J. 1966. Job's syndrome: Recurrent "cold" staphylococcal abscesses. *Lancet* 1: 1013–15

69. Donabedian, H., Gallin, J. I. 1983. The hyperimmunoglobulin E recurrent-infection (Job's) syndrome. *Medicine* 62: 195–208

70. Donabedian, H., Gallin, J. I. 1982. Mononuclear cells from patients with the hyperimmunoglobulin E-recurrent infection syndrome produce an inhibitor of leukocyte chemotaxis. *J. Clin. Invest.* 69: 1155–63

71. Donabedian, H., Gallin, J. I. 1983. Two inhibitors of neutrophil chemotaxis are produced by hyperimmunoglobulin E recurrent infection syndrome mononuclear cells exposed to heat-killed staphylococci. *Infect. Immun.* 40: 1030–37

72. Gallin, J. I., Wright, D. G., Malech, H. L., Davis, J. M., Klempner, M. S., Kirkpatrick, C. H. 1980. Disorders of phagocyte chemotaxis. *Ann. Intern. Med.* 92: 520–38

73. Dreskin, S. C., Goldsmith, P. K., Gallin, J. I. 1985. Immunoglobulins in the hyperimmunoglobulin E and recurrent infection (Job's) syndrome. *J. Clin. Invest.* 75: 26–34

74. Berger, M., Kirkpatrick, C. H., Goldsmith, P. K., Gallin, J. I. 1980. IgE antibodies to *Staphylococcus aureus* and *Candida albicans* in patients with the syndrome of hyperimmunoglobulin E and recurrent infections. *J. Immunol.* 125: 2437–43

75. Buckley, R. H., Sampson, H. A. 1981. The hyperimmunoglobulinemia E syndrome. In *Clinical Immunology Update*, ed. E. C. Franklin, pp. 148–167. Amsterdam: Elsevier/North Holland Biomedical

76. Dreskin, S. C., Goldsmith, P. K., Strober, W., Zech, L. A., Gallin, J. I.

1986. The rate of catabolism of polyclonal human IgE is decreased in patients with elevated IgE. *Clin. Res.* 34: 706A (Abstr.)

77. Van Dyke, T. E., Levine, M. J., Tabak, L. A., Genco, R. J. 1981. Reduced chemotactic peptide binding in juvenile periodontitis: A model for neutrophil function. *Biochem. Biophys. Res. Comm.* 100: 1278–84

78. Van Dyke, T. E. 1985. Role of the neutrophil in oral disease: Receptor deficiency in leukocytes from patients with juvenile periodontitis. *Rev. Infect. Dis.* 7: 419–25

79. Goldstein, I. M. 1984. Neutrophil degranulation. See Ref. 3, pp. 189–219

80. Baggiolini, M., Dewald, B. 1984. Exocytosis by neutrophils. See Ref. 3, pp. 221–46

81. Ernst, J. D., Meers, P., Hong. K., Duzgures, N., Papahadjopoulos, D., Goldstein, I. M. 1986. Human polymorphonuclear leukocytes contain synexin a calcium-binding protein that mediates membrane fusion. *Clin. Res.* 34: 722A (Abstr.)

82. Spitznagel, J. K., Shafer, W. M. 1985. Neutrophil killing of bacteria by oxygen-independent mechanisms: A historical summary. *Rev. Infect. Dis.* 7: 398–403

83. Weiss, J., Victor, M., Elsbach, P. 1983. Role of charge and hydrophobic interactions in the action of the bactericidal/permeability increasing protein of neutrophils on gram-negative bacteria. *J. Clin. Invest.* 71: 540–49

84. Ganz, T., Selsted, M. E., Szklarek, D., Harwig, S. S. L., Daher, K. Bainton, D. F., Lehrer, R. I. 1985. Defensins: Natural peptide antibiotics of human neutrophils. *J. Clin. Invest.* 76: 1427–35

85. Cohn, Z. A. 1986. Intracellular parasitism and the system of mononuclear phagocytes. In *Immunologic Diseases 14th Ed.* Ed. M. Samter, M. M. Frank, K. F. Austen, D. Talmadge, H. Claman. New York: Raven. In press

86. Cech, P., Lehrer, R. I. 1984. Heterogeneity of human neutrophil phagososomes: Functional consequences for candidacidal activity. *Blood* 64: 147–51

87. Klebanoff, S. J. 1980. Oxygen metabolism and the toxic properties of phagocytes. *Ann. Intern. Med.* 93: 480–89

88. Babior, B. M. 1984. The respiratory burst of phagocytes. *J. Clin. Invest.* 73: 599–601

89. Umei, T., Takeshige, K., Minakami, S. 1986. NADPH binding component of

neutrophil superoxide-generating oxidase. *J. Biol. Chem.* 261: 5229–32

90. Babior, B. M. 1983. The nature of NADPH oxidase. See ref. 98, pp. 91–114

91. Segal, A. W. 1983. Chronic granulomatous disease: A model for studying the role of cytochrome b_{245} in health and disease. See ref. 98, pp. 121–143

92. Tauber, A. I., Borregaard, N., Simons, E., Wright, J. 1983. Chronic granulomatous disease: A syndrome of phagocyte oxidase deficiencies. *Medicine* 62: 286–309

93. McPhail, L. C., Synderman, R. 1984. Mechanisms of regulating the respiratory burst in leukocytes. See Ref. 3, pp. 247–81

94. Lynch, R. E., Fridovich, I. 1978. Permeation of the erythrocyte stroma by superoxide radical. *J. Biol. Chem.* 253: 4697–99

95. Ohno, Y., Gallin, J. I. 1985. Diffusion of extracellular hydrogen peroxide into intracellular compartments of human neutrophils. *J. Biol. Chem.* 260: 8438–46

96. Henderson, W. R., Klebanoff, S. J. 1983. Leukotriene production and inactivation by normal, chronic granulomatous disease and myeloperoxidase-deficient neutrophils. *J. Biol. Chem.* 258: 13522–27

97. Clark, R. A., Klebanoff, S. J. 1979. Chemotactic factor inactivation by the myeloperoxidase-hydrogen peroxide-halide system. *J. Clin. Invest.* 64: 913–20

98. Gallin, J. I., Fauci, A. S. eds. 1983. *Advances in Host Defense Mechanisms, Vol. 3: Chronic Granulomatous Disease.* New York: Raven

99. Gallin, J. I., Buescher, E. S. 1983. Abnormal regulation of inflammatory skin responded in male patients with chronic granulomatous disease. *Inflammation* 7: 227–32

100. Hattori, H. 1961. Studies on the labile stable nadi oxidase and peroxidase staining reactions in the isolated particles of horse granulocytes. *Nagoya J. Med. Sci.* 23: 362–78

101. Ohno, Y., Buescher, E. S., Roberts, R., Metcalf, J. A., Gallin, J. I. 1986. Reevaluation of cytochrome *b* and flavin adenine dinucleotide in neutrophils from patients with chronic granulomatous disease and description of a family with probable autosomal recessive inheritance of cytochrome *b* deficiency. *Blood* 67: 1132–38

102. Weening, R. S., Corbeel, L., de Boer, M., Lutter, R., van Zwieten, R.,

Hamers, M. N., Roos, D. 1985. Cytochrome *b* deficiency in an autosomal form of chronic granulomatous disease. *J. Clin. Invest.* 75: 915–20

103. Royer-Pokora, B., Kunkel, L. M., Monaco, A. P., Goff, S. C., Newburger, P. E., Baehner, R. L., Cole, F. S., Curnutte, J. T., Orkin, S. H. 1986. Cloning the gene for an inherited human disorder-chronic granulomatous disease-on the basis of its chromosomal location. *Nature* 322: 32–38

104. Segal, A. W., Heyworth, P. G., Cockcroft, S., Barrowman, M. M. 1985. Stimulated neutrophils from patients with autosomal recessive chronic granulomatous disease fail to phosphorylate a M_r-44,000 protein. *Nature* 316: 547–49

105. Rossi, F., Bellavite, P., Papini, E. 1986. Respiratory response of phagocytes: Terminal NADPH oxidase and the mechanisms of its activation. In *Biochemistry of Macrophages.* 118: 172–95. London: Pitman

106. Harvath, L., Andersen, B. R. 1979. Defective initiation of oxidative metabolism in polymorphonuclear leukocytes. *New Engl. J. Med.* 300: 1130–35

107. Weening, R. S., Roos, D., Weemas, C. M. R., Homan-Muller, J. W. T., Schaik, M. L. J. 1976. Defective initiation of the metabolic stimulation in phagocytizing granulocytes: A new congential defect. *J. Lab. Clin. Med.* 88: 757–68

108. Lew, D. P., Southwick, F. S., Stossel, T. P., Whitin, J. C., Simons, E., Cohen, H. J. 1981. A variant of chronic granulomatous disease: Deficient oxidative metabolism due to a low-affinity NADPH oxidase. *New Engl. J. Med.* 305: 1329–33

109. Seger, R. A., Tiefenauer, L., Matsunaga, T., Wildfeuer, A., Newburger, P. E. 1983. Chronic granulomatous disease due to granulocytes with abnormal NADPH oxidase activity and deficient cytochrome-b. *Blood* 61: 423–28

110. Styrt, B., Klempner, M. S. 1984. Late-presenting variant of chronic granulomatous disease. *Prediatr. Infect. Dis.* 3: 456–59

111. Parry, M. F., Root, R. K., Metcalf, J. A., Delaney, K. K., Kaplow, L. S., Richar, W. J. 1981. Myeloperoxidase deficiency: Prevalence and clinical significance. *Ann. Intern. Med.* 95. 293–301

112. Stendahl, O., Coble, B. I., Dahlgren, C., Hed, J., Molin, L. 1984. Myeloperoxidase modulates the phagocytic

activity of polymorphonuclear neutrophil leukocytes. Studies with cells from a myeloperoxidase-deficient patient. *J. Clin. Invest.* 73: 366–73

113. Klebanoff, S. J., Clark, R. A. 1974. myeloperoxidase deficiency. See Ref. 62, pp. 711–33.

114. Nauseef, W. M., Root, R. K., Malech, H. L. 1983. Biochemical and immunologic analysis of hereditary myeloperoxidase deficiency. *J. Clin. Invest.* 71. 1297–1307

115. Nauseef, W. M. 1986. Myeloperoxidase biosynthesis by a human promyelocytic leukemia cell line: Insight into myeloperoxidase deficiency. *Blood* 67: 865–72

116. Koeffler, H. P., Ranyard, J., Pertcheck, M. 1985. Myeloperoxidase. Its structure and expression during myeloid differentiation. *Blood* 65. 484–91

117. Test, S. T., Lampert, M. B., Ossanna, P. J., Thoeno, J. G., Weiss, S. J. 1984. Generation of nitrogen-chlorine oxidants by human phagocytes. *J. Clin. Invest.* 74: 1341–49

Ann. Rev. Immunol. 1987. 5 : 151–74
Copyright © 1987 by Annual Reviews Inc. All rights reserved

TRANSGENIC MICE WITH IMMUNOGLOBULIN GENES

*Ursula Storb**

Department of Microbiology and Immunology, University of
Washington, Seattle, Washington 98195

INTRODUCTION

Immunoglobulin (Ig) genes encode antibody molecules expressed exclusively in the B-lymphocyte lineage. The most immature identifiable cells of the B-cell lineage, pre–B cells, give rise to B cells that differentiate terminally into antibody secreting plasma cells. Circulating antibody molecules are comprised of two identical light (L) chains and two identical heavy (H) chains. The expression of Ig genes is strictly regulated during development of B lymphocytes : At the pre-B-cell stage only μ chain genes are expressed, while L chains begin to be synthesized at the B-cell stage (1, 2). Ig genes are unusual in that they are arranged in separate subgenes. The limiting step required for correct transcription of Ig genes is the assembly of H genes from three subgenes, variable (V), diversity (D), and constant (C), and of L genes from V and C (reviewed in 3). Since each of the three Ig loci (H, κ, and λ) consists of multiple V, (D), and J subgenes, a large number of different H and L genes can be assembled by joining of different combinations of the subgenes (3). Somatic hypermutability contributes additional diversity (4, 5). Thus, it can be estimated that thousands of different H and L chains are produced by an individual. Since millions of different H-L combinations exist, B lymphocytes are extremely heterogenous with respect to the antibodies expressed and consequently to the antigens recognized. Lymphoid organs are a mixture of such highly diversified B cells, so it is impossible to study the development or immune response of a monoclonal B-cell lineage in the normal mouse. Malignant cell lines, such as myelomas in animals or cultured myeloma,

*Current address: Department of Molecular Genetics and Cell Biology, University of
Chicago, Chicago, Illinois 60637.

151

0732–0582/87/0410–0151$02.00

hybridoma, or pre–B cell lines, can provide homogeneous cell populations. These have been exploited extensively for the analysis of Ig structure and Ig gene organization (6). However, with the exception of a limited plasticity in certain pre–B cell lines, such transformed cells are frozen in a specific stage of differentiation. It was hoped that the technique of transgenic mice (7, 8) would provide B-lymphocyte populations, homogenously expressing Ig transgenes in a normal in vivo environment.

Transgenic mice can be produced in several ways, the most frequently used being microinjection of the cloned genes under study into one of the pronuclei of a fertilized egg (9). The microinjected embryos are developed to term in the uterus of a pseudopregnant mouse. About 25% of the resulting pups contain the injected gene, and they generally contain the foreign DNA stably integrated in all cells. This DNA is propagated in the germ line.

In most of the first transgenic mouse experiments, expression of the transgenes was either absent or aberrant (see 10 and 11). Good expression in a variety of cell types was only obtained with chimeric genes containing the mouse metallothionein (MT) promoter. The disadvantage of this system was the absence of tissue specificity due to the ubiquitous expression of the MT promoter. It is now known, however, that the absence of tissue-specific expression with a number of other genes was due to the presence of certain prokaryotic DNA sequences. Most introduced genes can be expressed correctly and at near normal endogenous levels under the direction of their own control regions, provided the plasmid or phage cloning vectors have been eliminated or restricted to a few hundred base pairs (11).

Based on the transgenic mouse data available in 1982, our expectations were that the MT promoter might be needed to obtain expression of Ig genes. It was, however, very desirable to express Ig genes exclusively in lymphoid cells and not ubiquitously, as are metallothionein genes. Therefore, the attempt was made first with a rearranged κ gene entirely under its own control regions (12). In this case, the κ gene was expressed correctly (see below). Other Ig genes that have now been introduced into the germline of mice are listed in Table 1.

TISSUE-SPECIFIC EXPRESSION OF IG TRANSGENES

Transgenic Mice With κ Genes

Three different κ genes have been introduced into transgenic mice. They are the functional genes of the myelomas MOPC-21 (12) and MOPC-167 (13) and of the hybridoma Sp6 (14).

The MPOC-21 κ transgene contains the complete rearranged gene with

Table 1 Summary of transgenic mice with immunoglobulin genes

	Gene injected	Source[a]	Antigen specificity	Enhancer	Transgene copy number	Expression	Reference number
1.	κ	Myeloma MOPC-21	Unknown	κ	16 to 64	B	12
2.	κ	Myeloma MOPC-167	PC	κ	1 to 41	B	13
3.	μ	Hybridoma 17.2.25	NP	H	17, 30	B, T	17
4.	μ	Myeloma MOPC-167-V_H spliced to $C\mu$	PC	H	1 to 70	B, T	13
5.	$\mu \Delta$ mem	Same as 4. without membrane terminus	PC	H	2 to 12	B, T	13
6.	$\gamma 1$	Human plasmacytoma	Unknown	None	~1 to 20	None	34
7.	$\gamma 1$	Same as 6.	Unknown	Human H	1 to 2	B	25
8.	ε	Mouse-$C\varepsilon$ gene	None	None	~1 to 10	None	35
9.	$\gamma 2b$	Mouse hybridoma VD93	Pseudomonas	H	~1 to 50	B, T	b
10.	$\mu + \kappa$	Hybridoma Sp6	TNP	H, κ	4	B, (T?)	14[d]
11.	$\mu + \kappa^c$ or $\mu \times \kappa^c$	μ: same as 4. κ: same as 2.	PC	H, κ	7 to 75	B, T	13
12.	$\mu \Delta$ mem$+\kappa^c$ or $\mu \Delta$ mem$\times \kappa^c$	μ: same as 5. κ: same as 2.	PC	H, κ	3 to 116	B, T	13
13.	Germline λ	Chicken	?	?	2 to 100	B	26
14.	Rearrangement test gene	Mouse κ	None	H	~1 to 20	?	e

[a] 1–12 the functional, rearranged gene of the respective cell type was used. In No. 13, the chicken λ genes were injected in the germline (unrearranged) configuration. In No. 14, a synthetic rearrangement test gene was injected containing small portions of the 3′ region of a mouse V_κ gene and the 5′ region of J_κ (82).
[b] H. Tsang, C. Pinkert, M. Lostrum, R. Brinster, U. Storb, unpublished.
[c] $\mu + \kappa$: Both genes were injected simultaneously into fertilized eggs. $\mu \times \kappa$: μ transgenic mice were bred with κ transgenic mice.
[d] Transgenic mice using the same gene were also produced by D. Solter (personal communication).
[e] P. Engler, C. Pinkert, R. Brinster, U. Storb, unpublished.

1.6 kb and 8.5 kb of flanking sequences 5′ of the leader and 3′ of C_κ respectively. Of eight different transgenic lines each had multiple tandem copies at a single chromosomal insertion site (15). As is the rule in transgenic mice, the insertion sites differed between mice, and there was no evidence for integration into the homologous κ locus. Despite the variation in integration sites, all the mice expressed the transgene. In the serum, high levels of κ chains with the mobility and isoelectric focusing pattern of MOPC-21 κ chains were seen (12, 15). The sources of these proteins were the lymphoid tissues. In general, elevated mRNA levels corresponding to transgene expression from nonlymphoid cells were not found in any nonlymphoid tissues tested [liver, kidney, heart, skeletal muscle, brain, thyroid, and testis (16)]. Normally, κ genes are only expressed in the B-cell lineage. When B and T lymphocytes from the κ-transgenic spleens were analyzed separately, κ-transgene expression was found to be restricted to B lymphocytes (16). Likewise, in the thymus of young mice transgenic κ-mRNA was not found beyond a low level attributed to infiltrating B lymphocytes (16).

The results with the MOPC-167 κ gene were very similar (13). The antibody produced by the myeloma MOPC-167 binds the antigen phosphorylcholine (PC). Sera from M-167 κ-transgenic mice were analyzed for PC binding by ELISA (C. Pinkert, J. Manz, R. O'Brien, R. Brinster, and U. Storb, unpublished). Normal mice raised in a conventional environment do have anti-PC antibodies in the serum. In some but not all of the M-167 κ-transgenic mice, the anti-PC titers were increased up to tenfold. With both of these κ transgenes, the level of expression of spleen κ-mRNA varied widely between mice (12, 13).

The third κ gene, derived from the hybridoma Sp6, was introduced into mice on one plasmid together with the μ gene from Sp6 (14). It is expressed in B cells, but the level of κ protein produced from the κ gene is one tenth the level of transgenic-μ protein (see below).

In conclusion, κ transgenes are expressed specifically in B lymphocytes and apparently independently of the chromosomal insertion site. Whether the differences in expression of different κ transgenes are due to environmental factors or are an effect of the insertion site still needs to be determined. As with most other transgenes (10, 11), the transgene copy number does not seem to influence the level of expression of any Ig transgene.

Transgenic Mice with μ Genes

Three different μ genes have been introduced into the germ line of mice (see Table 1). In two cases, μ 17.2.25 (17) and μ MOPC-167 (13), the μ genes were introduced alone. In the third case (14), referred to above, the μ gene and κ gene of the hybridoma Sp6 were present on the same plasmid.

All three μ genes are expressed in the serum of the transgenic mice at a high level. The 17.2.25 H gene is derived from a hybridoma with anti-NP activity. Serum anti-NP levels in the transgenic mice are elevated between 2- and 360-fold over control mice (17). The MOPC-167 H gene encodes anti-PC antibodies, and the serum of the μ-transgenic mice shows, on average, a tenfold increase of anti-PC antibodies (C. Pinkert, J. Manz, R. Brinster, and U. Storb, unpublished). In the case of the $\mu\kappa$ mice produced by Rusconi & Köhler (14), the serum levels of the Sp6 idiotype are increased 1000- to 4000-fold.

All three μ transgenes encode high levels of μmRNA in B cells; this RNA is of the correct size. Both the secreted and the membrane form of μRNA are found in spleen (13). There is more of the secreted form than of the membrane form (13) which suggests that the majority of the μ transcripts in the spleen are derived from plasma cells.

Nonlymphoid organs do not appear to transcribe the μ transgenes (13, 17), with the possible exception of the heart in one case (17). However, in contrast to κ transgenes, the μ transgenes are expressed in T cells (13, 17). Lyt-2^+ T cells from lymph nodes and thymus are enriched for transgenic μRNA (17). With fluorescent staining of transgenic thymus, about 60% of the cells are μ positive in the cytoplasm compared with less than 1% in normal thymus (13). A similar proportion of thymocytes express T-cell receptor (TCR) genes (18), and it is possible that transacting factors for activation of TCR genes also activate μ genes. The μRNA in the transgenic thymus is of the size of secreted and membrane RNA, with a slight preponderance of the membrane form (13). Thus, when a rearranged μ gene is present in certain T cells, it is transcribed and the transcripts are polyadenylated and processed correctly. The limitation for correct expression of H genes in normal T cells seems to be the absence of correct H-gene rearrangement. Abortive rearrangements of D-J occur in T cells and transcripts originating from a promoter associated with the D region as well as transcripts originating in the J-C intron are found (19–22). However, no V to D-J joints are made, perhaps because in contrast to B cells (23), unrearranged V_H genes do not become transcriptionally active in T cells.

The μ protein in transgenic thymuses seems to be restricted to an intracytoplasmic location (13). No μ is seen on the T-cell surface by immunofluorescence, and therefore it does not seem to associate with TCR α or β chains.

Transgenic mice with a μ gene which lacks membrane exons ($\mu\Delta$ mem) have also been produced (13). These mice express transgenic μ-mRNA at a high level in spleen and thymus. The implications of these mice for the regulation of B-cell triggering are discussed below.

When μ and κ genes are introduced into the same transgenic mouse, both μ and κ are expressed in most cases (13, 14). This seems to be independent of the mode by which the genes were introduced—i.e. whether both on one plasmid (14); or μ and κ on separate plasmids, but coinjected ($\mu + \kappa$) (13); or by mating a μ transgenic mouse with a κ transgenic mouse ($\mu \times \kappa$) (13). In the first and most likely the second case, the μ and κ genes are inserted at the same chromosomal site, whereas in the last case they are at different sites. It is interesting that in the $\mu + \kappa$ mice, both transgenic μ- and κ-mRNAs are expressed at a high level in the spleen, whereas in the thymus only μ-RNA is found (13). Thus, the close proximity of the expressed H gene does not lead to stable κ-mRNA transcription. This may be due to the inactivity of the κ transcriptional promoter in T cells (24).

In general, in transgenic mice with both μ and κ genes encoding anti-PC antibodies, the transgenic κ-mRNA levels are somewhat lower than the transgenic μ-mRNA levels in spleen (13). In hybridomas from anti-TNP transgenic mice, the transgenic κ-mRNA level is considerably lower than that of μ-mRNA, and κ protein is one tenth that of μ protein (14). The reason for these differences is not known.

Other Ig Genes in Transgenic Mice

Recently, transgenic mice were produced with a human $\gamma 1$ gene (25). The $\gamma 1$ gene is transcribed in B cells and not in brain, liver, or kidney cells of the transgenic mice; this suggests that the human promoter and enhancer sequences are recognized by transacting factors of mouse B cells. Evidence suggests that the $\gamma 1$ gene is not expressed at a high level in spleen T cells (25). However, a moderate level cannot be ruled out by the experiments, and it remains to be determined definitively, whether transacting factors in mouse T cells would interact with a human $\gamma 1$ gene.

A mouse $\gamma 2b$ gene has also been introduced into transgenic mouse lines (H. Tsang, C. Pinkert, R. Brinster, M. Lostrom, U. Storb, unpublished). It is expressed in spleen and thymus, but not in liver and kidney (U. Storb, unpublished).

Lastly, chicken λ genes were used to produce transgenic mice (26). The λ genes were not rearranged at the time of microinjection. These mice are discussed later.

The Molecular Basis for Tissue-Specific Expression of Ig Genes in Transgenic Mice

It is clear from the transgenic experiments that Ig genes, with relatively small additions of upstream and downstream flanking sequences, are sufficient for expression in B lymphocytes. Two features of this observation are immediately striking. First, specific expression in B cells is not a direct consequence of rearrangement. In the Ig transgenic mice, correctly rearranged Ig genes are present in all cell types, but the genes are not expressed in nonlymphoid cells.

Secondly, there does not seem to be a recognizable position effect. Random insertion into the genome is compatible with correct regulation. Of course, one would expect that certain genomic sites, such as the nucleolar organizer region, may not be accessible to polymerase-II transcriptional complexes. In our own experience, of 57 transgenic mice produced with H or L genes known to be functional, only 7 did not express the gene at a detectable level in the spleen (R. Brinster, R. Hammer, C. Pinkert, and U. Storb, unpublished). We have not determined the reason for the inactivity of these particular transgenes. In 27 of the mice, the complete plasmid vector was present in the microinjected genes; 3 of these did not express. Of 30 mice without the vector, 4 did not express. In the latter nonexpressors the transgenes were in low copy numbers, and it might be that deletions or disruptions of the transgene had occurred in some or all of them during the integration step in the microinjected embryos; that would explain the lack of expression. On the whole, the high success rate of Ig transgene expression suggests that strong tissue-specific control elements are present in Ig genes which may override negative effects of certain chromosomal sites. With some other genes, such as globin genes, the proportion of transgenic mice that express the gene appears to be lower (10, 11).

Certain sequences are required for correct transcription of Ig genes when transfected into cell lines in culture (reviewed by Calame—27). For H-chain genes, three control regions appear to be important (28). The first is a conserved octanucleotide ATGCAAAT upstream of the TATA box of the V_H gene that had been previously postulated to be involved in regulation of Ig genes because of its conserved nature 5' of all V genes (29). The second is the H-gene enhancer located in the intron between J-H and $C\mu$ (30). Finally, there seem to be intragenic sequences the exact position of which has not been defined and which are required for a high level of expression of H genes in B cells (28). Similarly, κ genes have two analogous regions, an upstream octanucleotide (the inverse of the H octamer) (29, 31) and an enhancer in the J_κ-C_κ intron (32), both of which are also required. Intragenic control regions for κ have not been explored. Lambda genes also contain a conserved octamer upstream of the V genes, the orientation of which is the same as that of V_κ genes (29). Despite the presence of an enhancer-"core"-like sequence in the J_λ-C_λ intron (30), enhancer activity could not be demonstrated in transfection experiments (33).

The three putative control regions, the upstream octamer, enhancer, and complete intragenic sequences were all present in the transgenes that have been shown to be active. We can probably assume that all are required in transgenic mice. Only two direct deletion tests have been done. In the case of the human $\gamma 1$ gene, no transcripts were found in transgenic mice if the microinjected gene lacked the enhancer (34). However, the same gene with enhancer was active in transgenic B cells (25). Of course, the human H enhancer (or other control regions) may not combine with mouse-B-cell transacting factors very efficiently: the amount of human $\gamma 1$ mRNA in the mouse cells is less than one tenth that produced in human cells (25). In a second case, a partial mouse gene containing the constant region domains of the ε gene was microinjected. The resulting transgenic mice did not express the transgene, but no positive control with the complete gene has been done (35).

Several variant B-cell lines with spontaneous deletion of the H-gene enhancer continue to produce H chains at the same rate as the parent cell line (36–39). However, these sequences are required for proper expression in transfection experiments. Based on these findings, Ig gene enhancers were postulated to be "transfection enhancers" only (27). They must also be "transgenic enhancers," if one can generalize from a single transgenic mouse; at least for a human H gene, the enhancer seems to be required. Perhaps, it is as others have postulated (27; 36–39): H enhancers may be required for establishing a stable transcriptional complex around an H gene when a pre–B cell first activates the normal H-gene locus or an H transgene, but the transcriptional competence may be stably propagated in the absence of the enhancer.

We have some preliminary evidence that DNA sequences located about 6 kb 3′ of C_κ may be required for high expression of κ transgenes (U. Storb, S. McKnight, C. Pinkert, and R. Brinster, unpublished). The nature and possible function of this region is unknown. Similarly, Rusconi (40) has reported that a κ gene which contains only 1 kb 3′ of C_κ is inactive in transgenic mice. However, no evidence exists for a similar region 3′ of $C\mu$. The mice with the $\mu\Delta$ mem gene express that stunted gene as highly as μ transgenic mice express the complete μ gene. This suggests that sequences beyond about 2 kb 3′ of the secreted terminus of the μ gene are nor required for activation of the μ locus or for stability and processing of secretory μmRNA.

In conclusion, Ig transgenes with their own control regions are correctly expressed in the B lymphocytes of transgenic mice. These mice provide therefore an ideal tool to investigate many parameters of the control of Ig synthesis, some of which are described below.

Despite the high levels of expression of the Ig transgenes, the transgenic mice produced so far appear to be healthy and of normal longevity. This would suggest that their immunity against a gamut of pathogens is not significantly impaired. The question arises, therefore, whether the Ig transgenes have any effect on the endogenous immune response. This question is the subject of the next two sections of this review.

ALLELIC AND ISOTYPIC EXCLUSION

More than 20 years ago, several laboratories observed in human (41), rabbit (42), and mouse (43) that plasma cells of Ig-allotype heterozygous individuals expressed only one allotype—not both. It was postulated that individual cells in the "phenotypically mosaic" (41) population had inactivated one of the alleles. Later it was determined that Ig genes have to rearrange before transcription of functional mRNAs occurs (44). However, it also became apparent that "allelic exclusion" could not be due to a priori inactivation of one allele, because both were found to be transcriptionally active in plasma cells, even if only one was rearranged (45). Two basic hypotheses were formulated to explain allelic exclusion. The *regulated model* proposed that when a correct H-gene or L-gene rearrangement had occurred the cell would receive a positive signal to stop further rearrangement (2). The *stochastic model* was based on the finding of a high proportion of incorrectly rearranged Ig genes. It proposed that the probability of two productive rearrangements in two alleles of a given cell was too low to result in the frequent occurrence of two different functional H or L chains (46, 47).

To distinguish between the two models of allelic exclusion transgenic mice were used that had a functionally rearranged Ig transgene present in every cell before any endogenous rearrangement could occur (48). If the feedback (regulated) model is correct, B cells which contain and express an Ig transgene should not rearrange the respective endogenous Ig genes.

Allelic Exclusion of κ Genes

The first series of experiments to distinguish between the two models of allelic exclusion in transgenic mice were carried out with κ-transgenic mice (48). These mice expressed the κ transgene at a high level in B lymphocytes, and it was expected that endogenous κ genes should not be rearranged if the feedback model was correct. To determine whether there was an effect of the κ transgene on the cellular level, hybridomas were produced from spleen B lymphocytes by fusion with the myeloma Sp2/0 (48). Analysis of the DNA, RNA, and Ig proteins produced by the transgenic hybridomas clearly showed that the κ transgene affected endogenous κ genes (48). In all the hybridomas where transgenic κ chains were present together with endogenous H chains, no rearrangement of endogenous κ genes had occurred. The effect was observed only when combinations of κ and H chains were present. In hybridomas without H chains, endogenous κ genes were rearranged despite the presence of transgenic κ chains.

Does this feedback by κ-H molecules occur with any combination of κ and H chains, or does preferential association play a role in allelic exclusion? Discovery of two full-size κ chains in the myeloma S107 and the association of only one of them with H chains (49, 50) suggested that preferential association may be important. However, one of the S107 L chains has a deletion of two amino acids, which makes it incapable of pairing with H chains (49, 50). When one considers immunochemical data concerning H-L association, it is apparent that most L and H chains can associate at random (51, 52, 53). (Specific association to bind a particular antigen is a different matter which probably does not have to concern us here.) The association constants for any H-L pair differ maximally by a factor of 10 (K. Dorrington, personal communication). This does not affect random associability except in competitive situations (54) where the chain combination with a higher association constant may be the only one to form. It remains to be determined for transgenic mice whether all κ-H associations can cause feedback inhibition of rearrangement. In the MOPC-21 κ-transgenic mice it appears that any H chain in combination with the transgenic-κ chain is able to exert the feedback. The endogenous H chains were of the μ or γ classes, and most likely their J, D, and V regions differed, since different restriction fragments were seen on Southern blots (55). However, the transgenic κ chain of these experiments may be unusual, in that it may be particularly capable of combining with any H chain. It is the same κ chain as that produced by the myeloma NS-1 (a derivative of MOPC-21). This myeloma has been widely used as a fusing line for hybridomas (51). In such hybridomas mixed Igs are produced in which the NS-1 κ chain is paired with H chains from the fusion partner. In addition, unlike most κ chains, this κ chain cannot be secreted on its own (56). Whether there are any quantitative or qualitative differences in feedback by different κ chains needs to be investigated with transgenic mice harboring various other κ transgenes.

One example has already resulted in a different outcome. In transgenic mice carrying a plasmid which combined μ and κ genes from an anti-TNP hybridoma, feedback inhibition of endogenous κ-gene rearragement could not be demonstrated (14). In these mice, however, a great imbalance existed between the μ and κ chains. The level of κ chains was only about one tenth of μ. In another case, in mice with μ and κ transgenes encoding anti-PC antibodies, where balanced amounts of μ and κ were produced (13), the endogenous κ-gene rearrangement appears to be suppressed (J. Manz, R. Brinster, and U. Storb, unpublished).

Thus, one may formulate a tentative rule for allelic exclusion of κ genes (Figure 1): In a developing B cell, the achievement of a level of κ chains sufficient to saturate the μ chains causes the stop of κ gene rearrangement.

The history of the B cell immortalized in a particular hybridoma is unknown. Such knowledge would be desirable in at least two respects. Firstly, in the case of mice with the MOPC-21 κ transgene, one would like to know whether the precursor cells ever produced H chains. Is it possible that those cells without evidence of feedback—i.e. cells which produce both transgenic and endogenous κ in the absence of H chains—previously produced H chains? If so, the postulated stop of κ-gene rearrangement would be reversible after loss of the H chain.

A second question concerning the history of the hybridomas asks in which developmental stage relative to endogenous Ig gene activation do the κ transgenes become transcriptionally activated. In μ-producing pre–B cell lines of Abelson murine leukemia virus (AMuLV), the κ transgene was not transcribed in cells from mice with a low copy number of transgenes (57). For κ transgenes or, in normal mice, endogenous rearranged κ genes to stop further rearrangement of κ genes in the same cell, a certain level of κ chains may be required. As we discussed above, perhaps the κ chains must saturate the μ chains. Thus, the timing and completeness of the feedback would depend upon the relative frequencies of correct and

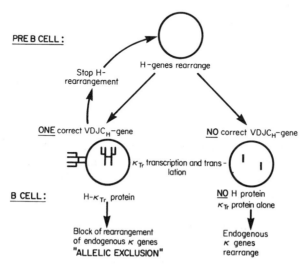

Figure 1   Scheme of allelic exclusion of H genes and κ genes.

aberrant κ-gene rearrangements and the rates of transcription and translation of the correctly rearranged κ as well as μ genes. Since mature B cells with two functional κ chains and H chains have not been observed in normal mice or in transgenic mice with balanced H- and L-gene expression, presumably the occurrence of a second correct κ-gene rearrangement before full expression of a first correct κ gene is a rare event. In addition, B cells that produce scrambled antibody molecules with two different κ chains may somehow be eliminated or not selectively expanded.

The molecular basis for the feedback inhibition of κ-gene rearrangement is not known. The κ-H unit most likely acts on the protein level. There is no known way that two mRNAs would act in a concerted fashion. The κH proteins could act within the cell or via the cell surface. The suggestion has been made that an H-chain-binding protein, BiP, may be replaced by L chain (58, 59). Free BiP or κ-H may directly interact with the genome. On the other hand, the initiation of L-chain production coincides with the appearance of Ig molecules on the surface of the B cell (60). Perhaps the accompanying perturbation of the membrane or an interaction of surface Ig with T lymphocytes leads to a second messenger response which shuts off rearrangement.

Other, more general changes occur in the transition from the pre-B to the B-cell stage. It is therefore possible that rearrangement is not the only target for the feedback. Perhaps the next step of differentiation is initiated and leads to the eventual expression of Fc and complement receptors, Ia and other surface markers, and to susceptibility to antigen stimulation, as well as cessation of gene rearrangement. Shutoff of κ-gene rearrangement may result when the genes are turned off that encode the specific nuclease-ligase complex involved in Ig gene rearrangement.

Allelic Exclusion of μ Genes

Does a functional μ transgene have any effect on the rearrangement of endogenous μ genes? To answer this question B cells from μ-transgenic mice were immortalized and studied individually (14, 61). Two methods of cell cloning were used. In AMuLV-transformed bone marrow cells, mainly pre–B cells are immortalized (2, 62), whereas in hybridomas from spleen cells, more mature cells of the B-cell lineage are obtained (51). AMuLV lines from the bone marrow of μ-transgenic mice all produced the transgenic μ protein (61). When their endogenous H genes were analyzed, it was found that 40% (10/25) of the cell lines had at least one H allele in the germline J-H configuration. In contrast, in AMuLV cell lines from the bone marrow of normal mice, no germline J-H genes were found (61). It had been previously shown that in more immature AMuLV lines from normal fetal liver, germline J-H genes are present (2). However, these show less than single gene–per-cell levels of band intensity on Southern blots, and this suggests ongoing rearrangement (2); eventually all germline H genes are lost in subsequent subclones owing to rearrangement. In the μ-transgenic AMuLV lines, the germline H genes are stable (61).

Was this effect of the μ transgene perhaps due to some targeting of the AMuLV to more immature cells? Several reasons made that seem unlikely. First, one of the lines with a germline H gene had rearranged κ genes, indicative of a more mature B-cell stage. Secondly, germline genes often appeared in hybridomas of mature B cells from μ-transgenic mice not derived by AMuLV transformation (14, 61). Furthermore 63% of the H alleles of μ-transgenic hybridomas were arrested at the D-J stage of rearrangement, in contrast to 23% in AMuLV-transformed bone marrow cells from normal mice (14).

Thus, the feedback on H-gene rearrangement may be due to a direct effect of the μ transgenes. It is a question, however, if this is a physiological effect. The μ transgenes were present in multiple copies of 4 (14), 17 (61) and 30 (61). It is possible that the transgenes competed out transacting factors that may be required for H-gene activation. This problem needs to be addressed directly by studying transgenic mice with incomplete μ genes, pure enhancer sequences, etc. However, based on experiments in which free enhancer sequences were transfected into B cells to compete out transacting factors (63), a larger number of the transgenes would probably be required for complete inhibition of endogenous H-gene expression. Also, several of the hybridomas from a $\mu\kappa$-transgenic mouse produced endogenous μ protein, in addition to the transgenic μ. These hybridomas included one cell with a D-J allele (14), indicating the presence of enough transacting factors to allow expression of a single endogenous μ gene.

Assuming then that the feedback effect reflects events in normal B-cell development, what may be its mechanism? It needs to be determined whether DNA, RNA, or protein is the agent. In most of the μ-transgenic AMuLV lines, some rearrangement of endogenous H genes occurs; this suggests a delay between the time of activation of the H locus and the feedback. Perhaps competence for rearrangement and transcription coincide, and a certain threshold of transcripts or protein encoded by the μ transgene has to be achieved before further H-gene rearrangement can be blocked.

The target for H-gene feedback is likely to be very specific for at least two reasons. Firstly, it cannot be the genes that encode the postulated recombinase, because the recombinase is not Ig locus specific; the recombinase even rearranges T cell–receptor genes transfected into B cells (64). Rearrangement of L-chain genes goes on after the postulated feedback on

H-gene rearrangement. Secondly, the feedback probably does not block the rearrangement competence of the H-locus completely. Rearrangement and transcriptional competence are very closely coupled (23, 27, 45, 65) and may involve the same transacting factors. The H locus, of course, continues to be transcribed at the functional V-D-J$^+$ allele and aberrantly rearranged V-D-J$^-$ or D-J or germline allele(s) in normal and transgenic B cells. Furthermore, H genes can switch C regions after V-D-J rearrangement is completed (66). Although the switching enzymology may differ from that of rearrangement, and active chromatin conformation of similar quality is presumably required. Thus, as a consequence of the feedback a change must occur in the early B cell; the change must not interfere with rearrangement of L genes or transcription of H genes but must specifically stop H-gene rearrangement. One would postulate tentatively that upstream regions of the H locus, namely, unrearranged V_H and D_H genes, become inactive and unable to rearrange. Alt and his colleagues have shown that unrearranged V_H genes are transcribed and are DNase-I sensitive in pre–B cells, but not in more mature B cells (23). Perhaps transacting factors that are specific for unrearranged V_H and D_H genes exist in pre–B cells competent to rearrange H genes, but the factors disappear after the feedback inhibition.

Antibody-producing cells normally synthesize either κ or λ L chains, but not both. Isotypic exclusion of κ chains is reflected in rearranged κ genes and germline λ genes in κ-producing cells. However, in almost all λ-producing cells, the κ genes are either aberrantly rearranged or deleted (67–69). These observations have led to development of a "sequential model" to explain λ isotypic exclusion (67). In this model, cells of the B lineage first become competent to rearrange κ genes after a productive H-gene rearrangement. Only if both κ alleles have not achieved a functional κ-gene rearrangement will the λ locus become activated and competent to rearrange (67).

In keeping with this model, one would assume that in κ-transgenic mice λ genes would not be rearranged in B cells that express the κ transgene. However, when λ-producing hybridomas were isolated after fusion of spleen cells from κ-transgenic mice, they were found to coexpress κ and λ chains (70). In addition, most of the hybridomas also produced H chains associated with both the κ and λ chains. Not only did these cells violate the rules of isotypic exclusion, they also did not obey allelic exclusion: Despite the presence of the transgenic κ together with H chains, endogenous κ genes were often rearranged or deleted. Finally, some of the hybridomas continued to rearrange their Ig genes.

Based on these findings a new model was formulated for L-chain gene expression (70) (Figure 2). It postulates two independent B-cell lineages—

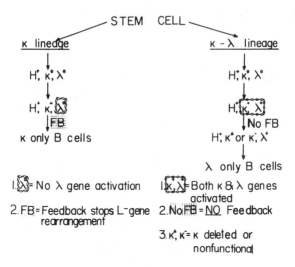

Figure 2   Model for B-cell differentiation: Scheme of κλ isotypic exclusion.

a κ lineage and a κλ lineage. The κ lineage is competent to rearrange only κ genes. The κ-gene rearrangement is controlled by feedback from κH protein. (These are the cells discussed earlier.) The κλ lineage becomes competent to rearrange κ and λ genes at the same time. In contrast to the κ lineage, the κλ one does not respond to feedback inhibition. Thus, κ- and λ-gene rearrangements continue until all sequences capable of rearrangement have been exhausted. In the κ locus, this may often lead to deletion of $C_κ$ by recombination of a sequence downstream of $C_κ$ with a region in the $J_κ$-$C_κ$ intron (71–73). There is no evidence for such a recombining sequence in the λ locus (E. Selsing, personal communication, concerning the mouse locus; K. Siminovitch, personal communication, for the human λ locus). Therefore, λ genes will be retained, and if a correct λ-gene rearrangement exists after exhaustive λ-gene rearrangements and κ-locus inactivation, the B cell will have become a pure λ producer.

This model does not explain λ-allelic exclusion or the presence of unrearranged Vλ genes in λ-producing cells. The former, i.e. the absence of two different λ chains in λ myelomas, may be due to a low frequency of antigen triggering of B cells that expose scrambled antibodies at the cell surface (because of different V-1 chains). The latter, i.e. the presence of germline λ genes in λ cells may be due to chance, or perhaps the κλ lineage rearrangement stops eventually, for example, at the plasma cell stage. These problems must and can be addressed experimentally.

OTHER IMMUNOREGULATORY INFLUENCES OF IG TRANSGENES

Although the presence of the Ig transgenes does not completely inhibit endogenous Ig production (14, 15, 17), it definitely modifies the expression of those endogenous Ig genes that respond to the same antigens as do the transgenic Ig's.

The anti-NP response is mainly λ_1 in normal Balb/c mice. Transgenic mice that possess the anti-NP μ gene without immunization show highly elevated levels of serum anti-NP antibodies, most of which contain endogenous λ chains (17).

Likewise, when transgenic mice have the MOPC-167 μ gene that encodes anti-PC H chains, they have a large increase of MOPC-167 κmRNA encoded by endogenous κ genes in spleen (13). Apparently, B cells that express the MOPC-167 μ transgene and happen to express an endogenous MOPC-167 κ gene are stimulated and expanded at a much higher rate than are B cells that express other endogenous κ genes. In the anti-NP and anti-PC μ-transgenic mice, it has not yet been determined whether the stimulus for B cells to express the transgenic μ and endogenous L chains is antigen (present in the environment) or an antiidiotype response.

Transgenic mice with the MOPC-167 μ gene without the exon encoding the transmembrane portion of μ chains ($\mu\Delta$ mem) had no increase in endogenous MOPC-167 κ-mRNA in spleen (13). These mice did, however, produce similar quantities of anti-PC antibodies and MOPC-167 μ-mRNA as did mice with the complete μ gene. The lack of stimulation of endogenous MOPC-167 κ-mRNA probably indicates that B cells that express the μ-Δ-mem gene and produce secreted μ chains are not specifically triggered by either PC or an antiidiotype response. Although secreted Ig molecules are released from the cell surface, apparently they do not dwell in the plasma membrane in a way that would permit cell activation.

Strong evidence for the elicitation of endogenous Ig's with the transgenic idiotype was obtained in transgenic mice with an anti-NP μ gene (74). Hybridomas were produced from lymph nodes and spleen of the unimmunized transgenic mice. The transgenic idiotype named 17.2.25 was found in the secreted Ig's of 68% of lymph-node and 28% of spleen hybridomas. In normal mice, less than 1% of unselected hybridomas produce this idiotype. In the majority of the hybridomas, the 17.2.25 idiotype was, for several reasons, most likely derived from endogenous genes, rather than from the transgene. First, the transgenic hybridoma 17.2.25 idiotype was present on H chains with an isotype other than μ, the isotype of the transgene. Second, transgenic hybridomas which produced μ often showed that the μ was of the recipient (μ^b, i.e. C5 7BL/6) rather than the donor (μ^a, i.e. Balb/c) allotype. Third, most of the hybridomas did not transcribe the transgene, according to S_1 nuclease protection assays of the 5' portion of the mRNA. Fourth, many of the hybridomas that produced the 17.2.25 idiotype had lost the transgenes.

The reason for the high level of endogenous expression of the transgene related idiotype is not clear. Two types of mechanisms either alone or together may explain the results. Perhaps portions of the transgene (or one of the multiple copies) recombined with endogenous genes. This could explain the hybridomas in which the V-D-J_H sequence was closely related to that of the transgene. In most of the hybridomas, however, the V-D-J_H sequence was very different from the transgene, often even containing a different V_H gene. In these cases, it seems likely that some idiotype regulation selected for cells that expressed the transgene-related idiotype.

These transgenic hybridomas were derived from the same mice that had bone marrow cells immortalized by AMuLV as in the experiments described above (61). Among the bone marrow pre–B cells, most expressed the μ transgene. The repertoire in the peripheral lymphoid organs is apparently quite different from that of bone marrow, since in the spleen and lymph node hybridomas only about 15% expressed the μ transgene (74). A related observation has been made in transgenic mice that have both μ and κ transgenes from MOPC-167 (anti-PC) (13). In a spleen focus forming assay, the spleen contained about 200 times fewer anti-PC precursor cells than did the bone marrow (N. Klineman, C. Pinkert, R. Brinster and U. Storb, unpublished).

Another unexplained phenomenon is the repression of transgenic κ or μ transcription (48, 74). Two hybridomas (8% of the total analyzed) from the MOPC-21 κ-transgenic mice that had retained the κ transgene did not produce transgenic κ chains but did produce endogenous μ and κ chains (48). In the 17.2.25 μ-transgenic mice, 19 out of 27 hybridomas (70%) that contained the μ transgene did not express it, although endogenous H chains were produced (74). In these cases the possibility has not been ruled out that the transgenes were permanently inactivated by some rearrangement event. For example, translocation of the total transgene insertion complex or more subtle deletions and insertions are a possibility. However, the very high frequency of transgene inactivation in the 17.2.25 μ hybridomas makes this somewhat unlikely. One should, however, not underestimate the selective forces operating in the immune system. Even if only a minute subpopulation of the total cells had inactivated all the transgenes, it may be selected by the demands of functional immunity. In addition, a mechanism, as yet unknown, may actively suppress the expression of unaltered transgenes in some cells, while permitting the expression of endogenous Ig genes.

REARRANGEMENT MECHANISM

The joining of V and J in L-gene rearrangement or of D-J and V-D in H-gene rearrangement is not understood in molecular terms. It is likely that "recognition sequences" containing a highly conserved 7-mer and 9-mer located downstream of V, upstream of J, and on both sides of D, are the sites for the rearrangement mechanism (reviewed in 3). It is not known whether, besides the 7-mer, 9-mer, and the spacers between them, other sequences are essential, or what changes in the recognition sequences can be tolerated. As far as the enzymology of the reaction is concerned, it can be expected that at least an endonuclease and ligase are involved. In addition, locus specific factors must exist, such that, for example, H genes alone are rearranged at the pre–B cell stage. Such factors may be stimulating or inhibitory.

Some experiments have been carried out in cell-free systems (75–77) or with cell transfection of rearrangement-test-genes (78–81), in attempts to define the DNA sequences and transacting factors involved in the rearrangement. Recently, transgenic mice with unrearranged genes were produced in two laboratories (26; and P. Engler, C. Pinkert, R. Brinster, U. Storb, unpublished), thus beginning study of the rearrangement process in the whole animal.

One of the experiments involved the chicken λ genes. In the chicken λ locus, the J and V_1 gene are very close (about 2 kb) to each other (82), in contrast to mammalian Ig genes, where the distances must be over 60 kb (J. Miller, S. Ogden and U. Storb, unpublished) and no V and J genes have been linked up. To determine if rearrangement recognition sequences of the chicken can be recognized in the mouse, transgenic mice were produced with an unrearranged 11.5-kb segment of the chicken DNA containing $V\lambda$, $J\lambda$, $C\lambda$, and flanking regions (26). Four transgenic mouse lines arose, all of which expressed chicken λ in the spleen but not in the thymus; this suggested the conservation of B cell–specific transcriptional control sequences between chicken and mouse. Rearrangement occurred in the spleen in approximately 2–5% of the chicken λ transgenes, indicating that the rearrangement mechanism is conserved between mouse and chicken at least as far as the target DNA sequences are concerned. In one mouse line, rearrangement of the chicken λ gene also occurred in the thymus, at about 30% of the level in the spleen. This was allegedly not due to contamination by B lymphocytes. Perhaps the integration of the chicken DNA was into a genomic site normally active in T cells. It will be interesting to determine, by sequencing the rearranged chicken transgenes, how accurate the rearrangement sites are with respect to the recognition sequences.

Another set of transgenic mice were produced by means of a rearrangement test gene of mouse (P. Engler, C. Pinkert, R. Brinster and U. Storb).
The test gene contains the rearrangement recognition sequences of a V_κ
and J_κ gene with very little additional flanking sequences separated by a
translational block (81). When transfected into a pre-B-cell line, about
90% of the transfectants rearrange the gene and translate a *gpt*-gene
located downstream of the J-recognition sequence (81). Thus, a very small
amount of DNA sequence is required as a target for the rearrangement
machinery. The experiment also shows formally that κ genes can rearrange
by deletion of DNA sequences between V and J. Transgenic mice containing this rearrangement test gene have been produced. It will be interesting to determine when in ontogeny rearrangement first occurs and what
sequences are required for B cell–specific rearrangement.

SOMATIC HYPERMUTATION OF κ GENES

One of the major modes of diversification of Ig genes is somatic hypermutation of V regions (reviewed in 3). In κ genes the mutations extend
about 1 kb to either side of the V-J joining site (83). Two major hypotheses
have been formulated to explain this specific hypermutation. A *rearrangement model* postulates that the mutations are introduced during Ig gene
rearrangement as a consequence of an error-prone DNA repair process
(84). This is postulated to operate after a site-specific endonuclease has
cleaved the DNA adjacent to V and J. In support of this model was the
finding that the mutated region is centered at the V-J joining site. An
accumulation model postulates that the mutations occur after rearrangement and after antigen stimulation. A special mutator is postulated to
recognize and mutate the rearranged V genes (85, 86). This model was
supported by the finding that the mutations "accumulated" with increasing
time after immunization. This observation may, however, be attributed to
affinity maturation, i.e. antigen selection of more highly mutated
sequences. To distinguish formally between these two models, we used
transgenic mice that harbor a rearranged κ gene (12, 13). If the rearrangement model of somatic mutation were correct, this already rearranged
transgene should not become mutated, and vice versa. At first, a random
collection of transgene clones isolated from a spleen cDNA library and a
genomic library of a hybridoma were sequenced (R. O'Brien, E. Selsing,
R. Brinster and U. Storb, unpublished). A possible candidate, not explainable by other mechanisms, for a mutation in the intron adjacent to JK5
was found in about 12 kb of sequenced DNA. This change, however, could
be due to some general instability of the transgene akin to that reported
in transfection. DNA sequences transfected into cells undergo a fairly high

rate of mutations (87). No data are available concerning transgene stability in mice.

To increase the chance for detecting mutations of the transgene if they existed, κ transgenes were selected from hyperimmunized mice (88). This would closely mimic the situation in normal mice where maximal rates of mutations have been found (85, 86). Transgenic mice with the MOPC-167 κ gene, which encodes κ chains of anti-PC antibodies, were hyperimmunized with PC; the anti-PC hybridomas were isolated from spleen B cells (88). In the response of mice to PC, one particular V_H gene (V1 of the T15 family) is used almost exclusively (89). Hybridomas were selected from the κ-transgenic mice whose endogenous V1 gene was muted. These cells were thus known to have undergone a mutation event; in the same cell as the endogenous V_H gene, the κ transgene would have had a chance to mutate. When the κ transgenes of such cells were sequenced, several mutations were found in the V regions (10 mutations in 7650 nucleotides). No mutations were found in 9500 nucleotides of C region. These results clearly show that somatic mutation of κ genes occurs in a stably rearranged gene. Apparently, a special mutator mechanism alters the gene. The results also show that insertion at genomic sites other than the κ locus is compatible with somatic mutation of κ and that about 15 kb of the κ gene and flanking regions is sufficient.

CONCLUSIONS

Transgenic mice have been an especially useful experimental model for the analysis of the control of expression of immunoglobulin genes. Beyond questions of tissue and development, specific expression relevant to all types of genes, immunoglobulin genes underlie special somatic processes unique to the immune system. Because of the extreme heterogeneity of the B-cell pool with respect to the specific Ig genes expressed, the ability to follow the fate of a single transgene in the majority of B cells is invaluable. Phenomena such as the control of allelic and isotypic exclusion, rearrangement, and somatic hypermutation can perhaps only be solved with transgenic mice harboring an Ig transgene whose sequence as it exists in the germ line is known. As this review indicates, major advances in these questions have already been made, and further answers can be expected.

Besides the problems of Ig gene expression, transgenic mice with Ig transgenes will probably also be very useful for the study of immune cell interactions. Because it appears that B cells which express a particular Ig transgene can be selected or suppressed, it is likely that T cells are involved which recognize the transgene-encoded Ig's at the B-cell surface.

ACKNOWLEDGMENTS

I am grateful to Ralph Brinster, Peter Engler, Rebecca O'Brien, and Benjamin Arp for critical reading of the manuscript and many valuable suggestions, to Jean-Claude Weill for unpublished results, and to Richard Palmiter for making his manuscript (11) available prior to publication. I also thank Jerry Vernier for typing the manuscript. Our work included in this review was supported by NIH grants CA/A1 25754, HD 19488, and CA 35979.

Literature Cited

1. Perry, R. P., Kelley, D. E., Coleclough, C., Kearney, J. F. 1981. Organization and expression of immunoglobulin genes in fetal liver hybridomas. *Proc. Natl. Acad. Sci. USA* 78: 247
2. Alt, F., Rosenberg, N., Lewis, S., Thomas, E., Baltimore, D. 1981. Organization and reorganization of immunoglobulin genes in A-MuLV-transformed cells: Rearrangement of heavy but not light chain genes. *Cell* 27: 381
3. Tonegawa, S. 1983. Somatic generation of antibody diversity. *Nature* 302: 575
4. Selsing, E., Storb, U. 1981. Somatic mutation of immunoglobulin light-chain variable-region genes. *Cell* 25: 47
5. Crews, S., Griffin, J., Huang, H., Calame, K., Hood, L. 1981. A single V_H gene segment encodes the immune response to phosphorylcholine: Somatic mutation is correlated with the class of the antibody. *Cell* 25: 59
6. Potter, M. 1967. The plasma cell tumors and myeloma proteins of mice. *Meth. Cancer Res.* 2: 105
7. Brinster, R. L., Chen, H. Y., Trumbauer, M. E. 1981. Mouse oocytes transcribe injected *Xenopus* 5S RNA gene. *Science* 211: 396
8. Brinster, R. L., Chen, H. Y., Trumbauer, M., Senear, W., Warren, R., Palmiter, R. 1981. Somatic expression and herpes thymidine kinase in mice following injection of a fusion gene into eggs. *Cell* 27: 223
9. Brinster, R. L., Chen, H. Y., Trumbauer, M., Yagle, M. K., Palmiter, R. D. 1985. Factors affecting the efficiency of introducing foreign DNA into mice by microinjecting eggs. *Proc. Natl. Acad. Sci. USA* 82: 4438
10. Brinster, R. L., Palmiter, R. D. 1986. Introduction of genes into the germ line of animals. *The Harvey Lectures, Series 80*, pp. 1–38
11. Palmiter, R. D., Brinster, R. L. 1986. Germline transformation of mice. *Ann. Rev. Genet.* 20: In press
12. Brinster, R. L., Ritchie, K. A., Hammer, R. E., O'Brien, R. L., Arp, B., Storb, U. 1983. Expression of a microinjected immunoglobulin gene in the spleen of transgenic mice. *Nature* 306: 332
13. Storb, U., Pinkert, C., Arp, B., Engler, P., Gollahon, K., Manz, J., Brady, W., Brinster, R. 1986. Transgenic mice with μ and κ genes encoding antiphosphorylcholine antibodies. *J. Exp. Med.* 164: 627
14. Rusconi, S., Köhler, G. 1985. Transmission and expression of a specific pair of rearranged immunoglobulin μ and κ genes in a transgenic mouse line. *Nature* 314: 330
15. Storb, U., Ritchie, K. A., Hammer, R. E., O'Brien, R. L., Manz, J. T., Arp, B., Brinster, R. L. 1985. Expression of a microinjected immunoglobulin kappa gene in transgenic mice. In *Banbury Report 20: Genetic Manipulation of the Early Mammalian Embryo*, ed. F. Constantini, R. Jaenisch, p. 197. Cold Spring Harbor Lab., Cold Spring Harbor, Me.
16. Storb, U., O'Brien, R. L., McMullen, M. D., Gollahon, K. A., Brinster, R. L. 1984. High expression of cloned immunoglobulin κ gene in transgenic mice is restricted to B lymphocytes. *Nature* 310: 238
17. Grosschedl, R., Weaver, D., Baltimore, D., Costantini, F. 1984. Introduction of a μ immunoglobulin gene into the mouse germ line: Specific expression in lymphoid cells and synthesis of functional antibody. *Cell* 38: 647
18. Roehm, N., Herron, L., Cambier, C., DiGuisto, D., Haskins, K., Kappler, J., Marrack, P. 1984. The major histocompatibility complex restricted antigen

receptor on T cells : Distribution on thymus and peripheral T cells. *Cell* 38 : 577

19. Kemp, D. J., Harris, A. W., Cory, S., Adams, J. M. 1980. Expression of the immunoglobulin C_μ gene in mouse T and B lymphoid and myeloid cell lines. *Proc. Natl. Acad. Sci. USA* 77 : 2876

20. Walker, I. D., Harris, A. W. 1980. Immunoglobulin C_μ RNA in T lymphoma cells is not translated. *Nature* 288 : 290

21. Kemp, D. J., Harris, A. W., Adams, J. M. 1980. Transcripts of the immunoglobulin C_μ gene vary in structure and splicing during lymphoid development. *Proc. Natl. Acad. Sci. USA* 77 : 7400

22. Lennon, G. G., Perry, R. P. 1985. C_μ-containing transcripts initiate heterogeneously within the IgH enhancer region and contain a novel 5'-nontranslatable exon. *Nature* 318 : 475

23. Yancopoulos, G. D., Alt, F. W. 1985. Developmentally controlled and tissue-specific expression of unrearranged V_H gene segments. *Cell* 40 : 271

24. Mason, J. O., Williams, G. T., Neuberger, M. S. 1985. Transcription cell type specificity is conferred by an immunoglobulin V_H gene promoter that includes a functional consensus sequence. *Cell* 41 : 479

25. Yamamura, K.-I., Kudo, A., Ebihara, T., Kamino, K., Araki, K., Kumahara, Y., Watanabe, T. 1986. Cell-type-specific and regulated expression of a human $\gamma 1$ heavy-chain immunoglobulin gene in transgenic mice. *Proc. Natl Acad. Sci. USA* 83 : 2152

26. Bruchini, D., Reynaud, C.-A., Ripoche, R.-A., Jamie, J., Weill, J.-C. 1987. Rearrangement of a chicken Ig gene occurs solely in the lymphocyte lineage of transgenic mice. Submitted

27. Calame, K. L. 1985. Mechanisms that regulate immunoglobulin gene expression. *Ann. Rev. Immunol.* 3 : 159

28. Grosschedl, R., Baltimore, D. 1985. Cell-type specificity of immunoglobulin gene expression is regulated by at least three DNA sequence elements. *Cell* 41 : 885

29. Parslow, T. G., Blair, D. L., Murphy, W. J., Granner, D. K. 1984. Structure of the 5' ends of immunoglobulin genes : A novel conserved sequence. *Proc. Natl. Acad. Sci. USA* 81 : 2650

30. Gillies, S. D., Morrison, S. L., Oi, V. T., Tonegawa, S. 1983. A tissue-specific transcription enhancer element is located in the major intron of a rearranged immunoglobulin heavy chain gene. *Cell* 33 : 717

31. Falkner, F. G., Neumann, E., Zachau, H. G. 1984. Tissue specificity of the initiation of immunoglobulin κ gene transcription. *Hoppe-Seyler's Z. Physiol. Chem.* 365 : 1331

32. Queen, C., Baltimore, D. 1983. Immunoglobulin gene transcription is activated by downstream sequence elements. *Cell* 33 : 741

33. Picard, D., Schaffner, W. 1984. A lymphocyte-specific enhancer in the mouse immunoglobulin κ gene. *Nature* 307 : 80

34. Yamamura, K.-I., Kikutani, H., Takahashi, N., Taga, T., Akira, S., Kawai, K., Fukuchi, K., Kumahara, Y., Honjo, Kishimoto, T. 1984. Introduction of human $\gamma 1$ immunoglobulin genes into fertilized mouse eggs. *J. Biochem. (Japan)* 96 : 357

35. Yamamura, K.-I., Miki, T., Suzuki, N., Ebihara, T., Kawai, K., Kumahara, Y., Honjo, T. 1985. Introduction of mouse C_c genes into cos-7 cells and fertilized mouse eggs. *J. Biochem.* (Japan) 97 : 333

36. Wabl, M. R., Burrows, P. D. 1984. Expression of immunoglobulin heavy chain at a high level in the absence of a proposed immunoglobulin enhancer element in *cis*. *Proc. Natl. Acad. Sci. USA* 81 : 2452

37. Klein, S., Sablitzky, F., Radbruch, A. 1984. Deletion of the IgH enhancer does not reduce immunoglobulin heavy chain production of a hybridoma IgD class switch variant. *EMBO J.* 3 : 2473

38. Zaller, D. M., Eckhardt, L. A. 1985. Deletion of a B-cell-specific enhancer affects transfected, but not endogenous, immunoglobulin heavy-chain gene expression. *Proc. Natl. Acad. Sci. USA* 82 : 5088

39. Klein, S., Gerster, T., Picard, D., Radbruch, A., Schaffner, W. 1985. Evidence for transient requirement of the IgH enhancer. *Nucl. Acids Res.* 13 : 8901

40. Rusconi, S. 1984. Gene transfer in living organisms. In *Colloquium-Mosbach 1984. The Impact of Gene Transfer Technique in Eukaryotic Cell Biology*, ed. P. Shell, P. Starlinger, p. 134. Berlin, Heidelberg : Springer-Verlag

41. Weiler, E. 1965. Differential activity of allelic γ-globulin genes in antibody-producing cells. *PNAS* 54 : 1765

42. Pernis, B., Chiappino, G., Kelus, A. S., Gell, P. G. H. 1965. Cellular localization of immunoglobulins with different allotypic specificities in rabbit lymphoid tissues. *J. Exp. Med.* 122 : 853

43. Bernier, G. M. 1964. Polypeptide chains of human gamma-globulin : Cellular localization by fluorescent antibody. *Science* 144 : 1590

44. Hozumi, N., Tonegawa, S. 1976. Evidence for somatic rearrangement of immunoglobulin genes coding for variable and constant regions. *Proc. Natl. Acad. Sci. USA* 73: 3628

45. Perry, R. P., Kelley, D. E., Coleclough, C., Seidman, J. G., Leder, P., Tonegawa, S., Matthyssens, G., Weigert, M. 1980. Transcription of mouse κ chain genes: Implications for allelic exclusion. *Proc. Natl. Acad. Sci. USA* 77: 1937

46. Coleclough, C., Perry, R. P., Karjalainen, K., Weigert, M. 1981. Aberrant rearrangements contribute significantly to the allelic exclusion of immunoglobulin gene expression. *Nature* 290: 372

47. Walfield, A., Selsing, E., Arp, B., Storb, U. 1981. Misalignment of V and J gene segments resulting in nonfunctional immunoglobulin gene. *Nucl. Acids Res.* 9: 1101

48. Ritchie, K. A., Brinster, R. L., Storb, U. 1984. Allelic exclusion and control of endogenous immunoglobulin gene rearrangement in κ transgenic mice. *Nature* 312: 5994

49. Bernard, O., Gough, N. M., Adams, J. M. 1981. Plasmacytomas with more than one immunoglobulin κ mRNA: Implications for allelic exclusion. *Proc. Natl. Acad. Sci. USA* 78: 5812

50. Kwan, S.-P., Max, E. E., Seidman, J. G., Leder, P., Scharff, M. D. 1981. Two kappa immunoglobulin genes are expressed in the myeloma S107. *Cell* 26: 57

51. Köhler, G., Milstein, C. 1976. Derivation of specific antibody-producing tissue culture and tumor lines by cell fusion. *Eur. J. Immunol.* 6: 511

52. Kranz, D. M., Voss, E. W. Jr. 1981. Restricted reassociation of heavy and light chains from hapten-specific monoclonal antibodies. *Proc. Natl. Acad. Sci. USA* 78: 5807

53. Hamel, P., Isenman, D., Klein, M., Luedtke, R., Dorrington, K. 1984. Structural basis for the preferential association of autologous immunoglobulin subunits: Role of the J region of the light chain. *Fed. Proc.* 43: 1429

54. Klein, M., Kortan, C., Kells, D. I. C., Dorrington, K. J. 1979. Equilibrium and kinetic aspects of the interaction of isolated variable and constant domains of light chain with the Fd' fragment of immunoglobulin G. *Biochem.* 18: 1473

55. Storb, U., Ritchie, K. A., O'Brien, R., Arp, B., Brinster, R. 1986. Control of immunoglobulin gene expression. *Immunol. Rev.* 89: 85

56. Cowan, N. J., Secher, D. S., Milstein, C. 1974. Intracellular immunoglobulin chain synthesis in non-secreting variants of a mouse myeloma: Detection of inactive light-chain messenger RNA. *J. Mol. Bio.* 90: 691

57. Storb, U., Denis, K. A., Brinster, R. L., Witte, O. N. 1985. Pre-B cells in κ-transgenic mice. *Nature* 316: 356

58. Wabl, M., Steinberg, C. 1982. A theory of allelic and isotypic exclusion for immunoglobulin genes. *Proc. Natl. Acad. Sci. USA* 79: 6976

59 Haas, I. G., Wabl, M. 1983. Immunoglobulin heavy chain binding protein. *Nature* 306: 387

60. Alt, F. W., Blackwell, T. K., Yancopoulos, G. D. 1985. Immunoglobulin genes in transgenic mice. *Trends Genet.* (August): 231

61. Weaver, D., Constantini, F., Imanishi-Kari, T., Baltimore, D. 1985. A transgenic immunoglobulin mu gene prevents rearrangement of endogenous genes. *Cell* 42: 117

62. Whitlock, C. A., Ziegler, S. F., Treiman, L. J., Stafford, J. I., Witte, O. N. 1983. Differentiation of cloned populations of immature B cells after transformation with Abelson murine leukemia virus. *Cell* 32: 903

63. Mercola, M., Goverman, J., Mirell, C., Calame, K. 1985. Immunoglobulin heavy-chain enhancer requires one or more tissue-specific factors. *Science* 227: 266

64. Yancopoulos, G. D., Blackwell, T. K., Suh, H., Hood, L., Alt, F. W. 1986. Introduced T cell receptor variable region gene segments recombine in pre-B cells: Evidence that B and T cells use a common recombinase. *Cell* 44: 251

65. Storb, U., Wilson, R., Selsing, E., Walfield, A. 1981. Rearranged and germline and immunoglobulin κ genes: Different states of DNase I sensitivity of constant κ genes in immunocompetent and nonimmune cells. *Biochem.* 20: 990

66. Shimizu, A., Honjo, T. 1984. Immunoglobulin class switching. *Cell* 36: 801

67. Hieter, P. A., Korsmeyer, S. J., Waldmann, T. A., Leder, P. 1981. Human immunoglobulin κ light-chain genes are deleted or rearranged in λ-producing B cells. *Nature* 290: 368

68. Korsmeyer, S. J., Hieter, P. A., Ravetch, J. V., Poplack, D. G., Waldmann, T. A., Leder, P. 1981. Developmental hierarchy of immunoglobulin gene rearrangements in human leukemic pre-B-cells. *Proc. Natl. Acad. Sci. USA* 78: 7096

69. Perry, R. P., Coleclough, C., Weigert,

174   STORB

M. 1981. Reorganization and expression of immunoglobulin genes: Status of allelic elements. In *Cold Spring Harbor Symposia and Quantitative Biology*, 45: 925

70. Gollahon, K. A., Brinster, R. L., Storb, U. 1986. Isotypic exclusion of κ and λ genes is regulated by a mechanism different from allelic exclusion of κ genes. Submitted

71. Durdik, J., Moore, M. W., Selsing, E. 1984. Novel κ light-chain gene rearrangements in mouse λ light chain-producing B lymphocytes. *Nature* 307: 749

72. Siminovitch, K. A., Bakhshi, A., Goldman, P., Korsmeyer, S. J. 1985. A uniform deleting element mediates the loss of κ genes in human B cells. *Nature* 316: 260

73. Moore, M. W., Durdik, J., Persiani, D. M., Selsing, E. 1985. Deletions of κ chain constant region genes in mouse λ chain-producing B cells involve intrachromosomal DNA recombinations similar to V-J joining. *Proc. Natl. Acad. Sci. USA* 82: 6211

74. Weaver, D., Reis, M. H., Albanese, C., Costantini, F., Baltimore, D., Imanishi-Kari, T. 1986. Altered repertoire of endogenous immunoglobulin gene expression in transgenic mice containing a rearranged mu heavy chain gene. *Cell* 45: 247

75. Desiderio, S., Baltimore, D. 1984. Double-stranded cleavage by cell extracts near recombinational signal sequences of immunoglobulin genes. *Nature* 308: 860

76. Kataoka, T., Kondo, S., Nishi, M., Kodaira, M., Honjo, T. 1984. Isolation and characterization of endonuclease J: A sequence-specific endonuclease cleaving immunoglobulin genes. *Nucl. Acids Res.* 12: 5995

77. Hope, T. J., Aguilera, R. J., Minie, M. E., Sakano, H. 1986. Endonucleolytic activity that cleaves immunoglobulin recomination sequences. *Science* 231: 1141

78. Lewis, S., Gifford, A., Baltimore, D. 1985. Joining of V_κ to J_κ gene segments

in a retroviral vector introduced into lymphoid cells. *Nature* 308: 425

79. Blackwell, T. K., Alt, F. W. 1984. Site-specific recombination between immuno-globulin D and J_H segments that were introduced into the genome of a murine pre-B cell line. *Cell* 37: 105

80. Lewis, S., Gifford, A., Baltimore, D. 1985. DNA elements are asymmetrically joined during the site-specific recombination of kappa immunoglobulin genes. *Science* 28: 677

81. Engler, P., Storb, U. 1987. Rearrangement of artificial immunoglobulin gene segments in cultured cells. In preparation. Submitted

82. Reynaud, C.-A., Anquez, V., Dahan, A., Weill, J.-C. 1985. A single rearrangement event generates most of the chicken immunoglobulin light chain diversity. *Cell* 40: 283

83. Gearhart, P. J., Bogenhagen, D. F. 1983. Clusters of point mutations are found exclusively around antibody variable genes. *Proc. Natl. Acad. Sci. USA* 80: 3439

84. Selsing, E., Storb, S. 1981. Somatic mutation of immunoglobulin light-chain variable-region genes. *Cell* 25: 47

85. Berek, C., Griffiths, G. M., Milstein, C. 1985. Molecular events during maturation of the immune response to oxazolone. *Nature* 316: 412–18

86. Clarke, S. H., Huppi, K., Ruezinsky, D., Staudt, L., Gerhard, W., Weigert, M. 1985. Inter- and intraclonal diversity in the immune response to influenza hemagglutinin. *J. Exp. Med.* 161: 687

87. Calos, M. P., Lebkowski, J. S., Botchan, M. R. 1983. High mutation frequency in DNA transfected into mammalian cells. *Proc. Natl. Acad. Sci. USA* 809: 3015

88. O'Brien, R., Brinster, R., Storb, U. 1987. Somatic hypermutation of an immuno-globulin transgene in kappa transgenic mice. Submitted

89. Crews, S., Griffin, J., Huang, H., Calame, K., Hood, L. 1981. A single V_H gene segment encodes the immune response to phosphorylcholine: Somatic mutation is correlated with the class of the antibody. *Cell* 25: 59

Ann. Rev. Immunol. 1987. 5 : 175–99

MOLECULAR MECHANISMS OF TRANSMEMBRANE SIGNALING IN B LYMPHOCYTES

*J. C. Cambier and J. T. Ransom**

Department of Medicine, National Jewish Center for Immunology and Respiratory Medicine, Denver, Colorado 80206, and Department of Microbiology, University of Colorado Health Sciences Center, Denver, Colorado 80262

INTRODUCTION

A large and growing body of evidence attests to the complexity of regulation of B-lymphocyte activation by physiologic mediators. In the mouse, the state of activation of quiescent, noncycling (G_0) B cells, which represent the majority of splenic B cells (~ 80–90%) (1), can be modulated by a variety of species, including antigen or surrogate anti-immunoglobulin antibodies (2, 3), IL-4 (formerly BSF1 ; 4, 5), gamma interferon (IFNγ ; 6), B-cell activating factor (BCAF ; 7, 8), insulin (9), lipopolysaccharide (LPS ; 10), and major histocompatibility class-II (Ia) molecule binding ligands (10–12). Cycling B cells may be regulated by these species and additional interleukins (for review see 14), including IL-5 (formerly BCGF2), IL-2, IL-1, and less well-defined B-cell differentiation factors. That each of these regulators affects the biology of a B cell differently suggests that each has a distinct receptor on the B-cell surface and also that respective receptors must be coupled to generation of distinct intracellular second messenger. To elicit a given biologic response, e.g. stimulation of a G_0 B cell to proliferate, may require the concerted or sequential generation of a particular combination of second messengers. The response

---
*Current address: DNAX Institute for Molecular Biology, 1454 Page Mill, Palo Alto, California 94306.

0732–0582/87/0410–0175$02.00

is clearly also determined by the quantity of given second messenger(s) generated. For example, the stimulation of G_0 B cells to proliferate generally requires both an antigenic or anti-mIg signal and an IL-4 and/or a BCAF signal (7, 8, 15). However, a very potent mIg cross-linking signal delivered via anti-Ig beads appears to generate sufficient second messenger, in this case diacylglycerol (DAG) and inositol trisphosphate (InsP3), to drive G_0 cells into S-phase (2, 16, 17). IL-4 and BCAF do not appear to induce the formation of these second messengers (J. Cambier, unpublished observations). That DAG and InsP3 are capable of mediating B-cell proliferation is suggested by the synergistic mitogenic effects of phorbol myristate acetate, a DAG analog, and calcium ionophores, which like InsP3 raise intracellular free calcium levels (18). Thus, it appears that there may be multiple pathways by which to initiate a given B-cell response. This redundancy has frustrated attempts to define the precise role of antigens, cytokines, and T cells in regulation of B-cell function; it illustrates the importance of defining the transmembrane signal transduction mechanisms operative following stimulation of B cells with specific immunoregulators. In fact, an in-depth understanding of the molecular bases of transmembrane signaling may be essential to our unraveling the complexities of B-cell immune regulation.

In the past few years, much progress has been made in understanding the molecular basis of transmembrane signaling in nonlymphoid tissues (for review, see 19–22). It appears that many of the mechanisms defined are operative in multiple tissues and are in many cases coupled to occupancy of different receptors. For example, signals generated following serotonin binding in blowfly salivary gland, f-methionyl-leucyl-phenylalanine binding to neutrophils, and acetylcholine binding to neuronal tissue, all appear to be transduced by very similar mechanisms. The groundwork laid by cell biologists studying nonlymphoid tissues has been instrumental in the recent rapid progress in defining transmembrane signaling mechanisms operative in T and B cells. This review is devoted to a discussion of current knowledge regarding transmembrane signaling mechanisms operative in regulation of quiescent B-lymphocyte function. We focus on signaling via membrane immunoglobulin (mIg), Ia, IL-4 receptors, and IFNγ receptors, and we begin with a review of the biology of G_0 B-cell regulation by their respective ligands.

THE BIOLOGY OF G_0 B-CELL IMMUNE REGULATION

B lymphocytes represent one of the major cellular components of the mammalian immune system. Immunogenic challenge prompts the activation, proliferation, and differentiation of G_0 B cells generating progeny that secrete antibody specific for the immunogen. The structural basis of selection for growth and differentiation of B cells whose progeny secrete specific antibody was identified by discoveries that B cells express surface immunoglobulin (mIg, 23–25) with antigen specificity identical to that of immunoglobulin secreted by their differentiated progeny (26–29). These findings suggest that membrane immunoglobulin plays an important role in antigen-induced activation of antigen-specific B cells. Consistent with this notion is the possibility that by binding antigen for subsequent processing (30), mIg may play a passive role in focusing mitogenic carrier moeities, e.g. LPS or antigen-specific helper T cells, which in turn activate the B cell (31). Alternatively or additionally, mIg may play an active role by transducing stimulatory signals across the plasma membrane (32). It was subsequently shown that anti-immunoglobulin antibodies stimulate B cells to increase expression of Ia molecules (33) and under certain conditions stimulate thymidine uptake by B cells (2, 23, 34, 35), leaving little doubt that mIg can act as a signal transducer. It should be noted that thymus-dependent antigens, while stimulating some cellular changes that include increased expression of Ia (36–39), do not stimulate proliferation of antigen-binding B cells in the absence of T-cell help (37, 40). Thus, it would appear that in physiologic situations, binding of classically defined thymus-dependent antigens to mIg leads to antigen processing and reexpression (30) and increased expression of Ia antigens, but not to B-cell growth. Both of these processes contribute to the efficient cognitive interaction of B cells with helper T cells that recognize processed antigen and Ia (30, 41, 42).

As stated, quiescent B cells appear to be regulated by species in addition to antigen. Recent studies have demonstrated that IL-4, formerly B-cell stimulatory factor 1 (15), is a potent stimulator of Ia expression by G_0 B cells (3, 4). Thus, G_0 B cells express functional receptors for IL-4. IL-4 released by activated T cells may amplify the immune response by further stimulating B-cell Ia expression, but IL-4 clearly also acts as a B-cell growth and differentiation factor (43, 44).

The possible role of Ia as a signal transducing molecule has been contested for the past decade. The necessity for Ia compatibility in T cell–dependent activation of quiescent B cells (45) may reflect a requirement for T cell–derived ligand binding to Ia for B-cell activation. Alternatively, this restriction may simply reflect Ia compatibility requirements for antigen presentation by B cells to T cells. Results of published studies using soluble anti-Ia antibodies to simulate the effects of helper T cells have been somewhat inconsistent but demonstrate generally that Ia binding ligands can suppress (11, 13, 46) B-cell proliferation and enhance (12, 47) B-cell differentiation. Clearly anti-Ia antibodies do affect the biology of small

B cells, indicating that Ia molecules can act as transmembrane signal transducers.

Finally, recent studies by Mond et al (6) indicate that interferon γ (IFNγ) suppresses soluble anti-IgD induction of B-cell Ia expression and proliferation. Further, IFNγ is only suppressive if present during the first seven hours of stimulation. Therefore, quiescent B cells appear to express functional IFNγ receptors, and those receptors transduce signals which antagonize mIg-mediated signaling.

Thus, four well-defined physiologic ligands, including antigen or surrogate anti-immunoglobulin antibodies, T cells or surrogate anti-Ia antibodies, and recombinant IL-4 and IFNγ, modulate the physiology of G_0 B cells. The cellular responses to these ligands appear to differ qualitatively, suggesting that the respective receptors may be coupled to their biologic responses by different mechanisms. In the succeeding paragraphs we discuss the extant mechanisms of transmembrane signal transduction which have been defined in nonlymphoid tissues and which are thus leading candidates for B cell receptor–mediated transduction. We also consider existing evidence regarding mechanisms utilized for transduction of signals generated upon ligation of mIg, mIa, IL-4 receptors, and IFNγ receptors.

EXTANT MECHANISMS OF TRANSMEMBRANE SIGNAL TRANSDUCTION: AN OVERVIEW

Cells, including lymphocytes, whose function is regulated by impermeant molecules in the extracellular space, face the problem of conveying information acquired via ligand–cell surface receptor interactions to the interior of the cell. This problem led Sutherland & Rall to propose in 1958 (48) the concept of "second messengers" based upon their studies of the effect of adrenaline on liver. In simplistic terms, they suggested that ligand binding on cell surface receptors leads to generation of a second messenger inside the cell that conveys the information. This concept has been adopted and expanded in a vast variety of stimulus-response coupling systems in a number of tissues. It is particularly attractive in explaining situations in which extracellular regulators evoke rapid changes in cell metabolism. Such responses must involve the functional modification of existing enzymes and/or other proteins and lipids. Such modifications are most notably achieved by phosphorylation. For example, adrenaline induces the formation of the second messenger cyclic AMP (cAMP) in the liver cell, which activates a cyclic AMP-dependent protein kinase. This enzyme, through protein phosphorylation, activates glycogen breakdown resulting in increased blood glucose.

A great diversity of proteins may be influenced by cycles of phosphorylation and dephosphorylation, and thus protein kinases represent an efficient means for regulation of cell metabolism. There are currently four well-defined second messenger–dependent protein kinases; they are cyclic AMP dependent, cyclic GMP dependent (49), calcium/calmodulin dependent (50), and calcium/phospholipid dependent (51). In addition, in some cases receptors themselves may exhibit ligand-activated kinase activity (52). All of these systems have been defined in stimulus response coupling models in nonlymphoid tissues, but evidence suggests that each may operate at some level in the regulation of lymphocyte function. For this reason, the salient features of each mechanism are discussed briefly here.

Perhaps the best-defined transmembrane signaling mechanisms are those activated by catecholamine interaction with β- and α_2-adrenergic receptors. There are at least three component membrane proteins in this system: (a) cell surface receptors such as the catecholamine receptor; (b) regulatory guanine nucleotide–binding proteins termed Ns and Ni; and (c) the adenylate cyclase catalytic subunit. In mammalian tissues, purified β-adrenergic receptors are single polypeptide chains in the 60,000–65,000 D range (53, 54). Guanine nucleotide binding proteins (N proteins) operative in this system mediate the stimulation (Ns) or inhibition (Ni) of adenylate cyclase activity. These proteins (55, 56) exist as heterotrimers with subunits of 42 kd (α_s) or 41 kd (α_i), 35 kd (β_s or β_i), and 5 kd (γ_s or γ_i). The adenylate cyclase catalytic subunit is less well characterized but has a reported molecular mass of ~ 150 kd (57).

The current hypothesis (for review see 20) for the sequence of events in this signaling cascade for activation of adenylate cyclase is as follows: Agonist binding to the adrenergic receptor initiates the interaction of this protein with Ns. Ns subsequently binds guanine nucleotide, and this is thought to cause the dissociation of the α subunit-GTP complex, from the β and γ subunits. Guanine nucleotide–associated α subunit then binds and activates previously inactive adenylate cyclase catalytic subunit. The complex is inactivated by GTPase activity which is associated with the α subunit (56, 58–60). Similarly, α_2 receptor activation causes dissociation of Ni into its α and β/γ subunits and results in inhibition of adenylate cyclase either by a direct inhibitory interaction of α_i with cyclase or perhaps by association of liberated β/γ with free α_s.

A somewhat analogous signaling mechanism has recently been described which involves similar structural components for the regulation of cellular cyclic-GMP levels (for review see 61). In the retinal rod model, rodopsin acts as a photon receptor, interacting with transducin, a GTP binding and hydrolysing N-protein. GTP-transducin modulates the activity of cGMP phosphodiesterase, controlling intracellular cGMP concentrations and thereby the activity of a c-GMP regulated protein kinase.

Recent studies, conducted most notably in the insulin receptor system have demonstrated that under certain circumstances ligation of receptors can lead to protein kinase activation in the absence of conventional second messengers (for review see 52). Specifically, ligation of the insulin or epidermal growth factor receptors results in the activation of tyrosine–specific protein kinase activity which is associated with the receptor itself. Ancillary protein molecules are apparently not required since purified receptors undergo autophosphorylation when stimulated with ligand in the presence of $[^{32}P]$ ATP.

The past 3–4 years have seen a very rapid evolution of our understanding of the molecular bases of transmembrane signaling by α_1 adrenergic type agonists. In 1953, ligand stimulation of $[^{32}P]$ PO$_4$ incorporation into phospholipids in pancreas was reported (62). Subsequent studies demonstrated that in many tissues, metabolism of inositol lipids (Figure 1) is activated by receptor ligation (for review see 19). Durell et al (63) proposed that the increased inositol lipid metabolism was intimately involved in receptor function. Observations that increased inositol lipid metabolism was virtually always associated with elevation of intracellular calcium led Michell (64) to propose that inositol lipid metabolism and calcium mobilization are closely linked if not physiologically coupled. An additional chapter of this story evolved in the early 1980s when Nishizuka and coworkers (for review see 51) described a calcium and diacylglycerol (DAG) regulated protein kinase, protein kinase C (PKC). Subsequently, this kinase was shown to translocate from cytosol to the plasma membrane following stimulation of cells with the DAG analogue phorbol myristate

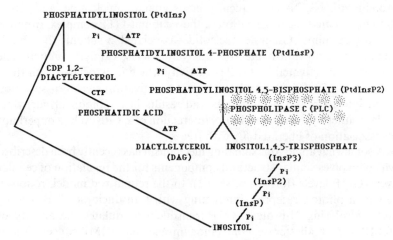

Figure 1   The phosphatidylinositol metabolic cycle.

acetate as well as with physiologic ligands which induce phosphoinositide metabolism (for review see 65). Durell (66) observed that phosphatidylinositol 4,5-bisphosphate (PtdInsP$_2$) is selectively hydrolyzed following ligand binding, yielding DAG and inositol trisphosphate (InsP$_3$). Finally, Streb et al (67) demonstrated that InsP$_3$ could induce the release of free calcium from intracellular stores, specifically, from the endoplasmic reticulum.

These observations provided the framework for the working model of α_1 adrenergic type receptor-mediated transmembrane signaling described below and pictured in Figure 2. Ligation of receptor leads to activation of PtdInsP$_2$ hydrolysis by phospholipase C, yielding DAG and InsP$_3$. In some systems this coupling appears to involve a GTP-binding protein (68–71). InsP$_3$ diffuses to the endoplasmic reticulum where it mediates the release of sequestered Ca^{++} into the cytosol. This Ca^{++} may act as a trigger to cause the influx of extracellular calcium detected in some systems. Mobilized Ca^{++} may act through calmodulin to activate calmodulin-dependent protein kinase(s), but it also appears to act in concert with DAG to cause high-affinity association of protein kinase C with the plasma membrane. Now active by virtue of its association with Ca^{++}, DAG, and phosphatidylserine, PKC alters by phosphorylation the activity of various ion transport mechanisms (72). The resulting changes in ion transport

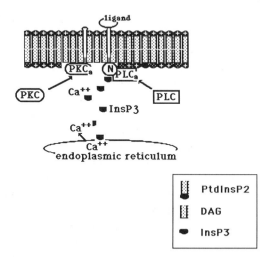

Figure 2 Diagram of α_1-adrenergic transmembrane signal transduction. Abbreviations: DAG, diacylglycerol; InsP3, inositol trisphosphate; N, GTP binding protein; PKC, protein kinase C; PKCa, active PKC; PLC, phospholipase C; PLCa, active PLC; PtdInsP2, phosphatidylinositol 4,5-bisphosphate.

appear to lead via unknown additional intermediary events to the ultimate biologic response, e.g. secretion or altered gene expression.

BIOCHEMICAL BASES OF TRANSMEMBRANE SIGNALING BY B-LYMPHOCYTE RECEPTORS

Membrane Immunoglobulin

PHOSPHOLIPID METABOLISM   The first studies to suggest the molecular basis of transmembrane signaling by B-cell mIg were conducted in 1975 by Maino et al (73). They demonstrated that anti-immunoglobulin antibodies induce increased incorporation of ^{32}P into phosphatidylinositol, indicating that phosphatidylinositol metabolic cycle (diagrammed in Figure 1) is activated following mIg ligation. This response is specific, with detectable increases in metabolism occurring only in the inositol lipids. It has very recently been demonstrated that antigens induce increased phosphatidylinositol metabolism in isolated antigen-binding cells (39). Phosphatidylinositol (PtdIns) and polyphosphoinositides (PtdInsP, $PtdInsP_2$) represent minor constituents of the membrane phospholipid pool. In B cells, PtdIns represents about 6% of total phospholipids, while PtdInsP and $PtdInsP_2$ are approximately 0.6% and 0.06% respectively (L. K. Harris, unpublished observations). Thus, the target of this response is a very minor portion of total membrane phospholipids.

The importance of increased PtdIns metabolism in the biologic response to anti-Ig was unknown until Monroe et al (74) and Lindsten et al (75) observed in 1984 that tumor-promoting phorbol diesters (PMA and 4βPDD) were, like anti-Ig antibodies, potent inducers of Ia expression on B cells. Phorbol diesters which neither promoted tumor formation nor activated PKC were ineffective Ia inducers. A variety of studies conducted at about this time (for review see 19, 51, 76) indicated that the potent biologic effects of phorbol esters documented in a number of systems were related to their ability to act as analogues of diacylglycerol (DAG) in the activation of protein kinase C. DAG is an intermediary in the PtdIns cycle (Figure 1) generated by phospholipase C–mediated hydrolysis of phosphoinositides. In view of our findings regarding the biological effects of phorbol diesters (74), we hypothesized that mIg-mediated signaling involves hydrolysis of PtdIns, PtdInsP, and/or $PtdInsP_2$ by phospholipase C (PLC), liberating DAG which in turn activates PKC. A prediction of this hypothesis was that anti-Ig antibodies should stimulate rapid formation of DAG or its phosphorylated product phosphatidic acid (PtdOH) and that this response should precede increased incorporation of ^{32}P into PtdIns. A further prediction was that inhibitors of the PtdIns cycle such as cAMP (77) should inhibit the biologic response to anti-Ig. Moreover, stimulators of PtdIns metabolism such as exogenous phospholipase C and exogenous DAG should stimulate the biologic response normally induced by anti-Ig (see Calcium Mobilization and Membrane Depolarization; 42). In 1984, Coggeshall & Cambier reported (78) that anti-Ig inducted rapid increases in ^{32}P incorporation into PtdOH. Rapid induction of DAG formation by anti-Ig was later reported by Bijsterbosch et al (79). The PtdOH response reported by Coggeshall & Cambier (78) was easily detectable within 30 sec to 1 min and was maximal 10 min following stimulation. The induction appears specific for the PtdIns cycle, since no increase in phosphatidylcholine metabolism was seen. The DAG appears to be derived from PtdIns, PtdInsP, or PtdInsP$_2$ because anti-Ig induced relatively little increase in incorporation of free [^3H]-glycerol into PtdOH (M. Coggeshall, unpublished observation). These studies support the hypothesis that ligation of mIg induces phospholipase-mediated hydrolysis of PtdIns, PtdInsP, and/or PtdInsP$_2$, generating DAG and leading to increased production of PtdOH and, somewhat later, PtdIns.

Ransom et al (80) and Bijsterbosch et al (79) subsequently addressed the source of DAG generated by analyzing the release of inositol phosphates InsP, InsP$_2$, and InsP$_3$ following anti-Ig stimulation of B cells. Bijsterbosch et al (79) found that cells prelabeled with [^3H]-inositol contained significantly increased levels of InsP$_3$ within 10 sec of exposure to anti-Ig; they reached a maximum level 60 sec after stimulation which then declined. Release of InsP$_2$ and InsP increased more slowly, reaching a maximum level 5 min after stimulation. In terms of total InsP release, the response was more pronounced in the presence of 5 mM LiCl which inhibits dephosphorylation of InsP. Our own results (80) were entirely consistent with those of Bijsterbosch except that anti-Ig-induced release of InsP$_2$ and InsP$_3$ was slower and more sustained. We believe that this discrepancy is explained by the fact that while Bijsterbosch et al (79) used supersaturating amounts of ligand, we used subsaturating doses. The observed sequential release of InsP$_3$, InsP$_2$, and InsP is consistent with the current view that receptor occupancy induces preferential hydrolysis of PtdInsP$_2$ and that the InsP$_3$ generated is converted by sequential dephosphorylation to InsP$_2$, InsP, and inositol before being reutilized in PtdIns synthesis.

Anti-Ig induction of PtdInsP$_2$ hydrolysis is also suggested by results (22) obtained using an isolated B lymphocyte membrane–transduction model adapted from a method described by Smith et al (70) using neutrophil membranes. In this system, membranes are isolated by differential centrifugation after B-cell disruption by sonication. Constituent PtdInsP, PtdInsP$_2$, and PtdOH are labeled selectively by incubation of membranes with $[^{32}P]$ ATP in pH 7.0 HEPES buffer containing 10 mM $MgCl_2$. Labeled membranes are washed and stimulated with anti-immunoglobulin antibodies before lipids are extracted and fractionated by thin-layer chromatography. Radioactivity associated with PtdInsP, PtdInsP$_2$, and PtdOH is counted to determine ligand effects on levels of labeled lipid species. Experiments indicate that anti-immunoglobulin induces the degradation of PtdInsP$_2$ ($\approx 20\%$) but has no effect on PtdInsP; this suggests that phospholipid hydrolysis induced by mIg cross-linking may be specific for PtdInsP$_2$. It is interesting that this response appears to be dependent on the presence of low levels of cytosolic protein, presumably providing phospholipase C, but is independent of Ca^{++}, GTP, and GDP concentration. This observation suggests that there may be no GTP binding protein involvement in anti-Ig activation of PLC-mediated PtdInsP$_2$ hydrolysis. This would be in contrast to GTP requirements recently defined for receptor coupling to PtdInsP$_2$ hydrolysis in a number of other systems (68–71).

CALCIUM MOBILIZATION   In virtually all systems in which ligand binding to receptor induces increased polyphosphoinositide hydrolysis, ligand binding also induces a rapid increase in intracellular free calcium levels (for review see 19 and 76). B-lymphocyte stimulation by mIg-cross-linking ligands is no exception to this rule. Braun et al (81) first documented anti-Ig induction of mobilization of intracellular calcium in B cells. These studies which utilized Ca^{45} indicated that the response is rapid, occurring within minutes of stimulation. Pozzan et al (82) later corroborated these findings and demonstrated using the Ca^{++} sensitive dye Quin 2 (83) that anti-Ig induced an increase in intracellular free Ca^{++}, from about 100 nM to 500–1000 nM. Quin 2 and the more recently developed fluorescent Ca^{++} indicator Indo 1 (84–86) can be used in conjunction with flow cytometry to conduct single cell analysis of Ca^{++} mobilization. Studies of Ransom et al (85) indicate that virtually all Ig-expressing B cells respond to anti-Ig by mobilizing intracellular free Ca^{++}. This is consistent with previous observations that all Ig-positive splenic B cells respond to anti-Ig by increasing expression of cell surface Ia molecules (4, 6). As with this later response, Ca^{++} mobilization requires stimulation by divalent antibodies (87). Thus, receptor cross-linking appears essential for PtdInsP$_2$ hydrolysis (78) and Ca^{++} mobilization as well as for the cell's biologic response, i.e. enhanced Ia expression.

Further studies which have addressed the molecular basis of the calcium mobilization response in B cells indicate that this response is not mediated by PKC activation since it is not induced by phorbol myristate acetate (PMA) or 1-oleyl-2-acetylglycerol (87). Furthermore, protein kinase C translocation and activation (see below) induced by anti-Ig occurs later following stimulation than Ca^{++} mobilization. The calcium mobilization response can be elicited by exposure of cells to exogenous phospholipase C (PLC) and is blocked by inhibitors of $PtdInsP_2$ metabolism (42, 87), suggesting that a product of PtdIns, PtdInsP, or $PtdinsP_2$ hydrolysis is responsible for anti-Ig induced Ca^{++} mobilization. The PLC effect was somewhat of a surprise since extracellular enzyme should not have access to $PtdInsP_2$ on the inner leaflet. We assume that some enzyme must gain entry into the cell where $PtdInsP_2$ hydrolysis occurs.

The fact that PLC but not exogenous DAG (87) induces Ca^{++} mobilization suggests that the inositol phosphate product of inositol lipid hydrolysis may be responsible for Ca^{++} mobilization. This would be consistent with observations of Streb et al (67) and others (for review see 19) that $InsP_3$ mediates the release of Ca^{++} from intracellular stores, specifically the endoplasmic reticulum (ER). We assessed the existence of an $InsP_3$-mobilizable ER Ca^{++} store in B cells by determining the ability of $InsP_3$ to induce release of Ca^{++} from saponin-permeabilized cells pre-loaded with $^{45}Ca^{++}$ (80). B cells were labeled with $^{45}Ca^{++}$ by incubation with 1 mM Mg^{++} ATP plus 5 mM azide for 10 min. Under these conditions, Ca^{++} is loaded only into nonmitochondrial stores. The endo-plasmic reticular component of this uptake was determined by loading in the presence of azide plus orthovanadate which blocks ER uptake. Based on these studies, we determined that 56% of the azide insensitive Ca^{++} uptake (21% of total uptake) by saponin permeabilized cells is by the endoplasmic reticulum. $InsP_3$ (10 μM) stimulated cells $^{45}Ca^{++}$ loaded in the presence of 1 mM Mg-ATP and azide to release $\sim 35\%$ of label or about 70% of the ER-associated Ca^{++}. Significant release of Ca^{++} was seen with 300 nM $InsP_3$, the lowest dose tested. Thus, the B-lymphocyte endoplasmic reticulum contains a significant calcium store, a large pro-portion of which can be liberated by the action of $InsP_3$. Considered in view of observations that $InsP_3$ is released following B-cell stimulation (79, 80) and that the kinetics of this release are similar to that of Ca^{++} mobilization in these cells, these findings are consistent with the possibility that at least a portion of the calcium mobilized is derived from intracellular stores by the action of $InsP_3$. In further support of this are recent findings that a portion of the anti-Ig-induced Ca^{++} response detected with either Quin 2 or Indo 1 is independent of extracellular Ca^{++} and insensitive to Ca^{++} channel blockade with competitive ions (88).

Cross-linking of mIg also causes a transient increase in the permeability of the plasma membrane to Ca^{++}, which is sensitive to membrane potential and negatively modulated by protein kinase C (80, 88). This permeability is sensitive to the K^+ channel blocker 4-aminopyridine and to organic Ca^{++} channel blockers such as nifedipine (90) at doses which have been shown to block voltage gated K^+ channels in T cells (91). However, these doses are in excess of those required to block well-characterized Ca^{++} channels in other tissues (92). These results suggest that permeability may be dependent on potential, because the membrane potential is primarily a function of K^+ permeability. A greater permeability to Sr^{++} than to Ba^{++} also develops in the anti-mIg-stimulated B cell, indicating a similarity to the voltage gated Ca^{++} channel described in a myeloma by Fukushima & Hagiwara (93, 94) using the patch-clamp technique. Whether the voltage gated channel is responsible for all, if any, of the Ca^{++} influx remains to be determined since no patch-clamp studies have been performed on normal B cells despite a rapidly growing volume of literature on application of the technique to T cells (95, 96). The inability to detect any further enhancement of Ca^{++} permeability in cells after anti-Ig stimulation for longer than 5 min and evidence that optimal activation of protein kinase C occurs within 2–4 minutes (97) are consistent with the possibility that protein kinase C may terminate the Ca^{++} influx caused by Ig cross-linking. Removal of elevated free cytoplasmic calcium (Ca_i^{++}) is most likely a function of the ATP-dependent Ca^{++} pumping activity such as that described in human T-cell plasma membrane preparations (98).

PROTEIN KINASE C ACTIVATION   In several diacylglycerol generating signaling systems, DAG release and Ca^{++}-mobilization have been shown to be accompanied by translocation of the Ca^{++}/phospholipid-dependent protein kinase, protein kinase C, to the plasma membrane (for review see 65). This translocation, which may simply represent transition of PKC from a loosely to a tightly membrane-associated state, presumably results from the high affinity binding of PKC to complexes of DAG, phosphatidylserine (PtdSer), and Ca^{++} (99). Translocation of PKC was first documented in EL-4 cells following stimulation with the DAG analog PMA (100, 101). We have demonstrated PKC translocation in normal B cells stimulated with PMA (97) and LPS (97) which do not stimulate $PtdInsP_2$ metabolism (79, 102) and with Ca^{++} ionophores (Z. Z. Chen, J. C. Cambier, submitted for publication). The ability of LPS to induce PKC translocation may be a function of its ability to act as a DAG or PtdSer analogue, as is suggested by its ability to activate PKC in cell-free systems (103). Thus, LPS may integrate into the plasma membrane, perhaps with the aid of a receptor, and act as a "sink" for intracellular PKC.

Chen et al (97) and Nel et al (104) have recently reported anti-immunoglobulin induction of PKC translocation in normal murine and human B cells respectively. This translocation is rapid, occurring within 1–2 min of stimulation and involving 80–90% of the cellular protein kinase C which can be activated in vitro. The response to anti-Ig is transient, with activity returning to the cytosol within 10–20 min. It is interesting that trans-location of PKC to the Triton-X100 soluble particulate fraction is followed by a brief translocation of about 20% of cellular PKC to a Triton X100 insoluble particulate compartment (97). We have recently identified this compartment as the nuclear envelope (Z. Z. Chen, J. C. Cambier, sub-mitted for publication). This translocation is detectable about 4 min after stimulation, and PKC begins to leave this compartment about 10 min later. Studies comparing the abilities of whole molecules and $F(ab')_2$ of rabbit antimouse immunoglobulin to induce translocation of PKC indicate that induction of nuclear translocation requires the presence of the Fc portion of the antibody molecule. Thus, this response may be mediated by the B-cell Fc receptor. Translocation to the plasma membrane is most efficient with $F(ab')_2$ fragments of anti-immunoglobulin antibodies.

The role played by mobilized Ca^{++} in the activation and in addition translocation of PKC is not clearly understood. Observations that the calcium ionophores stimulate exclusive phosphorylation of 20 kd myosin light chains in platelets while PMA induces phosphorylation of only a 44 kd protein (105) have been taken as evidence that in physiologic situations second messenger Ca^{++} and DAG function exclusively in activation of Ca^{++} calmodulin dependent protein kinase(s) and PKC respectively. How-ever, studies by Wolf et al (106) have demonstrated that both Ca^{++} and DAG promote binding of PKC to inside-out red cell ghosts. Similarly, we have recently shown that calcium ionophores induce translocation of B-cell PKC from the cytosol to the Triton-soluble particulate fraction, pre-sumably plasma membrane (Z. Z. Chen, J. C. Cambier, submitted for publication). These data are consistent with the possibility that in the physiologic response, mobilized calcium and DAG act in concert to modu-late PKC compartmentalization. Thus, both may be important in placing PKC in the appropriate subcellular site so that phosphorylation of sub-strates can in turn drive the cell's biologic response.

Based upon early studies of lipid and cation requirements for activation of purified PKC (for review see 76,105), one would predict that in B cells, DAG and free Ca^{++} elevated as a result of mIg cross-linking would induce activation of PKC. This possibility has been explored in the recent time period by assessing and comparing the ability of anti-Ig and DAG analogs to induce incorporation of ^{32}P into cytosolic and membrane proteins. Studies by Nel et al (104), Hornbeck & Paul (107), and Coggeshall et al (42) have documented anti-Ig and PMA induction of phosphorylation of many common substrates, consistent with the possibility that anti-Ig-induced phosphorylation is mediated by PKC. Nel et al (104) have cata-logued these substrates, defining major cytosolic substrates of 94, 66, 60, 56, 50, 43, 38, 35, 28–30, 20–23, and 15–18 kd and a membrane substrate of 29 kd. The function of these phosphoproteins is unknown. Apparently none are mIg subunits. Hornbeck & Paul (107) have demonstrated that down regulation of pKC by previous exposure of cells to PMA blocks the ability of anti-Ig to induce new protein phosphorylation, supporting the hypothesis that anti-Ig-induced protein phosphorylation is mediated by PKC. It is interesting that recent studies by Nel et al (108) indicate that anti-Ig induces phosphorylation of major 50- and 56-kd B-cell membrane proteins on tyrosine residues. This finding, like similar observations in fibroblasts (109), suggests that a tyrosine kinase may be activated following mIg cross-linking. This may occur through a kinase cascade involving PKC activation of a tyrosine kinase.

MEMBRANE DEPOLARIZATION : ALTERATIONS IN ION TRANSPORT   The cell physiologic events that intervene between translocation of PKC (2–4 min) and activation of transcription of genes encoding c-fos (15 min) c-myc (60 min) and Ia (120 min) (M. C. Klemsz, E. Palmer, J. C. Cambier, submitted for publication) are very poorly defined. In 1983, Monroe & Cambier (36, 37, 110) conducted a series of experiments using carbocyanine dyes which indicated that mIg cross-linking by anti-Ig antibodies or antigen induces B cells to undergo membrane depolarization. This response is quite slow in comparison to stimulus-coupled changes in ion transport in many other systems, being detectable within about 3 min and maximal by 60 min following stimulation. Membrane depolarization is inducible with an array of agents which directly or indirectly activate PKC including exogenous phospholipase C, DAG, PMA, and calcium ionophores (A23187 or ionomycin), suggesting that this response to anti-Ig is PKC mediated (42). Membrane depolarization appears in turn to be coupled to the B cell's subsequent biologic responses to anti-Ig, since valinomycin (a potassium ionophore) blocks anti-Ig induction of Ia, and passive membrane depolarization using 50 mM KCl stimulates increased cellular Ia expression (37). Taken together these data suggest that PKC may act by altering ion transport which in turn leads to increased Ia, c-myc, and c-fos encoding gene transcription.

The ionic basis of anti-Ig or antigen-induced B-cell membrane depolarization is unknown. Ions like any solute, diffuse down a concentration gradient across a biological membrane that is permeable to the ion, towards a state of thermodynamic equilibrium. Uneven distribution of charged particles (e.g. K^+ and Na^+) across the membrane establishes an electrical potential (normally inside negative) which exerts a strong influence on the movement of charged particles across the membrane. For example, the negativity of a cell's interior greatly contributes to the tendency of Ca^{++} to enter a cell down its inwardly directed concentration gradient ($[Ca^{++}]_i = 100$ nM, $[Ca^{++}]_o = 1$ mM) if the membrane is permeable to Ca^{++}. In addition, a rapid change in membrane permeability to a specific ion will result in a diffusion potential as long as the species crosses the membrane. Considering these simplified principles of ion movement, membrane potential may be a function of the contributions of a number of unevenly distributed ions, particularly Na^+ K^+, Ca^{++}, and Cl^- and of sudden changes in the permeability properties of the plasma membrane towards these ions. As such it is important to identify alterations in permeability of the plasma membrane to specific ions during activation to gain a clear understanding of how ion flux may regulate activation and proliferation.

K^+ Fluxes Electrophysiologic evidence shows the presence of voltage gated K^+ channels on a variety of cells of hematopoetic lineage including T cells; it is more thoroughly reviewed elsewhere (111–113). Currently, no published evidence indicates the presence of voltage gated K^+ channels on normal B cells, although the existence of such channels in human B cells has been alluded to recently (111). Measurements of total cell $86Rb^+$ (a frequently used substitute for $42K^+$) suggest that mIg cross-linking causes a loss of K^+, presumably due to K^+ efflux, which is maximal by 2 min and is followed by restoration of cell K^+ towards the level determined prior to stimulation (89). The loss of cell K^+ is blocked by concentrations of 4-aminopyridine and nifedipine identical to those that inhibit mIg-mediated Ca^{++} influx. This similarity between the pharmacologic sensitivity of mIg-mediated K^+ efflux and Ca^{++} influx suggests that the two phenomena may be related. Evidence that anti-Ig-stimulated murine B cells hyperpolarize only when Ca^{++} influx is minimized—i.e. when $[Ca^{++}]_o$ is < 10 uM—and that hyperpolarization is blocked by a high $[K^+]_o$ or by 50 μM nifedipine all suggest that the membrane potential of stimulated B cells is dependent upon both K^+ and Ca^{++} fluxes (114). Fukushima & Hagiwara (93, 94) have demonstrated that the voltage-gated Ca^{++} channel present in myelomas and hybridomas also conducts monovalent cations ($Na^+ > K^+ > Rb^+$) which, in light of their inability to detect any other voltage-gated ion-specific conductance, raises the possibility that the same transport pathway may be responsible for K^+ efflux and Ca^{++} influx. However, the existence of independent K^+ and Ca^{++} transport pathways has not been ruled out. Further characterization of specific ion transport pathways in individual normal B cells will greatly improve our understanding of the role of these ions in B-cell activation.

Na^+ and H^+ Fluxes Except for evidence that the voltage-gated Ca^{++} channel in myeloma cells can conduct Na^+ (94), there is no published evidence indicating the existence of voltage-gated Na^+ specific con-

ductance pathways or of Na^+/Ca^{++} exchange activity in B cells. There is strong evidence for the existence of a Na^+/H^+ exchange activity in a pre–B lymphocyte line which is activated by LPS or PMA to increase cytoplasmic pH and $[Na^+]_i$ slightly (115). Cytoplasmic alkalizinization and elevation of $[Na^+]_i$ are dependent upon $[Na^+]_o$ and can be blocked by amiloride, an inhibitor of Na^+ transport. Induction of pre–B cell differentiation by LPS is also inhibited by amiloride and can be mimicked by the Na^+/H^+ exchange ionophore monensin (116) suggesting that physiologic activation of the exchanger may be mediated by PKC: Whether activation of Na^+/H^+ exchange activity alone is sufficient to induce differentiation remains to be determined. Extensive work describing the presence and regulation of this activity in T cells has been recently reviewed by Grinstein et al (117). Preliminary evidence indicates that Na^+/H^+ exchange activity is inducible in quiescent B cells by anti-Ig antibodies (J. C. Cambier, M. H. Julius, D. F. Gerson, unpublished observations) and that cytoplasmic alkalinization mediated by this mechanism occurs with kinetics similar to membrane depolarization.

Cl⁻ Fluxes   Currently, no evidence supports a role for alterations in Cl^- permeability in B-cell (or T-cell) activation. In a series of elegant studies Gelfand and Grinstein and colleagues (118–122) have demonstrated that hypotonic treatment of lymphocytes causes an increase in the permeability of B and T cells to K^+ and Cl^-, although the effect on K^+ permeability is much more marked in T cells than B cells. These studies indicate that specific mechanisms which regulate Cl^- permeability are present and alterable in the B cell, raising the possibility that such pathways may be involved in B-cell activation.

Lipopolysaccharide

Lipopolysaccharide does not induce inositol lipid metabolism (79, 102) or Ca^{++} mobilization (79) in quiescent B cells. LPS does, however, induce protein kinase C translocation (97) as well as membrane depolarization (123). Recent evidence indicates that LPS also activates the GTP binding protein Ni (124) which in other systems leads to depression of cellular adenyl-cyclase activity and cAMP levels. Thus, LPS appears to act as a diacylglycerol analogue and as an α_2 adrenergic agonist.

Interleukin 4

IL-4 binding to its receptor does not induce inositol lipid metabolism, Ca^{++} mobilization, protein kinase C translocation, or membrane depolarization (125). Unlike anti-immunoglobulin or PMA, IL-4 does not induce significant protein phosphorylation in whole cells (P. Hornbeck, unpublished observation). However, IL-4 (but not IFNγ, anti-Ig, or Il-2) induces phosphorylation of 42 kd protein upon stimulation of isolated B-cell membranes in the presence of ^{32}P ATP (125; L. B. Justement, Z. Z. Chen, L. K. Harris, G. T. Ransom, V. S. Sandoval, C. Smith, D. Rennick, N. Roehm, J. C. Cambier, submitted for publication). Thus, Il-4 is similar to insulin and epidermal growth factor receptors (152) in inducing phosphorylation of proteins in isolated cell membrane. It is not known whether the 42-kd substrate is the receptor for IL-4. The amino acid specificity of this phosphorylation has not been determined.

Ia-Binding Ligands and Interferon γ

Anti-Ia antibodies and IFNγ have similar effects on the biology of quiescent B cells, antagonizing mitogen induction of Ia expression and proliferation (11, 13, 46, 126) but promoting differentiation to antibody secretion (12, 47). These agents also do not induce inositol lipid metabolism or Ca^{++} mobilization in B cells (J. Cambier, unpublished observation). We noted that the biologic effects are qualitatively similar to agents which elevate intracellular cAMP levels such as dbcAMP, forskolin, and cholera toxin; we therefore hypothesized that anti-Ia antibodies and IFNγ may signal B cells via elevation of cAMP levels. To address this possibility we analyzed effects of anti-Ia antibodies on cellular cAMP levels and found that anti-Ia antibodies induce a rapid rise, about three fold in 60 sec, in cAMP levels in B cells (M. K. Newell, J. C. Camber, unpublished observation). We also observed that both anti-Ia antibodies and dbcAMP induce recompartmentalization of protein kinase C to the nuclear envelope (125, 126). While the biologic significance of this phenomenon is unknown, the facts that both anti-Ia antibodies and cAMP agonists induce it and that anti-Ia antibodies induce an increase in intracellular cAMP levels are consistent with the possibility that B-cell Ia, and perhaps interferon receptors, transduce signals via a β-adrenergic-like mechanism.

CONCLUSIONS

In recent years, many laboratories have undertaken cell biological and biochemical studies of signal transduction in lymphocytes in hopes that the resultant information would contribute substantially to our understanding of immune regulation. Studies of some transduction mechanisms, e.g. that employed by membrane immunoglobulin, have progressed to the point that they are providing important insight into the role of antigen receptors in generation of immune responses. As recently as 10 years ago there was controversy about whether membrane immunoglobulin served any signal transducing role at all. Compelling molecular evidence now indicates that mIg molecules are signal transducers, and considerable evidence indicates that this transduction occurs via the biochemical cascade depicted in Figure 3. Specifically, cross-linking of mIgM or mIgD by antigen prompts the hydrolysis of polyphosphoinositides by phospholipase C yielding inositol trisphosphate and diacylglycerol. Newly formed inositol trisphosphate stimulates the release of Ca^{++} from the endoplasmic reticulum, and Ca^{++} influx; this Ca^{++} acts in concert with diacylglycerol to cause translocation of protein kinase C to the plasma membrane. Active, membrane-associated PKC phosphorylates its substrates, altering the activity of structures responsible for transport of various ions. In the succeeding minutes, changes in intracellular pH and membrane potential occur that reflect changes in transport of Na^+, H^+ and Ca^{++} and K^+. This cascade appears to lead via unknown intermediary events to increased transcription of genes encoding c-fos, c-myc, and Ia and to increased cell surface expression of Ia.

What are the physiologic implications of activation of this cascade? Data documenting the ability of phorbol diesters and calcium ionophores to synergize in induction of thymidine uptake by B cells (18) suggest that under appropriate circumstances activation of this cascade can lead to progression of quiescent B cells into S phase. This presumably depends on robust activation of the pathway for an extended period of time as

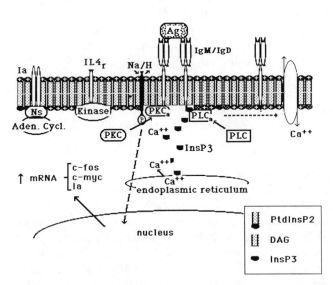

Figure 3 Working model for transmembrane signal transduction by membrane immunoglobulin (IgM and IgD), IL-4 receptors (IL4r) and Ia molecules. See legend of Figure 2 for abbreviations.

accomplished pharmacologically with these agents. Such a situation probably exists when B cells are stimulated with high doses of soluble heterologous anti-Ig antibodies, anti-Ig derivatized beads, or highly polyvalent thymus-independent type-II antigens. Thymus-dependent antigens, which are generally paucivalent, activate the cascade sufficiently to induce Ia expression, but they do not stimulate entry into cell cycle. Increased Ia expression with concomitant antigen processing and reexpression, confers upon the B cell the ability to present antigen to helper T cells. During the subsequent interaction, T cells may signal B cells directly by binding Ia, inducing an elevation in cellular cAMP levels and thereby promoting differentiation. This would provide a basis for MHC restriction of thymus-dependent plaque-forming cell responses by small B cells. T cells activated during conjugate formation release lymphokines including BCAF, IFNγ, IL-4, IL-5, and perhaps others, which promote B-cell growth and differentiation in an Ia unrestricted manner. With the exception of IL-4 receptors, which appear to transduce signals via activation of a membrane-associated kinase, we know virtually nothing about signaling mechanisms employed by receptors for lymphokines.

Clearly, much progress has been made in efforts to define the transmembrane signaling mechanisms operative in regulation of B-cell function. However, a very long and challenging road lies ahead.

ACKNOWLEDGMENTS

We thank Donna Thompson and Kelly Crumrine for their assistance in preparing this manuscript. We also thank Drs. Priscilla Campbell and Louis Justement for critical review of the manuscript. Original studies described here were supported by PHS grants AI 20519 and AI21768.

Literature Cited

1. Monroe, J. G., Cambier, J. C. 1983. Level of mIa expression on mitogen-stimulated murine B cells is dependent on position in the cell cycle. *J. Immunol.* 130: 626
2. Finkleman, F. D., Mond, J. J., Metcalf, E. S. 1986. Anti-immunoglobulin antibody induction of B lymphocyte activation and differentiation. In *B-Lymphocyte Differentiation*, ed. J. C. Cambier, pp. 41–61. Boca Raton: CRC Press
3. Cambier, J. C., Monroe, J. G., Coggeshall, K. M., Ransom, J. T. 1985. On the mechanisms of transmembrane signaling by membrane immunoglobulin. *Immunol. Today* 6: 218
4. Roehm, N. W., Leibson, J. L., Zlotnik, A., Kappler, J., Marrack, P., Cambier, J. C. 1984. Interleukin-induced increase in Ia expression by normal mouse B cells. *J. Exp. Med.* 160: 679
5. Noelle, R., Krammer, P. H., Ohara, T., Uhr, J. W., Vitetta, E. S. 1984. Increased expression of Ia antigens on resting B cells: An additional role for B cell growth factor. *Proct. Natl. Acad. Sci. USA* 81: 6149
6. Mond, J. J., Finkelman, F. D., Sarma, S., Ohara, J., Serrate, S. 1985. Recombinant interferon-γ inhibits the B cells proliferative response stimulated by soluble but not by sepharose-bound anti-immunoglobulin antibody. *J. Immunol.* 135: 2513
7. Leclercq, L., Bismuth, G., Theze, J. 1984. Antigen specific helper T-cell

clone supernatant is sufficient to induce both polyclonal proliferation and differentiation of small resting B lymphocytes. *Proc. Natl. Acad. Sci. USA* 81: 6491

8. Leclercq, L., Cambier, J. C., Mishel, Z., Julius, M. H., Theze, J. 1986. Supernatant from a cloned helper T cell stimulates most small resting B cells to undergo increased I-A expression, blastogenesis and progression through cell cycle. *J. Immunol.* 136: 539

9. Snow, E. C. 1985. Insulin and growth hormone function as minor growth factors which potentiate lymphocyte activation. *J. Immunol.* 135: 776s

10. Melchers, F., Andersson, J. 1984. B cell activation: Three steps and their variations. *Cell* 37: 715

11. Forsgren, S., Pobor, G., Coutinho, A., Pierres, M. 1984. Their role of I-A/E molecules in B lymphocyte activation 1. Inhibition of lipopolysaccharide-induced responses by monoclonal antibodies. *J. Immunol.* 133: 2104

12. Palacios, R., Martinez-Maza, O., Guy, K. 1983. Monoclonal antibodies against HLA-DR antigens replace T helper cells in activation of B lymphocytes. *Proc. Natl. Acad. Sci. USA* 80: 3456

13. Niederhuber, J. E., Frelinger, J. A., Dugan, E., Coutinho, A., Shreffler, D. C. 1975. Effects of anti-Ia serum on mitogenic responses. I. Inhibition of proliferative response to B cell mitogen, LPS, by specific anti-Ia sera. *J. Immunol.* 115. 1672

14. Howard, M., Nakanishi, K., Paul, W. 1984. B cell growth and differentiation factors. *Immunol. Rev.* 78: 185

15. Rabin, E. M., Ohara, J., Paul, W. E. 1985. B-cell stimulatory factor 1 activates resting B cells. *Proc. Natl. Acad. Sci. USA* 82: 2935

16. Bijsterbosch, M. K., Klaus, G. G. B. 1985. Crosslinking of surface immunoglobulin and Fc receptors on B lymphocytes inhibits stimulation of inositol phosphate breakdown via the antigen receptors. *J. Exp. Med.* 126: 1825

17. Leptin, M. 1985. Monoclonal antibodies specific for murine IgM. II. Activation of B lymphocytes by monoclonal antibodies specific for the four constant domains of IgM. *Eur. J. Immunol.* 15: 131

18. Klaus, G. G. B., O'Garra, A., Bijsterbosch, M. K., Holam, M. 1986. Activation and proliferation signals in mouse B cells. VIII. Induction of DNA synthesis in B cells by a combination of calcium ionophores and phorbol myristate acetate. *Eur. J. Immunol.* 16: 92

19. Berridge, M. J., Irvine, R. F. 1984. Inositol trisphosphate, a novel second messenger in cellular signal transduction. *Nature* 312: 315

20. Sibley, D. R., Lefkowitz, R. J. 1985. Molecular mechanisms of receptor desensitization using the β-adrenergic receptor-coupled adenylate cyclase system as a model. *Nature* 317: 124

21. Bell, R. M. 1986. Protein kinase C activation by diacylglycerol second messengers. *Cell* 45: 631

22. Cambier, J. C., Harris, L. K. 1986. Phosphoinositides and transmembrane signaling in the immune system. In *Receptor Biochemistry and Methodology*, ed. J. W. Putney, New York: Alan Liss. In press

23. Sell, S., Gell, P. 1965. Studies on rabbit lymphocytes *in vitro*. I. Stimulation of blast transformation with an anti-allotype serum. *J. Exp. Med.* 122: 423

24. Pernis, B., Forni, L., Amante, L. 1971. Immunoglobulins as cell receptors. *Ann. N.Y. Acad. Sci. USA* 190: 420

25. Raff, M. C., Steinberg, J., Taylor, R. 1970. Immunoglobulin determinants on the surface of mouse lymphoid cells. *Nature* 225: 553

26. Salsano, R., Frøland, S. S., Natvig, J. B., Michaelson, T. E. 1974. Same idiotype of B lymphocyte membrane IgD and IgM. Formal evidence for monoclonality of chronic lymphocytic leukemia cells. *Scand. J. Immunol.* 3: 841

27. Fu, S. M., Winchester, R. J., Kunkel, H. G. 1975. Similar idiotypic specificity for the membrane IgD and IgM of human B lymphocytes. *J. Immunol.* 114: 250

28. Pernis, B., Brouet, J. C., Seligmann, M. 1974. IgD and IgM on the membrane of lymphoid cells in the macroglobulinemia. Evidence of identity of membrane IgD and IgM antibody activity in a case with anti-IgG receptors. *Eur. J. Immunol.* 4: 776

29. Goding, J. W., Layton, J. E. 1975. Antigen-induced cocapping of IgM and IgD-like receptors on murine B cells. *J. Exp. Med.* 114: 852

30. Grey, H. M., Chestnut, R. 1985. Antigen processing and presentation to T cells. *Immunol. Today* 6: 1

31. Coutinho, A., Möller, G. 1974. Immune activation of B cells: Evidence for one non-specific triggering signal

not delivered by the Ig receptors. *Scand J. Immunol.* 3 : 133

32. Bretscher, P. A., Cohn, M. 1970. A theory of self-nonself discrimination. *Science* 169 : 1042

33. Mond, J. J., Seghal, E., Kung, J., Finkelman, F. D. 1981. Increased expression of I-region-associated antigen (Ia) on B cells after cross-linking of surface immunoglobulin. *J. Immunol.* 127 : 881

34. Weiner, H. L., Moorhead, J. W., Claman, H. N. 1976. Anti-immunoglobulin stimulation of murine lymphocytes. I. Age dependence of the proliferative response. *J. Immunol.* 116 : 1656

35. Parker, D. C. 1975. Stimulation of mouse lymphocytes by insoluble antimouse immunoglobulin. *Nature* 258 : 361

36. Monroe, J. G., Cambier, J. C. 1983. B cell activation. II. Receptor crosslinking by thymus-independent and thymus-dependent antigens induces a rapid decrease in the plasma membrane potential of antigen binding B lymphocytes. *J. Immunol.* 133 : 2641

37. Monroe, J. G., Cambier, J. C. 1983. B cell activation. III. B cell plasma membrane depolarization and hyper I-A antigen expression induced by receptor immunoglobulin crosslinking are coupled. *J. Exp. Med.* 158 : 1589

38. Snow, E. C., Fetherston, J. D., Zimmes, S. 1986. Induction of the *cMyc* after antigen binding to haptenspecific B cells. *J. Exp. Med.* 184 : 944

39. Grupp, S. H., Snow, E. C., Harmon, A. K. 1986. Phosphatidylinositol response is an early event in physiologically relevant activation of antigen specific B lymphocytes. *Proc. Natl. Acad. Sci. USA.* In press

40. Noelle, R. J., Snow, E. C., Uhr, J. W., Vitetta, E. S. 1983. Activation of antigen-specific B cells : The role of T cells, cytokines and antigens in the induction of growth and differentiation. *Proc. Natl. Acad. Sci.* 80 : 6628

41. Janeway, C. A., Bottomly, K. B., Babich, J., Conrad, P., Conzan, S., Jones, B., McVay, L., Murphy, D., Tite, J. 1984. Quantitative variation in Ia antigen expression plays a central role in immune regulation. *Immunol. Today* 5 : 99

42. Coggeshall, K. M., Monroe, J. G., Ransom, J. T., Cambier, J. C. 1985. Mechanisms of transmembrane signal transduction during B cell activation. In *B Lymphocyte Differentiation,* ed.

J. C. Cambier, pp. 1–22. Boca Raton, Florida : CRC Uniscience

43. Lee, F., Yokota, T., Otsuka, T., Meyerson, P., Villaret, D., Coffman, R., Mosmauh, T., Rennick, D., Roehm, N., Smith, C., Blatnik, A., Arai, K. 1986. Isolation and characterization of a mouse interleukin cDNA that expresses BSF1 activities and T cell and mast cell stimulating activities. *Proc. Natl. Acad. Sci. USA.* 83 : 2061

44. Noma, Y., Sideras, P., Naito, T., Bergstedt-Lindquist, S., Azuma, C., Severinson, E., Tanabe, T., Kinash, T., Matsuda, F., Yasitaff, O., Honjo, T. 1986. Cloning of cDNA encoding the murine IgG1 induction factor by a novel strategy using SPG promoter. *Nature* 319 : 640

45. Julius, M., Von Boehmer, H., Sidman, C. L. 1982. Dissociation of two signals required for activation of resting B cells. *Proc. Natl. Acad. Sci. USA* 79 : 1989

46. Clement, L. T., Tedder, T. F., Gartland, G. L. 1986. Antibodies reactive with class II antigens encoded for by the major histocompatibility complex inhibit human B cell activation. *J. Immunol.* 136 : 2375

47. Bishop, G. A., Haughton, G. 1986. Induction of differentiation of a transformed clone of Ly1+ B cells by clonal T cells and antibodies. *Proc. Natl. Acad. Sci. USA.* In press

48. Sutherland, E. W., Rall, T. W. 1958. Fractionation and characterization of a cyclic adenine ribonucleotide formed by tissue particles. *J. Biol. Chem.* 232 : 1077

49. Lohmann, S. M., Walter, U. 1984. Regulation of the cellular and subcellular concentrations and distribution of cyclic nucleotide-dependent protein kinases. *Adv. Cyclic. Nucleotide and Protein Phosphoryl. Res.* 18 : 116

50. Manalan, A. S., Klee, C. B. 1984. Calmodulin. *Adv. Cyclic nucleotide Protein Phosphoryl. Res.* 18 : 227

51. Takai, Y., Kaibuchi, K., Nishizuka, Y. 1984. Membrane phospholipid metabolism and signal transduction for protein phosphorylation. *Adv. Cyclic nucleotide Protein Phosphoryl. Res.* 18 : 119

52. Hunter, T., Cooper, J. A. 1985. Protein-tyrosine kinases. *Ann. Rev. Biochem.* 54 : 897

53. Benovic, J. L., Shorr, R. G. L., Caron, M. G., Lefkowitz, R. J. 1984. The Mammalian β_2-adrenergic receptor : Purification and characterization. *Biochemistry* 23 : 4510

54. Cubero, A., Malbon, C. C. 1984. The fat cell beta-adrenergic receptor. Purification and characterization of a mammalian beta 1-adrenergic receptor. *J. Biol. Chem.* 259: 1344

55. Codina, J. 1984. Mechanisms in the vectoral receptor-adenylate cyclase signal transduction. *Adv. Cyclic nucleotide Protein Phosphoryl. Res.* 17: 111

56. Gilman, A. G. 1984. G proteins and dual control of adenylate cyclase. *Cell* 36: 577

57. Pfeuffer, E., Dreher, R. M., Metzger, H., Pfeuffer, T. 1985. Catalytic unit of adenylate cyclase: Purification and identification by affinity crosslinking. *Proc. Natl. Acad. Sci. USA* 82: 3086

58. Stadel, J. M., Lefkowitz, R. J. 1983. The β-adrenergic receptor: Ligand binding studies illuminate the mechanisms of receptor-adenylate cyclase coupling. *Curr. Top. Membrane Transplant.* 18: 45

59. Schramm, M., Selinger, Z. 1984. Message transmission: Receptor controlled adenylate cyclase system. *Science* 255: 1350

60. Hildebrandt, J. D., Codina, J., Rosenthal, W., Sunger, T., Iyengar, R., Birnbaumer, L. 1985. Properties of human erythrocyte N_s and N_i, the regulatory components of adenylate cyclase, as purified without regulatory ligands. *Adv. Cyclic Nucleotide Protein Phosphoryl. Res.* 19: 87

61. Leibman, P. A., Sitaramayya, A. 1984. Role of G protein-receptor interaction in amplified phosphodiesterase activation of retinal rods. *Adv. Cyclic Nucleotide Protein Phosphoryl. Res.* 17: 215

62. Hokin, M. R., Hokin, L. E. 1953. Enzyme secretion and the incorporation of P^{32} into phospholipids of pancreas slices. *Biol. Chem.* 203: 967

63. Durell, J., Garland, J. T., Friedel, R. O. 1969. Acetylcholine action: Biochemical aspects. *Science* 165: 862

64. Michell, R. H. 1975. Inositol phospholipides and cell surface receptor function. *Biochem. Biophys. Acta.* 415: 81

65. Anderson, W. B., Estival, A., Tapiovaara, H., Gopalakrishna, R. 1985. Altered subcellular distribution of protein kinase C (a phorbol ester receptor). Possible role in tumor promotion and the regulation of cell growth: Relationship to changes in adenylate cyclase activity. *Adv. Cyclic Nucleotide Protein Phosphoryl. Res.* 19: 287

66. Durell, J., Sodd, M. A., Friedel, R. O. 1960. Acetylcholine stimulation of the phosphodiesteratic cleavage of guinea pig brain phosphoinositides. *Life Sci.* 7: 363

67. Streb, H., Irvine, R. F., Berridge, M. J., Schulz, I. 1983. Release of Ca^{++} from a nonmitochondrial intracellular store in pancreatic acinar cells by inositol-1,4,5-triphosphate. *Nature* 306: 67

68. Seyfred, M. A., Wells, W. W. 1984. Subcellular site and mechanism of vasopressin-stimulated hydrolysis of phosphoinositides in rat hepatocytes. *J. Biol. Chem.* 259: 7666

69. Litosch, I., Wallis, C., Fain, J. N. 1985. 5-hydroxytryptamine stimulates inositol phosphate production in a cell-free system from blowfly salivary glands. *J. Biol. Chem.* 260: 5464

70. Smith, C. D., Lane, B., Kusaka, I., Verghese, M. W., Snyderman, R. 1985. Chemoattractant receptor-induced hydrolysis of phosphatidyl-inositol 4,5-bisphosphate in human polymorphonuclear leukocyte membranes. *J. Biol. Chem.* 260: 5875

71. Jackowski, S., Rettenmier, C. W., Sherr, C. J., Rock, C. O. 1986. A guanine nucleotide-dependent phosphatidylinositol 4,5-diphosphate phospholipase C in cells transformed by the *v-fms* and *v-fes* oncogenes. *J. Biol. Chem.* 261: 4978

72. Moolenaar, W. H., Tertoolen, L. G. J., deLatt, S. W. 1984. Phorbol ester and diaclyglycerol mimic growth factors in raising cytoplasmic pH. *Nature* 312: 371

73. Maino, V. C., Hayman, M. J., Crumpton, M. J. 1975. Relationship between enhanced turnover of phosphatidylinositol and lymphocyte activation by mitogen. *Biochem. J.* 146: 247

74. Monroe, J. G., Neidel, J. E., Cambier, J. C. 1984. B cell activation. IV. Induction of cell membrane depolarization and hyper I-A expression by phorbol diesters suggests a role for protein kinase C in murine B lymphocyte activation. *J. Immunol.* 132: 1472

75. Lindsten, T., Thompson, C. B., Finkleman, F. D., Andersson, B., Scher, I. 1984. Changes in the expression of B cell surface markers on complement receptor–positive and complement receptor–negative B cells induced by phorbol myristate acetate. *J. Immunol.* 132: 235

76. Nishizuka, Y. 1984. Turnover of inositol phospholipids and signal transduction. *Science* 21: 1365

77. Billah, M. M., Lapetina, E. G., Cuatrecasas, P. 1979. Phosphatidylinositol-specific phospholipase C of platelets: Association with 1,2-diacylglycerol-kinase and inhibition by cyclic-AMP. *Biochem. Biophys. Res. Comm.* 90: 92

78. Coggeshall, K. M., Cambier, J. C. 1984. B cell activation. VIII. Membrane immunoglobulins transduce signals via activation of phosphatidylinositol hydrolysis. *J. Immunol.* 133: 3382

79. Bijsterbosch, M. K., Meade, J. C., Turner, G. A., Klaus, G. G. B. 1985. B lymphocyte receptors and polyphosphoinositide degradation. *Cell* 41: 999

80. Ransom, J. T., Harris, L. K., Cambier, J. C. 1986. Anti-Ig induces release of inositol 1,4,5 trisphosphate which mediates mobilization of intracellular Ca^{++} stores in B lymphocyte. *J. Immunol.* 137: 708

81. Braun, J., Sha'afi, R. I., Unanue, E. R. 1979. Crosslinking by ligands to surface immunoglobulin triggers mobilization of intracellular $^{45}Ca^{++}$ in B lymphocytes. *J. Cell Biol.* 82: 755

82. Pozzan, T., Arslan, P., Tsien, R. Y., Rink, T. J. 1982. Anti-immunoglobulin, cytoplasmic free calcium, and capping in B lymphocytes. *J. Cell. Biol.* 94: 335

83. Tsien, R. Y. 1980. New calcium indicators and buffers with high selectivity against magnesium and protons: Design, synthesis and properties of prototype structures. *Biochemistry* 19: 2396

84. Grynkiewicz, G., Poenie, M., Tsien, R. Y. 1985. A new generation of Ca^{++} indicators with greatly improved fluorescence properties. *J. Biol. Chem.* 260: 3440

85. Ransom, J. T., DiGiusto, D. L., Cambier, J. C. 1985. Single cell analysis of calcium mobilization in anti-receptor antibody stimulated B lymphocytes. *J. Immunol.* 136: 54

86. Ransom, J. T., DiGiusto, D. L., Cambier, J. C. 1986. Flow cytometric analysis of intracellular calcium mobilization. In *Cellular Regulators: Calcium and Calmodulin Binding Proteins, Methods in Enzymology*, ed. A. Means, P. M. Conn. New York: Academic Press. In press

87. Ransom, J. T., Cambier, J. C. 1986. B cell activation. VII. Independent and synergistic effects of mobilized calcium and diacylglycerol on membrane potential and I-A expression. *J. Immunol.* 136: 66

88. Ransom, J., Cambier, J. C. 1987. Submitted

89. Ransom, J., Cambier, J. C. Unpublished observation

90. Deleted in proof

91. Chandy, K. G., DeCoursey, T. E., Cahalan, M. D., McLaughlin, C., Gupta, S. 1984. Voltage-gated potassium channels are required for human lymphocyte activation. *J. Exp. Med.* 160: 369

92. Lee, K. S., Tsien, R. W. 1983. Mechanisms of calcium channel blockage by verapamil, D600, diltiazem and nitrendipine in single dialysed heart cells. *Nature* 302: 790

93. Fukushima, Y., Hagiwara, S. 1983. Voltage-gated Ca^{++} channel in mouse myeloma cells. *Proc. Natl. Acad. Sci. USA* 80: 2240

94. Fukushima, Y., Hagiwara, S. 1985. Currents carried by monovalent cations through calcium channels in mouse neoplastic B lymphocytes. *J. Physiol.* 358: 255

95. DeCoursey, T. E., Chandy, K. G., Gupta, S., Cahalan, M. D. 1984. Voltage-gated K^+ channels in human T lymphocytes: A role in mitogenesis? *Nature* 307: 465

96. Matteson, D. R., Duetsch, C. 1984. K channels in T lymphocytes: A patch clamp study using monoclonal antibody adhesion. *Nature* 307: 468

97. Chen, Z. Z., Coggeshall, K. M., Cambier, J. C. 1986. Translocation of protein kinase C during membrane immunoglobulin-mediated transmembrane signaling in B lymphocytes. *J. Immunol.* 136: 2300

98. Lichtman, A. H., Segel, G. B., Lichtman, M. A. 1981. Calcium transport and calcium-ATPase activity in human lymphocyte plasma membrane vesicles. *J. Biol. Chem.* 256: 6148

99. Ganong, B. R., Long, C. R., Hannun, Y. A., Bell, R. M. 1986. Specificity and mechanisms of protein kinase C activation by sn-1, 2-diacylglycerol. *Proc. Natl. Acad. Sci. USA* 83: 1184

100. Kraft, A. S., Anderson, W. B. 1983. Phorbol esters increase the amount of Ca^{++}, phospholipid-dependent protein kinase associated with plasma membrane. *Nature* 301: 621

101. Kraft, A. S., Anderson, W. B., Cooper, L., Sando, J. J. 1983. Decrease in cytosolic calcium/phospholipid-dependent

protein kinase activity following phorbol ester treatment of EL4 thymoma cells. *J. Biol. Chem.* 257: 13198

102. Grupp, S. A., Harmony, J. A. K. 1985. Increased phosphatidylinositol metabolism is an important but not an obligatory early event in B lymphocyte activation. *J. Immunol.* 134: 4087

103. Wightman, P., Raetz, C. R. H. 1984. The activation of protein kinase C by biologically active lipid moieties of lipopolysaccharide. *J. Biol. Chem.* 259: 10048

104. Nel, A. E., Wooten, M. W., Landreth, G. E., Goldshmidt-Clermont, P. J., Stevenson, H. C., Miller, P. G., Galbraith, R. M. 1986. Translocation of phospholipid/Ca^{++}-dependent protein kinase in B lymphocytes activated by phorbol ester or crosslinking of membrane immunoglobulin. *Biochem. J.* 233: 145

105. Nishizuka, Y. 1984. The role of protein kinase C in cell surface signal transduction and tumor promotion. *Nature* 308: 693

106. Wolf, M., Levine, H. III, May, W. S., Cautrecasas, P., Sahyoun, N. 1985. A model for intracellular translocation of protein kinase C involving synergism between Ca^{++} and phorbol ester. *Nature* 317: 546

107. Hornbeck, P., Paul, W. E. 1986. Anti-immunoglobulin and phorbol ester induce phosphorylation of proteins associated with plasma membrane and cytoskeleton in murine B lymphocytes. *J. Bio. Chem.* In press

108. Nel, A. E., Navailles, M., Rosberger, D. F., Landreth, G. E., Goldschmidt-Clermong, P. J., Baldwin, G. J., Galbraith, R. M. 1985. Phorbol ester induces tyrosine phosphorylation in normal and abnormal human B lymphocytes. *J. Immunol.* 135: 3448

109. Giulmore, T., Martin, G. S. 1983. Phorbol ester and diacylglycerol induce protein phosphorylation at tyrosine. *Nature* 306: 487

110. Monroe, J. C. Cambier, J. C. 1983. B cell activation. I. Receptor crosslinking by anti-immunoglobulin antibodies induces a rapid decrease in B cell plasma membrane potential. *J. Exp. Med.* 157: 2073

111. Chandy, K. G., DeCoursey, T. E., Cahalan, M. D., Gupta, S. 1985a. Ion channels in lymphocytes. *J. Clin. Immunol.* 5: 1

112. DeCoursey, T. E., Chandy, K. G., Gupta, S., Cahalan, M. D. 1986. Vol-

tage-dependent ion channels in T-lymphocytes. *J. Neuroimmunol.* In press

113. Gallin, E. K. 1986. Ionic channels in leukocytes. *J. Leukocyte Biol.* 39: 241

114. Ransom, J., Cambier, J. C. 1986. Loss of cell K^+ and Ca^{++} influx in anti-immunoglobulin stimulated B cells are interdependent. *J. Immunol.* In press

115. Rosoff, P. M., Stein, L. F., Cantley, L. C. 1984. Phorbol esters induce differentiation in a pre-B lymphocytic cell line by enhancing Na^+/H^+ exchange. *J. Biol. Chem.* 259: 7056

116. Rosoff, P. M., Cantley, L. C. 1983. Increasing the intracellular Na^+ concentration induces differentiation in a pre-B lymphocyte cell line. *Proc. Natl. Acad. Sci. USA* 80: 7547

117. Grinstein, S., Cohen, S., Goetz, J. D., Rothstein, A. 1985a. Na^+/H^+ exchange in volume regulation and cytoplasmic pH homeostasis in lymphocytes. *Fed. Proc.* 44: 2508

118. Cheung, R. K., Grinstein, S., Dosch, H.-M., Gelfand, E., W. 1982. Volume regulation by human lymphocytes: Characterization of the ionic basis for regulatory volume decrease. *J. Cell. Physiol.* 112: 189

119. Cheung, R. K., Grinstein, S., Gelfand, E. W. 1982. Volume regulation by human lymphocytes. Identification of differences between the two major lymphocyte subpopulations. *J. Clin. Invest.* 70: 632

120. Gelfand, E. W., Cheung, R. K., Grinstein, S. K. 1984. Volume regulation in lymphoid leukemia cells and assignment of cell lineage. *N. Engl. J. Med.* 311: 939

121. Grinstein, S., Clarke, C. A., DuPre, A., Rothstein, A. 1982. Volume-induced increase of anion permeability in human lymphocytes. *J. Gen. Phys.* 80: 801

122. Grinstein, S., Clarke, A., Rothstein, A., Gelfand, E. W. 1983. Volume-induced anion conductance in human B lymphocytes is cation independent. *Am. J. Physiol.* 245: C160

123. Kiefer, H., Blume, A. J., Kaback, H. R. 1980. Membrane potential changes during mitogenic stimulation of mouse spleen lymphocytes. *Proc. Natl. Acad. Sci. USA* 77: 2200

124. Jakway, J. P., Defranco, A. L., 1986. Biochemical signaling in B cell and macrophage cell lines stimulated with lipopolysaccharide. *Proc. 6th Intl. Cong. Immunol.* 2.65.36 (Abstr.)

125. Cambier, J. C., Ransom, J. T. Harris,

L. K., Coggeshall, K. M., Chen, Z. Z., Newell, M. K., Justement, L. B. 1986. Coupling of B cell surface Ig, Ia, and BSF1 receptors to intracellular "Second Messengers." In *Proceedings of the Conference, Lymphocyte Activation and Immune Regulation*, ed. S. Gupta, W. E. Paul, A. S. Fauci. New York: Plenum Press. In press

126. Chen, Z. Z., Cambier, J. C. 1986. Transmembrane signaling through B cell MHC class II molecules: Anti-Ia antibodies induce protein kinase C translocation to the nuclear fraction. *J. Immunol.* In press

Ann. Rev. Immunol. 1987. 5 : 201–222

SPECIFIC CELL-ADHESION MECHANISMS DETERMINING MIGRATION PATHWAYS OF RECIRCULATING LYMPHOCYTES

Judith J. Woodruff and Lorraine M. Clarke

Department of Pathology, State University of New York, Health Science Center at Brooklyn, Brooklyn, New York 11203

Yee Hon Chin

Department of Microbiology and Immunology, University of Miami School of Medicine, Miami, Florida 33101

INTRODUCTION

Recirculating lymphocytes traffic continuously between blood and lymph moving by a route that takes them through various lymphoid organs in the body. Though the cells circulate freely in the bloodstream and have opportunity for interaction elsewhere, they are not promiscuous but faithfully seek out those blood vessels that provide entry into these tissues (1–15).

Lymphocytes gain access to lymph nodes and mucosal-associated lymphoid tissues, e.g. Peyer's patches, via a system of specialized microvascular structures, the high endothelial cell venules (HEV) (4). Such interactions are crucial for recirculation and reassortment of lymphocytes because the vascular endothelium is the first barrier these cells must negotiate as they move from blood into tissue. Prime importance is therefore attached to the understanding of lymphocyte–high endothelial cell interactions. This review thus focuses on properties of lymphocytes and high endothelium that play a role in their affinity and specificity for each other.

201

0732–0582/87/0410–0201$02.00

Other aspects of lymphocyte recirculation have ben reviewed elsewhere (8, 10–15) and are only briefly considered here.

PHYSIOLOGIC CHARACTERISTICS OF LYMPHOCYTE RECIRCULATION

The lymphoid system is composed of fixed, distinct organs widely distributed throughout the body and circulating lymphocytes moving through blood, lymph, and interstitial fluids. Despite this diffuseness, the immune system is coordinated and functions with the precision of a single organ. The orderly movement of circulating lymphocytes between these compartments provides a means for functional integration of the system at the cellular level. This is brought about by the blood vascular and lymphatic systems which physically interconnect all peripheral lymphoid organs, i.e. lymph nodes, spleen, and mucosal-associated lymphoid tissues (gut and bronchial) (14, 16). At the same time, however, the endothelia of these vessels are barriers that effectively separate one organ from another. The bridging of this barrier by the migration of lymphocytes is a physiologic event which imparts to lymphocyte-endothelial cell interactions a crucial role in determining T- and B-cell distribution and consequently the immune responsiveness of the individual.

Analysis of mechanisms underlying lymphocyte recirculation must therefore focus on the cardiovascular system. Blood flow rates and the concentration of lymphocytes in blood determine the number of cells delivered to lymphoid organs. It is estimated that one in four of the lymphocytes delivered to a node in the bloodstream migrates across the endothlium into the tissue. This migration follows "first order" kinetics, that is, the rate of migration is a linear function of the concentration of lymphocytes in the blood perfusing the organs (10, 17–19). Since large-scale migration does not occur in other tissues, lymphoid organs must contain a specialized endothelium which enables lymphocytes to migrate with speed and precision across the vessel wall into the parenchyma. The vessel wall therefore serves the function of maintaining lymphocytes within the intravascular compartment and restraining their movement into non-lymphoid tissues. Direction to lymphocyte migration is imparted through construction of a modified endothelium in lymphoid tissues. In most mammals this modified endothelium is composed of polygonal cells with columnar or cuboidal morphology; these "high" endothelial cells are readily differentiated from the typical flat endothelia which line all other vessels (4, 20–22). However, the columnar/cuboidal morphology is not a requirement for this function of the modified endothelium; lymphocytes recirculate through lymph nodes in sheep even though high endothelial cell venules are not present in these organs (15, 23).

Early investigators noted that lymphocytes enter the blood from major lymphatic ducts, the largest being the thoracic duct, in numbers sufficient to replace the blood content several times daily (24, 25). Because the level of lymphocytes in the blood remains fairly constant, it was apparent that the lymphocytes must leave the blood in equal numbers. Lymphocyte destruction was initially thought to account for this observation (24, 26), and a very high rate of lymphocyte turnover was postulated. The first indication that lymphocyte recirculation accounts for these findings was provided by the work of Gowans (1–4). He demonstrated that chronic thoracic duct drainage in rats produces a severe reduction in the number of lymphocytes emerging in lymph. This depletion is prevented by rein-fusion of thoracic duct lymphocytes (TDL) intravenously into the can-nulated donor rat. It was then shown that the vast majority of TDL are not newly formed cells (6) and that radiolabeled TDL injected intravenously reappear several hours later in lymph (2) and lymphoid tissues (4) of recipient animals. By these findings, Gowans established the principle of lymphocyte recirculation and identified a major physiologic function of lymph circulation. It is now known that the recirculating pool of lymphocytes is composed of small T and B cells. They predominate in the circulation and comprise a substantial portion of cells in peripheral lymphoid tissues, but relatively few are in thymus and bone marrow (11, 12, 27–30).

The early work on lymphocyte recirculation in the rat led to the view that repeated passage of lymphocytes from blood to lymph through lymph nodes and Peyer's patch resulted in a complete mixing of the recirculating pool. This arrangement would mean that the composition of cells in thoracic duct lymph would be the same as that in efferent lymph draining each lymphoid tissue. By implication, small lymphocytes belonging to the recirculating pool would be uniform in their behavior. If this were the case, cells would not have an intrinsic preference for entering spleen or a particular group of nodes. Their distribution would depend on two factors: (a) delivery to an organ in the bloodstream and (b) capacity to cross the specialized endothelium. Analysis of recirculation in sheep, however, showed that the system is not entirely uniform. In these animals, efferent lymphatics draining peripheral lymph nodes (7, 31), the intestine (32), or granulomas (33) can be cannulated individually, and the circulation of lymphocytes through each tissue can be measured seperately and sim-ultaneously. Such studies have shown that there are three subsets of lym-phocytes that recirculate via different pathways. Lymphocytes from intestinal lymph recirculate preferentially through the intestine, whereas lymphocytes from efferent lymph of peripheral lymph nodes show preferential recirculation through these tissues (34). A third population exhibits preferential migration through inflammatory sites (33). These preferences, however, are not absolute, and many cells have the capacity to recirculate by all routes.

In rodents, lymph is collected by cannulating the abdominal portion of the thoracic duct (1). Initially most cells are derived from gut-associated tissues (particularly Peyer's patch and mesenteric lymph nodes), but within several hours, lymph collections also contain lymphocytes derived from peripheral lymph nodes. Lymph would therefore be expected to be a mixture of both subsets if such distinct populations, in fact, exist in rodents. It has not, however, been possible using adoptive transfer techniques to identify in these animals the populations that recirculate by different routes (29, 35). Recently, in studies described later in this review, experiments employing antibodies to high endothelial binding factors (HEBF) have shown that rat TDL can be fractionated into populations that exhibit affinity for either HEV of lymph nodes or HEV of Peyer's patch, thus suggesting that such subsets do exist in rodents (Y. H. Chin, J. J. Woodruff, unpublished observations). The predominant subset appears to be a population composed of cells expressing adhesion molecules for both lymph node and gut-associated high endothelial cell types which is therefore capable of recirculating via either organ.

Other experiments have examined whether spleen-seeking or lymph node-seeking subsets comprise the recirculating pool (29). In double transfer experiments in which cell localization was determined 30 min after intravenous injection, the TDL transferred from the spleen of an intermediate host showed no greater localization in the spleen of the final recipient than did TDL transferred from lymph nodes; corresponding TDL transferred from lymph node showed no preferential reentry into the nodes. Thus, these results provided no evidence of a spleen-seeking or lymph node-seeking population in the recirculating lymphocyte pool.

Interactions of lymphocyte and high endothelial cells in intact animals have been studied by determining the distribution of labeled lymphocytes injected intravenously into syngeneic animals (11, 12, 30). Donor lymphocytes rapidly leave the blood and begin to appear in lymph nodes and Peyer's patch within minutes after transfer. In peripheral nodes the arrival of intravenously injected lymphocytes is almost entirely via the blood and therefore involves migration across the high endothelium. In central nodes (mesenteric, celiac) substantial numbers of lymphocytes also enter via afferent lymphatics draining the intestine and liver. During the first 2 hours after transfer, however, donor cell entry into central nodes is almost entirely via HEV (30). Because of this arrangement, measurements are made soon after injection in order to assess specifically the donor cell entry into lymph node and Peyer's patch occurring through HEV; this results in minimal contributions from cells that had passaged through other organs, such as the spleen, or were delivered to the tissue by afferent lymph, in the case of central nodes.

Lymphocytes that physiologically interact with the vessel wall are those in the blood or about to enter the blood via lymph. Importance is thus attached to the use of lymphocytes from the blood or lymph streams for such studies (10, 36). When lymphocytes from teased lymph nodes and spleen are utilized, the suspensions are a mixture of both recirculating and nonrecirculating populations. These populations may reasonably be expected to have different affinities for specialized endothelium and may interact with the endothelium by different mechanisms. Furthermore, direct evidence now suggests that high endothelial cell–binding properties of recirculating lymphocytes are altered as the cells migrate across the vessel wall and enter lymph nodes and spleen (29). In double transfer studies, Ford showed that radiolabeled TDL, injected intravenously and reisolated from the lymph nodes and spleen of intermediate recipients, were markedly impaired in their ability to interact with HEV and to reenter lymph nodes of final recipients. No impairment was observed with donor TDL which has been reisolated from blood and lymph of intermediate hosts. Thus, even authentic recirculating lymphocytes show abnormal patterns of high endothelial reactivity when the cells are isolated from these organs. It appears from these results that recirculating lymphocytes temporarily lose their ability to adhere to high endothelium when the cells emigrate from the bloodstream.

Another technical consideration is that the results of experiments may be affected by methods used to prepare lymphocytes (37). Incubation of the cells for short periods in the cold has been shown to impair their ability to migrate across HEV. This treatment has no effect on the ability of lymphocytes to migrate into the spleen, and it does not promote localization of the cells in the liver. Because of such variables, a profile of the distribution of donor lymphocytes in all peripheral lymphoid organs and in highly vascularized tissues as well (i.e. liver, lung) is essential for critical analysis of results.

FUNCTION OF HIGH ENDOTHELIAL CELL VENULES IN LYMPHOCYTE RECIRCULATION

The first step in entry of blood lymphocytes into lymph nodes is the binding of these cells to venules lined by high endothelium. The cells enter only by crossing HEV and not by migrating through other vascular endothelia (4, 11). Moreover, only lymphocytes accumulate at HEV, whereas other cells in the bloodstream, such as granulocytes, do not normally migrate through these vessels. A tenfold increase in the number of lymphocytes passing through the lymph node per second can be induced by antigenic stimulation. Even under these conditions, specificity is maintained; granulocytes as well as other nonrecirculating cells in the bloodstream are excluded (38).

HEV are specialized microvascular structures which were first described in 1898 (20). The vessel wall is composed of one or two endothelial cell layers, a thick basement membrane, and a prominent perivascular sheath (22). Typically, large numbers of lymphocytes are present in the vessel lumen and within the various layers of the wall. HEV are found only in mammals and develop during the neonatal period in lymph nodes, tonsils, gut-associated (Peyer's patch and appendix) and bronchial-associated lymphoid tisues (23). HEV are found throughout the cortex and interfollicular regions, but they are not found in germinal centers or medulla (4, 22). These vessels may arise directly from postcapillary structures, but very commonly they are derived from an intervening segment where the venules are lined by flat-type endothelium (39). No HEV are present in thymus and bone marrow, and they are not normally found in nonlymphoid organs. They do appear in other sites but only in association with chronic lymphocytic infiltrates (40).

The high endothelium is formed by cuboidal/columnar cells which have distinct morphologic features, including abundant cytoplasm, large pale nuclei, and 'a dense nucleolus. The cells are linked together by macular tight junctions located near their luminal surfaces. Most high endothelial cells contain prominent Golgi, fine and clustered ribosomes, and mitochondria (41, 42). They have distinct histochemical properties including nonspecific esterase activity not found in flat-type endothelia (21). High endothelial cells also selectively incorporate ^{35}S-sulphate and secrete sulphated glycolipid in association with a glycoprotein (43–45).

Scanning electron microscopic studies have shown that lymphocytes within the HEV lumen have numerous microvilli and attach to endothelial cells via these projections (41, 42, 46). Microvilli are retracted as lymphocytes penetrate the vessel wall (41) by migrating between and not through endothelial cells (41, 47, 48).

It is not known what properties of the high endothelium are responsible for its function in lymphocyte migration. The height of the endothelium or the bulging of these cells into the lumen do not seem to be important for recognition events. Thus, the specialized vascular endothelium of sheep lymph nodes lacks high endothelial cells (15). Similarly, there is a marked decrease in the height of the endothelium of HEV in nude mice (49) and rats (50). In both situations there are no defects in lymphocyte migration into tissue (27, 28, 49, 50).

Cell recognition mechanisms mediating lymphocyte interaction with the high endothelium have been investigated in several laboratories using an in vitro adherence assay (51). The distinct morphology of high endothelial cells in rat and mouse lymph node sections made it possible to devise a system whereby their interaction with lymphocytes could be studied in vitro. The high endothelium is exposed in such sections, and lymphocytes deposited over the tissue can then be assessed for their capacity to bind these cells.

Adherence of Lymphocytes to High Endothelium In Vitro

Lymphocytes adhere specifically to high endothelial cells when overlaid onto frozen sections of peripheral and mesenteric lymph nodes, Peyer's patch, tonsils, and bronchial-associated tissues (51–67). Binding occurs rapidly at 7°C with maximal levels of adherence found in 15–30 min (52). The results are quantitated directly by counting either the number of HEV per section with bound lymphocytes (51, 52) or the number of lymphocytes bound per HEV (52, 57, 67). Under optimal conditions lymphocytes bind to 80% or more of HEV in each section. About 85% or more of lymphocytes adherent to the tissue section are bound to the high endothelium which comprises only about 1–2% of the total area of the tissue section (52). Lymphocytes bind to HEV in unfixed (56, 57), glutaraldehyde-fixed (51) and paraformaldehyde-fixed (68) sections. The same pattern of adherence is observed with these different preparations, but the sensitivity of the system is increased tenfold by use of fixed HEV (56). In vitro HEV adherence is a property of T and B cells and has been observed with rat (51, 68, 69), mouse (51, 56, 57), human (54, 59), guinea pig (60, 63), and rabbit lymphoid cells (63). Binding has also been observed with human (64) but not mouse (65) T-cell clones. Reciprocal binding studies (54; see below) have not supported the claim of species restriction in lymphocyte-HEV interactions (61).

The in vitro system exhibits the dual specificity with regard to vascular endothelium and lymphoid cell type which characterizes lymphocyte migration interactions occurring under physiological conditions in rodents (51, 61). Nearly all thoracic duct lymphocytes (TDL) are recirculating cells, and substantial numbers of these cells are also present in lymph nodes and spleen. Comparable levels of HEV adherence are observed in sections overlaid with cells from these compartments. In contrast, thymus and bone marrow are deficient in cells with this capability; the extent of HEV binding in sections overlaid with these cell populations is 10% or less of that observed with thoracic duct cells. With respect to endothelial cell specificity, adhesion is only to the high endothelium, and overlaid lymphocytes are rarely found in subcapsular sinuses. They do not adhere to capillaries, lymphatic channels, or nonspecialized venules in the cortex or medulla. We concluded from this work that: (a) High endothelial binding lymphocytes predominate in peripheral lymphoid tissues and are part of the recirculating pool of cells; (b) Acquisition of high endothelial binding properties appears to be a differentiation event; (c) Specific surface molecules of lymphocytes mediate their adherence to high endothelium (51).

Additional studies indicate, however, that lymphocyte binding is not a passive consequence of the interaction of these molecules with high endothelial cell determinants (53). Formation of a stable complex with the endothelium involves activities of certain cytoplasmic components of lymphocytes. Thus, binding requires viable lymphocytes, is dependent on calcium but not magnesium ions, and is inhibited by azide and cytochalasin B but not by colchicine. In addition, trypsin treatment of lymphocytes destroys the HEV binding ability of the cells. We interpret this to mean that microfilaments play a role in adhesion and that binding is mediated by cell surface molecules that are integral membrane proteins.

It is now clear that high endothelia of peripheral lymph nodes and Peyer's patch differ with respect to the specificity of lymphocyte adherence. Investigation of HEV binding properties of mouse lymphoma cells showed that certain lymphoma cell lines have the capacity to bind with nearly absolute specificity to HEV of peripheral lymph node (HEV_{LN}) or HEV of Peyer's patch (HEV_{PP}) (58). There is also evidence that some Peyer's patch and lymph node lymphocytes exhibit a preference for binding to HEV_{PP} and HEV_{LN}, respectively (58, 60), and that B cells exhibit a preference for HEV_{PP} and T cells, for HEV_{LN} (60, 62). Differences in recognition structures appear to mediate these observations; each interaction can be selectively blocked by antibodies against specific lymphocyte surface molecules. With rat lymphocytes, antibodies to cell surface $HEBF_{LN}$ block binding to HEV_{LN} but not HEV_{PP} (70, 71), whereas antibodies to cell surface $HEBF_{PP}$ block binding to HEV_{PP} but not HEV_{LN} (72). With mouse lymphocytes, the antibody designated MEL-14 blocks adhesion to HEV_{LN} but not HEV_{PP} (73). In a later section we present evidence with human lymphocytes that antibodies to cell surface $HEBF_{LN}$ block binding to HEV_{PLN} but not HEV_{MLN}. Further evidence of high endothelial cell heterogeneity involving lymphocyte attachment sites has been found in studies in which soluble HEBF exhibited specificity for rat HEV_{LN} or HEV_{PP} (71, 72, 74). Similarly, sialidase treatment destroyed lymphocyte binding sites of HEV_{LN} but had no effect on binding sites of HEV_{PP} (67).

LYMPHOCYTE–HIGH ENDOTHELIAL CELL RECOGNITION MECHANISMS

The idea that specific cell surface adhesion molecules mediate lymphocyte interaction with high endothelium originated from studies by Gesner (75–77) which examined the distribution of intravenously injected radiolabeled lymphocytes treated with enzymes known to alter surface membrane components. Thus, lymphocytes treated with glycosidase (75) or neuraminidase (76) show reduced entry into lymph nodes of recipient animals. This change occurs because donor cells are rapidly removed from the circulation in the liver; these enzymes have no direct effect on lymphocyte surface molecules for high endothelial cell adhesion (78). By contrast, trypsin treatment of lymphocytes (77) prevents their entry into lymph nodes in vivo and adhesion to HEV in vitro. This effect is specific; trypsin-treated lymphocytes accumulate normally in the spleen and show no increased retention in liver or lung. Thus, surface membrane changes can prevent lymphocyte entry into lymph nodes by two entirely different mechanisms: (a) alteration in specific adhesion molecules required for HEV binding or, (b) alterations which promote lymphocyte removal from the circulation and thereby decrease delivery of the cells to lymph nodes.

The evidence that treatment of lymphocytes with vasoactive intestinal peptides inhibits their migration into mesenteric lymph node and Peyer's patch but not into spleen raises the possibility that hormone receptors might exert a regulatory role in lymphocyte-HEV interactions, possibly by affecting expression or function of cell surface adhesion molecules (79).

It has been proposed (68, 69) that a lymphocyte cell surface carbohydrate-binding receptor with specificity for mannose-6-phosphate is involved in lymphocyte attachment. Two polysaccharides—a phosphomonoester mannan fragment containing only mannose and mannose-6-phosphate and fucoidin, a sulfated polysaccharide rich in L-fucose—are potent inhibitors of lymphocyte attachment to HEV_{LN}; other glycoconjugates have little or no inhibitory activity. Additional studies showed that fucoidin inhibits lymphocyte migration across HEV in vivo (80).

Two approaches have been used to develop antibodies against specific cell surface molecules of lymphocytes involved in HEV adhesion (70–73). One approach has involved production of a rat monoclonal antibody, MEL-14, by immunization with a cloned C3H/eb mouse B-cell lymphoma; these cells have the capacity to bind in vitro to HEV_{LN} but not to HEV_{PP} (73, 81). When mouse mesenteric node cells were pretreated with MEL-14, washed and overlaid onto tissue sections, adherence to HEV_{LN} was almost completely blocked, without significantly altering their binding to HEV$_{PP}$. Similarily, when labeled mesenteric lymph node lymphocytes treated with MEL-14 were injected intravenously, recovery in peripheral lymph nodes was inhibited, but recovery in Peyer's patch was essentially unaffected. These results suggest that MEL-14 blocks surface molecules for HEV adhesion, but it is not known if this change is the only relevant variable. Thus, it is not known if MEL-14 treatment promotes emigration of the cells into the liver or lung. In addition, it is yet to be determined if MEL-14 treatment inhibits lymphocyte migration into the spleen, a property not altered by reagents that specifically block membrane molecules involved in HEV adherence. MEL-14 recognizes a cell surface component of 85,000 to 95,000 daltons which is reported to be a ubiquitinated protein (82).

The approach undertaken in this laboratory was based on our hypothesis that surface molecules of lymphocytes interact with complementary components on the high endothelium. This lymphocyte surface recognition structure has been termed high endothelial binding factor or HEBF and its function studied through production of antibodies to soluble forms isolated from lymph (83, 84).

The strategy of these experiments was based on the idea that soluble HEBF (or fragments of these molecules) would have the capacity to bind to HEV when overlaid onto lymph node or Peyer's patch sections. As a result, the high endothelial cell structures that are the sites for lymphocyte attachment would be blocked, and HEV adhesion of lymphocytes overlaid onto these sections would be inhibited. This approach has led to the detection of 2 species of soluble HEBF (74). Material with specificity for HEV of lymph nodes is designated HEBF$_{LN}$ and that with specificity for HEV of Peyer's patch, HEBF$_{PP}$.

In preliminary experiments soluble HEBF$_{LN}$ was detected in supernatants of cultured rat TDL suggesting that the cells might spontaneously shed these adhesion molecules. This finding led to studies which showed that soluble HEBF is a component of rat thoracic duct lymph (74, 83, 84) and that rabbit antibodies to this material are directed against surface molecules of lymphocytes involved in HEV adhesion (70). Thus, rat TDL treated with rabbit anti-HEBF$_{LN}$ antibody, washed, and overlaid onto lymph node sections show an 80–90% reduction in HEV adhesion. Fab fragments of this antibody are also active, which rules out the possibility that the effect is Fc-mediated or is produced by cocapping of lymphocyte surface molecules. The antibody apparently does not act by blocking HEV sites, because binding is not reduced when lymph node sections are pretreated with anti-HEBF IgG, washed, and overlaid with untreated TDL. These in vitro findings are relevant to physiologic mechanisms of recirculation because treatment of radiolabeled TDL with rabbit anti-

HEBF$_{LN}$ Fab blocks entry of donor cells into lymph nodes (cervical and axillary) but not into Peyer's patches of recipient animals. Donor cells migrate normally into the spleen and show no increased tendency to accumulate in liver or lung. These results suggested that lymphocyte adhesion to lymph node and Peyer's patch high endothelium are mediated by distinct recognition structures. These findings also imply that migration into the spleen is mediated by an unrelated mechanism.

Properties of Soluble Rat HEBF$_{LN}$ and HEBF$_{PP}$

HEBF$_{LN}$ was originally isolated from lymph using a purification scheme in which the factor was identified by its ability to inhibit TDL binding to HEV$_{LN}$ (70, 74, 83, 84). Subsequent studies showed that HEBF$_{PP}$ is also a component of lymph (74) and that these factors can be physically separated from each other by affinity chromatography using rabbit anti-HEBF$_{LN}$ antibody (Table 1). Both factors appear to be glycoproteins (74, 84).

The effects of soluble HEBF$_{LN}$ and HEBF$_{PP}$ are produced by functional alterations in high endothelial cells and not by changes in adhesive properties of lymphocytes. Thus, when tissue sections are treated with factor, washed, and overlaid with untreated TDL, adhesion is inhibited. In contrast, HEV binding is not blocked when TDL treated with HEBF$_{LN}$ are washed and overlaid onto untreated sections (83; Y. H. Chin, J. J. Woodruff, unpublished observations). Our interpretation of these results is that the soluble factors achieve their effects by blocking high endothelial cell components which are lymphocyte attachment sites. It is unlikely that this occurs because HEBF enzymatically alters these sites; the effects are not

Table 1   Isolation of HEBF$_{LN}$ and HEBF$_{PP}$ from rat lymph proteins by affinity chromatography using rabbit anti-HEBF$_{LN}$ Sepharose 4B[a]

Fraction	Section pretreated	Inhibition HEV binding (%)	HEBF designation
Eluted	LN	85	HEBF$_{LN}$
	PP	0	
Unbound	LN	2	HEBF$_{PP}$
	PP	90	

[a] Lymph depleted of cells and chylomicra was fractionated with $(NH_4)_2SO_4$; material precipitating between 40% and 60% saturation was dialyzed against tris (0.05M, pH 8.3) and applied to the antibody column. Unbound and NaSCN-eluted fractions were obtained as described (74) and used for section pretreatment at 850 μg/ml.

temperature-dependent, and no increase in activity is produced by treatment of sections at 37°C rather than at 4°C (Y. H. Chin, J. J. Woodruff, unpublished observations). It is also unlikely that HEBF nonspecifically blocks these sites because $HEBF_{LN}$ does not inhibit TDL binding to HEV_{PP}, and $HEBF_{PP}$ does not inhibit TDL binding to HEV_{LN} (74). We interpret these results to mean that $HEBF_{LN}$ has affinity for lymphocyte attachment sites of lymph node high endothelium and that $HEBF_{PP}$ has affinity for attachment sites of Peyer's patch high endothelium. The implication of this finding is that endothelial cells of lymph node and Peyer's patch differ with respect to the specificity of surface molecules involved in lymphocyte migration.

Properties of Monoclonal Anti-Rat $HEBF_{LN}$ (A.11) and Anti-Rat $HEBF_{PP}$ (1.B2) Antibodies

Monoclonal antibodies to cell surface $HEBF_{LN}$ (A.11) and $HEBF_{PP}$ (1.B2) were developed by immunization of BALB/c mice with soluble factors isolated from rat lymph (71, 72). Indirect immunofluorescent studies showed that each antibody bound to 50–60% of rat TDL. The 1.B2 antibody immunoprecipitated a surface radioiodinated protein of TDL with an apparent molecular weight of 80,000 daltons (72) and the A.11 antibody, 3 polypeptides with apparent molecular weights of approximately 135,000, 60,000, and 40,000 (71). In addition, these antibodies detected antigen expressed on a large proportion of peripheral lymphocytes (35–50% lymph node, spleen, and Peyer's patch cells) but on relatively few thymus or bone marrow cells ($< 5\%$).

TDL surface HEBF function is specifically inhibited by these antibodies. When TDL are incubated for 30 min at 37°C with A.11 antibody, washed, and overlaid onto lymph node or Peyer's patch sections, lymphocyte binding to HEV_{LN} is inhibited, but adhesion to HEV_{PP} is not blocked (71). The opposite pattern of inhibition is produced by treatment of TDL with 1.B2 antibody. This antibody blocks TDL binding to HEV_{PP} but not to HEV_{LN} (72). Monoclonal antibodies to other cell surface determinants are not inhibitory (Figure 1).

The physiologic role of TDL surface molecules recognized by A.11 and 1.B2 antibodies was demonstrated in adoptive transfer experiments employing ^{51}Cr-labeled lymphocytes. The results showed that lymphocyte entry into lymph nodes or Peyer's patch could be blocked specifically by these monoclonal antibodies (Figure 2). Thus, treatment of radiolabeled TDL with A.11 antibody markedly reduced accumulation of injected cells into lymph nodes but did not interfere with lymphocyte entry into Peyer's patch. In addition the antibody had no effect on lymphocyte migration into spleen, liver, or lung. When TDL were treated with 1.B2 antibody,

TDL TREATMENT   TDL-HEV ADHERENCE

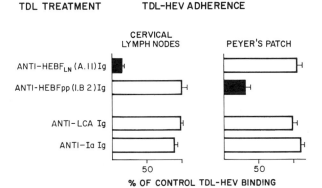

Figure 1 Role of TDL surface HEBF in HEV adhesion. TDL were treated with saturating amounts of the indicated monoclonal antibodies or nonimmune mouse Ig (37°C, 30 min), washed and overlaid onto tissue sections. Results were calculated as described (71).

the migration of injected cells into Peyer's patch was blocked, but accumulation of these cells in lymph nodes was unaffected. Further, 1.B2 antibody did not inhibit lymphocyte migration into the spleen or promote migration into the liver or lungs. Our interpretation of these results is that cell surface HEBF$_{LN}$ mediates entry into peripheral lymph nodes and cell surface HEBF$_{PP}$, entry into Peyer's patch. These cell surface molecules are not recognition structures for lymphocyte entry into spleen.

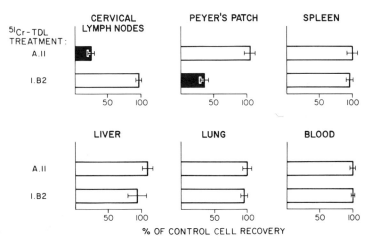

Figure 2 Role of TDL surface HEBF in migration. Accumulation of radioactivity in organs determined 2 hr after transfer of ^{51}Cr-labeled TDL treated in vitro with A.11, 1.B2, or nonimmune (control) IgG. Results calculated as in reference (70).

The properties of molecules recognized by these antibodies were examined in experiments where the A.11 and 1.B2 antigens were isolated and tested for the capacity to block lymphocyte binding sites of HEV_{LN} and HEV_{PP}, respectively (71, 72). The antigens were isolated from lymph proteins by affinity chromatography using A.11 or 1.B2 IgG coupled to Sepharose 4B beads. Lymph node or Peyer's patch sections were treated with A.11 or 1.B2 antigen, washed, and then overlaid with untreated TDL. The A.11 antigen blocked lymphocyte binding sites of HEV_{LN} and thereby caused a marked reduction in TDL-HEV adhesion as compared to sections treated with control antigen, i.e. material eluted from a nonimmune mouse-Ig Sepharose-4B column. The effect of the A.11 antigen was specific; it did not block binding sites of HEV_{PP}. In contrast to these results, pretreatment of sections with 1.B2 antigen prevented TDL binding to HEV_{PP} but not HEV_{LN}.

Further studies examined HEBF activity of A.11 and 1.B2 antigens isolated directly from detergent solubilized cells. For this purpose, lymphocytes were resuspended briefly in a lysis buffer, dialyzed against PBS to remove free detergent, and then chromatographed using anti-$HEBF_{LN}$ (A.11) or anti-$HEBF_{PP}$ (1.B2) antibody coupled to Sepharose 4B beads. Bound molecules were eluted with NaSCN, dialyzed against PBS, concentrated, and used to pretreat tissue sections. Earlier studies (85) had shown that biologically active HEBF could be recovered from detergent lysates of TDL by antibody affinity chromatography and that comparable amounts of biologically active HEBF are obtained from lysates of TDL, lymph node, and spleen cells (each at 10^8 cell equivalents per ml). With lysates of thymus or bone marrow, recovery of HEBF is at least tenfold less. No HEBF is recovered from lysates of trypsinized TDL, indicating that HEBF harvested from lysates of nontrypsinized cells is derived from molecules expressed at the cell surface. This approach was therefore used to characterize antigens recognized by the monoclonal antibodies. Results showed that TDL-derived A.11 and 1.B2 antigens blocked lymphocyte binding sites of HEV_{LN} and HEV_{PP}, respectively, and thereby selectively inhibited adhesion of overlaid TDL. The effects were specific because TDL-derived A.11 antigen did not block binding sites of HEV_{PP} and TDL-derived 1.B2 antigen did not block binding sites of HEV_{LN} (71, 72).

We conclude from these results that lymphocyte surface $HEBF_{LN}$ is recognized by the A.11 antibody, and surface $HEBF_{PP}$, by the 1.B2 antibody. Specific adhesion molecules on lymphocytes for recognition of high endothelium of either lymph nodes or Peyer's patch provide a molecular system with the potential to ensure that individual cell populations are distributed to either peripheral or intestinal lymphoid tissues, whereas the expression of both types of adhesion molecules would enable lymphocytes to recirculate through all of these organs. This problem has been studied directly by examining the HEV binding properties of TDL separated into subsets on the basis of A.11 and 1.B2 surface phenotype. TDL fractionation was achieved using the panning technique employing A.11 or 1.B2 IgG-coated petri dishes. Adherent and nonadherent lymphocytes were recovered from each dish and scored for surface A.11 and 1.B2 antigen expression by fluorescent microscopy. Each population was then tested for the ability to bind to HEV_{LN} and HEV_{PP}. The results of one experiment are summarized in Figure 3. The TDL subset deficient in A.11 positive cells (nonadherent cells from A.11 Ig-coated plates) exhibited little affinity for HEV_{LN}; this population contained 1.B2 positive cells and showed normal levels of HEV_{PP} binding. By contrast, the TDL subset deficient in 1.B2 positive cells (nonadherent cells from 1.B2 Ig-coated plates) exhibited little affinity for HEV_{PP}; these cells were A.11 positive and had normal HEV_{LN} binding capabilities. We interpret these findings to mean that expression of A.11 and 1.B2 surface molecules determines affinity of lymphocytes for high endothelium. This idea was tested in adoptive transfer experiments using ^{51}Cr-labeled TDL (70) fractionated on 1.B2 antibody-coated dishes. The nonadherent population (12% 1.B2$^+$, 48% A.11$^+$), the adherent population (90% 1.B2$^+$, 47% A.11$^+$), and unfractionated TDL (50% 1.B2$^+$, 60% A.11$^+$) were each labeled with ^{51}Cr and injected intravenously into recipient rats (5 per group) killed

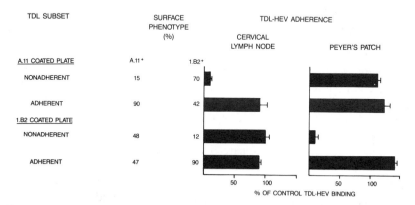

Figure 3 TDL subsets with the capacity to bind to either HEV_{LN} or HEV_{PP}. Subsets were isolated by panning on optilux plastic petri dishes coated with the indicated antibody. After nonadherent cells were collected, the dishes were thoroughly washed, incubated at 37°C for 30 min, and adherent cells removed by pipetting. Each subset was overlaid at 3×10^6 cells/ml on lymph node sections and at 15×10^6 cells/ml on Peyer's patch sections using unfractionated TDL as the control. Binding values calculated as described in (71).

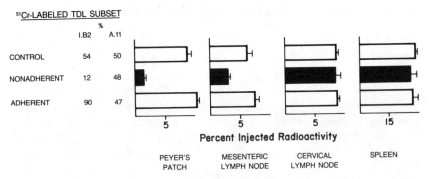

⁵¹Cr-LABELED TDL SUBSET

Figure 4   Restricted migration pathway of surface HEBF$_{PP}$-deficient TDL. Accumulation of radioactivity in organs determined 2 hr after transfer of ^{51}Cr-labeled TDL as described (70). Each recipient transfused with 5×10^6 cells containing approximately 50,000 cpm. Subsets were obtained by panning on antibody-coated plates (as described in Figure 2).

2 hr later; the distribution of donor TDL was determined by measuring radioactivity in organs (Figure 4). The results show that TDL deficient in 1.B2$^+$ surface expression exhibited little tendency to migrate into Peyer's patch and mesenteric lymph nodes. These lymphocytes nevertheless showed normal accumulation in cervical lymph nodes as well as in the spleen. Their recovery in other organs (liver, lung) was essentially the same as that observed with labeled unfractionated TDL. It appears that TDL lacking the 1.B2 molecule may migrate by a route restricted to peripheral lymph nodes and spleen.

Human HEBF: A Cell Surface Recognition Structure of Blood Lymphocytes

The above results indicate that antibodies which specifically block lymphocyte adhesion to HEV in vitro are directed against cell surface molecules that mediate lymphocyte migration into lymph nodes and Peyer's patches in vivo. The specificity of the in vitro HEV adherence reaction and its close functional relationship to cellular interaction involved in lymphocyte recirculation indicate that this system provides a means to analyze the interaction of human lymphocytes with high endothelial cells responsible for transport of the cells into lymph nodes and gut-associated tissues. Human peripheral blood lymphocytes (PBL) bind readily to HEV when overlaid onto sections of human tonsils and rat cervical lymph nodes (Table 2). Human and rat lymphocytes exhibit similar binding kinetics, specificities, and cell dose requirements in reciprocal assays with human and rat HEV. The results indicate that this recognition system is highly conserved and that both rat and human high endothelium express the target molecules for human PBL binding. We therefore considered the possibility that adherence molecules of human and rat recirculating lymphocytes are structurally similar. Initial experiments showed that rabbit–anti–rat $HEBF_{LN}$ antibody surface labeled 60–70% of PBL of healthy donors. However, attempts to block PBL-HEV binding with this reagent were not successful. Nevertheless, affinity chromatography studies using this polyclonal antibody led to the isolation of biologically active HEBF from detergent-solubilized PBL and soluble HEBF from sera of healthy adults (L. M. Clarke, J. J. Woodruff, unpublished observations). Anti-human HEBF antisera and monoclonal antibodies have been developed in BALB/c mice by immunization with soluble HEBF isolated from serum. Treatment of PBL with antiserum or monoclonal anti-human HEBF (3.A.7) antibodies blocked binding to rat HEV_{LN} (Table 3). In contrast, no inhibition was produced by several other monoclonal anti–human HEBF antibodies. Both the blocking and nonblocking monoclonal antibodies each surfaced labeled about 85% of PBL. Furthermore the inhibitory effects of 3.A.7 were specific because this antibody did not block PBL binding to HEV of mesenteric lymph nodes. Thus, human lymphocyte surface HEBF appears to mediate their interaction with high endothelial cells. The results provide further evidence that functionally distinct adhesion molecules play a role in lymphocyte binding to high endothelium of peripheral lymph nodes, on the one hand, and gut-associated tissues, on the other.

Table 2   Lack of species restriction in rat and human lymphocyte–HEV adhesion

| | Values per rat LN section | | | | | Values per human tonsillar section | | | | |
| | Number of HEV with indicated range of adherent cells | | | Number positive HEV | Specific binding (%) | Number of HEV with indicated range of adherent cells | | | Number positive HEV | Specific binding (%) |
	2–5	6–10	>10			2–5	6–10	>10		
Human PBL	41	26	19	86	94	34	12	6	52	85
Rat TDL	38	25	11	74	87	27	10	7	44	83

Cells overlaid at 3×10^6/ml.

Table 3 Anti–human HEBF polyclonal and monoclonal antibodies block human PBL binding to HEV_{LN}

Experiment number	PBL pretreatment[a]	Fluorescent positive cells (%)	Inhibition PBL-HEV_{LN} binding[b] (%)
<u>1</u>	*Polyclonal antibodies*		
	Rabbit anti–rat $HEBF_{LN}$	65	2
	Nonimmune rabbit serum	0	1
	Mouse anti–human HEBF	—	99
	Nonimmune mouse serum	—	0
	Monoclonal Antibodies		
	Anti–human HEBF:		
	3.A.7	85	89
	2.E.11	89	2
	4.C.6	93	0
	Control supernatant	—	1
	OKT4	63	2
	OKT8	17	0

		Inhibition PBL binding to:	
		HEV_{PLN} (%)	HEV_{MLN} (%)
<u>2</u>	Anti–human HEBF 3.A.7	92	3
	Control supernatant	2	0

[a] Performed with saturating amounts of antibody for 60 min at 24°C.
[b] Calculated as described (70) using PBL at 3×10^6/ml.

CONCLUSION

The evidence indicates that high endothelia of peripheral lymph nodes and Peyer's patch interact with distinct adhesion molecules of lymphocytes and that these recognition events play a role in transport of the cells into these tissues. Binding appears to be mediated by a class of molecules expressed on both T cells and B cells but not on most thymus and bone marrow cells. Biologic and chemical properties of the lymphocyte-HEV recognition system indicate that lymphocyte function–associated antigens which are known to be surface adhesive proteins of lymphocytes are not involved in this reaction.

We propose the designation HEBF (high endothelial binding factor) to denote this surface membrane molecular complex. Very little is known as to how this structure works, but the evidence summarized here suggests that at least operationally these cell surface molecules function as adhesive proteins. It is also possible that HEBF interaction with other cell surface–associated molecules might influence adhesive properties of the cells. In any event, the presence of soluble HEBF in the circulation raises the possibility that this material may exert a regulatory role in lymphocyte–high endothelial cell interactions. For example, it could be that changes in serum HEBF concentrations affect the rate of entry of lymphocytes into lymph nodes and Peyer's patch.

ACKNOWLEDGMENTS

This work was supported by NIH grant AI 10080. The authors wish to thank Ms. Dorothy McLean and Mr. James Kelly for excellent assistance in the preparation of this manuscript.

Literature Cited

1. Gowans, J. L. 1957. The effect of the continuous reinfusion of lymph and lymphocytes on the output of lymphocytes from the thoracic duct of unanaesthetized rats. *Br. J. Pathol.* 38: 67
2. Gowans, J. L. 1959. The recirculation of lymphocytes from blood to lymph in the rat. *J. Physiol.* 14: 654
3. Gowans, J. L. 1959. Lymphocyte recirculation. *Br. Med. Bull.* 15: 50
4. Gowans, J. L., Knight, E. J. 1964. The route of recirculation of lymphocytes in the rat. *Proc. R. Soc. Lond. Ser B* 159: 257
5. Marchesi, V. T., Gowans, J. L. 1964. The migration of lymphocytes through the endothelium of venules in lymph nodes: An electron microscopic study. *Proc. R. Soc. Lond. Ser B* 159: 283
6. Everett, N. B., Caffrey, R. W., Rieke, W. O. 1964. Recirculation of lymphocytes. *Ann. N.Y. Acad. Sci.* 113: 887
7. Hall, J. G., Morris, B. 1965. The origin of cells in the efferent lymph from a single lymph node. *Br. J. Exp. Pathol.* 46: 450
8. Gowans, J. L., McGregor, D. D. 1965. The immunological activities of lymphocytes. *Prog. Allergy* 9: 1
9. Goldschneider, I., McGregor, D. D. 1968. Migration of lymphocytes and thymocytes in the rat. I. The route of migration from blood to spleen and lymph nodes. *J. Exp. Med.* 127: 1
10. Ford, W. L., Gowans, J. L. 1969. The traffic of lymphocytes. *Semin. Hematol.* 6: 67
11. Ford, W. L. 1975. Lymphocyte migration and immune response. *Prog. Allergy.* 19: 1
12. Sprent, J. 1977. Recirculating lymphocytes. In *Lymphocyte Structure and Function, Part 1*, ed. J. J. Marchalonis, p. 43. New York: Marcel-Dekker
13. Parrott, D. M. V., Wilkinson, P. C. 1981. Lymphocyte locomotion and migration. *Prog. Allergy* 28: 193
14. Hall, J. G. 1980. An essay on lymphocyte circulation and the gut. *Monogr. Allergy* 16: 100
15. Trevella, W., Morris, B. 1980. Reassortment of cell populations within the lymphoid apparatus of the sheep. In *Blood Cell and Vessel Walls: Functional Interactions.* Ciba Found. 71: 127
16. Bienenstock, J., Befus, A. D., McDermott, M. 1980. Mucosal immunity. *Monogr. Allergy* 16: 1
17. Hay, J. B., Hobbs, B. B. 1977. The flow of blood to lymph nodes and its relation to lymphocyte traffic and the immune response. *J. Exp. Med.* 145: 31
18. Hay, J. B., Johnston, M. G., Vadas, P., Chia, W., Issekutz, T., Movat, H. Z. 1979. Relationships between changes in blood flow and lymphocyte migration induced by antigen. *Monogr. Allergy* 16: 100
19. Bjerkness, M., Cheng, H., Ottaway, C. A. 1986. Dynamics of lymphocyte-endothelial interactions in vivo. *Science* 231: 402
20. Thome, R. 1898. Endothelien als phagocyten (aus den Lymphdrussen von Macacus cynomalgus). *Arch. Microsk. Anat.* 52: 820

21. Smith, C., Hernon, B. K. 1959. Histological and histochemical study of high endothelium of post-capillary veins of the lymph node. *Anat. Recirc.* 135: 207

22. Cleaesson, M. O., Jorgenson, O., Ropke, C. 1971. Light and electron microscopic studies of the paracortical post-capillary high endothelial venules. *Z. Zellforsch. Mikrosk. Anat.* 119: 195

23. Miller, J. J. III. 1969. Studies of the phylogeny and ontogeny of the specialized lymphatic venules. *Lab. Invest.* 21: 284

24. Yoffey, J. M. 1936. Variation in lymphocyte production. *J. Anat.* 70: 507

25. Sanders, A. G., Florey, H. W., Barnes, J. M. 1940. The output of lymphocytes from the thoracic duct in cats and rabbits. *Br. J. Exp. Pathol.* 21: 254

26. Hughes, R., May, A. J., Widdicombe, J. G. 1956. The output of lymphocytes from the lymphatic system of the rabbit. *J. Physiol.* 132: 384

27. Fossum, S., Smith, M. E., Ford, W. L. 1983. The migration of lymphocytes across specialized vascular endothelium. VII. The migration of T and B lymphocytes from the blood of the athymic, nude rat. *Scand. J. Immunol.* 17: 539

28. Fossum, S., Smith, M. E., Ford, W. L. 1983. The recirculation of T and B lymphocytes in the athymic nude rat. *Scand. J. Immunol.* 17: 551

29. Smith, M. E., Ford, W. L. 1983. The migration of lymphocytes across specialized vascular endothelium. VI. The migratory behaviour of rat thoracic duct lymphocytes re-transferred from the lymph nodes, spleen, blood or lymph of a primary recipient. *Cell Immunol.* 78: 161

30. Smith, M. E., Ford, W. L. 1983. The recirculating lymphocyte pool of the rat: A systematic description of the migratory behaviour of recirculating lymphocytes. *Immunology* 49: 83

31. Hall, J. G., Scollay, R. G., Smith, M. E. 1976. Studies on the lymphocytes of sheep. I. Recirculation of lymphocytes through peripheral nodes and tissues. *Eur. J. Immunol.* 6: 17

32. Scollay, R. G., Hopkins, J., Hall, J. G. 1976. Possible role of surface Ig in non-random recirculation of small lymphocytes. *Nature* 260: 528

33. Chin, W., Hay, J. B. 1980. A comparison of lymphocyte migration through intestinal lymph nodes, subcutaneous lymph nodes, and chronic inflammatory sites of sheep. *Gastroenterology* 79: 1231

34. Hill, R. N. P., Poskitt, D. C., Frost, H., Trnka, Z. 1977. Two distinct pools of recirculating T lymphocytes: Migratory characteristics in intestinal T lymphocytes. *J. Exp. Med.* 145: 420

35. deFreitas, A. A., Rose, M. L., Parrott, D. M. V. 1977. Mesenteric and peripheral lymph nodes in the mouse: A common pool of small T lymphocytes. *Nature* 230: 731

36. Hall, J. 1985. The study of circulating lymphocytes in vivo: A personal view of artifice and artifact. *Immunol. Today* 6: 149

37. Ford, W. L., Allen, T. D., Pitt, M. E., Smith, M. E., Stoddart, R. W. 1984. The migration of lymphocytes across specialized vascular endothelium. VIII. Physical and chemical conditions influencing the surface morphology of lymphocytes and their ability to enter lymph nodes. *Am. J. Anat.* 170: 377

38. Cahill, R. N. P., Frost, H., Trnka, Z. 1976. The effects of antigen on the migration of recirculating lymphocytes through single lymph nodes. *J. Exp. Med.* 143: 870

39. Herman, P. G., Yamamoto, I., Mellins, H. Z. 1972. Blood microcirculation in the lymph node, during the primary immune response. *J. Exp. Med.* 136: 697

40. Freemont, A. J. 1983. A possible route for lymphocyte migration into diseased tissues. *J. Clin. Pathol.* 36: 161

41. Anderson, N. D., Anderson, A. O., Wyllie, R. G. 1976. Specialized structure and metabolic activities of high endothelial venules in rat lymphatic tissues. *Immunology* 31: 455

42. Anderson, A. O., Anderson, N. D. 1976. Lymphocyte immigration from high endothelial venules in rat lymph nodes. *Immunology* 31: 731

43. Andrews, P., Ford, W. L., Stoddart, R. W. 1980. Metabolic studies of high-walled endothelium of post-capillary venules in rat lymph nodes. See Ref. 15, 71: 211

44. Andrews, P., Milsom, D. W., Ford, W. L. 1982. Migration of lymphocytes across specialized vascular endothelium. V. Production of a sulphated macromolecule by high endothelial cells in lymph nodes. *J. Cell Sci.* 57: 277

45. Andrews, P., Milsom, D. W., Stoddart, R. W. 1983. Glycoconjugates from high endothelial cells. I. Partial characterization of a sulphated glycoconjugate from the high endothelial cells of rat lymph nodes. *J. Cell Sci.* 59: 231

46. Van Ewijk, W., Brons, N. H. C., Rozing, J. 1975. Scanning electron microscopy of homing and recirculating lymphocyte populations. *Cell Immunol.* 19: 245

47. Sugimura, M., Furuhata, K., Kudo, N., Takahata, K., Mifune, Y. 1964. Fine

structure of post-capillary venules in mouse lymph nodes. *Jpn. J. Vet. Res.* 12: 38

48. Schoefl, G. I. 1972. The migration of lymphocytes across the vascular endothelium in lymphoid tissue: A re-examination. *J. Exp. Med.* 136: 568

49. de Sousa, M. A. B., Parrott, D. M. V., Pantelouris, E. M. 1969. The lymphoid tissues in mice with congenital aplasia of the thymus. *Clin. Exp. Immunol.* 4: 637

50. Fossum, S., Smith, M. E., Bell, E. B., Ford, W. L. 1980. The architecture of rat lymph nodes. III. The lymph nodes and lymph bone cells of the congenitally athymic nude rat (rnu). *Scand. J. Immunol.* 12: 421

51. Stamper, H. B. Jr., Woodruff, J. J. 1976. Lymphocyte homing into lymph nodes: In vitro demonstration of the selective affinity of recirculating lymphocytes for high endothelial venules. *J. Exp. Med.* 144: 828

52. Stamper, H. B. Jr., Woodruff, J. J. 1977. An in vitro model of lymphocyte homing. I. Characterization of the interaction between thoracic duct lymphocytes and specialized high endothelial venules of lymph nodes. *J. Immunol.* 119: 772

53. Woodruff, J. J., Katz, I. M., Lucas, L. E., Stamper, H. B. Jr. 1977. An in vitro model of lymphocyte homing. II. Membrane and cytoplasmic events involved in lymphocyte adherence to specialized high-endothelial venules of lymph nodes. *J. Immunol.* 119: 1603

54. Woodruff, J. J., Kuttner, B. J. 1980. Adherence of lymphocytes to the high endothelium of lymph nodes in vitro. See Ref. 15, 71: 243

55. Kuttner, B. J., Woodruff, J. J. 1979. Adherence of recirculating T and B lymphocytes to high endothelium of lymph nodes in vitro. *J. Immunol.* 123: 1421

56. Woodruff, J. J., Rasmussen, R. A. 1979. In vitro adherence of lymphocytes to unfixed and fixed high endothelial cells of lymph nodes. *J. immunol.* 123: 2369

57. Butcher, E. C., Scollay, R. G., Weissman, I. L. 1979. Lymphocyte adherence to high endothelial venules: Characterization of a modified in vitro assay, and examination of the binding of syngeneic and allogeneic lymphocyte populations. *J. Immunol.* 123: 1996

58. Butcher, E. C., Scollay, R., Weissman, I. 1980. Organ specificity of lymphocyte interaction with organ-specific determinants on high endothelial venules. *Eur. J. Immunol.* 10: 556

59. Jalkanen, S. T., Butcher, E. C. 1985. In vitro analysis of the homing properties of human lymphocytes. Developmental regulation of functional receptors for high endothelial venules. *Blood* 66: 577

60. van der Brugge-Gamelkoorn, G. J., Kraal, G. 1985. The specificity of the high endothelial venule in bronchus-associated lymphoid tissue (BALT). *J. Immunol.* 134: 3746

61. Butcher, E., Scollay, R., Weissman, I. 1979. Evidence of continuous evolutionary change in structures mediating adherence of lymphocytes to specialized venules. *Nature* 280: 496

62. Stevens, S. K., Weissman, I. L., Butcher, E. C. 1982. Differences in the migration of B and T lymphocytes: Organ-selectivity and the role of lymphocyte endothelial cell recognition. *J. Immunol.* 128: 844

63. Braaten, B. A., Spangrude, G. J., Daynes, R. A. 1984. Molecular mechanisms of in vitro lymphocyte adherence to high endothelial venules. *J. Immunol.* 133: 117

64. Navarro, R. F., Jalkanen, S. T., Hsu, M., Soenderstrup-Hansen, G., Goronzy, J., Weyand, C., Fathman, C. G., Clayberger, C., Krensky, A. M., Butcher, E. C., 1985. Human T cell clones express functional homing receptors required for normal lymphocyte trafficking. *J. Exp. Med.* 66: 577

65. Dailey, M. O., Fathman, C. G., Butcher, E. C., Pillemer, E., Weissman, I. 1982. Abnormal migration of T lymphocyte clones. *J. Immunol.* 128: 2134

66. Jalkanen, S., Steere, A. C., Fox, R. I., Butcher, E. C. 1986. A distinct endothelial cell recognition system that controls lymphocyte traffic into inflamed synovium. *Science* 233: 556

67. Rosen, S. D., Singer, M. S., Yednock, T. A., Stoolman, L. M. 1985. Involvement of sialic acid on endothelial cells in organ-specific lymphocyte recirculation. *Science* 228: 1005

68. Stoolman, L. M., Rosen, S. D. 1983. Possible role for cell-surface carbohydrate-binding molecules in lymphocyte recirculation. *J. Cell Biol.* 96: 722

69. Stoolman, L. M., Tenforde, T. S., Rosen, S. D. 1984. Phosphomannosyl receptors may participate in the adhesive interaction between lymphocytes and high endothelial venules. *J. Cell Biol.* 99: 1535

70. Chin, Y. H., Carey, G. D., Woodruff, J. J. 1982. Lymphocyte recognition of lymph node high endothelium. IV. Cell surface structures mediating entry into lymph nodes. *J. Immunol.* 129: 1911

71. Rasmussen, R. A., Chin, Y. H., Woodruff, J. J., Easton, T. G. 1985. Lympho-

cyte recognition of lymph node high endothelium. VII. Cell surface proteins involved in adhesion defined by monoclonal anti-HEBF$_{LN}$ (A.11) antibody. *J. Immunol.* 135: 19

72. Chin, Y. H., Rasmussen, R. A., Woodruff, J. J., Easton, T. G. 1986. A monoclonal anti-HEBF$_{PP}$ antibody with specificity for lymphocyte surface molecules mediating adhesion to Peyer's patch high endothelium of the rat. *J. Immunol.* 136: 2556

73. Gallatin, W. M., Weissman, I. L., Butcher, E. C. 1983. A lymphoid cell-surface molecule involved in organ-specific homing of lymphocytes. *Nature* 304: 30

74. Chin, Y. H., Rasmussen, R., Cakiroglu, A. K., Woodruff, J. J. 1984. Lymphocyte recognition of lymph node high endothelium. VI. Evidence of distinct structures mediating binding to high endothelial cells of lymph nodes and Peyer's patches. *J. Immunol.* 133: 2961

75. Gesner, B. M., Ginsburg, V. 1964. Effect of glycosidases on the fate of transfused lymphocytes. *Proc. Natl. Acad. Sci. USA* 52: 750

76. Woodruff, J. J., Gesner, B. M. 1969. The effect of neuraminidase on the fates of transfused lymphocytes. *J. Exp. Med.* 129: 551

77. Woodruff, J. J., Gesner, B. M. 1968. Lymphocyte recirculation altered by trypsin. *Science* 161: 176

78. Ford, W. L., Sedgley, M., Sparshott, S. M., Smith, M. E. 1976. The migration of lymphocytes across specialized vascular endothelium. II. The contrasting consequences of treating lymphocytes with trypsin or neuraminidase. *Cell Tissue Kinet.* 9: 351

79. Ottaway, C. A. 1984. In vitro alteration of receptors for vasoactive peptide changes the in vitro localization of mouse T cells. *J. Exp. Med.* 160: 1054

80. Spangrude, G. J., Braaten, B. A., Daynes, R. A. 1984. Molecular mechanisms of lymphocyte extravasation. I. Studies of two selective inhibitors of lymphocyte recirculation. *J. Immunol.* 132: 354

81. Rouse, R. V., Reichert, R. A., Gallatin, W. M., Weissman, I. L., Butcher, E. C. 1984. Localization of lymphocyte subpopulations in peripheral lymphoid organs: Directed migration and segregation into specific microenvironments. *Am. J. Anat.* 170: 391

82. Siegelman, M., Bond, M. W., Gallatin, W. M., St. John, T., Smith, H. T., Fried, V. A., Weissman, I. L. 1986. Cell surface molecule associated with lymphocyte homing is a ubiquitinated branched-chain glycoprotein. *Science* 231: 823

83. Chin, Y. H., Carey, G. D., Woodruff, J. J. 1980. Lymphocyte recognition of lymph node high endothelium. I. Inhibition of in vitro binding by a component of thoracic duct lymph. *J. Immunol.* 125: 1764

84. Chin, Y. H., Carey, G. D., Woodruff, J. J. 1980. Lymphocyte recognition of lymph node high endothelium. II. Characterization of an in vitro inhibitory factor isolated by antibody affinity chromatography. *J. Immunol.* 125: 1770

85. Chin, Y. H., Carey, G. D., Woodruff, J. J. 1983. Lymphocyte recognition of lymph node high endothelium. V. Isolation of adhesion molecules from lysates of rat lymphocytes. *J. Immunol.* 131: 1368

Ann. Rev. Immunol. 1987. 5 : 223–52

THE LYMPHOCYTE FUNCTION-ASSOCIATED LFA-1, CD2, and LFA-3 MOLECULES: Cell Adhesion Receptors of the Immune System

Timothy A. Springer, Michael L. Dustin, Takashi K. Kishimoto, and Steven D. Marlin

Department of Pathology, Harvard Medical School and Laboratory of Membrane Immunochemistry, Dana-Farber Cancer Institute, Boston, MA 02115

INTRODUCTION

Cell adhesion molecules are thought to play an important role in guiding cell migration and localization in the development of the embryo and in organogenesis. In the immune system, cell adhesion molecules enhance the efficiency of specific receptor-dependent lymphocyte-accessory cell and lymphocyte-target cell interactions; they are also important in leukocyte-endothelial cell interactions and lymphocyte recirculation. Recent studies with monoclonal antibodies (MAb) that perturb antigen-receptor-dependent T-lymphocyte functions have defined a number of cell surface molecules that are associated with lymphocyte function (lymphocyte function–associated or LFA antigens) (Table 1). The antigens LFA-1, CD2, LFA-3, CD8, and CD4 appear to enhance antigen-specific functions by acting as cell adhesion molecules. Further studies have shown that the LFA-1, CD2, and LFA-3 molecules are also important in antigen-independent T-lymphocyte adherence and function and that the LFA-1 molecule is important in the adherence and function of essentially all leukocyte cell types.

This review focuses on LFA-1, CD2, and LFA-3. The role of CD4 and CD8 is reviewed by Littman in this volume. We discuss (*a*) the contributions of LFA-1, CD2, and LFA-3 to antigen-dependent and antigen-

223

0732–0582/87/0410–0223$02.00

Table 1 Cell surface molecules that regulate T-cell interactions

Name	T Cells M_r	Cell distribution	Name	Other Cells M_r	Cell distribution
LFA-1	$\alpha = 180K, \beta = 95K$	Thymocytes, T and B lymphocytes, LGL, monocytes, activated macrophages, neutrophils	ICAM-1	90–110K	Wide, regulated by IL-1, TNF, and IFN-γ
CD2 (LFA-2/T11)	50–58K	Thymocytes, T lymphocytes, LGL	LFA-3	55–70K	Wide
Antigen Receptor-CD3 (T3) Complex	$\alpha = 50K, \beta = 40K$ $\gamma = 28K\ \delta = 22K$	Mature thymocytes T lymphocytes	Antigen in association with major histocompatibility complex (MHC) molecules		
CD8 (T8)	32K disulfide-linked dimer	Subset of thymocytes and T lymphocytes, LGL	MHC class I	$\alpha = 44K, \beta = 12K$	Wide, regulated by IFN-α, β, γ
CD4 (T4)	55K	Subset of thymocytes and T lymphocytes, monocytes/ macrophages	MHC class II	$\alpha = 34K, \beta = 29K$	Wide, regulated by IFN-γ

independent adhesiveness, (*b*) a putative ligand for LFA-1, designated intercellular adhesion molecule 1 (ICAM-1), (*c*) the direct molecular interaction of CD2 with its ligand, LFA-3, and (*d*) the dual function of CD2 in T lymphocyte adhesion and triggering. The reader is also referred to previous reviews (1–4), an excellent recent review by Martz on accessory (LFA) molecules (5), and a concurrent review by Anderson & Springer on inherited leukocyte adhesion deficiency disease (6).

LFA-1

Mouse and Human LFA-1

LFA-1 has been defined in the mouse with rat MAb (7, 8) and in the human with mouse MAb (9, 10). The tissue distribution, structure, and function of murine and human LFA-1 are highly similar. In hybrid cells, the α and β subunits of mouse and human LFA-1 promiscuously coassociate in interspecies $\alpha\beta$ complexes, further suggesting homology (11). Results from the mouse and human are described interchangeably below.

Tissue Distribution

LFA-1 is expressed by all leukocytes, with the exception of some macrophages (Table 1) (12, 13). There are 15,000 to 40,000 LFA-1 surface sites per peripheral lymphocyte, with more abundant expression on T than B lymphocytes and increased expression on T blasts (12, 14). LFA-1 is present on $\sim 50\%$ of bone marrow cells. In B and myeloid lineages, LFA-1 is first seen at the pre–B cell (cytoplasmic μ chain positive) and late myeloblast stages, respectively (15). LFA-1 is absent or low on myeloid and erythroid precursor cells (15, 16) and is absent from nonhematopoietic cells.

Structure and Biosynthesis of LFA-1

LFA-1 is a heterodimer consisting of an α subunit of 180 kd and a noncovalently associated β subunit of 95 kd (17, 18). Crosslinking experiments show the presence of only one α and one β subunit per complex. The α and β subunits are synthesized as separate precursors of 170 kd and 87 kd respectively (18). The precursors contain N-glycoside high mannose carbohydrate groups linked to polypeptide chain backbones of 130 kd (α) and 72 kd (β) (19). The α and β precursors must associate intracellularly before conversion of the high mannose carbohydrates to a complex form occurs in the Golgi apparatus (18–20). The $\alpha\beta$ complex is then expressed on the cell surface. The N-linked carbohydrates of LFA-1 are sulfated in thymocytes and splenic T cells but not on macrophages, splenic B cells, or bone marrow cells (21).

Functional Studies

LFA-1 was initially defined on human and murine lymphocytes by monoclonal antibody–mediated inhibition of killing by cytotoxic T lymphocytes (CTL) and natural killer cells (7–10). Since then, MAb, $F(ab')_2$, and Fab fragments to both the α and β subunits of LFA-1 have been shown to inhibit a wide variety of adhesion-dependent leukocyte functions. The pattern of inhibition is highly discreet, and MAb binding to many other antigens present at higher density on the cell surface has no effect (1, 2, 22, 23). Furthermore, normal cells treated with LFA-1 MAb exhibit the same pattern of defects as genetically LFA-1-deficient cells (see below).

LFA-1 MAb inhibit CTL-mediated lysis of allogeneic (2, 7–9, 13, 24–30), xenogeneic (7, 31–33), virus-infected (10), and hapten modified targets (25) by both cloned CTL and bulk populations. In addition to T cell–mediated cytotoxicity, LFA-1 is involved in natural killer (NK) cell–mediated cytotoxicity and antibody-dependent cytotoxicity mediated by granulocytes or peripheral blood mononuclear cells (10, 13, 24, 26, 34–39).

Two steps in cytolytic T lymphocyte–mediated killing have clearly been distinguished: adhesion and lethal hit delivery (40). These steps are Mg^{+2} and Ca^{+2} dependent, respectively. Cytolytic T lymphocytes can be distinguished from target cells by size or by means of fluorescent dyes. Adhesion of CTL to target cells (conjugate formation) can be quantitated microscopically or with flow microfluorometry, while killing can be measured as the release of label from the target cell. Adhesion to target cells clearly precedes and is required for lethal hit delivery. LFA-1 MAb block CTL-mediated killing by acting at the Mg^{+2}-dependent adhesion stage rather than the Ca^{+2}-dependent lethal hit delivery step (1). LFA-1 MAb inhibit conjugate formation between CTL and target cell, and preformed conjugates are reversed (1, 25, 29, 41, 42).

LFA-1 is also involved in helper-T-cell functions. Anti-LFA-1 MAb inhibit the proliferation of T cells in response to soluble antigens, viruses, alloantigen, xenoantigen, and mitogens (8, 13, 24–26, 43–45). LFA-1 MAb block only if added before or within the first few hours of initiation of these assays, before proliferation begins. Thus, induction of proliferation rather than proliferation itself is inhibited. Responses of cell lines such as CTLL2 that require only IL-2 for proliferation are not inhibited. These results suggest that the T cell–antigen presenting cell interaction is blocked, but this remains to be demonstrated.

In contrast to conventional LFA-1 MAb, a MAb reactive with an activation determinant present on LFA-1 and additional surface molecules stimulates proliferation and IFN-γ release by T-cell clones and inhibits cytolysis by the same cells (46).

Antibody responses by B cells are inhibited by anti-LFA-1 MAb, apparently by affecting interactions with T cells or antigen-presenting cells. T cell–dependent antibody responses to antigen or mitogens are inhibited, while T-independent responses are unaffected (8, 25, 35, 37, 44, 45). Pretreatment of either T cells or monocytes, but not B cells, inhibits the in vitro antibody response to influenza virus (43).

Inhibition of adhesion in helper-T-lymphocyte and B-cell responses is consistent with results cited above but has thus far been assessed in only one report which showed that LFA-1 MAb inhibit conjugate-formation of hapten-specific B cells with carrier-specific T cells (47). Other reports show a differential effect of LFA-1 MAb on cell-cell interactions. Anti-LFA-1 MAb block IL-2 production by T-cell hybrids when stimulated by antigen-presenting or allogeneic cells (48–50) but not when stimulated with anti-T cell receptor antibody linked to Sepharose (48). Anti-CD3 induced cytolysis (51) and lysis by human CTL of murine hybridomas bearing surface membrane anti-CD3 immunoglobulin (32) are inhibited.

LFA-1 is absent from resident or thioglycollate-elicited peritoneal macrophages but is present on activated, tumoricidal macrophages. LPS and IFN-γ induce LFA-1 on thioglycollate-elicited macrophages in vitro (52). Pretreatment with LFA-1 F(ab')$_2$ inhibits selective binding of activated murine macrophages to tumor cells and prevents development of weak into strong binding as shown by the centrifugal force required for dissociation (53).

Inherited LFA-1, Mac-1, and p150,95 Deficiency

A novel immunodeficiency disease has been defined in which expression of LFA-1 and the related Mac-1 and p150,95 glycoproteins is selectively defective (54–57). Mac-1, p150,95, and LFA-1 are $\alpha\beta$ heterodimers that have identical β subunits. The α subunits are distinct but are 33–50% identical in amino acid sequence (58) (L. J. Miller, M. Wiebe, T. A. Springer, submitted). LFA-1, Mac-1, and p150,95 thus constitute a family of related $\alpha\beta$ complexes. Mac-1 (and p150,95) mediate 'nonspecific' adhesion of granulocytes and monocytes to endothelial cells and other substrates and also function as the complement receptor type 3 (CR3), binding to the complement component iC3b. In common with LFA-1, adhesion reactions mediated by Mac-1 and p150,95 require Mg^{+2}. Mac-1 and p150,95 are stored in intracellular pools in circulating monocytes and granulocytes; binding of chemoattractants to specific receptors results in translocation of Mac-1 and p150,95 to the cell surface. Adhesiveness of these cells is thereby increased, and this appears to mediate binding to endothelial cells and localization in inflammatory sites. A detailed discussion and pertinent references to the Mac-1 and p150,95 glycoproteins and the inherited disease is presented in a review by Anderson & Springer (6).

Patients deficient in LFA-1, Mac-1, and p150,95 are characterized by recurrent life-threatening bacterial and fungal infections, progressive periodontitis, lack of pus formation, and leukocytosis. Granulocytes, monocytes, and lymphocytes from patients display profound defects in both in vivo and in vitro adherence-dependent immune functions. We have suggested the designation leukocyte adhesion deficiency (LAD) for this disease, which has now been characterized in 30 patients worldwide. Quantitative analysis of Mac-1 and LFA-1 surface expression by flow cytometry indicates that all leukocytes are affected. There are two patient phenotypes, designated severe ($<0.3\%$ of normal expression) and moderate (5–10% of normal expression) deficiency (20, 54). For a given patient, the three α subunits and the common β subunit are deficient to similar extents, as shown with subunit-specific MAbs. The severity of the clinical complications correlates directly with the degree of LFA-1 deficiency. That patients with severe deficiency rarely survive beyond childhood underscores the importance of this family of molecules in vivo.

The deficiency of LFA-1, Mac-1, and p150,95 appears due to a defect in the common β subunit. Biosynthetic labeling of LFA 1–deficient T-cell blasts and EBV-transformed B cells shows that a normal amount of the LFA-1 α-chain precursor is made, but little is processed or transported to the cell surface (20, 59). Patient cells lack both α and β subunits on the surface. In mouse × human hybrids formed with patient cells (11), the patient LFA-1 α subunit complexes with the mouse β subunit, and the interspecies complex is expressed on the cell surface. Hybrids expressing the human β subunit complexed with the mouse α subunit can be derived from healthy human but not from patient cells. These results show that the α subunit is competent for surface expression if a functional β subunit is present and imply that the β subunit is defective. Immunoprecipitation with a rabbit antiserum produced against purified, denatured β subunit has recently confirmed that the β subunit is defective and has revealed that in some patients the β-chain precursor is of an aberrant size (T. K. Kishimoto, D. C. Anderson, and T. A. Springer, manuscript in preparation).

Chromosomal analysis of human × mouse hybrids shows that the LFA-1 α-chain gene is encoded on human chromosome 16 (11). The β-chain gene, and hence the genetic defect, is on chromosome 21 (11, 60). This agrees with the autosomal recessive inheritance of LAD (6).

Functional Consequences of LFA-1 Deficiency

Defects in adhesion-dependent lymphocyte functions have been observed in patients with LAD. Moderately to profoundly impaired proliferative responses to mitogens, allogeneic cells, and antigen were found in all (37, 55, 56, 61, 62) but one (63) study. Proliferation was most impaired at suboptimal mitogen doses ($\bar{x}$ = 12% of normal) (55, 56, 61, 62) and the dose-response curve was shifted (62). Mitogen proliferative responses of lymphocytes of LFA 1–deficient patients were further depressed when LFA-1 MAb was added (37, 55, 56). This agrees with the finding that deficiency in most patients is quantitative rather than absolute and shows that small amounts of LFA-1 present on patient lymphocytes can be functionally important. IFN-α and IFN-γ production by MLR or mitogen stimulated patient lymphocytes was severely deficient (62, 63).

After primary mixed lymphocyte culture, cytolytic T lymphocyte–mediated killing was 8–40% of normal ($\bar{x}$ = 18% for 4 patients) and was more depressed in the severe than moderate phenotypes (61, 63). Natural killing ranged from strikingly deficient ($\sim$10% of normal) (34, 61–63) to normal (37, 64). Antibody-dependent cytotoxicity by K cells and polymorphonuclear leukocytes was markedly depressed (34) or normal (37). These differences may be related to the extent of LFA-1 deficiency (34, 61). After repeated restimulation with allogeneic cells, CTL lines could be established from patient lymphocytes that showed cytolytic activity somewhat lower (61) or comparable (37, 55) to normal. The improvement in patient T-cell function after secondary stimulation (61) suggests that T lymphocytes can partially adapt to LFA-1 deficiency, perhaps by clonal selection of T lymphocytes with high-affinity antigen receptors.

For 5 of 6 patients, cytolysis by CTL, NK cells, and ADCC effectors was further diminished by LFA-1 MAb (37, 55, 61), and killing by patient cells was inhibited by much lower than normal concentrations of anti–LFA 1 MAb (37, 61). The above studies used LFA-1$^+$ target cells. Killing by LFA-1–deficient patients' cells was more markedly depressed with LFA-1$^-$ target cells (65). With normal CTL, LFA-1 MAb have been found to exert their inhibitory effect by binding to the CTL (8, 13, 25, 61), although pretreatment with MAb of both target cells and effectors has been shown to be more inhibitory than pretreatment of effector cells only (8). In comparable studies with patient CTL, LFA-1 MAb pretreatment of LFA-1$^+$ target cells only (severe deficiency) or of both CTL and LFA-1$^+$ target cells (moderate deficiency) was found to be inhibitory (61). This shows that LFA-1 on both the CTL and target cell can be functionally important.

Antibody production by B lymphocytes in LAD patients was found to be abnormal in vitro (37, 43) and abnormal in vivo in some patients (37, 43, 66). Strikingly, repeated immunization with tetanus and diphtheria toxoids, *Bordetella pertussis*, and influenza virus produced no response, but anti-mannose antibodies were produced in response to chronic candida infection and total Ig levels were elevated (43). In other patients, antibody production, T lymphocyte–dependent DTH, and recovery from viral infection occur, but very little quantitative data has been published.

Lymphocytes are present in normal amounts and other lymphocyte surface antigens are normal in LAD. T- and B-lymphocyte defects occur in LAD, but it is not surprising that they are milder than in a disease in which lymphocytes are missing altogether, as in severe combined immunodeficiency disease. Other accessory molecules, levels of LFA-1 still present on patient cells, and antigen-receptor-affinity maturation may contribute to patient lymphocyte responsiveness. The dramatic defects in granulocyte and monocyte mobilization in LAD have clinically overshadowed lymphocyte defects. However, lymphocyte defects in antibody production to bacteria (66) may contribute to the recurring bacterial infections. In addition T lymphocyte defects may be related to characteristic chronic mucous membrane and cutaneous candidal infections, and to the death of one patient due to picorna virus infection [54].

Prevention of Graft Rejection by LFA-1 MAb

The poor prognosis of LAD patients prompted Griscelli, Fischer, and coworkers (67) in France to attempt bone marrow transplants to correct the deficiency in two patients. Although only HLA-mismatched bone marrow was available, both transplants were accepted and the recipients are disease-free. Fischer et al (68) observed that LAD patients, all of whom did not mount allogeneic mixed lymphocyte responses, did not reject grafts. In their previous experience with bone marrow transplantation, this group found that T cell-depleted, HLA-mismatched bone marrow could be accepted by patients with severe combined immunodeficiency but was rapidly rejected by patients with other immune disorders who could mount allogeneic mixed lymphocyte responses.

The acceptance of HLA-mismatched bone marrow by LFA-1 deficient patients suggested that LFA-1 may be important in graft rejection. Since (*a*) graft rejection can be mediated by both T and non-T cells, (*b*) LFA-1 MAb inhibit both T cell and NK immune functions in vitro and (*c*) LFA-1 is low or absent on hematopoietic stem cells (15, 16), Fischer et al (68) treated graft recipients with 0.1 mg kg^{-1} anti-LFA-1 α subunit MAb from 3 days before to 5 days after transplantation. Recipients had a variety of inherited diseases such as Wiskott-Aldrich syndrome and osteopetrosis, and all received HLA-mismatched transplants. The use of LFA-1 MAb resulted in 7/7 successful engraftments, a dramatic improvement over previous experience. Thus, the clinical experience with LAD, and concepts based on the functional effects of LFA-1 MAb in vitro, have led to new treatment modalities in the therapy of other diseases.

LFA-1 in Antigen-Independent Adhesion

Lymphocyte activation is accompanied by increased adhesiveness and motility. Although specific antigen may be used to stimulate increased adhesiveness, stimulated lymphocytes show a generalized increased adherence to cells lacking specific antigen. Cells cultured in the MLR acquire the ability to adhere to a wide variety of tumor cell types. There is no MHC restriction in this adhesion, although species specificity has been shown (69). Adhesion of lymphocytes to one another can also be measured as cluster formation, i.e. aggregation. After autologous MLR or periodate stimulation, 5–35% of the viable lymphocytes are found in clusters (70). Lymphocytes isolated from clusters by vigorous vortexing readily reaggregate. Aggregation is also induced by phorbol esters, perhaps by bypassing specific activation mechanisms through direct stimulation of protein kinase C (71). Within 15 min, human PBL show uropod formation, and within 30 min, exhibit hairy surface projections, ruffled membranes, and the beginnings of aggregation (72). Similar aggregation is seen with monocytes (72) and some leukocyte cell lines, including EBV-transformed B lymphocytes (73).

Antigen-independent aggregation of a single cell type (homotypic adhesion) has recently been found to be an excellent model system for studying the cell biology of LFA 1–dependent adhesion. Phorbol ester–induced homotypic aggregation of T, B, and myeloid lineage cells (22, 30) and peripheral blood lymphocytes (22, 23) is inhibited by anti-LFA-1 MAb (Figure 1A,B). Clustering of MLR, autologous MLR, lectin, periodate, and lipopolysaccharide-activated lymphocytes (70), spontaneous clustering of EBV B-lymphoblastoid cells (74), and clustering of monocytes cultured in IFN-γ [75] are also LFA-1 dependent.

The importance of LFA-1 in homotypic adhesion has been further demonstrated by the finding that phorbol ester–activated lymphocytes from LFA 1–deficient patients fail to aggregate (22). LFA-1$^-$ lymphocytes, however, are able to coaggregate with LFA-1$^+$ lymphocytes (22). This demonstrates that LFA 1–dependent adhesion is not mediated by homophilic interactions whereby LFA-1 molecules on one cell interact with those on another, and it is consistent with observations of LFA 1–dependent CTL and NK interactions with LFA-1$^-$ targets (8, 48, 65, 76–78).

The characteristics of phorbol ester–stimulated lymphocyte aggregation and the adhesion step in CTL-mediated killing are similar. LFA 1–depen-

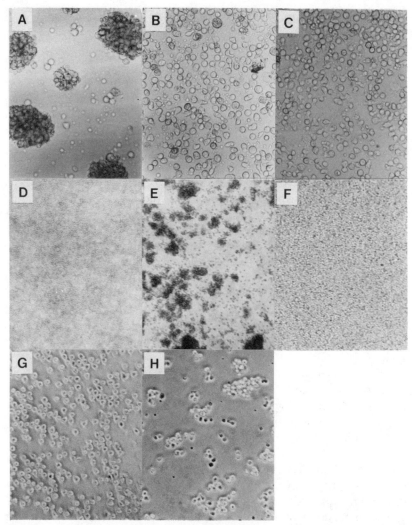

Figure 1 Visualization of LFA molecule–dependent adherence in homotypic aggregation assays [22, 82, 104]. A-C: LFA-1— and ICAM-1-dependent adherence of phorbol ester-stimulated JY lymphoblastoid cells at 37°C. JY cells aggregated in the presence of control MAb (A), but not in the presence of LFA-1 MAb (B), or ICAM-1 MAb (C). D-F: LFA-3-dependent aggregation of human erythrocytes in the presence of added CD2 at 4°C. Human erythrocytes were aggregated at 4°C by 100 μg ml¹ purified, native CD2 protein (E) but not by heat denatured CD2 (D). Aggregation was reversed by addition of 5 μg ml⁻¹ LFA-3 MAb (F). G, H: LFA-3-dependent aggregation of JY lymphoblastoid cells in the presence of added CD2 at 4°C. JY cells do not aggregate at 4°C (not shown) except in the presence of purified CD2 (H). Aggregation is reversed by LFA-3 MAb (G).

dent aggregation requires Mg^{+2}, and Ca^{+2} has a synergistic effect with suboptimal concentrations of Mg^{+2} (5, 22). A cytoskeleton with functional microfilaments is required for aggregation as shown by inhibition with cytochalasin B (22, 79). Aggregation is energy and temperature dependent (5, 80). The similar requirements for Mg^{+2}, a functional cytoskeleton, energy, and temperature in both CTL adhesion (5, 40) and phorbol ester–stimulated aggregation may all be intimately related to the involvement of LFA-1 in these processes.

Lymphocyte activation appears to be required for the LFA 1–dependent, antigen-independent adhesion system to become operative. $LFA-1^{+}$ T and B lymphocytes and thymocytes do not aggregate significantly in the absence of activation by culture, antigen, mitogen, phorbol ester, or EBV transformation (22, 23, 70, 80). Since phorbol-ester activation does not increase the amount of cell surface LFA-1 and protein synthesis is not required (22, 79, 81, 82), some other mechanism must be responsible for the enhanced LFA 1–dependent adhesion after activation. Phorbol esters are pleiotropic, inducing pseudopod formation, motility, and more rapid capping in blood lymphocytes (72, 83) (but not EBV-transformed cells: 22). Some phosphorylation of the LFA-1 β subunit is induced by phorbol esters (84). However, the molecular mechanism(s) which regulate LFA 1–dependent adherence remain unknown.

Antigen-Independent Interactions with Nonhematopoietic Cells

The interaction of T cells with vascular endothelium is a prerequisite to the migration of lymphocytes into sites of inflammation and is important in the pathophysiology of graft rejection. T-lymphocyte adherence to endothelial cells, augmented by phorbol-ester stimulation of the T cells, is inhibited by LFA-1 MAb (81). T lymphoblast adherence to endothelial cells and fibroblasts is also LFA-1 dependent (85, 86). Binding of lymphocytes to high endothelial venules in vitro and homing of lymphocytes to peripheral lymph nodes in vivo is specifically inhibited 50% by LFA-1 MAb; MAb to a lymphocyte homing receptor inhibits more completely (87).

LFA-1 in Antigen-Dependent Adhesion to Nonhematopoietic Cells

Elegant studies with antigen-specific T-cell hybridomas have shown that LFA-1 MAb block antigen presentation by lymphoid cells but do not block presentation by Ia-transfected fibroblasts or artificial Ia-containing membranes (48, 78, 88–90). Similarly, anti–LFA 1 MAb can inhibit the killing of hematopoietic target cells but not nonhematopoietic targets by the same CTL line (77, 90, 91). However, other studies have shown LFA-1 MAb can block killing of transfected fibroblasts even across a species barrier (92) and can moderately inhibit killing of endothelial cells (76, 91).

These discrepancies are not understood, and the idea that a ligand for LFA-1 is absent from nonhematopoietic cells is one of several possible explanations. Some feature of the surface of nonhematopoietic cells, which correlates with their tendency to grow as adherent cells, may facilitate stronger and hence LFA 1–independent adherence by lymphocytes. Both LFA 1–dependent and –independent mechanisms of adherence to endothelial cells have been defined (81). Endothelial cells and fibroblasts secrete extracellular matrix components such as fibronectin; perhaps extracellular matrix receptors (which appear related to LFA-1, see below) can substitute for LFA-1. Variable expression of the putative cell surface ligand for LFA-1, ICAM-1 (see below) might also explain some of the conflicting results.

A Unifying Hypothesis on the Involvement of LFA-1 in Specific Receptor-Dependent and -Independent Adhesion

The T-lymphocyte antigen receptor was held early on to have three functions: to dictate antigen specificity, to stabilize adhesion to the antigen-bearing cell, and to trigger delivery of effector functions. However, the function of the antigen receptor in stabilizing adhesion has received varying experimental support.

The antigen specificity of adhesion can be measured by comparing CTL conjugation with target cells bearing or lacking specific antigen. This topic has recently been reviewed in more detail by Martz (5). The ratio of specific:nonspecific adhesion for binding of mouse CTL, generated in vivo, to allogeneic tumor target cells is typically five-fold, and ranges from two- to fifteen-fold in different studies (5). In contrast, studies with cloned human CTL stimulated in vitro have shown equally strong conjugation with specific alloantigen positive and negative tumor target cells (42, 93). Thus, while the antigen receptor can contribute to target cell adhesion, antigen-independent conjugates are always found at significant levels, and sometimes they predominate.

In both mouse and human studies, the specificity of target cell killing is much ($\sim$thirty-fold) higher than that of conjugate formation (5). Thus, much or all of the specificity contributed by the antigen receptor appears to be in triggering effector function, while the contribution of the antigen receptor to adhesive specificity is variable. Antibodies to framework determinants of the antigen receptor and to the associated T3 molecule have been reported to inhibit killing but not conjugate formation, while anti–LFA 1 and Lyt-2 (CD8) MAb inhibited adhesion in parallel experiments

(42, 94). However, these results are from systems in which conjugation was not antigen specific (42) or in which specificity was not tested (94). It has been suggested that accessory molecule–dependent adhesion precedes antigen receptor ligation (42); however, results of experiments in the mouse system demonstrating antigen-specific adhesion argue that the antigen receptor can contribute to adhesion. While antigen-independent adhesion may procede antigen receptor engagement in systems in which conjugation is not antigen specific (42, 93), there is no kinetic evidence bearing on this question. The existence of both antigen-dependent and -independent conjugates suggests variations in the contribution of antigen-dependent and -independent mechanisms.

How is LFA 1–dependent adhesion activated physiologically? Reasoning by analogy to phorbol ester–stimulated homotypic adhesion, we hypothesize that binding of the antigen receptor during initial T cell–antigen presenting cell or effector cell–target cell contact releases diacyl glycerol (the physiologic activator of protein kinase C) and stimulates LFA 1–dependent adherence. Thus, binding of specific receptors to their ligands would provide only a small part of the decrease in free energy required to stabilize cell-cell adhesion; most would be provided by LFA-1. LFA-1 thus is hypothesized to provide a mechanism for amplifying adherence. Triggering of adhesion strengthening mechanisms by antigen receptor ligation would make antigen recognition more sensitive. We believe that the difference between CTL which show antigen-specific conjugate formation and those which show antigen-independent conjugate formation is that in vitro propagation and repeated stimulation with antigen of the latter cells has already activated LFA 1–dependent adherence to such an extent that it cannot be further elevated by antigen recognition.

Relation to Extracellular Matrix Receptors

Like LFA-1, the cell surface receptors for extracellular matrix components such as fibronectin are $\alpha\beta$ heterodimers (95). The N-terminal sequences of the α subunits of the vitronectin receptor and platelet gpIIb IIIa protein (a receptor for fibronectin, fibrinogen, and vitronectin) are homologous to the LFA-1 and Mac-1 α subunits (96a,b). The cDNA sequence of the human LFA-1 β subunit (Kishimoto, O'Connor, Lee, Roberts, Springer, 96c) shows $\sim 45\%$ homology to the chicken integrin (fibronectin receptor) β subunit (97). These results suggest that the LFA-1, Mac-1, and p150,95 family and the extracellular matrix receptor family constitutes a supergene family of adhesion molecules. Position-specific proteins that appear to control cell migration and localization in the developing Drosophila embryo may also belong to this supergene family (95). Although it has been proposed that the α subunits of Mac-1, LFA-1, and gpIIb IIIa are derived by differential splicing (98), the Northern blots with a Mac-1 DNA clone (101) and the absence of shared peptides (99, 100) do not support this idea.

Structural similarities between extracellular matrix receptors and the LFA-1, Mac-1, and p150,95 leukocyte adhesion proteins strongly suggest functional similarities. Extracellular matrix receptors recognize a core sequence of *arg-gly-asp* within ligands, and additional ligand sequences can modify specificity (95). Although a synthetic peptide containing the fibronectin recognition sequence failed to inhibit phorbol ester–stimulated lymphocyte aggregation (R. Rothlein, M. Pierschbacher, T. A. Springer, unpublished), it remains possible that LFA-1 recognizes ligand(s) containing similar sequences.

Matrix receptors form a link between the extracellular matrix and the cytoskeleton and are localized at adhesion plaques (97). It is interesting that LFA-1 has been found co-localized with actin at the site of adhesion between NK cells and targets (102). Modulation by lymphocyte activation of the interaction between LFA-1 and the cytoskeleton would be an attractive mechanism for regulating LFA 1–dependent adherence.

ICAM-1: A Putative LFA-1 Ligand

Coaggregation of LFA-1$^+$ cells with LFA-1$^-$ cells and LFA 1–dependent interactions of CTL and NK cells with LFA-1$^-$ targets suggest that at least one other molecule, perhaps a ligand for LFA-1, is involved in LFA 1–dependent leukocyte adhesion. To detect such molecules, MAb were elicited to lymphocytes from LFA-1 deficiency patients and screened for inhibition of phorbol ester–induced aggregation of LFA-1$^+$ cells (82). One MAb inhibited aggregation by reacting with a novel antigen distinct from LFA-1. This antigen, intercellular adhesion molecule-1 (ICAM-1), is widely distributed on cells of both hematopoietic and nonhematopoietic origin (86). ICAM-1 is expressed in low levels on peripheral blood cells and in higher levels on mitogen-activated T lymphoblasts, EBV-transformed B cells, and some cell lines of T cell and myeloid lineage (82, 86). Immunohistochemical staining of thin sections has shown that ICAM-1 is expressed on most vascular endothelial cells, tissue macrophages, germinal center dendritic cells, and thymic and mucosal epithelial cells (86). Expression on vascular endothelium is greatest in inflammation. ICAM-1 is a heavily glycosylated protein with a heterogeneous weight ranging from 90 kd to 114 kd. The deglycosylated precursor is 55 kd (86).

The ICAM-1 MAb inhibits the LFA 1–dependent phorbol ester–induced aggregation of T-lymphoblasts, B-lymphoblastoid, and myeloid cell lines, and also the binding of T lymphocytes to fibroblasts (Figure 1A, C) (82, 86). The binding of T cells to fibroblasts can be inhibited by either anti–ICAM 1 treatment of fibroblasts or by anti–LFA 1 treatment of the T cells. The inhibitory effects of anti–LFA 1 and anti–ICAM 1 are not additive. These results suggest a possible receptor-ligand interaction. ICAM-1 expression on dermal fibroblasts is increased several fold over a 4–10 hour period after treatment with interleukin-1 (IL-1) or interferon-γ (IFN-γ), and increased expression directly correlates with increased ICAM 1–dependent T-cell binding (86). Similarly, IL-1 and IFN-γ, as well as tumor necrosis factor, induce a rapid rise in ICAM-1 expression on endothelial cells (103).

The expression of ICAM-1 on cells at inflammatory sites and its induction on fibroblasts and endothelial cells by cytokines suggest that ICAM-1 may regulate adherence during inflammatory and immune responses. Up-modulation of ICAM-1 may facilitate margination and subsequent movement of lymphocytes into inflammatory regions and may potentiate the immune response. ICAM-1 expression may also regulate adhesion of monocytes. When monoblastoid U937 cells are induced to differentiate along the monocytic pathway, ICAM-1 expression increases almost twenty fold and correlates with the induction of LFA-1 dependent aggregation (82, 86).

The distribution of ICAM-1 and its role in many LFA 1–dependent adhesion systems make it an attractive candidate for the ligand of LFA-1. However, ICAM-1 MAb fails to inhibit the LFA 1–dependent aggregation of one T-cell line (82), and this argues against a hypothesis that ICAM-1 is the sole ligand for LFA-1. ICAM-1 may be a member of a family of related LFA-1 ligands, only some of which bear the determinant defined by the ICAM-1 MAb.

CD2 AND LFA-3

CD2 Is a Receptor for the Cell Surface Ligand LFA-3

The widely distributed surface molecule LFA-3 has recently been shown to be the ligand for the T-lymphocyte surface molecule CD2 (93, 104, 105). Inhibition of CTL-mediated killing and helper-T-lymphocyte functions by anti-CD2 and anti-LFA-3 appears to be due to the ability of these MAb to inhibit binding of CD2 to its ligand, LFA-3. The emphasis of the following section is to discuss the data supporting this hypothesis. Another mechanism involving negative signaling via CD2 or LFA-3 is also considered. Finally, the concept of stimulation of T-lymphocyte function through CD2 and the potential role of LFA-3 in this signaling is examined.

CD2 is a glycoprotein of 45–50 kd found on all T lymphocytes, large

238   SPRINGER ET AL

granular lymphocytes, and thymocytes (9, 13, 106–109). CD2 (cluster of differentiation 2) is the internationally accepted nomenclature (110) for the antigen variously referred to in earlier work as OKT11, T11, LFA-2, Leu5, Tp50, and the sheep erythrocyte receptor (9, 13, 106–109, 111). MAb to CD2 inhibit a variety of T-lymphocyte functions including antigen-specific T lymphocyte–mediated cytolysis and T lymphocyte–proliferative responses to lectins, alloantigens, and soluble antigens (9, 13, 109, 112, 113). CD2 MAb inhibit CTL-mediated killing by binding to the CTL rather than the target cell (which is often CD2$^-$) (13). CD2 MAb inhibit conjugation of CTL to target cells (29). Some (109, 114, 115) but not other (13, 109) CD2 MAb partially inhibit NK activity. CD2 MAb also inhibit rosetting of sheep erythrocytes with human T lymphocytes (106–108), as described in more detail below, and inhibit antigen-independent conjugation of thymocytes, T lymphoblasts, and CTL to B lymphoblast and K562 cells (93, 116, 117).

Inhibition of proliferation of peripheral blood lymphocytes and T-cell lines by anti-CD2 MAb is accompanied by a failure to induce IL-2 mRNA accumulation and IL-2 secretion, and a failure to express IL-2 receptor mRNA and protein (113, 118–122). The effects of anti-CD2 MAb are overcome in some systems by addition of exogenous IL-2, suggesting that the failure to express IL-2 receptor is secondary to the failure to secrete IL-2 (119, 121).

An MAb to a target cell structure (LFA-3) involved in CTL activity was described concurrently with anti-human LFA-1 and LFA-2 (CD2) MAb (9, 13). Since the LFA-3 MAb inhibited CTL-mediated killing by binding to the target cell, it was speculated that LFA-3 might be a ligand for either LFA-1 or CD2; it now appears that the latter is correct. This is discussed below. LFA-3 has a weight of 55–70 kd and has a broad tissue distribution including expression on endothelial, epithelial, and connective tissue cells in most organs studied and on most blood cells including erythrocytes (13, 104). LFA-3 has been mapped to chromosome 1 (123). LFA-3 MAb, like CD2 MAb, also inhibits a number of T helper lymphocyte–dependent functions (13) and inhibits conjugate formation between CTL and target cells (29, 93, 117).

Studies on antigen-independent conjugation of CTL to B-lymphoblastoid target cells have clarified the relationships of the LFA-1, CD2, and LFA-3 avidity-enhancing mechanisms (93). MAb to each antigen partially ($\sim 50\%$) inhibit conjugate formation. Combinations of saturating concentrations of LFA-1 MAb and CD2 MAb or LFA-1 MAb and LFA-3 Mab inhibit conjugate formation totally and thus are additive, while the combination of CD2 MAb and LFA-3 MAb is not any more effective than either MAb alone (13, 93). The LFA-1 dependent and the CD2/LFA-3 dependent pathways were further resolved by the dependence of the LFA-1 pathway but not the CD2/LFA-3 pathway on Mg^{+2} and temperature (93). Studies with purified CD2 and autologous E-rosetting have recently directly demonstrated an interaction between CD2 and LFA-3 (104).

Binding of human T lymphocytes to sheep (E)rythrocytes is an antigen-independent adhesion termed "E rosetting." E-rosetting has long been used as a technique for the purification of human T cells. Since some MAb against CD2 inhibit E-rosetting, CD2 has been inferred to be the "sheep E rosette receptor" (106–108, 112, 124). CD2 has recently been purified to homogeneity from the tumor cell line Jurkat (125). Purified CD2 inhibits T-lymphocyte E-rosetting and absorbs specifically to sheep E (125), confirming that CD2 interacts directly with a ligand on sheep E.

Human E, like sheep E, rosette with thymocytes, activated T lymphocytes, and some T-cell tumors; however, human E do not rosette with resting T cells (126). Thus, human E may express a ligand for CD2 analogous to that on sheep E. Since CD2 and LFA-3 are involved in the same functional adhesion pathway in CTL-target conjugation, the role of LFA-3 in rosetting of human T lymphocytes with human E was examined (104). Rosetting could be abolished by pretreating T lymphocytes with CD2 MAb or by pretreating E with LFA-3 MAb, suggesting a parallel between the CD2/LFA-3 functional pathway in CTL adhesion and E-rosetting. Experiments with purified CD2 suggest that it binds to LFA-3 (104a,b). Saturable binding of iodinated CD2 to human erythrocytes and to the B-lymphoblastoid cell line JT, a good target in CTL-mediated killing, is inhibited by LFA-3 MAb (104a,b). Reciprocally, preincubation of cells with purified CD2 inhibits LFA-3 MAb binding. High concentrations of purified CD2 aggregate sheep and human erythrocytes and the B lymphoblastoid cell line JY. Aggregation of human E and JY cells by purified CD2 protein is inhibited by anti-LFA-3 MAb (Figure 1D-H). This is the first demonstration that a purified lymphocyte protein can mediate adhesion and suggests that purified CD2 and its membrane bound counterpart bind directly to LFA-3 on human cells. Reciprocal results have recently been obtained with purified LFA-3 (104c), and cell surface CD2 shown to mediate binding of T lymphocytes to purified LFA-3 reconstituted into planar membranes.

An MAb to a sheep erythrocyte cell surface determinant that inhibits sheep E-rosetting with human T-lymphocytes has recently been reported (127). This MAb recognizes a 42-kd glycoprotein which may be the sheep erythrocyte equivalent of LFA-3; it does not cross-react with human erythrocytes. The purified antigen blocks sheep E-rosetting and reduces the staining of human T-lymphocytes by anti-CD2 MAb in immunofluorescence flow cytometry, suggesting that the antigen binds to CD2

(127). Hence this antigen was called the T11 target structure (T11TS). T11TS MAb inhibits the sheep mixed lymphocyte reaction, and the pattern of expression of T11TS on sheep peripheral blood lymphocytes and activated T-lymphocytes (128) is very similar to that of LFA-3 on human cells (13).

The Negative Signal Hypothesis

Negative signal transduction has been suggested as an alternative mechanism for inhibition of function by anti-LFA-3 and anti-CD2 MAb. In this hypothesis, the CD2 or LFA-3 molecules would be critical membrane transport proteins, channels, etc of general importance but not specifically involved in immune interactions; MAb bound to these molecules would inhibit many functions including adherence. It was suggested that anti-CD2 MAb may elicit a negative signal since removal of the CD2 epitope with trypsin eliminated the ability of anti-CD2 to inhibit the response to mitogenic anti-CD3 MAb (113). Anti-CD2 MAb also inhibited proliferation induced by the Ca^{+2} ionophore A23187 suggesting that the 'negative signal' occurred subsequent to the rise in intracellular Ca^{+2} (113). In a study on CTL-mediated killing, removal of the LFA-3 determinant by trypsin treatment of target cells did not affect their susceptibility to lysis, suggesting that LFA-3 did not participate in an adhesion strengthening interaction (129). Similar conclusions were reached in studies with human × mouse hybrids. The presence or absence of human LFA-3 on hybrid cells failed to correlate with their susceptibility to lysis by human CTL, while anti-LFA-3 could inhibit CTL-mediated killing of LFA-3$^+$ but not LFA-3$^-$ hybrid cells (123). The studies with trypsinization and somatic cell hybrids (113, 123, 129), are difficult to interpret, however, since the effects of CD2 and LFA-3 MAb were never examined in the same experiment. It would be interesting to extend the above studies by examining the effects of CD2 MAb and LFA-3 MAb in parallel. The possibility has not been examined that some of the inhibitory effects of anti-CD2 MAb on helper-T-lymphocyte functions (13, 112, 113, 118–122) may be mediated by abrogation of cell-cell contact and adhesion. Therefore, some data that have been interpreted as indicating negative signaling via CD2 may be due to inhibition of critical cell-cell interactions. Indeed, one study noted the failure of mitogen-stimulated lymphocytes to cluster in the presence of anti-CD2 (113). Because our conclusion that CD2 interacts with LFA-3 is supported by direct observations with purified molecules, we believe that it is more convincing than the negative signal hypothesis, which is based on complicated functional experiments.

Stimulation of Function by CD2 MAb

Activation-related epitopes have been defined on CD2 which are strongly expressed on thymocytes and activated mature lymphocytes but are absent or weakly expressed on resting peripheral blood lymphocytes (111, 130–132). Specific combinations of MAb to certain of these CD2 epitopes can result in proliferation which is IL-2 dependent (111, 133). This has been termed an 'alternative pathway' of T-cell activation, in contrast to the antigen receptor-CD3 complex "classical pathway." In one case an activation epitope (called $T11_3$) was induced on peripheral blood T lymphocytes by MAb to another non-E-rosette blocking epitope ($T11_2$); a combination of anti-$T11_2$ and anti-$T11_3$ led to proliferation in the absence of accessory cells (111). The $T11_2$/$T11_3$ MAb combination was also capable of inducing T-helper activity for antibody responses (111), antigen-independent killing by CTL clones, and killing of inappropriate targets by NK clones (134). Expression of the CD2 activation epitope D66 was increased after incubation with D66 MAb and a Fab fragment of rabbit anti-mouse IgG at 4°C (130). It was subsequently demonstrated that combinations of E-rosette blocking anti-CD2 MAb and D66 MAb could also stimulate proliferation, but only in the presence of accessory cells (133). The accessory cells (monocytes) in this system were contributing an Fc receptor-dependent interaction (133). With thymocytes, the combination of anti-CD2 MAb to the $T11_2$ and $T11_3$ epitopes failed to induce proliferation but did induce expression of IL-2 receptors and addition of exogenous IL-2 resulted in proliferation (135).

It has been suggested that the ability of MAb binding to specific epitopes on CD2 to induce or augment expression of the $T11_3$- and D66-type epitopes may be due to a conformational change in the CD2 molecule, because these effects occur rapidly and at 0°C (111, 130). However, it is also possible that the MAb to the activation epitopes bind with low avidity and that clustering of CD2 by a specific second anti-CD2 (resulting in a specific cluster geometry) enhances binding to the activation epitope by allowing bivalent interaction.

Do components of the 'classical pathway' (the antigen receptor–CD3 complex) regulate the 'alternative pathway' of T-cell activation? The modulation of CD3 (and the antigen receptor) with anti-CD3 prior to exposure to mitogenic combinations of anti-CD2 MAb abrogates subsequent proliferation in response to otherwise mitogenic combinations of anti-CD2 MAb (111, 133). However, CD3 modulation results in a generalized refractory state of T lymphocytes to signals evoking $[Ca^{+2}]^i$ increases (136). Some work suggests that while CD3 may regulate signaling through CD2, CD3 is not required. IL-2 receptor expression is induced by combinations of CD2 MAb on both $CD3^+$ and $CD3^-$ thymocytes (135). Binding of CD3 MAb to T cells at 4°C has been shown to induce the 9–1 CD2 activation epitope. However, in this case the CD3 MAb synergizes with the 9-1 CD2 MAb in inducing proliferation in the absence of accessory cells (137).

Stimulation via mitogenic combinations of anti-CD2 MAb results in rapid increases in $[Ca^{+2}]_i$ in T-lymphocyte clones and peripheral blood T lymphocytes (138–140). Furthermore, stimulation of a population of $Fc\gamma R^+$, $CD3^-$ lymphocytes with a single MAb to a D66-like epitope induces a small but significant increase in $[Ca^{+2}]_i$, again suggesting that surface CD3 expression is not required for signaling via CD2. It has also been shown that increases in $[Ca^{+2}]_i$ in response to PHA-P are specifically blocked by anti-CD2 MAb, suggesting that the PHA-P may activate a Ca^{+2} flux in T lymphocytes via CD2 (136).

CD2 and LFA-3 in Thymic Ontogeny

The regulation of proliferation and differentiation of immature T lymphocytes in the thymus is very likely a property of the thymic microenvironment which includes a number of cell types in addition to thymocytes (141). A major role has been proposed for the thymic epithelium in this regulation based on observations in pathological and normal states. Immunohistochemical staining of thymus has demonstrated a close association between thymocytes and thymic epithelial cells, particularly in the cortex where the most immature thymocytes are localized and where expression of T-cell antigen receptor first occurs. Recent advances in culture techniques for thymic epithelium have allowed their interactions with thymocytes to be studied in vitro with enriched epithelial cell populations obtained after serial passage (142). The mechanism by which thymocytes adhere to thymic epithelial cells depends largely on CD2 and LFA-3, based on the ability of CD2 MAb and LFA-3 MAb to block rosetting of thymic epithelial cells with thymocytes (105). Furthermore, CD2 MAb and LFA-3 MAb inhibit the accessory cell function of thymic epithelial cells for PHA stimulation of macrophage-depleted thymocytes (143, 144). Thymocyte IL-2 receptor expression is inhibited by CD2 and LFA-3 MAb. Mitogen responses of cells in their native microenvironment (4-mm thymus chunks) are also inhibited.

How immature thymocyte proliferation is triggered and regulated is of key importance in understanding thymocyte ontogeny. Purified thymic epithelial cells have been shown to provide accessory cell support for mitogen-induced proliferation of mature or $CD3^+$, and $CD2^+$ thymocytes and this proliferation is inhibited by CD2 and LFA-3 MAb (143). CD2 antibodies inhibit by binding to thymocytes and LFA-3 antibody inhibits by binding to TE cells. Moreover, purified thymic epithelial cells have been shown as well to induce spontaneous proliferation of the most immature or CD3$^-$, T4$^-$, T8$^-$, CD2$^+$, CD7$^+$ thymocytes. These results suggest that LFA-3 is an endogenous ligand for binding and activating thymocytes through CD2.

CD2 and LFA-3 in Mature T-Lymphocyte Function

Since CD2 has a cell surface ligand on human and sheep erythrocytes (LFA-3 and T11TS, respectively) and has the ability to transduce activation signals, it would be plausible for E-rosetting to affect the activation state of T lymphocytes. It has long been known that E-rosetted T lymphocytes are functionally altered. Sheep E-rosetting has been reported to result in acquisition of responsiveness to crude activated lymphocyte supernatants by resting human T lymphocytes (145). We have recently reproduced these results and found that CD2 MAb and LFA-3 MAb inhibit E-stimulated proliferation (M. Plunkett, T. A. Springer, unpublished). It will be interesting to determine whether isolated LFA-3 can duplicate this effect and to determine the effects of LFA-3 in other thymocyte and lymphocyte functional assays.

CD2- and LFA 3–dependent, antigen-independent, CTL-target conujugation does not result in increased $[Ca^{+2}]_i$, while the antigen-dependent interaction does (117). Thus, CD2 and LFA-3 appear to act strictly as an avidity enhancing mechanism in this CTL system. It will be of interest to determine if ligation of CD2 by LFA-3 in other systems can modulate $[Ca^{+2}]_i$ or T-cell function. Localization of T cells in the skin, which is often associated with T-cell activation, is of particular interest since epithelium is rich in LFA-3 (13).

CONCLUDING PERSPECTIVES

Studies on LFA-1, CD2, and LFA-3 have established the functional importance of these molecules in a wide variety of cell-cell interactions of the immune system. They may also be important in vivo in controlling lymphocyte migration and localization in specialized microenvironments. The expression of LFA-3 on thymic epithelial cells, of ICAM-1 on follicular dendritic cells, and the regulated expression of ICAM-1 on endothelial and epithelial cells may be particularly relevant to localization in vivo.

The importance of LFA-1 in the increased adhesiveness accompanying lymphocyte activation has been established, and a model was proposed in which regulation of LFA 1–dependent adhesiveness by specific receptors amplifies adherence. CD2 and LFA 3–dependent adherence may also be regulated by lymphocyte activation, as is suggested by the finding that thymocytes and T lymphoblasts, but not resting lymphocytes, show CD2 and LFA 3–dependent adherence to human erythrocytes. Considerable variation in SDS-PAGE mobility of CD2 and LFA-3 on different cell types suggests heterogenous glycosylation; this, increased expression of CD2 on activated lymphocytes, and CD2 activation epitopes are among many possible mechanisms for regulating CD2 and LFA 3–dependent adherence.

Demonstration that CD2 is a receptor for LFA-3 and the identification of ICAM-1 as a putative ligand for LFA-1 have advanced our understanding of how these molecules function. Much remains to be learned at the molecular level about possible additional ligands, receptor and ligand binding sites, regulation of receptor activity, interaction with the cytoskeleton, and signal transduction. The homologies discovered between the LFA-1 family of leukocyte adhesion proteins and extracellular matrix receptors suggest many new concepts concerning functional mechanisms which can now be tested. The complete structure of the LFA-1, ICAM-1, CD2, and LFA-3 proteins will soon be known from cloned genes and these should provide rich insights for future studies on the molecular basis of lymphocyte adhesion and signal transduction.

ACKNOWLEDGMENTS

We thank E. Martz, B. Haynes, A. Fischer, W. Golde, A. Hamann, G. Janossy, F. Takei, E. Vitetta, M. Pierschbacher, and J. Hansen for sharing prepublication manuscripts. Work from this lab was supported by NIH grant CA31798 and an American Cancer Society Faculty Award to T. A. Springer.

Literature Cited

1. Springer, T. A., Davignon, D., Ho, M. K., Kürzinger, K., Martz, E., Sanchez-Madrid, F. 1982. LFA-1 and Lyt-2,3 molecules associated with T lymphocyte-mediated killing; and Mac-1, an LFA-1 homologue associated with complement receptor function. *Immunol. Rev.* 68: 111–135
2. Golstein, P., Goridis, C., Schmitt-Verhulst, A. M., Hayot, B., Pierres, A., Van Agthoven, A., Kaufmann, Y., Eshhar, Z., Pierres, M. 1982. Lymphoid cell surface interaction structures detected using cytolysis-inhibiting monoclonal antibodies. *Immunol. Rev.* 68: 5–42
3. Martz, E., Heagy, W., Gromkowski,

S. H. 1983. The mechanism of CTL-mediated killing: Monoclonal antibody analysis of the roles of killer and target cell membrane proteins. *Immunol. Rev.* 72: 73–96
4. Burakoff, S. J., Weinberger, O., Krensky, A. M., Reiss, C. S. 1984. A molecular analysis of the cytolytic T lymphocyte response. In *Advances in Immunology, Vol. 36*, ed. F. J. Dixon. New York: Academic Press
5. Martz, E. 1986. LFA-1 and other accessory molecules functioning in adhesions of T and B lymphocytes. *Human Immunol.* In press
6. Anderson, D. C., Springer, T. A. 1986. Leukocyte adhesion deficiency: An

inherited defect in the Mac-1, LFA-1, and p150,95 glycoproteins. *Ann. Rev. Med.* In press

7. Davignon, D., Martz, E., Reynolds, T., Kürzinger, K., Springer, T. A. 1981. Lymphocyte function-associated antigen 1 (LFA-1): A surface antigen distinct from Lyt-2,3 that participates in T lymphocyte-mediated killing. *Proc. Nat. Acad. Sci. USA* 78: 4535–39

8. Pierres, M., Goridis, C., Golstein, P. 1982. Inhibition of murine T cell-mediated cytolysis and T cell proliferation by a rat monoclonal antibody immunoprecipitating two lymphoid cell surface polypeptides of 94,000 and 180,000 molecular weight. *Eur. J. Immunol.* 12: 60–69

9. Sanchez-Madrid, F., Krensky, A. M., Ware, C. F., Robbins, E., Strominger, J. L., Burakoff, S. J., Springer, T. A. 1982. Three distinct antigens associated with human T lymphocyte-mediated cytolysis: LFA-1, LFA-2, and LFA-3. *Proc. Natl. Acad. Sci. USA* 79: 7489–93

10. Hildreth, J. E. K., Gotch, F. M., Hildreth, P. D. K., McMichael, A. J. 1983. A human lymphocyte-associated antigen involved in cell-mediated lympholysis. *Eur. J. Immunol.* 13: 202–8

11. Marlin, S. D:, Morton, C. C., Anderson, D. C., Springer, T. A. 1986. LFA-1 immunodeficiency disease: Definition of the genetic defect and chromosomal mapping of alpha and beta subunits by complementation in hybrid cells. *J. Exp. Med.* 164: 855–67

12. Kürzinger, K., Reynolds, T., Germain, R. N., Davignon, D., Martz, E., Springer, T. A. 1981. A novel lymphocyte function-associated antigen (LFA-1): Cellular distribution, quantitative expression, and structure. *J. Immunol.* 127: 596–602

13. Krensky, A. M., Sanchez-Madrid, F., Robbins, E., Nagy, J., Springer, T. A., Burakoff, S. J. 1983. The functional significance, distribution, and structure of LFA-1, LFA-2, and LFA-3: cell surface antigens associated with CTL-target interactions. *J. Immunol.* 131: 611–16

14. Van Agthoven, A. J., Truneh, A. 1985. Lymphocyte function-associated antigens one (LFA-1) on B and on T lymphocytes bind a monoclonal antibody with different affinities. *Cell. Immunol.* 91: 255–62

15. Campana, D., Sheridan, B., Tidman, N., Hoffbrand, A. V., Janossy, G. 1986. Human leukocyte function-associated antigens on lymphohemopoietic precursor cells. *Eur. J. Immunol.* 16: 537–42

16. Miller, B. A., Antognetti, G., Springer, T. 1985. Identification of cell surface antigens present on murine hematopoietic stem cells. *J. Immunol.* 134: 3286–90

17. Kürzinger, K., Springer, T. A. 1982. Purification and structural characterization of LFA-1, a lymphocyte function-associated antigen, and Mac-1, a related macrophage differentiation antigen. *J. Biol. Chem.* 257: 12412–18

18. Sanchez-Madrid, F., Nagy, J., Robbins, E., Simon, P., Springer, T. A. 1983. A human leukocyte differentiation antigen family with distinct alpha subunits and a common beta subunit: The lymphocyte function-associated antigen (LFA-1), the C3bi complement receptor (OKM1/Mac-1), and the p150,95 molecule. *J. Exp. Med.* 158: 1785–1803

19. Sastre, L., Kishimoto, T. K., Gee, C., Roberts, T., Springer, T. A. (1986). The mouse leukocyte adhesion proteins Mac-1 and LFA-1: Studies on mRNA translation and protein glycosylation with emphasis on Mac-1. *J. Immunol.* 137: 1060–65

20. Springer, T. A., Thompson, W. S., Miller, L. J., Schmalstieg, F. C., Anderson, D. C. 1984. Inherited deficiency of the Mac-1, LFA-1, p150,95 glycoprotein family and its molecular basis. *J. Exp. Med.* 160: 1901–18

21. Dahms, N. M., Hart, G. . 1985. Lymphocyte function-associated antigen 1 (LFA-1) contains sulfated N-linked oligosaccharides. *J. Immunol.* 134: 3978–86

22. Rothlein, R., Springer, T. A. 1986. The requirement for lymphocyte function-associated antigen 1 in homotypic leukocyte adhesion stimulated by phorbol ester. *J. Exp. Med.* 163: 1132–49

23. Patarroyo, M., Beatty, P. G., Fabre, J. W., Gahmberg, C. G. 1985. Identification of a cell surface protein complex mediating phorbol ester-induced adhesion (binding) among human mononuclear leukocytes. *Scand. J. Immunol.* 22: 171–82

24. Beatty, P. G., Ledbetter, J. A., Martin, P. J., Price, T. H., Hansen, J. A. 1983. Definition of a common leukocyte cell-surface antigen (Lp95-150) associated with diverse cell-mediated immune functions. *J. Immunol.* 131: 2913–18

25. Davignon, D., Martz, E., Reynolds, T., Kürzinger, K., Springer, T. A. 1981. Monoclonal antibody to a novel lymphocyte function-associated antigen (LFA-1): Mechanism of blocking

of T lymphocyte-mediated killing and effects on other T and B lymphocyte functions. *J. Immunol.* 127 : 590–95

26. Hildreth, J. E. K., August, J. T. 1985. The human lymphocyte function-associated (HLFA) antigen and a related macrophage differentiation antigen (HMac-1): Functional effects of subunit-specific monoclonal antibodies. *J. Immunol.* 134 : 3272–80

27. Kaufman, Y., Golstein, P., Pierres, M., Springer, T. A., Eshhar, Z. 1982. LFA-1 but not Lyt-2 is associated with killing activity of cytotoxic T lymphocyte hybridomas. *Nature* 300 : 357–60

28. Sarmiento, M., Loken, M. R., Trowbridge, I., Coffman, R. L., Fitch, F. W. 1982. High molecular weight lymphocyte surface proteins are structurally related and are expressed on different cell populations at different times during lymphocyte maturation and differentiation. *J. Immunol.* 128 : 1676–84

29. Krensky, A. M., Robbins, E., Springer, T. A., Burakoff, S. J. 1984. LFA-1, LFA-2 and LFA-3 antigens are involved in CTL-target conjugation. *J. Immunol.* 132 : 218082.

30. Springer, T. A., Rothlein, R., Anderson, D. C., Burakoff, S. J., Krensky, A. M. 1985. The function of LFA-1 in cell-mediated killing and adhesion : Studies on heritable LFA-1, Mac-1 deficiency and on lymphoid cell self-aggregation. In *Mechanisms of Cell-Mediated Cytotoxicity II* ed. P. Henkart, E. Martz, pp. 311–20. New York : Plenum

31. Greenstein, J. L., Foran, J. A., Gorga, J. C., Burakoff, S. J. 1986. The role of T cell accessory molecules in the generation of class II-specific xenogeneic cytolytic T cells. *J. Immunol.* 136 : 2358–63

32. Hoffman, R. W., Bluestone, J. A., Oberdan, L., Shaw, S. 1985. Lysis of anti-T3-bearing murine hybridoma cells by human allospecific cytotoxic T cell clones and inhibition of that lysis by anti-T3 and anti-LFA-1 antibodies. *J. Immunol.* 135 : 5–8

33. Sanchez-Madrid, F., Davignon, D., Martz, E., Springer, T. A. 1982. Antigens involved in mouse cytolytic T-lymphocyte (CTL)-mediated killing : Functional screening and topographic relationship. *Cell. Immunol.* 73 : 1–11

34. Kohl, S., Springer, T. A., Schmalstieg, F. C., Loo, L. S., Anderson, D. C. 1984. Defective natural killer cytotoxicity and polymorphonuclear leukocyte antibody-dependent cellular cytotoxicity in patients with LFA-1/OKM-

1 deficiency. *J. Immunol.* 133 : 2972–78

35. Miedema, F., Terpstra, F. G., Melief, C. J. M. 1986. Functional studies with monoclonal antibodies against function-associated leukocyte antigens. In *Leukocyte Typing II, Vol. 3 : Human Myeloid and Hematopoietic Cells* ed. E. L. Reinherz, B. F. Haynes, L. M. Nadler, I. D: Bernstein, pp. 55–68. New York : Springer-Verlag

36. Miedema, F., Tetteroo, P. A. T., Hesselink, W. G., Werner, G., Spits, H., Melief, C. J. M. 1984. Both Fc receptors and LFA-1 on human Tγ lymphocytes are required for antibody-dependent cellular cytotoxicity (K-cell activity). *Eur. J. Immunol.* 14 : 518–23

37. Miedema, F., Tetteroo, P. A. T., Terpstra, F. G., Keizer, G., Roos, M., Weening, R. S., Weemaes, C. M. R., Roos, D., Melief, C. J. M. 1985. Immunologic studies with LFA-1 and Mol-deficient lymphocytes from a patient with recurrent bacterial infections. *J. Immunol.* 134 : 3075–81

38. Schmidt, R. E., Bartley, G., Levine, H., Schlossman, S. F., Ritz, J. 1985. Functional characterization of LFA-1 antigens in the interaction of human NK clones and target cells. *J. Immunol.* 135 : 1020–25

39. Schmidt, R. E., Bartley, G., Hercend, T., Schlossman, S. F., Ritz, J. 1986. NK-associated and LFA-1 antigens : Phenotypic and functional studies utilizing human NK clones. See Ref. 35 p. 133–44

40. Martz, E. 1977. Mechanism of specific tumor cell lysis by alloimmune T-lymphocytes: Resolution and characterization of discrete steps in the cellular interaction. *Contemp. Topics Immunobiol.* 7 : 301–61

41. Bongrand, P., Pierres, M., Golstein, P. 1983. T-cell mediated cytolysis: On the strength of effector-target cell interaction. *Eur. J. Immunol.* 13 : 424–29

42. Spits, H., van Schooten, W., Keizer, H., van Seventer, G., van de Rijn, M., Terhorst, C., de Vries, J. E. 1986. Alloantigen recognition is preceded by nonspecific adhesion of cytotoxic T cells and target cells. *Science* 232 : 403–5

43. Fischer, A., Durandy, A., Sterkers, G., Griscelli, C. 1986. Role of the LFA-1 molecule in cellular interactions required for antibody production in humans. *J. Immunol.* 136 : 3198–3203

44. Howard, D. R., Eaves, A. C., Takei, F,. 1986. Lymphocyte function-associated antigen (LFA-1) is involved in B cell activation. *J. Immunol.* 136 : In press

45. Keizer, G. D., Borst, J., Figdor, C. G., Spits, H., Miedema, F., Terhorst, C., De Vries, J. E. 1985. Biochemical and functional characteristics of the human leukocyte membrane antigen family LFA-1, Mo-1 and p150,95. *Eur. J. Immunol.* 15: 1142–47

46. Pircher, H., Groscurth, P., Baumhütter, S., Auget, M., Zinkernagel, R. M., Hengartner, H. 1986. A monoclonal antibody against altered LFA-1 induces proliferation and lymphokine release of cloned T cells. *Eur. J. Immunol.* 16: 172–81

47. Sanders, V. M., Snyder, J. M., Uhr, J. W., and Vitetta, E. S. 1986. Characterization of the physical interaction between antigen-specific B and T cells. *J. Immunol.* 137: 2395–404

48. Golde, W. T., Kappler, J. W., Greenstein, J., Malissen, B., Hood, L., Marrack, P. 1985. Major histocompatibility complex-restricted antigen receptor on T cells. VIII. Role of the LFA-1 molecule. *J. Exp. Med.* 161: 635–40

49. Kaufman, Y., Berke, G. 1983. Monoclonal cytotoxic T lymphocyte hybridomas capable of specific killing activity, antigenic responsiveness, and inducible interleukin secretion. *J. Immunol.* 131: 50–56

50. Golde, W. T., Gay, D., Kappler, J., Marrack, P. 1986. The role of LFA-1 in class II restricted, antigen specific T cell responses. Manuscript submitted

51. Spits, H., Yssel, H., Leeuwenberg, J., de Vries, J. E. 1985. Antigen-specific cytotoxic T cell and antigen-specific proliferating T cell clones can be induced to cytolytic activity by monoclonal antibodies against T3. *Eur. J. Immunol.* 15: 88–91

52. Strassmann, G., Springer, T. A., Adams, D. O. 1985. Studies on antigens associated with the activation of murine mononuclear phagocytes: Kinetics of and requirements for induction of Lymphocyte Function-Associated (LFA)-1 Antigen in vitro. *J. Immunol.* 135: 147–51

53. Strassmann, G., Springer, T. A., Somers, S. D., Adams, D. O. 1986. Mechanisms of tumor cell capture by activated macrophages: Evidence for involvement of lymphocyte function associated (LFA)-1 antigen. *J. Immunol.* 136: 4328–33

54. Anderson, D. C., Schmalstieg, F. C., Finegold, M. J., Hughes, B. J., Rothlein, R., Miller, L. J., Kohl, S., Tosi, M. F., Jacobs, R. L., Waldrop, T. C., Goldman, A. S., Shearer, W. T., Springer, T. A. 1985. The severe and moderate phenotypes of heritable Mac-1, LFA-1 deficiency: Their quantitative definition and relation to leukocyte dysfunction and clinical features. *J. Infect. Dis.* 152: 668–89

55. Arnaout, M. A., Spits, H., Terhorst, C., Pitt, J., Todd, R. F. I. 1984. Deficiency of a leukocyte surface glycoprotein (LFA-1) in two patients with Mol deficiency. *J. Clin. Invest.* 74: 1291–1300

56. Beatty, P. G., Harlan, J. M., Rosen, H., Hansen, J. A., Ochs, H. D., Price, T. D., Taylor, R. F., Klebanoff, S. J. 1984. Absence of monoclonal-antibody-defined protein complex in boy with abnormal leucocyte function. *Lancet* I: 535–37

57. Buescher, E. S., Gaither, T., Nath, J., Gallin, J. I. 1985. Abnormal adherence-related functions of neutrophils, monocytes, and Epstein-Barr virus-transformed B cells in a patient with C3bi receptor deficiency. *Blood* 65: 1382–90

58. Springer, T. A., Teplow, D. B., Dreyer, W. J. 1985. Sequence homology of the LFA-1 and Mac-1 leukocyte adhesion glycoproteins and unexpected relation to leukocyte interferon. *Nature* 314: 540–42

59. Lisowska-Grospierre, B., Bohler, M. C,. Fischer, A., Mawas, C., Springer, T. A., Griscelli, C. 1986. Defective membrane expression of the LFA-1 complex may be secondary to the absence of the beta chain in a child with recurrent bacterial infection. *Mol. Immunol.* In press

60. Suomalainen, H. A., Gahmberg, C. G., Patarroyo, M., Beatty, P. G., Schröder, J. 1986. Genetic assignment of gp90, leukocyte adhesion glycoprotein to human chromosome 21. *Somat. Cell Mol. Genet.* 12: 297–302

61. Krensky, A. M., Mentzer, S. J., Clayberger, C., Anderson, D. C., Schmalstieg, F. C., Burakoff, S. J., Springer, T. A. 1985. Heritable lymphocyte function-associated antigen-1 deficiency: Abnormalities of cytotoxicity and proliferation associated with abnormal expression of LFA-1. *J. Immunol.* 135: 3102–08

62. Davies, E. G., Isaacs, D., Levinsky, R. J. 1982. Defective immune interferon production and natural killer activity associated with poor neutrophil mobility and delayed umbilical cord separation. *Clin. Exp. Immunol.* 50: 454–60

63. Fischer, A., Seger, R., Durandy, A., Grospierre, B., Virelizier, J. L., Le

Deist, F., Griscelli, C., Fischer, E., Kazatchkine, M., Bohler, M. C., Descamps-Latscha, B., Trung, P. H., Springer, T. A. Olive, D., Mawas, C. 1985. Deficiency of the adhesive protein complex lymphocyte function antigen 1, complement receptor type 3, glycoprotein p150,95 in a girl with recurrent bacterial infections. *J. Clin. Invest.* 76: 2385–92

64. Dana, N., Todd, R. F. III, Pitt, J., Springer, T. A., Arnaout, M. A. 1984. Deficiency of a surface membrane glycoprotein (Mol) in man. *J. Clin. Invest.* 73: 153–59

65. Mentzer, S. J., Bierer, B. E., Anderson, D. C., Springer, T. A., Burakoff, S. J. 1986. Abnormal cytolytic activity of lymphocyte function associated antigen 1 deficient human cytolytic T lymphocyte clones. *J. Clin. Invest.* 78: 1387–91

66. Bissenden, J. G., Haeney, M. R., Tarlow, M. J., Thompson, R. A. 1981. Delayed separation of the umbilical cord, severe widespread infections, and immunodefiency. *Arch. Dis. Child.* 56: 397–99

67. Fischer, A., Descamps-Latscha, B., Gerota, I., Scheinmetzler, C., Virelizier, J. L., Trung, P. H., Lisowska-Grospierre, B., Perez, N., Durandy, A., Griscelli, C. 1983. Bone-marrow transplantation for inborn error of phagocytic cells associated with defective adherence, chemotaxis, and oxidative response during opsonised particle phagocytosis. *Lancet* II : 473–76

68. Fischer, A., Blanche, S., Veber, F., LeDeist, F., Gerota, I., Lopez, M., Durandy, A., Griscelli, C. 1986. Correction of immune disorders by HLA matched and mismatched bone marrow transplantation. In *Recent Advances in Bone Marrow Transplantation* ed. R. P. Gale, A. R. Liss, p. 000–00. New York : In press

69. Galili, U., Galili, N., Vanky, F., Klein, E. 1978. Natural species-restricted attachment of human and murine T lymphocytes to various cells. *Proc. Natl. Acad. Sci. USA* 75 : 2396–2400

70. Hamann, A., Jablonski-Westrich, D., Thiele, H. G. 1986. Contact interaction between lymphocytes is a general event following activation and is mediated by LFA-1. *Eur. J. Immunol.* 16 : 847–50

71. Erard, F., Nabholz, M., Dupuy-D'Angeac, A., MacDonald, H. R. 1985. Differential requirements for the induction of interleukin 2 responsiveness in L3T4+ and Lyt-2+ T cell subsets. *J. Exp. Med.* 162 : 1738–43

72. Patarroyo, M., Yogeeswaran, G., Biberfeld, P., Klein, E., Klein, G. 1982. Morphological changes, cell aggregation and cell membrane alterations caused by phorbol 12,13-dibutyrate in human blood lymphocytes. *Int. J. Cancer* 30 : 707–17

73. Hoshino, H., Miwa, M., Fujiki, H., Sugimura, T. 1980. Aggregation of human lymphoblastoid cells by tumor-promoting phorbol esters and dihydroteleocidin B. *Biochem. Biophys. Res. Commun.* 95 : 842–48

74. Mentzer, S. J., Gromkowski, S. H., Krensky, A. M., Burakoff, S. J., Martz, E. 1985. LFA-1 membrane molecule in the regulation of homotypic adhesions of human B lymphocytes. *J. Immunol.* 135 : 9–11

75. Mentzer, S. J., Faller, D. v., Burakoff, S. J. 1986. Interferon-gamma induction of LFA-1-mediated homotypic adhesion of human monocytes. *J. Immunol.* 137 : 108–113

76. Collins, T., Krensky, A. M., Clayberger, C., Fiers, W., Gimbrone, M. A. J., Burakoff, S. J., Pober, J. S. 1984. Human cytolytic T lymphocyte interactions with vascular endothelium and fibroblasts. Role of effector and target cell molecules. *J. Immunol.* 133 : 1878–84

77. Shimonkevitz, R., Cerottini, J. C., MacDonald, H. R. 1985. Variable requirement for murine lymphocyte function-associated antigen-1 (LFA-1) in T cell-mediated lysis depending upon the tissue origin of the target cells. *J. Immunol.* 135 : 1555–57

78. Watts, T. H., Brian, A. A., Kappler, J. W., Marrack, P., McConnell, H. M. 1984. Antigen presentation by supported planar membranes containing affinity-purified I-A^d. *Proc. Natl. Acad. Sci. USA* 81 : 7564–68

79. Patarroyo, M., Jondal, M., Gordon, J., Klein, E. 1983. Characterization of the phorbol 12,13-dibutyrate(P(Bu2)) induced binding between human lymphocytes. *Cell. Immunol.* 81 : 373–83

80. Patarroyo, M., Biberfeld, P., Klein, E., Klein, G. 1983. Phorbol 12,13-dibutyrate (P(Bu2))-treated human blood mononuclear cells bind to each other. *Cell. Immunol.* 75 : 144–53

81. Haskard, D., Cavender, D., Beatty, P., Springer, T., Ziff, M. 1986. T. Lymphocyte adhesion to endothelial cells : Mechanisms demonstrated by anti-LFA-1 monoclonal antibodies. *J. Immunol.* 137 : 2901–6

82. Rothlein, R., Dustin, M. L., Marlin, S. D., Springer, T. A. 1986. A human

intercellular adhesion molecule (ICAM-1) distinct from LFA-1 *J. Immunol.* 137: 1270–74

83. Patarroyo, M., Gahmberg, C. G. 1984. Phorbol 12,13-dibutyrate enhances lateral redistribution of membrane glycoproteins in human blood lymphocytes. *Eur. J. Immunol.* 14: 781–87

84. Hara, T., Fu, S. M. 1986. Phosphorylation of alpha, beta subunits of 180/100-Kd polypeptides (LFA-1) and related antigens. See Ref. 35, pp. 77–89

85. Mentzer, S. J., Burakoff, S. J., Faller, D. V. 1986. Adhesion of T lymphocytes to human endothelial cells is regulated by the LFA-1 membrane molecule. *J. Cell. Physiol.* 126: 285–90

86. Dustin, M. L., Rothlein, R., Bhan, A. K., Dinarello, C. A., Springer, T. A. 1986. A natural adherence molecule (ICAM-1): Induction by IL 1 and interferon-gamma, tissue distribution, biochemistry, and function. *J. Immunol.* 137: 245–54

87. Hamann, A., Jablonski-Westrich, D., Duijvestijn, A., Butcher, E. C., Harder, R., Thiele, H. G. 1986. LFA-1 is involved in binding of lymphocytes to high endothelial venules during homing. Manuscript submitted

88. Gay, D., Coeshott, C., Golde, W., Kappler, J., Marrack, P. 1986. The major histocompatibility complex-restricted antigen receptor on T cells. IX. Role of accessory molecules in recognition of antigen plus isolated IA. *J. Immunol.* 136: 2026–32

89. Lechler, R. I., Norcross, M. A., Germain, R. N. 1985. Qualitative and quantitative studies of antigen-presenting cell function by using I-A-expressing L cells. *J. Immunol.* 135: 2914–22

90. Naquet, P., Malissen, B., Bekkhoucha, F., Pont, S., Pierres, A., Hood, L., Pierres, M. 1985. L3T4 but not LFA-1 participates in antigen presentation by A^k-positive L-cell transformants. *Immunogenetics* 22: 247–56

91. Clayberger, C., Uyehara, T., Hardy, B., Eaton, J., Karasek, M., Krensky, A. M. 1985. Target specificity and cell surface structures involved in the human cytolytic T lymphocyte response to endothelial cells. *J. Immunol.* 135: 12–18

92. Cowan, E. P., Coligan, J. E., Biddison, W. E. 1985. Human cytotoxic T-lymphocyte recognition of an HLA-A3 gene product expressed on murine L cells: The only human gene product required on the target cells for lysis is the class I heavy chain. *Proc. Natl. Acad. Sci. USA* 82: 4490–94

93. Shaw, S., Luce, G. E. G., Quinones, R., Gress, R. E., Springer, T. A., Sanders, M. E. 1986. Two antigen-independent adhesion pathways used by human cytotoxic T cell clones. *Nature* 323: 262

94. Landegren, U., Ramstedt, U., Axberg, I., Ullberg, M., Jondal, M., Wigzell, H. 1982. Selective inhibition of human T cell cytotoxicity at levels of target recognition or initiation of lysis by monoclonal OKT3 and Leu-2a antibodies. *J. Exp. Med.* 155: 1579–84

95. Leptin, M. 1986. The fibronectin receptor family. *Nature* 321: 728

96a. Suzuki, S., Pytela, R., Arai, H., Argraves, W. S., Krusius, T., Pierschbacher, M. D., Ruoslahti, E. 1986. cDNA and amino acid sequences of the cell adhesion receptor recognizing vitronectin reveal a transmembrane domain and homologies with other adhesion receptors. *Proc. Natl. Acad. Sci. USA.* In press

96b. Charo, I. F., Fitzgerald, L. A., Steiner, B., Rall, S. C. Jr., Bekeart, L. S., Phillips, D. R. 1986. Platelet glycoproteins IIb and IIIa: Evidence for a family of immunologically and structurally related glycoproteins in mammalian cells. *Proc. Natl. Acad. Sci. USA* 83: 8351–55

96c. Kishimoto, T. K., O'Connor, K., Lee, A., Roberts, T. M., Springer, T. A. 1987. Cloning of the beta subunit of the leukocyte adhesion proteins: Homology to an extracellular matrix receptor defines a novel supergene family. *Nature.* Manuscript submitted

97. Tamkun, J. W., DeSimone, D. W., Fonda, D., Patel, R. S., Buck, C., Horwitz, A. F., Hynes, R. O. 1986. Structures of integrin, a glycoprotein involved in the transmembrane linkage between fibronectin and actin. *Cell* 46: 271–82

98. Cosgrove, L. J., Sandrin, M. S., Rajasekariah, P., McKenzie, I. F. C. 1986. A genomic clone encoding the alpha chain of the OKM1, LFA-1, and platelet glycoprotein IIb-IIIa molecules. *Proc. Natl. Acad. Sci. USA* 83: 752–56

99. Trowbridge, I. S., Omary, M. B. 1981. Molecular complexity of leukocyte surface glycoproteins related to the macrophage differentiation antigen Mac-1. *J. Exp. Med.* 154: 1517–24

100. Kürzinger, K., Ho, M. K., Springer, T. A. 1982. Structural homology of a macrophage differentiation antigen and an antigen involved in T-cell-mediated killing. *Nature* 296: 668–70

101. Sastre, L., Roman, J., Teplow, D., Dreyer, W., Gee, C., Larson, R.,

Roberts, T., Springer, T. A. 1986. A partial genomic DNA clone for the alpha subunit of the mouse complement receptor type 3 and cellular adhesion molecule Mac-1. *Proc. Natl. Acad. Sci. USA.* 83: 5644–48

102. Carpen, O., Keiser, G., Saksela, E. 1986. LFA-1 and actin filaments codistribute at the contact area in lytic NK-cell conjugates. *6th International Congress of Immunology, Toronto,* 572 (Abstract)

103. Pober, J. S., Gimbrone, M. A. Jr., Lapierre, L. A., Mendrick, D. L., Fiers, W., Rothlein, R., Springer, T. A. 1986. Overlapping patterns of activation of human endothelial cells by interleukin 1, tumor necrosis factor and immune interferon. *J. Immunol.* 137: 1893–96

104a. Plunkett, M. L., Sanders, M. E., Selvaraj, P., Dustin, M. L., Shaw S., Springer, T. A. 1986. Rosetting of activated T lymphocytes with autologous erythrocytes: Definition of the receptor and ligand molecules as CD2 and lymphocyte function-associated antigen-3 (LFA-3). *J. Exp. Med.* In press

104b. Selvaraj, P., Plunkett, M. L., Dustin, M., Sanders, M. E., Shaw, S., Springer, T. A. 1986. The T lymphocyte glycoprotein CD2 (LFA-2/T11/E-rosette receptor) binds the cell surface ligand LFA-3. *Nature.* Submitted

104c. Dustin, M. L., Sanders, M. E., Shaw, S., Springer, T. A. 1986. Purified lymphocyte function associated antigen-3 (LFA-3) binds to CD2 and mediates T lymphocytes adhesion. *J. Exp. Med.* Manuscript submitted

105. Wolf, L. S., Tuck, D. T., Springer, T. A., Haynes, B. F., Singer, K. H. 1986. Thymocyte binding to human thymic epithelial cells is inhibited by monoclonal antibodies to CD-2 and LFA-3 antigens. *J. Immunol.* In press

106. Howard, F. D., Ledbetter, J. A., Wong, J., Bieber, C. P., Stinson, E. B., Herzenberg, L. A. 1981. A human T lymphocyte differentiation marker defined by monoclonal antibodies that block E-rosette formation. *J. Immunol.* 126: 2117–22

107. Kamoun, M., Martin, P. J., Hansen, J. A., Brown, M. A., Siadak, A. W., Nowinski, R. C. 1981. Identification of a human T lymphocyte surface protein associated with the E-rosette receptor. *J. Exp. Med.* 153: 207–12

108. Verbi, W., Greaves, M. F., Schneider, C., Koubek, K., Janossey, G., Stein, H., Kung, P. C., Goldstein, G. 1982. Monoclonal antibodies OKT11 and OKT11a have pan-T reactivity and block sheep erythrocyte receptors. *Eur. J. Immunol.* 12: 81–86

109. Martin, P. J., Longton, G., Ledbetter, J. A., Newman, W., Braun, M. P., Beatty, P. G., Hansen, J. A. 1983. Identification and functional characterization of two distinct epitopes on the human T cell surface protein Tp50. *J. Immunol.* 131: 180–85

110. Haynes, B. F. 1986. Summary of T cell studies performed during the Second International Workshop and Conference on Human Leukocyte Differentiation Antigens. In *Leukocyte Typing II, Vol. 1: Human T Lymphocytes* ed. E. L. Reinherz, B. F. Haynes, L. M. Nadler, I. D. Bernstein, pp. 3–30. New York: Springer-Verlag

111. Meuer, S. C., Hussey, R. E., Fabbi, M., Fox, D., Acuto, O., Fitzgerald, K. A., Hodgdon, J. C., Protentis, J. P., Schlossman, S. F., Reinherz, E. L. 1984. An alternative pathway of T-cell activation: A functional role for the 50 kd T11 sheep erythrocyte receptor protein. *Cell* 36: 897–906

112. Van Wauwe, J., Goossens, J., Decock, W., Kung, P., Goldstein, G. 1981. Suppression of human T-cell mitogenesis and E-rosette formation by the monoclonal antibody OKT11A. *Immunology* 44: 865–71

113. Palacios, R.,, Martinez-Maza, O. 1982. Is the E receptor on human T lymphocytes a "negative signal receptor"? *J. Immunol.* 129: 2479–85

114. Fast, L. D., Hansen, J. A., Newman, W. 1981. Evidence for T cell nature and heterogeneity within natural killer (NK) and antibody-dependent cellular cytotoxicity (ADCC) effectors: A comparison with cytolytic T lymphocytes (CTL). *J. Immunol.* 127: 448–52

115. Bolhuis, R. L. H., Roozemond, R. C., van de Griend, R. J. 1986. Induction and blocking of cytolysis in CD2+, CD3− NK and CD2+, CD3+ cytotoxic T lymphocytes via CD2 50 kD sheep erythrocyte receptor. *J. Immunol.* 136: 3939–44

116. Schlesinger, M., Levy, J., Laskov, R., Hadar, R., Weinstock, J., Ben Bassat, H., Rabinowitz, R. 1983. The role of E receptors in the attachment of thymocytes and T lymphocytes to human target cells. *Clin. Immunol. Immunopathol.* 29: 349–58

117. Mentzer, S., Barbosa, J., Crimmins, M., Bierer, B., Strominger, J., Burakoff, S. 1986. Human CTL-target cell interactions involve antigen nonspecific adhesion but antigen specific activation *Fed. Proc.* 45: 1100 (Abstract)

118. Reed, J. C., Tadmori, W., Kamoun, M., Koretzky, G., Nowell, P. C. 1985. Suppression of interleukin 2 receptor acquisition by monoclonal antibodies recognizing the 50 kD protein associated with the sheep erythrocyte receptor on human T lymphocytes. *J. Immunol.* 134: 1631–39

119. Tadmori, W., Reed, J. C., Nowell, P. C., Kamoun, M. 1985. Functional properties of the 50 kd protein associated with the E-receptor on human T lymphocytes: Suppression of IL 2 production by anti-p50 monoclonal antibodies. *J. Immunol.* 134: 1709–16

120. Tadmori, W., Kant, J. A., Kamoun, M. 1986. Down regulation of IL-2 mRNA by antibody to the 50-kd protein associated with E receptors on human T lymphocyte. *J. Immunol.* 136: 1155–60

121. Reed, J. C., Greene, W. C., Hoover, R. G., Nowell, P. C. 1985. Monoclonal antibody OKT11A inhibits and recombinant interleukin 2 (IL 2) augments expression of IL 2 receptors at a pretranslational level. *J. Immunol.* 135: 2478–82

122. Wilkinson, M., Morris, A. 1984. The E receptor regulates interferon-gamma production: Four-receptor model for human lymphocyte activation. *Eur. J. Immunol.* 14: 708–13

123. Barbosa, J. A., Mentzer, S. J., Kamarck, M. E. Hart, J., Biro, P. A., Strominger, J. L., Burakoff, S. J. 1986. Gene mapping and somatic cell hybrid analysis of the role of human lymphocyte function-associated antigen-3 (LFA-3) in CTL-target cell interactions. *J. Immunol.* 136: 3085–91

124. Haynes, B. F. 1981. Human T lymphocyte antigens as defined by monoclonal antibodies. *Immunol. Rev.* 57: 127–61

125. Plunkett, M. L., Springer, T. A. 1986. Purification and characterization of the lymphocyte function-associated-2 (LFA-2) molecule. *J. Immunol.* 136: 4181–87

126. Baxley, G., Bishop, G. B., Cooper, A. G., Wortis, H. H. 1973. Rosetting of human red blood cells to thymocytes and thymus-derived cells. *Clin. Exp. Immunol.* 15: 385–92

127. Hünig, T. 1985. The cell surface molecule recognized by the erythrocyte receptor of T lymphocytes. *J. Exp. Med.* 162: 890–901

128. Hünig, T. R. 1986. The ligand of the erythrocyte receptor of T lymphocytes: Expression on white blood cells and possible involvement in T cell activation. *J. Immunol.* 136: 2103–8

129. Gromkowski, S. H., Krensky, A. M., Martz, E., Burakoff, S. J. 1985. Functional distinctions between the LFA-1, LFA-2, and LFA-3 membrane proteins on human CTL are revealed with trypsin-pretreated target cells. *J. Immunol.* 134: 244–49

130. Bernard, A., Gelin, C., Raynal, B., Pham, D., Gosse, C., Boumsell, L. 1982. Phenomenon of human T cells rosetting with sheep erythrocytes analyzed with monoclonal antibodies. "Modulation" of a partially hidden epitope determining the conditions of interaction between T cells and erythrocytes. *J. Exp. Med.* 155: 1317–33

131. Bernard, A., Brottier, P., Georget, E., Lepage, V., Boumsell, L. 1986. The epitopic dissection of the CD2 defined molecule: Relationship of the Second Workshop antibodies in terms of reactivities with leukocytes, rosette blocking properties, induction of positive modulation of the molecule, and triggering T cell activation. See ref. 110, pp. 53–66

132. Holter, W., Majdic, O., Liszka, K., Stockinger, H., Knapp, W. 1985. Kinetics of activation antigen expression by in vitro-stimulated human T lymphocytes. *Cellular Immunol.* 90: 322–30

133. Brottier, P., Boumsell, L., Gelin, C., Bernard, A. 1985. T cell activation via CD2 (T,gp50) molecules: Accessory cells are required to trigger T cell activation via CD2-D66 plus CD2-9.6/T11(sub 1) epitopes. *J. Immunol.* 135: 1624–31

134. Siliciano, R. F., Pratt, J. C., Schmidt, R. E., Ritz, J., Reinherz, E. L. 1985. Activation of cytolytic T lymphocyte and natural killer cell function through the T11 sheep erythrocyte binding protein. *Nature* 317: 428–30

135. Fox, D. A., Hussey, R. E., Fitzgerald, K. A., Bensussan, A., Daley, J. F., Schlossman, S. F., Reinherz, E. L. 1985. Activation of human thymocytes via the 50KD T11 sheep erythrocyte binding protein induces the expression of interleukin 2 receptors on both T3+ and T3− populations. *J. Immunol.* 134: 330–35

136. O'Flynn, K., Russul-Saib, M., Ando, I., Wallace, D. L., Beverley, P. C. L., Boylston, A. W., Linch, D. C. 1986. Different pathways of human T-cell activation revealed by PHA-P and PHA-M. *Immunology* 57: 55–60

137. Yang, S. Y., Chouaib, S., Dupont, B. 1986. A common pathway for T lymphocyte activation involving both the

CD3-Ti complex and CD2 sheep erythrocyte receptor determinants. *J. Immunol.* 137: 10971–1100

138. Alcover, A., Weiss, M. J., Daley, J. F., Reinherz, E. L. 1986. The T11 glycoprotein is functionally linked to a calcium channel in precursor and mature T-lineage cells. *Proc. Natl. Acad. Sci. USA* 83: 2614–18

139. June, C. H., Ledbetter, J. A., Rabinovitch, P. S., Martin, P. J., Beatty, P. G., Hansen, J. A. 1986. Distinct patterns of transmembrane calcium flux and intracellular calcium mobilization after differentiation antigen cluster 2 (E rosette receptor) or 3 (T3) stimulation of human lymphocytes. *J. Clin. Invest.* 77: 1224–32

140. O'Flynn, K., Knott, L. J., Russul-Saib, M., Abdul-Gaffar, R., Morgan, G., Beverley, P. C. L., Linch, D. C. 1986. CD2 and CD3 antigens mobilize Ca(2+) independently. *Eur. J. Immunol.* 16: 580–84

141. Haynes, B. F. 1984. The human thymic microenvironment. In *Advances in Immunology, Vol. 36.* New York: Academic Press

142. Singer, K. H., Harden, E. A., Robertson, A. L., Lobach, D. F., Haynes, B. F. 1985. In vitro growth and phenotypic characterization of mesodermal-derived and epithelial components of normal and abnormal human thymus. *Human Immunol.* 13: 161–76

143. Haynes, B. F. 1986. The role of the thymic microenvironment in promotion of early stages of human T cell maturation. *Clin. Res.* 34: 422–31

144. Denning, S. M., Tuck, D. T., Wolf, L. S., Springer, T. A., Singer, K. H., Haynes, B. F. 1986. Monoclonal antibodies to LFA-1, CD-2, and LFA-3 antigens inhibit human thymic epithelial accessory cell function for mature thymocyte activation. *J. Immunol.* In press

145. Larsson, E. L., Andersson, J., Coutinho, A. 1978. Functional consequences of sheep red blood cell rosetting for human T cells: Gain of reactivity to mitogenic factors. *Eur. J. Immunol.* 8: 693–96

Ann. Rev. Immunol. 1987. 5 : 253–77

THE ROLE OF CHROMOSOMAL TRANSLOCATIONS IN B- AND T-CELL NEOPLASIA[1]

Louise C. Showe and Carlo M. Croce

The Wistar Institute of Anatomy and Biology, 36th Street at Spruce, Philadelphia, Pennsylvania 19104

TRANSLOCATIONS AND CANCER

At the turn of the century, Boveri (1) suggested that abnormal chromosomal patterns in tumor samples reflect the processes responsible for the development of the malignant phenotype. However, the first demonstrated association of a consistent chromosomal rearrangement with a specific disease did not occur until 1960 when a marker chromosome, the Philadelphia chromosome (Ph[+]), was discovered to be consistently associated with chronic myelogenous leukemia (CML) (2). The Ph[+] chromosome was subsequently shown to result from a reciprocal translocation between chromosomes 9 and 22 (3).

Due to the introduction of new techniques for "high resolution" banding-analysis of chromosomes, a large majority of human cancer cells are now known to carry clonal cytogenetic changes (4). This has been most clearly demonstrated for the hematopoietic malignancies and has resulted in a renewed focus on chromosomal rearrangements as a causative factor in malignant transformation. While this view was originally not widely accepted outside of the discipline of cytogenetics, the painstaking efforts to observe and catalogue these rearrangements have now been justified. The application of the new technologies of molecular biology and somatic cell genetics have allowed the characterization of a variety of chromosomal abnormalities, and in particular chromosomal translocations, at the mol-

---

[1] The US Government has the right to retain a nonexclusive royalty-free license in and to any copyright covering this paper.

ecular level. As a result of these studies, chromosomal translocation is now generally accepted as a mechanism for protooncogene activation and malignant transformation. This acceptance is based primarily on the characterization of the translocations involving the *c-myc* gene in Burkitt lymphomas and the translocations involving the *c-abl* gene associated with the Ph+ chromosome in CML. In both cases, oncogene activation as a result of translocation has been demonstrated, although the mechanisms in the two cases are very different. In the case of *c-abl*, the translocation results in the production of a chimeric protein (derived from sequences contributed by both involved chromosomes) which has an acquired tyrosine kinase activity. In the case of *c-myc*, while mutations sometimes occur in the protein coding sequences, the translocated genes are for the most part normal, eliminating an altered protein as the mechanism of protooncogene activation. Transcriptional studies show instead that inappropriate expression of the *c-myc* gene as a function of the new chromosomal environment is a likely mechanism for protooncogene activation.

Observation of additional nonrandom translocations that associate with specific malignancies (5) suggest that protooncogene activation as a result of chromosomal translocations may be a more general mechanism than previously thought. In that event, the analysis of nonrandom chromosomal translocations involving regions where no known oncogenes have been previously mapped should result in the identification of new potential oncogenes that may also be important for cellular regulation and/or differentiation. In at least one example, a new putative protooncogene has been identified by the isolation of such a translocation breakpoint.

PROPERTIES OF TRANSLOCATIONS IN B-CELL AND T-CELL NEOPLASMS

Burkitt lymphoma (BL) is an extremely aggressive B-cell neoplasm primarily affecting children. While this tumor is only occasionally seen throughout most of the world, BL is endemic in equatorial Africa. These tumors are operationally referred to as sporadic BL (sBL) and endemic BL (eBL). As early as 1972, Manolov & Manolova (6) had detected a marker chromosome 14 (14q+) associated with a large majority of Burkitt lymphomas. Zech et al (7) subsequently showed that the 14q+ chromosome results from the translocation of a small segment of the long arm of chromosome 8 to chromosome 14. This t(8; 14) (q24; q32) translocation is present in 75% of the described cases of BL. A minority of the cases carry a variant class of translocations involving the same region on chromosome 8 and either chromosome 22 (16%) or chromosome 2 (9%)

(reviewed in 8, 9). The involvement of chromosomes 14, 22, and 2 was significant, as the immunoglobulin heavy-chain genes had been mapped to q32 on chromosome 14 (10), and the λ and κ light-chain genes had been mapped to q11 on chromosome 22 (11, 12) and p11 on chromosome 2, respectively (12, 13). It was suggested that an oncogene located at q24 on chromosome 8 was activated as a result of the translocation by juxtaposition to a locus that is specifically and highly expressed in B cells (14, 15). These speculations were strengthened by the localization of the cellular homologue (c-myc) of the avian myelocytomatosis virus v-myc gene to chromosome 8 at band q24 (16). The v-myc gene has been shown to be responsible for the production of avian bursal lymphomas (17).

Mineral oil–induced mouse plasmacytomas also contain chromosomal translocations similar to those involved in BL. The c-myc translocations associated with murine plasmacytomas have been extensively characterized. We draw on those characterizations where they relate specifically to the topics discussed here.

Translocations involving q32 on chromosome 14 are also frequently associated with non-Burkitt B-cell neoplasms (18). A chromosomal translocation involving q32 on chromosome 14 and q21 on chromosome 18 [t(14;18) (q32;q21)] is observed in more than 80% of follicular lymphomas, the most common of the B-cell malignancies (18). In addition, a t(11;14) (q13;q32) is frequently associated with B-cell diffuse lymphoma, multiple myeloma, and chronic lymphocytic leukemia (B-CLL) (5). No known protooncogenes had been previously mapped to either band q21 of chromosome 18 or band q13 of chromosome 11; these two translocations might thus mark the positions of two new protooncogenes.

Indeed, a transcript has been identified by Northern analyses which is encoded by sequences adjacent to the breakpoint on chromosome 18 in the t(14; 18) translocation (19, 19a). This gene has been named bcl-2 and codes for three different transcripts of 8.5, 5.5, and 3.5 kb. The gene consists of a 3.5 kb 5′ exon and a 5 kb 3′ exon, separated by a >65 kb intron. The 3.5 kb transcript is encoded entirely in the 5′ exon. Two overlapping reading frames have been identified with the potential of encoding polypeptides of 239 and 205 amino acids (19a). The function of these polypeptides and their roles in the development of follicular lymphoma are presently under study.

Translocations in T-cell neoplasias are also frequently associated with chromosome 14, but these usually involve 14q11.2 and not q32. The localization of the α-chain for the T-cell receptor (Tcr-α) to this region (20) has suggested that these translocations might involve mechanisms of translocation and oncogene activation similar to those involving the immunoglobulin loci in B-cell neoplasias.

The c-myc *Model System*

The human *c-myc* gene is composed of three exons separated by two introns (21, 22). The first exon is noncoding and contains two promoters separated by 160 base pairs (22). The second and third exons encode a protein of 439 amino acids with a calculated molecular weight of 48,812 (21). Southern blot analysis of DNA derived from a variety of Burkitt cell lines carrying a t(8;14) translocation with a *c-myc* probe indicated that the gene was rearranged and perhaps translocated in many cases (23, 24). This suggested that the *c-myc* gene was directly involved in the chromosomal translocations associated with BL. A direct demonstration that this was the case came from the analysis of somatic cell hybrids of BL cells (reviewed in 8) and by cloning of the translocation breakpoints from BL cell lines (reviewed in 9). Somatic cell hybrids, between murine myeloma and Burkitt lymphoma cells, which segregated the involved normal and translocated chromosomes were selected by Southern analysis of DNA derived from the hybrids, with probes for both *c-myc* and the immunoglobulin genes. Analysis of the t(8;14) (q24;q32) chromosomal translocations in the Daudi eBL cell line showed that the IgH locus is split by the translocation, with some variable region genes being translocated to the involved chromosome 8 (8q−). The genes for the immunoglobulin heavy chain (IgH) constant regions remain on chromosome 14 (14q+), thus providing a chromosomal orientation for the IgH locus (23a). The c-myc gene at 8q24, which is not detectably rearranged in the Daudi cell line (23), is translocated to chromosome 14. Analysis of the sBL ST486 cell line by somatic cell genetics showed that in this case the *c-myc* gene is split as a result of the translocation. The second and third coding exons are translocated to chromosome 14 while the first exon and the two normal promoter regions remain on chromosome 8 (23b). Transcription of the truncated, translocated gene occurs by the activation of cryptic promoters in the first intron (23c–24b). Therefore, the 5′ end of the *c-myc* gene is more proximal to the centromere on chromosome 8. It now appears that the *c-myc* rearrangement associated with the ST486 translocation is characteristic of the translocated gene in sBL while *c-myc* genes associated with eBL are characteristically intact. This difference is discussed more fully in the next section.

Somatic cell hybrids from variant Burkitt lymphomas with t(8; 22) and t(2; 8) have also been analyzed. In these translocations involving the Ig light-chain genes, *c-myc* always remains on chromosome 8 and the λ or κ genes translocate to a region 3′ of *c-myc* (25–26a). In both the t(8; 14) and the variant translocations, *c-myc* is always upstream of an immunoglobulin constant region. Figure 1 illustrates the chromosomal structure which

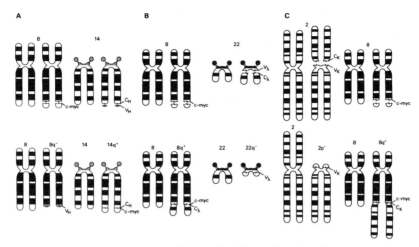

Figure 1 In Burkitt's lymphomas with the t(8; 14) translocation, the *c-myc* oncogene translocates to the heavy chain locus (A), and a portion of the immunoglobulin locus (V_H) is translocated to chromosome 8. In Burkitt's lymphoma with the less frequent t(8; 22) (B) and t(2; 8) (C) translocations, the *c-myc* oncogene remains on the involved chromosome 8, but the genes for the immunoglobulin light-chain constant regions (C_κ and C_λ) translocated to a region 3′ (distal) to the *c-myc* oncogene on the involved chromosome 8 (8q+). Again with these translocations, the immunoglobulin loci are split so that sequences encoding the variable portion of the immunoglobulin molecule (V_κ or V_λ) remain on chromosome 2 or 22, respectively.

results from these translocations and the basic orientation of both the *c-myc* and Ig loci on the marker chromosomes.

Differences Between Sporadic and Endemic Burkitt Lymphomas

While the breakpoint positions that map within the IgH region are distributed over perhaps hundreds of kilobases, they appear to fall into two general classes. Most African or endemic BL carry t(8; 14) translocations that associate *c-myc* with the VH or JH regions. The translocations associated with the sporadic BL resemble more closely those in mouse plasmacytomas and primarily involve the heavy-chain switch regions (8, 27). The characterization of a panel of BL cell lines and fresh patient samples shows that the breakpoints relative to the *c-myc* gene for the eBL and the sBL lymphomas also fall into two general classes. In 16 of 18 examples of sBL, the translocations truncate the *c-myc* gene within the first intron or exon, or in the near 5′ flanking region, as shown by Southern analysis. In 14 of 14 examples of eBL, the break-points are located far 5′ of the first exon (27), in most cases greater than 15 kb 5′ of exon 1 (24, 27). The differences in the character of the translocations in eBL as compared to sBL may be a function of the stage of B-cell differentiation represented by the two types of neoplasms (8, 27). The African BL express only cytoplasmic or membrane-bound immunoglobulins, indicative of early B cells, whereas the sporadic type secretes immunoglobulins, indicating a more advanced stage of B-cell differentiation (8, 27). If the translocation event is facilitated by the normal mechanisms involved in immunoglobulin V-D-J joining or in isotype switching, then the positions of the translocations within the IgH locus for eBL and sBL are consistent with the recombination systems active at that stage of B-cell differentiation where the translocation occurred. In this model the translocations in the African BL must have occurred in cells undergoing V-D-J rearrangements, whereas in the sporadic BL, they were undergoing isotype switching.

Environments for Translocations

Some indication of the environment that may lead to chromosomal translocations has come from the analysis of the regulation of B-cell division in populations where BL is frequently seen. Children in equatorial Africa, where BL is endemic, are widely infected by Epstein-Barr virus (EBV) and suffer repeated bouts of malaria. However, although more than 80% of the population is EBV-positive, eBL is restricted to certain regions of equatorial Africa and New Guinea (28). In these two areas, malaria is hyperendemic. It has been widely suggested that malarial infections can affect the normal function of the immune system through the mitogenic stimulation of B-cells by a malaria antigen (reviewed in 29). Alternatively, the primary role of malaria in eBL might be to alter T-cell function, as suggested by the recent demonstration that infection with *Plasmodium falciparum* causes a state of immunodeficiency characterized by a decrease in the T-cell population, in particular helper T cells, thus upsetting the T-cell regulation of B-cell expansion (30). Repeated bouts of malaria infection might result in repeated rounds of unregulated B-cell division, thereby increasing the possibility for a chromosomal translocation (30). EBV-infected B-cells would appear to have a selective advantage in the expansion process, as the eBL are almost always positive for EBV antigens and are characteristically early B-cells. Cells that happen to carry *c-myc* translocations and as a result unregulated *c-myc* transcription, could continue to divide, eventually leading to BL.

The frequent occurrence of BL in patients with acquired immune deficiency syndrome (AIDS) (31) may result from a physiological environment that directly parallels that produced by malaria infections in the development of eBl. In this scenario, the HTLV III/LAV infection of T-cells rather than malaria infection results in a reduced T-cell population and severe immunodepression (32). These conditions provide the environment for unregulated B-cell division and a probable chromosomal translocation, consistent with the model for the generation of translocations in eBL in association with malaria infection. Similar events might be involved in the generation of sBL that are not AIDS-related, such as the BL in transplant patients who are immunosuppressed (29). These observations suggest that repeated or protracted events of immunosuppression could provide the unregulated B-cell division that increases the probability of a translocation event. While the truly sporadic BL are usually negative for EBV antigens, the BL associated with AIDS, like the African BL, are frequently EBV-positive. A systematic comparison of the types of translocations and a determination of the level in the B-cell hierarchy represented in the BL associated with AIDS and the eBL could provide important clues to the greater susceptibility of these groups of patients to BL.

MECHANISMS FOR TRANSLOCATION

The linkage between B-cell proliferation, Ig gene rearrangement and expression, and the confined association of the translocations with the three Ig loci all suggested that translocations associated with BL may be facilitated by the recombination enzymes active in immunoglobulin gene rearrangements. In the simplest model, this would require that the appropriate recombination signals are also associated with the c-myc locus.

Translocations that Involve Ig Joining Signals

While the eBL make up the largest class of c-myc translocation in BL, very few translocation breakpoints in eBL have been characterized. Until recently, most of these characterizations have come through the analysis of somatic cell hybrids. Southern blot analysis of DNA from somatic cell hybrids for the eBL cell lines P3HR1 and Daudi indicated that the P3HR1 translocation involves the J_H region, while the Daudi translocation occurs into V_H genes, as variable region gene sequences are detected on both the $8q^+$ and the $14q^-$ chromosomes (8, 24). It now appears that the majority of these translocations occur primarily into V_H or J_H regions rather than Ig switch regions.

The first sequence data for breakpoints in nonswitch regions, however, did not come from the analysis of Burkitt-associated translocations but rather from the characterization of two additional chromosomal translocations associated with the IgH region on chromosome 14. The t(14; 18) is very frequently associated with follicular lymphomas and much more rarely with pre–B ALL. The t(11; 14) translocation has been found in cases of diffuse B-cell lymphoma, multiple myeloma, and chronic lymphocytic leukemia. These two translocations appeared analogous to the t(8; 14), although no known protooncogenes had been mapped to the translocation regions on either chromosome 11 or 18. The characterization of the breakpoints associated with these two translocations provides the basis for the hypothesis that the enzymatic systems which carry out Ig-joining reactions are involved in these chromosomal translocations. These observations have now been expanded to include the translocations in eBL as well as T-cell neoplasms.

Taking advantage of the association with the IgH locus, for which multiple probes were available, Tsujimoto et al (33, 34) have cloned the associated sequences on chromosomes 11 and 18 and sequenced a number of the breakpoints associated with both translocations (35, 36). In both cases, the translocations relative to the IgH locus are within the J regions (one example may associate with a V region). Most breakpoints on chromosome 18 are clustered in a 4-kb restriction fragment (36, 37), and a similar clustering of breakpoints is found on chromosome 11 (35). In several examples, a precise rearrangement with a J region is consistent with a D-J recombination. Based on this finding, it was suggested that the enzymes involved in the immunoglobulin joining reactions may also be involved in the translocations (36). In support of this hypothesis, sequences with homologies to the heptamer-nonamer joining signals have been identified within a region on chromosome 18 where at least five t(14; 18) translocations have been mapped (36–38, 19), as well as on chromosome 11 where two t(11; 14) breakpoints map. Based on the significant occurrence of the joining signals in chromosomal regions to which a number of translocations have been mapped, and the association with regions on chromosome 14 known to take part in gene rearrangements, it has been suggested that both the t(11; 14) and t(14; 18) translocations result from "mistakes" in V-D-J joining (35, 36). If this is the case, the translocations must have occurred in B-cells where the enzymes involved in immunoglobulin gene rearrangement are active, i.e. in pre–B cells. Several extra nucleotides, not derived from either chromosome 11, 18, or 14, are found at breakpoints. These extra nucleotides resemble the N regions thought to be added by terminal deoxytransferase (TdT) at the recombination junction during heavy-chain and Tcr gene rearrangement (38–39). Since TdT is a marker for pre–B and pre–T cells, the presence of N-like regions at the breakpoints supports the idea that the translocation occurs as a pre–B cell event. Characterization of the t(14; 18) and t(11; 14) breakpoints suggested that the translocations in African BL that typically occur into J_H and V_H regions may also involve joining signals on chromosome 8.

The breakpoint region for the P3HR1 cell line derived from an African

Burkitt lymphoma has now been cloned and sequenced (40). The break-point region on chromosome 8 in these cells is very close to the breakpoint region in cell line 380 derived from a pre–B ALL that harbors a t(8 ; 14) in addition to a t(14 ; 18) translocation (41). The t(14 ; 18) may occur by a joining reaction (36). The breakpoint regions for the t(8 ; 14) in P3HR1 and 380 occur within 63 basepairs of each other on chromosome 8. The 380 rearrangement directly involves the J_6 region on chromosome 14, while the P3HR1 breakpoint is upstream of J_5. Sequences with loose homologies to heptamer-nonamer signals are present on chromosome 8, and extra nucleotides homologous to N-regions are located at the breakpoint junctions (40). The breakpoint in the eBL Daudi cell line has been mapped by Southern blotting to this same general region on chromosome 8, which may be analogous to that of the clustered breakpoints on chromosome 18 which is frequently rearranged in follicular lymphomas. Using probes from the 5′ flanking region of c-myc and 3′ of the cloned P3HR1 breakpoint on chromosome 8, we can estimate that the breakpoints are greater than 35 kb upstream of the c-myc exon 1 (40). The position of the translocations in relationship to J or V regions, the identification of what appears to be N-region sequences at the breakpoint junctions, and the identification of joining signals on chromosome 8, 11, and 18 all strongly suggest that these translocations take place as a function of V-D-J recombination.

In the variant Burkitt cell lines, the translocations involve the c-myc gene and the λ and κ light-chain genes. Many of the variant BL were isolated as sporadic lymphomas, which primarily involve translocations into switch regions when the heavy-chain locus is involved. It was not expected that the translocations involving the λ or κ loci would utilize switch regions within the c-myc or light-chain loci, because the light-chain genes do not undergo isotype switching. Somatic cell hybrids had been used to characterize one translocation that occurs within $V\kappa$ genes (26) and another which occurs close to a λ constant region gene (25). The breakpoints in relation to the c-myc gene all occur within the 3′ flanking region of c-myc and range from 400 bp to greater than 50 kb 3′ of the third exon (42, 43). A small cluster of rearrangements map within the 4 kb immediately 3′ of exon 3 (Figure 2), and some sequence data are available for the translocations in this region (reviewed in 43). The second class of translocations occur beyond the 50 kb shown in the figure and may represent a class of translocations similar to those described for the variant translocations in mouse plasmacytomas (44). Seven variant translocations associated with mouse plasmacytomas have been mapped to a 4-kb region more than 75 kb downstream of the murine c-myc gene. This region has been designated "plasmacytoma variant translocation" (pvt-1) (44). Although it was suggested that a gene involved in c-myc

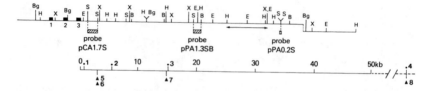

Figure 2 Restriction map of the *c-myc* locus. The restriction map was generated from overlapping bacteriophage λ clones derived from the genomic DNA library of Burkitt's lymphoma cells PA682 (PB). Two probes used in the library screening were pCA1.7S containing a 1.7-kb SstI fragment and pPA1.3SB containing a 1.3-kb SstI/BamHI fragment. The three exons of *c-myc* are indicated by numbered boxes 1, 2, and 3. Distances in kb from the 3′ end of *c-myc* are marked on the bottom line. Restriction sites beyond 38 kb from the 3′ end of *c-myc*, as shown on a discontinued line, were obtained by Southern blot analysis with probe pPAO.2S. The two-way arrow below the map indicates the region within which the breakpoint is located in the Burkitt's lymphoma cell line JI. Restriction sites are abbreviated as follows: K, KPNI; E, EcoRI; H, HindIII; B, BamHI; S, SstI; Bg, BglII. SstI, KPNI and BglII sites are not complete. Triangles identify published breakpoint positions in T-cell neoplasias, and the circles the breakpoint positions in B-cell neoplasias (reviewed in 43).

regulation might be located in this region, a transcription unit has not been identified. All of the translocations in this study involve the murine κ gene and the breakpoint positions in relation to κ all map 5′ of the κ enhancer region [the translocations in the t(8;14) eBL all map 5′ of the heavy-chain enhancer]. The pvt-1 region has now been shown by sequencing homology to be the murine homologue of a proviral integration site in rat thymomas Mis-1 (45), and has also been shown to contain proviral inserts in some murine T lymphomas. This raises the question of whether the same enzymatic mechanisms and the same recognition sequences are used in proviral insertion and chromosomal translocation into these regions.

Some of the best characterized translocation points are those associated with the lymphomas carrying the t(8; 14), where translocations of *c-myc* into heavy-chain variable, joining, and switch regions have been described. Translocations into constant region coding sequences have not yet been described.

The majority of the translocations associated with the sBL and the mouse plasmacytomas occur in IgH switch regions. A number of these have breakpoints within the first intron of *c-myc* (reviewed in 9, 43). Three of these breakpoints have been mapped within a 20-bp region situated between two groups of pentameric repeats characteristic of heavy-chain switch regions (46). Two additional breakpoints in the JD38 (L. Showe, unpublished observations) and the Ly65 (47) cell lines occur approximately 100 bp further 5′ within intron 1 and also within a region that contains a number of switch-like repeats (43, 47). Switch pentanucleotides are also found near the breakpoint in the BL22 cell line, which is 5′ of the first exon of c-myc (48).

The most complete characterization of the translocations into switch regions comes from the analysis of the breakpoints in mouse plasmacytomas where translocations occur primarily into the Sα region. Piccoli et al (49) compared the sequences associated with 12 different c-myc breakpoints and noted that a large number of these sequences included a GAGG tetranucleotide situated 10–12 bp from the breakpoint. They suggest that this sequence, which meets the minimum requirements for a switch consensus (49), may play an important part in the translocation. This tetranucleotide occurs 8 times within the first intron of the murine c-myc gene (49) and 24 times within a 125-bp sequence (reviewed in 50) of the murine S_a region where the majority of the translocations associated with plasmacytomas occur.

Although the spacing of the GAGG sequence in relation to the breakpoints in the human BL is not so precise as for the plasmacytomas, this tetranucleotide is also frequently associated with these breakpoints. In fact, the GAGG tetranucleotide is associated with the murine and human Ig genes, T-cell receptor genes and the c-myc genes at a statistically significant frequency (reviewed in 50). As the requirements for isotype switching are not well defined, it is not clear whether the short arrays of pentanucleotide repeats and the GAGG sequences are sufficient to direct this type of recombination. However, simple homologous sequences which probably represent enzyme binding sites (such as for topoisomerase I) have been identified at regions of nonhomologous recombination frequently associated with viral insertion or excision (51). The frequent occurrence of the GAGG tetranucleotide within introns and flanking regions of genes where switching events do not take place suggests that they are not involved in the switch recombination per se. The periodic occurrence of this sequence within these loci may play a more general role, perhaps the maintenance of these regions in a configuration favorable for recombination by interaction with specific proteins involved in recombination or gene regulation. A diagram of the various translocations involving c-myc and the Ig loci is shown in Figure 3.

The similarities between the Ig genes and the Tcr genes with respect to the mechanisms of gene assembly and tissue-specific expression make the Tcr genes prime candidates for involvement in chromosomal translocations. Thus, it was not surprising to find that the gene for the α-chain of the Tcr maps to a region on chromosome 14 (band q11) (20, 20a) frequently involved with translocations associated with T-cell neoplasias (52, 53). The α locus for the T-cell receptor may play a role in T-cell leukemias similar to the role of the Ig gene in B-cell malignancies (20, 53).

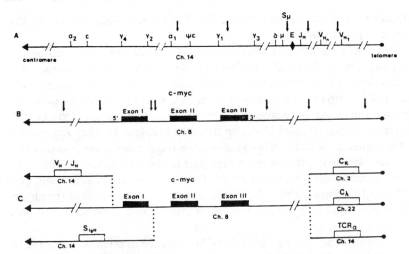

Figure 3 Translocation points within the *c-myc*, Ig and Tcr loci. Diagrammatic representation of the different classes of *c-myc* translocations involving the immunoglobulin genes and the gene for the α-chain of the T-cell receptor. (A) is the normal IgH locus at 14q32; (B), the normal *c-myc* gene on 8q24; (C), representation of the region 5' and 3' of *c-myc* which are utilized in translocations involving *c-myc* and the Ig heavy- and light-chain genes or the Tcr-α gene. Diagrams are not to scale, and vertical breaks indicate distances are not known.

Somatic cell hybrids between leukemic cells carrying a t(11; 14) (q13q11) from two patients with T-cell acute lymphocytic leukemia and mouse leukemic T-cells have been analyzed. The Tcr-α locus is split in the translocation carried by these cells, with the Cα genes translocated to chromosome 11 and at least some Vα genes remaining on chromosome 14 (54). These studies demonstrate the direct involvement of the Tcr-α locus in these translocations. No known protooncogene maps to the vicinity of this breakpoint on chromosome 11. It should be possible to isolate the t(11; 14) breakpoint region using genomic probes derived from the Tcr-α locus. This could lead to the identification of a new protooncogene and provide probes for the q13 region on chromosome 11. The latter might be useful in studies of Wilms tumor and aniridia, which characteristically have a deletion in the 11q13 region.

A further analogy with B-cell translocation has come from the characterization of chromosomal translocations involving the *c-myc* gene and the gene for the α-chain of the T-cell receptor (55, 56) which have now been described in human T-cell leukemias and lymphomas. These translocations are directly analogous to the Burkitt variant translocations as the *c-myc* gene remains on chromosome 8 and the Tcr-α locus translocates into the 3′ flanking region. The rearrangements 3′ of *c-myc* separate into two classes: those that cluster within a 4-kb region just 3′ of exon 1 and those that map at some position > 50 kb 3′ (55, L. Showe, unpublished results). The positions of some translocations in T-cell neoplasms in relation to the *c-myc* gene are indicated in Figures 2 and 3. The sequence through the breakpoint for one of these translocations carried by the SKW3 cell line which maps within 3 kb of *c-myc* is shown in Figure 4 (57). The SKW3 cell line was derived from a patient with T-ALL, and the cells carry a t(8; 14) (q24; q11.2) chromosomal translocation (55, 56). The breakpoint for this translocation maps to a region 3′ of *c-myc* where several other B- and T-cell translocations cluster on chromosome 8 (43). The breakpoint on chromosome 14 is in the region of a J_α gene which maps 35 kb upstream of the Tcr-α constant region (57). The translocation is a precise rearrangement with a Jα region. A heptamer-like joining signal is located just 3′ of the breakpoint on chromosome 8, and a number of nonamer-like sequences are indicated further downstream. The spacings between the heptamers and nonamers on chromosome 8 do not conform to the 23 or 12 bp spacer rules. This translocation could not have occurred by a classical joining reaction. We have previously suggested a similar illegitimate use of joining signals at a recombination breakpoint for a variant BL which occurs within this region of chromosome 8 (50). Therefore, translocations within the 3′ flanking region of *c-myc* in at least some cases appear to occur by mechanisms similar to those that have been suggested for the eBL and the t(14;18) and t(11;14) chromosomal translocations. Is there a similar region of clustered recombination signals defined by the far 3′ breaks in human B- and T-cell translocations, and is this far 3′ region homologous to the pvt-1 locus in the mouse system? The answers to these questions must await cloning and characterization of the far 3′ breakpoints.

The characterization of a breakpoint on the inverted chromosome 14 carried by the SupT1 cell line, derived from a patient with a T-ALL, provides further evidence for the involvement of the joining enzymes in chromosomal translocations associated with B- and T-cell neoplasias (58, 58a). This chromosomal inversion involves q32, where the IgH locus is located, and q11.2 where the Tcr-α locus is located. The characterization of the cloned breakpoint region shows a recombination between a V_H gene and a J gene of Tcr-a (58, 58a). Analysis of somatic cell hybrids which carry only the inverted chromosome 14 with a variety of probes derived from 5′ of the Jα gene involved in this recombination indicates that a large segment of the Tcr-α locus including both Jα and Vα sequences has been deleted from this chromosome (L. Showe, unpublished results). This suggests that the inversion event occurred at some position upstream of the

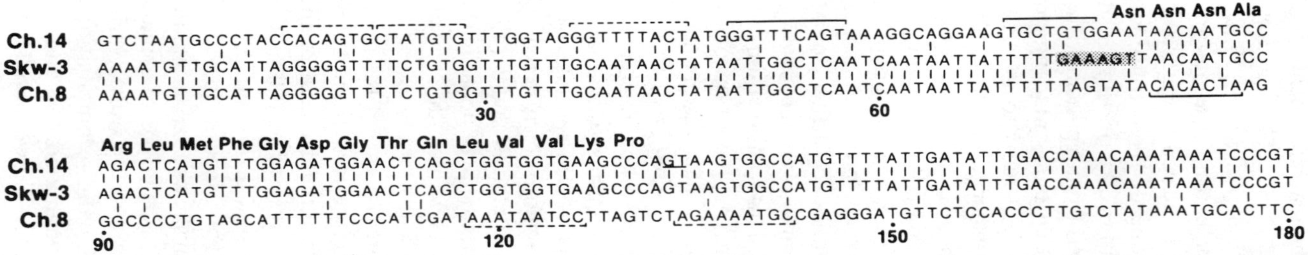
Asn Asn Asn Ala
Ch.14 GTCTAATGCCCTACCACAGTGCTATGTGTTTGGTAGGGTTTTACTATGGGTTTCAGTAAAGGCAGGAAGTGCTGTGGAATAACAATGCC
Skw-3 AAAATGTTGCATTAGGGGGTTTTCTGTGGTTTGTTTGCAATAACTATAATTGGCTCAATCAATAATTATTTTTGAAAGTTAACAATGCC
Ch.8 AAAATGTTGCATTAGGGGGTTTTCTGTGGTTTGTTTGCAATAACTATAATTGGCTCAATCAATAATTATTTTTTAGTATACACACTAAG
30
60
Arg Leu Met Phe Gly Asp Gly Thr Gln Leu Val Val Lys Pro
Ch.14 AGACTCATGTTTGGAGATGGAACTCAGCTGGTGGTGAAGCCCAGTAAGTGGCCATGTTTTATTGATATTTGACCAAACAAATAAATCCCGT
Skw-3 AGACTCATGTTTGGAGATGGAACTCAGCTGGTGGTGAAGCCCAGTAAGTGGCCATGTTTTATTGATATTTGACCAAACAAATAAATCCCGT
Ch.8 GGCCCCTGTAGCATTTTTTCCCATCGATAAATAATCCTTAGTCTAGAAAATGCCGAGGGATGTTCTCCACCCTTGTCTATAAATGCACTTC
90
120
150
180

Figure 4 DNA sequences of the breakpoint in SKW-3 and the sequences for the corresponding regions on the normal chromosome 8 and chromosome 14. The predicted protein sequence for the Jα coding segment is shown. Brackets indicate the heptamer and nonamer signal sequences adjacent to the breakpoints on chromosomes 8 and 14. Additional sequences with homologies to heptamers and nonamers are indicated by dashed brackets. The RNA donor splice site 3′ of the Jα coding region is underlined.

involved Jα gene and that the VH/Jα joining occurred as a subsequent joining event. Regardless of whether the VH/Jα joining is a direct result of the inversion or occurred as a second event, it suggests that the Ig joining enzymes were active at the time of the inversion event and provides further evidence that the chromosomal rearrangements have occurred in an early T-cell environment.

Joining reactions that appear to violate the one turn–two turn joining rule have been suggested to account for a V–J rearrangements in the β locus of the T-cell receptor gene (Tcr-β) (59, 60). Illegitimate recombinations may also account for the high percentage of inactive lymphocytes produced in the thymus. Thus, abberant "joining" events are not unusual within the Ig and Tcr loci. Single heptamer-like sequences that occur near the deletion sites for heavy-chain enhancer regions have also been suggested to play a role in that deletion process (61). The presence of joining signals in regions where multiple breakpoints occur, which are not associated with the Ig loci (chromosomes 8, 11, and 18) regions, strongly suggests that Ig joining enzymes are active participants in some classes of chromosomal translocations.

C-MYC REGULATION

The normal c-myc gene is not expressed in terminally differentiated cells. The observation that the normal c-myc is turned off in BL and mouse plasmacytomas, while the gene associated with the translocation is inappropriately transcribed, is consistent with this (61a,b). Two different hypotheses have been proposed to explain the expression of the translocation associated c-myc gene in BL and plasmacytomas. The first invokes a negative regulatory system mediated by trans acting factors and their interactions with putative repressor binding sites 5' of exon 1 (62); the other postulates the deregulation to be a function of the association of the myc gene with a highly expressed, tissue-specific locus (Ig genes for B-cell neoplasias, TCR genes for T-cell neoplasias) by cis activation (8, 61a,b).

Expression of the Truncated c-myc

The characterization of the translocations associated with the sBL provides the foundation for the first model. These translocations frequently involve the separation of the first exon and putative 5'-regulatory regions (62) from the second and third coding exons. Transcription of the translocated c-myc gene is initiated from cryptic promoters in the first intron (24, 24b). It has been suggested that these translocations interfere with a negative regulatory region in the 5' flanking region of exon 1 (62). While it is reasonable to suggest that rearrangements or mutations surrounding the first exon might affect the regulation of *c-myc* transcription, a variety of observations suggest that decapitation per se is not sufficient to explain deregulation.

In the murine system, plasmacytoma cells transfected with a normal *c-myc* gene from which the first exon has been removed (63) or with a similar defective gene cloned from a BL cell line do not express the decapitated gene construct (64). The only exception is the truncated *c-myc* gene cloned from the Manca cell line where the IgH enhancer is present in the gene construction used for the transfections (24, 64). Even so, the level of expression of the transfected Manca gene is only a fraction of that in the gene in Manca cells (24, 64). These results suggest that the association with the Ig locus is necessary to drive the transcription of the truncated gene, regardless of the effects of truncation on transcriptional regulatory elements in the 5' flanking region. While expression of transfected, truncated *c-myc* genes that are not associated with the IgH enhancer has been reported, these constructions always include a viral enhancer and/or a selectable marker that could function to stimulate *c-myc* expression from these constructs.

While transfection studies with cloned genes are an important tool for unraveling the mechanisms of *c-myc* deregulation, ideally one would like to compare the levels of *c-myc* expression from a truncated gene that remains in the normal position on chromosome 8 with one translocated to the IgH locus. A patient carrying a thyroid carcinoma has been described whose normal lymphocytes carry a *c-myc* gene with a deletion that removes exon 1 and a part of intron 1 (65). In situ hybridization detects *c-myc* sequences only on the long arm of chromosome 8, eliminating a possible translocation. Mapping data indicate that the cryptic promoters within the first intron are not deleted. The level of *c-myc* transcription in peripheral blood cells of this patient is equivalent to that in normal lymphocytes, and no elevation of *c-myc* transcripts was detected in the patient's thyroid cells (65). These results suggest that decapitation of the *c-myc* in situ is not sufficient to activate constitutive transcription of this gene, and so these data support the results of transfection studies which demonstrate that the additional association of the Ig locus is required for *c-myc* activation.

c-myc *and Ig Expression*

The strongest evidence for the influence of the Ig loci on translocated *c-myc* expression has come from the analysis of somatic cell hybrids with BL cell lines. S1 protection studies of RNA derived from somatic cell hybrids between sBL and mouse plasmacytomas show that the truncated *c-myc* genes derived from the t(8 ; 14) sBL as well as the nontruncated genes derived from the variant sBL are efficiently transcribed in this background (61a,b). Similar sBL hybrids with a lymphoblastoid fusion partner show that the truncated gene is turned off in this background (23b) even though the normal endogenous *c-myc* gene is transcribed. These results imply that the expression of the truncated *c-myc* gene requires a positive regulatory element. They do not differentiate between a lack of expression due to the absence of a binding site for a *trans* acting regulator or because of the inactivity of a *cis*-regulator in the lymphoblastoid environment. However, based on the results described below, we suggest that in the proper environment (BL or plasmacytoma), the associated Ig locus is capable of positively regulating the *c-myc* gene by *cis* activation. This regulation must in some cases be independent of the presence of the IgH enhancer, as that regulatory element is translocated to the reciprocal marker chromosome in most of the sBL.

Expression of the nontruncated c-myc

Analysis of *c-myc* expression in somatic cell hybrids between the eBL Daudi cell line and a lymphoblastoid fusion partner shows that the translocated nontruncated *c-myc* gene is expressed in the lymphoblastoid background (68), in contrast to the hybrids with the sBL. Most of the eBL translocations do result in the placement of *c-myc* upstream of the IgH enhancer (8, 27), and the expression of the Daudi *c-myc* gene in the lymphoblastoid background may be a function of the presence of that enhancer upstream of the translocated *c-myc* gene (8, 66). By contrast the translocations in the sBL, which represent a more differentiated B-cell stage, consistently result in the *c-myc* gene and the IgH enhancer residing on different marker chromosomes. Therefore, the presence of the IgH enhancer cannot be responsible for the induction of transcription of the translocated *c-myc* in these tumors.

The presence of *c-myc* and the IgH enhancer on the same marker chromosome in the eBL may reflect the need for this enhancer in the early B cell to achieve the efficient expression of the IgH locus and thus for the associated *c-myc* gene. While the IgH and IgK enhancers are thought to play a major part in determining the tissue-specific expression of the Ig genes in early B-cells, they do not appear to be required for continued Ig expression in more differentiated cells, as the levels of Ig expression are unaffected in a number of B-cell lines that have spontaneously deleted the enhancer (69). Therefore, while the onset of Ig transcription presumably requires the organization of the locus into a stable transcription complex, a process which is orchestrated by the enhancer element, the enhancer is not required to maintain this activated structure (67). We would suggest that eBL cells require the presence of the enhancer to maintain both Ig and *c-myc* transcription at least at the time of translocation, whereas the Ig loci in the more differentiated sBL cells have achieved a stage of activation that no longer requires the continued presence of the enhancer element. While the effect of this activated state may be transmitted to the associated *c-myc* gene in the sBL or plasmacytoma background, the Ig locus is not able to maintain this activated state when placed in the lymphoblastoid background in somatic cell hybrids. This observation suggests that *trans*-acting factors expressed in mature B-cells but not in lymphoblastoid cells are required to maintain the Ig locus in this activated state.

The *c-myc* genes that are translocated 5′ of the IgH enhancer and may be transcriptionally activated by it are in some cases located >35 kb upstream of this sequence (40). This requires that the effect on gene transcription can be transmitted over long distances. Enhancer sequences capable of *cis*-activation of gene transcription were originally thought to be able to exert this influence over a distance of a few hundred base pairs up to a few kb. These estimates of the range of enhancer influence are primarily derived from studies of transcriptional activation of cloned genes. Recent reports have shown that the endogenous murine IgH enhancer is capable of activating a promoter 17 kb away and separated from the enhancer sequence by an intervening promoter which is also activated (68). A similar result has been reported for the IgK enhancer which has been demonstrated to stimulate equally two promoters 7 kb apart. These studies further indicate that the enhancer sequence does not function as a bidirectional RNA polymerase entry site, but the results suggest instead that transcriptional enhancement follows a change in the local chromatin structure (68, 69). How then can these sequences transmit this effect on chromatin structure over the long stretches of DNA which would be required to activate the translocated Daudi *c-myc* gene? One suggestion is that while the linear distances along the chromatin are large, loop structures might bring the two sequences closer together thereby eliminating the requirement for long range effects (a *trans-cis* activation?) (44). Alternatively, the effect on chromatic structure which is initiated by the enhancer sequences might be transmitted along the chromatin by sequences common to both the Ig and *c-myc* loci (68). Either the GAGG sequences or the heptamer-nonamer signal sequences associated with the Ig, *c-myc*, *bcl*-1, *bcl*-2, as well as the Tcr loci might be reasonable candidates for such a function.

The primary event of *c-myc* deregulation which imparts a selective advantage for growth to a specific cell may be the association of *c-myc* with one of the three immunoglobulin loci. This allows *c-myc* transcription under the influence of a constitutively expressed B-cell specific locus in a cellular background where *c-myc* is usually silent. The deregulation of *c-myc* genes associated with the translocated Tcr-α locus has also been demonstrated. The many similarities between the Ig and Tcr loci suggest that the mechanisms of deregulation as well as translocation are the same in the two systems. Other oncogenes that come under the influence of either the Ig or the Tcr loci may be similarly deregulated as demonstrated for the *bcl*-2 gene (19a). Recent studies have demonstrated that m-RNA from truncated or mutated *c-myc* genes has a longer half-life than the normal *c-myc* m-RNA. These observations may account for the high steady state levels of *c-myc* RNA in some plasmacytomas (70) and BL (71, 72). However, although the steady state level of *c-myc* RNA is high in a variety of the cell lines, both plasmacytoma and BL cell lines have been described with levels of *c-myc* RNA which are equivalent to and even lower than normal controls (42, 73, 74). Thus, additional changes or mutations that accumulate in these genes with progression of the neoplasia can clearly affect the level of this expression posttranscriptionally (70–72) and perhaps even at the level of translation (24) or protein stability. However, the primary event in the malignant transformation is that which turns the translocation-associated *c-myc* on when the normal gene is silent.

TRANSLOCATIONS AND CHRONIC MYELOGENOUS LEUKEMIA (CML)

Most CML cells carry a Philadelphia chromosome. This chromosome arises from a t(9; 22) (q24; q11) translocation resulting in the transfer of the *c-abl* protooncogene from chromosome 9 to chromosome 22 (75). Most of these translocations have been shown to occur within a short segment of DNA on chromosome 22, designated "breakpoint cluster region" or bcr (75). Studies of transcription using probes derived from the bcr and *c-abl* loci indicate that CML cells express a novel hybrid transcript which has sequences derived from both loci (76); the normal 5′ terminus of *c-abl* is replaced by bcr sequences. The chimeric protein is a highly active tyrosine kinase (77).

Approximately 8% of CML patients are Ph⁻, and it has recently been reported that in several cases these Ph⁻ patients carry a "masked Ph⁺ chromosome" which has both bcr and *c-abl* rearrangements (78). Several different classes of Ph⁻ CML have been described. In one patient with a t(9;12) translocation, both *c-abl* and bcr sequences were detected on the 12q⁻ chromosome (78). In another patient, bcr and *c-abl* rearrangements were detected, but no translocation was evident; hybridization in situ indicated the presence of *c-abl* sequences on the normal appearing chromosome 22 (79). Thus, it seems that *c-abl* sequences can be transferred to chromosome 22 by mechanisms other than reciprocal translocation. A third class is indicated by the analysis of Ph$^+$ acute lymphocytic leukemia (ALL) patients, who lack the usual bcr and *c-abl* rearrangements associated with CML and make no detectable large chimeric transcript (80). This result indicates there can be heterogeneity of breakpoints that generate the Ph$^+$ chromosome and suggests that the genetic events leading to Ph$^+$ ALL may be different from those responsible for CML (80).

MULTIPLE TRANSLOCATIONS AND LEUKEMIA

There are some human B-cell leukemias and lymphomas that carry two chromosomal translocations. The 380 cell line is a pre-B leukemia which carries a t(14; 18), the hallmark of follicular lymphoma (FL), as well as a typical t(8; 14), indicative of BL (41). FL is typically a low-grade malignancy, while BL is extremely aggressive. We have speculated that the two translocations did not occur simultaneously, but rather sequentially. The t(14; 18), which is a high-probability translocation (FL is one of the most common neoplasms), probably occurred first, inducing the clonal expansion of a transformed B-cell population and also increasing the probability for the second translocation. Cell lines with similar double translocations are found in association with both B- and T-cell neoplasias, and the involvement of *c-myc* in the second event is not unusual. In one case of Ph+ chronic granulocytic leukemia with typical *c-abl* and bcr rearrangements, progression from a chronic to an acute phase was demonstrated to coincide with amplification and rearrangement of *c-myc* not present in the chronic phase (81). While one can only speculate on the temporal sequence of these events from the analysis of established cell lines, the patient studies suggest a similar picture of tumor progression. Both the t(9; 22) that gives rise to the Ph+ chromosome and the t(14; 18) associated with FL primarily give rise to low-grade malignancies. The development to more acute malignancies may require a second event, and a rearrangement of *c-myc* appears to be one possible pathway.

Secondary rearrangements of *c-myc* are also found in association with T-cell neoplasias. In addition to the inversion of chromosome 14, the SupT1 cell line carries a *c-myc* gene with a rearrangement in the 3' flanking region, and the level of *c-myc* transcripts is elevated compared to a non-rearranged control cell line (L. Finger, C. M. Croce, unpublished results). No translocation involving 8q24 is evident in this cell line, and thus it is similar to the examples of the masked rearrangements of the *c-abl* gene. The secondary involvement of *c-myc* rearrangements in T-cell as well as B-cell neoplasias suggests that the activation of *c-myc* may be an important event in the progression of a variety of malignancies. This would be consistent with the experiments that demonstrate the requirement of two cooperating oncogenes in tumorigenic conversion of primary embryo fibroblasts (82).

The observation of oncogene rearrangement in the absence of chromosomal translocation suggests that alternate routes may also exist for *c-myc* activation in addition to the chromosome translocations and the gene amplifications already described in a variety of neoplasias.

ACKNOWLEDGEMENTS

We would like to thank the Wistar editorial staff for preparing this manuscript, Marina Hoffman for excellent editorial assistance, and Dr. Donna George and Dr. M. K. Showe for helpful suggestions. The work is supported by a grant from the National Institutes of Health CA-39860 (C. M. Croce) and American Cancer Society CD-274 (L. C. Showe).

Literature Cited

1. Boveri, R. 1914. *Zur Frage der Enstehung Maligner Tumoren.* (Transl. M. Boveri, 1929). Baltimore, Md: Williams & Williams
2. Nowell, P. C., Hungerford, D. A. 1960. A minute chromosome in chronic granulocytic leukemia. *Science* 132: 1497
3. Rowley, J. D. 1973. Identification of a translocation with quinacrine fluorescence in a patient with acute leukemia. *Ann. Genet.* 16: 109–12
4. Yunis, J. J., Oken, M. M., Kaplan, M. E., Ensrud, K. M., Howe, R. R., Theologides, A. 1982. Distinctive chromosomal abnormalities in histologic subtypes of non-Hodgkin's lymphoma. *N. Engl. J. Med.* 307: 1231–36
5. Yunis, J. J. 1983. The chromosomal basis of human neoplasia. *Science* 221: 227–36
6. Manolov, G., Manolova, Y. 1972. Marker band in one chromosome 14 from Burkitt lymphoma. *Nature* 237: 33–34
7. Zech, L., Haglund, V., Nilson, N., Klein, G. 1976. Characteristic chromosomal abnormalities in biopsies and lymphoid cell lines from patients with Burkitt and non-Burkitt lymphomas. *Int. J. Cancer* 17: 47–56
8. Croce, C. M., Nowell, P. C. 1985. Molecular basis of human B cell neoplasia. *Blood* 65: 1–7
9. Leder, P., Battey, J., Lenoir, G., Moulding, C., Murphy, W., Potter, H., Stewart, T., Taub, R. 1983. Transloca-

tions among antibody genes in human cancer. *Science* 222: 765–71
10. Croce, C. M., Shander, M., Martinis, J., Cicurel, L., D'Ancona, G. G., Dolby, T. W., Koprowski, H. 1979. Chromosomal location of the human immunoglobulin heavy chain genes. *Proc. Natl. Acad. Sci. USA* 76: 3416–19
11. Erikson, J., Martinis, J., Croce, C. M. 1981. Assignment of the human genes for lambda immunoglobulin chains to chromosome 22. *Nature* 294: 173–75
12. McBride, D. W., Heiter, P. A., Hollis, G. F., Swan, D., Otey, M. C., Leder, P. 1982. Chromosomal location of human kappa and lambda immunoglobulin light chain constant region genes. *J. Exp. Med.* 155: 1480–90
13. Malcolm, S., Barton, P., Murphy, C., Ferguson-Smith, M. A., Bentley, D. L., Rabbitts, T. H. 1982. Localization of human immunoglobulin kappa light chain variable region genes to the short arm of chromosome 2 by in situ hybridization. *Proc. Natl. Acad. Sci. USA* 79: 4957–61
14. Klein, G. 1981. The role of gene dosage and genetic transpositions in carcinogenesis. *Nature* 294: 313–16
15. Rowley, J. 1982. Identification of the constant chromosome regions involved in human hematologic malignant disease. *Science* 216: 749–51
16. Dalla Favera, R., Bregni, M., Erikson, J., Patterson, D., Gallo, R. C., Croce, C.

M. 1982. Assignment of the human c-myc oncogene to the region of chromosome 8 which is translocated in Burkitt lymphoma cells. *Proc. Natl. Acad. Sci. USA* 79 : 7824–27

17. Hayward, W. S., Neel, B. J., Astrin, S. M. 1981. Activation of a cellular onc gene by promoter insertion in ALV induced lymphoid leucosin. *Nature* 290 : 475–80

18. Fukuhara, S., Rowley, J. D., Variakojis, D., Golomb, H. M. 1979. Chromosome abnormalities in poorly differentiated lymphocytic lymphoma. *Cancer Res.* 39 : 3119–28

19. Bakshi, A., Jensen, J. P., Goldman, P., Weight, J. J., McBride, J. W., Epstein, A. L., Korsmeyer, S. J. 1985. Cloning of the chromosomal breakpoint of t(14 ; 18) human lymphomas : Clustering around J4H5 on chromosome 14 and near a transcriptional unit on 18. *Cell* 41 : 899–906

19a. Tsujimoto, Y., Croce, C. M. 1986. Analysis of the structure, transcripts and protein products of bcl-2, the gene involved in follicular lymphoma. *Proc. Natl. Acad. Sci. USA* 83 : 5214–18

20. Croce, C. M., Isobe, M., Palumbo, A., Puck, J., Ming, J., Tweardy, D., Erikson, J., Davis, M., Rovera, G. 1985. Gene for Ca-chain of human T-cell receptor : Location on chromosome 14 region involved in T-cell neoplasms. *Science* 227 : 1044–47

21. Colby, W. W., Cheny, E. Y., Seth, D. H., Levinson, A. D. 1983. Identification and nucleotide sequence of a human locus homologous to the v-myc oncogene of avian myelocytomatosis virus MC29. *Nature* 301 : 722–25

21a. Watt, R., Stanton, L. W., Marcu, K. B., Gallo, R. C., Croce, C. M., Rovera, G. 1983. Nucleotide sequence of cloned cDNA of the human c-myc gene. *Nature* 303 : 725–28

22. Watt, R., Nishikura, K., Sorrentino, J., ar-Rushdi, A., Croce, C. M., Rovera, G. 1983. The structure and nucleotide sequence of the 5' end of the human c-myc gene. *Proc. Natl. Acad. Sci. USA* 80 : 6307–11

23. Dalla Favera, R., Martinotti, S., Gallo, R. C., Erikson, J., Croce, C. M. 1983. Translocation and rearrangements of the c-myc oncogene in human differentiated B-cell lymphomas. *Science* 219 : 963–67

23a. Erickson, J., Finan, J., Nowell, P. C., Croce, C. M. 1982. Translocation of immunoglobulin V4H genes in Burkitt lymphoma. *Proc. Natl. Acad. Sci. USA* 80 : 810–24

23b. Croce, C. M., Erikson, J., ar-Rushdi, A., Aden, D., Nishikura, K. 1984. Translocated c-myc oncogene of Burkitt lymphoma is transcribed in plasma cells and repressed in lymphoblastoid cells. *Proc. Natl. Acad. Sci. USA* 81 : 3170–74

23c. Taub, R., Kirsch, I., Morton, C., Lenoir, G., Swan, D., Tronick, S., Aaronson, S., Leder, P. 1982. Translocation of the c-myc gene into the immunoglobulin heavy chain locus in human Burkitt lymphoma and mouse plasmacytome cells. *Proc. Natl. Acad. Sci. USA* 79 : 7837–41

24. Saito, H., Hayday, A. C., Wiman, K., Hayward, W. S., Tonegawa, S. 1983. Activation of the c-myc gene by translocation : A model for translational control. *Proc. Natl. Acad. Sci. USA* 80 : 7476–80

24a. Shen-Ong, G. L. C., Keath, E. J., Piccoli, S. P., Cole, M. D. 1982. Novel myc oncogene RNA from abortive immunoglobulin gene recombination in mouse plasmacytomas. *Cell* 31 : 443–52

24b. Stanton, L. W., Watt, R., Marcell, K. B. 1983. Translocation, breakage and truncated transcripts of c-myc oncogene in murine plasmacytomas. *Nature* 303 : 401–6

25. Hollis, G. F., Mitchell, K. F., Battey, J., Potter, H., Taub, R. A., Lenoir, G., Leder, P. 1984. A variant translocation places the immunoglobulin genes 3' to the c-myc oncogene in Burkitt lymphoma. *Nature* 307 : 752–55

25a. Croce, C. M., Theirfelder, W., Erikson, J., Nishikura, K., Finan, J., Lenoir, G., Nowell, P. C. 1983. Transcriptional activation of an unrearranged and untranslocated of a *c-myc* oncogene by translocation of C lambda locus in Burkitt lymphoma. *Proc. Natl. Acad. Sci. USA* 80 : 6922–26

26. Erikson, J., Nishikura, K., ar-Rushdi, A., Finan, J., Emanuel, B., Lenoir, G., Nowell, P. C., Croce, C. M. 1983. Translocation of a kappa immunoglobulin locus to a region 3' of an unrearranged c-myc oncogene enhances c-myc transcription. *Proc. Natl. Acad. Sci. USA* 80 : 7581–85

26a. Taub, R., Kelley, K., Battey, J., Latt, S., Lenoir, G. M., Tantravahi, U., Tu, Z., Leder, P. 1984. A novel alteration in the structure of an activated c-myc gene in a variant t(2 ; 8) Burkitt lymphoma. *Cell* 37 : 511–20

27. Pelicci, P. G., Knowles, D. M., II, Magrath, I., Dalla Favera, R. 1986. Chromosomal breakpoints and structural rearrangements of the c-myc locus differ in endemic and sporadic forms of Bur-

kitt lymphoma. *Proc. Natl. Acad. Sci. USA* 83: 2984–88

28. Burkitt, D. P. 1970. Geographical distribution. In *Burkitt's Lymphoma*, ed. D. P. Burkitt, D. H. Wright. London: Livingstone. 186 pp.

29. Ablashi, D. V., Levine, P. H., Papas, T., Pearson, G. R., Kottaridis, S. D. 1985. *First International Symposium on Epstein-Barr virus and associated malignant diseases. Cancer Res.* 45: 3981–84

30. Whittle, H. C., Brown, J., Marsh, K., Greenwood, B. M., Seidelin, P., Tighe, H., Wedderburn, L. 1984. T-cell control of Epstein-Barr virus infected B-cells is lost during *P. falciparum* malaria. *Nature* 312: 449–50

31. Zeigler, J. L., Miner, R., Rosenbaum, E., Lenette, E. T., Shillitoe, E., Casavant, C., Drew, W. L., Mintz, L., Gershow, J., Greenspan, J., Beckstead, J., Yamamoto, K. 1982. Outbreak of Burkitt's-like lymphoma in homosexual men. *Lancet* 2: 631–33

32. Groopman, J. E., Sullivan, J. L., Mulder, C., Ginsberg, D., Orkin, S. H., O'Hara, C. J., Falchuk, K., Wong-Staal, F., Gallo, R. C. 1986. Pathogenesis of B-cell lymphoma in a patient with AIDS. *Blood* 67: 612–14

33. Tsujimoto, Y., Yunis, J., Onorato-Showe, L., Erikson, J., Nowell, P. C., Croce, C. M. 1984. Molecular cloning of the chromosomal breakpoint of B-cell lymphomas and leukemias with the t(11; 14) chromosome translocation. *Science* 224: 1403–6

34. Tsujimoto, Y., Cossman, J., Jaffe, E., Croce, C. M. 1985. Involvement of the bcl-2 gene in human follicular lymphoma. *Science* 228: 1440–43

35. Tsujimoto, Y., Jaffe, E., Cossman, J., Gorham, J., Nowell, P. C., Croce, C. M. 1985. Clustering of breakpoints on chromosome 11 in human B cell neoplasms with the t(11; 14) chromosome translocation. *Nature* 315: 340–43

36. Tsujimoto, Y., Gorham, J., Cossman, J., Jaffe, E., Croce, C. M. 1984. The t(14; 18) chromosome translocations involved in B cell neoplasms result from mistakes in VDJ joining. *Science* 229: 1390–93

37. Cleary, M. L., Sklar, J. 1985. Nucleotide sequence of a t(14; 18) chromosome breakpoint in follicular lymphoma, and demonstration of a breakpoint cluster region near a transcriptionally active locus on chromosome 18. *Proc. Natl. Acad. Sci. USA* 82: 7439–43

38. Quertermous, T., Strauss, W., Murre, C., Dialynas, D. P., Strominger, J. L., Seidman, J. G. 1986. Human T-cell gamma genes contain N segments and have marked junctional viability. *Nature* 322: 184–87

39. Desiderio, S. V., Yancopoulos, D. G., Paskind, M., Thomas, E., Boss, M. A., Landau, N., Alt, F. W., Baltimore, D. 1984. Insertion of N regions into heavy-chain genes is correlated with expression of terminal deoxytransferase in B-cells. *Nature* 311: 752–55

40. Haluska, F. G., Finver, S., Tsujimoto, Y., Croce, C. M. 1986. The t(8; 14) chromosome translocation occurring in African Burkitt lymphoma and in acute pre-B-cell leukemia results from mistakes in V-D-J joining. *Nature.* In press

41. Pegoraro, L., Palumbo, A., Erikson, J., Falda, M., Giovanoizzo, B., Emanuel, B. S., Rovera, G., Nowell, P. C., Croce, C. M. 1984. A 14; 18 and an 8; 14 chromosome translocation in a cell line derived from an acute B-cell leukemia. *Proc. Natl. Acad. Sci. USA* 81: 7166–70

42. Sun, L. K., Showe, L. C., Croce, C. M. 1986. Analysis of the 3′ flanking region of the human c-myc gene in lymphomas with the t(8; 22) and t(2; 8) chromosomal translocations. *Nucleic Acids Res.* 14: 4037–50

43. Showe, L. C., Croce, C. M. 1986. Translocations in B- and T-cell neoplasms. *Seminars in Haemotology.* 23: 237–44

44. Cory, S., Graham, M., Webb, E., Corcoran, L., Adams, J. 1985. Variant (6; 15) translocations in murine plasmacytomas involve a chromosome 15 locus at least 72kb from the c-myc oncogene. *EMBO J.* 4: 675–81

45. Villeneuve, L., Rassart, E., Jolicoeur, P., Graham, M., Adams, J. 1986. Proviral integration site *mis-1* in rat thymomas corresponds to the pvt-1 translocation breakpoint in murine plasmacytomas. *Mol. Cell. Biol.* 6: 1834–37

46. Showe, L. C., Ballantine, M., Nishikura, K., Erikson, J., Kaji, H., Croce, C. M. 1985. Cloning and sequencing of a c-myc oncogene in a Burkitt's lymphoma cell line that is translocated to a germ line alpha switch region. *Mol. Cell. Biol.* 5: 501–509

47. Murphy, W., Sarid, J., Taub, R., Vasicek, T., Battey, J., Lenoir, G., Leder, P. 1986. Translocated human c-myc oncogene is altered in a conserved coding sequence. *Proc. Natl. Acad. Sci. USA* 83: 2939–43

48. Battey, J., Moulding, C., Taub, R., Murphy, W., Stewart, T., Potter, H., Lenoir, G., Leder, P. 1983. The human c-myc oncogene: Structural consequences of translocation into the IgH locus in Burkitt lymphoma. *Cell* 34: 779–87

49. Piccoli, S. P., Caimi, P. G., Cole, M. D. 1984. A conserved sequence at the c-myc oncogene translocation breakpoints in plasmacytomas. *Nature* 310: 327–30

50. Showe, L. C., Croce, C. M. 1986. Mechanisms involved in chromosomal translocations in B- and T-cell neoplasms. *Cancer Rev.* 2: 18–23

51. Bullock, P., Champoux, J. J., Botchan, M. 1985. Association of crossover points with topoisomerase I cleavage sites: A model for non-homologous recombination. *Science* 230: 954

52. Hecht, F., Morgan, R., Hecht, B. K-M., Smith, S. D. 1984. Common region on chromosome 14 in T-cell leukemia and lymphoma. *Science* 226: 1445–47

53. Williams, D. L., Look, A. T., Melvin, S. L., Roberson, P. K., Dahl, G., Flake, T., Stass, S. 1984. New chromosomal translocations correlate with specific immunophenotypes of childhood acute lymphoblastic leukemia. *Cell* 36: 101–9

54. Erikson, J., Williams, D. L., Finan, J., Nowell, P. C., Croce, C. M. 1985. Locus of the Ca chain of the T-cell receptor is split by chromosome translocation in T-cell leukemias. *Science* 229: 784–86

54a. Lewis, W. H., Michalopoulos, E. E., Williams, D. L., Minden, M. D., Mak, T. W. 1985. Breakpoints in the human T-cell antigen receptor α-chain locus in two T-cell leukemia patients with chromosomal translocations. *Nature* 317: 544–45

55. Erikson, J., Finger, L., Sun, L., ar-Rushdi, A., Nishikura, K., Minowada, J., Finan, J., Emanuel, B. S., Nowell, P. C., Croce, C. M. 1985. C-myc deregulation by translocation of the alpha-locus of the T-cell receptor in T-cell leukemia. *Science* 232: 884–86

55a. Mathieu-Mahu, D., Caubet, J. F., Bernheim, A., Mauchauffé, M., Palmer, E., Berger, R., Larsen, C. J. 1985. Molecular cloning of a fragment from human chromosome 14(14q11) involved in T-cell malignancies. *EMBO J.* 4: 3427–33

56. Shima, E. A., LeBeau, M. M., McKeitnon, T. W., Minowada, J., Showe, L. C., Mak, T. W., Minden, M. D., Rowley, J. D., Diaz, M. O. 1985. The gene encoding the α-chain of the T-cell receptor is moved immediately downstream of c-myc in a chromosomal 8; 14 translocation in a cell line from a human T-cell leukemia. *Proc. Natl. Acad. Sci. USA* 83: 3439–44

57. Finger, L. R., Harvey, R., Moore, R. C. A., Showe, L. C., Croce, C. M. 1986. Nucleotide sequence of the breakpoint of a t(8; 14) translocation in T-cell leukemia indicates a common mechanism of translocation in T-cell and B-cell neoplasia. *Science.* 243: 982–85

58. Baer, R., Smith, K. C., Rabbitts, T. 1985. Fusion of an immunoglobulin variable gene and a T-cell receptor constant region gene in the chromosome 14 inversion associated with a T-cell tumor. *Cell* 43: 705–13

58a. Denny, C. T., Yoshikai, Y., Mak, T. W., Smith, S. D., Hollis, G. F., Kirsch, I. 1986. A chromosome 14 inversion in a T-cell lymphoma is caused by site-specific recombination between immunoglobulin and T-cell receptor loci. *Nature* 320: 549–51

59. Ikuta, K., Ogura, T., Shimizu, A., Honjo, T. 1986. A joining-diversity-joining complex generated by inversion mechanism and a variable-diversity complex in the B-chain gene of the human t-cell receptor. *Nucl. Acids Res.* 14: 4899–4909

60. Duby, A. D., Seidman, J. G. 1986. Abnormal recombination products result from aberrant DNA rearrangement of the human T-cell antigen receptor B-chain. *Proc. Natl. Acad. Sci. USA* 83: 4890–94

61. Aguilera, R. J., Hope, T. J., Sakano, H. 1985. Characterization of immunoglobulin enhancer deletions in murine plasmacytomas. *EMBO J.* 4: 3689–93

61a. ar-Rushdi, A., Nishikura, K., Erikson, J., Watt, R., Rovera, G., Croce, C. 1983. Differential expression of the translocated and untranslocated c-myc oncogene in Burkitt lymphoma. *Science* 222: 390–93

61b. Nishikura, K., ar-Rushdi, A., Erikson, J., Watt, R., Rovera, G., Croce, C. M. 1983. Differential expression of the normal and of the translocated human c-myc oncogenes in B-cells. *Proc. Natl. Acad. Sci. USA* 80: 4822–26

62. Siebenlist, U., Hennighausen, L., Battey, J., Leder, P. 1984. Chromatin structure and protein binding in the putative regulatory region of the c-myc gene in Burkitt lymphoma. *Cell* 37: 381–91

63. Prehn, J., Mercola, M., Calame, K. 1984. Translocation affects normal c-myc promoter usage and activates fifteen cryptic c-myc transcription starts in plasmacytoma M603. *Nucl. Acids Res.* 12: 8987–9007

64. Feo, S., Harvey, R., Showe, L., Croce, C. M. 1986. Regulation of the translocated c-myc genes transfected into plasmacytoma cells. *Proc. Natl. Acad. Sci. USA* 83: 706–9

65. Del Senno, L., Umberti, E., Rossi, M., Buzzoni, D., Barbieri, R., Rossi, P.,

Patracchini, P., Bernardi, F., Marchetti, G., Comconi, F., Gambari, R. 1986. Identification of a c-myc oncogene lacking the exon-1 in the normal cells of a patient carrying a thyroid carcinoma. *FEBS Lett.* 196: 296–300

66. Croce, C. M., Erikson, J., Huebner, K., Nishikura, K. 1985. Co-expression of translocated and normal c-myc oncogene in hybrids between Daudi and lymphoblastoid cells. *Science* 227: 1235–38

67. Klein, S., Gerster, T., Picard, D., Radbruch, A., Schaffner, W. 1985. Evidence for transient requirement for the IgH enhancer. *Nucl. Acids Res.* 13: 8901–12

68. Wang, X. F., Calame, K. 1985. The endogenous immunoglobulin heavy chain enhancer can activate tandem V_H promoters separated by a large distance. *Cell* 43: 659–65

69. Atchison, M. L., Perry, R. 1986. Tandem kappa immunoglobulin promoters are equally active in the presence of the kappa enhancer: Implications for models of enhancer function. *Cell* 46: 253–62

70. Piechaczyk, M., Yang, J. O., Blanchard, J. M., Jeanteur, P., Marcu, K. 1985. Posttranscriptional mechanisms are responsible for accumulation of truncated c-myc RNAs in murine plasma cell tumors. *Cell* 42: 589–97

71. Eick, D., Piechaczyk, M., Henglein, B., Blanchard, J. M., Traub, B., Kofler, E., Wiest, S., Lenoir, G. M., Bornkamm, G. W. 1985. Aberrant c-myc RNA's of Burkitt's lymphoma cells have longer half-lives. *EMBO J.* 4: 3717–25

72. Rabbits, P. H., Forster, A., Stinson, M. A., Rabbits, T. H. 1985. Truncation of exon 1 from c-myc gene results in prolonged c-myc m-RNA stability. *EMBO J.* 4: 3727–33

73. Keath, E. J., Kelekar, A., Cole, M. D. 1984. Transcriptional activation of the translocated c-myc oncogene in mouse plasmacytomas: Similar RNA levels in tumor and proliferating normal cells. *Cell* 37: 521–28

74. Taub, R., Moulding, C., Battey, J., Murphy, W., Vasicek, T., Lenoir, G. M., Leder, P. 1984. Activation and somatic mutation of the translocated c-myc gene in Burkitt lymphoma cells. *Cell* 36: 339–48

75. Groffen, J., Stephenson, J. R., Heisterkamp, N., deKlein, A., Bartram, C. R., Grosveld, G. 1984. Philadelphia chromosomal breakpoints are clustered within a limited region, bcr, on chromosome 22. *Cell* 36: 93–99

76. Gale, R., Canaani, E. 1985. An 8-kilobase abl RNA transcript in chronic myelogenomic leukemia. *Proc. Natl. Acad. Sci. USA* 81: 5648–52

77. Kanopka, J. B., Watanabe, S. M., Witte, O. N. 1984. An alteration of the human c-abl protein in K562 leukemia cells unmasks associated tyrosine kinase activity. *Cell* 37: 1035–42

78. Bartram, C. R., Kleihauer, E., deKlein, A., Grosveld, G., Teyssier, J. R., Heisterkamp, N., Groffen, J. 1985. c-abl and bcr are rearranged in a Ph^1-negative CML patient. *EMBO J.* 4: 683–86

79. Morris, C. M., Reeve, A. E., Fitzgerald, P. H., Hollings, P. E., Beard, M. E. J., Heaton, D. C. 1986. Genomic diversity correlates with clinical variation in Ph$^-$-negative chronic myeloid leukaemia. *Nature* 320: 281–82

80. Erikson, J., Griffin, C., ar-Rushdi, A., Valtieri, M., Hoxie, J., Finan, J., Emanuel, B. S., Rovera, G., Nowell, P. C., Croce, C. M. 1986. Heterogeneity of chromosome 22 breakpoint in Ph-positive acute lymphocytic leukemia. *Proc. Natl. Acad. Sci. USA* 183: 1403–7

81. McCarthy, D. M., Rassool, F. V., Goldman, J. M., Graham, S. V., Birnie, G. D. 1984. Genomic alterations involving the c-myc proto-oncogene locus during the evolution of a case of chronic granulocytic leukemia. *Lancet* 15: 1362–65

82. Land, H., Parada, L. F., Weinberg, R. A. 1983. Tumorgenic conversion of primary embryo fibroblasts requires at least two co-operating oncogenes. *Nature* 304: 602–5

Ann. Rev. Immunol. 1987. 5 : 279–304

VIRUSES PERTURB LYMPHOCYTE FUNCTIONS: Selected Principles Characterizing Virus-Induced Immunosuppression

Michael B. McChesney and Michael B. A. Oldstone

Department of Immunology, Scripps Clinic and Research Foundation, La Jolla, California 92037

INTRODUCTION

Viruses can perturb the host immune system in general by two basic mechanisms. In the first case, disordered immune regulation may be a direct consequence of viral replication in immunocompetent cells. For example, cell dysfunction due to infection of suppressor T lymphocytes could result in heightened or aberrant immune responses and auto-immunity; infection of helper T lymphocytes could lead to immu-nosuppression; or the establishment of persistent virus infection could involve virus-specific cytotoxic T lymphocytes (CTL). The second general mechanism by which viruses interfere with the immune system is an indirect one. In this case the infection of a lymphocyte, macrophage, or other cell would result in the release of lymphokines (monokines), lymphokine inhibitors, or other soluble factors that affect many other uninfected cells. In this way, a marked amplification can occur whereby the infection of a small number of cells results in the dysfunction of the whole system.

Another consideration in the study of virus-lymphocyte interactions is the variable cytopathic effects of different viruses in cells. A given virus infection may be productive (replication of infectious virus), or non-productive, i.e. restricted (virus replication blocked, but viral genetic infor-mation persists). As illustrated in Figure 1, productive infection may be lytic due to inhibition of cellular protein synthesis or altered cell membrane integrity, for example. However, some viruses are relatively noncytopathic.

279

0732–0582/87/0410–0279$02.00

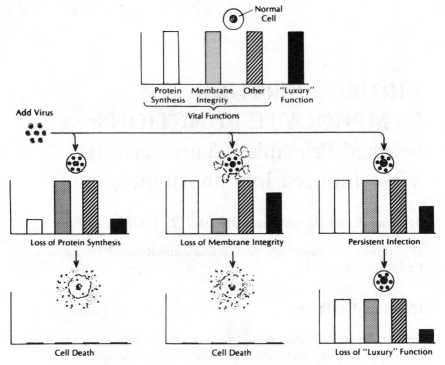

Figure 1 Mechanisms by which viruses cause disease. A viral infection alters cell function, disturbs homeostasis, and leads to disease in several ways. One mechanism (left and center) involves the loss of vital cell functions resulting in cell lysis. A more subtle mechanism (right) does not perturb vital functions, and cell morphology is normal, but differentiated functions are altered.

Vital cellular functions are not affected, but differentiated or luxury functions may be inhibited. This type of cytopathology is associated with several persistent virus infections (reviewed in 1). Restricted or latent infection usually results in no apparent cytopathology until viral replication is activated. These principles are illustrated in Figure 2, together with the assays required to detect virus during different stages of replication.

The literature of virus-lymphocyte interactions has expanded greatly in the last 25 years. We are indebted to the reviews of Notkins, Mergenhagen & Howard (2), Wheelock & Toy (3), and Woodruff & Woodruff (4) for providing a firm foundation. The current scope of virus-lymphocyte interactions is indicated in Table 1 by the list of viruses that infect lymphocytes. This list is not comprehensive but rather indicates representative

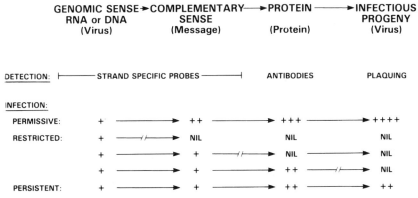

Figure 2 Methods for detection of virus during productive or restricted infection. During productive, i.e. permissive or persistent, infection, viral replication proceeds through the stages of transcription and replication of the viral genome, translation of viral proteins, then maturation and assembly of infectious virus progeny. During restricted infection, viral replication may be blocked at one of these stages and the presence of virus in cells may only be detected by a specific method. Viral nucleic acids can be detected by hybridization with cDNA or RNA probes. Strand-specific RNA probes can distinguish genomic and message sense viral nucleic acids. Specific viral proteins are detected by monoclonal antibody or antibody to synthetic peptides. Infectious virus can be quantitated by direct plaque assay on permissive cell monolayers. If viral replication is blocked at the final stage of maturation, cocultivation of infected cells on permissive cell monolayers can recover infectious virus.

viruses from different taxonomic groups. Viruses with every type of genomic nucleic acid, encompassing divergent replication strategies, are now known to infect lymphocytes. With few exceptions, immunologic dysfunction has been associated with the infections. The majority of viruses, e.g. measles, undergo restricted infection in unstimulated lymphocytes. Productive infection frequently follows mitogenic stimulation (3). With some viruses, e.g. lymphocytic choriomeningitis virus (LCMV), replication is restricted so that recovery of infectious virus requires cocultivation with permissive cells. The replication of other viruses, notably human cytomegalic virus (HCMV), is more restricted, and infectious virus progeny have not been found. The percentage of cells harboring virus in a population may be small, e.g. in HCMV and LCMV, or large, e.g. in measles. Many viruses that replicate in lymphocytes are not lytic, and cytopathic effects are not often defined.

In this chapter we discuss four different models of virus-lymphocyte interaction that demonstrate specific principles for understanding virus-induced immunosuppression. It is not our intention to review the entire literature concerning the effect of these viruses on the immune system, but rather to point to the novel concepts that have emerged. Both measles

Table 1 Viruses that infect lymphocytes and monocytes

Virus[a]	Host	Infected Cells	Reference
Double Strand DNA Viruses			
Hepatitis B virus	Human, monkey	PBMC,[b] T and B lymph	(5)
Papovavirus	Human, monkey	PBMC	(6)
Group C adenoviruses	Human	T, B and null lymph	(7, 8)
Herpes simplex virus	Human	T Lymph	(9)
Epstein-Barr virus	Human	B lymph	(10)
Cytomegalovirus	Human	Lymph, mono	(11)
Pox virus	Rabbit	Spleen cells	(12)
Single Strand DNA Viruses			
Porcine parvovirus	Pig	Spleen cells	(13)
Minute virus of mice	Mouse	Lymph	(14)
Positive Strand RNA Viruses			
Poliovirus	Human	Lymph, mono	(15)
Rubella	Human	T and B lymph	(16)
Negative Strand RNA Viruses			
Measles	Human	T and B lymph, mono	(17, 18)
Mumps	Human	T and B lymph	(19)
Respiratory syncytial virus	Human	Lmph, mono	(20)
Vesicular stomatitis virus	Human, mouse	T lymph	(21)
Influenza A	Human	Lymph, mono	(22)
Parainfluenza	Human	Lymph, mono	(23)
Ambisense RNA Viruses			
Lymphocytic choriomeningitis virus	Mouse	T and B lymph, mono	(24, 25)
Junin virus	Human	PBMC	(26)
Retroviruses			
Murine leukemia virus	Mouse	B lymph	(27)
Feline leukemia virus	Cat	T and B lymph, mono	(28)
HTLV I, II	Human	T, B and null lymph	(29, 30)
HTLV III	Human	T and B lymph, mono	(30–32)
Endogenous C-type virus	Mouse	Spleen cells	(33)

[a] Viruses that infect monocytes but not lymphocytes are not listed.
[b] Abbreviations: PBMC—peripheral blood mononuclear cells, lymph—lymphocyte, mono—monocyte or macrophage.

virus and HCMV cause antigen nonspecific immunosuppression (Figure 3). The general suppression of differentiated functions that require lymphocyte activation and proliferation is described in the measles virus model. A different mechanism of general immunosuppression mediated by specific viral structural proteins is presented in the retrovirus model. The recent evidence that HCMV infects lymphocytes and monocytes and a proposed mechanism whereby lymphocyte-monocyte interaction results in lymphocyte dysfunction are discussed. The LCMV model presents a

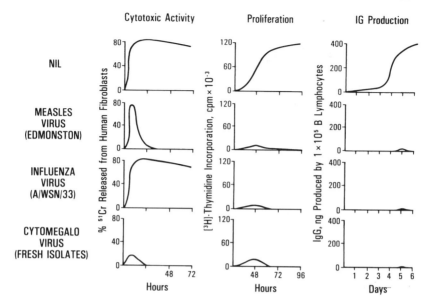

Figure 3   In vitro alteration of human lymphocyte functions by different viruses. Shown schematically are the effects of virus infection of lymphocytes on selected immunologic functions. Cytotoxic activity refers either to natural cytotoxicity or to virus-specific HLA-restricted cytotoxic-T-lymphocyte killing. Proliferation refers to the response to mitogen or to an antigen unrelated to the infecting virus. IG production indicates polyclonal immunoglobulin secretion in PWM cultures. Data from Casali et al 1984; Rice et al 1985; Schrier & Oldstone 1986.

mechanism of specific suppression of virus-specific cytotoxic-T-lymphocyte (CTL) function by viral variants selected in vivo during persistent infection. In this system the adoptive transfer of uninfected immune lymphocytes terminates the infection by clearing virus materials.

MEASLES VIRUS

Measles virus infection has been recognized for centuries. Complications of measles including the development of chronic pulmonary disease and active tuberculosis were described by Heberden (34), Osler (35), and Holt (36). These authors speculated that measles infection could reactivate dormant tuberculosis or allow opportunistic infection by other common pathogens to occur with greater virulence. Clement von Pirquet (37) made the first scientific observation of virus-induced immunosuppression when he demonstrated that the tuberculin skin test response of immune individuals became negative transiently during the course of acute measles infection. This observation was followed by the demonstration that neither lymphocytes from tuberculosis-immune donors isolated during the course of acute measles infection nor lymphocytes cultured in vitro with measles virus could proliferate in the presence of tuberculin antigen (38) or a mitogen (38, 39). Live attenuated measles virus vaccine suppressed the tuberculin skin test response, but inactivated virus vaccine did not (40).

Measles virus is a relatively noncytopathic virus; it does not dramatically suppress the macromolecular synthesis of cells that it infects. Destruction of infected cells is due to the positioning in the cell membrane of cleaved fusion protein of the virus. Infected adjacent cells form syncytia that cannot survive. This process can be blocked, and viral replication then proceeds normally (41). During acute infection, measles virus replicates in lymphoid tissues (reviewed in 42). Both T and B lymphocytes taken from the peripheral blood during natural infection will express viral antigen after mitogenic stimulation in vitro (43). Monocytes and T-helper, T-suppressor, and B lymphocytes can be infected in vitro, but not poly-morphonuclear cells (17, 18, 44). Viral replication is restricted in unstimu-lated cells, but if lymphocytes are stimulated by mitogen either before or after infection, they become fully permissive, and the majority of cells in culture express viral antigen (18, 44, 45). Infected unstimulated lym-phocytes do not express viral antigen after day 3 or 4 of culture, but viral RNA persists and subsequent mitogenic stimulation leads to productive infection (44, 45). The cellular requirements for viral replication and the nature of the block in replication in unstimulated cells are not known.

As mentioned above, measles virus suppresses lymphocyte proliferation (38, 39). However, inactivation of virus by heat, ultraviolet rays, or neu-tralizing antibody abrogates the suppression, and increased proliferation may occur (46, 47, 48). Purified glycoproteins of the virus are mitogenic (49) and activate natural killer (NK) cells (48). Some discrepancies in the literature about suppression by measles virus could be explained by the nature of the virus preparation used—by whether it is infectious or inac-tivated. A product of measles virus infection of HeLa cells or Vero cells suppresses cell proliferation. The factor is not a component of the virus but is a heat stable protein that elutes in multiple molecular weight fractions by gel filtration (50). Although alpha interferon and prostaglandin E are secreted by infected mononuclear cell cultures, neither these agents nor other monocyte products could account for the suppression of cell pro-liferation by measles virus (51, 52). In contrast, experiments designed to assay T-suppressor-cell function in measles-virus-infected lymphocyte cultures demonstrated that infected T or B lymphocytes inhibited the proliferative response of uninfected lymphocytes to phytohemagglutinin (PHA). This cell-associated suppression was resistant to radiation but was abrogated by incubation of the infected cells with antibody to measles virus (52). This suggests that suppression is mediated directly by virus, whether free or cell associated, although a role of suppressor cells has not been formally excluded.

The generation of several lymphocyte functions in vitro is inhibited by measles virus (Figure 3), but already established effector functions are not affected (53, 54). The addition of virus to mixed lymphocyte (MLR) cultures resulted in decreased proliferation and reduced CTL activity. However, if virus was added to MLR cultures on day 5 or 7, no effect on cytotoxicity was seen (53). Infected, unstimulated lymphocytes were able to mediate antibody-dependent cellular cytotoxicity (ADCC) even after 4 days of culture, but there was no increase in lytic units after PHA stimulation (53). Addition of virus to 7–9-day pokeweed mitogen (PWM) cultures inhibited immunoglobulin (Ig) secretion if virus was added on day 0 or day 3, but the inhibition was reduced if virus was added on day 5 or later (53, 54). The infection of purified NK cells had no effect on cell viability, but infected cells could not lyse NK targets in chromium release assays (54). In comparison, infection of lymphocytes with influenza A virus also inhibited cell proliferation and PWM-driven Ig secretion but had no effect on NK cell activation (Figure 3). PWM-driven Ig secretion was further studied as a model of cell-cell interactions that result in differentiation in vitro. The target cell of measles-virus-induced suppression could be OKT4 helper or inducer cells, OKT8 suppressor cells, monocytes, or NK or B cells in this system. In the simplest model of cellular interaction in PWM cultures, T cells plus a small percentage of monocytes cultured in PWM generate conditioned medium containing B-cell growth and differentiating factors. This conditioned medium added to purified B cells with PWM replaces the accessory cell requirement for B-cell differentiation (55). The effect of measles virus infection of T cells or B cells can then be seen without the spread of virus from one to the other cell in culture (Figure 4; 56). Uninfected B cells cultured in conditioned medium from uninfected T cells supplemented with PWM secrete IgG and IgM. Conditioned medium from measles-virus-infected T cells, cleared of infectious virus by ultracentrifugation, also supported IgG and IgM secretion. In some experiments uninfected B cells secreted more Ig when cultured in conditioned medium harvested from infected cells than from uninfected T cells. However, infected B cells secreted little Ig in any conditioned medium. Possible suppression by NK cells was excluded by depletion of Leu 7 and Leu 11 positive cells from the B-cell population. Proliferation of infected B cells was blocked when measured on day 3 or 5 of culture (56). Thus, measles infection of B cells but not T cells is required for suppression of PWM-driven Ig secretion. Infected T lymphocytes (and monocytes) are

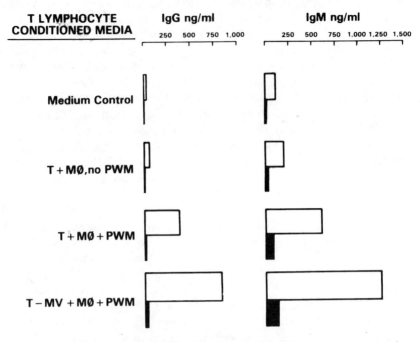

Figure 4 Measles virus suppresses B lymphocytes, not T lymphocytes or monocytes, in PWM-driven Ig synthesis. B lymphocytes were cultured in conditioned medium from uninfected (T+mΦ) or measles virus infected (T−MV+mΦ) T lymphocyte monocyte cultures. Open bars, Ig secretion by mock infected B lymphocytes; closed bars, Ig secretion by virus-infected B lymphocytes. From McChesney et al 1986.

able to secrete B-cell growth and differentiation factors. That helper-T-cell function is not inhibited by virus is not surprising. This T-cell function in PWM culture is resistant to radiation and thus does not require cell proliferation. Specific antigen induced secretion of IL-2 by a human T-cell line was not suppressed by measles virus infection (57). In summary, measles virus suppresses T- and B-lymphocyte responses that require cellular activation and proliferation prior to expression of differentiated functions.

In order to characterize further the cytopathic effect of measles virus in lymphocytes, events occurring during T- and B-lymphocyte activation were examined in infected cells (M. McChesney, J. Kehrl, A. Altman, unpublished data). Infected T cells were able to secrete IL-2 and gamma interferon and to express IL-2 receptors at 24 and 48 hr after mitogenic stimulation. And yet proliferation was inhibited by 50–90% at 72 hours. Exogenous IL-2 could not reverse the inhibition. B-lymphocyte proliferation was also inhibited by 80–90% at 72 hours. However, infected B cells underwent a normal increase in cell volume at 24 and 48 hr post-stimulation. Infected T and B lymphocytes responded to mitogens with the expected increase in cellular RNA synthesis at 24 and 48 hr, but RNA synthesis was reduced by 50% of controls in infected T lymphocytes at 72 hr. Infected T and B lymphocytes expressed normal levels of cell surface activation antigen 4F2 and transferin receptor. In parallel cultures, measles virus replicated to high titers in T and B lymphocytes at 48 and 72 hr poststimulation. These observations suggest that (*a*) infected lymphocytes can undergo activation by mitogens, but the cells are blocked in late G1 of the cell cycle, (*b*) lymphokine secretion during lymphocyte activation is not inhibited by measles virus, and (*c*) infected cells cannot respond to lymphokines by proliferation or terminal differentiation.

ENVELOPE PROTEINS OF RETROVIRUSES

Retroviruses of murine, avian, feline, and human origin are immunosuppressive as well as oncogenic in their hosts (30, 58, 59). The evidence of depressed cellular and humoral immune responses in most cases precedes leukemogenesis and is independent of the transforming function of the virus. Animals genetically resistant to leukemogenesis are also resistant to immunosuppression (58). In several experimental models with C-type retroviruses, both UV inactivated virus and disrupted virions cause immunosuppression in a dose dependent manner. There is no interspecies restriction, i.e. both murine leukemia viruses (MuLV) and feline leukemia virus (FeLV) can suppress mouse and human lymphocyte proliferation in vitro. Thus, a component of the virion structure, but not replicating virus or some other product of infection, is required.

A partially purified 15-kd structural protein of FeLV inhibited the proliferation of feline lymphocytes induced by concanavalin A (Con A) (60). The inhibition was dose dependent and occurred when the protein was added as late as day 3 of a 4-day culture. In contrast, another structural protein of the virus, p27, was not inhibitory. Both UV inactivated FeLV (UV-FeLV) and the 15-kd envelope protein of FeLV (p15E) suppressed the proliferation of human lymphocytes to Con A (61). Suppression was not mediated by monocytes, as pretreated monocytes were able to secrete IL-1 and serve as accessory cells when cultured with untreated T lymphocytes. In fact, when monocytes were added back to treated T lymphocytes, suppression was partially abrogated, suggesting that monocytes could prevent the suppression. Suppressor T lymphocytes could not be detected. There was no decrease in cell number or viability in lymphocyte cultures incubated with UL-FeLV. Other experiments suggested that T-lymphocytes cultured with UV-FeLV could not secrete IL-2. The response of T lymphocytes to exogenous IL-2 (i.e. stimulated lymphocyte culture supernates) was also impaired (61). This latter result was confirmed by experiments using the IL-2-dependent murine T-cell line (CTLL-20) and EL4 cell supernates as a source of IL-2 (62). UV-FeLV or P15E inhibited the response of CTLL-20 cells to IL-2 after brief exposure. The inhibition was reversible, and inhibited cells were able to absorb IL-2 activity as well as did untreated cells. The virus preparations did not absorb IL-2. This is in contrast to another model of immunosuppression by several viruses in which viral particles nonspecifically absorb IL-2 or block IL-2 receptors (63). In the present system, suppression is due to a direct effect on the metabolism of T lymphocytes independent of the requirement of IL-2. Murine spleen cells and cloned T-cell lines could not secrete IL-2 after early exposure of mitogen stimulated cells to UV-FeLv (64). This inhibition was reversible. Cells incubated with UV-FeLV for 24 hr, then washed and recultured with mitogen, were able to secrete IL-2.

In MLR cultures both the proliferative response of murine spleen cells to alloantigen and the generation of alloreactive CTL were blocked by UV-FeLV, but target cell lysis was not inhibited when UV-FeLV was added to the CTL assay (62). An alloreactive CTL clone was impaired in secretion of macrophage activating factor in the presence of UV-FeLV but was not impaired in target cell lysis (64). This suggests that FeLV suppresses lymphocyte functions requiring activation of proliferation but does not suppress established effector functions.

In addition to an inhibitory effect on lymphocyte functions, retroviral proteins, particularly p15E, suppress some functions of mononuclear phagocytes. Low molecular weight extracts of Friend, Moloney, and Rauscher leukemia viruses, and a protein fraction enriched for p15E, when injected subcutaneously into mice, inhibited the accumulation of peritoneal macrophages following an intraperitoneal injection of PHA (65); they directly inhibited the polarization of human monocytes in vitro to a chemoattractant stimulus (66).

The p15E proteins of C-type retroviruses are hydrophobic transmembrane proteins that are thought to function in anchoring viral nucleocapsids to regions of the host cell plasma membrane prior to budding and final maturation of the virion (67). Viral typing sera revealed broadly cross-reactive determinants among the envelope proteins of both C- and D-type retroviruses (68). The deduced amino acid sequences of p15E proteins of MuLV and FeLV, gp21 proteins of HTLV-I and -II, and gp20 of Mason-Pfizer monkey virus all have regions of homology spanning up to 26 amino acid residues (69, 70), indicating a highly conserved structure in the evolution of retroviruses. Cianciolo et al (71) have made a synthetic

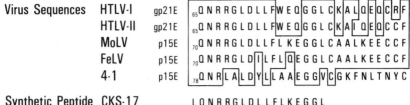

Virus Sequences

HTLV-I	gp21E	$_{65}$Q N R R G L D L L F W E Q G G L C K A L Q E Q C R F
HTLV-II	gp21E	$_{65}$Q N R R G L D L L F W E Q G G L C K A I Q E Q C C F
MoLV	p15E	$_{70}$Q N R R G L D L L F L K E G G L C A A L K E E C C F
FeLV	p15E	$_{70}$Q N R R G L D I L F L Q E G G L C A A L K E E C C F
4-1	p15E	$_{78}$Q N R L A L D Y L L A A E G G V C G K F N L T N Y C

Synthetic Peptide CKS-17   L Q N R R G L D L L F L K E G G L

Figure 5 Amino acid sequences for the conserved region of retrovirus envelope proteins gp21E of human T lymphotropic viruses I and II, and gp15E of Moloney leukemia virus (MoLV), feline leukemia virus (FeLV), and a human endogenous C-retrovirus type (4-1). The number at the lower left of each sequence respresents the residue at which the sequence begins. Modified from Cianciolo et al 1985.

peptide encompassing a portion of this region of homology that suppresses the proliferation of mouse and human lymphocytes in vitro (Figure 5).

We can extend the observations of immunosuppression by p15E-like proteins to tumor systems in which retroviral particles have not been detected. Here, low molecular weight proteins in human cancer effusions (66) and murine tumor cells (72) inhibit the monocyte chemoattractant response, and this activity was immunoprecipitated by a monoclonal antibody directed to the immunosuppressive region of retroviral p15E (72). Snyderman & Cianciolo (73) have proposed that p15E-like proteins present in a wide variety of viral and nonviral tumor systems may be important cofactors in the neoplastic process because they could suppress the host immune response to transformed cells.

HUMAN CYTOMEGALOVIRUS

There has been a growing awareness that HCMV is associated with immunosuppression in diverse clinical settings: congenital infection, organ transplantation, leukemia and lymphoma, and the acquired immunodeficiency syndrome (74, 75). For several of these settings, it is difficult to know whether the virus is a cause of immunosuppression and mortality or an opportunistic pathogen that is permitted to replicate in an immunosuppressed host; however, these possibilities are not mutually exclusive. Like the Epstein Barr virus, HCMV can cause a mononucleosis syndrome associated with transient acquired immunosuppression (76). HCMV, like the other herpes viruses, infects a majority of the population, most often asymptomatically, and establishes latent infection which can be reactivated. Lymphocytes and monocytes are now known to be sites of infection (11).

Attempts to demonstrate infection of peripheral blood mononuclear cells (PBMC) with HCMV either in vivo or in vitro were unsuccessful until recently. Then two groups independently showed that the immediate-early and early antigens of HCMV replication but not the late antigens (i.e. HCMV structural proteins) were expressed in a small percentage of cells exposed to virus in vitro (77, 78). Immunofluorescent staining of cells with polyclonal (77) or monoclonal (78) antibody to these nonstructural antigens was positive in PBMC, although infectious virus was not rescued. More cells were immunofluorescence positive when infected with recent clinical isolates of HCMV than with the laboratory strain AD169 (78). PBMC are less permissive for replication of AD169 than for low passage recent isolates. Laboratory strains of HCMV have been adapted to long-term passage in fibroblasts, and this makes it likely that tropism for lymphocytes and monocytes is lost. HCMV replication proceeds through three stages (74). Immediate-early or α genes are transcribed within the first 30 min of infection, and their proteins regulate the transcription of early or β genes which function about 6 hr after the initiation of infection. Early genes encode the viral polymerase and regulate late or γ genes that are turned on about 24 hr after infection. The late genes encode the viral structural proteins. The detection of immediate-early and early proteins but not structural proteins or infectious virus means that virus replication is blocked at an intermediate stage of replication in PBMC. This observation was extended by Schrier et al (11) in two ways: (a) RNA transcripts of the immediate-early genes were detected, and (b) the transcripts were found in PBMC from normal HCMV seropositive donors indicating a natural infection in vivo with HCMV. Use of the technique of in situ hybridization with a radiolabelled cDNA probe revealed that viral RNA containing the major immediate-early transcript was present in a small percentage of lymphocytes (range 0.3 to 2%) and monocytes (5%). OKT4+ cells were the most common lymphocyte subset involved. These results have recently been confirmed (C. Jordin, personal communication). In a murine CMV model that provided the initial evidence of CMV replication in lymphocytes, restricted infection of such cells could be converted to productive infection by stimulation in MLR (79). Attempts to induce productive infection of HCMV in human lymphocytes by mitogens or alloantigens have so far been unsuccessful (R. Schrier, unpublished data).

Infection of PBMC with low passage recent clinical isolates of HCMV resulted in depression of the following lymphocyte functions: T cell proliferation to antigen or mitogen (78, 80), NK activity but not ADCC (78, 80), antibody production (G. Rice, unpublished), and generation of CMV-specific CTL (81) (Figure 6). The laboratory strain AD169, in contrast,

LYMPHOCYTE ASSAY	VIRUS SOURCE	EXPERIMENTAL DATA	% SUPRESSION
PROLIFERATION	mock	29,158	
(^{3}HT incorporation cpm)	AD 169	44,478	0
	I-G	3,081	89
MITOGEN	I-V	6,129	79
	mock	8,215	
ANTIGEN	AD 169	5,070	38
SPECIFIC	I-G	3,581	56
IMMUNOGLOBULIN	mock	303	
PRODUCTION	AD 169	367	0
(ng Ig made/1×10^5 B lymphocytes)	I-G	8	99
CYTOTOXICITY			
(specific ^{51}Cr released)	AD 169	30	0
	I-S	nil	100
CTL	AD 169+I-S	1	99
	mock	11	
NK	AD 169	24	0
	I-R	2	82
	I-P	1	91
	mock	60	
	AD 169	60	0
ADCC	I-M	58	0
	I-R	71	0
	I-S	72	0

Figure 6 Recent clinical isolates of HCMV alter several lymphocyte functions. For virus source, mock means uninfected. AD169 is a laboratory strain of HCMV. The different clinical isolates are designated I-G, I-V, I-S, I-R, I-F and I-M. From Rice et al 1984; Schrier et al 1986; and Schrier & Oldstone 1986.

has little or no effect on these functions. Further experiments demonstrated that the degree of suppression varied with the strain of virus, the multiplicity of infection, the mode of infection (cell-free virus is less suppressive than virus grown in autologous fibroblasts), and the duration of virus-lymphocyte culture prior to assay. Suppression of both lymphocyte proliferation and NK activity requires live virus. The amount of clinical isolate virus needed to effect suppression is typically 2 $\log_{10}$ less than for AD169. Suppression of proliferation could not be abrogated by exogenous IL-2. The suppression of NK cell function required at least 72 hr of culture with virus. Lytic activity could not be restored by IL-2, but there was partial restoration with alpha interferon. Monocyte depletion prior to culture with virus prevented the suppression, and infected monocytes could suppress uninfected lymphocyte cultures (80). The role of monocytes in suppression of lymphocyte proliferation has been described in HCMV mononucleosis (76).

It was reported that monocyte-mediated suppression of lymphocyte proliferation is due to an inhibitor of IL-1 released from HCMV-infected macrophage cultures (82). Further analysis indicated that IL-1 inhibitor was a protein with a mol wt of about 90,000 and that it required active HCMV for its release. The infected macrophage still releases IL-1, but the released IL-1 is inhibited by the virus-induced inhibitor in the supernatant fluids. Other evidence indicates that monocytes infected with a recent clinical isolate of CMV and stimulated with LPS are unable to secrete IL-1, but the subsequent suppression of lymphocyte proliferation can be restored by exogenous IL-1 (83).

HCMV-specific, MHC class-I restricted OKT8+ CTL can be induced by culture of seropositive PBMC with AD169 virus (81, 84). However, no CTL activity occurred when lymphocytes were cultured with recent clinical isolates of HCMV (81). Generation of HCMV-specific CTL was suppressed by the clinical isolates. Experiments combining clinical isolate and AD169 virus demonstrated that live virus was required for suppression. However, CTL generated by stimulation with AD169 were able to lyse target cells infected with clinical isolate virus as well as targets infected with AD169. Thus, suppressive strains of HCMV are still antigenic and cross-reactive with AD169 for CTL recognition.

LYMPHOCYTIC CHORIOMENINGITIS VIRUS

The cardinal principles by which the immune system clears infectious virus from a host during an acute infection but fails to do so during persistent infection has come, in large part, from the study of lymphocytic choriomeningitis virus (LCMV). This virus induces either an acute or persistent infection in its natural host, the mouse. The virus is not lytic and does not cause cytopathology in those cells that it infects. Rather, lysis of infected cells and tissue injury is immunopathologic, i.e. resulting from the generation of antiviral immune responses that react with virus infected cells.

Infection of an immunocompetent host with LCMV leads to the generation of several antiviral effector mechanisms (reviewed in 85). However, the primary mechanism of both virus clearance and virus-induced immunopathology is the activity of LCMV-specific and H-2-restricted CTL (86, 87). This point is best emphasized through two observations. First, failure to generate CTL or depletion of these cells allows the progression from an acute to a persistent infection (85). Second, adoptive transfer of cloned CTL rapidly and effectively clears virus from acutely infected mice (87). Once established, virus persists throughout the animal's lifespan, and viral materials (nucleic acid sequences, proteins, and infectious virus) are maintained in many tissues (88). During persistence, antibody to all LCMV proteins are made, but the LCMV-specific CTL response is decreased (85). This diminished CTL response is apparently selective for LCMV in that CTL responses to several other viruses can be generated (89). Mice persistently infected with LCMV do show a limited number of other immune response disorders when challenged with unrelated antigens (90, 91). Most intriguing in this regard is their inability to be tolerized to human or bovine gamma globulins (91, 92).

Persistent infection by LCMV is likely caused by a specific defect, i.e. the inability to generate virus-specific H-2-restricted CTL. Support of this notion comes from experiments showing that persistently infected mice reconstituted with LCMV-specific H-2-restricted Thyl.2^+, Lyt 2.2^+, L3T4$^-$ lymphocytes clear infectious virus and viral materials from blood and other tissues (Figure 7; 93, 94). The adoptive transfer of cloned CTL or of long lived LCMV-immune memory cells into persistently infected recipients causes acute immunopathologic injury and death (95, 96). However, transfer of early immune lymphocytes (harvested 30–80 days post-infection) does not produce immunopathology but clears virus materials (94). Clearance of virus from persistently infected recipients is dramatic: 10^4 to 10^5 plaque-forming units of virus are cleared from blood, liver, and spleen by day 15. Virus is cleared more slowly from the kidney and brain, but by 60–120 days after adoptive transfer, greater than 99.9% of virus has been removed from these organs as well. Treated mice remain free of virus a year later (94; A. Tishon, M. Oldstone, unpublished). As shown in Figure 7, viral nucleic acids are also cleared. This is an important point because during persistent LCMV infection there is a continuous accumulation of viral nucleic acid sequences associated with a 3–4 $\log_{10}$ decrease in infectious virus titer, indicating the generation of defective or incomplete virus particles (88). Defective or incomplete virus may have an important role in certain diseases that can occur during persistent infection (1).

LCMV infects mononuclear phagocytes and lymphocytes in vivo during both the acute and persistent infection, but virus does not kill these cells (24, 25). The percentage of infected cells in peripheral blood is small (3%, range 0.5–10%), as detected by infectious center assay (Figure 8; 24, 25) or by in situ hybridization (P. Schwimmbeck, A. Tishon, M. Oldstone, unpublished). The infection is restricted in that virus has only been recovered by cocultivation of infected lymphocytes with a fully permissive

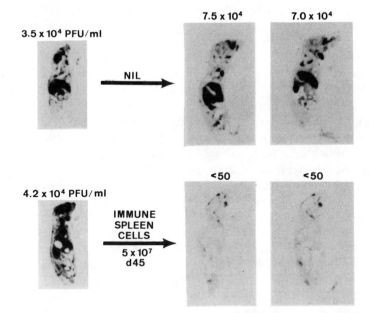

4-5 WEEKS   16-20 WEEKS

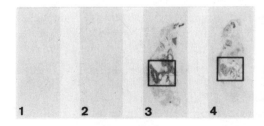

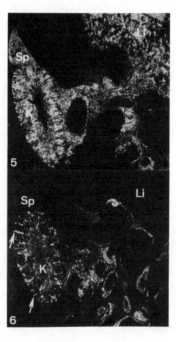

PFU/gm or ml

	d0	d60	d120
SERUM	4.5	<1.6	<1.6
SPLEEN	5.9	<1.6	<1.6
LIVER	5.2	<1.6	<1.6
LUNG	4.8	<1.6	<1.6
BRAIN	5.2	3.0	1.9
KIDNEY	6.2	4.1	3.1

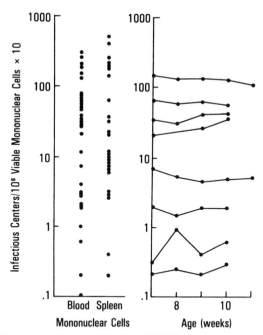

Figure 8 Lymphocytic choriomeningitis virus (LCMV) persists in lymphocytes in vivo. Panel on left shows the numbers of infected lymphocytes from blood or spleen of individual 6–8-week-old persistently infected mice. Panel on right shows multiple samples of blood lymphocytes at different times post-infection of 8 individual mice, indicating that a constant number of cells are infected for a prolonged period. Infected lymphocytes were scored by an infectious center assay. From Doyle & Oldstone 1978.

cell line. The percentage of infected lymphocytes in the peripheral blood remains relatively constant throughout the animal's lifespan (Figure 8; 24). Adoptive transfer of virus-infected lymphocytes from F_1 offspring into an uninfected parent demonstrated that the transferred cells remain

---

Figure 7 Adoptive transfer of LCMV-immune H-2-restricted lymphocytes clears virus from persistently infected mice. In the top figure, clearance of viral nucleic acids from two persistently infected 6–10-week-old mice after adoptive transfer of day 45 immune spleen cells is shown by whole body sections, hybridized with a ^{32}P-labeled nick-translated probe and developed by autoradiography. Numbers above the photographs indicate virus titer by plaque assay. In the bottom figure, clearance of virus and viral proteins is shown. The table indicates the $\log_{10}$ of viral titers by plaque assay of serum or organs from persistently infected mice on the day of transfer or 60 and 120 days after transfer of immune spleen cells. Panels 1–6 show expression of viral proteins in whole animal sections. Panel 1, control uninfected mouse section reacted with antibody reagents; Panel 2, infected mouse section reacted with second antibody alone; Panels 3 & 5, and Panels 4 & 6, viral nucleoprotein antigen prior to lymphocyte transfer and 60 days after transfer, respectively. Arrows point to rows of glomeruli. Abbreviations: Sp—spleen, Li—liver, K—kidney. From Oldstone et al 1986.

viable and retain virus (24). Virus was not found in lymphocytes of the uninfected recipient. From these studies we can conclude that virus perists in infected lymphocytes but does not destroy them. Further, input virus is not released in vivo, or if it is, it is unable to infect the lymphocytes of the adult recipient. Other studies using cell surface markers and infectious center assay indicate that Thyl.2$^+$ cells are the major lymphocytes infected (24), and of these cells, the L3T4$^+$ subset (24a).

Recently the biologic mechanism by which CTLs become suppressed during LCMV persistent infection has been uncovered (89). The principles involved are cartooned in Figure 9. Inoculation of newborn mice with plaque purified LCMV (Armstrong strain) results in a persistent infection associated with the failure to generate LCMV-specific H-2-restricted CTL. In contrast, inoculation of the same cloned virus into immunocompetent adult mice leads to generation of a brisk CTL response. When spleen lymphocytes are harvested from two- or three-month-old persistently infected mice and then adoptively transferred to immunocompetent adult mice, they actively suppress the expected LCMV-specific CTL response of those animals when challenged with LCMV. Further analysis indicated that the suppression is not mediated by suppressor cells but rather by genetic variants of LCMV found in spleen lymphocytes of persistently infected mice. Spleen variants of the parental Armstrong strain of LCMV

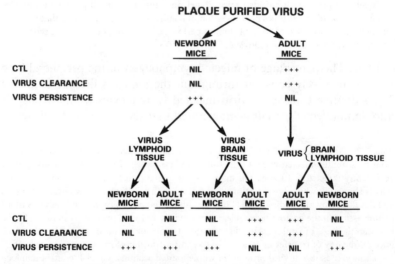

Figure 9 Selection of LCMV variants by replication in lymphoid tissue during persistent virus infection. These variants, unlike parental virus, do not induce LCMV-specific CTL, but do establish persistent infection in adult immunocompetent mice. Schematic presentation of data reported in Ahmed et al 1984.

inoculated into adult immunocompetent mice prevent the generation of virus-specific CTLs and establish persistent infection (89, 97). Thus, the evidence on hand indicates that a viral variant emerging during infection in vivo plays a critical role in the suppression of virus-specific CTL and the establishment of persistent infection. Virus-induced suppression of virus-specific CTL function has been described in two other systems in vitro (98, 99).

EPILOGUE

Viruses are obligate intracellular parasites that require host cell metabolic processes and substrates for their replication. Virus infection of lymphocytes can be regarded as a special case of virus-cell interaction. The variety of phenomena describing these interactions reflects the replication strategy of a particular virus and the metabolic potential of the infected cell to support viral replication.

Measles is a negative strand RNA virus. That is, the genomic RNA is not messenger RNA; instead, a complementary positive strand is transcribed (100). The block in viral replication in unstimulated lymphocytes could be at the level of transcription or replication of the genomic strand. When a viral genome was sequenced recently, a nonstructural viral protein was discovered that migrates to the cell nucleus early during productive infection (101). Suppression of lymphocyte function could be mediated by viral RNA or protein intermediates during replication. The possibility is being addressed in several laboratories with strand-specific RNA probes in hybridization to infected cell RNAs. The study of measles virus during persistent infection in vivo has provided other leads. Measles RNA has been detected by in situ hybridization in peripheral blood lymphocytes during SSPE (102) and in mononuclear cells present in Paget's disease bone tissue (103). These viruses are not immunosuppressive in vivo (42), and this suggests that defective replication aborts suppression or selection of non-suppressive virus occurs during persistent infection.

Evidence suggests that functional genes encoding retroviral p15E-like proteins may be present in the germline DNA of mammals including man. Recently, the nucleotide sequence of a full-length endogenous C-type human retrovirus has been reported (104), with *gag* and *pol* gene homologies to Moloney MuLV, but an *env* gene sequence that is unique. However, within the *env* open reading frame, extensive homology in deduced amino acid sequence to the immunosuppressive region of infectious retrovirus p15E was seen (clone 4-1 in Figure 5). This human provirus is considered to be replication defective because stop codons or changes in reading frame are present within the open reading frames. But polyadenylated RNAs hybridizing to *gag*, *pol*, and *env* region probes have been detected in both normal and neoplastic human cells (105) indicating the transcription of some provirus genes. Mitogen-stimulated human peripheral blood mononuclear cells were positive by indirect immunofluorescence with monoclonal antibody to p15E, but unstimulated cells were not (106). Similar proteins could not be detected in murine spleen cells by immunoprecipitation (72). It is not yet clear whether a suppressive p15E-like protein is synthesized during normal lymphocyte activation. Yet these observations pose a potentially important set of questions about virus-lymphocyte interactions: the effect of endogenous retrovirus elements on physiologic functions of lymphocytes (reviewed in 107).

Elucidation of the mechanism(s) of latency and reactivation of HCMV will allow an understanding of the suppression of lymphocyte biology by this virus. The virus is restricted in undifferentiated cells. Chemical factors that induce differentiation (108, 109) and allogeneic cell interactions (79) play a role in activating viral replication. Studies on the molecular basis of HCMV latency show that the block in replication occurs at the level of transcription of the major immediate-early gene (110). DNase I hypersensitivity analysis of chromatin structure in nonpermissive and permissive cells and deletion analysis of an upstream regulatory region suggest that there is differential binding of positive and negative regulatory proteins to regions upstream of the gene. Recent evidence suggests that there is a negative regulatory protein in nonpermissive cells (J. Nelson, unpublished).

In order to define the genetic basis for the different pathologic effects of LCMV parental virus and the spleen variants, cloning and sequencing of the viral genome are needed. Special attention is focused on the open reading frame containing the glycoprotein gene. Experiments using reassortants of LCMV genes (111) as well as recombinant truncated LCMV genes in vaccinia vectors (112) have shown that the induction of and recognition by LCMV-specific CTL map to the glycoprotein region. The selective pressure for generation of immunosuppressive viral variants in lymphocytes is not known. Parental virus and several variants isolated from nonlymphoid organs induce CTL. Work to define the tropism of these viruses at the molecular level is in progress.

Future studies of the effects of viruses in lymphocytes will complement ongoing work in the area of lymphocyte activation and differentiation (reviewed in 113, 114). The evidence already obtained from the models discussed in this chapter suggests that there may be several interesting pathways for the effects of different viruses on lymphocytes. The increasing observations that viruses infect lymphocytes (Table 1) and can persist in them suggest that viral infections may be associated with diseases not traditionally known to be of viral etiology. Persistent viral infections may play a role in diseases of immune regulation, i.e. autoimmune and immunosuppressive disorders. Understanding the principles and specific defects in a particular virus-induced disorder of immune regulation should allow for its correction with specific immunotherapy or perhaps a needed growth factor.

ACKNOWLEDGMENTS

This is Publication Number 4527-IMM from the Department of Immunology, Scripps Clinic and Research Foundation, La Jolla, California. This work was supported in part by USPHS grants AI-09484, AI-07007, NS-12428 and AG-04342. M. B. McChesney has a postdoctoral fellowship award from the National Arthritis Foundation.

We are grateful to R. D. Schrier, J. A. Nelson, P. J. Southern, and R. Ahmed for helpful discussions. We are indebted to Mrs. Gay L. Schilling for manuscript preparation.

Literature Cited

1. Oldstone, M. B. A. 1984. Virus can alter cell function without causing cell pathology: Disordered function leads to imbalance of homeostasis and disease. In *Concepts in Viral Pathogenesis*, ed. A. L. Notkins, M. B. A. Oldstone, pp. 269–76. New York: Springer-Verlag
2. Notkins, A. L., Mergenhagen, S. E., Howard, R. J. 1970. Effect of virus infections on the function of the immune system. *Ann. Rev. Micro.* 24: 525
3. Wheelock, E. F., Toy, S. 1973. Participation of lymphocytes in viral infections. *Adv. Immunol.* 16: 123
4. Woodruff, J. F., Woodruff, J. J. 1975. T lymphocyte interaction with viruses and virus-infected tissues. *Prog. Med. Virol.* 19: 120
5. Korba, B. E., Wells, F., Tennant, B. C., Yoakum, G. H., Purcell, R. H., Gerin, J. L. 1986. Hepadnavirus infection of peripheral blood lymphocytes *in vivo*: Woodchuck and chimpanzee models of viral hepatitis. *J. Virol.* 58: 1
6. zur Hausen, H., Gissmann, L. 1979. Lymphotropic papovaviruses isolated from African green monkey and human cells. *Med. Microbiol. Immunol.* 167: 137
7. Lambriex, M., van der Veen, J. 1976. Comparison of replication of adenovirus type 2 and type 4 in human lymphocyte cultures. *Infect. Immun.* 14: 618
8. Horvath, J., Palkonyay, L., Weber, J.

1986. Group C adenovirus DNA sequences in human lymphoid cells. *J. Virol.* 59: 189
9. Pelton, B. K., Imrie, R. C., Denman, A. M. 1977. Susceptibility of human lymphocyte populations to infection by herpes simplex virus. *Immunology* 32: 803
10. Greaves, M. F., Brown, G., Rickinson, A. B. 1975. Epstein-Barr virus binding sites on lymphocyte subpopulations and the origin of lymphoblasts in cultured lymphoid cell lines and in the blood of patients with infectious mononucleosis. *Clin. Immunol. Immunopathol.* 3: 514
11. Schrier, R. D., Nelson, J. A., Oldstone, M. B. A. 1985. Detection of human cytomegalovirus in peripheral blood lymphocytes in a natural infection. *Science* 230: 1048
12. Strayer, D. S., Skaletsky, E., Leibowitz, J. L. 1985. *In vitro* growth of two related leporipoxviruses in lymphoid cells. *Virology* 145: 330
13. Paul, P. S., Mengeling, W. L., Brown, T. T. 1979. Replication of procine parvovirus in peripheral blood lymphocytes, monocytes and peritoneal macrophages. *Infect. Immun.* 25: 1003
14. Kimsey, P. B., Engers, H. D., Hirt, B., Jongeneel, C. V. 1986. Pathogenicity of fibroblast- and lymphocyte-specific variants of minute virus of mice. *J. Virol.* 59: 8
15. Willems, F. T. C., Melnick, J. L.,

Rawls, W. E. 1969. Replication of poliovirus in phytohemagglutinin-stimulated human lymphocytes. *J. Virol.* 3: 451

16. Chantler, J. K., Tingle, A. J. 1982. Isolation of rubella virus from human lymphocytes after acute natural infection. *J. Infect. Dis.* 145: 673

17. Sullivan, J. L., Barry, D. W., Lucas, S. J., Albrecht, P. 1975. Measles infection of human mononuclear cells. I. Acute infection of peripheral blood lymphocytes and monocytes. *J. Exp. Med.* 142: 773

18. Joseph, B. S., Lampert, P. W., Oldstone, M. B. A. 1975. Replication and persistence of measles virus in defined subpopulations of human leukocytes. *J. Virol.* 16: 1638

19. Fleischer, B., Kreth, H. W. 1982. Mumps virus replication in human lymphoid cell lines and in peripheral blood lymphocytes: Preference for T cells. *Infect. Immun.* 35: 25

20. Domurat, F., Roberts, N. J. Jr., Walsh, E. E., Dagan, R. 1985. Respiratory syncytial virus infection of human mononuclear leukocytes *in vitro* and *in vivo*. *J. Infect. Dis.* 152: 895

21. Bloom, B. R., Senik, A., Stoner, G., Ju, G., Nowakowski, M., Kano, S., Jiminez, L. 1976. Studies on the interactions between viruses and lymphocytes. *Cold Spring Harbor Symp. Quant. Biol.* 41: 73

22. Roberts, N. J. Jr., Horan, P. K. 1985. Expression of viral antigens after infection of human lymphocytes, monocytes and macrophages with influenza virus. *J. Infect. Dis.* 151: 308

23. Verini, M. A., Lief, F. S. 1979. Interaction between 6/94 virus, a parainfluenza type 1 strain, and human leukocytes. *Infect. Immun.* 24: 734

24. Doyle, M. V., Oldstone, M. B. A. 1978. Interactions between viruses and lymphocytes. I. *In vivo* replication of lymphocytic choriomeningitis virus in mononuclear cells during both chronic and acute viral infections. *J. Immunol.* 121: 1262

24a. Ahmed, R., King, C. C., Oldstone, M. B. A. 1986. Virus-lymphocyte interaction: T cells of the helper subset are infected with lymphocytic choriomeningitis during persistent infection in vivo. *J. Virol.* Submitted

25. Popescu, M., Lohler, J., Lehmann-Grube, F. 1979. Infectious lymphocytes in lymphocytic choriomeningitis virus carrier mice. *J. Gen. Virol.* 42: 481

26. Ambrosio, A. M., Enria, D. A., Maiztegui, J. I. 1986. Junin virus isolation from lympho-mononuclear cells of patients with Argentine hemorrhagic fever. *Intervirology* 25: 97

27. Cerny, J., Hensgen, P. A., Fistel, S. H., Demler, L. M. 1976. Interactions of murine leukemia virus (MuLV) with isolated lymphocytes. II. Infection of B and T cells with Friend virus complex in diffusion chambers and *in vitro*: Effect of polyclonal mitogens. *Int. J. Cancer* 18: 189

28. Rojko, J. L., Hoover, E. A., Finn, B. L., Olsen, R. G. 1981. Determinants of susceptibility and resistance to feline leukemia virus infection. II. Susceptibility of feline lymphocytes to productive feline leukemia virus infection. *J. Natl. Cancer Inst.* 67: 899

29. Hinuma, Y. 1984. Retrovirus in adult T cell leukemia. *Prog. Med. Virol.* 30: 156

30. Wong-Staal, F., Gallo, R. C. 1985. Human T lymphotropic retroviruses. *Nature* 317: 395

31. Hoxie, J. A., Haggarty, B. S., Rackowski, J. L., Pillsbury, N., Levy, J. A. 1985. Persistent noncytopathic infection of normal human T lymphocytes with AIDS-associated retrovirus. *Science* 229: 1400

32. McDougal, J. S., Mawle, A., Cort, S. P., Nicholson, J. K. A., Cross, G. D., Scheppler-Campbell, J. A., Hicks, D., Sligh, J. 1985. Cellular tropism of the human retrovirus HTLV-III/LAV. I. Role of T cell activation and expression of the T4 antigen. *J. Immunol.* 135: 3151

33. Hirsch, M. S., Phillips, S. M., Solnik, C., Black, P. H., Schwartz, R. S., Carpenter, C. B. 1972. Activation of leukemia viruses by graft-versus-host and mixed lymphocyte reactions *in vitro*. *Proc. Natl. Acad. Sci. USA* 69: 1069

34. Heberden, W. 1802. *Commentaries on the History and Cure of Diseases*, pp. 313–22. London: T. Payne

35. Osler, W. 1982. *The Principles and Practice of Medicine*, pp. 77–81. New York: Appleton

36. Holt, L. E. 1897. *The Diseases of Infancy and Childhood*, pp. 910–26. New York: Appleton

37. von Pirquet, C. 1908. Das verhalten der kutanen tuberkulin-reaktion wahrend der massern. *D. Med. Wochenschr.* 30: 1297

38. Smithwick, E. M., Berkovich, S. 1969. The effect of measles virus on the *in vitro* lymphocyte response to tuberculin. In *Cellular Recognition*, ed. R. T. Smith, R. A. Good, p. 131. New York: Appleton Century Crofts

39. Zweiman, B. 1971. *In vitro* effects of measles virus on proliferating human lymphocytes. *J. Immunol.* 106: 1154

40. Fireman, P., Friday, G., Kumate, J. 1969. Effect of measles vaccine on immunologic responsiveness. *Pediatrics* 43: 264

41. Norrby, E. 1971. The effect of a carbenzoxy tripeptide on the biological activities of measles virus. *Virology* 44: 599

42. ter Meulen, V., Stephenson, J. R., Kreth, H. W. 1983. Subacute sclerosing panencephalitis. In *Comprehensive Virology*, ed. H. Fraenkel-Conrat, R. R. Wagner, 18: 105. New York: Plenum

43. Whittle, H. C., Dossetor, J., Oduloju, A., Bryceson, A. D. M., Greenwood, B. M. 1978. Cell-mediated immunity during natural measles infection. *J. Clin. Invest.* 62: 678

44. Huddlestone, J. R., Lampert, P. W., Oldstone, M. B. A. 1980. Virus-lymphocyte interactions: Infection of Tg and Tm subsets by measles virus. *Clin. Immunol. Immunopathol.* 15: 502

45. Lucas, C. J., Ubels-Postma, J. C., Rezee, A., Galama, J. M. D. 1978. Activation of measles virus from silently infected human lymphocytes. *J. Exp. Med.* 148: 940

46. Sullivan, J. L., Barry, D. W., Albrecht, P., Lucas, S. J. 1975. Inhibition of lymphocyte stimulation by measles virus. *J. Immunol.* 114: 1458

47. Lucas, C. J., Galama, J. M. D., Ubels-Postma, J. C. 1977. Measles virus induced suppression of lymphocyte reactivity *in vitro*. *Cell. Immunol.* 32: 70

48. Casali, P., Sissons, J. G. P., Buchmeier, M. J., Oldstone, M. B. A. 1981. Generation of human cytotoxic lymphocytes by virus: Viral glycoproteins induce nonspecific cell-mediated cytotoxicity without release of interferon *in vitro*. *J. Exp. Med.* 154: 840

49. Rose, J. W., Bellini, W. J., McFarlin, D. E., McFarland, H. F. 1984. Human cellular immune response to measles virus polypeptides. *J. Virol.* 49: 988

50. Minagawa, T., Nakaya, C., Ida, H. 1974. Host DNA synthesis-suppressing factor in culture fluid of tissue cultures infected with measles virus. *J. Virol.* 13: 1118

51. Lucas, C. J., Ubels-Postma, J., Galama, J. M. D., Rezee, A. 1978. Studies on the mechanism of measles virus-induced suppression of lymphocyte functions *in vitro*. Lack of a role for interferon and monocytes. *Cell. Immunol.* 37: 448

52. Lanier, M. S. 1986. *Possible mechanisms of measles virus-induced immunosuppression*. PhD thesis. Univ. N. Mex. Albuquerque, N. Mex.

53. Galama, J. M. D., Ubels-Postma, J., Vos, A., Lucas, C. J. 1980. Measles virus inhibits acquisition of lymphocyte functions but not established effector functions. *Cell. Immunol.* 50: 405

54. Casali, P., Rice, G. P. A., Oldstone, M. B. A. 1984. Viruses disrupt functions of human lymphocytes. Effects of measles virus and influenza virus on lymphocyte-mediated killing and antibody production. *J. Exp. Med.* 159: 1322

55. Hirano, T., Kuritani, T., Kishimoto, T., Yamamura, Y. 1977. *In vitro* immune response of human peripheral blood lymphocytes. I. The mechanism(s) involved in T cell helper functions in the pokeweed mitogen-induced differentiation and proliferation of B cells. *J. Immunol.* 119: 1235

56. McChesney, M. B., Fujinami, R. S., Lampert, P. W., Oldstone, M. B. A. 1986. Viruses disrupt functions of human lymphocytes. II. Measles virus suppresses antibody production by acting on B lymphocytes. *J. Exp. Med.* 163: 1331

57. Borysiewicz, L. K., Casali, P., Rogers, B., Morris, S., Sissons, J. G. P. 1985. The immunosuppressive effects of measles virus on T cell function-failure to affect IL-2 release or cytotoxic T cell activity *in vitro*. *Clin. Exp. Immunol.* 59: 29

58. Dent, P. B. 1972. Immunodepression by oncogenic viruses. *Prog. Med. Virol.* 14: 1

59. Essex, M. 1975. Horizontally and vertically transmitted oncornaviruses of cats. *Adv. Cancer Res.* 21: 175

60. Mathes, L. E., Olsen, R. G., Hebebrand, L. C., Hoover, E. A., Schaller, J. P. 1978. Abrogation of lymphocyte blastogenesis by a feline leukemia virus protein. *Nature* 274: 687

61. Copelan, E. A., Rinehart, J. J., Lewis, M., Mathes, L., Olsen, R., Sagone, A. 1983. The mechanism of retrovirus suppression of human T cell proliferation *in vitro*. *J. Immunol.* 131: 2017

62. Orosz, C. G., Zinn, N. E., Olsen, R. G., Mathes, L. E. 1985. Retrovirus mediated immunosuppression. I. FeLV-UV and specific FeLV proteins alter T lymphocyte behavior by inducing hyporesponsiveness to lymphokines. *J. Immunol.* 134: 3396

63. Wainberg, M. A., Vydelingum, S., Boushira, M., Legace-Simard, J., Margolese, R. G., Spira, B., Mendelson, J. 1984. Reversible interference with TCGF activity by virus particles. *Clin. Exp. Immunol.* 57: 663

64. Orosz, C. G., Zinn, N. E., Olsen, R. G., Mathes, L. E. 1985. Retrovirus mediated immunosuppression. II. FeLV-UV alters *in vitro* murine T lymphocyte behavior by reversibly impairing lymphokine secretion. *J. Immunol.* 135: 583

65. Cianciolo, G. J., Matthews, T. J., Bolognesi, D. P., Snyderman, R. 1980. Macrophage accumulation in mice is inhibited by low molecular weight products from murine leukemia viruses. *J. Immunol.* 124: 2900

66. Cianciolo, G., Hunter, J., Silva, J., Haskill, J. S., Snyderman, R. 1981. Inhibitors of monocyte responses to chemotaxins are present in human cancerous effusions and react with monoclonal antibodies to the P15(E) structural protein of retroviruses. *J. Clin. Invest.* 68: 831

67. Bolognesi, D. P., Montelaro, R. C., Frank, H., Schafer, W. 1978. Assembly of Type C oncornaviruses: A model. *Science* 199: 183

68. Thiel, H. J., Broughton, E. M., Matthews, T. J., Schafer, W., Bolognesi, D. P. 1981. Interspecies reactivity of Type C and D retroviruses p15E and p15C proteins. *Virology* 111: 270

69. Cianciolo, G. J., Kipnis, R. J., Snyderman, R. 1984. Similarity between p15E of murine and feline leukemia viruses and p21 of HTLV. *Nature* 311: 515

70. Sonigo, P., Barker, C., Hunter, E., Wain-Hobson, S. 1986. Nucleotide sequence of Mason-Pfizer monkey virus: An immunosuppressive D-Type retrovirus. *Cell* 45: 375

71. Cianciolo, G. J., Copeland, T. D., Oroszlan, S., Snyderman, R. 1985. Inhibition of lymphocyte proliferation by a synthetic peptide homologous to retroviral envelope proteins. *Science* 230: 453

72. Cianciolo, G. J., Lostrom, M. E., Tam, M., Snyderman, R. 1983. Murine malignant cells synthesize a 19,000-dalton protein that is physicochemically and antigenically related to the immunosuppressive retroviral protein, p15E. *J. Exp. Med.* 158: 885

73. Snyderman, R., Cianciolo, G. J. 1984. Immunosuppressive activity of the retroviral envelope protein P15E and its possible relationship to neoplasia. *Immunol. Today* 5: 240

74. Ho, M. 1982. *Cytomegalovirus, Biology and Infection.* New York: Plenum. 309 pp.

75. Blaser, M. J., Cohn, D. L. 1986. Opportunistic infections in patients with AIDS: Clues to the epidemiology of AIDS and the relative virulence of pathogens. *Rev. Infect. Dis.* 8: 21

76. Carney, W. P., Hirsch, M. S. 1981. Mechanisms of immunosuppression in cytomegalovirus mononucleosis. II. Virus-monocyte interactions. *J. Infect. Dis.* 144: 47

77. Einhorn, L., Ost, A. 1984. Cytomegalovirus infection of human blood cells. *J. Infect. Dis.* 149: 207

78. Rice, G. P. A., Schrier, R. D., Oldstone, M. B. A. 1984. Cytomegalovirus infects human lymphocytes and monocytes: Virus expression is restricted to immediate-early gene products. *Proc. Natl. Acad. Sci. USA* 81: 6134

79. Olding, L. B., Jensen, F. C., Oldstone, M. B. A. 1975. Pathogenesis of cytomegalovirus infection. I. Activation of virus from bone marrow derived lymphocytes by *in vitro* allogenic reaction. *J. Exp. Med.* 141: 561

80. Schrier, R. D., Rice, G. P. A., Oldstone, M. B. A. 1986. Suppression of natural killer cell activity and T cell proliferation by fresh isolates of human cytomegalovirus. *J. Infect. Dis.* 153: 1084

81. Schrier, R. D., Oldstone, M. B. A. 1986. Recent clinical isolates of cytomegalovirus suppress human leukocyte antigen-restricted cytotoxic T lymphocyte activity. *J. Virol.* 59: 127

82. Rodgers, B. C., Scott, D. M., Mundin, J., Sissons, J. G. P. 1985. Monocyte derived inhibitor of interleukin 1 induced by human cytomegalovirus. *J. Virol.* 55: 527

83. Dudding, L. R., Garnett, H. M. 1986. Interaction of AD169 and a clinical isolate of cytomegalovirus with peripheral monocytes: The effect of LPS stimulation. *J. Infect. Dis.* In press

84. Boryscewicz, L. K., Morris, S., Page, J. D., Sissons, J. G. P. 1983. Human cytomegalovirus-specific T lymphocytes: Requirements for *in vitro* generation and specificity. *Eur. J. Immunol.* 13: 804

85. Buchmeier, M. J., Welsh, R. M., Dutko, F. J., Oldstone, M. B. A. 1980. The virology and immunobiology of lymphocytic choriomeningitis virus infection. *Adv. Immunol.* 30: 275

86. Zinkernagel, R. M., Doherty, P. C. 1974. Restriction of *in vitro* T cell mediated cytotoxicity in lymphocytic choriomeningitis within a syngeneic or semiallogeneic system. *Nature* 248: 701

87. Byrne, J. A., Ahmed, R., Oldstone, M. B. A. 1984. Biology of cloned cytotoxic T lymphocytes specific for lymphocytic choriomeningitis virus. I. Generation

and recognition of virus strains and H-2b mutants. *J. Immunol.* 133 : 433

88. Southern, P. J., Blount, P., Oldstone, M. B. A. 1984. Analysis of persistent virus infections by *in situ* hybridization to whole-mouse sections. *Nature* 312 : 555

89. Ahmed, R., Salmi, A., Butler, L. D., Chiller, J. M., Oldstone, M. B. A. 1984. Selection of genetic variants of lymphocytic choriomeningitis virus in spleens of persistently infected mice. Role in suppression of cytotoxic T lymphocyte response and viral persistence. *J. Exp. Med.* 160 : 521

90. Mims, C. A., Wainwright, S. 1968. The immunosuppressive action of lymphocytic choriomeningitis virus in mice. *J. Immunol.* 101 : 717

91. Oldstone, M. B. A., Tishon, A., Chiller, J. M., Weigle, W. O., Dixon, F. J. 1973. Effect of chronic viral infection on the immune system. I. Comparison of the immune responsiveness of mice chronically infected with LCM virus with that of noninfected mice. *J. Immunol.* 110 : 1268

92. Pincus, T., Rowe, W. P., Staples, P. J., Talal, N. 1970. Inability to induce tolerance to bovine gamma globulin in mice infected at birth with lymphocytic choriomeningitis virus. *Proc. Soc. Exp. Biol. Med.* 133 : 986

93. Volkert, M. 1963. Studies on immunologic tolerance to LCM virus. 2. Treatment of virus carrier mice by adoptive immunization. *Acta. Pathol. Microbiol. Scand.* 57 : 465

94. Oldstone, M. B. A., Blount, P., Southern, P. J., Lampert, P. W. 1986. Cytoimmunotherapy for persistent virus infection reveals a unique clearance pattern from the central nervous system. *Nature* 321 : 239

95. Byrne, J. A., Oldstone, M. B. A. 1986. Biology of cloned cytotoxic T lymphocytes specific for lymphocytic choriomeningitis virus. VI. Migration and activity *in vivo* in acute and persistent infection. *J. Immunol.* 136 : 698

96. Baenziger, J., Hengartner, H., Zinkernagel, R. M., Cole, G. A. 1986. Induction or prevention of immunopathologic disease by cloned cytotoxic T cell lines specific for lymphocytic choriomeningitis virus. *Eur. J. Immunol.* 16 : 387

97. Tishon, A., Oldstone, M. B. A. 1986. Persistent virus infection associated with chemical manifestations of diabetes. II. Role of viral strain, environmental insult and host genetics. *Am. J. Pathol.* In press

98. Lamb, J. R., Skidmore, B. J., Green, N., Chiller, J. M., Feldman, M. 1983. Induction of tolerance in influenza virus-immune T lymphocyte clones with synthetic peptides of influenza hemagglutinin. *J. Exp. Med.* 157 : 1434

99. Mitsuya, H., Guo, H., Megson, M., Trainer, C., Reitz, M. S. Jr., Broder, S. 1984. Transformation and cytopathic effect in an immune human T cell clone infected by HTLV-I. *Science* 223 : 1293

100. Norrby, E. 1986. Measles. In *Virology*, ed. B. N. Fields, Pp. 1305–21. New York : Raven

101. Bellini, W. J., Englund, G., Rozenblatt, S., Arnheitter, H., Richardson, C. D. 1985. Measles virus P gene codes for two proteins. *J. Virol.* 53 : 908

102. Fournier, J. G., Tardieu, M., Lebon, P., Robain, O., Pousot, G., Rozenblatt, S., Bouteille, M. 1985. Detection of measles virus RNA in lymphocytes from peripheral blood and brain perivascular infiltrates of patients with subacute sclerosing panencephalitis. *N. Engl. J. Med.* 313 : 910

103. Basle, M. F., Fournier, J. G., Rozenblatt, S., Rebel, A., Bouteille, M. 1986. Measles virus RNA detected in Paget's disease bone tissue by *in situ* hybridization. *J. Gen. Virol.* 67 : 907

104. Repaske, R., Steele, P. E., O'Neill, R. R., Rabson, A., Martin, M. A. 1985. Nucleotide sequence of a full-length human endogenous retroviral segment. *J. Virol.* 54 : 764

105. Rabson, A. B., Steele, P. E., Garon, C. F., Martin, M. A. 1983. mRNA transcripts related to full-length endogenous retroviral DNA in human cells. *Nature* 306 : 604

106. Cianciolo, G. J., Phipps, D., Snyderman, R. 1984. Human malignant and mitogen-transformed cells contain retroviral P15E-related antigen. *J. Exp. Med.* 159 : 964

107. Wecker, E., Horak, I. 1982. Expression of endogenous viral genes in mouse lymphocytes. *Curr. Top. Microbiol. Immunol.* 98 : 27

108. Dutko, F. J., Oldstone, M. B. A. 1981. Cytomegalovirus causes a latent infection in undifferentiated cells and is activated by induction of cell differentiation. *J. Exp. Med.* 154 : 1636

109. Gonczol, E., Andrew, P. W., Plotkin, S. A. 1984. Cytomegalovirus replicates in differentiated but not in undifferentiated human embryonal carcinoma cells. *Science* 224 : 159

110. Nelson, J. A., Groudine, M. 1986. Transcriptional regulation of the human cytomegalovirus major immediate-early gene is associated with in-

duction of DNaseI-hypersensitive sites. *Mol. Cell. Biol.* 6 : 452

111. Riviere, Y., Southern, P. J., Ahmed, R., Oldstone, M. B. A. 1986. Biology of cloned cytotoxic T lymphocytes specific for lymphocytic choriomeningitis virus. V. Recognition is restricted to gene products encoded by the viral S RNA segment. *J. Immunol.* 136 : 304

112. Whitton, J. L., Southern, P., Tishon, A., Oldstone, M. B. A. 1986. LCMV glycoprotein expressed in vaccinia virus allows H-2 restricted LCMV-specific CTL recognition/lysis. *Fed. Proc.* 45 : 979

113. Howard, M., Paul, W. E. 1983. Regulation of B-cell growth and differentiation by soluble factors. *Ann. Rev. Immunol.* 1 : 307

114. Weiss, A., Imboden, J., Hardy, K., Manger, B., Terhoust, C., Stobo, J. 1986. The role of the T3/antigen receptor complex in T-cell activation. *Ann. Rev. Immunol.* 4 : 593

Ann. Rev. Immunol. 1987. 5 : 305–24

VACCINIA VIRUS EXPRESSION VECTORS[1]

Bernard Moss and Charles Flexner

Laboratory of Viral Diseases, National Institute of Allergy and Infectious Diseases, National Institutes of Health, Bethesda, Maryland 20892

INTRODUCTION

We depend on a variety of specific and nonspecific mechanisms for defense against microbial invaders. Although the value of vaccination for protection against infectious disease is well established, neither the significant target antigens nor the relative roles of humoral and cell mediated immunity have been adequately defined in most cases. Nevertheless, a thorough understanding of these parameters is critical for the engineering of new vaccines. Recent advances in recombinant DNA technology, peptide synthesis, and hybridoma production are providing important tools for analyzing the protective immune response. In this review we focus on one new approach, namely the use of vaccinia virus as a general expression vector for stimulation of humoral and cell mediated immunity to specified proteins and for determination of target antigens of cytotoxic T lymphocytes (CTL). The protective effect of vaccination with live recombinant vaccinia viruses and their potential as vaccines for veterinary and medical purposes are discussed.

TYPES OF EXPRESSION VECTORS

Recombinant DNA technology has made it possible to synthesize proteins in cells from diverse sources including bacteria, yeast, and mammals. Generally speaking, homologous systems are best for producing proteins that are properly folded and processed. Continuous high level expression of mammalian proteins can be obtained by gene amplification in stably

---

[1] The US Government has the right to retain a nonexclusive, royalty-free license in and to any copyright covering this paper.

transformed cell lines. The use of viruses as expression vectors, however, provides advantages of rapid construction and portability to different cell types and even whole animals. For technical reasons, DNA viruses and retroviruses have been developed as vectors. The small and intermediate sized viruses, including SV40, bovine papilloma virus, retroviruses, and adenoviruses offer simplicity of vector construction and relatively high level expression. As a trade-off, however, these viruses have a limited host range and/or can accommodate little extra genetic information before becoming defective in replication. The large DNA viruses, such as vaccinia virus and herpes simplex virus, are intrinsically more difficult to engineer genetically than the viruses listed above, but they have a wider host range and can accommodate substantial amounts of foreign DNA while retaining infectivity. Since vaccinia virus has been used extensively for gene expression, identification of targets for humoral and cell mediated immunity, and protection of experimental animals against viral diseases, further discussion is limited to this vector. Although vaccinia virus has some unique characteristics, many of the approaches described here could be readily adapted to other poxviruses as well as to members of the herpesvirus and adenovirus families.

HISTORICAL PERSPECTIVES

The distinctive scars left by smallpox and the severity of the disease led to the early recognition that recurrences were rare. This observation was put to practical use, perhaps first in China, by inoculating susceptible individuals with infectious material from mild cases of smallpox. Although this process, known as variolation, had a mortality as high as 1%, it otherwise effectively prevented even more severe natural infections.

The recognition, in England, that milkmaids who had contracted cowpox also were protected against smallpox led to the next advance. Edward Jenner, a country physician and noted naturalist, tested the validity of this folklore by inoculating several children with cowpox and then challenging them by variolation (1). The absence of smallpox lesions in these children convinced Jenner that vaccination would protect against naturally transmitted smallpox, and he correctly predicted that this means of prophylaxis would eventually eradicate the disease. The value of vaccination was quickly and generally accepted and fostered a burgeoning interest in immunization by Pasteur and others in the 1800s. In this sense, smallpox vaccination played a major role in the inception of the science of immunology.

In Jenner's time, poxvirus infections occurred naturally in cattle and horses, and more than one virus type may have been prevalent. Jenner's vaccine was passaged initially in humans and subsequently in cattle and sheep. In some places it was even mixed with smallpox virus. For these reasons the pedigree of vaccinia virus, as it came to be known, is somewhat clouded. Comparisons of the restriction endonuclease maps of the genomes of vaccinia virus isolates from all over the world, however, revealed a remarkable uniformity indicating a common origin. It is interesting that vaccinia virus DNA maps are similar to but distinct from those of modern cowpox and variola (smallpox) viruses, confirming their classification as three separate species within the orthopoxvirus genus (2).

The success of the vaccination program in eradicating smallpox was due both to the properties of the vaccine and the characteristics of the disease (3). With regard to the latter, the simplicity of diagnosis, antigenic stability of variola virus, and absence of animal reservoirs are of special note. Important properties of the live vaccinia virus vaccine include its close antigenic relationship to variola virus, its high immunogenicity, economy of manufacture in animal skin or tissue culture, stability in freeze-dried form (making continuous refrigeration unnecessary), and simplicity of administration—by skin scratch with a bifurcated needle or jet gun. It is important that many of these advantages could also apply to a recombinant vaccinia virus designed to protect against other diseases.

BIOLOGY OF VACCINIA VIRUS

Poxviruses comprise a large family with members that infect both vertebrate and invertebrate hosts (4). Unlike other DNA viruses, they replicate in the cytoplasm of infected cells. The best known members of this family, vaccinia virus and variola virus, are in the orthopoxvirus genus. All orthopoxviruses, including cowpox and ectromelia (mousepox) are very closely related, as judged by electron micrographic appearance, DNA analysis, and immunological cross-reactivity.

The genome of vaccinia virus consists of a long linear double-stranded DNA molecule of about 185,000 base pairs. Hairpin structures at the two ends of the molecule link the DNA strands together. The DNA is packaged within the virus core, which also contains a complete transcription system. The presence of a virus-specific RNA polymerase, as well as modifying enzymes for capping, methylation, and polyadenylation of mRNA, are unique features of poxviruses that are intimately tied to their cytoplasmic life-style. The outer portion of the poxvirus particle is composed of lipoprotein membranes.

Following entry into a cell, the viral transcription system is activated and approximately 100 early genes are expressed. The viral proteins synthesized include a DNA polymerase and other factors needed for replication of the vaccinia genome. Replication then signals the onset of expression of the late class of genes, which also number about 100, as well as the decline of early gene expression. The sequences that regulate early and late RNA formation are located just upstream of the coding regions and differ significantly from eukaryotic promoters (5).

The products of the late genes include the majority of structural proteins assembled into virus particles within the cytoplasm. Some virions remain intracellular while others become wrapped in golgi and are extruded through the plasma membrane with an additional envelope. Both the intracellular and extracellular forms of vaccinia virus are infectious.

FORMATION OF RECOMBINANT VIRUSES

Insertion of foreign DNA into nonessential regions of the vaccinia virus genome has been carried out by homologous recombination (6, 7). Expression of foreign genes within the DNA may occur because of transcriptional regulatory elements at or near the site of insertion or by more precise genetic engineering. Plasmid vectors that greatly facilitate insertion and expression of foreign genes have been constructed (8, 9). These vectors contain an expression site, composed of a vaccinia transcriptional promoter and one or more unique restriction endonuclease sites for insertion of the foreign coding sequence, flanked by DNA from a nonessential region of the vaccinia genome. The choice of promoter determines both the time (e.g. early or late) and level of expression, whereas the flanking DNA sequence determines the site of homologous recombination.

Only about one in a thousand virus particles produced by this procedure is a recombinant. Although recombinant virus plaques can be identified by DNA hybridization, efficient selection procedures have been developed. By using segments of the nonessential vaccinia virus thymidine kinase (TK) gene as flanking sequences, the foreign gene recombines into the TK locus and by insertion inactivates the TK gene. Selection of TK^- virus is achieved by carrying out the virus plaque assay in TK^- cells in the presence of 5-bromodeoxyuridine. Phosphorylation of this nucleoside analogue and consequent lethal incorporation into viral DNA occurs only in cells infected with TK^+ parental virus. Depending on the efficiency of the transfection and recombination, up to 80% of the plaques are desired recombinants, and the rest are spontaneous TK^- mutants.

Plasmid vectors that contain the *E. coli* β-galactosidase gene, as well as an expression site for a second gene, permit an altenative method of distinguishing recombinant from parental virus (10). Plaques formed by such recombinants can be positively identified by the blue color that forms upon addition of an appropriate indicator. By combining both TK⁻ selection and β-galactosidase expression, recombinant virus is readily and quickly isolated. The recombinants are then amplified by propagation in suitable cell lines and expression of the inserted gene is checked by appropriate enzymological, immunological, or physical procedures.

An upper limit to the amount of genetic information that can be added to the vaccinia virus genome is not yet known. However, the addition of nearly 25,000 base pairs of foreign DNA had no apparent deleterious effect on virus yield (11). Were it necessary, large segments of the vaccinia virus genome could be deleted to provide additional capacity (12, 13).

SYNTHESIS, PROCESSING, AND TRANSPORT OF PROTEINS

Apparently any continuous open reading frame, whether derived from prokaryotic or eukaryotic sources, can be expressed by vaccinia virus vectors. The kinetics of protein synthesis depends on the vaccinia promoter chosen. When early transcriptional signals are used, expression occurs within the first 6 hours; when late signals are used expression occurs from 6 hours on (8, 14). The most widely used promoter, referred to as P7.5, contains both early and late signals and leads to continuous expression for 1–2 days depending on the conditions (8, 15). Significantly longer periods of protein synthesis are not possible since vaccinia virus infection leads to cell death. Nevertheless, there is little lysis, and the majority of cells remain intact. Production of 1–2 mg of foreign protein per liter of cell culture has been achieved, although the levels of gene expression have not yet been optimized. Synthesis of multiple foreign proteins by a single recombinant virus is also possible (10, 16).

When the proteins expressed by vaccinia virus vectors are of mammalian or mammalian virus origin, posttranslational modifications appear to occur faithfully. Table 1 lists examples of proteins that have been shown to be proteolytically processed, glycosylated, secreted, and inserted into the plasma membrane. Perhaps most dramatic was the demonstration of correct polarity of transport: The influenza hemagglutinin was directed to the apical surface of infected cells, whereas murine leukemia and vesicular stomatitis virus envelope glycoproteins were directed to the basolateral surface (17). Because of the extraordinarily wide host range of vaccinia virus, the cell-type specificity of processing can be analyzed. Variation in the efficiency of proteolytic cleavage of the envelope protein of human immunodeficiency virus type 1, the causative agent of AIDS (18), and preproenkephalin (19) was noted in different cell lines.

Table 1 Proteins expressed by recombinant vaccinia viruses

Origin	Protein	Glycosylation	Cleavage	Cell location or biological activity	Reference
DNA Viruses					
Epstein-Barr	Membrane antigen (gp 340)	+	Signal[a]	Cell surface	22
Hepatitis B	Surface antigen (S, MS, LS)	+		Secreted particle	15, 24–27
Herpes simplex 1	Glycoprotein D	+	Signal	Cell surface	24, 28
	Thymidine kinase			Active enzyme	6, 7
RNA Viruses					
Friend murine leukemia	Envelope protein	+	Signal, subunit	Basolateral cell surface	17, 42
Human immunodeficiency (HTLV-III, LAV)	Envelope protein	+	Signal, subunit	Cell surface	18, 43
Influenza A	Hemagglutinin	+	Signal	Apical cell surface	17, 20, 30, 31
	Nucleoprotein			Internal	33, 34
Rabies	Glycoprotein	+	Signal	Cell surface	39, 40
Respiratory syncytial	Glycoprotein	+		Cell surface	36, 37
	Fusion protein	+	Signal, subunit	Cell surface	38a

Sindbis	Structural proteins (C, E1, E2)	+	Signal, polyprotein	Cell surface, virion	45
Transmissible gastroenteritis	Spike protein (gp 195)	+	Signal		44
Vesicular stomatitis	Glycoprotein	+	Signal	Basolateral cell surface	17, 21
	Nucleoprotein			Internal	21
Protozoa					
P. knowlesi	Circumsporozoite			Internal	47
P. falciparum	Circumsporozoite			Internal	48
	S antigen		Signal	Secreted	49
Bacteria					
E. coli	β-galactosidase			Active enzyme	10
	Chloramphenicol acetyltransferase			Active enzyme	8
Mammalian					
Human	Factor IX[b]	+	Signal, precursor	Secreted, coagulant	56
	Interleukin 2	+	Signal	Secreted, lymphokine	52
	Preproenkephalin		Signal, precursor	Secreted	19
Bovine	γ-interferon		Signal	Secreted, antiviral	51

[a] Cleavage of signal peptide assumed because of translocation.
[b] Evidence for γ-carboxylation.

STIMULATION OF NEUTRALIZING ANTIBODY

A broad spectrum of antibodies directed against multiple protein components of a virus frequently appears following natural infection. Determination of which antibodies are responsible for neutralizing virus infectivity, however, is an important task. Immunization of animals with vaccinia virus vectors that express only one gene of an infectious agent provides a direct approach to the resolution of this problem. When possible, sera from such vaccinated animals can be analyzed for neutralizing ability in vitro and in vivo. Although similar experiments can be carried out with subunit proteins, requirements for extensive purification of the native form of the antigen may be technically difficult. Vaccinia virus vectors have particular advantages when the neutralizing antigens are membrane proteins, since they will be presented in a manner that closely mimics the natural infection.

Antibody production has been elicited with live recombinant vaccinia viruses in a variety of animals including rabbits, mice, hamsters, cotton rats, and cattle (Table 2). Usually intradermal inoculations have been used, but in some cases the route was intravenous, intraperitoneal, or intranasal. In the latter case secretory IgA as well as circulating IgG was demonstrated (20).

In most instances, intradermal inoculation leads to the production of a pox lesion that lasts one to two weeks; there is no evidence for persistent infection of animals. Antibody is frequently detected by the end of the second week after vaccination and remains elevated for extended periods. Since the experimental animal also is immunized against the vaccinia virus vector, subsequent vaccinations may produce a mild or no discernible skin lesion, depending in part on the dose and interval of time between inoculations. Nevertheless, significant boosting of antibody production to a recombinant protein has been reported in several cases and should be tried if higher levels of immunity are required (16, 21, 22).

STIMULATION OF CELL MEDIATED IMMUNITY AND IDENTIFICATION OF TARGET ANTIGENS

Natural virus infections stimulate specific cell mediated as well as humoral immune mechanisms. Cytotoxic T cells (CTL) are effectors in cell mediated immune reactions that specifically recognize a foreign antigen in conjunction with a syngenic MHC protein on the surface of target cells and cause their lysis. The identification of viral antigens that prime T lymphocytes in vivo and serve as CTL targets, however, may be quite difficult. Although the most direct approach involves the use of isolated proteins, proper protein presentation is an important consideration. Virus vectors consequently appear to be most suitable for such purposes.

Virus-specific and MHC-restricted CTL effectors have been demonstrated following in vitro stimulation of spleen cells from mice that had been vaccinated with recombinant vaccinia virus (23). In this paradigm, the recombinant vaccinia virus used to immunize the animal expressed only a single candidate immunogen of the test virus (e.g. the hemagglutinin of influenza virus). Spleen cells were than isolated, and effectors were stimulated in vitro with cells infected with the recombinant vaccinia or the test virus itself. Target cells infected in vitro with the test virus and loaded with ^{51}Cr were incubated with the effector cells, and lysis was measured by ^{51}Cr release. Several examples in which recombinant vaccinia viruses have been successfully used to demonstrate CTL priming are shown in Table 2.

A complementary approach is to obtain effector cells from animals naturally infected with the test virus and to prepare target cells in vitro with a recombinant vaccinia virus that expresses a single test virus protein (23). Although target cells can also be prepared by other recombinant DNA techniques, vaccinia virus vectors have several advantages. It is most important that the same recombinant vaccinia virus can be used to prime animals and prepare targets (in different experiments), to determine the influence of MHC by infecting several inbred mouse strains or cell lines, and to prepare target cells from individuals of outbred animals such as humans.

PROTECTIVE IMMUNIZATION

Protection of animals against infection may rely on humoral and cell mediated immune mechanisms depending to a large extent on the nature of the microbial agent. Although many different approaches are required to understand fully the complex host reactions and their protective roles, recombinant vaccinia viruses can be used as one facet in such investigations. In several cases, immunization with live recombinant vaccinia expressing a single protein of a pathogenic virus has protected experimental animals against subsequent challenge with that virus. Examples include both RNA and DNA viruses, viruses that cause local or systemic infections, and viruses that produce acute or latent infections (Table 2). A synopsis of information obtained in this manner is presented below.

HEPATITIS B VIRUS

Hepatitis B virus (HBV) is an important cause of acute and chronic liver disease and a factor in the development of hepatocellular carcinoma.

Table 2 Immune response to vaccinia vectors and disease protection

Infectious agent	Protein expressed	Antibody response	T-cell response	Protection	Reference
DNA Viruses					
Epstein-Barr	Membrane antigen (gp 340)	Neutralizing (rabbit)			22
Hepatitis B	Surface antigens (pre-S and S)	anti-S and pre-S (rabbit), priming (chimpanzee)		Hepatitis (chimpanzee)	15, 24–27
Herpes simplex 1	Glycoprotein D	Neutralizing (mouse, rabbit, cow)	Proliferative (mouse), CTL target (human)	Lethal and latent infections (mouse)	24, 28, 29
RNA Viruses					
Friend murine leukemia	Envelope	Priming, neutralizing (mouse)	Proliferative, CTL priming (mouse)	Leukemia (mouse)	42
Influenza A	Hemagglutinin (HA)	Anti-HA, neutralizing (hamster, mouse)	Type specific: CTL priming, target (mouse)	Lower respiratory infection (hamster, mouse)	17, 20, 30, 31, 34
	Nucleoprotein		Cross-reactive: priming, target (mouse, man)	Lethal infection (partial)	33, 34
Rabies	Glycoprotein	Neutralizing (mouse, rabbit, fox)	CTL priming (mouse)	Encephalitis (mouse, rabbit, fox)	39, 40, 41

Respiratory syncytial	Glycoprotein	Neutralizing (cotton rat)	Lower respiratory infection (cotton rat)	37
	Fusion	Neutralizing (cotton rat)	Lower respiratory infection (cotton rat)	38a
Sindbis	Structural (C, E1, E2)	Neutralizing (cow)		46
Transmissible gastroenteritis	Spike (gp 195)	Neutralizing (mouse)		44
Vesicular stomatitis	Glycoprotein	Neutralizing (mouse, cow)	Type specific: CTL priming, target (mouse)	21, 38b
	Nucleoprotein	Anti-N (mouse, man)	Cross-reactive; CTL priming, target (mouse)	38b, 57
Protozoan				
P. knowlesi	Circumsporozoite antigen	Anti-sporozoite (rabbit, mouse)		47
P. falciparum	Circumsporozoite antigen	Anti-sporozoite (rabbit, mouse)		48
	Blood stage secretory (S) antigen	Anti-S (rabbit)		49

Worldwide, the number of cases of chronic hepatitis B is believed to be about 200 million. The development of effective plasma-derived and recombinant yeast subunit vaccines has considerably brightened the prospect for reducing the incidence of hepatitis B and perhaps eliminating the disease entirely. Studies with recombinant vaccinia viruses have been undertaken to learn more about the expression and immunogenicity of HBV proteins and to provide an economical mass delivery system for a hepatitis B virus eradication campaign similar to that successfully used for smallpox.

Recombinant vaccinia viruses that express the major HBsAg gene have been described (15, 24). The hepatitis protein is glycosylated and secreted as particles that are indistinguishable by sedimentation or electron microscopy from human plasma–derived material. Moreover, vaccinated rabbits mounted a high and sustained antibody response. Although two chimpanzees that were inoculated intradermally with recombinant vaccinia virus had little or no detectable circulating anti-HBsAg, they were protected against any signs of hepatitis following intravenous challenge with HBV (25). Since the HBV challenge was promptly followed by a vigorous anti-HBsAg response, which was not seen for several months in control animals, it was evident that the vaccination had a priming effect. Most likely the challenge HBV caused a mild inapparent infection which produced a protective anamnestic neutralizing antibody response. A role for cell mediated immunity in prevention of HBV infections has not been determined. Despite the qualified success of this experiment, monitoring the efficacy of vaccination would be difficult without antibody to HBsAg. Use of recombinants that express higher amounts of HBsAg or improvement in the vaccination protocol may be necessary.

Another alternative is the use of recombinant vaccinia viruses that express the larger forms of HBsAg containing the highly immunogenic pre-S region that has been implicated in binding of HBV to hepatocytes. Vaccinia recombinants that selectively express the large and middle forms of HBsAg have both been constructed. It is interesting that the middle protein is secreted from infected tissue culture cells in a form similar to particles made of the major small HBsAg (26). By contrast, the large form of the protein (not yet selectively made by other recombinant DNA methods) is not secreted as a particle and even interferes with secretion of the major HBsAg when cells are infected simultaneously with two recombinants (27). Rabbits vaccinated with recombinants that express the middle or large forms of HBsAg produce antibodies to both pre-S and S epitopes. The immunogenicity of these pre-S recombinant vaccinia viruses in chimpanzees, however, has not yet been tested.

HERPES SIMPLEX VIRUS (HSV)

Like other members of the herpesvirus family, HSV is a large complex DNA virus. The two serotypes of HSV cause acute infections at the portal of entry and latent infections in nerve ganglia. Recurrent infections can occur despite the presence of antibody. Nevertheless, experimental animals immunized with purified glycoprotein D (gD), one of several glycoproteins in the HSV envelope, display considerable protection against subsequent challenge.

Recombinant vaccinia viruses that express HSV-1 gD protect mice against lethal and latent infections with HSV-1 (24, 28). The high degree of immunity was indicated by cross-protection against HSV-2 (28), and the presence of neutralizing antibody in mice following vaccination suggested that this was an important component of protection (28). It is surprising that no evidence of MHC class-1 restricted CTL priming was obtained, although spleen cells did exhibit a proliferative response (J. Bennink, personal communication). This was consistent with the failure of cells infected with the gD recombinant virus to serve as targets for CTL from HSV-infected mice. Thus, the target antigens of mouse CTL remain to be determined. On the other hand, human CTL clones that lyse autologous cells infected with HSV-1 or HSV-2 were generated by stimulating lymphocytes with the recombinant vaccinia virus that expresses HSV-1 gD (29). Furthermore, CTL clones generated with HSV-1 or cloned HSV-1 gD lysed autologous cells infected with the vaccinia recombinant.

INFLUENZA VIRUS

Influenza viruses contribute significantly to respiratory diseases in man and domesticated animals. Protection against influenza virus is type specific and is mediated primarily by antibody to the hemagglutinin (HA), one of the two surface proteins.

Changes in the antigenicity of the HA, and to a lesser extent the neuraminidase, allow influenza virus to elude host defenses and make long-term immunization difficult. In contrast to the type specificity of neutralizing antibody, cytotoxic T cells are cross-reactive. Until recently, it was not known whether the cross-reactivity was due to conserved regions of the HA or to other more highly conserved internal proteins of influenza virus. Direct evidence for the latter has been obtained by using recombinant DNA–transformed and recombinant vaccinia virus–infected cells as targets.

Recombinant vaccinia viruses that express the influenza HA have been constructed (30, 31). The protein is properly glycosylated, is transported to the plasma membrane where it is cleavable into two subunits, and is reactive with a panel of monoclonal antibodies (31, 32). Antibody production has been demonstrated in experimental animals that have been vaccinated (30, 31). Moreover, intradermal vaccinations of hamsters and mice were shown to prevent lower respiratory infections in a type-specific manner (31). Intranasal vaccination of mice, however, led to production of secretory IgA antibodies to HA and prevented both upper and lower respiratory infections (20).

Vaccination of mice with the influenza HA recombinant vaccinia virus also produced a CTL priming effect that was predominantly type specific (23, 32). An even greater CTL priming effect occurred after vaccination with an influenza nucleoprotein (NP) recombinant vaccinia virus, and CTLs from these animals had cross-type specificity (33, 34). Vaccination with the NP recombinant, however, did not prevent lower respiratory infection, although partial protection against lethality has been reported (34). This result confirms the primary defensive role of neutralizing antibody against influenza virus. Nevertheless, efforts to enhance the immunogenicity of NP are being pursued because of the anticipated cross-type specificity.

HA and NP are not the only targets of influenza virus CTL. In mice, CTL directed against the nonstructural and polymerase proteins also have been demonstrated (J. R. Bennink, J. W. Yewdell, personal communication). The latter studies also revealed that CTL recognition of individual viral antigens occurs in conjunction with only a limited number of available class-I MHC restriction elements.

Determination of the target specificity of human CTL is difficult for several reasons. The HLA specificity of the CTL response poses a particular problem. By using recombinant vaccinia virus to prepare target cells in vitro, however, both target and effector cells can be derived from the same individual. With this type of protocol, evidence has been obtained that NP (35) and other internal proteins (F. Gotch, A. McMichael, personal communications) are human CTL targets.

RESPIRATORY SYNCYTIAL VIRUS

Respiratory syncytial virus (RSV) is the most important pediatric lower respiratory pathogen. At present an effective vaccine against this disease has not been developed, although there is data from animal models that neutralizing antibody can be protective. The two surface glycoproteins, G and F, are considered to be the most likely targets of a successful vaccine.

Recombinant vaccinia viruses that express either G (36, 37) or F (38a) have been constructed. The proteins produced by the recombinant viruses are glycosylated, processed, and transported to the plasma membrane of infected tissue culture cells. Intradermal vaccination of cotton rats, an experimental model for RSV, with either G or F recombinant produced neutralizing antibodies and protected against lower respiratory infection (37, 38a). In the case of the F recombinant, the level of neutralizing antibody was even higher than that which occurred after intranasal RSV infection, and protection against lower respiratory infection was as good. The intradermal route of inoculation with the F recombinant vaccinia virus diminished but did not prevent upper respiratory infection. The studies with influenza virus, summarized in the previous section, suggest the likelihood that intranasal inoculation would prevent both upper and lower respiratory infections. Continued studies with these and other RSV/vaccinia recombinants should provide valuable information regarding the targets and role of cell mediated immunity.

VESICULAR STOMATITIS VIRUS AND RABIES VIRUS

Vesicular stomatitis virus (VSV) and rabies virus are both members of the rhabdovirus family. VSV causes significant infection of horses, cattle, and pigs whereas rabies virus infects a variety of wild and domesticated animals as well as man. Both viruses contain a surface glycoprotein (G) that can induce neutralizing antibody, and inactivated and attenuated rabies virus vaccines have been developed.

The VSV G protein produced by recombinant vaccinia virus is glycosylated and expressed on the surface of infected cells (21). When polarized cells were used, transport to the basalolateral surface of the plasma membrane was demonstrated as in natural VSV infection (17). Cattle that were vaccinated intradermally with the recombinant virus produced neutralizing antibody, and the majority were protected against direct lingual injection of VSV (21).

CTL studies similar to those carried out with influenza virus were extended to VSV. The results were entirely consistent: CTL recognition of G was type specific, whereas recognition of the internal nucleoprotein was cross-reactive (38b).

Recombinant vaccinia viruses that express the rabies virus G protein induce high levels of neutralizing antibody and protect a variety of animals against direct intracerebral inoculation with lethal doses of rabies virus (39, 40). Of potential practical importance, foxes inoculated orally were also protected against rabies (41). This result raises the possibility of using bait vaccines to immunize wildlife.

FRIEND MURINE LEUKEMIA VIRUS (FMuLV)

The recent finding that certain human leukemias and acquired immune deficiency syndrome (AIDS) are caused by retroviruses has greatly increased the interest in immune defenses against this class of viruses. While murine retrovirus infection should not be considered as a model for AIDS, considerable useful information may be derived. Friend virus complex, which consists of replication-competent FMuLV and replication-defective spleen-focus-forming virus, causes erythroleukemia and splenomegaly in newborn and adult mice. As with other retroviruses, FMuLV contains two envelope glycoproteins that are generated by cleavage of a precursor. Evidence that repeated immunization with the envelope proteins can protect against murine leukemia has been obtained.

The FMuLV envelope protein made by a recombinant vaccinia virus was glycosylated, transported to the plasma membrane, and correctly processed into large and small subunits (42). In polarized cells, the protein appeared on the basolateral cell surface (17). Mice that were vaccinated intradermally with the recombinant virus were protected against splenomegaly and leukemia when challenged with Friend virus complex several weeks later (42). In order to understand the nature of the protection, several immune parameters were measured both before and after challenge. Before challenge, neutralizing antibody and CTL were difficult to detect, but spleen cells exhibited a specific proliferative response. After challenge, both neutralizing antibody and CTLs rapidly appeared, indicating that the animals had been primed. By contrast, control animals had a more delayed neutralizing antibody response and never developed a measurable CTL response. A second interesting finding was that seeding of the spleen with FMuLV was not entirely prevented by vaccination, but the vaccinated animals were able largely to clear their spleens of virus. Whether protection and recovery are due to CTL or antibody or both may be determined by further investigations.

EXPRESSION OF GENES OF OTHER VIRUSES AND PARASITES

We have not described in detail reports of expression of genes from certain other viruses because of the absence thus far of protection data. Nevertheless, it is worth mentioning the expression and processing of the envelope protein of the virus causing AIDS (18, 43), a surface glycoprotein of Epstein-Barr virus (22), the spike glycoprotein of transmissible gastroenteritis virus of swine (44) and Sindbis virus structural proteins (45, 46). The approaches described here may also find use for studies of other classes of infectious agents, as demonstrated by expression of malarial circumsporozoite (47, 48) and blood stage secretory (49) antigens.

FUTURE PROSPECTS

It seems likely that vaccinia virus expression vectors will continue to be improved both with regard to efficiency of expression and convenience of construction. Increased expression may be achieved by using stronger natural or synthetic promoters, altering the site of gene insertion, and introducing multiple gene copies. One novel approach being explored involves the integration of a bacteriophage RNA polymerase gene into the genome of vaccinia virus. This allows the efficient and selective expression of coding segments containing bacteriophage promoters in transfected plasmids (50) or in a second recombinant vaccinia virus (T. Fuerst, B. Moss, personal communication).

Enhancement of the immunological response either through antigen presentation or immune modulation also has been considered. An increased antibody response which resulted from the placing of a membrane anchor sequence on the end of a plasmodial secretory protein is an example of the former (49). The placement of genes for interferon gamma (51) and interleukin-2 (52) into the vaccinia virus genome represent attempts at the latter approach.

A third active area of research is directed toward altering the biological properties of the vaccinia virus vector. By inactivating the TK gene or deleting other gene segments, the pathogenicity of vaccinia virus can be sharply reduced (53). Even the host range of vaccinia virus can be restricted by mutation (54).

The value of recombinant vaccinia viruses for experimental immunology has been established, and new uses will undoubtedly be found. Serious consideration is also being given to the development of recombinant vaccinia viruses as vaccines for veterinary and medical purposes. It is a fact that the most widely used viral vaccines, e.g. for smallpox, poliomyelitis, measles, rubella, mumps, and yellow fever, are live attenuated viruses. Recombinant vaccinia viruses maintain many of the advantages of live virus vaccines, particularly with regard to stimulation of cell mediated immunity, economical production, and simple delivery for mass immunization, while eliminating certain hazards such as reversion to virulence. The impressive degree of protection obtained in immunizing foxes with a rabies G protein recombinant opens up the possibility of wildlife vaccination (41). Similarly, the resistance of cotton rats to RSV induced by recombinant vaccinia viruses that express the RSV G or F proteins suggests possible human application (37, 38). Development of effective and safe recombinant vaccinia virus vaccines for hepatitis B, Epstein-Barr, and malaria would surely benefit underdeveloped areas of the world. Individual and environmental safety is, of course, a major concern with all live vaccines, whether developed by attenuation or by recombinant DNA methods (55). Experience with smallpox vaccination indicates that serious complications varied with the vaccine strain and were relatively rare with some. Even more attenuated strains were developed but were not used on a large scale because of the overriding necessity of eradicating smallpox. Either these attenuated strains or new ones developed by genetic engineering may provide vectors with low risk relative to potential benefits.

Literature Cited

1. Jenner, E. 1978. *An Inquiry into the Causes and Effects of the Variolae Vaccinae, a Disease Discovered in Some of the Western Counties of England, Particularly Gloucestershire, and Known by the Name of the Cow Pox*. London: Sampson Low
2. Mackett, M., Archard, L. C. 1979. Conservation and variation in orthopoxvirus genome structure. *J. Gen. Virol.* 47: 683–701
3. Fenner, F. 1985. Poxviruses. In *Virology*, ed. B. N. Fields, D. M. Knipe, R. M. Chanock, J. L. Melnick, B. Roizman, R. E. Shope, pp. 661–84. New York: Raven. 1614 pp.
4. Moss, B. 1985. Replication of poxviruses. See Ref. 3, pp. 685–704
5. Cochran, M. A., Puckett, C., Moss, B. 1985. In vitro mutagenesis of the promoter region for a vaccinia virus gene: Evidence for tandem early and late regulatory signals. *J. Virol.* 54: 30–37
6. Panicali, D., Paoletti, E. 1982. Construction of poxviruses as cloning vectors: Insertion of the thymidine kinase gene from herpes simplex virus into the DNA of infectious vaccinia virus. *Proc. Natl. Acad. Sci. USA* 79: 4927–31
7. Mackett, M., Smith, G. L., Moss, B. 1982. Vaccinia virus: A selectable eukaryotic cloning and expression vector. *Proc. Natl. Acad. Sci. USA* 79: 7415–19
8. Mackett, M., Smith, G. L., Moss, B. 1982. A general method for the production and selection of infectious vaccinia virus recombinants expressing foreign genes. *J. Virol.* 49: 857–64
9. Mackett, M., Smith, G. L., Moss, B. 1985. The construction and characterization of vaccinia virus recombinants expressing foreign genes. In *DNA Cloning*, vol. II, ed. D. M. Glover, pp. 191–212. Oxford: IRL Press. 245 pp.
10. Chakrabarti, S., Brechling, K., Moss, B. 1985. Vaccinia virus expression vector: Coexpression of β-galactosidase provides visual screening of recombinant virus plaques. *Mol. Cell. Biol.* 5: 3403–9
11. Smith, G. L., Moss, B. 1983. Infectious poxvirus vectors have capacity for at least 25,000 base pairs of foreign DNA. *Gene* 25: 21–28
12. Panicali, D., Davis, S. W., Mercer, S. R., Paoletti, E. 1981. Two major DNA variants present in serially propagated stocks of the WR strain of vaccinia virus. *J. Virol.* 40: 1000–10
13. Moss, B., Winters, E., Cooper, J. A. 1981. Deletion of a 9,000 base pair segment of the vaccinia virus genome that encodes non-essential polypeptides. *J. Virol.* 40: 387–95
14. Weir, J. P., Moss, B. 1984. Regulation of expression and nucleotide sequence of a late vaccinia virus gene. *J. Virol.* 51: 662–69
15. Smith, G. L., Mackett, M., Moss, B. 1983. Infectious vaccinia virus recombinants that express hepatitis B virus surface antigen. *Nature* 302: 490–95
16. Perkus, M. E., Piccini, A., Lipinskas, B. R., Paoletti, E. 1985. Recombinant vaccinia virus: Immunization against multiple pathogens. *Science* 229: 981–84
17. Stephens, E. B., Compans, R. W., Earl, P., Moss, B. 1986. Surface expression of viral glycoproteins is polarized in epithelial cells infected with recombinant vaccinia viral vectors. *EMBO J.* 5: 237–45
18. Chakrabarti, S., Robert-Guroff, M., Wong-Stall, F., Gallo, R. C., Moss, B. 1986. Expression of the HTLV-III envelope gene by a recombinant vaccinia virus. *Nature* 320: 535–37

19. Thomas, G., Herbert, E., Hruby, D. E. 1986. Expression and cell-type-specific processing of human preproenkephalin with a vaccinia recombinant. *Science* 232: 1641–46

20. Small, P. A., Smith, G. L., Moss, B. 1985. Intranasal vaccination with a recombinant vaccinia virus containing influenza hemagglutinin prevents both influenza virus pneumonia and nasal infection: Intradermal vaccination prevents only viral pneumonia. In *Vaccines 85. Molecular and Chemical Basis of Resistance to Parasitic, Bacterial and Viral Diseases*, ed. Lerner, R. A., Chanock, R. M., Brown, F., pp. 175–76. Cold Spring Harbor, NY: Cold Spring Harbor Laboratory

21. Mackett, M., Yilma, T., Rose, J. K., Moss, B. 1985. Vaccinia virus recombinants: Expression of VSV genes and protective immunization of mice and cattle. *Science* 227: 433–35

22. Mackett, M., Arrand, J. R. 1985. Recombinant vaccinia virus induces neutralizing antibodies in rabbits against Epstein-Barr virus membrane antigen gp 340. *EMBO J.* 4: 3229–34

23. Bennink, J. R., Yewdell, J. W., Smith, G. L., Moller, C., Moss, B. 1984. Recombinant vaccinia virus primes and stimulates influenza virus HA-specific CTL. *Nature* 311: 578–7

24. Paoletti, E., Lipinskas, B. R., Samsonoff, C., Mercer, S. R., Panicali, D. 1984. Construction of live vaccines using genetically engineered poxviruses: Biological activity of vaccinia virus recombinants expressing the hepatitis B virus surface antigen and the herpes simplex virus glycoprotein D. *Proc. Natl. Acad. Sci. USA* 81: 193–97

25. Moss, B., Smith, G. L., Gerin, J. L., Purcell, R. H. 1984. Live recombinant vaccinia virus protects chimpanzees against hepatitis B. *Nature* 311: 67–69

26. Cheng, K.-C., Moss, B. 1987. Selective synthesis and secretion of particles composed of the hepatitis B virus middle surface protein directed by a recombinant vaccinia virus; induction of antibodies to pre-S and S epitopes. *J. Virol.* In press

27. Cheng, K.-C., Smith, G. L., Moss, B. 1986. Hepatitis B virus large surface protein is not secreted but is immunogenic when selectively expressed by recombinant vaccinia virus. *J. Virol.* 60: 337

28. Cremer, K. J., Mackett, M., Wohlenberg, C., Notkins, A. L., Moss, B. 1985. Vaccinia virus recombinant expressing herpes simplex virus type 1 glycoprotein D prevents latent herpes in mice. *Science* 228: 737–40

29. Zarling, J. M., Moran, P. A., Lasky, L. A., Moss, B. 1986. Herpes simplex virus (HSV)-specific human T-cell clones recognize HSV glycoprotein D expressed by a recombinant vaccinia virus. *J. Virol.* 59: 506–9

30. Panicali, D., Davis, S. W., Weinberg, R. L., Paoletti, E. 1983. Construction of live vaccines by using genetically engineered poxviruses: Biological activity of recombinant vaccinia virus expressing influenza virus hemagglutinin. *Proc. Natl. Acad. Sci. USA* 80: 5364–68

31. Smith, G. L., Murphy, B. R., Moss, B. 1983. Construction and characterization of an infectious vaccinia virus recombinant that expresses the influenza hemagglutinin gene and induces resistance to influenza virus infection in hamsters. *Proc. Natl. Acad. Sci. USA* 80: 7155–59

32. Bennink, J. R., Yewdell, J. W., Smith, G. L., Moss, B. 1986. Recognition of cloned influenza virus hemagglutinin gene products by cytotoxic T lymphocytes. *J. Virol.* 57: 786–91

33. Yewdell, J. W., Bennink, J. R., Smith, G. L., Moss, B. 1985. Influenza A virus nucleoprotein is a major target antigen for cross-reactive anti-influenza A virus cytotoxic T lymphocytes. *Proc. Natl. Acad. Sci. USA* 82: 1785–89

34. Andrew, M. E., Coupar, B. E. H., Ada, G. L., Boyle, D. B. 1986. Cell-mediated immune response to influenza virus antigens expressed by vaccinia virus recombinants. *Microb. Path.* 1: 443–52

35. McMichael, A. J., Michie, C. A., Gotch, F. M., Smith, G. L., Moss, B. 1986. Recognition of influenza A virus nucleoprotein by human cytotoxic T lymphocytes. *J. Gen. Virol.* 67: 719–26

36. Ball, L. A., Young, K. K. Y., Anderson, K., Collins, P. L., Wertz, G. W. 1986. Expression of the major glycoprotein G of human respiratory syncytial virus from recombinant vaccinia virus vectors. *Proc. Natl. Acad. Sci. USA* 83: 246–50

37. Elango, N., Prince, G. A., Murphy, B. R., Venkatesan, S., Chanock, R. M., Moss, B. 1986. Resistance to human respiratory syncytial virus (RSV) infection induced by immunization of cotton rats with a recombinant vaccinia virus expressing the RSV G glycoprotein. *Proc. Natl. Acad. Sci. USA* 83: 1906–10

38a. Olmsted, R. A., Elango, N., Prince, G. A., Murphy, B. R., Johnson, P. R., Moss, B., Chanock, R. M., Collins, P. L. 1986. Expression of the F glycoprotein of respiratory syncytial virus by a recombinant vaccinia virus: Comparison of the individual contributions of the F and G glycoproteins to host immunity. *Proc. Natl. Acad. Sci. USA.* 83: 7462

38b. Yewdell, J. W., Bennink, J. R., Mackett, M., LeFrancois, L., Lyles, D. S., Moss, B. 1986. Recognition of cloned vesicular stomatitis virus internal and external gene products by cytotoxic T lymphocytes. *J. Exp. Med.* 163: 1529–39

39. Kieny, M. P., Lathe, R., Drillien, R., Spehner, D., Skory, S., Schmitt, D., Wiktor, T., Koprowski, H., Lecocq, J. P. 1984. Expression of rabies virus glycoprotein from a recombinant vaccinia virus. *Nature* 312: 163–66

40. Wiktor, T. J., MacFarlan, R. I., Reagan, K. J., Dietzschold, B., Curtis, P. J., Wunner, W. H., Kieny, M.-P., Lathe, R., Lecocq, J.-P., Hockett, M., Moss, B., Koprowski, H. 1984. Protection from rabies by a vaccinia virus recombinant containing the rabies virus glycoprotein gene. *Proc. Natl. Acad. Sci. USA* 81: 7194–98

41. Blancou, J., Kieny, M. P., Lathe, R., Lecocq, J. P., Pastoret, P. P., Soulebot, J. P., Desmettre, P. 1986. Oral vaccination of the fox against rabies using a live recombinant vaccinia virus. *Nature* 322: 373–75

42. Earl, P. L., Moss, B., Wehrly, K., Nishio, J., Chesebro, B. 1986. T cell priming and protection against Friend murine leukemia by a recombinant vaccinia virus expressing env gene. *Science.* 234: 728–31

43. Hu, S., Kosowski, S. G., Dalrymple, J. M. 1986. Expression of AIDS virus envelope gene in recombinant vaccinia viruses. *Nature* 320: 537–40

44. Hu, S., Bruszewski, J., Smalling, R., Browne, J. K. 1985. Studies of TGEV spike protein gp 195 expressed in *E. coli* and by a TGE-vaccinia virus recombinant. *Adv. Exp. Med. Biol.* 185: 63–82

45. Rice, C. M., Franke, C. A., Strauss, J. H., Hruby, D. E. 1985. Expression of sindbis virus structural proteins via recombinant vaccinia virus: Synthesis, processing, and incorporation into mature sindbis virions. *J. Virol.* 56: 227–39

46. Franke, C. A., Berry, E. S., Smith, A. W., Hruby, D. E. 1985. Immunization of cattle with a recombinant togavirus–vaccinia virus strain. *Res. Vet. Sci.* 39: 113–15

47. Smith, G. L., Godson, G. N., Nussenzweig, V., Nussenzweig, R. S., Barnwell, J., Moss, B. 1984. *Plasmodium knowlesi* sporozoite antigen: Expression by infectious recombinant vaccinia virus. *Science* 224: 397–99

48. Cheng, K. C., Smith, G. L., Moss, B., Zavala, F., Nussenzweig, R., Nussenzweig, V. 1986. Expression of malaria circumsporozoite protein and hepatitis-B virus surface antigen by infectious vaccinia virus. In *Vaccines 86*, eds. F. Brown, R. M., Chanock, R. A. Lerner, pp. 165–68. Cold Spring Harbor, NY: Cold Spring Harbor Laboratory. 418 pp.

49. Langford, C. J., Edwards, S. J., Smith, G. L., Moss, B., Kemp, D. J., Anders, R. F., Mitchell, G. F. 1986. Anchoring a secreted plasmodium antigen on the surface of recombinant vaccinia virus-infected cells increases its immunogenicity. *Mol. Cell. Biol.* 6: 3191–99

50. Fuerst, T. R., Niles, E. G., Studier, F. W., Moss, B. 1986. Eukaryotic transient expression system based on recombinant vaccinia virus that synthesizes bacteriophage T7 RNA polymerase. *Proc. Natl. Acad. Sci. USA.* 83: 8122

51. Yilma, T., Anderson, K., Ristow, S., Brechling, K., Chakrabarti, S., Moss, B. 1986. *An infectious vaccinia virus recombinant expresses the bovine interferon-gamma gene.* Presented at Modern Approaches to New Vaccines Including Prevention of AIDS, Cold Spring Harbor, NY

52. Flexner, C., Hügin, A., Moss, B. 1986. *Expression of human interleukin-2 by live recombinant vaccinia virus.* Presented at Modern Approaches to New Vaccines Including Prevention of AIDS, Cold Spring Harbor, NY

53. Buller, R. M. L., Smith, G. L., Cremer, K., Notkins, A. L., Moss, B. 1985. Decreased virulence of recombinant vaccinia virus expression vectors is associated with a thymidine kinase-negative phenotype. *Nature* 317: 813–15

54. Gillard, S., Spehner, D., Drillien, R., Kirn, A. 1986. Localization and sequence of a vaccinia virus gene required for multiplication in human cells. *Proc. Natl. Acad. Sci. USA* 83: 5573–77

55. Brown, F., Schild, G. C., Ada, G. L. 1986. Recombinant vaccinia viruses as vaccines. *Nature* 319: 549–50

56. De LaSalle, Altenburger, W., Elkaim, R., Dott, K., Dieterlé, A., Drillien, R., Cazenav, J.-P., Tolstoshev, P., Lecocq, J.-P. 1985. Active γ-carboxylated human factor IX expressed using recombinant DNA techniques. *Nature* 316: 268–70

57. Jones, L. Ristow, S., Yilma, T., Moss, B. 1986. Accidental human vaccination with vaccinia virus expressing nucleoprotein gene. *Nature* 316: 543

Ann. Rev. Immunol. 1987. 5 : 325–65

EARLY EVENTS IN T-CELL MATURATION

Becky Adkins, Christoph Mueller, Craig Y. Okada, Roger A. Reichert, Irving L. Weissman, and Gerald J. Spangrude

Department of Pathology, Stanford University School of Medicine, Stanford, California 94305

INTRODUCTION

The thymus is the major if not the sole site of maturation of T lymphocytes from committed hematopoietic progenitors. The end products of this maturation—the T lymphocytes—are well equipped to survey the entire lymphoid system for the appearance of foreign antigens; within the lymphoid organs they initiate, regulate, and play a direct role in at least a part of the immune response to antigens. T lymphocytes are highly diverse in terms of their immune functions, their surface phenotypes, their MHC recognition or corecognition capabilities, and their migration behaviors. Like B lymphocytes they are clonally precommitted to recognize only a small fraction of the antigenic shapes the universe may present. Unlike B cells, T cells are precommitted primarily to recognize those shapes which appear to be MHC restricted in their presentation. The hematopoietic precursors of thymocytes and T cells are apparently not clonally restricted as to the types and specificity of antigen receptors they may express or they are not yet capable of immune function; they do not express surface markers characteristic of T lymphocytes and do not yet have the kinds of homing receptors that direct the migration of peripheral T and B lymphocytes. It is even questionable whether these T cell precursors are capable of self/non-self recognition. Thus, the functions of the thymus must include a series of biochemical and biological commitments induced in or selected in developing thymocyte and T cell populations. The selective and/or inductive processes in the thymus must be stringent. It has been estimated that 99% or more of the progeny of thymus lymphocyte divisions die in the thymus and thus do not leave the thymus to participate in

325

0732–0582/87/0410–0325$02.00

peripheral immune responses. The normal rate of thymic precursor entry into the steady state thymus is extremely low compared to the proliferative rate of cells within the thymus, and even extremely low compared to the rate of emigration of successful progeny from the thymus to the periphery. Thus, a major portion of thymic proliferation clearly is via intrathymic stem cells, or progenitors, with a significant self-renewal capacity. Because of the diversity of outcomes of thymic lymphocyte maturation, and because of the requirement for precise regulation of thymic lymphocyte production and function, the study of the developmental biology of T lymphocytes has become one of the central models for the study of cellular differentiation in complex vertebrate systems.

Despite the intensity of research on thymic lymphocyte maturation, and despite the major clinical importance for understanding T-cell maturation and function, it is surprising and disappointing that we know so little about T-cell development and T-cell developmental lineages. Within this review we attempt to catalogue the various developmental steps that thymic-produced T lymphocytes must undergo on their pathway from hematopoietic progenitors to functional virgin T cells. We attempt to approach this from the viewpoint of developmental biologists wishing to understand the lineage relationships between these virgin immuno-competent products of T-lymphocyte maturation and their various pre-cursors within the thymus and the hematopoietic tissues. We hope eventu-ally to understand the stages at which developing thymic lymphocytes commit to phenotypes, functions, intrathymic developmental micro-environments, self/non-self recognition, MHC restriction, antigen rec-ognition, and homing receptor expression. To do so we begin with a review of mature T-cell types and what is known of the phenotypically defined subsets of their progenitors within the thymus.

THYMOCYTE SUBPOPULATIONS DEFINED
BY CELL SURFACE PHENOTYPE

Intrathymic differentiation of T-progenitor cells produces mature T lym-phocytes which possess the cellular machinery necessary for carrying out various effector functions, such as target cell lysis and lymphokine secretion (1–3). The ability to perform these tasks is, in part, dependent upon the expression of a variety of cell surface antigens, including (a) the T-cell receptor (TCR) for antigen, (b) growth factor receptors, (c) the associative recognition antigens Lyt-2 and L3T4 which are thought to participate in interactions with class-I and class-II MHC structures, respectively (4, 5), (d) homing receptors involved in the homing of thymus emigrants to peripheral lymphoid organs (6, 7), and (*e*) an abundance of molecules which have been characterized but do not, as yet, have assigned functions (e.g. Thy-1 and Ly-1) (8). Thymocytes are remarkably heterogeneous with respect to the expression of different combinations of these cell surface markers. Since several excellent reviews have covered the full spectrum of thymocyte phenotypes (8–12), we focus here on thymocyte subpopulations of major significance.

The major markers used to divide thymocytes into phenotypically distinct subsets are Lyt-2, L3T4, Thy-1, Ly-1, H-2K, expression of binding sites for the plant lectin peanut agglutinin (PNA), and expression of the homing receptor identified by the monoclonal antibody MEL-14 (7–12). Thymocytes express most of these markers to a greater or lesser degree rather in an all-or-none fashion, and the degree of staining intensity above background may be critical in distinguishing subpopulations with different levels of immunocompetence (7, 9). Thus, for many markers, the designations "lo," "med," and "hi" are more useful than simply using the " + " symbol to indicate staining above background.

Mature Cells Within the Thymus

By comparison with peripheral T cells, a proportion of thymocytes can be identified as phenotypically mature (8–12). These cells are homogeneously Thy-1^{med}, PNA^{lo}, H-2K^{hi}, and Ly-1^{hi} and express either L3T4 or Lyt-2, but not both. Mature phenotype cells of the L3T4$^-$ Lyt-2$^+$ (mostly cytotoxic/suppressor precursors) and L3T4$^+$ Lyt-2$^-$ (generally helper/inducer precursor) types comprise 3–8% and 8–15% of total thymocytes, respectively. Nearly all functionally mature thymocytes are contained within this phenotypically mature population, and thymus emigrants which seed peripheral lymphoid organs are culled from this pool of phenotypically mature thymocytes (13–16).

The compartments in which these mature cells reside and their sites of exit from the thymus are points of some controversy. Largely from studies on cortisone resistant thymocytes (CRT) (13, 17, 18), it was widely accepted that properly selected cortical cells give rise to mature thymocytes in the medulla, which then exports thymus emigrants to the periphery. Recently, studies with the monoclonal antibody MEL-14 have led to a different version of intrathymic maturation. Approximately 1–3% of normal adult thymocytes express high levels of the MEL-14-defined homing receptor (7), a molecule that enables mature lymphocytes to migrate from the bloodstream into peripheral lymph nodes via recognition of specific postcapillary high-walled endothelial cells (HEV) (6). The MEL-14^{hi} subset of thymocytes is greatly enriched in cells of mature surface phenotype and function (19, 20; A. Wilson et al, manuscript in preparation). Furthermore,

MEL-14^{hi} thymocytes appear to be a major, although not the sole, source of thymus emigrants (7). Conventional wisdom would predict a medullary location for these cells. However, immunohistochemical analysis reveals that MEL-14^{hi} thymocytes are scattered throughout *cortical* regions and are virtually undetectable in the medulla (7). The question that arises, therefore, is, "Can I believe my eyes?" In one view, one cannot believe one's eyes; that is, in this case, immunohistochemical localization of cells is in error, and all mature thymocytes reside in the medulla with their MEL-14 epitopes occluded (21). If this were true, all PNA^{lo}, L3T4$^+$ Lyt-2$^-$, or L3T4$^-$ Lyt-2$^+$ "mature cells" would, by suspension staining, be MEL-14^{hi}. In fact, the PNA^{lo} subpopulation which is virtually all L3T4$^+$ Lyt-2$^-$ or Lyt-2$^+$ L3T4$^-$ is divided, with approximately one third in the MEL-14^{hi} subset and two thirds in the MEL-14^{lo} subset (20). Both subsets respond to lectin stimulation in the presence of IL-2 by proliferation (PTL-P), whereas both allo- and lectin-stimulated cytotoxic precursors (CTL-P) are concentrated in the MEL-14$^{med\,to\,hi}$, Ly2$^+$ L3T4$^-$ subpopulation (19; A. Wilson et al, manuscript in preparation). We propose that most MEL-14^{hi} PNA^{lo} cells are cortical, while most MEL-14^{lo} PNA^{lo} cells are medullary in location. The location of the immediate and ultimate precursors of the thymic cortical MEL-14^{hi} cells is not yet known and could include populations in both cortex and medulla. Since the MEL-14 population is enriched in but does not contain all of the mature cells in the thymus or thymic emigrants, we are *not* suggesting that all mature cells reside in the cortex. Rather, considering all the data, it seems reasonable to propose that mature cells may reside in either the cortex or the medulla, perhaps exiting via common vessels in the juxtamedullary cortex. As more markers defining distinct thymocyte cell subsets become available (e.g. other organ-specific homing receptors), we may find that different types of mature thymocytes locate exclusively to cortical or medullary regions.

As mentioned above, studies with CRT have contributed significantly to the conventional wisdom that the immunocompetent thymus emigrants exit from medullary sites (17, 18, 22). Thymocytes that remain following cortisone treatment represent approximately 5% of the population, are located in the medulla 2–5 days after treatment, are phenotypically mature, and are immunocompetent in terms of their migratory capabilities and their ability to serve as precursors of helper and cytotoxic T cells (13, 17, 18, 22). In contrast to medullary thymocytes in untreated hosts, a significant fraction of medullary cells 2–5 days after cortisone treatment are MEL-14^{hi} and express functional homing receptors as assayed by their ability to bind to high endothelial venules in vitro (7). Frozen section analysis and dual immunofluorescence studies suggest that cortisone administration may lead to the selective survival of cortisone-resistant MEL-14^{hi} cortical thymocytes, with their subsequent relocation in the medulla where they join the cortisone resistant medullary PNA^{lo} $MEL\text{-}14^{lo}$ "mature" cells (7, 23). It therefore seems likely that CRT may actually represent a complex mixture of cortisone-resistant subgroups of medullary and $MEL\text{-}14^{hi}$ cortical thymocytes. In any case, true medullary thymocytes differ from many CRT in their level of expression of the MEL-14-defined homing receptor. This difference alone renders data derived from studies with CRT inapplicable to the medullary population and necessitates a reevaluation of the conventional concepts of the role of the thymic medulla in T-cell maturation.

Nonmature Cells Within the Thymus

Much evidence indicates that the majority of thymocytes never exit and, thus, are destined to die within the thymus (24). The most popular candidates for this fate are small cells located almost exclusively in cortical areas. These small cortical cells are $Thy\text{-}1^{hi}$, PNA^{hi}, $H\text{-}2K^{lo}$, $Ly\text{-}1^{med}$ and, notably, are positive for both the L3T4 and Lyt-2 antigens (8–12, 25). We refer to these cells as small $L3T4^{+}$ $Lyt\text{-}2^{+}$ cells to distinguish them from large cortical blasts with the same phenotype but different functional properties (see below). The small $L3T4^{+}$ $Lyt\text{-}2^{+}$ cells do not appear to be cycling in vivo and have a short lifespan of 3–5 days, within the thymus. In addition, these cells are functionally inactive in vitro in a variety of assay systems (reviewed in 25) and are not represented in cells emigrated from the thymus, even very recently (24). These properties, coupled with the abundance (60–70%) of these cells among total thymocytes, strongly imply that the generation and loss of small $L3T4^{+}$ $Lyt\text{-}2^{+}$ cells account for most of the cell death in the thymus. Since small $L3T4^{+}$ $Lyt\text{-}2^{+}$ cells are not mature by any criteria and, in addition, appear to die without further proliferation or differentiation, we have adopted the term "nonmature" to describe this subset.

Immature Cells Within the Thymus

Although the cortex contains many quiescent thymocytes, it also houses at least 95% of the proliferating cells in the thymus (26). Up to 20% of cortical thymocytes, many of which are located in the subcapsular blast zone of the outer cortex, divide with a cell cycle time of 7–9 hours. These cycling cells are of two phenotypic types, those that are large $L3T4^{+}$ $Lyt\text{-}2^{+}$ blasts and those with a $L3T4^{-}$ $Lyt\text{-}2^{-}$ phenotype.

The large $L3T4^{+}$ $Lyt\text{-}2^{+}$ cells comprise 10–15% of total thymocytes and, as noted above, are phenotypically indistinguishable from small cortical cells (8–12, 25). Because of this phenotypic identity, it has been proposed that these $L3T4^{+}$ $Lyt\text{-}2^{+}$ blasts give rise to the nondividing, small cells. It has also been suggested by some that large L3T4$^+$ Lyt-2$^+$ cells are functional intermediates between less mature cells and fully mature thymocytes (B. J. Mathieson, personal communication). Consistent with that hypothesis is the finding that a significant fraction of these cells are MEL-14^{hi}, whereas the vast majority of nonmature cells are MEL-14^{lo} (20). If some L3T4$^+$ Lyt-2$^+$ blasts are precursors of mature cells, L3T4 or Lyt-2 expression would have to be shut down in order to generate mature cells expressing only one of these markers. Unfortunately there is, as yet, no evidence that directly identifies the progeny of large double positive cells.

The L3T4$^-$ Lyt-2$^-$ subset of thymocytes, most of which are mitotically active, constitutes 3% of the total thymocyte population. L3T4$^-$ Lyt-2$^-$ cells are Thy-1^{hi}, H-2K^{hi}, and mainly Ly-1^{lo} (about 15% are Ly-1^{hi}) (8–12, 25) and are unique among all thymocytes in the expression of several surface antigens. Notably, about 50% of these cells express receptors for the lymphokine, interleukin 2 (IL-2) (27, 28). The IL-2 receptor is also found on mature T cells and on at least some mature B cells shortly after mitogen or antigen activation but is only transiently expressed (reviewed in 29). Although some L3T4$^-$Lyt-2$^-$ thymocytes constitutively express the IL-2 receptor, IL-2 alone does not stimulate these cells to proliferate in vitro (the mitogen, con A, must also be added) (28, 30). In addition to the IL-2 receptor, a proportion of L3T4$^-$ Lyt-2$^-$ cells are positive for the Pgp-1 marker, an antigen found on most cells in the bone marrow but not on mature or nonmature thymocytes (31, 32). Intriguingly, although high level expression is not unique to this thymocyte subset, 10–15% of L3T4$^-$ Lyt-2$^-$ cells are MEL–14^{hi} (20). This phenotypic heterogeneity may reflect important functional diversity in this cell population; for example, perhaps the presence or absence of MEL-14 determinants on these precursors determines whether their progeny express the MEL-14 determinant, and whether they are scheduled for imminent emigration. Since Ly-1^{lo} Lyt-2$^-$ L3T4$^-$ thymocytes: (a) contain cells that have significant self-renewal capacity (9, 33), (b) are the predominant cell type in the mouse fetal thymus at day 15 and day 16 of gestation (9, 34), and (c) are mitotically active and phenotypically immature (9, 10), these cells are thought to represent at least one type of intrathymic stem cell.

We consider in greater detail below the lineage relationships between these thymic subsets and what is known of the stages at which they rearrange and express their T-cell receptor genes.

T-CELL ANTIGEN RECEPTOR

Both T cells and B cells recognize foreign antigens via cell surface receptor proteins. Several important functional differences exist, however, between the T-cell antigen receptor (TCR) and the B-cell antigen receptor (surface bound immunoglobulin) (35, 36). Unlike B cells, which recognize free antigens, T cells recognize surface bound antigens almost exclusively when they are associated with molecules encoded by the major histocompatibility complex (MHC). This phenomenon is called MHC restriction (37, 38). Different subsets of mature T cells are restricted by different classes of MHC molecules. In general, Lyt-2$^+$ L3T4$^-$ cells, usually cytotoxic T cells, recognize antigens in association with class-I (H-2) MHC gene products; whereas L3T4$^+$ Lyt-2$^-$ cells, usually helper T cells, recognize antigens in the context of class-II (Ia) antigens (4, 5). The information for MHC class restriction as well as specific antigen recognition is largely contained within the TCR (39-41) and is thought to be acquired during the intrathymic phase of T-cell development (1–3). Therefore, a clear definition of the structure and expression of the TCR is important for a basic understanding of T-cell maturation and function.

The development of T-cell clones and hybridomas (reviewed in 42) greatly facilitated identification and characterization of the TCR. Clones or hybridomas with distinct antigen or MHC restrictions were used to generate monoclonal antibodies (MAb) specific for the immunizing cells (43–46). Many of these antibodies were capable of mimicking antigen-specific activation, which suggests that they recognized the antigen receptor. Immunoprecipitations with these serological reagents revealed that the predominant class of TCR is composed of a glycosylated disulfide-linked heterodimer with an apparent molecular weight of 80–90 kd. The heterodimer consists of a 48–54 kd acidic α-chain and a neutral to basic 40–44 kd β-chain (42, 47). Peptide mapping studies of the α- and β-chains demonstrated that each chain contained constant and variable domains expected for an antigen receptor (47).

The TCR $\alpha\beta$-heterodimer is associated on the cell surface with several other molecules that may be involved in signal transduction. This has been best demonstrated in human T cells and subsequently in mouse T cells where the TCR is associated with a protein complex termed T3 (reviewed in 48). In humans the T3 polypeptide structure is composed of at least three chains: a 20 kd glycoprotein δ-chain, a 25-kd glycoprotein γ-chain, and a 20-kd nonglycosylated ε-chain. Several lines of evidence support the concept that the TCR is intimately associated in a complex with T3 proteins and that the presence of both TCR and T3 is required for either to appear on the cell surface (49). Recently, a cDNA clone for the murine T3 δ-chain has been isolated using a human T3 δ-chain clone as a probe (50). Immunoprecipitation with anti-TCR antibodies has revealed four candidate T3-like chains in murine T cells (51–53). The total number of components which constitute the TCR/T3 complex is not known, and it is possible that more protein chains will be identified. Obviously, a subset of L3T4 or Lyt-2 molecules in mice (CD4 and CD8, respectively, in humans) is likely to be involved.

T-Cell Receptor β-Chain

The first TCR gene was identified independently by two separate groups using subtractive cDNA libraries from a mouse T-cell hybridoma (54) and from a human T-cell leukemia (55). The protein sequences predicted from these cDNA clones showed domains analogous to those found in immunoglobulins: a hydrophobic leader (L) sequence (18–29 amino acids); a variable (V) segment (102–119 amino acids); a constant (C) segment (87–113 amino acids); a transmembrane region (20–24 amino acids); and a short cytoplasmic tail (5–12 amino acids).

The β-chain locus in mice, located on chromosome 6 (56, 57), consists of a relatively small number (less than 30) of variable (V_β) gene segments (58–61) and two constant (C_β) regions, each with a cluster of one diversity (D_β) and six joining (J_β) gene segments (62–65). The V_β, D_β, and J_β gene segments code for the variable segment of the β-chain protein. The V_β gene segments, unlike immunoglobulin V genes, are limited in number and mostly belong to single member families (58–60). The 20 mouse V_β sequences cloned thus far constitute 16 different families (58–60). Although there are few V_β gene segments, there does not appear to be any correlation between a particular V_β gene expression and MHC or class restriction (66, 67). In addition, a deletion of 10 V_β gene segments (one half of the known V_β gene segments) in the SJL mouse strain does not appear to immunocompromise the animal grossly (61).

Northern analysis of thymocytes using a β-chain constant region probe reveals a 1.3-kb and a 1.0-kb transcript. The 1.3-kb transcript is derived from a VDJ_β rearrangement and codes for the functional β-chain. The 1.0-kb transcript appears to be derived from rearrangements that join a D_β to a J_β gene segment (without a V_β gene segment rearrangement) (64, 65). The presence of this smaller transcript implies that there is a promoter 5′ to the D_β regions which becomes activated upon rearrangement. It is not known if a polypeptide is translated from the 1.0 kb message, but an open reading frame with an in-frame start codon suggests that a truncated protein could be made. Analogous D-J transcripts have been found for immunoglobulins, and the translation of these transcripts gives rise to truncated polypeptides (68).

T-Cell Receptor α-Chain

cDNA clones for the TCR α-chain have also been isolated by several groups (69, 70). Similar to the β-chain, α-chain mRNA codes for a protein with L, V, C, transmembrane, and cytoplasmic domains. The genomic locus for the mouse α-chain has been mapped in chromosome 14 (71, 72). Although the overall organization of the α-chain locus is similar to that of the β-chain gene locus, there are some major differences. The α-chain locus contains only one constant region gene segment, and there are at least 50 J_α gene segments spread out over more than 65 kb of DNA 5' to C_α (73, 74). There are many more V_α gene segments (greater than 100) than V_β gene segments, and they form families with 1–10 members (69, 75, 76). It is not known if D_α gene segments exist, nor if N-nucleotides (see below) are found at V-J junctions.

There are two α-chain transcripts of 1.7 kb and 1.4 kb. Analogous to β-chain transcripts, the 1.7-kb transcript codes for the functional α-chain polypeptide while the 1.4-kb message lacks V_α region sequences. It is not known if the shorter 1.4-kb transcripts are initiated by a promoter 5' to the J_α segments or by a promoter 5' to as-yet-unidentified D_α gene segment(s).

T-Cell Receptor γ-Chain

During efforts to isolate a cDNA for the α-chain, another cDNA clone with TCR-like properties was unexpectedly found. This clone hybridizes to T-cell-specific mRNA and to genomic DNA which is rearranged only in T cells (77). The predicted protein, termed the TCR γ-chain, contains L, V, C transmembrane and cytoplasmic domains similar to the α- and β-chains. The γ-chain genomic locus has been mapped to mouse chromosome 13 (72). It consists of three cross-hybridizing and one non-cross-hybridizing C_γ regions, at least four J_γ and six V_γ gene segments (78–81). It is possible that more gene segments may yet be found.

The γ-chain transcript is 1.3–1.5 kb and contains sequences coding for V_γ, D_γ, J_γ, and C_γ. A truncated transcript lacking V_γ sequences has not been identified. The γ-chain gene is apparently not expressed in some mature T cells (82, 83), but γ-mRNA can be found in most cytotoxic T cells (79) as well as in fetal thymocytes (84, 85). An increase of up to ten-fold in the level of γ-chain transcripts can be induced in resting spleen cells by alloantigen stimulation. It is interesting that the increased γ-mRNA levels are only detected in $Ly2^+$ $L3T4^-$ cells (79).

Evidence for a protein product for the γ-chain gene comes from experiments in human cells involving immunoprecipitation with antibodies against T3. When T3 is immunoprecipitated from immature T cells or from T lymphomas negative for TCR α- and β-chains, two polypeptides coprecipitate (86, 87). One of these peptides (55 kd) also precipitates with antisera against C_γ and V_γ. The γ-chain may be a part of a putative second T-cell receptor family expressed on a distinct subset of T cells. The associated 40–45-kd glycoprotein has tentatively been called the TCR δ chain. cDNA clones encoding this chain have not yet been reported.

Generation of Receptor Diversity

Like B cells, T cells use several mechanisms for generating diversity in the antigen receptor: (a) utilization of several different gene segments (V, D, and J) for variable region formation; (b) random combinatorial joining of different germline gene segments; (c) addition at the junction between the joined gene segments of N nucleotides not encoded in the genome; and (d) use of the D gene segments in all three translational reading frames. Unlike immunoglobulin genes, however, the expressed V gene segments are identical in sequence to their germline gene segment counterparts, suggesting a very low rate of somatic mutation (66, 88). N-nucleotides are found in regions encoding likely contact regions for antigen. Thus, by analogy to immunoglobulin heavy chains, it is likely they play an important role in TCR diversity generation. Although the contribution to generating diversity by each of these mechanisms differs between B and T cells (e.g. the number of V and J region genes differs for immunoglobulins and TCR genes), estimates for the overall diversity of their antigen recognition regions are similar (35).

Recognition of MHC and Antigen by the T-Cell Receptor

As mentioned earlier, T-cell antigen recognition is mainly MHC restricted. This observation has led to the formulation of two basic models of recognition. In the single receptor model, the TCR recognizes both antigen and MHC either as a combinatorial determinant or as two individual determinants recognized by distinct combining sites on the same receptor. Alternative models propose that there are two receptors, one which recognizes antigenic determinants and a second receptor which recognizes a polymorphic site of MHC. Recently a "one and a half" receptor model has been proposed where one chain is shared between two receptor complexes (89).

Several experiments have directly tested the hypothesis that the TCR $\alpha\beta$ heterodimer is responsible for both antigen and MHC recognition. In experiments by Yague et al (39), a hybridoma specific for chicken ovalbumin (cOVA) was isolated. Variants of this hybridoma, which had lost either α- or β-chain expression, had also lost the ability to respond to antigen plus MHC. Fusion of an α-loss with a β-loss variant restored the ability to respond to cOVA and MHC, which suggests that the α- and β-chains are responsible for both antigen and MHC recognition. However, since the α-loss and β-loss variants were due to loss of chromosomes, it was formally possible that other gene products from these chromosomes contributed to the restoration of reactivity. In addition, combinatorial association of nonvariable proteins such as LFA1, Lyt-2, or L3T4 on multimeric functional TCR complexes might be involved in specificity generation. Experiments by Ohashi et al (40) provided evidence that loss of other contributing products from the α- and/or β-chain chromosomes was probably not the case. A variant of a human T-cell leukemia line (Jurkat) that had lost the ability to express the α-chain was isolated. Transfection of a Jurkat β-chain gene into this variant cell line restored the appearance of the TCR on the cell surface as well as responsiveness to antiidiotype antibodies and PHA (phytohaemagglutinin). Dembic et al (41) extended the transfection analysis to include both α- and β-chain genes. The α- and β-chain genes were isolated from a CTL clone specific for fluorescein (FL) and I-A^d. These genes were transfected into another CTL clone, the specificity of which was against a different hapten and MHC type. The resultant transfectants obtained both the antigen and MHC restriction characteristics of the donor CTL clone. These experiments provide strong evidence that the αβ-heterodimer is critical in the formation of a TCR complex which recognizes both antigen and MHC and that a second receptor is not necessary for MHC-restricted recognition of cell bound antigen by T cells. Nevertheless, until it is determined whether γδ-chain heterodimers can coexist on T cells with αβ-heterodimers and what functions they possess, one cannot exclude a role for other proteins in MHC and/or antigen recognition.

Ontogeny of the TCR

An examination of mass populations of thymocytes during fetal development indicates that rearrangement of the TCR genes appears to proceed in an ordered fashion: γ- and β-segments rearrange first, followed by rearrangements at the α-locus. Indications of γ-chain gene rearrangements are first seen in day 14–15 fetal thymocytes (90). From fetal day-15 onward, new γ-chain gene rearrangements appear (90). Some of the fetal γ-chain gene rearrangements are similar to those found in adult thymocytes, but three types of rearrangements are apparently unique to fetal thymocytes (79). Rearrangements in the β-chain gene are first observed in DNA from day-14 fetal thymocytes (90). At day 15 in gestation, there is a dramatic increase in the number of rearrangements at the β-chain locus (92). By analogy with the sequence of immunoglobulin segment joining in B cells, the day-14 rearrangements may represent $D_\beta J_\beta$-rearrangements which precede $V_\beta D_\beta J_\beta$ rearrangements by one day (92). The majority of β-chain alleles are rearranged in fetal thymocytes by day 17 of gestation (92, 93). Due to the large separation between the different segments of the α-chain gene, it is difficult to detect rearrangements of the α-chain locus with standard techniques. However, based on the late appearance of α-chain transcripts in ontogeny (see below), it seems likely that α-chain gene rearrangements occur after γ- and β-chain loci are rearranged.

The onset of transcription of the different TCR-genes during ontogeny follows essentially the same sequence as the DNA rearrangements. Transcripts of the γ-chain gene are seen as early as day 14 in gestation (90). Expression of the γ-chain gene rapidly increases, reaching maximum levels in day-15 fetal thymocytes. Thereafter, the quantity of γ-chain RNA on a per cell basis declines slowly until day 18–19 of gestation (84). At birth (day-20 fetal gestation), transcripts of the γ-chain gene decrease dramatically to the very low level found in adult thymocytes (90). Beta-chain gene expression is first detected at day 14 of fetal life as a 1.0-kb RNA transcript derived from a $D_\beta J_\beta$ rearrangement (90). One day later, some full-length transcripts ($V_\beta D_\beta J_\beta$) of 1.3-kb can be found. By 17 days of gestation, the 1.3-kb species is the predominant β-chain RNA (84, 90, 93). Transcripts of the α-chain gene first appear in day-16 fetal thymocytes (90). At this stage, most of the α-chain transcripts are smaller (1.4 kb) than the 1.7-kb α-chain mRNA found in adult thymocytes. The exact structure of the 1.4-kb α-chain transcript is unknown. Like the early 1.0-kb β-chain RNA, the 1.4-kb α-chain transcript may represent $D_\alpha J_\alpha$-rearrangements, if there is a D_α gene pool. The level of the 1.7-kb α-chain mRNA, presumably encoding a functional α-chain polypeptide, increases steadily from day 16 onwards and reaches maximal levels in adult thymocytes (84). Information about expression of different T3-genes in the mouse is, at the time of this writing, incomplete, as only the mouse T3 δ-gene has been cloned to date. The kinetics of the expression of the T3 δ-gene during fetal development are quite similar to those observed for the TCR β-gene (90).

Shortly after the appearance of full-length transcripts for both the α- and β-chain genes, cell surface heterodimer structures are detected. A small proportion of day-17 fetal thymocytes react with the Mab KJ16-133 (95) which recognizes an epitope found on gene products from the $V_\beta 8$ subfamily (61, 96). The percentage of thymocytes reacting with this Mab increases rapidly from day 17 of gestation until birth. At birth, the percentage of KJ16-133 positive thymocytes is approximately half of the proportion of peripheral T cells reacting with this antibody (roughly 20% of peripheral T cells are KJ16-133 positive) (95).

TCR Rearrangements and Expression in the Adult Thymus

Greater than 95% of adult thymocyte β-chain alleles and approximately 50% of their γ-chain alleles are arranged from the germline configuration (84, 92). However, certain subpopulations of thymocytes contain cells with unrearranged β-chain loci. Trowbridge et al (97) isolated thymocytes that were resistant to treatment with anti-Thy-1 Mab and complement. This subpopulation of thymocytes was highly enriched in cells expressing the Pgp-1 marker and in T-progenitor activity (see below). An estimated 80% of these cells showed the germline configuration of the β-chain gene. Another population of immature thymocytes, those with the L3T4$^-$ Lyt-2$^-$ Ly-1^{lo} phenotype, is also enriched in cells with unrearranged β-chain genes (94). As many as one third of the cells in the total L3T4$^-$ Lyt-2$^-$ Ly-1^{lo} population may have the β-chain gene in the germline configuration, since 4 out of the 12 hybridomas made with L3T4$^-$ Lyt-2$^-$ Ly-1^{lo} cells showed only unrearranged β-chain gene sequences (94). The L3T4$^-$ Lyt-2$^-$ Ly-1^{lo} population as a whole contains both 1.0- and 1.3-kb β-chain transcripts at levels equivalent to those found in unfractionated thymocytes (94). Alpha-chain transcripts, however, appear to be completely absent (94). Thus, it has been suggested that the L3T4$^-$ Lyt-2$^-$ Ly-1^{lo} population contains at least two types of cells: those that have rearranged neither the α- nor β-chain gene, and those that have rearranged and express only the β-chain gene.

In the L3T4$^-$ Lyt-2$^-$ thymocyte population, which includes Ly-1^{lo} and Ly-1^{hi} cells, low levels of α-chain mRNA have been detected (84). The levels of β-chain transcripts in L3T4$^-$ Lyt-2$^-$ cells are approximately the same as in unfractionated thymocytes (84); γ-chain transcripts, however, are twenty-fold more abundant in L3T4$^-$ Lyt-2$^-$ cells than in unfractionated populations of adult thymocytes (84). How or why this occurs is unknown since the proportion of rearranged γ-chain alleles is roughly the same in L3T4$^-$ Lyt-2$^-$ cells and unfractionated thymocytes (79). Like total thymocytes, adult L3T4$^-$ Lyt-2$^-$ cells show limited rearrangement at the γ-chain locus with one type of rearrangement predominating (79).

The major subset of adult thymocytes consists of nonmature cells that are positive for both the L3T4 and Lyt-2 antigens (8–12, 25). By enriching for thymocytes expressing low levels of H-2K, Snodgrass et al (93), obtained a population consisting mostly ($>98\%$) of cells with the L3T4$^+$ Lyt-2$^+$ phenotype. At least 95% of these cells had rearranged their β-chain genes and contained levels of β-chain transcripts equivalent to those found in unfractionated thymocytes. Evidence that some L3T4$^+$ Lyt-2$^+$ cells have rearranged and transcribe α- as well as β-chain genes comes from analyses of primary thymoma cells. Richie et al (98) found that all of the 11 examined thymomas with the L3T4$^+$ Lyt-2$^+$ phenotype contained both α- and β-chain transcripts.

Thymocyte subpopulations containing PNA^{hi}, PNA^{lo}, MEL-14^{hi}, and MEL-14^{lo} cells have also been analyzed for TCR gene rearrangements. In all these subsets, β-chain genes appear to be nearly entirely rearranged on both chromosomal homologues (90).

$\alpha\beta$-Heterodimer Expression on the Cell Surface

Fractionation of adult thymocytes into cortisone resistant or PNA^{lo} cells (both fractions representing mainly mature phenotype thymocytes) reveals the same frequency of KJ16-133 positive cells as found among peripheral T cells (95). In the PNA^{hi} fraction of adult thymocytes (mainly cortical small thymocytes), cells reacting with KJ16-133 are about half as frequent, and the staining intensity is much reduced as compared to PNA^{lo} cells. This could indicate that, in general, mature phenotype thymocytes represent a more advanced or independent stage in T-cell differentiation than do nonmature thymocytes or that mature phenotype thymocytes are generated independently from the remaining thymocyte pool. Other explanations such as a differing frequency of $V_{\beta}8$ usage in mature vs nonmature thymocytes cannot be ruled out formally. Ultrastructural analysis revealed that a subset of cortical thymocytes showed cytoplasmic staining, or cytoplasmic with a faint cell surface staining, or cell surface staining alone with KJ16-133 (99). In contrast, medullary (presumably mature) thymocytes reacting with the same Mab showed cell surface staining exclusively. The explanation for this reaction pattern is not clear. One can speculate, however, that cells with cytoplasmic staining lack either a functional T3 complex, or α-polypeptide, or some other protein required for $\alpha\beta$-heterodimer expression on the cell surface.

It is noteworthy that at least some of the L3T4$^+$ Lyt-2$^+$ thymocytes appear to have $\alpha\beta$-heterodimer structures on the cell surface. As discussed above, H-2K^{lo} thymocytes are primarily of the L3T4$^+$ Lyt-2$^+$ type. Diagonal gel electrophoresis of lysates of H-2K^{lo} thymocytes indicates that an estimated 20% of these cells contain a cell surface molecule with TCR-like properties (93). Furthermore, surface-labeled $\alpha\beta$-heterodimer structures have been immunoprecipitated from L3T4$^+$ Lyt-2$^+$ primary thymoma cells (98).

THYMIC MICROENVIRONMENT

The thymus is organized into two distinct compartments: an outer, cortical region of densely packed cells and a more sparsely populated, inner medullary area. The two compartments differ in their organization of lymphoid cells (see above) as well as of nonlymphoid, stromal components (reviewed in 100). The majority of stromal cells in the cortex are epithelial cells with long processes which reticulate among cortical thymocytes. The cortex also contains a small number of bone marrow–derived macrophages, mainly in the region adjacent to the medulla. In the medulla, likewise, both epithelial and bone marrow–derived stroma are present. Medullary epithelial cells have truncated processes and thus are morphologically distinguishable from their cortical counterparts. The bone marrow–derived cells found in medullary regions include large numbers of macrophages as well as interdigitating, dendritic cells.

Cell Surface Antigens on Thymic Stroma

Since thymic stromal components are thought to be important in the acquisition of mature T-cell functions, much attention has been focused on surface antigens expressed by these cells. Perhaps of particular importance to the acquisition of MHC restriction and self-tolerance by maturing T cells is the expression of class-I (H-2) and class-II (Ia) MHC antigens on thymic stroma. Medullary stromal cells express high levels of both H-2K and H-2D antigens, whereas cortical epithelial cells express class-I antigenic determinants at levels ranging from undetectable to moderate in intensity; H-2K allodeterminants are undetectable, whereas H-2D allodeterminants are expressed (101, 102; R. V. Rouse, personal communication). Ia is expressed at very high levels on cortical epithelium, medullary epithelium, and medullary dendritic cells (101–103). Notably, neither cortical nor medullary macrophages express detectable levels of surface Ia antigens (104). With this exception, therefore, maturing thymocytes have the opportunity to encounter Ia and H-2D antigens on stromal cells in the cortex, and all MHC antigens in the medulla.

Although most cortical and medullary stromal cells share the characteristic of MHC antigen expression, subsets of these cells can be selectively identified with specific monoclonal antibodies (MAbs). We have recently found that an antigen (6C3Ag), first found as a marker for pre–B cell neoplasms induced by diverse conditions (105; H. C. Morse III et al, manuscript submitted), is expressed in the thymus exclusively on cortical epithelial cells (B. Adkins et al, manuscript submitted). In normal bone marrow, this antigen is expressed at high levels by the stromal cell subset which supports exclusively pre–B cell proliferation and differentiation (C. A. Whitlock et al, manuscript submitted). Since the thymic cortex is the major region for proliferation and perhaps also for differentiation within the thymus, it is tempting to speculate that the 6C3Ag is a marker for cells mediating early events in both B and T lymphopoiesis. Several other groups have also raised MAbs which, like 6C3 MAb, recognize exclusively cortical epithelial cells (106, 107; R. V. Rouse, manuscript in preparation). Cell surface antigens which specifically mark medullary epithelia and other medullary stromal cells have also been described (106, 107; R. V. Rouse, manuscript in preparation). One type of monoclonal antibody recognizes only thymic subcapsular epithelial cells plus medullary epithelial cells (reviewed in 108). Apart from the suggestive association between the 6C3Ag and lymphopoiesis, it is not known whether these cell surface antigens are important in mediating the proliferation or differentiation of thymocytes. Nevertheless, MAbs against these surface antigens are useful tools for identifying specific types of stroma in the normal adult thymus, during ontogeny and under various experimental or disease conditions. Furthermore, in vivo treatment with MAbs to eliminate either the antigen or the cell type expressing that antigen may help to define the impact of cortical vs medullary stromal components on the generation of specific mature T-cell functions.

Thymocyte-Stromal Cell Complexes

Within the thymus, interactions between maturing thymocytes and stromal elements are postulated to result in self-restricted and self-tolerant mature T cells (1, 3). The question is: Which thymocytes interact with which stromal cells and in what compartment? Gentle protease digestion of the thymus has been used to liberate complexes of thymocytes and stromal cells. One type of complex, the thymic nurse cell (TNC) complex, contains a single, cortical epithelial cell (the TNC) which completely surrounds a core of up to 200 thymocytes (109–111). Located in the subcapsular region TNCs are Ia^+ and express the 6C3Ag, along with other markers typical of cortical epithelial cells (109–112; B. Adkins et al, manuscript submitted). Much experimental evidence suggests that TNC complexes are not an artifact of the digestion process (111, 113). However, it is not yet clear whether the associated thymocytes are in intimate association with or are actually enclosed within the TNC in situ. Since the outer cortex contains the majority of proliferating thymocytes, it has been proposed that TNC complexes in vivo are specialized sites for the division, and perhaps education, of immature thymocytes. Were that the case, the expectation is that thymocytes contained in TNC complexes would be phenotypically and functionally distinct from the free cortical thymocyte pool. Evidence for and against this hypothesis exists. Thymocytes in TNC complexes are not significantly different phenotypically from the remaining outer cortical thymocytes (111, 112). In addition, like most cortical cells, thymocytes in TNC complexes are cortisone and radiation sensitive. On a functional level, however, TNC-associated thymocytes are enriched in cycling cells and are slightly enriched in both helper and cytotoxic activity relative to the bulk of cortical thymocytes (110, 114–116). It is difficult to prove experimentally that these activities are *not* due to the presence of a small proportion of contaminating, phenotypically mature cells. Nevertheless, these observations raise the possibility that association of thymocytes with

TNC leads to a stage of intrathymic differentiation in which, for some cells, functional maturity is reached prior to phenotypic maturity.

Another type of thymocyte-stromal cell complex is the thymus rosette (T-ROS). T-ROS consist of a single, central macrophage or dendritic cell (DC), surrounded by 10–15 thymocytes (117). Approximately half of the T-ROS are formed with a central Ia$^-$ macrophage, whereas the other half contain Ia$^+$ DC. Since thymic dendritic cells are found exclusively in the medulla (100), the latter type of T-ROS almost certainly is formed in medullary regions. While a substantial proportion of T-ROS-associated thymocytes are phenotypically immature or nonmature, and are cortisone and radiation sensitive, the frequency of cytotoxic precursors among these cells is at least twice as great as among TNC-associated thymocytes (114). Here again, the precise role(s) of T-ROS in intrathymic differentiation is a matter of conjecture. It is interesting, however, that the subset of T-ROS with Ia$^+$ DCs has been shown to be active at antigen presentation in vitro (118, 119). These DCs also appear to process and present at least some circulating antigens in vivo (118, 119), suggesting that some of the T-ROS may be sites for the recognition of antigen either by recently maturing thymocytes or by the rare subset of peripheral T cells that, upon activation in the periphery, return to the thymus (120, 121). It will be important to clarify the role of such antigen presentation in the thymus. It is conceivable that it plays an important role in tolerance (or suppression) to circulating self antigens; or that it is a selective step in amplifying immunocompetent thymocytes to self plus foreign antigens just prior to the release of their progeny to the periphery.

Ontogeny of Thymic Microenvironment

The organization of the mouse thymus into cortical and medullary stroma, the phenotypic maturation of the thymic microenvironment, and the formation of thymocyte-stromal cell complexes all occur early in fetal development. At 10–11 days in ontogeny, prior to colonization by T progenitor cells, the epithelial components are formed from the ectoderm and endoderm of the third pharyngeal cleft and pouch, respectively (122). The stromal microenvironment is well organized and maturing phenotypically as early as 14 days in fetal development. Desmosomes and tonofilaments characteristic of adult thymic epithelium appear at this time (123, 124), and cells expressing Ia antigens are present in focal clusters within medullary regions (125, 126). Class-I MHC expression is first detected at day 16 of gestation, by which time Ia expression has expanded throughout the thymic lobe showing a distribution similar to that seen in the adult (125, 126). The formation of thymocyte-stromal cell complexes is temporally linked to the maturation of the thymic microenvironment. T-ROS, some of which contain Ia$^+$ medullary dendritic cells, can be isolated first at day 14 in ontogeny (117). TNC-complex formation, involving Ia$^+$ cortical epithelial cells, lags behind somewhat, becoming evident by day 17. T-ROS and TNC complexes, therefore, appear in ontogeny concomitant with the expression of Ia antigens by the associated medullary and cortical stromal cells, respectively.

Many, but not all, of the antigenic markers specific for cortical or medullary stromal cells appear by day 14–15 of fetal development (126). At this stage, cortical epithelial cells express the ER-TR4 antigen found on adult cortical epithelium, and cells positive for medullary stromal cell markers are also apparent, providing a clear definition of cortex and medulla. Notably, the 6C3Ag which, like ER-TR4 Ag, is present on adult cortical epithelial cells, does not appear at any time during fetal life. 6C3Ag$^+$ cells are not present until one week following birth (B. Adkins et al, manuscript submitted); this indicates distinct and probably important differences between the fetal and the adult thymic microenvironments.

What types of changes might follow from the appearance of the 6C3Ag on thymic cortical epithelial cells? Elsewhere we have demonstrated that the 6C3Ag first appears at high levels on all AbLV-induced malignant early B-lineage lymphocytes at about the time that they escape the need for bone marrow stromal cell dependent growth in vitro (C. A. Whitlock et al, manuscript submitted), and at the time of the development of tumors outside the bone marrow microenvironment in vivo (105). The cloned stromal cell lines from the Whitlock-Witte mixed stroma feeder layers

Table 1 Antigenic markers on lymphoid and nonlymphoid cells in the thymus

Cell type	Location	Phenotype[a]	Size
Lymphoid			
Mature thymocytes	Cortex (MEL-14^{hi}) Medulla (MEL-14^{lo})	L3T4$^+$Lyt-2$^-$ PNA^{lo}H-2K^{hi}Ly-1^{hi} or L3T4$^-$Lyt-2$^+$	
Nonmature thymocytes	Cortex	PNA^{hi}H-2K^{lo}Ly-1^{lo}L3T4$^+$Lyt-2$^+$	Small
Immature thymocytes	Cortex	PNA^{hi}H-2K^{lo}Ly-1^{lo}L3T4$^+$Lyt-2$^+$ or PNA^{med-hi}H-2K^{hi}Ly-1^{lo}L3T4$^-$Lyt-2$^-$	Large
Nonlymphoid			
Epithelial cells	Cortex	Ia$^+$, H-2$^\pm$, ER-TR4$^+$, 6C3-Aq$^+$	
Epithelial cells	Medulla	Ia$^+$, H-2$^+$, ER-TR5$^+$	
Dendritic cells	Medulla	Ia$^+$, H-2$^+$, ER-TR6$^+$	
Macrophages	Medulla and Juxtamedullary Cortex	Ia$^-$, H-2$^+$, ER-TR6$^+$	

[a] The phenotype of the majority (>80%) of cells in each category is listed. Thymocyte phenotypes determined by FACS analysis; stromal cell phenotypes by immunohistochemical staining. Referenced in text.

which support normal and neoplastic pre–B cell maturation are all 6C3Ag^{hi}, while other cloned lines from the same cultures which fail to support pre–B cell proliferation are 6C3Ag$^-$ (C. A. Whitlock et al, manuscript submitted). We have proposed that the gp160-6C3Ag might represent a cell surface growth factor, or its precursor, for which both normal and neoplastic pre–B cells bear a specific receptor (C. A. Whitlock et al, manuscript submitted). If this speculation proves correct, it is reasonable to propose that the appearance of 6C3Ag at high levels on thymic cortical epithelial cells heralds the possible switch to, or addition of, interactions between stromal cells and developing thymocytes which express the 6C3Ag receptor, whether or not that receptor is coupled to growth regulation in these cells.

Alterations in the Thymic Microenvironment

The importance of the establishment and maintenance of a normal thymic microenvironment is perhaps best illustrated by athymic nude mice. As adults, these mice are indeed athymic in that they lack detectable thymic tissue. However, during ontogeny, limited development of a thymus primordium does occur, forming an incomplete structure which is subsequently resorbed. Epithelia in this fetal thymic rudiment lack Ia expression; and the ER-TR-4 and -5 antigens, characteristic, respectively, of cortical and medullary epithelial cells in the normal adult thymus, are undetectable (124–126). The inadequacy of this thymic microenvironment, present only transiently in development, results in adult mice severely (but not completely) depleted in phenotypically and functionally mature T cells. This depletion is clearly due to the absence of a thymus since nude mouse bone marrow contains T-cell precursors capable of differentiating in normal thymic grafts (reviewed in 127). Although both cytotoxic and helper activities have been detected in spleens from nude mice, the frequency of cells showing either of these activities is orders of magnitude lower than that found in normal mouse spleen (reviewed in 128). Similarly, complete removal of thymic stroma by neonatal thymectomy of otherwise normal animals leads to a drastic depletion of the peripheral T-cell pool (129).

While it is clear that removal of the entire thymic stroma severely affects T lymphopoiesis, the role(s) of individual stromal cell types in T-cell maturation are not well understood. Included in a long list of unanswered questions are these: (a) Do bone marrow–derived and epithelial stromal cells mediate the same or different aspects of intrathymic differentiation; (b) What are the relative functions of cortical and medullary stroma; and (c) What is the importance of stromal cell antigens in the proliferation and differentiation of thymocytes? To address these questions, experimental systems for the depletion of specific stromal cell subsets or antigens have been (and are being) developed.

Several methods are available for depleting the thymus of virtually all bone marrow–derived cells, leaving only a framework of epithelial cells. Since thymic epithelial cells do not proliferate or only do so extremely slowly, these cells are not grossly affected by treatments that eliminate proliferating bone marrow–derived cells. Conditions of this sort include lethal irradiation of adult tissue and exposure of fetal thymuses in vitro to low temperatures or deoxyguanosine (130–132). If the resultant, mainly epithelial, cell framework is grafted to another animal, both the thymocyte and the bone marrow–derived stromal cell compartments are repopulated by host cells. As discussed in greater detail below, these procedures have been used to address the relative importance of bone marrow–derived vs epithelial cells in MHC restriction and tolerance induction.

An experimental protocol for generating abnormalities almost exclusively in the medullary subpopulation of stromal cells was recently discovered in a collaboration between our laboratory and that of Dr. Samuel Strober. This protocol, called total lymphoid irradiation (TLI), is a fractionated radiation regimen used in the treatment of certain human lymphomas and autoimmune diseases (133). In this irradiation regimen, the major lymphoid organs, including the thymus, are exposed while the rest of the body is shielded from the radiation source. Mice given TLI receive 17 doses of 200 rads each over a 3–4 week period, for a cumulative total dose of 3400 rads. Immediately following TLI, the thymus shows major disruptions in overall architecture as well as in thymocyte and stromal cell subpopulations (B. Adkins et al, manuscript submitted). Although most of these effects are transient, stromal cells in the medulla show persistent abnormalities as long as 4 months after TLI. Medullary stromal cells reacting with the medullary stromal cell-specific Mab (MD-1) (R. V. Rouse, manuscript in preparation) as well as endogenous peroxidase-containing macrophages, both abundant in the normal medulla, are severely reduced or absent following TLI. Since cortical epithelial cells appear antigenically normal after TLI, this regimen could prove useful for assessing the importance of these medullary stromal cells in the acquisition of particular T-cell phenotypes and functions.

The importance of stromal cell antigens in intrathymic differentiation has, quite naturally, focused on MHC antigens. To test the impact of Ia expression by thymic stromal cells on T-cell generation, Kruisbeek et al (134, 135) injected neonates multiple times with anti-Ia antibodies. The effects of this treatment within the thymus were dramatic. Ia antigens on thymic stroma were either modulated off the cell surface (of interdigitating DC) or masked (on epithelial cells) by the injected antibody. Furthermore, cells with the L3T4$^+$ Lyt-2$^-$ phenotype (i.e. those restricted to interaction with Ia antigens) were severely depleted both in the thymus and in the periphery, suggesting that Ia expression by thymic stromal cells is crucial for the generation and perhaps survival of the correspondingly restricted T-cell subset. The success of these experiments suggests that similar in vivo treatments can be performed with a wide variety of antibodies. Since multiple reagents for marking specific subpopulations of thymic stromal cells are now available, this approach may be very useful for assigning functions to particular antigens or to the cells that bear them. Because of the complexity of the architecture of the thymic stroma and the difficulty in isolating and maintaining in tissue culture a representative sampling of these microenvironments, most of what we expect to learn in the near future will depend on in vivo experimental manipulations of these types. But this, we hope, will soon change.

FUNCTIONS ASCRIBED TO THE THYMUS

The thymus has long been postulated to be a controlling element in the selective maturation of T lymphocytes with specific reactivity patterns. Mature T cells recognize antigens only in the context of host (self) class-I or class-II MHC determinants on target cells. Further, T cells exhibit a tolerance to host (self) determinants, while they respond vigorously against polymorphisms of foreign cell surface molecules. These phenomena have been extensively studied by employing thymectomy, lethal irradiation, and bone marrow reconstitution, with or without simultaneous thymic grafting, to establish the role of the thymus in the maturation of restriction patterns.

Restriction to Class-I Molecules

Several groups, using various experimental systems, have shown that radio-resistant elements in the thymus are responsible for 'educating' maturing thymocytes to recognize preferentially foreign antigens in the context of thymic class-I MHC molecules (1, 136). This appeared initially to occur without regard to the class-I haplotype of the maturing thymocytes themselves. However, Zinkernagel (137) has shown that there is a preferential restriction for class-I haplotypes that are shared by the injected bone marrow cells and the stromal elements of the thymus. Thus, adult thymectomized, lethally irradiated mice of strain A that were reconstituted with bone marrow of (AxB)F$_1$ origin and given thymic grafts of (BxC)F$_1$ origin utilized restriction to B but not to C class-I molecules in a virus-specific cytotoxic-T-cell assay (137).

Restriction to Class-II Molecules

Longo and coworkers have shown in an extremely complicated experimental model that proliferative responses restricted to recognition of class-II molecules may be dictated by bone marrow–derived cells in the thymuses of radiation chimeras (138, 139). This was demonstrated by constructing chimeras of F_1 into parental strains and immunizing with soluble antigens to which the parental strain was a nonresponder and the F_1 hybrid was a responder. Under certain conditions, the experiment showed that bone marrow–derived cells dictate the class-II restriction of the immune responses. In contrast, Lo & Sprent (140) have utilized in vitro deoxyguanosine (dGuo) treatment of thymus explants, a procedure that depletes the thymus of nonepithelial cells; they have shown that class-II restriction of proliferative responses to antigen is dictated by the dGuo-resistant (epithelial?) component of the thymus. The contradiction of these two results is undoubtedly related to differing experimental approaches, most notably the use by Longo et al of antigens under immune response gene control and of cortisone and antilymphocyte serum treatment as part of their experimental protocol. Sprent and coworkers (140a) have attempted to resolve this issue by utilizing a second irradiation and reconstitution of radiation chimeras to ensure the dominance of donor-derived macrophages and dendritic cells in the thymus prior to testing for restriction patterns. Their results (140a) agree with the dGuo experiment in that T cells from twice-irradiated chimeras were predominantly restricted to host (thymic) H-2 determinants and not to donor (bone marrow–derived macrophage and dendritic cell) H-2 determinants. Therefore, the role of bone marrow–derived cells in dictating restriction to class-II molecules remains to be clarified.

The Questions of Self Tolerance

Organ culture of thymic explants at a suboptimal temperature produces thymic 'epithelial' cell frameworks similar to those generated by dGuo treatment (141). Use of these two methods has provided some clues to answer the question of how self-tolerance might be generated. Ready and coworkers (142) showed that dGuo-treated embryonic thymus lobes could be transplanted across major histocompatibility barriers without being rejected, in spite of their continued expression of foreign class-I and class-II antigens. However, both the intragraft and mature peripheral T-cell populations proliferated in vitro in response to the foreign MHC antigens of the graft, indicating that tolerance had failed to develop. Cytotoxic-T-cell activity against the MHC type of the graft has also been demonstrated in thymocytes within the grafted tissue (143). In contrast, minor histocompatibility antigens of the grafted thymus epithelium are not recognized in the context of recipient MHC antigens, indicating that tolerance has developed to these antigens (144). These results have been interpreted as evidence that tolerance induction to foreign MHC antigens is dependent upon a nonepithelial component of the thymus, probably the bone marrow–derived stromal elements (142). More correctly, treatment of embryonic thymuses with dGuo results in a structure apparently incapable of imposing self-tolerance. Two major points are of special interest here. First, in spite of the presence of cytotoxic and proliferating T-cell precursors within the thymic graft that are specific for the foreign antigens of the graft, no rejection response occurs. Second, as previously mentioned, MHC restriction to thymic class-II antigens of graft origin does develop in these animals (144), indicating a clear separation between the generation of self-tolerance and of MHC restriction. This is important evidence in support of theories that suggest that MHC restriction is learned by self-reactive clones in the thymus.

The mechanism of MHC restriction has been thought to involve positive selection within the thymus of maturing thymocytes that express low affinity receptors for self-MHC molecules on stromal elements of the thymus. While this mechanism explains restriction to self, it is not clear how the association of antigen and self changes the recognition events. The role of antigens in these interactions is difficult to envision, since every possible antigen–MHC complex would have to be available for intrathymic education of developing T cells. An alternative hypothesis states that self antigens may play a role in shaping the repertoire of maturing T cells. Singer and colleagues (145) have recently presented evidence that T cells that can respond to a foreign class-I antigen, e.g. the mutant K^{bm6} antigen, may be I^b restricted T helper cells selected during ontogeny by recognition of K^b plus I^b, but not by recognition of K^{bm1} plus I^b or by K^b plus I^{bm12}. Thus, it seems that self antigens can play a role in the development of responses for non-self antigens.

Extrathymic Influences on Restriction

While the repertoire of intrathymic lymphoid cells seems to be guided by the expression of MHC antigens on thymic stroma, it is clear that there is a strong extrathymic influence which, under some experimental conditions, can modify the repertoire of thymic emigrants. This effect is readily apparent in class-I restricted responses, which have been shown in both radiation chimeras (146) and nude recipients of allogeneic thymic grafts (147, 148) to be influenced by extrathymic as well as intrathymic elements. Class-II restricted responses, on the other hand, seem to be selected only by the MHC of the thymic stroma (146). Stutman (149) has presented a strong case for further differentiation and diversification of postthymic lymphocytes which contrasts with the more widely accepted view that only antigen driven proliferation can occur, in the absence of further diversification in repertoire, following the exit of cells from the thymus. As with issues of self restriction, tolerance, and repertoire diversification, the answers remain to be determined.

In summary, the data presently at hand would seem to indicate that bone marrow–derived stroma in the thymus might prove to be the primary elements for inducing tolerance. The origin of MHC restriction is less clear, since various groups have presented evidence that thymic macrophages and dendritic cells (138, 139) and thymic epithelial cells (140) determine class-II restriction and that intrathymic as well as extrathymic events occur in the generation of class-I restriction (146–148).

THYMIC COLONIZATION BY BONE MARROW–DERIVED PROGENITOR CELLS

The issue of cellular migration into, within, and out of the thymus has been investigated extensively, particularly in the period following the clarification of the role of the thymus in T-cell maturation. Several excellent reviews have recently been published (24, 149, 150), and only the major points of interest are summarized here.

Thymic Colonization During Ontogeny

In the normal animal, a large-scale migration of stem cells into the thymus occurs only in fetal life. In the mouse, this occurs at approximately day 11 of gestation (151). The lymphoid stem cells are presumably derived from the yolk sac or fetal liver, since the bone marrow does not develop into a significant site of hematopoiesis until shortly before birth (151, 152). Thymic colonization during avian ontogeny presents a much more complicated picture, in which three successive waves of lymphocyte precursor cells enter the embryonic thymus, each wave being separated by a refractory period of roughly 5 days during which no significant seeding occurs (153). Recent evidence indicates that the quail embryonic thymus produces peptides that induce bone marrow cells to migrate in a directed fashion (154). The observation that these chemotactic peptides are produced by the avian thymus only during the receptive period of thymic development (155) fits well with data which indicate that it is the developing thymus, not the availability of stem cells in the bloodstream, which dictates the cyclic colonization of the avian thymus (153). Although it is difficult, in light of the rapid movement of blood within vessels, to imagine a soluble chemoattractant functioning to attract blood-borne stem cells, one can easily imagine the chemotactic effect of peptides expressed on the luminal cell surface of endothelium within the thymic microvasculature. In a formal sense, these would be ligands displayed on the cell surface for specific lymphocyte homing receptors (156). Studies of interactions between stem cells and thymic endothelium may provide valuable information on the regulation of thymic colonization by stem cells.

Thymic Colonization in Adult Animals

In marked contrast to the fetal thymus, the thymus of the normal adult animal does not receive a large-scale input of precursor cells from hematopoietic sources; apparently it relies primarily upon an intrinsic intrathymic pool of self-renewing precursor cells to maintain its lymphoid components. This conclusion is drawn from the observation that large doses of bone marrow (2×10^7 cells) given intravenously do not contribute significantly either to the intrathymic population or to the mature T-cell pool in peripheral tissues (150, 157; and our own unpublished observations). This is of course a nonphysiological approach to the question, as are most other experimental systems used to analyze cellular migration into the adult thymus. Hence, accurate quantitation of the normal level of cellular influx into the thymus remains an open issue.

Colonization of the Thymus Under Experimental Conditions

In contrast to the paucity of cellular migration to the thymus under normal conditions, experimental systems allow a large scale influx of bone marrow–derived cells into the thymus. The most commonly utilized system involves whole body irradiation followed by an intravenous transfer of syngeneic, congenic, or allogeneic bone marrow cells. The cellular influx to the thymus is dependent upon the dose of radiation (158), but it is interesting that the thymus need not be irradiated since a considerable cellular influx eventually results following irradiation of the lower half of the body (159). The effect of whole body irradiation is immediate, since migrant cells can be detected in the thymus within three hours of irradiation and reconstitution (160). In spite of widespread application of the irradiation and reconstitution technique, the mechanism by which the thymus is rendered receptive to colonization remains a mystery. The answer does not seem to be as simple as the elimination of thymocytes to create space within the thymus, since the receptiveness occurs before thymus depletion is evident (160) and since hydrocortisone-depleted thymuses remain unreceptive to intravenously transferred bone marrow cells (150, and our unpublished observations).

A second experimental approach to the question of thymic colonization has been by surgical parabiosis of two adult mice, the lymphoid cells of which could be distinguished from each other by means of a chromosomal marker (161). This experiment is a natural extension of the observation that dizygotic twin chimeras share stem cells and, hence, thymic populations of cells (151, 161). If a free exchange of cellular elements in the blood is established between two karyotypically distinct mice, one would predict that the percentage of cells derived from each partner should equilibrate with time to about 50% of the total cells in each organ. This situation does in fact result within the lymph nodes and spleens of parabiotic mice, demonstrating that a free cellular interchange is established, and the 50% equilibrium is attained between 12 and 30 weeks (161). The thymus and bone marrow, however, reach an exchange level of 5–10% within the first 4 weeks of parabiosis, but very little interchange occurs subsequently. Scollay & Shortman (24) have suggested that the initial 5–10% colonization of the partner thymus in these experiments may have been facilitated by stress-induced depletion of the thymus following the initial surgery, and that the failure of further exchange of thymocytes may reflect a very low level of thymic seeding from bone marrow under steady state conditions. This point is of interest in light of data from part-body irradiation experiments (159), which seemed to indicate that the level of bone marrow replacement was roughly equivalent to the eventual level of thymic replacement. As Scollay & Shortman (24) noted, however, the results of experiments which utilize parabiosis do not fit with the part-body irradiation experiments, since different estimates of steady state seeding are implied by the two techniques. The problem that low thymic seeding efficiency poses in the study of early maturational events within the thymus has been overcome to some extent by the development of techniques for intrathymic injections of bone marrow and thymocytes (158). The results of these experiments have demonstrated that thymic seeding efficiency is not the only factor that limits the external contribution to thymic populations, since bone marrow cells injected directly into the thymus will not produce a population of thymocytes in the absence of preirradiation of the recipient animal. This obligatory radiation conditioning can be avoided at least to some extent by utilizing newborn animals rather than adults as bone marrow recipients (B. Adkins et al, manuscript submitted). A clear definition of the spectrum of radiation effects on the thymus is sorely needed, especially in regard to experimental systems that utilize radiation and bone marrow reconstitution to study the phenomenon of MHC restriction in the thymus.

Phenotype of Pre–T Cells in Bone Marrow

The isolation and characterization of T-cell precursors, as well as pluri-potent hematopoietic precursors, from normal bone marrow cells have been of considerable interest for many years. Investigators have utilized separation techniques such as density gradients (162), antibodies plus complement (163), and flow cytometry (164) in attempts to enrich for precursor activity in bone marrow. By far, the assays of choice for moni-toring enrichment of precursor activity have been splenic colony formation (CFU-S) (165) and thymic repopulating activity, both assays being per-formed in lethally irradiated mice. In the most recent of such investigations, Muller-Sieburg et al (166) utilized flow cytometry to characterize B-cell precursors in mouse bone marrow, as assayed in an in vitro bone marrow culture system capable of supporting B-cell maturation (167). It is sur-prising that the pre–B cell activity can be traced to a population of bone marrow cells that expresses the T-cell antigen Thy-1 as a surface marker, albeit at levels one-tenth of those found on peripheral T cells. Apart from the low level of Thy-1 expression, this population of cells does not express surface markers normally found on mature T cells, and B cells, granulo-cytes, or macrophages, and thus is refered to as Thy-1^{lo} T$^-$B$^-$G$^-$M$^-$ (166). CFU-S activity enriches to an equal extent in the same fraction of cells, roughly 100–200 fold over unseparated bone marrow cells. Recently, we tested this fraction of bone marrow for pre–T cell activity and found a similar level of enrichment (G. Spangrude et al, manuscript submitted), demonstrating that the Thy-1^{lo}T$^-$B$^-$G$^-$M$^-$ population contains the pre-cursors for every hematopoietic lineage. Since this fraction of cells also reconstitutes lethally irradiated animals following a transfer of 300–500 cells, it contains self-renewing precursor activity as well. Whether these activities are due to single pluripotent of several types of multipotent hematopoietic stem cells is currently under investigation. If individual pluripotent stem cells exist, clearly they are contained within the Thy-1^{lo}T$^-$B$^-$G$^-$M$^-$ fraction of bone marrow and as such are contained within only 0.1–0.2% of the total population of bone marrow cells.

Commitment of Pre–T Cells in Bone Marrow

With the identification of a very small fraction of bone marrow which is highly enriched in thymus homing pre–T cell activity, it becomes interesting and important to ask whether precommitment to T-cell lineages or antigen receptor expression occurs prior to thymus homing. Several lines of evi-dence argue that some antigen receptor specificities are already expressed in pre–T populations in the bone marrow. Scott and coworkers (168) showed that bone marrow precursors of T helper cells could be rendered unresponsive to specific hapten carrier complexes, and this suggests the existence of antigen specific receptors on some bone marrow pre–T cells, although the transfer of tolerogen and/or suppressive cells to irradiated hosts is difficult to rule out in such experiments. Experiments in two other systems have demonstrated the same basic observation for anti-self class-I and class-II responses within the thymus (169, 170) and for anti-dinitrophenyl-modified self responses in peripheral tissues (171). In each case, the evidence presented suggested the expression of antigen-specific receptors by pre–T cells in the bone marrow. Although mRNA coding for the alpha and beta subunits of the T-cell receptor is not found in bone marrow cells depleted of mature T cells, the recent demonstration of a cell surface protein coded for by the gamma gene (86) may prove to be the key to this puzzle, since evidence exists for gamma rearrangement and expression prior to thymic homing (90).

Another clue to precommitment of pre–T cells in the bone marrow may come from a system which allows repopulation of the thymus with bone marrow precursors in a limiting dilution situation (172). In these experiments, clones of thymocytes which are the progeny of individual bone marrow–derived precursors can be characterized. The clones occupy discrete areas of the thymus, confined to cortical or medullary regions, and give rise to peripheral migrants. We have used this system to characterize the diversity of the progeny of a single bone marrow–derived precursor with regard to Lyt-2 and L3T4 expression. From our studies it seems clear that a bias for expression of either mature marker may occur in the progeny of a single clone, but in all cases so far mature lymphocytes of both L3T4$^+$ Lyt-2$^-$ and L3T4$^-$ Lyt-2$^+$ subclasses are detected (G. Spangrude and I. Weissman, in preparation). Therefore, it seems likely that commitments to these lineages occurs only intrathymically.

PRECURSOR-PROGENY RELATIONSHIPS AMONG THYMOCYTES

Based largely on phenotypic analyses, numerous models for intrathymic differentiation have been proposed. The major questions these proposals address are (a) Do single or multiple progenitor cells give rise to the L3T4$^+$ Lyt-2$^+$, L3T4$^+$ Lyt-2$^-$, and L3T4$^-$ Lyt-2$^+$ subpopulations? (b) Are one- or multiple-step pathways taken to generate each of these subpopulations? And (c) does the maturation of specific T-cell subsets occur in the cortex or medulla or both? Since the answers to these questions are still unknown, we do not attempt to give an overview here and refer the reader to several excellent reviews covering these areas (8–10, 12, 25). Instead, we focus on what has recently been learned about the phenotypic characteristics and proliferative and differentiative capacities of the intrathymic progenitor cell populations.

Differentiation of Fetal Intrathymic Progenitor Cells

Some of the first indications that the thymus contained a distinct population of progenitor cells capable of giving rise to the entire spectrum of thymocyte subsets came from analyses of thymic ontogeny. At day 14 of fetal development, the predominant cell type in the thymus is Thy-1$^+$, and it expresses low levels of the Ly-1 antigen (Ly-1^{lo} or dull Ly-1) (34, 173, 174). Cells expressing either L3T4 or Lyt-2 are undetectable at this stage. Within 2–3 days, however, cells of the Thy-1$^+$ Ly-1^{lo} L3T4$^-$ Lyt-2$^-$ phenotype decline dramatically in proportion concomitant with the appearance of cells expressing one or both of the L3T4 and Lyt-2 markers. By day 19 of fetal gestation, the proportions of thymocytes in each of the L3T4/Lyt-2 phenotypic classes have nearly attained adult levels. The development of mature function in the fetal thymus parallels this phenotypic progression. Helper activity, assayed by IL-2 secretion, is first evident at 19 gestational days (34). Cytotoxic activity lags somewhat behind, appearing in the thymus around birth (20 gestational days) (175, 176). Although the development of mature phenotype and function occurs over a very short time span (3–4 gestational days), it might be argued that the mature (and nonmature) cells arise not from the resident Thy-1$^+$ Ly-1^{lo} L3T4$^-$ Lyt-2$^-$ cells but from other cells entering the thymus during this period. Experiments with fetal thymus cultures make that possibility unlikely. Organ cultures initiated with 14-day fetal thymus lobes show essentially the same patterns of phenotypic and functional maturation as observed in vivo over a corresponding developmental time span (174, 177, 178). Since cell influx does not occur in these cultures, the Thy-1$^+$ Ly-1^{lo} L3T4$^-$ Lyt-2$^-$ population is a leading candidate to give rise to all subsequent populations.

That single, 14-day fetal thymocytes contain full progenitor capacity has been demonstrated elegantly by Kingston et al (179). In their experiments, individual fetal thymocytes were isolated and placed into hanging drop cultures together with portions of deoxyguanosine treated fetal thymus lobes. Over a period of 12 days, 2–3% of the cultures proliferated and produced cells expressing the L3T4 and Lyt-2 antigens. In all cultures, cells of the L3T4$^+$ Lyt-2$^+$ phenotype were present. Approximately half of the cultures also contained L3T4$^+$ Lyt-2$^-$ and L3T4$^-$ Lyt-2$^+$ cells. The remaining cultures showed L3T4$^-$ Lyt-2$^+$ cells but completely lacked the L3T4$^+$ Lyt-2$^-$ subpopulation. This observation might be due simply to a deficit in the culture conditions. Such an explanation would be consistent with the common observation that L3T4$^+$ Lyt-2$^-$ cells are more difficult to culture than L3T4$^-$ Lyt-2$^+$ cells. An alternative possibility, and one which we favor (see below), is that the cultures were initiated by different types of intrathymic progenitor cells—those which give rise to L3T4$^+$ Lyt-2$^+$ and L3T4$^-$ Lyt-2$^+$ but not L3T4$^+$ Lyt-2$^-$ cells, and those which give rise to all three phenotypic types.

Fetal-Like Progenitor Cells in the Adult Thymus

The adult thymus contains a small population of cells which bear a striking phenotypic resemblance to day-14 fetal thymocytes. Approximately 3% of adult thymocytes are L3T4$^-$ Lyt-2$^-$, and most of these are also Ly-1^{lo} (8–10, 12, 25). This small population of adult thymocytes also resembles fetal thymocytes in the expression of three other antigens. As discussed above, some of the L3T4$^-$ Lyt-2$^-$ cells express the Pgp-1 marker and/or the IL-2 receptor and/or bear high levels of the MEL-14 antigen (20, 27, 28, 180–182). Whether an individual L3T4$^-$ Lyt-2$^-$ cell has only one or a combination of these antigens on the cell surface is not known. All three of these antigens are expressed on the majority of fetal thymocytes between day 14 and 16 of gestation, subsequently declining to their low adult levels (1–5% of total thymocytes) by day 19 (27, 28, 30, 32, 180, 182, 183, 185). In addition to these phenotypic similarities, both fetal thymocytes and IL-2 receptor bearing L3T4$^-$ Lyt-2$^-$ adult thymocytes can proliferate in vitro, but require stimulation with IL-2 plus con A (28, 30, 184). Therefore, the thymus appears to retain into adulthood a small population of cells that, by most measures, are strikingly similar to 14-day fetal thymocytes.

Differentiation of Adult Intrathymic Progenitor Cells

The similarity of L3T4$^-$ Lyt-2$^-$ thymocytes to fetal thymocytes suggested that this population might contain at least some of the intrathymic progenitors in the adult thymus. The methods for testing this hypothesis directly were adopted from earlier reports which showed that the thymuses of irradiated animals were transiently repopulated by donor cells following intravenous (i.v.) injection of large numbers of total thymocytes (162, 186). The obvious question was: Among total thymocytes, are the repopulating cells contained within the L3T4$^-$ Lyt-2$^-$ population? Using a similar assay system, Lesley et al (181) showed that the transient repopulation of the thymus was inhibited by pretreatment of the injected thymocytes with Pgp-1-specific antibody plus complement. Since Pgp-1$^+$ thymocytes are almost entirely L3T4$^-$ Lyt-2$^-$, these experiments suggested that at least some of the L3T4$^-$ Lyt-2$^-$ cells had both thymus-homing and progenitor activity. In fact, within the L3T4$^-$ Lyt-2$^-$ population, Pgp-1 may mark the most immature cells; Pgp-1$^+$ cells are enriched for cells with a non-rearranged TCR beta-chain gene (97) whereas the L3T4$^-$ Lyt-2$^-$ population as a whole apparently contains some cells which have rearranged the β-chain locus (92, 94).

In a more direct approach, Fowlkes et al (33) prepared thymocytes enriched for L3T4$^-$ Lyt-2$^-$ Lyt-1^{lo} cells, injected these cells i.v. into lethally irradiated, congenic hosts, and analyzed the colonization of the thymus by these cells and their progeny. Within 7 days, donor cells were detected in the thymus, and by 14 days, these cells had proliferated and differentiated to all the L3T4/Lyt-2 phenotypic types. While these experiments are important for clearly demonstrating intrathymic progenitor activity among L3T4$^-$ Lyt-2$^-$ Ly-1^{lo} cells, the interpretations are limited by the experimental design. For example, since the donor cells had to home to the thymus, it is not known what proportion (or which subset) of the injected cells contributed to the repopulation, or whether the sequence of repopulation and differentiation was affected by this constraint.

In our laboratory, experiments were designed to follow the intrathymic differentiation of L3T4$^-$ Lyt-2$^-$ cells independent of their ability to home to the thymus. In these experiments, highly purified L3T4$^-$ Lyt-2$^-$ cells were injected directly into the thymuses of nonirradiated, congenic hosts and donor cells and their progeny were analyzed at various times after injection (B. Adkins, I. L. Weissman, manuscript submitted). Even in the unirradiated host, L3T4$^-$ Lyt-2$^-$ cells proliferate vigorously, attaining levels 40- to 50-fold higher than the actual injected number within 5 days of injection. Strikingly, the acquisition of the L3T4 and Lyt-2 antigens is very rapid. Donor cells expressing these antigens are readily detected at 24 hr, and by 6 days after injection, the proportions of donor cells in each L3T4/Lyt-2 subpopulation attain normal, steady state levels. In the unirradiated thymus, therefore, the differentiation of L3T4$^-$ Lyt-2$^-$ adult thymocytes occurs over a short time span, similar to that required for fetal thymocyte differentiation.

In our experiments, donor L3T4$^-$ Lyt-2$^-$ cells showed several patterns of L3T4 and Lyt-2 expression at early times after intrathymic injection. Specifically, a high proportion of the cells appearing 1 day after injection were of "mature" phenotype, L3T4$^+$ Ly2$^-$ or Ly2$^+$ L3T4$^-$. The number of Ly2$^+$ L3T4$^+$ cells in these mice was so low that it is unlikely that "mature" phenotype cells obligatorily pass through the L3T4$^+$ Ly2$^+$ stage, although it is conceivable that the single positive state represents a precursor of double positive cells. For a number of reasons, different patterns seem not to be due to variations in experimental regimen but, rather, appear to reflect the range of short-term differentiation events by the L3T4$^-$ Lyt-2$^-$ population as a whole. The interpretation we favor is that the variety of early outcomes is representative of a heterogeneous L3T4$^-$ Lyt-2$^-$ progenitor cell pool. That is, contained within the L3T4$^-$ Lyt-2$^-$

population are several different types of progenitor cells whose relative proliferative and differentiative advantages may differ at early times after injection. Within 4–6 days of injection, the normal ratios of each of these progenitor cell types have been restored, and constant patterns of L3T4 and Lyt-2 expression appear. The possibility that the L3T4$^-$ Lyt-2$^-$ population is functionally heterogeneous is certainly consistent with the heterogeneous expression by this population of other cell surface markers (20, 27, 28, 180–182). The idea that more than one type of progenitor cell is found in the L3T4$^-$ Lyt-2$^-$ population might be tested with several approaches. Further fractionation of the L3T4$^-$ Lyt-2$^-$ population into e.g. Pgp-1$^+$, IL-2 receptor positive, or MEL-14^{hi} subpopulations or, alternatively, the cloning of the total population in vivo by intrathymic injection may help to determine whether commitment to particular T-cell function or phenotype has already occurred among subsets of L3T4$^-$ Lyt-2$^-$ cells.

In summary, the L3T4$^-$ Lyt-2$^-$ thymocyte population, as a whole, contains both thymus-homing and repopulating activities. However, the proportion of cells within this population which contributes to each of these activities is not known. It is also not clear whether a single L3T4$^-$ Lyt-2$^-$ cell gives rise to all of the phenotypic subsets in the thymus or whether individual L3T4$^-$ Lyt-2$^-$ cells are precommitted to produce progeny of some, but not all, phenotypes. Nonetheless, the identification of L3T4$^-$ Lyt-2$^-$ cells as at least one type of intrathymic progenitor cell population is an important step toward defining the pathway(s) of intrathymic maturation.

EPILOGUE: MOVING TOWARD CERTAINTY

What we do is colored by the framework in which we do it. For over 100 years research on the thymus has been directed, and often misdirected, by the scientific expectations and standards of contemporary relevant science. We have passed from the time that the thymus was felt to give rise to the lower limbs or was the imminent cause of crib death. The recent era of the study of the thymus has drawn largely from contemporary immunobiology, with its emphasis on the genetics and tissue-specific expression of cell-surface proteins. The power of these genetic tools, coupled with precise technological tools such as the fluorescence-activated cell sorter, has allowed the description of many distinguishable subpopulations of thymic lymphocytes, prethymic bone marrow cells, and postthymic T cells. Unfortunately, correlations between phenotypes of cells in different locations has often substituted for a precise and unequivocal demonstration of progenitor-progeny relationships within a cell lineage, much in the way that early histologists used morphological similarities to make (often incorrect) determinations of broader developmental lineages. The need to establish unequivocal lineages in lymphopoiesis from experiment, not correlation, is immediately apparent when one considers that in the primary sites of generation of B and T cells—the bone marrow and thymus— $\sim 99\%$ of the lymphoid cells are destined to die without maturing.

How does one establish unequivocal lineages? To answer this question we must take our standards from the best of contemporary developmental and cell biology. First, one must obtain defined subsets of independently marked cells of sufficient purity and viability that, when placed in appropriate microenvironments, their short- and long-term lineal transitions can be recorded. Independent markers such as genetic allotypes, non-reutilizable labels, allotypic DNA restriction polymorphisms, inserted retroviral or other transfected genetic elements, and clonal antigen receptor gene rearrangements have provided abundant tools to identify cells. Second, one must be able to identify clonogenic cells at each stage of lineal progression, in order to define the developmental commitments that have occurred during that progression. Third, we must identify and be able to use in vitro these microenvironments and/or cloned growth and differentiation factors that impact on lymphocyte development. Our studies on the thymus remain necessarily vague and phenomenological due to our inability to reconstruct and dissect the thymic microenvironment in vitro, and therefore the second and third requirements have remained elusive goals. But not much longer. The identification of thymic stromal elements is at hand; their transfer to culture and their subsequent cloning must surely follow. Then we can hope to move from models toward certainty.

Literature Cited

1. Fink, P. J., Bevan, M. J. 1978. H-2 antigen of the thymus determines lymphocyte specificity. *J. Exp. Med.* 148 : 766
2. Sprent, J. 1978. Role of H-2 gene products in the function of T helper cells from normal and chimeric mice *in vivo*. *Immunol. Rev.* 42 : 108
3. Zinkernagel, R. M. 1978. Thymus and lymphohaemopoietic cells: Their role in T-cell maturation in selection of T cells' H-2 restriction specificity and H-2 linked gene control. *Immunol. Rev.* 42 : 224
4. Dialynas, D. P., Wilde, D. B., Marrack, P., Pierres, A., Wall, K. A., Havran, W., Otten, G., Loken, M. R., Pierres, M., Kappler, J., Fitch, F. W. 1983. Characterization of the murine antigenic determinant, designated L3T4a, recognized by monoclonal antibody GK1.5: Expression of L3T4a by functional T cell clones appears to correlate primarily with class II MHC antigen-reactivity. *Immunol. Rev.* 74 : 29
5. Swain, S. L. 1983. T cell subsets and the recognition of MHC class. *Immunol. Rev.* 74 : 129
6. Gallatin, W. M., Weissman, I. L., Butcher, E. C. 1983. A cell-surface molecule involved in organ-specific homing of lymphocytes. *Nature* 304 : 30
7. Reichert, R. A., Gallatin, W. M., Butcher, E. C., Weissman, I. L. 1984. A homing receptor-bearing cortical thymocyte subset: Implications for thymus cell migration and the nature of cortisone-resistant thymocytes. *Cell* 38 : 89
8. Scollay, R., Shortman, K. 1983.

Thymocyte subpopulations: An experimental review, including flow cytometric cross-correlations between the major murine thymocyte markers. *Thymus* 5: 245

9. Mathieson, B. J., Fowlkes, B. J. 1984. Cell surface antigen expression on thymocytes: Development and phenotypic differentiation of intrathymic subsets. *Immunol. Rev.* 82: 141

10. Scollay, R., Bartlett, P., Shortman, K. 1984. T cell development in the adult murine thymus: Changes in the expression of the surface antigens Ly2, L3T4, and B2A2 during development from early precursor cells to emigrants. *Immunol. Rev.* 82: 79

11. Ceredig, R., MacDonald, H. R. 1985. Intrathymic differentiation: Some unanswered questions. *Surv. Immunol. Res.* 4: 87

12. Rothenberg, E., Lugo, J. P. 1985. Differentiation and cell division in the mammalian thymus. *Dev. Biol.* 112: 1

13. Ceredig, R., Glasebrook, A. L., MacDonald, H. R. 1982. Phenotypic and functional properties of murine thymocytes. I. Precursors of cytotoxic T lymphocytes and Interleukin-2 producing cells are all contained within a subpopulation of "mature" thymocytes as analyzed by monoclonal antibodies and flow micro-fluorometry. *J. Exp. Med.* 155: 358

14. Chen, W. F., Scollay, R., Shortman, K. 1982. The functional capacity of thymus subpopulations: Limit-dilution analysis of all precursors of cytotoxic lymphocytes and of all T cells capable of proliferation in subpopulations separated by the use of peanut agglutinin. *J. Immunol.* 129: 18

15. Scollay, R. 1982. Thymus cell migration: Cells migrating from thymus to peripheral lymphoid organs have a "mature" phenotype. *J. Immunol.* 128: 1566

16. Scollay, R., Chen, W.-F., Shortman, K. 1984. The functional capabilities of cells leaving the thymus. *J. Immunol.* 132: 25

17. Blomgren, H., Andersson, B. 1971. Characteristics of the immunocompetent cells in the mouse thymus: Cell population changes during cortisone-induced atrophy and subsequent regeneration. *Cell. Immunol.* 1: 545

18. Blomgren, H., Andersson, B. 1972. Recirculating lymphocytes in the mouse thymus are part of the relatively cortisone resistant cell population. *Clin. Exp. Immunol.* 10: 297

19. Fink, P. J., Gallatin, W. M., Reichert, R. A., Butcher, E. C., Weissman, I. L. 1985. Homing receptor-bearing thymocytes, an immunocompetent cortical subpopulation. *Nature* 313: 233

20. Reichert, R. A., Weissman, I. L., Butcher, E. C. 1986. Phenotypic analysis of thymocytes that express homing receptors for peripheral lymph nodes. *J. Immunol.* 136: 3521

21. Shortman, K., Mandel, T., Andrews, P., Scollay, R. 1985. Are any functionally mature cells of medullary phenotype located in the thymus cortex? *Cell. Immunol.* 93: 350

22. Blomgren, H., Andersson, B. 1969. Evidence for a small pool of immunocompetent cells in the mouse thymus. *Exp. Cell. Res.* 57: 185

23. Reichert, R. A., Weissman, I. L., Butcher, E. C. 1986. Dual immunofluorescence studies of cortisone-induced thymic involution: Evidence for a major cortical component to cortisone-resistant thymocytes. *J. Immunol.* 136: 3529

24. Scollay, R., Shortman, K. 1984. Cell traffic in the adult thymus. Cell entry and exit, cell birth and death. In *Recognition and Regulation in Cell-Mediated Immunity*, ed. J. D. Watson, J. Marbrook, p. 3. New York/Basel: Dekker

25. Shortman, K., Scollay, R. 1984. Cortical and medullary thymocytes. See Ref. 24, p. 31.

26. Metcalf, D. 1966. *The Thymus: Its Role in Immune Responses, Leukemia Development and Carcinogenesis*. New York: Springer-Verlag

27. Ceredig, R., Lowenthal, J. W., Nabholz, M., MacDonald, H. R. 1985. Expression of Interleukin-2 receptors as a differentiation marker on intrathymic stem cells. *Nature* 314: 98

28. Raulet, D. H. 1985. Expression and function of Interleukin-2 receptors on immature thymocytes. *Nature* 314: 101

29. Smith, K. A. 1984. Interleukin 2. *Ann. Rev. Immunol.* 2: 319

30. von Boehmer, H., Crisanti, A., Kisielow, P., Haas, W. 1985. Absence of growth by most receptor expressing fetal thymocytes in the presence of Interleukin-2. *Nature* 314: 539

31. Trowbridge, I. S., Lesley, J., Schulte, J., Hyman, R., Trotter, J. 1982. Biochemical characterization and cellular distribution of a polymorphic, murine cell-surface glycoprotein expressed on lymphoid tissues. *Immunogenetics* 15: 299

32. Lesley, J., Trotter, J., Hyman, R. 1985. The Pgp-1 antigen is expressed on early

fetal thymocytes. *Immunogenetics* 22: 149

33. Fowlkes, B. J., Edison, L., Mathieson, B. J., Chused, T. M. 1985. Early T lymphocytes. Differentiation *in vivo* of adult intrathymic precursor cells. *J. Exp. Med.* 162: 802

34. Ceredig, R., Dialynas, D. P., Fitch, F. W., MacDonald, M. R. 1983. Precursors of T cell growth factor producing cells in the thymus. Ontogeny, frequency and quantitative recovery in a subpopulation of functionally mature thymocytes defined by monoclonal antibody GK1.5. *J. Exp. Med.* 158: 1654

35. Kronenberg, M., Siu, G., Hood, L., Shastri, N. 1986. The molecular genetics of the T cell antigen receptor and T cell antigen recognition. *Ann. Rev. Immunol.* 4: 529

36. Marrack, P., Kappler, J. 1986. The antigen-specific, major histocompatibility complex-restricted receptor on T cells. *Adv. Immunol.* 38: 1

37. Kindred, B., Shreffler, D. C. 1972. H-2 dependence of cooperation between T and B cell in vivo. *J. Immunol.* 109: 940

38. Katz, D. H., Hamaoka, T., Dorf, M. E., Maurer, P. H., Benacerraf, B. 1973. Cell interactions between histoincompatible T and B lymphocytes. IV. Involvement of the immune response (Ir) gene in the control of lymphocyte interactions in responses controlled by the gene. *J. Exp. Med.* 1: 138: 734

39. Yague, J., White, J., Coleclough, C., Kappler, J., Palmer, E., Marrack, P. 1985. The T cell receptor: The alpha and beta chains define idiotype, and MHC specificity. *Cell* 42: 81

40. Ohashi, P. S., Mak, T. W., Van den Elsen, P., Yanagi, Y., Yoshikai, Y., Calman, A. F., Terhorst, C., Stobo, J. D., Weiss, A. 1985. Reconstitution of an active surface T3/T-cell antigen receptor by DNA transfer. *Nature* 316: 606

41. Dembic, Z., Werner, H., Weiss, S., McCubrey, J., Kiefer, H., von Boehmer, H., Steinmetz, M. 1986. Transfer of specificity by murine alpha and beta T cell receptor genes. *Nature* 320: 232

42. Fathman, C. G., Frelinger, J. G. 1983. T-lymphocyte clones. *Ann. Rev. Immunol.* 1: 633

43. Allison, J., McIntyre, B., Block, D. 1982. Tumor-specific antigen of murine T lymphoma defined with monoclonal antibody. *J. Immunol.* 129: 2293

44. Haskins, K., Kubo, R., White, J., Pigeon, M., Kappler, J., Marrack, P. 1983. The major histocompatibility complex-restricted antigen receptor on T cells. I. Isolation with a monoclonal antibody. *J. Exp. Med.* 157: 1149

45. Samelson, L. E., Germain, R. N., Schwartz, R. H. 1983. Monoclonal antibodies against the antigen receptor on a cloned T-cell hybrid. *Proc. Natl. Acad. Sci. USA* 80: 6972

46. Kaye, J., Porcelli, S., Tite, J., Jones, B., Janeway, C. A. Jr. 1983. Both a monoclonal antibody and antisera specific for determinants unique to individual cloned helper T cell lines can substitute for antigen and antigen-presenting cells in the activation of T cells. *J. Exp. Med.* 158: 836

47. Kappler, J., Kubo, R., Haskins, K., Hannum, C., Marrack, P., Pigeon, M., McIntyre, B., Allison, J., Trowbridge, I. 1983. The major histocompatibility complex-restricted antigen receptor on T cells in mouse and man: Identification of constant and variable peptides. *Cell* 35: 295

48. Weiss, A., Imboden, J., Hardy, K., Manger, B., Terhorst, C., Stobo, J. 1986. The role of the T3/antigen receptor complex in T-cell activation. *Ann. Rev. Immunol.* 4: 593

49. Weiss, A., Stobo, J. D. 1984. Requirement for the coexpression of T3 and the T cell antigen receptor on a malignant human T cell line. *J. Exp. Med.* 160: 1284

50. van den Elsen, P., Shepley, B. A., Cho, M., Terhorst, C. 1985. Isolation and characterization of a cDNA clone encoding the murine homologue of the human 20K T3/T-cell receptor glycoprotein. *Nature* 314: 542

51. Allison, J. P., Lanier, L. L. 1985. Identification of antigen receptor-associated structures on murine T cells. *Nature* 314: 107

52. Samelson, L. E., Harford, J. B., Klausner, R. D. 1985. Identification of the components of the murine T cell antigen receptor complex. *Cell* 43: 223

53. Oettgen, H. C., Pettey, C. L., Maloy, W. L., Terhost, C. 1986. A T3-like protein complex associated with the antigen receptor on murine T cells. *Nature* 320: 272

54. Hedrick, S. M., Cohen, D. I., Nielsen, E. A., Davis, M. M. 1984. Isolation of cDNA clones encoding T cell-specific membrane-associated proteins. *Nature* 308: 149

55. Yanagi, Y., Yoshikai, Y., Leggett, K., Clark, S. P., Aleksander, I., Mak, T. W. 1984. A human T cell-specific cDNA clone encodes a protein having exten-

360 ADKINS ET AL

sive homology to immunoglobulin chains. *Nature* 308 : 145

56. Lee, N. E., D'Eustachio, P., Pravtcheva, D., Ruddle, F. H., Hedrick, S. M., Davis, M. M. 1984. Murine T cell receptor beta chain is encoded on chromosome 6. *J. Exp. Med.* 160 : 905

57. Caccia, N., Kronenberg, M., Saxe, D., Haars, R., Bruns, G. A., Goverman, J., Malissen, M., Willard, H., Yoshikai, Y., Simon, M., Hood, L., Mak, T. W. 1984. The T cell receptor beta chain genes are located on chromosome 6 in mice and chromosome 7 in humans. *Cell* 37 : 1091

58. Patten, P., Yokota, T., Rothbard, J., Chien, Y., Arai, K., Davis, M. M. 1984. Structure, expression and divergence of T-cell receptor beta-chain variable regions. *Nature* 312 : 40

59. Barth, R. K., Kim, B. S., Lan, N. C., Hunkapiller, T., Sobieck, N., Winoto, A., Gershenfeld, H., Okada, C., Hansburg, D., Weissman, I. L., Hood, L. 1985. The murine T-cell receptor uses a limited repertoire of expressed V beta gene segments. *Nature* 316 : 517

60. Behlke, M. A., Spinella, D. G., Chou, H. S., Sha, W., Hartl, D. L., Loh, D. Y. 1985. T-cell receptor beta-chain expression : Dependence on relatively few variable region genes. *Science* 229 : 566

61. Behlke, M. A., Chou, H. S., Huppi, K., Loh, D. Y. 1986. Murine T-cell receptor mutants with deletions of betachain variable region genes. *Proc. Natl. Acad. Sci. USA* 83 : 767

62. Gascoigne, N. R., Chien, Y., Becker, D. M., Kavaler, J., Davis, M. M. 1984. Genomic organization and sequence of T-cell receptor beta-chain constant-and joining-region genes. *Nature* 310 : 387

63. Malissen, M., Minard, K., Mjolsness, S., Kronenberg, M., Goverman, J. Hunkapiller, T., Prystowsky, M. B., Yoshikai, Y., Fitch, F., Mak, T. W., Hood, L. 1984. Mouse T cell antigen receptor : Structure and organization of constant and joining gene segments encoding the beta polypeptide. *Cell* 37 : 1101

64. Siu, G., Kronenberg, M., Strauss, E., Haars, R., Mak, T. W., Hood, L. 1984. The structure, rearrangement and expression of D beta gene segments of the murine T-cell antigen receptor. *Nature* 311 : 344

65. Kavalar, J., Davis, M. M., Chien, Y. 1984. Localization of a T-cell receptor diversity-region element. *Nature* 310 : 421

66. Goverman, J., Minard, K., Shastri, N.,

Hunkapiller, T., Hansburg, D., Sercarz, E., Hood, L. 1985. Rearranged beta T cell receptor genes in a helper T cell clone specific for lysozyme : No correlation between V beta and MHC restriction. *Cell* 40 : 859

67. Rupp, F., Acha-Orbea, H., Hengartner, H., Zinkernagel, R., Joho, R. 1985. Identical V beta T-cell receptor genes used in alloreactive cytotoxic and antigen plus I-A specific helper T cells. *Nature* 315 : 425

68. Reth, M. G., Alt, F. W. 1984. Novel immunoglobulin heavy chains are produced from DJ$_H$ gene segment rearrangements in lymphoid cells. *Nature* 319 : 418

69. Chien, Y., Becker, D. M., Lindsten, T., Okamura, M., Cohen, D. I., Davis, M. M. 1984. A third type of murine T-cell receptor gene. *Nature* 312 : 31

70. Saito, H., Kranz, D. M., Takagaki, Y., Hayday, A. C., Eisen, H. N., Tonegawa, S. 1984. A third rearranged and expressed gene in a clone of cytotoxic T lymphocytes. *Nature* 312 : 36

71. Dembic, Z., Bannwarth, W., Taylor, B. A., Steinmetz, M. 1985. The gene encoding the T-cell receptor alpha-chain maps close to the Np-2 locus on mouse chromosome 14. *Nature* 314 : 271

72. Kranz, D. M., Saito, H., Disteche, C. M., Swisshelm, K., Pravtcheva, D., Ruddle, F. H., Eisen, H. N., Tonegawa, S. 1985. Chromosomal locations of the murine T-cell receptor alpha-chain gene and the T-cell gamma gene. *Science* 227 : 941

73. Hayday, A. C., Diamond, D. J., Tanigawa, G., Heilig, J. S., Folsom, V., Saito, H., Tonegawa, S. 1985. Unusual organization and diversity of T-cell receptor alpha-chain genes. *Nature* 316 : 828

74. Winoto, A., Mjolsness, S., Hood, L. 1985. Genomic organization of the genes encoding mouse T-cell receptor alpha chain. *Nature* 316 : 832

75. Arden, B., Klotz, J. L., Siu, G., Hood, L. E. 1985. Diversity and structure of genes of the alpha family of mouse T-cell antigen receptor. *Nature* 316 : 783

76. Becker, D. M., Patten, P., Chien, Y., Yokota, T., Eshhar, Z., Giedlin, M., Gascoigne, N. R., Goodnow, C., Wolf, R., Arai, K., Davis, M. M. 1985. Variability and repertoire size of T-cell receptor V alpha gene segments. *Nature* 317 : 430

77. Saito, H., Kranz, D. M., Takagaki, Y., Hayday, A. C., Eisen, H. N., Tonegawa, S. 1984. Complete primary structure of a heterodimeric T-cell re-

ceptor deduced from cDNA sequences. *Nature* 309: 757

78. Hayday, A. C., Saito, H., Gillies, S. D., Kranz, D. M., Tanigawa, G., Eisen, H. N., Tonegawa, S. 1985. Structure, organization, and somatic rearrangement of T cell gamma genes. *Cell* 40: 259

79. Garman, R. D., Doherty, P. J., Raulet, D. H. 1986. Diversity, rearrangement and expression of murine T cell gamma genes. *Cell* 45: 733

80. Heilig, J. S., Tonegawa, S. 1986. Diversity of the murine gamma genes and their expression in fetal and adult T lymphocytes. *Nature* 322: 836

81. Iwamoto, A., Rupp, F., Ohashi, P. S., Walker, C. L., Pircher, H., Joho, R., Hengartner, H., Mak, T. W. 1986. T cell specific gamma genes in C57Bl/10 mice. *J. Exp. Med.* 163: 1203

82. Heilig, J. S., Glimcher, L. H., Kranz, D. M., Clayton, L. K., Greenstein, J. L., Saito, H., Maxam, A. M., Burakoff, S. J., Eisen, H. N., Tonegawa, S. 1985. Expression of the T cell specific gamma gene is unnecessary in T cells recognizing class II MHC determinants. *Nature* 317: 68

83. Kranz, D. M., Saito, H., Heller, M., Takagaki, Y., Haas, W., Eisen, H. N., Tonegawa, S. 1985. Limited diversity of the rearranged T-cell gamma gene. *Nature* 313: 752

84. Raulet, D. H., Garman, R. D., Saito, H., Tonegawa, S. 1985. Developmental regulation of T-cell receptor gene expression. *Nature* 314: 103

85. Snodgrass, H. R., Dembi, C. Z., Steinmetz, M., von Boehmer, H. 1985. Expression of T-cell antigen receptor genes during fetal development in the thymus. *Nature* 315: 232

86. Brenner, M. B., McLean, J., Deno, D. P., Strominger, J. L., Smith, J. A., Owen, F. L., Seidman, J. G., Ip, S., Rosen, F., Krangel, M. S. 1986. Identification of a putative second T-cell receptor. *Nature* 322: 145

87. Bank, I., DePinho, R. A., Brenner, M. B., Cassimeris, J., Alt, F. W., Chess, L. 1986. A functional T3 molecule associated with a novel heterodimer on the surface of immature human thymocytes. *Nature* 322: 179

88. Chien, Y. H., Gascoigne, N. R., Kavaler, J., Lee, N. E., Davis, M. M. 1984. Somatic recombination in a murine T-cell receptor gene. *Nature* 309: 322

89. Pernis, B., Axel, R. 1985. A one and a half receptor model for the MHC-restricted antigen recognition by T lymphocytes. *Cell* 41: 13

90. Haars, R., Kronenberg, M., Gallatin, W. M., Weissman, I. L., Owen, F. L., Hood, L. 1986. Rearrangement and expression of T-cell antigen receptor and gamma genes during thymic development. *J. Exp. Med.* 164: 1

91. Deleted in proof

92. Born, W., Yague, J., Palmer, E., Kappler, J., Marrack, P. 1985. Rearrangement of T-cell receptor beta-chain genes during T-cell development. *Proc. Natl. Acad. Sci. USA* 82: 2925

93. Snodgrass, H. R., Kisielow, P., Kiefer, M., Steinmetz, M., von Boehmer, H. 1985. Ontogeny of the T-cell antigen receptor within the thymus. *Nature* 313: 592

94. Samelson, L. E., Lindsten, T., Fowlkes, B. J., van den Elsen, P., Terhorst, C., Davis, M. M., Germain, R. N., Schwartz, R. H. 1985. Expression of genes of the T-cell antigen receptor complex in precursor thymocytes. *Nature* 315: 765

95. Roehm, N., Herron, L., Cambier, J., DiGuisto, D., Haskins, K., Kappler, J., Marrack, P. 1984. The major histocompatibility complex-restricted antigen receptor on T cells: Distribution on thymus and peripheral T cells. *Cell* 38: 577

96. Sim, G. K., Augustin, A. A. 1985. V beta gene polymorphism and a major polyclonal T cell receptor idiotype. *Cell* 42: 89

97. Trowbridge, I. S., Lesley, J., Trotter, J., Hyman, R. 1985. Thymocyte subpopulation enriched for progenitors with an unrearranged T-cell receptor beta-chain gene. *Nature* 315: 666

98. Richie, E. R., McIntire, B., Guyden, J., Lanier, L., Allison, J. P. Expression of T-cell receptor α and β-chain genes in phenotypically characterized murine T-lymphomas: Implications for thymocyte differentiation. *J. Exp. Med.* Submitted

99. Farr, A. G., Anderson, S. K., Marrack, P., Kappler, J. 1985. Expression of antigen-specific, major histocompatibility complex-restricted receptors by cortical and medullary thymocytes *in situ*. *Cell* 43: 543

100. ven Ewijk, W. 1984. Immunohistology of lymphoid and non-lymphoid cells in the thymus in relation to T lymphocyte differentiation. *Am. J. Anat.* 170: 311

101. Rouse, R. V., van Ewijk, W., Jones, P. P., Weissman, I. L. 1979. Expression of MHC antigens by mouse thymic dendritic cells. *J. Immunol.* 122: 2508

102. van Ewijk, W., Rouse, R. V., Weiss-

man, I. L. 1980. Distribution of H-2 microenvironments in the mouse thymus. *J. Histochem. Cytochem.* 28 : 1089

103. Barclay, N. A., Mayrhofer, G. 1981. Bone marrow origin of Ia-positive cells in the medulla of rat thymus. *J. Exp. Med.* 153 : 1666

104. Duijvestijn, A. M., Schutte, R., Kohler, Y. G., Korn, G., Hoefsmit, E. C. M. 1983. Characterization of the population of phagocytic cells in thymic cell suspensions. A morphological and cytochemical study. *Cell Tissue Res.* 231 : 313

105. Tidmarsh, G. F., Dailey, M. O., Whitlock, C. A., Pillemer, E., Weissman, I. L. 1985. Transformed lymphocytes from Abelson diseased mice express levels of a B lineage transformation-associated antigen elevated from that found on normal lymphocytes. *J. Exp. Med.* 162 : 1421

106. Van Vliet, E., Mells, M., van Ewijk, W. 1984. Monoclonal antibodies to stromal cell types of the mouse thymus. *Eur. J. Immunol.* 14 : 524

107. de Maagd, R. A., Mackenzie, W. A., Schuurman, H.-J., Ritter, M. A., Price, K. M., Broekhuizen, R., Kater, L. 1985. The human thymus microenvironment: Heterogeneity detected by monoclonal anti-epithelial cell antibodies. *Immunol.* 54 : 745

108. Kaneshima, H., Asai, J., Hiai, H., Nishizuka, Y. 1985. The structural organization of thymic microenvironments. *Clin. Immunol.* 17 : 439

109. Wekerle, H., Ketelsen, Y.-P. 1980. Thymic nurse cells. Ia-bearing epithelium involved in T-lymphocyte differentiation. *Nature* 283 : 402

110. Wekerle, H., Ketelsen, U.-P., Ernst, M. 1980. Thymic nurse cells: Lymphoepithelial cell complexes in murine thymuses: Morphological and serological characterization. *J. Exp. Med.* 151 : 925

111. Kyewski, B. A., Kaplan, H. S. 1982. Lymphoepithelial interactions in the mouse thymus: Phenotypic and kinetic studies on thymic nurse cells. *J. Immunol.* 128 : 2287

112. Van Vliet, E., Mells, M., van Ewijk, W. 1984. Immunohistology of thymic nurse cells. *Cell. Immunol.* 87 : 101

113. Andrews, P., Boyd, R. L. 1985. The murine thymic nurse cell: An isolated thymic microenvironment. *Eur. J. Immunol.* 15 : 36

114. Fink, P. J., Weissman, I. L., Kaplan, H. S., Kyewski, B. A. 1984. The immunocompetence of murine stromal cell associated thymocytes. *J. Immunol.* 132 : 2266

115. Vakharia, D. D., Mitchison, N. A. 1984. Helper T cell activity demonstrated by thymic nurse T cells (TNC-T). *Immunology* 51 : 269

116. Andrews, P., Boyd, R. L., Shortman, K. 1985. The limited immunocompetence of thymocytes within murine thymic nurse cells. *Eur. J. Immunol.* 15 : 1043

117. Kyewski, B. A., Rouse, R. V., Kaplan, H. S. 1982. Thymocyte rosettes: Multicellular complexes of lymphocytes and bone marrow-derived stromal cells in the mouse thymus. *Proc. Natl. Acad. Sci. USA* 79 : 5646

118. Kyewski, B. A., Fathman, C. G., Kaplan, H. S. 1984. Intrathymic presentation of circulating non-major histocompatibility complex antigens. *Nature* 308 : 196

119. Kyewski, B. A., Fathman, C. G., Rouse, R. V. 1986. Intrathymic presentation of circulating non-MHC antigens by medullary dendritic cells. An antigen-dependent microenvironment for T cell differentiation. *J. Exp. Med.* 163 : 231

120. Fink, P. J., Bevan, M. J., Weissman, I. L. 1984. Thymic cytotoxic T lymphocytes are primed *in vivo* to minor histocompatibility antigens. *J. Exp. Med.* 159 : 436

121. Naparstek, Y., Holoshitz, J., Eisenstein, S., Reshef, T., Rappaport, S., Chemke, J., Ben-Nun, A., Cohen, I. 1982. Effector T lymphocyte line cells migrate to the thymus and persist there. *Nature* 300 : 262

122. Owen, J. J. T., Jenkinson, E. J. 1984. Early events in T lymphocyte genesis in the fetal thymus. *Am. J. Anat.* 170 : 301

123. Clark, S. L. 1963. The thymus in mice of strain 129/J, studied with the electron microscope. *Am. J. Anat.* 112 : 1

124. Kingston, R., Jenkinson, E. J., Owen, J. J. T. 1984. Characterization of stromal cell populations in the developing thymus of normal and nude mice. *Eur. J. Immunol.* 14 : 1052

125. Jenkinson, E. J., van Ewijk, W., Owen, J. J. T. 1981. Major histocompatibility antigen expression on the epithelium of the developing thymus in normal and nude mice. *J. Exp. Med.* 153 : 280

126. Van Vliet, E., Jenkinson, E. J., Kingston, R., Owen, J. J. T., van Ewijk, W. 1985. Stromal cell types in the developing thymus of the normal and nude mouse embryo. *Eur. J. Immunol.* 15 : 675

127. Kindred, B. 1979. Nude mice in immunology. *Prog. Allergy* 26 : 137

128. Chen, W.-F., Scollay, R., Shortman, K., Skinner, M., Marbrook, J. 1984. T-cell development in the absence of a thymus: The number, the phenotype, and the functional capacity of T lymphocytes in nude mice. *Am. J. Anat.* 170: 339

129. Doenhoff, M. J., Leuchars, E., Kerbel, R. S., Wallis, V., Davies, A. J. S. 1979. Adult and pre-adult thymectomy of mice: Contrasting effects on immune responsiveness, and on numbers of mitogen-responsive and Thy-1$^+$ lymphocytes. *Immunology* 37: 397

130. Jenkinson, E. J., Franchi, L. L., Kingston, R., Owen, J. J. T. 1982. Effect of deoxyguanosine on lymphopoiesis in the developing thymus rudiment *in vitro*: application in the production of chimeric thymus rudiments. *Eur. J. Immunol.* 12: 583

131. Jordan, R. K., Bentley, A. L., Perry, G. A., Crouse, D. A. 1985. Thymic epithelium. I. Lymphoid-free organ cultures grafted in syngeneic intact mice. *J. Immunol.* 134: 2155

132. Pyke, K. W., Bartlett, P. G., Mandel, T. E. 1983. Lymphocyte depletion of murine fetal thymus and subsequent chimeric recolonization *in vitro*. *Thymus* 5: 95

133. Kaplan, H. S. 1972. *Hodgkin's Disease*. Cambridge, Massachusetts: Harvard Univ. Press

134. Kruisbeek, A. M., Fultz, M. J., Sharrow, S. O., Singer, A., Mond, J. J. 1983. Early development of the T cell repertoire. *In vivo* treatment of neonatal mice with anti-Ia antibodies interferes with differentiation of I-restricted T cells but not K/D-restricted T cells. *J. Exp. Med.* 157: 1932

135. Kruisbeek, A. M., Mond, J. J., Fowles, B. J., Bridges, S., Longo, D. L. 1985. *In vivo* treatment of neonatal mice with anti-Ia antibodies: Development of the L3T4$^+$ Lyt-2$^-$ T cell subset is dependent on thymic Ia-bearing antigen-presenting function. *J. Exp. Med.* In press

136. Zinkernagel, R. M., Callahan, G. N., Klein, J., Dennert, G. 1978. Cytotoxic T cells learn specificity for self H-2 during differentiation in the thymus. *Nature* 271: 251

137. Zinkernagel, R. M. 1981. Restriction specificity of virus-specific cytotoxic T cells from thymectomised irradiated bone marrow chimeras reconstituted with thymus grafts. *Thymus* 2: 321

138. Longo, D. L., Schwartz, R. H. 1980. T-cell specificity for H-2 and Ir gene phenotype correlates with the phenotype of thymic antigen-presenting cells. *Nature* 287: 44

139. Longo, D. L., Davis, M. L. 1983. Early appearance of donor-type antigen-presenting cells in the thymuses of 1200 R radiation-induced bone marrow chimeras correlates with self-recognition of donor I region gene products. *J. Immunol.* 130: 2525

140. Lo, D., Sprent, J. 1986. Identity of cells that imprint H-2-restricted T-cell specificity in the thymus. *Nature* 319: 672

141. Jordan, R. K., Robinson, J. H., Hopkinson, N. A., House, K. C., Bentley, A. L. 1985. Thymic epithelium and the induction of transplantation tolerance in nude mice. *Nature* 314: 454

142. Ready, A. R., Jenkinson, E. J., Kingston, R., Owen, J. J. T. 1984. Successful transplantation across major histocompatibility barrier of deoxyguanosine-treated embryonic thymus expressing class II antigens. *Nature* 310: 231

143. Von Boehmer, H., Schubiger, K. 1984. Thymocytes appear to ignore class I major histocompatibility complex antigens expressed on thymus epithelial cells. *Eur. J. Immunol.* 14: 1048

144. Von Boehmer, H., Hafen, K. 1986. Minor but not major histocompatibility antigens of thymus epithelium tolerize precursors of cytolytic T cells. *Nature* 320: 626

145. Singer, A., Mizuochi, T., Munitz, T. I., Gress, R. E. 1986. Role of self antigens in the selection of the developing T cell repertoire. In *Prog. Immunol. vi*, ed. B. Cinader, R. G. Miller. New York: Academic

146. Bradley, S. M., Kruisbeek, A. M., Singer, A. 1982. Cytotoxic T lymphocyte responses in allogeneic radiation bone marrow chimeras. *J. Exp. Med.* 156: 1650

147. Lake, J. P., Andrew, M. E., Pierce, C. W., Braciale, T. J. 1980. Sendai virus-specific, H-2-restricted cytotoxic T lymphocyte responses of nude mice grafted with allogeneic or semi-allogeneic thymus glands. *J. Exp. Med.* 152: 1805

148. Kruisbeek, A. M., Sharrow, S. O., Mathieson, B. J., Singer, A. 1981. The H-2 phenotype of the thymus dictates the self-specificity expressed by thymic but not splenic cytotoxic T lymphocyte precursors in thymus-engrafted nude mice. *J. Immunol.* 127: 2168

149. Stutman, O. 1986. Postthymic T-cell development. *Immunol. Rev.* 91: 159

150. Scollay, R., Smith, J., Stauffer, V. 1986. Dynamics of early T cells: Prothy-

mocyte migration and proliferation in the adult mouse thymus. *Immunol. Rev.* 91 : 129

151. Metcalf, D., Moore, M. A. S. 1971. *Haemopoietic Cells.* Amsterdam/-London: North-Holland Publishing Company

152. Moore, M. A. S., Metcalf, D. 1970. Ontogeny of the haemopoietic system: Yolk sac origin of *in vivo* and *in vitro* colony forming cells in the developing mouse embryo. *Br. J. Haematol.* 18 : 279

153. Jotereau, F. V., Le Douarin, N. M. 1982. Demonstration of a cyclic renewal of the lymphocyte precursor cells in the quail thymus during embryonic and perinatal life. *J. Immunol.* 129 : 1869

154. Champion, S., Imhof, B. A., Savagner, P., Thiery, J. P. 1986. The embryonic thymus produces chemotactic peptides involved in the homing of hemopoietic precursors. *Cell* 44 : 781

156. Gallatin, M., St. John, T. P., Siegelman, M., Reichert, R., Butcher, E. C., Weissman, I. L. 1986. Lymphocyte homing receptors. *Cell* 44 : 673

157. Micklem, H. S., Clarke, C. M., Evans, E. P., Ford, C. E. 1968. Fate of chromosome-marked mouse bone marrow cells transfused into normal syngeneic recipients. *Transplantation* 6 : 299

158. Goldschneider, I., Komschlies, K. L., Greiner, D. L. 1986. Studies of thymocytopoiesis in rats and mice. I. Kinetics of appearance of thymocytes using a direct intrathymic adoptive transfer assay for thymocyte precursors. *J. Exp. Med.* 163 : 1

159. Ford, C. E., Micklem, H. S., Evans, E. P., Gray, J. G., Ogden, D. A. 1966. The inflow of bone marrow cells to the thymus: Studies with part-body irradiated mice injected with chromosome-marked bone marrow and subjected to antigenic stimulation. *Ann. N.Y. Acad. Sci.* 129 : 283

160. Lepault, F., Weissman, I. L. 1981. An *in vivo* assay for thymus homing bone marrow cells. *Nature* 293 : 151

161. Wolstenholme, G. E. W., Porter, R., eds. 1966. *The Thymus: Experimental and Clinical Studies.* Boston: Little, Brown

162. Kadish, J. L., Basch, R. S. 1977. Hematopoietic thymocyte precursors. III. A population of thymocytes with capacity to return ("home") to the thymus. *Cell. Immunol.* 30 : 12

163. Watt, S. M., Gilmore, D. J., Clark, M. R., Davis, J. M., Swirsky, D. M., Waldmann, H. 1984. Haemopoietic progeni-

tor cell heterogeneity revealed by a single monoclonal antibody, YW 13.1.1. *Mol. Biol. Med.* 2 : 351

164. Basch, R. S., Berman, J. W. 1982. Thy-1 determinants are present on many murine hematopoietic cells other than T cells. *Eur. J. Immunol.* 12 : 359

165. Till, J. E., McCulloch, E. A. 1961. A direct measurement of the radiation sensitivity of normal mouse bone marrow cells. *Radiat. Res.* 14 : 213

166. Muller-Sieburg, C. E., Whitlock, C. A., Weissman, I. L. 1986. Isolation of two early B lymphocyte progenitors from mouse marrow: A committed pre-pre-B cell and a clonogenic Thy-1^{lo} hematopoietic stem cell. *Cell* 44 : 653

167. Whitlock, C. A., Witte, O. N. 1982. Long-term culture of B lymphocytes and their precursors from murine bone marrow. *Proc. Natl. Acad. Sci. USA* 79 : 3608

168. Cohn, M. L., Scott, D. W. 1979. Functional differentiation of T cell precursors. I. Parameters of carrier-specific tolerance in murine helper T cell precursors. *J. Immunol.* 123 : 2083

169. Morrissey, P. J., Kruisbeek, A. M., Sharrow, S. O., Singer, A. 1982. Tolerance of thymic cytotoxic T lymphocytes to allogeneic H-2 determinants encountered prethymically: Evidence for expression of anti-H-2 receptors prior to entry into the thymus. *Proc. Natl. Acad. Sci. USA* 79 : 2003

170. Bradley, S. M., Morrissey, P. J., Sharrow, S. O., Singer, A. 1982. Tolerance of thymocytes to allogeneic I region determinants encountered prethymically. Evidence for expression of anti-Ia receptors by T cell precursors before their entry into the thymus. *J. Exp. Med.* 155 : 1638

171. Chervenak, R., Moorhead, J. W., Cohen, J. J. 1985. Prethymic T cell precursors express receptors for antigen. *J. Immunol.* 134 : 695

172. Ezine, S., Weissman, I., Rouse, R. 1984. Bone marrow cells give rise to distinct cell clones within the thymus. *Nature* 309 : 629

173. Mathieson, B., Sharrow, S., Rosenberg, Y., Hammerling, U. 1981. Ly1$^+$23$^-$ cells appear in the thymus before Lyt123$^+$ cells. *Nature* 289 : 179

174. van Ewijk, W., Jenkinson, E., Owen, J. 1982. Detection of Thy-1, T200, Lyt-1, and Lyt-2 bearing cells in the developing lymphoid organs of the mouse thymus *in vivo* and *in vitro*. *Eur. J. Immunol.* 12 : 262

175. Ceredig, R. 1979. Frequency of alloreactive cytotoxic T cell precursors in the

mouse thymus and spleen during ontogeny. *Transplantation* 28 : 377

176. Widmer, M. B., MacDonald, H. R., Cerottini, J.-C. 1981. Limiting dilution analysis of alloantigen-reactive T lymphocytes. VI. Ontogeny of cytolytic T lymphocyte precursors in the thymus. *Thymus* 2 : 245

177. Kamarck, M., Gottlieb, P. 1977. Expression of thymocyte surface alloantigens in the fetal mouse thymus *in vivo* and in organ culture. *J. Immunol.* 119 : 407

178. Kisielow, P., Leiserson, W., von Boehmer, H. 1984. Differentiation of thymocytes in fetal organ culture : Analysis of phenotypic changes accompanying the appearance of cytolytic and interleukin 2-producing cells. *J. Immunol.* 133 : 1117

179. Kingston, R., Jenkinson, E. J., Owen, J. J. T. 1985. A single stem cell can recolonize an embryonic thymus, producing phenotypically distinct T-cell populations. *Nature* 317 : 811

180. Habu, S., Okumura, K., Diamantstein, T., Shevach, E. M. 1985. Expression of interleukin 2 receptor on murine fetal thymocytes. *Eur. J. Immunol.* 15 : 456

181. Lesley, J., Hyman, R., Schulte, R. 1985. Evidence that Pgp-1 glycoprotein is expressed on thymus-homing progenitor cells of the thymus. *Cell. Immunol.* 91 : 397

182. Lugo, J. P., Krishnan, S. N., Sailor, R. D., Koen, P., Malek, T., Rothenberg, E. 1985. Proliferation of thymic stem cells with and without receptors for interleukin 2. Implications for intrathymic antigen recognition. *J. Exp.* *Med.* 161 : 1048

183. Takas, L., Osawa, H., Diamantstein, T. 1984. Detection and localization by the monoclonal anti-interleukin 2 receptor antibody AMT-13 of IL-2 receptor-bearing cells in the developing thymus of the mouse embryo and in the thymus of cortisone-treated mice. *Eur. J. Immunol.* 14 : 1152

184. Deleted in proof

185. Reichert, R. A., Jerabek, L., Gallatin, W. M., Butcher, E. C., Weissman, I. L. 1986. Ontogeny of lymphocyte homing receptor expression in the mouse thymus. *J. Immunol.* 136 : 3535

186. Takada, A., Takada, Y. 1973. Proliferation of donor marrow and thymus cells in the myeloid and lymphoid organs of irradiated syngeneic host mice. *J. Exp. Med.* 137 : 543

References Added in Proof

140a. Ron, Y., Lo, D., Sprent, J. 1986. T cell specificity in twice-irradiated F_1-into-parent bone marrow chimeras : Failure to detect a role for immigrant marrow-derived cells in imprinting intrathymic H-2 restriction. *J. Immunol.* 137 : 1764

145. Singer, A., Mizuochi, T., Munitz, T. Y., Gress, R. E. 1986. Role of self antigens in the selection of the developing T cell repertoire. *Progr. Immunol. VI*, ed. B. Cinader, R. G. Miller. New York : Academic Press

155. Ben Slinane, S., Houllier, F., Tucker, G., Thiery, J. P. 1983. In vitro migration of avian hemopoietic cells to the thymus : Preliminary characterization of a chemotactice mechanism. *Cell Diff.* 13 : 1

NOTE ADDED IN PROOF

This work was supported by National Institutes of Health grant AI 09072 to I.L.W. ; NIH Cancer Biology Training Grant CA 09302 to B.A. and R.A.R. ; NIH General Medical Sciences Training Grant GM 07365 to C.Y.O. ; a Swiss National Foundation fellowship to C.M. ; and American Cancer Society (National) fellowship PF-2598 to G.J.S.

Ann. Rev. Immunol. 1987. 5 : 367–403

GENETICALLY DETERMINED MURINE MODELS OF IMMUNODEFICIENCY

Leonard D. Shultz and Charles L. Sidman

The Jackson Laboratory, Bar Harbor, Maine 04609

INTRODUCTION

Numerous investigations of the mammalian immune system in normal and pathologic states have been facilitated by the study of genetically determined immunologic dysfunctions in experimental animals. Heritable immunologic defects in mice have been studied extensively as models for immunodeficiency and autoimmune diseases of humans. This article focuses on mutations of the mouse that cause impairments in the development or regulation of the immune system. Since spontaneous murine autoimmune disease has been the subject of numerous reviews (1–4), genetically determined murine immunodeficiency disorders are stressed.

Investigations into basic mechanisms underlying the function of the mammalian immune system have benefited greatly by the development of inbred strains. However, the complex developmental and functional relationships within the immune system have made it difficult to elucidate those mechanisms that underlie genetically determined immunologic dysfunction in humans. Early experiments using heritable defects in mice to learn about normal immunologic processes focused on NZB mice and the (NZB × NZW)F_1 hybrid. These mice have functional defects in their T cells, B cells, and macrophages which yield a wide spectrum of immunologic abnormalities, including polyclonal immunoglobulin (Ig) production, autoantibodies, and immune complex disease (1–4). Unfortunately, the genetic bases of immunologic dysfunction in NZB mice and its F_1 hybrids are exceedingly complex (reviewed in 4). The difficulties inherent in determining the genetic bases for immunologic diseases in a polygenic model such as the NZB strain are evident when one compares

367

0732–0582/87/0410–0367$02.00

the allelic differences between these mice and mice of a "normal" strain such as C57BL/6J. Calculations based on known allelic differences suggest that these two strains differ at more than 10,000 loci. Thus, it is not surprising that determination of the genetic control of immunologic dysfunction in this polygenic model has proved to be a formidable task.

Our understanding of immunologic diseases has advanced through study of spontaneous mutations that cause defects in the development or regulation of the immune system. Although the congenitally athymic nude mouse is the most widely studied of the immunological mutants, there are many euthymic mutants of the mouse with marked immunological deficits. In contrast to the imposing problems inherent in polygenic models, it is much less difficult to manipulate single mutant genes by appropriate genetic crosses and to study the interaction of these deleterious alleles with different background modifying genes. Approximately 900 genes of the mouse have been assigned to specific chromosomal locations (reviewed in 5). Many of these genes were first recognized after deleterious mutations occurred. Although several of these mutations cause primary defects in the immune system, most of the immunological mutants were first recognized on the bases of phenotypic abnormalities in the skin, skeleton, endocrine system, and elsewhere. Although certain immunological mutations have served as models for specific human diseases, their chief value is as tools to dissect the immune system. The published literature on these mutants varies from a few descriptive reports to thousands of published studies. It is beyond the scope of this article to discuss each mutation in detail. However, their major phenotypic characteristics and the deleterious effects of these mutant genes on the development and regulation of the immune system are addressed here.

MUTATIONS THAT CAUSE SEVERE DEFECTS IN IMMUNOLOGIC FUNCTION

Congenitally Athymic Nude Mice

Homozygotes for the recessive "nude" (*nu*) mutation on chromosome 11 were first described as hairless mice with low body weight and short life span associated with degeneration and atrophy of the liver (6). The subsequent report (7) that this mutant was congenitally athymic heralded the widespread use of *nu/nu* mice in biomedical research. Nude mice are used extensively in studies of the role of T cells in host resistance to pathogenic microorganisms and in immune responses to a variety of antigens. Such investigations have been the subject of numerous workshops, symposia, and reviews (8–11). The extensive use of *nu/nu* mice over the past 10 years is attested to by approximately 5000 published papers reporting investigations with this mutant. Findings that certain human neoplasms can grow in *nu/nu* mice have stimulated work on the mechanisms of transplantable tumor growth, including metastasis and the response to experimental therapeutic protocols. Absence of the thymus is traceable to a developmental failure of the thymic anlage which arises from the third pharyngeal pouch. The thymic rudiment remains small and cystic, whereas other derivatives of the third pharyngeal pouch are unaffected. Although the nude mouse has been considered as an animal model for DiGeorge syndrome in humans (12), the mutant shows neither parathyroidism nor the anomalies of great vessels which are characteristic of infants with DiGeorge syndrome (13).

The developing thymus in normal mice incorporates ectodermal and endodermal tissue by 9–10 days of gestation. By day 14, thymic epithelial cells contain tonofilaments and desmosomes characteristic of the adult gland. At this stage, the lymphocytic invasion of the thymus has begun (14). The underlying basis for the failure of the stromal thymic elements in *nu/nu* mice to interact with lymphocyte precursors may be attributable to abnormal expression of Ia antigens by epithelial cells. While normal thymic epithelial cells express such class-II major histocompatibility complex (MHC) antigens by gestational day 13, there is no measurable Ia antigen expression in the 13-day thymic rudiment of *nu/nu* mice (15). Immunofluorescence analyses of antigens expressed on normal thymic stromal cells have shown that only a small percentage of epithelial cells in the thymic rudiment of *nu/nu* mice have detectable levels of stromal cell–specific antigens (16).

Since the thymus plays a central role in determining the maturation of functional T cells, it is not surprising that *nu/nu* mice have a severe T-cell deficit. However, by six months of age, functional T cells appear in lymphoid tissues (17–20). It has been shown recently that the majority of peripheral Thy-1$^+$ cells in *nu/nu* mice lack both L3T4 and Ly-2 differentiation antigens (21). Nude mice have elevated macrophage and natural killer (NK) cell activities. Macrophages from these athymic mice can kill tumor cells (22) and bacteria (23) more efficiently than macrophages from euthymic littermates. The heightened macrophage activity may be due both to increased levels of lipopolysaccharide associated with the resident gut flora and to a lack of T cell–mediated suppression (24). Increased NK-cell activity appears to be due to elevated levels of interferon (25, 26).

Because the original mutation to *nu* occurred in a noninbred mouse colony, much of the early work with these animals was subject to variation associated with heterogeneity in the genetic background. Although *nu* has been backcrossed onto a variety of inbred strains, the possibility of contaminating genes must always be considered. However, there has been a second spontaneous mutation at this locus on an inbred strain background. This mutation, called "streaker" (nu^{str}) occurred in a colony of AKR/J mice (27). Streaker mice resemble nu/nu mice, the difference between the two alleles being no greater than previously reported variations among nu/nu mice. Although nearly all AKR/J mice develop thymic lymphomas by 1 year of age (28), AKR/J-nu^{str}/nu^{str} mice do not develop T-cell neoplasms (27). However, thymic lymphomas have been induced experimentally by the grafting of thymic implants under the kidney capsule (L. D. Shultz, unpublished data).

Autosomal recessive mutations causing absence of the thymus and hairlessness have been reported also in the rat (29) and guinea pig (30). This developmental relationship between the thymus and the skin emphasizes the role of the skin as an immunological organ. The immunologic function of the skin has been reviewed recently in detail (31–33). Demonstration of Ia expression by Langerhans cells (34) and keratinocytes (35) suggests that these epidermal cell populations may play a role in the development of extrathymic T cells in nu/nu mice. This possibility is further supported by the findings that (a) coculture of bone marrow cells with keratinocytes induces expression of terminal deoxynucleotidyl transferase (TdT) which is a marker for immature T cells and that (b) keratinocytes contain a substance reactive with antibody to thymopoietin, a hormone known to support T-cell differentiation (31).

Severe Combined Immune Deficiency

Mice homozygous for the recessive mutation "severe combined immunodeficiency" (scid) on chromosome 16 have greatly reduced numbers of T cells and B cells; consequently, they are deficient in both humoral and cell-mediated immune function and lack detectable levels of circulating Ig. This mutation, first described by Bosma et al (36), occurred in the C.B-17 strain, an Igh congenic partner of BALB/cAnIcr. Despite the lack of functional lymphocytes and hypogammaglobulinemia, scid/scid mice survive up to one year of age under specific pathogen-free (SPF) conditions (36). They are suitable for investigations in vivo, but by two years homozygotes often die with Pneumocystis carinii pneumonia. This mutation does not appear to seriously impair the hematopoietic microenvironment necessary for lymphoid differentiation because homozygotes can be repopulated with functional T cells and B cells by transplanting bone marrow from normal syngeneic mice (36, 37), or from long-term normal bone marrow cultures (38). However, reconstitution may be incomplete unless the bone marrow recipients receive low-dose irradiation prior to injection (39). The presence of B-cell precursors in scid/scid mice has been investigated by Fulop et al (40), who found that Abelson murine leukemia virus (A-MuLV) infection of bone marrow cells from these mice results in normal numbers of transformed foci. Cell lines obtained from such foci express B cell–specific antigens that are undetectable on nontransformed bone marrow cells from *scid/scid* mice. Since *scid/scid* mice have normal numbers of A-MuLV target cells whose transformation leads to cell lines which express pre–B cell differentiation antigens, it is evident that this mutant is not deficient in early B-cell precursors.

In an investigation of the molecular basis for defective lymphoid differentiation, Schuler et al (41) found that the *scid* mutation is associated with a defect in the rearrangement of genes that code for antigen-specific receptors on B cells and T cells. Expression of these antigen receptor genes is dependent on the appropriate recombination of the V, D, and J elements. Bone marrow cells and thymocytes from *scid/scid* mice fail to show antigen receptor gene rearrangements. Although such rearrangements were observed in A-MuLV-transformed bone marrow cells and in thymic lymphoma cells, the majority of rearranged alleles were abnormal and showed deletions of the entire Jh and Jβ2 regions respectively. It should be noted that some productive rearrangements do occur in so called "leaky" *scid/ scid* mice. Approximately 15% of these mice have detectable levels of serum Ig with restricted heterogeneity (36). Furthermore, such "leaky" mice appear to have functional T cells, as is shown by their ability to reject skin allografts (G. Bosma, personal communication). It has been postulated that homozygosity for *scid* results in an abnormal recombinase system for the assemblage of antigen receptor genes in developing lymphocytes (41). Recent evidence suggests that B cells and T cells use a common recombinase (42). A mechanism underlying defective J-associated deletions might be an abnormality in TdT. This enzyme catalyzes the removal of nucleotides at D-J junctions and the replacement of these nucleotides with additional nucleotides not encoded by either gene segment (43). However, *scid/scid* mice have normal TdT activity (G. Bosma, personal communication). Likewise, they have normal levels of adenosine deaminase (ADA) (36); a deficiency of this purine salvage pathway enzyme has been shown to underlie one form of severe combined immunodeficiency disease in humans (44).

Numbers of myeloid progenitors (45), antigen-presenting cell (APC) function (46), and NK-cell activity (47) are normal in *scid/scid* mice. Since macrophages are affected by soluble factors released as a consequence of T-cell activation (48), the absence of functional T cells in *scid/scid* mice has provided an in vivo model to examine macrophage development and function in the absence of lymphocytes. Although *scid/scid* mice have near normal numbers of Ia-bearing macrophages in the peritoneal cavity, spleen, and liver, Ia expression in macrophages from these mice is not enhanced following secondary challenge with a T cell–dependent antigen (49). Following infection with *Listeria monocytogenes*, these mice show lymphocyte-independent macrophage activation. However, levels of activation sufficient to eliminate the bacterial load are not achieved (49). Investigation with *scid/scid* mice has shown that NK progenitor cells are distinct from the transplantable progenitors of lymphocytes. Bone marrow cells from these mice contain normal numbers of progenitors that generate mature NK cells following transfer into NK cell–depleted and lethally irradiated recipients (50).

The *scid* mutation may be a homologue of the form of severe combined immunodeficiency disease in humans characterized by a dearth of functional lymphocytes and normal ADA levels (51). It is clear that *scid/scid* mice will be utilized extensively in many areas of basic and applied research. Xenogeneic as well as allogeneic hybridomas grow well as ascites tumors in these mice and produce high yields of monoclonal antibodies in the absence of host Ig. This provides an ideal system for large-scale Ig production (52).

Deleterious Alleles at the Motheaten Locus

Mice homozygous for the recessive allelic genes "motheaten" (*me*) or "viable motheaten" (*me^v*) on chromosome 6 are severely immunodeficient and develop autoimmune disease early in life. Homozygotes (*me/me*) and (*me^v/me^v*) have a mean lifespan of 22 and 61 days, respectively (53). Unlike the *scid* mutation which selectively impairs lymphocyte development, mutations at the motheaten locus cause severe defects in nonlymphoid cell populations as well as in lymphocytes. The first reported mutation at the *me* locus occurred spontaneously in the C57BL/6J strain in 1965. Homozygotes for this autosomal recessive mutation are recognizable at 1–2 days of age by neutrophilic aggregates in the epidermis. Subepidermal accumulations of neutrophils displace hair follicles and result in patchy absence of pigment (54). The cutaneous abnormalities for which this mutant is named occur even under germfree conditions (55) and may be associated with autoimmunity rather than with infection.

Motheaten mice show an impaired response to thymic-dependent and thymic-independent antigens, have reduced proliferative responses to B-cell and T-cell mitogens (56–58), and lack cytotoxic-T-cell (57) as well as NK-cell (59) activity. Immunodeficiency in these mice is not thought to be due to any primary defect in IL-1 or IL-2 production or utilization (60). The functional impairment of the immune system in *me/me* mice is accompanied by production of abnormally high levels of serum Ig and expression of multiple autoantibodies (56, 61). The proximate cause of death depends on focal intra-alveolar hemorrhages with accumulations of alveolar macrophages and neutrophils in the lower respiratory tract (54, 62, 63). Such pneumonitis may be associated with an abnormal proliferative capacity of macrophages since these cells show increased reproductive capacity in vitro (64).

The short lifespan of *me/me* mice has discouraged experimentation and has made it difficult to distinguish basic defects in the immune system from those associated with incomplete maturation and from pathologic changes associated with poor health. The longer life span of mice homozygous for the *me^v* mutation stimulated additional recent investigations. Although *me^v/me^v* mice have severe defects in many immunological responses and develop hyperimmunoglobulinemia with multiple autoantibodies, we have recently found that these mice show significant humoral responses to certain T cell–dependent antigens in vivo and in vitro (C. L. Sidman, L. D. Shultz, unpublished). A dramatic overproduction of two B-cell maturation–promoting lymphokines, which drive normal resting B cells to differentiate into antibody-secreting cells, may underlie the observed polyclonal B-cell activation in *me/me* and *me^v/me^v* mice (65, 66). Almost all the B cells in *me^v/me^v* mice express the Ly-1 cell surface antigen (67) which marks a minor population of B cells in normal mice and the autoantibody-producing B cells in NZB mice (68). Concentrations of IgM and IgG3 in serum from *me^v/me^v* mice are 50 and 20 times greater than those in age-matched control animals (67). The role of autoimmunity in the disease process of these mice is underscored by the observation that animals simultaneously homozygous for *scid* and *me^v* show a threefold increase in longevity compared with *me^v/me^v* mice (unpublished). As *me^v/me^v* mice reach 5 weeks of age, their lymphoid tissues show large numbers of atypical plasma cells with discrete Ig inclusions (69). Such plasmacytoid cells have been termed Mott cells, and the inclusions termed Russell bodies. Mott cells have been described in association with malignancy (70), autoimmune disease (71), and acquired immunodeficiency disease (72) in man. These cells are defective in Ig secretion and may represent a mechanism by which spontaneous autoantibody secretion is reduced after chronic B-cell stimulation.

By 4 weeks of age *me/me* and *me^v/me^v* mice show loss of cells from the thymus (53, 54). Such thymic involution is associated with the inability of prothymocytes to home effectively to the thymus following intravenous injection (73). Bone marrow from *me/me* and *me^v/me^v* mice is deficient in TdT$^+$ cells (73, 74). Although it has been suggested that defective production of TdT underlies pathologic changes in these mice (74), this is unlikely to be true since levels of TdT enzymatic activity in thymocytes from *me/me* and *me^v/me^v* mice are not significantly different from levels measured in littermate control animals (73). Phenotypic abnormalities caused by deleterious alleles at the *me* locus are not identical to a recognized specific human disease. However, determination of mechanisms underlying their marked effects on the immune system early in life may help elucidate the bases of severe immunologic dysfunctions in man.

SINGLE GENE MODELS OF LESS SEVERE IMMUNODEFICIENCIES

X-Linked Immunodeficiency

The critical roles that X-linked genes play in the development of the immune system are evidenced by the occurrence of seven distinct heredi-table human immunologic dysfunctions which map to this chromosome (75). The best studied murine X-linked immunologic mutation is "X-linked immunodeficiency" (*xid*). This recessive mutation was originally found in CBA/N mice obtained from Harwell in 1966 by investigators at the National Institutes of Health (76). Hemizygous (*xid/Y*) males and homo-zygous (*xid/xid*) females are unable to generate humoral responses to certain antigens, subsequently termed thymus-independent type-2 antigens (77); and they show reduced responses to thymic-dependent antigens. These mice exhibit reduced serum IgM and IgG3 levels as well as defective proliferative responses to B-cell mitogens (reviewed in 78). Immuno-deficiency in *xid* mutant mice is associated with the absence of a specific population of mature B cells. Studies of B-cell surface marker expression, heterogeneity, and activation requirements have indicated that B cells in these mice resemble the B cells of young (1–3 week old) normal mice. B cells from *xid* mutant mice fail to express a variety of differentiation antigens, including Lyb-3, Lyb-5, Lyb-7, and Mls, usually found on normal B cells (79–81). Although it was initially thought that the *xid* mutation resulted in the absence of a particular B-cell subset, recent findings suggest that the paucity of mature B cells may represent an overall impairment in B-cell maturation (82). Mature B cells can be isolated from Peyer's patches of young adult *xid* mice. Such B cells express Lyb-5 and Mls antigens (83) and respond to TI-2 antigens (84). Recently, the speculation that B cells in *xid* mice are simply immature normal B cells has been disputed (85, 86). Rather, it has been suggested that the B cells in these mice are idiosyncratic and do not represent immature normal B cells. Development of B cells in *xid* mice appears to be thymus-dependent, since mice simultaneously homozygous for *nu* and homozygous or hemizygous for *xid* have a severe deficit in both B cells and T cells (87, 88). B-cell development in these athymic *xid* mice is blocked at the early pre–B cell stage (89).

Recently, Cohen et al (90) sought *xid* mRNA by the "subtraction" cloning method. They prepared a lymphocyte-specific cDNA clone which recognizes an X-linked gene family with 15–20 members. The "X-linked lymphocyte regulated" (XLR) gene family apparently includes the locus defined by the *xid* mutation. This conclusion is based on the finding that *xid* congenic mice have a restriction fragment length polymorphism of one of the members of the XLR gene family and also on the finding that plasmacytomas from *xid* mice do not contain XLR transcripts (91). Since the X chromosome of eutherian mammals has been conserved throughout evolution (92), it seems likely that this gene family is present on the X-chromosome of man and may be important in regulating immune functions that are disturbed in specific X-linked immunodeficiency diseases. Although the X-linked Wiscott-Aldrich syndrome in man is similar to the *xid* mutation in that both conditions cause defective antibody responses to certain polysaccharides, defects in T-cell and platelet function usually observed in patients with Wiscott-Aldrich syndrome (51, 93) are not characteristic of *xid* mutant mice.

Takatsu and Hamaoka reported that an X-linked recessive gene determined the defective responses of B cells from DBA/2Ha mice to a family of lymphokines termed T-cell replacing factors (reviewed in 94). However, recent investigation of the inability of B cells from these mice to respond to a similar family of lymphokines, termed B-cell maturation factors, found that the defect was due to the combined action of alleles at two autosomal loci, termed "BMF responsiveness-1" (*Bmfr-1*) on chromosome 4 and "BMF responsiveness-2" (*Bmfr-2*) on chromosome 9 (95). Resolution of whether DBA/2Ha mice carry both autosomal and X-linked deleterious genes will depend on further genetic studies and a more complete understanding of the different lymphokines and cellular systems used in these various studies.

Impaired Responses to Bacterial Lipopolysaccharide

Bacterial lipopolysaccharide (LPS) is a major component of the outer membrane of gram-negative bacteria. It is composed of a lipid moiety, termed lipid A, and a polysaccharide component responsible for the serologic distinction of LPS molecules. LPS is an important regulator of the immune system, displaying a wide spectrum of biological activities including toxicity, macrophage activation, B-cell mitogenicity and polyclonal activation, enhancement of T-cell responses, and complement activation (96, 97). Although early attempts at selective breeding for resistance to the toxic effects of LPS indicated that these effects were under genetic control (98), our current understanding of the mechanisms underlying biologic responses to LPS is due in large part to the presence of a spontaneous codominant mutation at the "lipopolysaccharide" *Lps* locus in

C3H/HeJ mice. This locus on chromosome 4 controls responsiveness to the lipid A moiety of endotoxin derived from bacteria belonging to the family Enterobacteriaceae. The allele for defective response (Lps^d) to LPS originated by mutation in this strain between 1960 and 1968 (99). A second spontaneous mutation to the Lps^d allele is responsible for the defective LPS responsiveness of C57BL/10ScN and C57BL/10ScCR mice (reviewed in 100). Most inbred strains and other sublines of C3H and C57BL/10 mice express the allele for normal response, Lps^n. Defective LPS responses in C3H/HeJ mice encompass all known lipid A–induced effects on the immune system and include: reduced proliferation, maturation, and poly-clonal activation of B cells; impaired LPS-induced phenotypic conversion of T-cell precursors; reduced LPS-enhancement of T-cell stimulation by lectins; resistance of macrophages to LPS cytotoxicity; and reduced sen-sitivity to the inhibitory effect of LPS on phagocytosis (reviewed in 96, 100). In addition to impaired responses to LPS, the Lps^d allele causes defects in the immune system unrelated to the exogenous administration of LPS. For example, macrophages from C3H/HeJ mice exhibit reduced tumoricidal activity (101), fail to respond to the lymphokine migration inhibitory factor (102), and do not retain Fc receptor–mediated phagocytic capacity after in vitro culture (103). It has recently been shown that macrophages from C3H/HeJ mice exhibit impaired endogenous interferon (IFN) production and that such production in normal macrophages acts as an autostimulatory differentiation signal (104). Correction of the Fc receptor deficit in C3H/HeJ mice after administration of exogenous IFN supports the hypothesis that impaired Fc receptor expression is associated with low endogenous IFN production (105, 106).

The mechanism by which the Lps^d allele exerts its effects on many diverse cell populations remains elusive. Recent experiments to determine the molecular basis for LPS nonresponsiveness have concentrated on the effects of LPS on B cells. Defective LPS responses of B cells in C3H/HeJ mice are intrinsic to this lineage and are not due to abnormalities in accessory cells or T-helper cells (107). Although it has been suggested that the Lps^d allele determines the absence of a cell surface receptor for LPS (108), B cells from Lps^d/Lps^d mice exhibit normal binding of radiolabeled lipid A (109). Alternatively it has been postulated that the Lps^d defect may result from an impairment in membrane signal transduction following LPS binding. This postulate is supported by the finding that microinjection of cytosol fractions from LPS-stimulated normal lymphocytes increased LPS responsiveness of C3H/HeJ lymphocytes to normal levels (110). In addition, B cells from C3H/HeJ mice show nearly normal proliferative responses following incubation with a lipid A precursor molecule derived from a mutant of Salmonella typhimurium (111). Direct evidence that the

Lps^d defect is at the level of the plasma membrane comes from the findings that fusion of B cells from C3H/HeJ mice with plasma membranes from C3HeB/FeJ responder mice results in the acquisition of partial LPS mitogenic responses (112) and that nuclei from C3H/HeJ B cells can respond to LPS after transfer into LPS-responsive cells (113). It has recently been suggested that the absence of a protease or the inability of LPS to activate a protease might underlie the LPS defect in C3H/HeJ mice, since culture with trypsin of splenic B cells from these mice resulted in partial restoration of the LPS mitogenic response (114).

The C3H/HeJ mouse has long been used for investigating the biology of mammary tumors because these tumors occur at high incidence with short latent periods in breeding females of this strain (115, 116). However, at about the time the *Lps^d* allele was fixed in the C3H/HeJ strain of mice at the Jackson Laboratory, a radical decrease in mammary tumor incidence occurred. This decrease is associated with an attenuation of the exogenous milk-transmitted murine mammary tumor virus (MuMTV) (117). Although the mechanism by which the *Lps^d* allele might provide the genetic pressure to select for an attenuated MuMTV is unknown, it is possible that the selection is related to the effect of this allele on lymphocytes since the MuMTV is transmitted to the target organ by lymphoid cells (118, 119).

Hemolytic Complement Deficiency

The complement system is composed of approximately 20 protein molecules that circulate in unactivated plasma. Activation of this complex enzyme cascade is critical for a number of host defense mechanisms and is responsible for many aspects of the inflammatory response (120, 121). Many inherited complement deficiency disorders have been described in experimental animals and in man.

C5 DEFICIENCY   The most severe, genetically determined complement deficiency state in mice occurs in homozygotes for the *Hc^0* allele at the "hemolytic complement" (*Hc*) locus on chromosome 2. Homozygotes (*Hc^0/Hc^0*) lack detectable hemolytic complement activity in serum because of an absence of C5. Approximately 30% of inbred strains lack C5 (122). Complement-sufficient normal strains of mice are homozygous for the *Hc^1* allele. Genetic polymorphisms in sera of inbred strains of mice were originally used to describe C5 as the MuB1 antigen (reviewed in 123). In normal mice, C5 is produced in macrophages as a single-chain precursor termed pro-C5. The precursor is secreted and then converted to the two chain C5 protein. In *Hc^0/Hc^0* mice, macrophages produce pro-C5. However, pro-C5 is not secreted (124). Although C5 deficiency prohibits the formation of the C5b-9 membrane attack complex (120), C5-deficient mice exhibit normal viability. Nevertheless, their absence of hemolytic complement is associated with increased susceptibility to a variety of pathogenic microorganisms, impaired chemotactic responses of neutrophils, prolonged rejection of skin allografts, decreased ability to reject transplantable tumor cells and increased susceptibility to autoimmunity (reviewed in 125).

C4 DEFICIENCY   The fourth component of complement is composed of three disulphide-linked subunits, termed alpha, beta, and gamma. C4 is initially synthesized as a single-chain precursor (pro-C4) which is glycosylated and cleaved to yield the three-subunit molecule (reviewed in 126). The structural genes for complement components C4, C2, factor B, and Slp map within the S region of the *H-2* complex on chromosome 17. Mice bearing the S region of the *H-2*^{k} and the wild-derived *H-2*^{w7}, *H-2*^{w16}, and *H-2*^{19} haplotypes have a three- to tenfold reduction in hemolytic C4 activity compared with mice congenic for the *H-2*^{d} haplotype (121, 127, 128). Reduced hemolytic C4 efficiency in mice with the *H-2*^{w7}, *H-2*^{w16}, and *H-2*^{19} haplotypes is probably due to the absence of carbohydrate on the alpha chain (128).

Genetic defects in man have been reported for nearly all the complement components, including C5 and C4. These defects are all inherited as autosomal recessive traits. C5 deficiency is associated with recurrent bacterial infections and an SLE-like syndrome, while C4 deficiency causes an SLE-like syndrome but is not associated with decreased resistance to infection (129).

IMMUNODEFICIENCY ASSOCIATED WITH SYSTEMIC AUTOIMMUNITY

Depressed immune function often occurs in association with systemic autoimmune disease in man (75, 130, 131). Likewise, functional abnormalities of T cells, B cells, and macrophages have been described in mice bearing mutations that either induce or accelerate autoimmunity. In addition to deleterious alleles at the motheaten locus described earlier, single gene mutants that have been studied as models for SLE in man include lymphoproliferation (*lpr*), generalized lymphoproliferative disease (*gld*), and Y chromosome–linked autoimmune accelerator (*Yaa*). These loci were first described by Murphy & Roths at the Jackson Laboratory (132–135). Strain MRL/Mp mice homozygous for the unmapped autosomal recessive mutation *lpr* develop lymphadenopathy accompanied by multiple autoantibodies, glomerulonephritis, vasculitis, and arthritic lesions (133–136). Homozygotes (MRL/Mp-*lpr/lpr*) die by 6 months of age. The hyperplastic lymph nodes are composed largely of dull Ly-1$^+$, L3T4$^-$, Ly-2$^-$ T cells. However, the majority of these T cells aberrantly express the Ly-5 (B220) antigen normally detected on cells of the B-cell lineage (137). C3H/HeJ mice that are homozygous for the recessive mutation "generalized lymphoproliferative disease" (*gld*), which maps on chromosome 1, develop a nonmalignant lymphoproliferative disease which results from the expansion of an aberrant lymphocyte subset closely resembling that seen in *lpr/lpr* mice. Aberrant lymphoid cells in both *lpr/lpr* and *gld/gld* mice are similar in their expression of cell surface differentiation antigens, functional abnormalities, and increased expression of the *myb* proto-oncogene (138, 139). C3H/HeJ-*gld/gld* mice however, survive to 13 months and die with interstitial pneumonitis (135).

The Y chromosome–linked *Yaa* gene was first described in strain BXSB mice. Males of this recombinant inbred strain develop B-cell hyperplasia, express multiple autoantibodies, and die by 6 months of age with immune complex glomerulonephritis (132–134, 136). Female BXSB mice develop a more chronic form of autoimmune disease and die by 16 months of age. Although *Yaa* accelerates autoimmunity in strains of mice genetically predisposed to this condition, analysis of Y-consomic stocks developed by backcrossing the Y-chromosome of BXSB mice to other strain backgrounds revealed that the *Yaa* mutation has little deleterious effect on normal mouse strains (140).

Since these models for murine SLE have been the subject of numerous reviews (1–4), we discuss briefly defects in immunologic function and do not examine features of autoimmune disease. Functional abnormalities of T cells, B cells, and macrophages of autoimmune mutant mice have recently been reviewed in detail (4). Although depressed immune function has been described following clinical manifestations of murine autoimmune disease, it is noteworthy that most studies have shown normal immune function when the mice were examined prior to the age of clinical disease, after which time immunocompetent cells are greatly diluted by the massive increase of lymphoproliferative cells. Limiting dilution analyses were conducted of precursor frequencies of cytotoxic T cells and Con-A responsive cells (141) as well as of cells that produce IL-2 and colony-stimulating factor (142); these analyses revealed that the absolute numbers of immunocompetent T cells in aged MRL/Mp-*lpr/lpr* mice are normal. Functional analyses were carried out on the abnormal population of lymphoproliferative cells in MRL/Mp-*lpr/lpr* mice following complement-mediated lysis of "normal" Lyt-2$^+$ and L3T4$^+$ cells; these studies showed that the major remaining subset of *lpr*-induced T cells produce neither IL-2 nor gamma interferon and have little potential for antigenic stimulation (143). Thus, it appears that the decreased T-cell function is a consequence of progressive dilution of immunocompetent cells by abnormal cells. Less is known about the etiology of macrophage defects in genetically autoimmune mice. MRL/Mp-*lpr*/*lpr* mice exhibit increased frequency of Ia$^+$ resident peritoneal macrophages (144–146). This increased Ia$^+$ macrophage recruitment appears to be mediated by macrophage stimulatory factors released by proliferating T cells (144). A functional defect in cell-mediated immunity in *lpr*/*lpr* mice is indicated by their increased susceptibility to infection with *L. monocytogenes* (147). While increased Ia expression in resident peritoneal macrophages may be a consequence of the lymphoproliferative disease in *lpr*/*lpr* mice, increased surface Ia expression on cells from spleen, thymus, lymph nodes, and bone marrow is evident by 1 month of age, prior to clinical autoimmune disease (148).

In BXSB mice carrying the *Yaa* gene, development of atypical peripheral blood monocytes may be involved in abnormal immune function. Wofsy (149) has reported that these mice exhibit a progressive increase in the percentage of blood monocytes by 2 months of age. The atypical cells had the morphological characteristics of macrophage lineage cells; they were Mac-1$^+$ and expressed Fc receptors; however, they lacked detectable levels of myeloperoxidase and nonspecific esterase and were Ia$^-$.

IMMUNODEFICIENCY ASSOCIATED WITH MAJOR CELLULAR OR ORGAN SYSTEM ABNORMALITIES

Congenitally Asplenic Dominant Hemimelia Mice

The mutation known as "Dominant hemimelia" (*Dh*) on chromosome 1 is the only mutation other than *nu* which results in absence of a lymphoid organ. This mutation was discovered in 1954 at the Institute of Animal Genetics in Edinburgh. In 1959, Searle reported that homozygotes (*Dh*/*Dh*) and heterozygotes (*Dh*/+) were asplenic (150). Homozygotes die within a few days after birth with severe pathologic changes in the urogenital and skeletal systems. The absence of the spleen as well as the widespread visceral and skeletal abnormalities were found to be associated in *Dh*/+ and *Dh*/*Dh* mice with a defect at the level of the splanchnic mesoderm in early embryos (151). Although the spleen is not essential for survival, it is an important site for the interaction of immunocompetent cell populations. Since the spleen is the major site of antibody synthesis in mice up to 6 months of age (152), it is not surprising that *Dh*/+ mice have reduced levels of serum IgM (153) and specific IgM antibody following immunization with sheep erythrocytes (SRBC) (154). In contrast, levels of IgG1 and IgE are normal in $Dh/+$ mice (155). Evidently the spleen is not essential for the development of cell-mediated immunity. Normal in vivo cell-mediated immunity in $Dh/+$ and Dh/Dh mice is expressed in their ability to reject allografts (153) and their response to contact allergens (156). Likewise, lymph node cells from congenitally asplenic mice show normal proliferative responses in vitro to allogeneic cells and increased proliferative responses to phytohemagglutinin (PHA) (157). Host defense against malaria in mice appears to require an intact spleen since $Dh/+$ mice exhibit increased mortality, compared with eusplenic control animals, following infection with *Plasmodium yoelli* (158). A reduction in absolute numbers of lymph node mast cells in $Dh/+$ mice suggests that the spleen is involved in regulating these cells (159). Most instances of congenital asplenia in humans appear to be sporadic, but familial occurrences have been reported (160). The spleen appears to play an important role in resistance to infectious disease in childhood since the hereditary absence of this organ leads to an increased incidence of bacterial and viral infections early in life (161).

Murine Pigmentation Mutations

Mutations at more than 60 loci are known to affect the coat color of mice. Several of these pigmentation mutations, described below, also cause deleterious changes in the immune system.

THE BEIGE GENE AND OTHER LOCI AFFECTING DEVELOPMENT OF LYSO-SOMES   The first reported mutation at the "beige" (bg) locus on chromosome 13 occurred in the offspring of an irradiated mouse at the Oak Ridge National Laboratory. Since the initial description in 1957, there have been at least nine additional independent mutations at this locus in a variety of inbred strains. Homozygotes show an increased incidence of spontaneous pneumonitis (162) and an increased susceptibility to bacterial (163), parasitic (164), and viral (165) infections. Reports that bg/bg mice have a selective defect in endogenous NK-cell activity stimulated further research (166, 167). The low NK-cell activity, due to a defect in a post-recognition step in the lytic cycle, is associated with decreased resistance to transplantable syngeneic leukemia (168) and melanoma (169) cells. Although it had been suggested that the bg mutation selectively impairs NK-cell activity (166, 167), homozygosity for this mutation also causes impaired neutrophil chemotaxis and bactericidal activity (170), delayed development of macrophage-mediated anti-tumor killing (171), defective cytotoxic-T-cell responses (172), decreased graft-vs-host (GVH) reactivity of lymphoid cells (173), and impaired humoral immune responses to allogeneic tumor cells (174). The decreased bactericidal capacity of leukocytes from *bg/bg* mice may be associated with a deficiency of serotonin storage granules in platelets (175), since addition of normal platelets to *bg/bg* leukocytes in cultures restores the bactericidal activity to normal levels. This restorative effect can be mimicked by the addition of serotonin to the culture (176).

Abnormal lysosomal granules are present in all granule-containing cells of this mutant. Subcellular organelles in *bg/bg* mice that exhibit altered morphology or function include lysosomes, melanosomes, platelet dense granules, and mast cell granules (177, 178). The *bg* mutation affects lysosomal biogenesis or processing. Although the underlying basis of the lysosomal membrane abnormality is not well understood, a high rate of fusion among preexisting normal lysosomal granules suggests either an abnormality facilitating fusion in the lysosomal membrane or a defective control in the size of the lysozomes (179). Altered lysosomal function is associated with decreased lysosomal enzyme secretion from the kidney (180). The light coat color of *bg/bg* mice is thought to result from reduced synthesis of pigment granules in the melanocytes and from the fusion of these granules into progressively larger bodies (178). The *bg* mutation has been considered as a homologue for the Chediak-Higashi syndrome in humans. This is a rare autosomal recessive disease characterized by abnormal leukocyte granulation, partial oculocutaneous albinism, and severe recurrent infections associated with functional defects in polymorphonuclear leukocytes (181). The impaired NK-cell activity in *bg/bg* mice led to the subsequent finding of low NK-cell activity in patients with Chediak-Higashi disease (182, 183). The Chediak-Higashi syndrome, as defined by reduced pigmentation and increased size of lysosomal granules, has also been reported in Aleutian mink (184), Hereford cows (184), Persian cats (185), and a killer whale (186).

The original study of spontaneous pneumonitis in *bg/bg* mice focused on the SB/Le strain, doubly homozygous for *bg* and the recessive mutation "satin" (*sa*) on chromosome 13 (162). The possibility that *sa* contributed to the increased susceptibility to pulmonary infection in SB/Le mice is supported by the report that mice doubly homozygous for *sa* and *bg* show a more severe depression in cytolytic-T-cell and NK-cell activity than do homozygotes for *bg* alone (187). Two autosomal recessive mutations, "pale ear" (*ep*) on chromosome 19 and "pallid" (*pa*) on chromosome 2, share similarities with *bg* in that they produce abnormalities in lysosome function (180) and may cause reduced NK-cell activity (188). The association of impaired NK-cell activity with lysosomal abnormalities in three independent nonallelic mutations suggests that this activity requires the normal function of lysosomal membranes.

MUTATIONS AT THE AGOUTI LOCUS   The "agouti" (A) locus on chromosome 2 is a determinant of coat color that acts via the hair follicle. There are 17 known alleles at this locus (5). Two of these mutations have been of interest because of their effects on tumorigenesis. For example, heterozygosity for the dominant alleles called lethal yellow (A^y) and viable yellow (A^{vy}) is associated with an increased incidence of a variety of spontaneous and induced neoplasms (189, 190). In a study aimed at determining whether the increased tumorigenesis was associated with decreased immunologic function, Gasser & Fischgrund (191) found that spleen cells from both A^y/a and A^{vy}/a mice exhibited reduced GVH reactivity. More recently, it has been reported that A^{vy}/a mice show reduced humoral immune responses to the T cell–dependent antigen tetanus toxoid and also decreased rates of carbon clearance (192). The A^y allele is associated with an ecotropic murine leukemia virus (MuLV) genome, suggesting that this mutation may have been caused by virus integration (193). A^{vy}/a mice show variability in phenotype. Their coat color varies from clear yellow, through mottling with dark patches, to an agouti-like coat. Both A^y/a and A^{vy}/a mice become obese, the degree of obesity being correlated with the amount of yellow in the coat (194). Reduced immune responses of these mice are probably secondary effects of hyperinsulinemia, since the severity of immune dysfunction is associated with circulating insulin levels and with the degree of obesity (194).

MUTATIONS CAUSING PLEIOTROPIC DEFECTS IN PIGMENT FORMING CELLS, ERYTHROCYTES, AND MAST CELLS   Mutations at the "dominant spotting" (W) and "steel" (Sl) loci, on chromosomes 5 and 10 respectively, cause deficiencies in pigment forming cells, germ cells, red blood cells, and mast cells. Experimentation with W-locus mutants has made extensive use of the W and W^v alleles and the compound W/W^v mutant mouse, while investigation with Sl-locus mutants has concentrated on the Sl and Sl^d-alleles and the compound Sl/Sl^d mutant mouse. Homozygotes or compounds for alleles at these loci are white, sterile, have macrocytic anemia, and lack tissue mast cells (195). While erythrocyte and melanocyte abnormalities in W-locus mutants are due to a defect at the level of progenitor cells, mutants at the Sl locus possess normal progenitor cells but provide an impaired microenvironment for the proper development of these cell types. Since W- (196) and Sl- (197) locus mutants have less than 1% of the normal number of tissue mast cells, recent investigations have focused on the effects of hereditary mast cell depletion on immunologic mechanisms thought to be mediated by mast cells. Immediate-type hypersensitivity responses, as measured by the ability to develop systemic anaphylaxis, are expressed in W/W^v mice (198). Peripheral basophils, which are present in normal numbers, may mediate systemic anaphylaxis (198). Although W/W^v mice are clearly deficient in the ability to manifest passive cutaneous anaphylaxis reactions (199, 200), controversy has occurred over the capacity of mast cell–deficient mice to mount delayed-type hypersensitivity (DTH) responses. Several investigations have suggested that T cell–dependent activation of cutaneous mast cells is required for the elicitation of DTH responses (201, 202). These findings would suggest an impairment of DTH responses in mast cell–deficient mice. Consequently, W/W^v and Sl/Sl^d mice were found to be deficient in the elicitation of sheep erythrocyte (SRBC)–induced DTH responses and of contact sensitivity responses to picryl chloride (199). However, this deficiency in DTH responses has not been confirmed by recent investigations (203–205). The reported differences in DTH responses may be dependent in part on the age of the animals tested, since responses of this kind are strongly age-dependent (206). Because mastocytosis occurs during certain parasitic infections, mucosal mast cells are thought to play a role in the rejection of intestinal parasites (207). In many but not all investigations of host-parasite relationships in W/W^v mice, expulsion of parasites was delayed (reviewed in 205). Restoration of mast cells by bone marrow injection reversed deficient responses to certain parasites (208). Primary and secondary humoral immune responses are normal in W/W^v mice (205, 207–209). Investigations in which normal bone marrow cells correct defective mast cell–mediated responses must be interpreted with caution because injection of marrow results in partial population by donor cells of the lymphoid tissues of W-anemic mice (210), curing of the recipient's anemia (195), and possibly other effects that might influence experimental results.

Immunodeficiency and Lymphomagenesis in Hairless Mice

Unlike nude mice, which are essentially hairless throughout life, mice homozygous for the "hairless" (hr) mutation on chromosome 14 develop and maintain a normal pelage up to 10 days of age, at which time they lose their hair (211, 212). Although hr/hr mice are euthymic, they undergo a marked thymic cortical atrophy by 6 months of age (213). Immunological studies were stimulated by the finding that HRS/J-hr/hr mice have a 45% incidence of thymic lymphoma by 10 months of age, the incidence of lymphoma in +/hr controls being only 1% (214). Lymphomagenesis in hr/hr mice is associated with the expression of polytropic retroviruses during the leukemic and preleukemic periods (215). An association between defective immune function and lymphomagenesis is suggested by impaired immune responsiveness to syngeneic lymphoma cells and purified murine leukemia viruses (216). Altered immune function in hr/hr mice may be associated with abnormalities at the level of T cells. Although splenic T cells from this mutant generate a strong cytotoxic-T-cell response, such cells have depressed proliferative response to I-region alloantigens (217). Since mixed lymphocyte reactions (MLR) are mediated by helper T cells, it has been suggested that the impairment in alloantigen response reflects a selective defect in this T-cell subset. Splenocytes from hr/hr mice show an inversion in the normal proportions of Ly-1$^+$ and Ly-123$^+$ T cells (218), decreased proliferative responses to B-cell mitogens, and impaired PFC responses to T-independent antigens (100).

Homozygosity for the "rhino" (hr^{rh}) mutation, a second recessive allele at the hr locus, also results in premature thymic involution and impaired helper-T-cell function (219). It has been suggested that T-cell abnormalities in rhino mice may be associated with autoimmunity, in that high levels of circulating antinuclear antibodies and other autoimmune changes are evident by 3–5 months of age (220). A recently described semidominant mutation at the hairless locus, called "near naked" (Hr^n), has not been examined for immunological abnormalities (221).

Mutant Genes Affecting Endocrine Glands

HYPOPITUITARY DWARF MUTANTS   Three nonallelic recessive mutations in the mouse cause dwarfism as a result of insufficient growth hormone production due to a primary defect in the anterior lobe of the pituitary gland. These mutant genes are "dwarf" (dw) on chromosome 16, "Ames dwarf" (df) on chromosome 1, and "little" (lit) on chromosome 6. While lit/lit mice have isolated growth hormone (GH) deficiency due to resistance to hypothalamic GH releasing hormone (222), dw/dw and df/df mice lack acidophilic cells and are deficient in other anterior pituitary hormones, including prolactin and thyroid stimulating hormone, as well as GH (223). Both dw/dw and df/df mice have been extensively studied as models for immunodeficiency secondary to anterior pituitary defects; however, there have been no published studies on immunologic function in lit/lit mice. Although a large number of investigations of the immune system of dw/dw and df/df mice indicated that these hypopituitary mutants exhibit defective cell-mediated immune function associated with early thymic involution (reviewed in 224), more recent investigations have failed to confirm the existence of such defects. Schneider (225) found that with the exception of a transient loss of cortical thymocytes at the time of weaning, dw/dw mice had normal lymphoid architecture and thymic-dependent immune responses. Similarly, Dumont et al (226) reported that dw/dw mice do not have major alterations in either T-cell or B-cell function. Although several explanations, including differences in genetic background, might contribute to the discrepant findings on the immunocompetence of dwarf mice, it is probable that animal husbandry conditions played a major role in the conflicting results. While early studies showed the mean lifespan of *dw/dw* mice to be only 2–4 months (227–229), recent investigations report a mean lifespan of 18 months (225) to 26 months (230). Hormonal reconstitution of *dw/dw* mice with GH and thyroxine corrects any deficiency in thymic-dependent immune function (231).

DIABETES-OBESITY MUTANTS   Several single gene mutations causing obesity are available for studies on the development of diabetes. Diabetes development in a given inbred mouse strain requires a complex interaction between background modifiers and the obesity-producing mutation. In most models, common features include hyperinsulinemia, hyperphagia, obesity, and pancreatic β-cell necrosis (reviewed in 232). The best understood of these single gene mutations are those designated as "diabetes" (*db*) on chromosome 4 and "obese" (*ob*) on chromosome 6. Homozygosity for these recessive obesity-causing mutations produces severe insulin resistance, a characteristic feature of many humans who have the non-insulin-dependent type of diabetes (233). Reports of an increased prevalence of infections and decreased cell-mediated immune function in patients with diabetes mellitus (234) stimulated investigations of the immune system in genetically diabetic mice. Although impairment in cell-mediated immune function in vivo in *db/db* and *ob/ob* mice has been widely reported, these mice have normal T-cell responses in vitro to mitogens and to allogeneic cells (235–241). Discrepancies between immunocompetence in vivo and in vitro of these genetically diabetic mice suggest that metabolic changes in vivo contribute to impaired immune function. In both *db/db* and *ob/ob* mice, corticosterone levels are elevated by 10–12 weeks of age (242–244). Such elevation may contribute to the observed immunological defects since glucocorticosteroids are known to have a wide range of immunosuppressive effects on a variety of lymphocyte-mediated functions in man and experimental animals (reviewed in 245). Necrosis of islet cells in *db/db* mice may be associated with autoimmunity. Anti-islet antibodies (246) and macrophages cytotoxic for cultured islet cells (247) have been reported. It has been suggested that T-cell dysfunction and autoimmunity in *db/db* mice are associated with thymic abnormalities because these mice show an accelerated age-dependent decline in numbers of thymic lymphocytes (248) and in levels of serum thymic factor (246). Similarly, impaired T-cell function may be due to hyperinsulinemia since increased insulin levels in *ob/ob* mice are associated with decreased numbers of insulin receptors on thymocytes, as well as with the impaired ability of thymocytes to transport amino acids after insulin stimulation (249).

Neurological-Neuromuscular Mutants

More than 100 neurological and neuromuscular mutations of the mouse have been described (5, 250). Recent investigations have shown a developmental and functional association between the mammalian nervous system and the immune system (251–253). Therefore, it is not surprising that several mutations first recognized because of neurological abnormalities also cause immunological defects.

WASTED   Homozygotes for the recessive mutation "wasted" (wst) on chromosome 2 can be recognized by 20 days of age by uncoordinated body movement and tremor. Affected mice develop progressive paralysis and die by 30 days of age with degeneration of neurons in the brain and spinal cord (254, 255). Immunological abnormalities in these mice include marked lymphoid hypoplasia affecting both the thymic-dependent and thymic-independent areas of the lymphoid organs (254, 256, 257), diminished DTH responses (254), reduced numbers of IgA plasma cells in the gut (258), and increased frequency of B cells that bear a large amount of surface Ig per cell (255). Abbott et al (259) have reported that wst/wst mice have a marked deficiency in levels of adenosine deaminase (ADA). They also observed an alteration in the apparent K_m of ADA for adenosine. Although ADA enzymatic activity in this mutant may be decreased, antibody studies have indicated normal levels of ADA protein (260). These findings support the suggestion that wst is a mutation in the structural gene for ADA. However, decreased ADA activity in wst/wst mice has not been confirmed in other studies (D. A. Carson, personal communication). Since the wst mutation shares some features with the human disease ataxia telangiectasia (AT), these mice have been examined for increases in chromosomal breakage and defects in DNA repair, features commonly found in cells from AT patients (261). Wasted mice have increased levels of spontaneous and gamma ray–induced chromosomal damage in bone marrow cells (254, 262). Although fibroblasts from wst/wst mice show normal DNA repair following gamma-irradiation or treatment with radiomimetic drugs (263, 264), spleen cells exhibit abnormal responses to agents that damage DNA (264). Although wst is evidently not a homologue of any known human immunodeficiency disease, it provides a valuable tool for investigating the relationships between neuropathology, immunodeficiency, chromosomal aberrations, and DNA repair.

LETHARGIC   Mice homozygous for the recessive mutation "lethargic" (lh) on chromosome 2 show an instability of gait at 15 days associated with peripheral neuropathy. No histological changes have been found in the central nervous system or in muscles (265). However, electrophysiologic

studies of homozygotes have shown peripheral neuropathies (266). Homozygotes (*lh/lh*) exhibit a generalized lymphoid hypoplasia at 2–4 weeks of
age (265) and show deficient cell-mediated immune function between 3
and 4 weeks (267). Defects in the immune system diminish with age, and
those mutant mice that survive to 45 days have normal immune function
(267). It has been suggested that immunologic abnormalities in young *lh/lh*
mice might be due to adrenocortical hypersecretion because unilaterally
adrenalectomized homozygotes show decreased mortality and improved
thymic cellularity (268).

MOTOR END-PLATE DISEASE   Homozygotes for the recessive mutation
"motor end-plate disease" (*med*) on chromosome 15 exhibit progressive
neuromuscular weakness due to a functional denervation of skeletal muscle
(269, 270). Three allelic forms of this mutation are called *med*, *med^J*, and
med^{jo}, in decreasing order of severity (271). Immunological studies have
been limited to the *med^J* allele. Since the majority of homozygotes die by
4 weeks of age, immunological analyses have focused on homozygotes at
2–3 weeks of age and on the few affected animals that survive to 4–5
months of age. Homozygotes (*med^J/med^J*) show hypoplasia of the thymus
and spleen at 2–4 weeks, decreased splenic PFC responses to SRBC at all
ages tested, reduced cytotoxic-T-cell responses to allogeneic cells at 2 weeks
of age, and decreased NK-cell activity at 4 months of age (272). Assessment
of T-cell subsets have shown increased ratios of $Ly1^+/Ly2^+$ cells compared
with $+/+$ controls (273).

DYSTROPHIA MUSCULARIS   There have been two spontaneous mutations
at the "dystrophia muscularis" (*dy*) locus on chromosome 10. Homozygotes for the recessive alleles *dy* or *dy^{2J}* develop a progressive deterio-
ration of striated muscle. Weakness and paralysis are evident by 3–4 weeks,
and homozygotes do not survive beyond 6 months (274–276). Degenerative
changes in skeletal muscle are associated with defective myelination in
the dorsal and ventral roots of the peripheral nerves, delayed onset of
myelination (277), and fewer myelinated axons in peripheral nerves (278,
279). Although defective myelination appears to be due to an intrinsic
defect at the level of Schwann cells, an extra–Schwann cell deficiency has
been suggested (280). In 1976, two reports of thymic abnormalities in *dy/dy*
mice described a transient thymic hypoplasia at 3–5 weeks of age (281,
282). Although it was suggested that *dy/dy* mice might have a defect in
immunologic function, a subsequent investigation of T-cell and B-cell
function showed only minor alterations in cell surface antigen expression
and no evidence of immunodeficiency (283). It appears that the initial
findings of thymic hypoplasia in young *dy/dy* mice may have been due to

the general poor health of these animals rather than to a direct effect of the mutation on the immune system.

Osteopetrosis mutants   Congenital osteopetrosis is a metabolic bone disease characterized by a general increase in skeletal mass. Excessive accumulation of bone results largely from a failure in bone remodeling processes. Reduced resorption of bone is thought to underlie congenital osteopetrosis in man and in experimental mammals (reviewed in 284). Investigation of possible abnormalities in the immune system of osteopetrotic patients and laboratory animals was prompted by observations of increased incidence of infections in affected patients (285) and by the finding that osteoclasts originate from mononuclear hematopoietic precursors (286). Four autosomal recessive mutations in the mouse cause a generalized reduction in bone resorption.

MICROPHTHALMIA   Alleles at the "microphthalmia" (*mi*) locus on chromosome 6 affect capacity for secondary bone resorption, eye size, and pigmentation (287–289). Although nine alleles at this locus have been described (5), immunological studies have focused on homozygotes for the (*mi*) allele. Osteoclasts in *mi/mi* mice are predominantly mononuclear and lack ruffled borders characteristic of normal osteoclasts (290). Osteoclasts are formed by the fusion of cells of the monocyte-macrophage lineage (reviewed in 287). It has been postulated that osteoclast precursor cells in *mi/mi* mice are defective in fusion; the large multinucleated osteoclasts of normal mice are present in reduced numbers at birth in mutant animals (291). Reports that normal bone resorption could be restored in *mi/mi* mice by transfer of hematopoietic cells from normal littermates and that transplantation of bone marrow cells from *mi/mi* mice could cause osteopetrosis in normal recipients (292, 293) stimulated further research. Homozygotes (*mi/mi*) show decreased chemotactic responses of macrophages (294), reduced proliferative responses of spleen cells to T-cell and B-cell mitogens, diminished PFC responses to T-dependent and T-independent antigens in vitro (295), and reduced NK-cell activity (296). However, ficoll-hypaque separated splenic lymphoid cells from *mi/mi* mice showed normal blastogenic responses to PHA and LPS (297). It has been suggested that reduced immune responses of whole spleen cell populations might be due to dilution of immunocompetent cells by stem cells and immature hematopoietic cells. This suggestion is supported by observations of increased levels of erythropoietic tissue relative to lymphoid tissue in spleens of *mi/mi* mice (297) and by the finding of a tenfold increase in the frequency of hematopoietic stem cells in the spleens of these animals (298).

GREY-LETHAL   Homozygotes for the "grey-lethal" (*gl*) mutation on chromosome 10 develop severe osteopetrosis and die at 3–5 weeks of age (287, 299, 300). As with *mi/mi* mice, the defect in bone resorption is at the level of the osteoclast. Transplantation of precursor cells from the bone marrow or spleen of normal mice reversed the defect in bone resorption (293). In 1938, Grüneberg (300) reported depletion of cortical thymocytes in *gl/gl* mice at 2–3 weeks of age. Since the teeth of these mutant mice fail to erupt (299), it is possible that the observed thymic defect was attributable to starvation. The thymic atrophy of *gl/gl* mice was reexamined by Wiktor-Jedrzejczak et al (301), who reported normal thymic cellularity at 3 weeks of age. Thymic involution, however, was observed at 5 weeks. The thymic atrophy was evidently associated with the ill health of *gl/gl* mice. Mutant thymic stromal cells in culture showed a relative increase in the frequency of cells forming colonies in vitro but no major qualitative differences in the cellular composition of stromal colonies (301).

OSTEOPETROSIS   The "osteopetrosis" (*op*) mutation maps to a linkage group initially assigned to chromosome 12 (302), but this locus is now known to be on chromosome 3. Homozygotes (*op/op*) have retarded skeletal growth and excessive bone accumulation until about 6 weeks of age. After 6 weeks, there is a significant reduction in bone matrix formation and skeletal defects slowly disappear. Thus, *op/op* mice are the longest lived of the osteopetrotic mouse mutants, with homozygotes surviving to 1–2 years of age. Although numbers of osteoclasts in *op/op* mice are greatly reduced (302), these mice are not cured by transplants of normal bone marrow or spleen cells (303). Thus, the abnormality in osteoclast development in *op/op* mice may be due to an environmental defect, reminiscent of mutations at the *Sl* locus, rather than an intrinsic impairment at the level of the hematopoietic precursor cell. Homozygotes have markedly reduced numbers of peritoneal macrophages and a monocytopenia (304). The postulate that *op/op* mice have a defective hematopoietic microenvironment is supported by the finding that their monocytes develop normally in vitro and in syngeneic +/+ hosts (304). Although the reduction of blastogenic responses of *op/op* spleen cells to T-cell and B-cell mitogens was reported in a preliminary study (305), these abnormalities were not confirmed in a subsequent investigation (298).

OSTEOSCLEROSIS   The "osteosclerosis" (*oc*) mutation on chromosome 19 was first observed in 1966. Most of the affected mice die by 3 weeks of age with a generalized increase in skeletal density and an absence of marrow cavities (306). Morphological evidence of rachitic changes in *oc/oc* mice distinguishes this mutant from the osteopetrotic mutants described above (306, 307). As in *mi/mi* mice, osteoclasts from *oc/oc* mice do not exhibit the ruffled borders characteristic of normal bone resorbing cells (287). Homozygotes (*oc/oc*) exhibit increased splenic erythropoiesis and show normal cellularity of the thymus at 3 weeks of age (298).

There is considerable heterogeneity among the human osteopetrosic syndromes (284, 287). Limited investigation of immunological abnormalities in several patients has revealed depressed bactericidal activity of peripheral blood monocytes, impaired NK-cell activity, and reduced thymic hormone levels (285, 308, 309). Because osteopetrosis was cured in *mi/mi* and *gl/gl* mice by treatment with bone marrow and spleen cells (292, 293), attempts to use bone marrow therapy in humans have been initiated. Although such treatment has met with only limited success, there are reports of correction of osteopetrosis following bone marrow therapy (308–310). Moreover, in several of these young patients, the immunological deficits were reversed (308, 309).

SUMMARY

We have discussed more than 30 mutant genes known to cause abnormalities in the development and regulation of the immune system. The loci defined by these deleterious alleles have been assigned to 13 different autosomal chromosomes in addition to X and Y. It is important to note that these single genes do not act alone but function in concert with the background genome. Studies of these mutations on different inbred strain backgrounds are contributing important information on the influence of background modifying genes. The development of stocks of mice carrying multiple mutations on an inbred strain background enables the use of a well-characterized mutation to explore a less-well-understood genetic model. Investigators are urged to assure proper conditions for studies with immunological mutants by using the appropriate methods of animal husbandry. A detailed guide for maintaining immunologically compromised rodents has been prepared (311). These experiments performed by nature provide a valuable resource for investigating the immune system in normal and pathologic states. As the gene products of the loci defined by these mutations become known, the information obtained will provide additional insight into mechanisms underlying normal immune function as well as immunologic disease processes in man.

ACKNOWLEDGMENTS

This work was supported by United States Public Health Service Grants CA 20408, CA 35845, and AI 20232. We thank Dale Rex Coman, Margaret Green, and Derry Roopenian for critical reading of the manuscript. We are grateful to colleagues who sent us preprints of their work.

Literature Cited

1. Steinberg, A. D., Huston, D. P., Taurog, J. D., Cowdery, J. S., Raveche, E. S. 1981. The cellular and genetic basis of murine lupus. *Immunol. Rev.* 55: 121

2. Theofilopoulos, A. N., Dixon, F. J. 1981. Etiopathogenesis of murine systemic lupus erythematosus. *Immunol. Rev.* 55: 179

3. Smith, H. R., Steinberg, A. D. 1983. Autoimmunity—A perspective. *Ann. Rev. Immunol.* 1: 175

4. Theofilopoulos, A. N., Dixon, F. J. 1985. Murine models of systemic lupus erythematosus. *Adv. Immunol.* 37: 269

5. Green, M. C. 1981. *Genetic Variants and Strains of Laboratory Animals.* New York: Fisher

6. Flanagan, S. P. 1966. "Nude," a new hairless gene with pleiotropic effects in the mouse. *Genet. Res.* 8: 295

7. Pantelouris, E. M. 1968. Absence of thymus in a mouse mutant. *Nature* 217: 370

8. Reed, N. D., ed. 1982. *Proceedings of the Third International Workshop on Nude Mice, Vols. 1 and 2.* New York: Gustav Fischer

9. Sordat, B., ed. 1984. *Immune-Deficient Animals.* New York: Karger

10. Fogh, J., Giovanella, B. C., eds. 1982. *The Nude Mouse in Experimental and Clinical Research, Vols. 1 and 2.* New York: Academic Press

11. Kindred, B. 1981. Deficient and sufficient systems in the nude mouse. In *Immunologic Defects in Laboratory Animals, Vol. 1,* ed. M. E. Gershwin, B. Merchant, pp. 215–65. New York: Plenum

12. Gershwin, M. E. 1977. DiGeorge Syndrome: Congenital thymic hypoplasia. *Am. J. Pathol.* 89: 809

13. Asherson, G. L., Webster, A. D. P., eds. 1980. *Diagnosis and Treatment of Immunodeficiency Diseases.* London: Blackwell Sci. Publ.

14. Mandel, T. 1970. Differentiation of epithelial cells in the mouse thymus. *Z. Zellforsch.* 106: 498

15. Jenkinson, E. J., Van Ewijk, W., Owen, J. J. T. 1981. Major histocompatibility antigen complex expression on the epithelium of the developing thymus in normal and nude mice. *J. Exp. Med.* 153: 280

16. Van Vliet, E., Jenkinson, E. J., Kingston, R., Owen, J. J. T., Van Ewijk, W. V. 1985. Stromal cell types in the developing thymus of the normal and nude mouse embryo. *Eur. J. Immunol.* 15: 675

17. Hunig, T. 1983. T-cell function and specificity in athymic mice. *Immunol. Today* 4: 84

18. MacDonald, H. R. 1984. Phenotypic and functional characteristics of "T-like" cells in nude mice. In *Immune-Deficient Animals,* ed. B. Sordat, pp. 2–6. Basel: Karger

19. Ikehara, S., Pahwa, R. N., Fernandes, G., Hansen, C. T., Good, R. A. 1984. Functional T cells in athymic nude mice. *Proc. Natl. Acad. Sci. USA* 81: 886

20. Chen, W. F., Scollay, R., Shortman, K., Skinner, M. K., Marbrook, J. 1984. T-cell development in the absence of a thymus: The number, the phenotype, and the functional capacity of T lymphocytes in nude mice. *Am. J. Anat.* 170: 339

21. MacDonald, H. R., Blanc, C., Lees, R. K., Sordat, B. 1986. Abnormal distribution of T cell subsets in athymic mice. *J. Immunol.* 136: 4337

22. Meltzer, M. S. 1976. Tumoricidal responses in vitro of peritoneal macrophages from conventionally housed and germ-free nude mice. *Cell. Immunol.* 22: 176–81

23. Cheers, C., Waller, R. 1975. Activated macrophages in congenitally athymic "nude" mice and in lethally irradiated mice. *J. Immunol.* 115: 844

24. Sharp, A. K., Colson, M. J. 1984. Elevated macrophage activity in nude mice. See Ref. 18, pp. 44–47

25. Herberman, R. B., Holden, H. T. 1978. Natural cell-mediated immunity. *Adv. Cancer Res.* 27: 305

26. Clark, E. A., Harmon, R. C. 1980. Genetic control of natural cytotoxicity and hybrid resistance. *Adv. Cancer Res.* 31: 227

27. Shultz, L. D., Bedigian, H. G., Heiniger, H. J., Eicher, E. M. 1982. The congenitally athymic streaker mouse. See Ref. 8, pp. 33–39

28. Hartley, J. W., Wolford, N. K., Old, L. J., Rowe, W. P. 1977. A new class of murine leukemia virus associated with development of spontaneous lymphomas. *Proc. Natl. Acad. Sci. USA* 74: 789

29. Festing, M. W. 1982. Characteristics of nude rats. See Ref. 8, pp. 41–49

30. Reed, C., O'Donoghue, J. L. 1982. The hairless athymic guinea pig. See Ref. 8, pp. 51–57

31. Edelson, R. L., Fink, J. M. 1985. The immunologic function of skin. *Sci. Am.* 252: 46

32. Streilein, J. W. 1983. Skin-associated lymphoid tissues (SALT): Origins and functions. *J. Invest. Dermatol.* 80(6): 12s (Suppl.)

33. Shultz, L. D. 1987. Pleiotropic mutations causing abnormalities in the murine immune system and the skin. *Curr. Probl. Dermatol.* 17: In press

34. Stingl, G., Tamaki, K., Katz, S. 1980. Origin and function of epidermal Langerhans cells. *Immunol. Rev.* 53: 149

35. Roberts, L. K., Spangrude, G. J., Daynes, R. A., Krueger, G. G. 1985. Correlation between keratinocyte expression of Ia and the intensity and duration of contact hypersensitivity responses in mice. *J. Immunol.* 135: 2929

36. Bosma, G. C., Custer, R. P., Bosma, M. J. 1983. A severe combined immunodeficiency mutation in the mouse. *Nature* 301: 527

37. Custer, R. P., Bosma, G. C., Bosma, M. J. 1985. Severe combined immunodeficiency (SCID) in the mouse. Pathology, reconstitution, neoplasms. *Am. J. Pathol.* 120: 464

38. Phillips, R. A., Bosma, M., Dorschkind, K. 1984. Reconstitution of immune-deficient mice with cells from long-term bone marrow cultures. In *Long-Term Bone Marrow Cultures*, ed. J. S. Greenberger, G. J. Nemo, D. G. Wright, pp. 309–21. New York: Liss

39. Fulop, G. M., Phillips, R. A. 1986. Full reconstitution of the immunodeficiency in *scid* mice with normal stem cells requires low dose irradiation of the recipients. *J. Immunol.* 136: 4438

40. Fulop, G. M., Bosma, G. C., Bosma, M. J., Phillips, R. A. 1987. Evidence for normal numbers of early B cell precursors in *Scid* mice. Submitted for publication

41. Schuler, W., Weiler, I. J., Schuler, A., Phillips, R. A., Rosenberg, N., Mak, T. W., Kearney, J. F., Perry, R. P., Bosma, M. J. 1986. Rearrangement of antigen receptor genes is defective in mice with severe combined immune deficiency. *Cell* 46: 963

42. Yancopoulos, G. D., Blackwell, T. K., Suh, H., Hood, L., Alt, F. W. 1986. Introduced T cell receptor variable region gene segments recombine in Pre-B cells: Evidence that T and B cells use a common recombinase. *Cell* 44: 251

43. Desiderio, S. V., Yancopoulos, G. D., Paskind, M., Thomas, E., Boss, M. A., Landau, N., Alt, F. W., Baltimore, D. 1984. Insertions of N-regions into heavy chain genes is correlated with

expression of terminal deoxytransferase in B cells. *Nature* 311: 752

44. Hirschhorn, R., Vawter, G. F., Kirkpatrick, J. A., Rosen, F. S. 1979. Adenosine deaminase deficiency: Frequency and comparative pathology in autosomally recessive severe combined immunodeficiency. *Clin. Immunol. Immunopathol.* 14: 107

45. Dorshkind, K., Keller, G. M., Phillips, R. A., Miller, R. G., Bosma, G. C., O'Toole, M., Bosma, M. J. 1984. Functional status of cells from lymphoid and myeloid tissues in mice with severe combined immunodeficiency disease. *J. Immunol.* 132: 1804

46. Czitrom, A. A., Edwards, S., Phillips, R. A., Bosma, M. J., Marrack, P., Kappler, J. W. 1985. The function of antigen-presenting cells in mice with severe combined immunodeficiency. *J. Immunol.* 134: 2276

47. Dorshkind, K., Pollack, S. B., Bosma, M. J., Phillips, R. A. 1985. Natural killer (NK) cells are present in mice with severe combined immunodeficiency. *J. Immunol.* 134: 3798

48. Unanue, E. R. 1981. The regulatory role of macrophages in antigenic stimulation. II. Symbiotic relationship between lymphocytes and macrophages. *J. Immunol.* 108: 1110

49. Bancroft, G. J., Bosma, M. J., Bosma, G. C., Unanue, E. R. 1986. Regulation of macrophage Ia expression in mice with severe combined immunodeficiency: Induction of Ia expression by a T-cell independent mechanism. *J. Immunol.* 137: 4

50. Hackett, J., Bosma, G. C., Bosma, M. J., Bennet, M., Kumar, V. 1986. Transplantable progenitors of natural killer cells are distinct from those of T and B lymphocytes. *Proc. Natl. Acad. Sci. USA* 83: 3427

51. Rosen, F. J., Cooper, M. D., Wedgewood, R. J. P. 1984. The primary immunodeficiencies (second of two parts). *New Engl. J. Med.* 311: 300

52. Ware, C. F., Donato, N. J., Dorschkind, K. 1985. Human, rat or mouse hybridomas secrete high levels of monoclonal antibodies following transplantation into mice with severe combined immunodeficiency disease (SCID). *J. Immunol. Meth.* 85: 353

53. Shultz, L. D., Coman, D. R., Bailey, C. L., Beamer, W. G., Sidman, C. L. 1984. "Viable motheaten," a new allele at the motheaten locus. I. Pathology. *Am. J. Pathol.* 116: 179

54. Green, M. C., Shultz, L. D. 1975. Motheaten, an immunodeficient

mutant of the mouse. I. Genetics and Pathology. *J. Hered.* 66: 250

55. Lutzner, M. A., Hansen, C. T. 1976. Motheaten: An immunodeficient mouse with markedly less ability to survive than the nude mouse in a germfree environment. *J. Immunol.* 116: 1496

56. Shultz, L. D., Green, M. C. 1976. Motheaten, an immunodeficient mutant of the mouse. II. Depressed immune competence and elevated serum immunoglobulins. *J. Immunol.* 116: 936

57. Sidman, C. L., Shultz, L. D., Unanue, E. R. 1978. The mouse mutant motheaten. II. Functional studies of the immune system. *J. Immunol.* 121: 2399

58. Davidson, W. F., Morse, H. C. III, Sharrow, S. O., Chused, T. M. 1979. Phenotypic and functional effects of the motheaten gene on murine B and T lymphocytes. *J. Immunol.* 122: 884

59. Clark, E. A., Shultz, L. D., Pollack, S. B. 1981. Mutations in mice that influence natural killer (NK) cell activity. *Immunogenetics* 12: 601

60. Stanton, T. H., Tubbs, C., Clagett, J. 1985. Cytokine production and utilization by the motheaten mouse. *J. Immunol.* 135: 4021

61. Shultz, L. D., Zurier, R. B. 1978. Motheaten, a single gene model for stem cell dysfunction and early onset autoimmunity. In *Genetic Control of Autoimmune Disease,* ed. N. R. Rose, P. E. Bigazzi, N. L. Warner, pp. 229–40. Amsterdam: Elsevier/North Holland

62. Ward, J. M. 1978. Pulmonary pathology of the motheaten mouse. *Vet. Pathol.* 15: 170

63. Rossi, G. A., Hunninghake, G. W., Kawanami, O., Ferrans, V. J., Hansen, C. T., Crystal, R. G. 1985. Motheaten mice—an animal model with an inherited form of interstitial lung disease. *Am. Rev. Respir. Dis.* 131: 150

64. McCoy, K. L., Chi, E., Engel, D., Clagett, J. 1983. Accelerated rate of mononuclear phagocyte production in vitro by splenocytes from autoimmune motheaten mice. *Am. J. Pathol.* 112: 18

65. Sidman, C. L., Marshall, J. D., Masiello, N. C., Roths, J. B., Shultz, L. D. 1984. Novel B-cell maturation factor from spontaneously autoimmune viable motheaten mice. *Proc. Natl. Acad. Sci. USA* 81: 7199

66. Sidman, C. L., Shultz, L. D., Evans, R. 1985. A serum-derived molecule from autoimmune viable motheaten mice potentiates the action of a B cell maturation factor. *J. Immunol.* 135: 870

67. Sidman, C. L., Shultz, L. D., Hardy, R. R., Hayakawa, K., Herzenberg, L. A. 1986. Production of immunoglobulin isotypes by Ly-1$^+$ B cells in viable motheaten and normal mice. *Science* 232: 1423

68. Herzenberg, L. A., Stall, A. M., Lalor, P. A., Sidman, C. L., Moore, W. A., Parks, D. R., Herzenberg, L. A. 1986. The Ly-1 B cell lineage. *Immunol. Rev.* 93: 81

69. Shultz, L. D., Coman, D. R., Sidman, C. L., Lyons, B. 1986. Development of plasmacytoid cells with Russell bodies in autoimmune viable motheaten mice. *Am. J. Pathol.* In press

70. Blom, J., Mansa, B., Wiik, A. 1976. A study of Russell bodies in human monoclonal plasma cells by means of immunofluorescence and electron microscopy. *Acta Pathol. Microbiol. Scand. Sect. A* 84: 335

71. Gebbers, J. O., Otto, H. F. 1976. Plasma cell alterations in ulcerative colitis. An electron microscopic study. *Pathol. Eur.* 11: 27

72. Armstrong, J. A., Dawkins, R. L., Horne, R. 1985. Retroviral infection of accessory cells and the immunological paradox in AIDS. *Immunol. Today* 6: 121

73. Greiner, D. L., Goldschneider, I., Komschlies, K. K., Medlock, E. S., Bollum, F. J., Shultz, L. D. 1986. Defective lymphopoiesis in the bone marrow of motheaten (*me/me*) and viable motheaten (*me^v/me^v*) mutant mice. I. Analysis of the development of prothymocytes, early lineage B cells and terminal deoxynucleotidyl transferase-positive cells. *J. Exp. Med.* 164: 1129

74. Landreth, K. S., McCoy, K., Clagett, J., Bollum, F. J., Rosse, C. 1981. Deficiency in cells expressing terminal transferase in autoimmune (motheaten) mice. *Nature* 290: 409

75. Rosen, F. S., Wedgewood, R. J., Aiuti, F., Cooper, M. D., Good, R. A., Hanson, L. A., Hitzig, W. H., Matsumoto, S., Seligman, M., Soothill, J. F., Waldman, T. A. 1983. Meeting report. Primary immunodeficiency diseases. *Clin. Immunol. Immunopathol.* 28: 450

76. Amsbaugh, D. F., Hansen, C. T., Prescott, B., Stashak, P. W., Barthold, D. R., Baker, P. J. 1972. Genetic control of the antibody response to type III pneumococcal polysaccharide in mice. *J. Exp. Med.* 136: 931

77. Mosier, D. E., Mond, J. J., Goldings, E. A. 1977. The ontogeny of thymic

independent antibody responses in vitro in normal mice and mice with an X-linked B cell defect. *J. Immunol.* 119: 1874

78. Scher, I. 1982. CBA/N immune defective mice: Evidence for the failure of a B cell subpopulation to be expressed. *Immunol. Rev.* 64: 117

79. Ahmed, A., Scher, I., Sharrow, S. D., Smith, A. H., Paul, W. E., Sachs, D. H., Sell, K. W. 1977. B-lymphocyte heterogeneity: Development and characterization of an alloantiserum which distinguishes B-lymphocyte differentiation alloantigens. *J. Exp. Med.* 145: 101

80. Huber, B., Gershon, R. K., Cantor, H. 1977. Identification of a B cell surface structure involved in antigen-dependent triggering: Absence of this structure on B cells from CBA/N mutant mice. *J. Exp. Med.* 145: 10

81. Subbarao, B., Ahmed, A., Paul, W. E., Scher, I., Lieberman, R., Mosier, D. E. 1979. Lyb-7, a new B cell alloantigen controlled by genes linked to the IgC_H locus. *J. Immunol.* 122: 2279

82. Smith, H. R., Yaffe, L. J., Kastner, D. L., Steinberg, A. D. 1986. Evidence that Lyb-5 is a differentiation antigen in normal and *xid* mice. *J. Immunol.* 136: 1194

83. Eldridge, J. H., Kiyono, H., Michalek, S. M., McGhee, J. R. 1983. Evidence for a mature B cell subpopulation in Peyer's patches of young adult *xid* mice. *J. Exp. Med.* 157: 789

84. Eldridge, J. H., Yaffe, L. J., Ryan, J. J., Kiyono, H., Scher, I., McGhee, J. R. 1984. Expression of Lyb-5 and Mls determinants by Peyer's patch B-lymphocytes of X-linked immunodeficient mice. *J. Immunol.* 133: 2308

85. Hardy, R. R., Hayakawa, K., Parks, D. R., Herzenberg, L. A. 1983. Demonstration of B cell maturation in X-linked immunodeficient mice by simultaneous three-colour immunofluorescence. *Nature* 306: 270

86. Mosier, D. E. 1985. Are *xid* B lymphocytes representative of any normal B cell population? *Mol. Cell. Immunol.* 2: 170

87. Wortis, H. H., Burkley, L., Hughes, D., Roschelle, S., Waneck, G. 1982. Lack of mature B cells in nude mice with X-linked immune deficiency. *J. Exp. Med.* 155: 903

88. Mond, J. J., Scher, I., Cossman, J., Kessler, S., Mongini, A., Hansen, C., Finkelman, F. D., Paul, W. E. 1982. Role of the thymus in directing the development of a subset of B lympho

cytes. *J. Exp. Med.* 155: 924

89. Karagogeos, D., Rosenberg, N., Wortis, H. H. 1986. Early arrest of B cell development in nude, X-linked immune deficient mice. *Eur. J. Immunol.* 16: 1125

90. Cohen, D. I., Hedrick, S. M., Nielsen, E. A., D'Eustachio, P., Ruddle, F., Steinberg, A. D., Paul, W. E., Davis, M. M. 1985. Isolation of a cDNA clone corresponding to an X-linked gene family (XLR) closely linked to the murine immunodeficiency disorder (*xid*). *Nature* 314: 369

91. Cohen, D. I., Steinberg, A. D., Paul, W. E., Davis, M. M. 1985. Expression of an X-linked gene family (XLR) in late stage B cells and its alteration by the *xid* mutation. *Nature* 314: 372

92. Ohno, S. 1967. *Sex Chromosomes and Sex-Linked Genes.* New York: Springer-Verlag

93. Spitler, L. N., Kevin, A. S., Stites, D. P., Fudenberg, H., Huber, H. 1975. The Wiscott-Aldrich syndrome. Immunologic studies in nine patients and selected family members. *Cell. Immunol.* 19: 201

94. Takatsu, K., Hamaoka, T. 1982. DBA/2H mice as a model of an X-linked immunodeficiency which is defective in the expression of TRF-acceptor site(s) on B lymphocytes. *Immunol. Rev.* 64: 25

95. Sidman, C. L., Marshall, J. D., Beamer, W. G., Nadeau, J. H., Unanue, E. R. 1986. Two loci affecting B cell responses to B cell maturation factors. *J. Exp. Med.* 163: 116

96. Scibienski, R. J. 1981. Defects in murine responses to bacterial lipopolysaccharide. See Ref. 11, Vol. 2, pp. 241–58

97. Morrison, D. C., Ryan, J. L. 1979. Bacterial endotoxins and host immune responses. *Adv. Immunol.* 28: 293

98. Hill, A. B., Hatswell, J. M., Topley, W. W. C. 1940. The inheritance of resistance, demonstrated by the development of a strain of mice resistant to experimental inoculation with a bacterial endotoxin. *J. Hyg.* 40: 538

99. Glode, L. M., Rosenstreich, D. L. 1976. Genetic control of B cell activation by bacterial lipopolysaccharide is mediated by multiple distinct genes or alleles. *J. Immunol.* 117: 2061

100. Vogel, S. N., Weinblatt, A. C., Rosenstreich, D. L. 1981. Inherent macrophage defects in mice. See Ref. 11, Vol. 1, pp. 327–57

101. Ruco, L. P., Meltzer, M. S., Rosenstreich, D. L. 1978. Macrophage activa-

tion for tumor cytotoxicity: Control of macrophage tumoricidal capacity by the LPS gene. *J. Immunol.* 121 : 543

102. Tagliabue, A., McCoy, J. L., Herberman, R. B. 1978. Refractoriness to migration inhibitory factor of macrophages of LPS nonresponder mouse strains. *J. Immunol.* 121 : 1223

103. Vogel, S. N., Rosenstreich, D. L. 1979. Defective Fc receptor-mediated phagocytosis in C3H/HeJ macrophages. I. Correction by lymphokine-induced stimulation. *J. Immunol.* 123 : 2842

104. Vogel, S. N., Fertsch, D. 1984. Endogenous interferon production by endotoxin-responsive macrophages provides an autostimulatory differentiation signal. *Infect. Immun.* 45 : 417

105. Fertsch, D., Vogel, S. N. 1984. Recombinant interferons increase macrophage Fc receptor capacity. *J. Immunol.* 132 : 2436

106. Vogel, S. N., Fertsch, D., Falk, L. A. 1986. Effects of endogenous and exogenous interferon signals on macrophages. *UCLA Symposium on Molecular and Cellular Biology, New Series, Vol. 50, Interferons as Cell Growth Inhibitors and Antitumor Factors*, ed. R. Freedman, T. Meregan, T. Sreevalsan. New York: Liss

107. Glode, L. M., Scher, I., Osborne, B., Rosenstreich, D. L. 1976. Cellular mechanism of endotoxin unresponsiveness in C3H/HeJ mice. *J. Immunol.* 116 : 454

108. Coutinho, A., Forni, L., Watanabe, T. 1978. Genetic and functional characterization of an antiserum to the lipid A-specific triggering receptor on murine B lymphocytes. *Eur. J. Immunol.* 8 : 63

109. Gregory, S. H., Zimmerman, D. H., Kern, M. 1980. The lipid A moiety of lipopolysaccharide is specifically bound to B cell subpopulations of responder and nonresponder animals. *J. Immunol.* 125 : 102

110. Eda, Y., Ohara, J., Watanabe, T. 1983. Restoration of LPS responsiveness of C3H/HeJ mouse lymphocytes by microinjection of cytoplasmic factors from LPS-stimulated normal lymphocytes. *J. Immunol.* 131 : 1294

111. Vogel, S. N., Madonna, G. S., Wahl, L. M., Rick, P. D. 1984. Stimulation of spleen cells and macrophages of C3H/HeJ mice by a lipid A precursor derived from *Salmonella typhimurium*. *Rev. Infect. Dis.* 6 : 535

112. Jakobovits, A., Sharon, N., Zan-bar, I. 1982. Acquisition of mitogenic responsiveness by nonresponding lymphocytes upon insertion of appropriate membrane components. *J. Exp. Med.* 156 : 1274

113. Watanabe, T., Ohara, J. 1981. Functional nuclei of LPS-nonresponder C3H/HeJ mice after transfer into LPS-responder C3H/HeN cells by cell fusion. *Nature* 290 : 58

114. Kuus-Reichel, K., Ulevitch, R. J. 1986. Partial restoration of the lipopolysaccharide-induced proliferative response in splenic B cells from C3H/HeJ mice. *J. Immunol.* 137 : 472

115. Moore, D. H., Long, C. A., Vaidya, A. B. 1979. Mammary tumor viruses. *Adv. Cancer Res.* 29 : 347

116. Hilgers, J., Bentvelzen, P. 1978. Interaction between viral and genetic factors in murine mammary cancer. *Adv. Cancer Res.* 26 : 143

117. Outzen, H. C., Corrow, D., Shultz, L. D. 1985. Attenuation of exogenous murine mammary tumor virus in the C3H/HeJ substrain bearing the *Lps* mutation. *J.N.C.I.* 75 : 917

118. Bentvelzen, P., Brinkhof, J. 1977. Organ distribution of exogenous murine mammary tumor virus as determined by bioassay. *Eur. J. Cancer* 13 : 241

119. Gillette, R. W., Robertson, S., Brown, R., Blackman, K. E. 1974. Expression of mammary tumor virus antigen on the membranes of lymphoid cells. *J.N.C.I.* 53 : 499

120. Müller-Eberhard, H. J. 1986. The membrane attack complex of complement. *Ann. Rev. Immunol.* 4 : 503

121. Perlmutter, D. H., Colten, H. R. 1986. Molecular immunology of complement biosynthesis. *Ann. Rev. Immunol.* 4 : 231

122. Shreffler, D. 1979. Variants of complement and other proteins: Mouse. In *Inbred and Genetically Defined Strains of Laboratory Animals. Part 1. Mouse and Rat*, ed. P. L. Altman, D. L. Katz, p. 103. Bethesda: Fed. Am. Soc. Exp. Biol.

123. Alper, C. A., Rosen, F. S. 1971. Genetic aspects of the complement system. *Adv. Immunol.* 14 : 251

124. Ooi, Y. M., Colten, H. R. 1979. Genetic defect in secretion of complement C5 in mice. *Nature* 282 : 207

125. Hammer, C. H., Gaither, T., Frank, M. M. 1981. Complement deficiencies of laboratory animals. See Ref. 11, Vol. 2, pp. 207–40

126. Chan, A. C., Karp, D. R., Shreffler, D. C., Atkinson, J. P. 1984. The 20 faces of the fourth component of complement. *Immunol. Today* 5 : 200

127. Atkinson, J. P., McGinnis, K., Brown, L., Peterein, J., Shreffler, D. 1980. A murine C4 molecule with reduced hemolytic efficiency. *J. Exp. Med.* 151: 492

128. Karp, D. R., Atkinson, J. P., Shreffler, D. C. 1982. Genetic variation of the fourth component of murine complement. *J. Biol. Chem.* 257: 7330

129. Day, N. K., Good, R. A. 1975. Deficiencies of the complement system in man. *Birth Defects Orig. Art. Ser.* 11: 306

130. Amman, A. J. 1977. Immunodeficiency and autoimmunity. In *Autoimmunity*, ed. N. Talal, pp. 479–512. New York: Academic Press

131. Touraine, J.-L. 1976. Autoimmunity in children and animals with a primary immunodeficiency. *Paediatrician* 5: 18

132. Murphy, E. D., Roths, J. B. 1978. Autoimmunity and lymphoproliferation: Induction by mutant gene *lpr* and acceleration by a male-associated factor in strain BXSB mice. See Ref. 61, pp. 207–21

133. Murphy, E. D. 1981. Lymphoproliferation (*lpr*) and other single locus models for murine lupus. See Ref. 11, Vol. 2, pp. 143–73

134. Murphy, E. D., Roths, J. B. 1979. A Y chromosome associated factor in strain BXSB producing accelerated autoimmunity and lymphoproliferation. *Arthritis Rheum.* 22: 1188

135. Roths, J. B., Murphy, E. D., Eicher, E. M. 1984. A new mutation, *gld* that produces lymphoproliferation and autoimmunity in C3H/HeJ mice. *J. Exp. Med.* 159: 1

136. Andrews, B. S., Eisenberg, R. A., Theofilopoulos, A. N., Izui, S., Wilson, C. B., McConahey, P. J., Murphy, E. D., Roths, J. B., Dixon, F. J. 1978. Spontaneous murine lupus-like syndromes. Clinical and immunological manifestations in several strains. *J. Exp. Med.* 148: 1198

137. Morse, H. C., Davidson, W. F., Yetter, R. A., Murphy, E. D., Roths, J. B., Coffman, R. L. 1982. Abnormalities induced by the mutant gene *lpr*: Expansion of a unique lymphocyte subset. *J. Immunol.* 129: 2612

138. Davidson, W. F., Holmes, K. L., Roths, J. B., Morse, H. C. III. 1985. Immunologic abnormalities of mice bearing the *gld* mutation suggest a common pathway for murine nonmalignant lymphoproliferative disorders with autoimmunity. *Proc. Natl. Acad. Sci. USA* 82: 1219

139. Mountz, J. D., Mushinski, F., Smith, H. R., Klinman, D. M., Steinberg, A. D. 1985. Modulation of *C-myb* transcription in autoimmune disease by cyclophosphamide. *J. Immunol.* 135: 2417

140. Hudgins, C. C., Steinberg, R. T., Klinman, D. M., Reeves, M. J. P., Steinberg, A. D. 1985. Studies of consomic mice bearing the Y chromosome of the BXSB mouse. *J. Immunol.* 134: 3849

141. Simon, M. M., Prester, M., Nerz, G., Kuppers, R. C. 1984. Quantitative studies on T-cell functions in MRL/Mp-*lpr/lpr* mice. 1. Frequency analysis of precursor cells of proliferating, cytotoxic, and T-cell growth factor-secreting T lymphocytes reveals an increase in absolute numbers, with a concomitant decrease in percentage of immunocompetent cells. *Cell Immunol.* 33: 39

142. Hefeneider, S. H., Conlon, P. J., Dower, S. K., Henney, C. S., Gillis, S. 1984. Limiting dilution analysis of interleukin 2 and colony-stimulating factor producer cells in normal and autoimmune mice. *J. Immunol.* 132: 1863

143. Davignon, J. L., Budd, R. C., Ceredig, R., Piguet, P. F., Macdonald, H. R., Cerottini, J. C., Vassalli, P., Izui, S. 1985. Functional analysis of T cell subsets from mice bearing the *lpr* gene. *J. Immunol.* 135: 2423

144. Lu, C., Unanue, E. R. 1982. Spontaneous T-cell lymphokine production and enhanced macrophage Ia expression and tumoricidal activity in MRL-lpr mice. *Clin. Immunol. Immunopathol.* 25: 213

145. Kelley, V., Roths, J. B. 1982. Increase in macrophage Ia expression in autoimmune mice. *J. Immunol.* 129: 923

146. Kofler, R., Schreiber, R. D., Dixon, F. J., Theofilopoulos, A. N. 1984. Macrophage l-A/l-E expression and macrophage-stimulating lymphokines in murine lupus. *Cell. Immunol.* 87: 92

147. Kelley, V. E., Wing, E. 1982. Loss of resistance to Listeria infection in autoimmune MRL/lpr mice: Protection by prostaglandin E₁. *Clin. Immunol. Immunopathol.* 23: 705

148. Dauphinee, M. J., Talal, N. 1984. Increased Ia expression, T lymphocyte subset abnormalities and autoimmunity in murine strains bearing the *lpr* gene. *Clin. Exp. Immunol.* 58: 145

149. Wofsy, D., Kerger, C. E., Seamen, W. E. 1984. Monocytosis in the BXSB model for systemic lupus erythematosis. *J. Exp. Med.* 159: 629

150. Searle, A. G. 1959. Hereditary absence

of spleen in the mouse. *Nature* 184: 1419

151. Green, M. C. 1967. A defect of the splanchnic mesoderm caused by the mutant gene dominant hemimelia in the mouse. *Dev. Biol.* 15: 62

152. Haaijman, J. J., Schuit, H. R. E., Hijmans, W. 1977. Immunoglobulin-containing cells in different lymphoid organs of the CBA mouse during its lifespan. *Immunology* 32: 427

153. Fletcher, M. P., Ikeda, R. M., Gershwin, M. E. 1977. Splenic influences on T cell function. The immunobiology of the inbred hereditarily asplenic mouse. *J. Immunol.* 119: 110

154. Lozzio, B. B., Wargon, L. B. 1974. Immune competence of hereditarily asplenic mice. *Immunology* 27: 167

155. Barnett, J. B., Wust, C. J. 1978. Levels of homocytotropic antibody in hereditarily asplenic, splenectomized, and normal mice. *Int. Arch. Allergy Appl. Immunol.* 56: 558

156. Welles, W. L., Battisto, J. R. 1976. Splenic input on immune capability of lymphoid cells. In *Immuno-Aspects of the Spleen*, ed. J. R. Battisto, J. W. Streilein, pp. 157–70. Amsterdam: Elsevier/North Holland

157. Welles, W. F., Battisto, J. R. 1981. The significance of hereditary asplenia for immunologic competence. See Ref. 11, pp. 191–212. New York: Plenum

158. Oster, O. N., Koontz, L. C., Wyler, D. J. 1980. Malaria in asplenic mice: Effects of splenectomy, congenital asplenia, and splenic reconstitution on the course of infection. *Am. J. Trop. Med. Hyg.* 29: 1138

159. Wlodarski, K., Morrison, K., Michowski, D. 1982. The effect of Dominant hemimelia genes on the number of mast cells in the lymph nodes. *Folia biologia (Praha)* 28: 255

160. McKusick, V. 1983. *Mendelian Inheritance in Man*. Baltimore: Johns Hopkins Univ. Press. 6th ed.

161. Kevy, S. V., Tefft, M., Vawier, G. F., Rosen, F. S. 1968. Hereditary splenic hypoplasia. *Pediatrics* 42: 752

162. Lane, P. W., Murphy, E. D. 1972. Susceptibility to spontaneous pneumonitis in an inbred strain of beige and satin mice. *Genetics* 72: 451

163. Elin, R. J., Edelin, J. B., Wolff, S. M. 1974. Infections and immunoglobulin concentrations in Chediak-Higashi mice. *Infect. Immunol.* 10: 88

164. Kirkpatrick, C. E., Farrell, J. P. 1982. Leishmaniasis in beige mice. *Infect. Immunol.* 38: 1208

165. Shellam, G. R., Allan, J. E., Papadi-mitriou, J. M., Bancroft, G. J. 1981. Increased susceptibility to cytomegalovirus infection in beige mutant mice. *Proc. Natl. Acad. Sci. USA* 78: 5104

166. Roder, J., Duwe, A. 1979. The beige mutation in the mouse selectively impairs natural killer cell function. *Nature* 278: 451

167. Roder, J. C., Lohmann-Matthes, M. L., Domzig, W., Wigzell, H. 1979. The beige mutation in the mouse. II. Selectivity of the natural killer (NK) cell defect. *J. Immunol.* 123: 2174

168. Karre, K., Klein, G. O., Kiessling, R., Klein, G., Roder, J. C. 1980. Low natural in vivo resistance to syngeneic leukemias in natural killer-deficient mice. *Nature* 284: 624

169. Talmadge, J. E., Meyers, K. M., Prieur, D. J., Starkey, J. R. 1980. Role of NK cells in tumour growth and metastasis in beige mice. *Nature* 284: 622

170. Gallin, J. I., Bujak, J. S., Patten, E., Wolff, S. M. 1974. Granulocyte function in the Chediak-Higashi syndrome of mice. *Blood* 43: 201

171. Mahoney, K. H., Morse, S. S., Morahan, P. S. 1980. Macrophage functions in beige (Chediak-Higashi syndrome) mice. *Cancer Res.* 40: 3934

172. Saxena, R. K., Saxena, Q. B., Adler, W. H. 1982. Defective T-cell response in beige mutant mice. *Nature* 295: 240

173. Halle-Pannenko, O., Bruley-Rosset, M. 1985. Decreased graft-versus-host reaction and T cell cytolytic potential of beige mice. *Transplantation* 39: 85

174. Carlson, G. A., Marshall, S. T., Truesdale, A. T. 1984. Adaptive immune defects and delayed rejection of allogeneic tumor cells in beige mice. *Cell Immunol.* 87: 348

175. Holland, J. M. 1976. Serotonin deficiency and prolonged bleeding in beige mice. *Proc. Soc. Exp. Biol. Med.* 151: 32

176. Kaplan, S. S., Boggs, S. S., Nardi, M. A., Basford, R. E., Holland, J. M. 1978. Leukocyte-platelet interactions in a murine model of Chediak-Higashi syndrome. *Blood* 52: 719

177. Brandt, E. J., Swank, R. T., Novak, E. K. 1981. The murine Chediak-Higashi mutation and other pigmentation mutations. See Ref. 11, Vol. 1, pp. 99–117

178. Silvers, W. K. 1979. *The Coat Colors of Mice*. New York: Springer-Verlag

179. Oliver, C., Essner, E. 1975. Formation of anomalous lysosomes in monocytes, neutrophils, and eosinophils from bone marrow of mice with Chediak-Higashi syndrome. *Lab. Invest.* 32: 17

180. Novak, E. K., Swank, R. T. 1979. Lysosomal dysfunctions associated with mutations at mouse pigment genes. *Genetics* 92: 189

181. Oliver, J. M. 1978. Cell biology of leukocyte abnormalities—membrane and cytoskeletal function in normal and defective cells. *Am. J. Pathol.* 93: 221

182. Roder, J. C., Haliotis, T., Klein, M., Korec, S., Jett, J. R., Ortaldo, J., Herberman, R. B., Katz, P., Fauci, A. S. 1980. A new immunodeficiency disorder in humans involving NK cells. *Nature* 284: 553

183. Haliotis, T., Roder, J., Klein, M., Ortaldo, J., Fauci, A. S., Herberman, R. B. 1980. Chediak-Higashi gene in humans. I. Impairment in natural-killer function. *J. Exp. Med.* 151: 1039

184. Padgett, G. A., Leader, R. W., Gorham, J. R., O'Mary, C. C. 1964. The familial occurrence of Chediak-Higashi syndrome in mink and cattle. *Genetics* 49: 505

185. Kramer, J. W., Davis, W. C., Prieur, D. J. 1977. The Chediak-Higashi syndrome of cats. *Lab. Invest.* 36: 554

186. Taylor, R. F., Farrel, R. K. 1973. Light and electron microscopy of peripheral blood neutrophils in a killer whale affected with Chediak-Higashi syndrome. *Fed. Proc.* 32: 822A

187. McGarry, R. C., Walker, R., Roder, J. C. 1984. The cooperative effect of the *satin* and *beige* mutations in the suppression of NK and CTL activities in mice. *Immunogenetics* 20: 527

188. Orn, A., Hakansson, E., Gidlund, M., Ramstedt, U., Axberg, I., Wigzell, H., Lundin, L. G. 1982. Pigment mutations in the mouse which also affect lysosomal functions lead to suppressed natural killer cell activity. *Scand. J. Immunol.* 15: 305

189. Heston, W. E., Vlahakis, G. 1961. Influence of the A^y gene on mammary-gland tumors, hepatomas, and normal growth in mice. *J. Natl. Cancer Inst.* 26: 969

190. Heston, W. E., Vlahakis, G. 1968. C3H-A^{vy-} A high hepatoma and high mammary tumor strain of mice. *J. Natl. Cancer Inst.* 40: 1161

191. Gasser, D. L., Fischgrund, T. 1973. Genetic control of the immune response in mice. IV. Relationship between graft vs host reactivity and possession of the high tumor genotypes A^ya and $A^{vy}a$. *J. Immunol.* 110: 305

192. Roberts, D. W., Wolff, G. L., Campbell, W. L. 1984. Differential effects of the mottled yellow and pseudoagouti phenotypes on immunocompetence in A^{vy}/a mice. *Proc. Natl. Acad. Sci. USA* 81: 2152

193. Copeland, N. G., Jenkins, N. A., Lee, B. K. 1983. Association of the lethal yellow (A^y) coat color mutation with an ecotropic murine leukemia virus genome. *Proc. Natl. Acad. Sci. USA* 80: 247

194. Wolff, G. L., Roberts, D. W., Galbraith, D. B. 1986. Prenatal determination of obesity, tumor susceptibility, and coat color pattern in viable yellow (A^{vy}/a) mice. *J. Hered.* 77: 151

195. Russell, E. S. 1979. Hereditary anemias of the mouse: A review for geneticists. *Adv. Gen.* 20: 357

196. Kitamura, Y., Go, S., Hatanaka, K. 1978. Decrease of mast cells in W/W^v mice and their increase by bone marrow transplantation. *Blood* 52: 447

197. Kitamura, Y., Go, S. 1979. Decreased production of mast cells in Sl/Sl^d anemic mice. *Blood* 53: 492

198. Jacoby, W., Cammarata, P. V., Findlay, S., Pincus, S. H. 1984. Anaphylaxis in mast cell-deficient mice. *J. Invest. Dermatol.* 83: 302

199. Askenase, P. W., Loveren, H. V., Kraeuter-Kops, S., Ron, Y., Meade, R., Theoharides, T. C., Norlund, J. J., Scovern, H., Gershon, M. D., Ptak, W. 1983. Defective elicitation of delayed-type hypersensitivity in W/W^v and Sl/Sl^d mast cell-deficient mice. *J. Immunol.* 131: 2687

200. Thomas, W. R., Schrader, J. W. 1983. Delayed hypersensitivity in mast-cell-deficient mice. *J. Immunol.* 130: 2565

201. Askenase, P. W., Bursztajn, S., Gershon, M. D., Gershon, R. K. 1980. T cell-dependent mast cell degranulation and release of serotonin in murine delayed-type hypersensitivity. *J. Exp. Med.* 152: 1358

202. Askenase, P. W., Meltzer, C. M., Gershon, R. K. 1982. Localization of leukocytes in sites of delayed-type hypersensitivity and in lymph nodes: Dependence of vasoactive amines on the elicitation of delayed-type hypersensitivity. *Immunology* 47: 239

203. Galli, S., Dvorak, A. M. 1984. What do mast cells have to do with delayed hypersensitivity? *Lab. Invest.* 50: 365

204. Galli, S. J., Hammel, I. 1984. Unequivocal delayed hypersensitivity in mast cell-deficient and beige mice. *Science* 226: 710

205. Ha, T., Reed, N., Crowle, P. K. 1986. Immune response potential in mast

cell-deficient mice. *Int. Arch. Allergy Appl. Immunol.* 80 : 85

206. Shultz, L. D., Bailey, D. W. 1975. Genetic control of contact sensitivity in mice: Effect of H-2 and non H-2 loci. *Immunogenetics* 1 : 570

207. Ferguson, A., Miller, H. R. P. 1979. Role of the mast cell in the defence against gut parasites. In *The Mast Cell*, ed. J. Pepys, A. M. Edward, p. 159. London: Pitman Medical

208. Oku, Y., Itayama, H., Kamiya, M. 1984. Expulsion of *Trichinella spiralis* from the intestine of *W/W*^{v} mice reconstituted with haematopoietic and lymphopoietic cells and origin of mucosal mast cells. *Immunology* 53 : 337

209. Mekori, T., Phillips, R. A. 1969. The immune response in mice of genotypes *W/W*^{v} and *Sl/Sl*^{d}. *Proc. Soc. Exp. Biol. Med.* 132 : 115

210. Harrison, D. E., Astle, C. M. 1976. Population of lymphoid tissues in cured *W*-anemic mice by donor cells. *Transplantation* 22 : 42

211. Brooke, H. C. 1926. Hairless mice. *J. Hered.* 17 : 173

212. Mann, S. J. 1971. Hair loss and cyst formation in hairless and rhino mutant mice. *Anat. Rec.* 170 : 845

213. Heiniger, H. J., Meier, H., Kaliss, N., Cherry, M., Chen, H., Stoner, R. 1974. Hereditary immunodeficiency and leukemogenesis in HRS/J mice. *Cancer Res.* 34 : 201

214. Meier, H., Huebner, J. 1969. Genetic control by the *hr*-locus of susceptibility and resistance to leukemia. *Proc. Natl. Acad. Sci. USA* 63 : 759

215. Green, N., Hiai, H., Elder, J. H., Schwartz, R. S., Khiroya, R. H., Thomas, C. Y., Tsichlis, P. N., Coffin, J. M. 1980. Expression of leukemogenic retroviruses associated with a recessive gene in HRS/J mice. *J. Exp. Med.* 152 : 249

216. Johnson, D. A., Meier, H. 1981. Immune responsiveness of HRS/J mice to syngeneic lymphoma cells. *J. Immunol.* 127 : 461

217. Morrisey, P. J., Parkinson, D. R., Schwartz, R. S., Waskal, S. D. 1980. Immunologic abnormalities in HRS/J mice. I. Specific defect in T lymphocyte helper T cell function in a mutant mouse. *J. Immunol.* 125 : 1558

218. Reske-Kunz, A. B., Scheid, M. P., Boyse, E. A. 1979. Disproportion in T cells subpopulations in immuno-deficient mutant *hr/hr* mice. *J. Exp. Med.* 149 : 228

219. Takaoki, M., Kawaji, H. 1980. Impaired antibody response against T-dependent antigens in Rhino mice. *Immunology* 40 : 27

220. Kawaji, H., Tsukuda, R., Nakaguchi, T. 1980. Immunopathology of rhino mouse, an autosomal recessive mutant with murine lupus-like disease. *Acta Pathol. Jpn.* 30: 515

221. Stelzner, K. F. 1983. Four dominant autosomal mutations affecting skin and hair development in the mouse. *J. Hered.* 74 : 193

222. Jansson, J.-O., Downs, T. R., Beamer, W. G., Frohman, L. A. 1986. Receptor-associated resistance to growth hormone-releasing factor in dwarf "little" mice. *Science* 232 : 511

223. Cheng, T. C., Beamer, W. G., Phillips, J. A., Bartke, A., Mallonee, L., Dowling, C. 1983. Etiology of growth hormone deficiency in little, Ames, and Snell dwarf mice. *Endocrinology* 113 : 1669

224. Duquesnoy, R. J., Pederson, G. M. 1981. Immunologic and hematologic deficiencies of the hypopituitary dwarf mouse. See Ref. 11, Vol. 1, pp. 309–24

225. Schneider, G. B. 1976. Immunological competence in Snell-Bagg pituitary dwarf mice: Response to the contact-sensitizing agent oxazolone. *Am. J. Anat.* 145 : 371

226. Dumont, F., Robert, F., Bischoff, P. 1979. T and B lymphocytes in pituitary dwarf Snell-Bagg mice. *Immunology* 38 : 23

227. Baroni, C. 1967. Mouse thymus in hereditary pituitary dwarfism. *Acta Anat.* 68 : 361

228. Duquesnoy, R. J., Kalpaktsoglou, P. K., Good, R. A. 1970. Immunological studies of the Snell-Bagg pituitary dwarf mouse. *Proc. Soc. Exp. Biol. Med.* 133 : 201

229. Fabris, N., Pierpaoli, W., Sorkin, E. 1972. Lymphocytes, hormones, and aging. *Nature* 240 : 557

230. Eicher, E. M., Beamer, W. G. 1980. New mouse *dw* allele: Genetic location and effects of lifespan and growth hormone levels. *J. Hered.* 71 : 187

231. Fabris, N., Pierpaoli, W., Sorkin, E. 1971. Hormones and the immunological capacity. IV. Restorative effect of developmental hormones or of lymphocytes on the immunodeficiency syndrome of the dwarf mouse. *Clin. Exp. Immunol.* 9 : 227

232. Coleman, D. L. 1982. Diabetes-obesity syndromes in mice. *Diabetes* 31 (Suppl.) : 1

233. Coleman, D. L. 1978. Obese and diabetes: Two mutant genes causing dia-

betes-obesity syndromes in mice. *Diabetologia* 14: 141

234. Plouffe, J. F., Silva, J., Fekety, R., Allen, J. L. 1978. Cell-mediated immunity in diabetes mellitus. *Infect. Immun.* 21: 425

235. Fernandes, G., Handwerger, B. S., Yunis, E. J., Brown, D. M. 1978. Immune response in the mutant diabetic C57BL/Ks-*db*+ mouse. Discrepancies between in vitro and in vivo immunological assays. *J. Clin. Invest.* 61: 243

236. Mandel, M. A., Mahmoud, A. A. F. 1978. Impairment of cell-mediated immunity in mutation diabetic mice (*db/db*). *J. Immunol.* 120: 1375

237. Sheena, J., Meade, C. 1978. Mice bearing the *ob/ob* mutation have impaired immunity. *Int. Arch. Allergy Appl. Immunol.* 57: 263

238. Meade, C. J., Sheena, J., Mertin, J. 1979. Effects of the obese (*ob/ob*) genotype on spleen cell immune function. *Int. Arch. Allergy Appl. Immunol.* 58: 121

239. Chandra, R. K., Au, B. 1980. Spleen hemolytic plaque-forming cell response and generation of cytotoxic cells in genetically obese (C57Bl/6J *ob/ob*) mice. *Int. Arch. Allergy Appl. Immunol.* 62: 94

240. Chandra, R. K. 1980. Cell-mediated immunity in genetically obese (C57BL/6J *ob/ob*) mice. 33: 13

241. Meade, C. J., Sheena, J. 1979. Immunity in genetically obese rodents. In *Animal Models of Obesity*, ed. M. J. Festing, pp. 205–20. London: Oxford Univ. Press

242. Dubuc, P. U. 1976. Basal cortisone levels in young *ob/ob* mice. *Horm. Metab. Res.* 9: 95

243. Garthwaite, T. L., Martinson, D. R., Tseng, L. F., Hagen, T. C., Menahan, L. A. 1980. A longitudinal hormone profile of the genetically obese mouse. *Endocrinology* 107: 671

244. Coleman, D. L., Burkart, D. L. 1977. Plasma cortisone concentrations in diabetic (*db*) mice. *Diabetologia* 13: 25

245. Cupps, T. R., Fauci, A. S. 1982. Corticosteroid-mediated immunoregulation in man. *Immunol. Rev.* 65: 133

246. Debray-Sachs, M., Dardenne, M., Sai, P., Savino, W., Quiniou, M.-C., Boillot, D., Gepts, W., Assan, R. 1983. Anti-islet immunity and thymic dysfunction in the mutant diabetic C57BL/KsJ *db/db* mouse. *Diabetes* 32: 1048

247. Schwizer, R. W., Leiter, E. H., Evans, R. 1984. Macrophage-mediated cytotoxicity against cultured pancreatic islet cells. *Transplantation* 37: 539

248. Boillot, D., Assan, R., Dardenne, M., Debray-Sachs, M., Bach, J. F. 1986. T-lymphopenia and T-cell imbalance in diabetic *db/db* mice. *Diabetes* 35: 198

249. Soll, A. H., Goldfine, I. D., Roth, J., Kahn, C. R., Neville, D. M. Jr. 1974. Thymic lymphocytes in obese (*ob/ob*) mice. A mirror of the insulin receptor defect in liver and fat. *J. Biol. Chem.* 249: 4127

250. Leiter, E. H., Beamer, W. G., Shultz, L. D., Barker, J. E., Lane, P. W. 1987. Mouse models of genetic diseases. In *Medical and Experimental Mammalian Genetics: A Perspective*, ed. V. A. McKusick, T. H. Roderick, March of Dimes Birth Defects Ser. New York: Liss

251. Bockman, D. E., Kirby, M. L. 1984. Dependence of thymic development on derivatives of the neural crest. *Science* 223: 498

252. Besedovsky, H. O., del Rey, A. E., Sorkin, E. 1983. What do the immune system and the brain know about each other? *Immunol. Today* 4: 342

253. Blalock, J. E., Smith, E. M. 1985. The immune system: Our mobile brain? *Immunol. Today* 6: 115

254. Shultz, L. D., Sweet, H. O., Davisson, M. T., Coman, D. R. 1982. "Wasted," a new mutant of the mouse with abnormalities characteristic of ataxia telangiectasia. *Nature* 297: 5865

255. Woloschak, G. E., Rodriguez, M., Krco, J. 1986. Characterization of immunologic and neuropathologic abnormalities in "wasted" mice. *Proc. 6th International Congress Immunological Abstracts*, p. 1. Ottawa, Quebec

256. Goldowitz, D., Shipman, P. M., Porter, J. F., Schmidt, R. R. 1985. Longitudinal assessment of immunological abnormalities of mice with the autosomal recessive mutation, "*wasted.*" *J. Immunol.* 135: 1806

257. Kaiserlianian, D., Savino, W., Uriel, J., Hassid, J., Dardenne, M., Bach, J. F. 1986. The wasted mutant mouse. II. Immunological abnormalities in a mouse described as a model of ataxia-telangiectasia. *Clin. Exp. Immunol.* 63: 562

258. Kaiserlianian, D., Delacroix, D., Bach, F. 1985. The wasted mutant mouse. I. An animal model of secretory IgA deficiency with normal serum IgA. *J. Immunol.* 135: 1126

259. Abbott, C. M., Skidmore, C. J., Searle, A. G., Peters, J. 1986. Deficiency of adenosine deaminase in the wasted mouse. *Proc. Natl. Acad. Sci. USA* 83: 693

260. Abbott, C. M., Skidmore, C. J., Peters,

J. 1986. Adenosine deaminase deficiency in wasted mice. *Mouse News Letter*. In press

261. Bridges, B. A., Harnden, D. G., eds. 1982. *Ataxia Telangiectasia*. Chichester: Wiley

262. Tezuka, H., Inoue, T., Noguti, T., Kada, T., Shultz, L. D. 1986. Evaluation of the mouse mutant "wasted" as an animal model for ataxia telangiectasia. I. Age-dependent and tissue-specific effects. *Mutation Res.* 161: 83

263. Nordeen, S. K., Schaefer, V. G., Edgell, M. H., Hutchison, C. A., Shultz, L. D., Swift, M. 1984. Evaluations of wasted mouse fibroblasts and SV-40 transformed human fibroblasts as models of ataxia telangiectasia in vitro. *Mutation Res.* 140: 219

264. Inoue, T., Aikawa, K., Tezuka, H., Kada, T., Shultz, L. D. 1986. Effect of DNA-damaging agents on isolated spleen cells and lung fibroblasts from the mouse mutant "wasted," a putative animal model for ataxia telangiectasia. *Cancer Res.* 46: 3979

265. Dung, H. C., Swigart, R. H. 1972. Histo-pathologic observations of the nervous and lymphoid tissues of "lethargic" mutant mice. *Texas Rep. Biol. Med.* 30: 23

266. Dung, H. C. 1981. Lethargic mice. See Ref. 11, Vol. 2, pp. 17–37

267. Dung, H. C. 1977. Deficiency in the thymus-dependent immunity in "lethargic" mutant mice. *Transplantation* 23: 39

268. Dung, H. C. 1976. Relationship between the adrenal cortex and thymic involution in "lethargic" mutant mice. *Am. J. Anat.* 147: 255

269. Duchen, L. W. 1970. Hereditary motor end-plate disease in the mouse: Light and electron microscopic studies. *J. Neurol. Neurosurg. Psychiat.* 33: 238

270. Rieger, F., Pincon-Raymond, M., Lombert, A., Ponzio, G., Lazdunski, M., Sidman, R. L. 1984. Paranodal dysmyelination and increase in tetrodoxin binding sites in the sciatic nerve of the motor end-plate disease (*med/med*) mouse during postnatal development. *Dev. Biol.* 101: 401

271. Sidman, R. L., Cowen, J. S., Eicher, E. M. 1979. Inherited muscle and nerve diseases in mice: A tabulation with commentary. *Ann. N.Y. Acad. Sci.* 317: 497

272. Papiernik, M., Rieger, F., Ezine, S., Pincon-Raymond, M. 1982. Impairment of T lymphocyte functions in mice with motor end-plate disease. *Clin. Exp. Immunol.* 48: 429

273. Ezine, S., Papiernik, M., Rieger, F.,

Pincon-Raymond, M. 1983. Modification of helper and suppressor/cytotoxic lymphocyte subsets in mice with motor end-plate disease. *Clin. Exp. Immunol.* 51: 475

274. Michelson, A. M., Russell, E. S., Harman, P. J. 1955. Dystrophia muscularis: A hereditary primary myopathy in the house mouse. *Proc. Natl. Acad. Sci. USA* 41: 1079

275. Meier, H., Southard, J. L. 1970. Muscular dystrophy in the mouse caused by an allele at the *dy* locus. *Life Sci.* 9: 137

276. Bradley, W. G., Jenkinson, M. 1973. Abnormalities of peripheral nerves in murine muscular dystrophy. *J. Neurol. Sci.* 18: 227

277. Tsuji, S., Matsushita, H. 1985. Evidence on hypomyelination of central nervous system in murine muscular dystrophy. *J. Neurol. Sci.* 68: 175

278. Montgomery, A., Swenarchuk, L. 1977. Dystrophic mice show age related muscle fibre and myelinated axon losses. *Nature* 267: 167

279. Montgomery, A., Swenarchuk, L. 1978. Further observations on myelinated axon numbers in normal and dystrophic mice. *J. Neurol. Sci.* 38: 77

280. Peterson, A. C., Bray, G. M. 1984. Normal basal laminas are realized in dystrophic Schwann cells in dystrophic ≪-----≫ shiverer chimera nerve. *J. Cell Biol.* 99: 1831

281. De Krester, T. A., Livett, B. G. 1976. Evidence of a thymic abnormality in murine muscular dystrophy. *Nature* 263: 682

282. Karmali, R. A., Horrobin, D. F. 1976. Abnormalities of thymus growth in dystrophic mice. *Nature* 263: 684

283. Ludwig, C. L., Kanellopoulos-Langevin, C., Jin Kim, K., Mathieson, B. J., Morse, H. C. III. 1981. Immunologic function and cell surface antigen expression of lymphocytes of dystrophic mice. *Cell. Immunol.* 59: 138

284. Marks, S. C. Jr., McGuire, J. L. 1987. Primary bone cell dysfunction. II. Osteopetrosis. In *Metabolic Bone Disease: A Pathogenetic Approach Based on Cellular Mechanisms*, ed. C. S. Tam, J. N. M. Heersche, T. M. Murray. Boca Raton: CRC Press. In press

285. Reeves, J. D., August, C. S., Humberg, J. R., Weston, W. L. 1979. Host defense in infantile osteopetrosis. *Pediatrics* 64: 202

286. Ash, P., Loutit, J. F., Townsend, K. M. S. 1980. Osteoclasts derived from haematopoietic stem cells. *Nature* 283: 669

287. Marks, S. C. Jr. 1984. Congenital osteopetrotic mutations as probes of the origin, structure, and function of osteoclasts. *Clin. Orthop. Rel. Res.* 189 : 239

288. Hollander, W. F. 1968. Complementary alleles at the *mi*-locus in the mouse. *Genetics* 60 : 189 (Abstr.)

289. Konyukhov, B. V., Osipov, V. V. 1968. Interallelic complementation of microphthalmia and white genes in mice. *Soviet Genet.* 4 : 1457

290. Holtrop, M. E., Cox, K. A., Eilon, G., Simmons, H. A., Raisz, L. G. 1981. The ultrastructure of osteoclasts in microphthalmic mice. *Metab. Bone Dis. Rel. Res.* 3 : 123

291. Thesingh, C. W., Scherft, J. P. 1985. Fusion disability of embryonic osteoclast precursor cells and macrophages in the microphthalmic osteopetrotic mouse. *Bone* 6 : 43

292. Walker, D. G. 1975. Control of bone resorption by hematopoietic tissue. The induction and reversal of congenital osteopetrosis in mice through use of bone marrow and splenic transplants. *J. Exp. Med.* 142 : 651

293. Walker, D. G. 1975. Bone resorption restored in osteopetrotic mice by transplants of normal bone marrow and spleen cells. *Science* 190 : 784

294. Minkin, C. 1981. Defective macrophage chemotaxis in osteopetrotic mice. *Calcif. Tissue Int.* 33 : 677

295. Minkin, C., Trump, G., Stohlman, S. A. 1982. Immune function in congenital osteopetrosis: Defective lymphocyte function in microphthalmic mice. *Develop. Comp. Immunol.* 6 : 151

296. Seamen, W. E., Gindhart, T. D., Greenspan, J. S., Blackman, M. A., Talal, N. 1979. Natural killer cells, bone, and the bone marrow: Studies in estrogen-treated and in congenitally osteopetrotic (*mi/mi*) mice. *J. Immunol.* 122 : 2541

297. Schneider, G. B., Marks, S. C. 1983. Immunological competence in osteopetrotic (microphthalmic) mice. *Expl. Cell Biol.* 51 : 327

298. Wiktor-Jedrzejczak, W., Skelly, R. R., Ahmed, A. 1981. Hematopoietic stem cell differentiation and its role in osteopetrosis. See Ref. 11, Vol. 1, pp. 51–77

299. Grüneberg, H. 1936. Grey-lethal, a new mutation in the house mouse. *J. Hered.* 27 : 105

300. Grüneberg, H. 1938. Some new data on the grey-lethal mouse. *J. Genet.* 36 : 153

301. Wiktor-Jedrzejczak, W., Grzybowski, J., Ahmed, A., Kaczmarek, L. 1983. Osteopetrosis associated with premature thymic involution in grey-lethal mice. In vitro studies of thymic microenvironment. *Clin. Exp. Immunol.* 52 : 465

302. Marks, S. C. Jr., Lane, P. W. 1976. Osteopetrosis, a new recessive skeletal mutation on chromosome 12 of the mouse. *J. Hered.* 67 : 11

303. Marks, S. C. Jr., Seifert, M. F., McGuire, J. L. 1984. Congenitally osteopetrotic (*op/op*) mice are not cured by transplants of spleen or bone marrow cells from normal littermates. *Metab. Bone Dis. Rel. Res.* 5 : 183

304. Wiktor-Jedrzejczak, W., Ahmed, A., Szcylik, C., Skelly, R. R. 1982. Hematological characterization of congenital osteopetrosis in *op/op* mouse. *J. Exp. Med.* 156 : 1516

305. Olsen, C. E., Wahl, S. M., Wahl, L. M., Sandberg, A. L., Mergenhagen, S. E. 1977. Immunological defects in osteopetrotic mice. In *Mechanisms of Localized Bone Loss*, ed. J. E. Horton, T. M. Tarpley, W. F. Davis, pp. 389–398. Washington, DC: IRI

306. Marks, S. C. Jr., Seifert, M. F., Lane, P. W. 1985. Osteosclerosis, a recessive skeletal mutation on chromosome 19 in the mouse. *J. Hered.* 76 : 171

307. Banco, R., Seifert, M. F., Marks, S. C. Jr., McGuire, J. L. 1985. Rickets and osteopetrosis: The osteosclerotic (*oc*) mouse. *Clin. Orthop.* 201 : 238

308. Coccia, P. F., Krivit, W., Cervenka, J., Clawson, C., Kersey, J. H., Kim, T. H., Nesbit, M. E., Ramsay, N. K. C., Warkentin, P. I., Teitelbaum, S. L., Kahn, A. J., Brown, D. M. 1980. Successful bone marrow transplantation for infantile malignant osteopetrosis. *New Engl. J. Med.* 302 : 701

309. Sorell, M., Kapoor, N., Kirkpatrick, D., Rosen, J. F., Chaganti, R. S. K., Lopez, C., Dupont, B., Pollack, M. S., Terrin, B. N., Harris, M. B., Vine, D., Rose, J. S., Goossen, C., Lane, J., Good, R. A., O'Reilly, R. J. 1981. Marrow transplantation for juvenile osteopetrosis. *Am. J. Med.* 70 : 1280

310. Ballet, J. J., Griscelli, C., Coutris, C., Milhaud, G., Maroteaux, P. 1977. Bone marrow transplantation in osteopetrosis. *Lancet* 2 : 1137

311. National Research Council. 1987. *Immunodeficient Rodents: A guide To Their Immunobiology, Husbandry, and Use. A Report of the Institute of Laboratory Animal Resources Committee on Immunologically Compromised Rodents*, Washington, DC: Natl. Acad. Press. In press

Ann. Rev. Immunol. 1987. 5 : 405–427
Copyright © 1987 by Annual Reviews Inc. All rights reserved

THE I-J PUZZLE*

Donal B. Murphy

Department of Pathology (Immunology), Yale Medical School,
310 Cedar Street, New Haven, Connecticut 06510

Overview

In 1976, several laboratories independently demonstrated that antibodies produced in major histocompatibility complex (MHC) I region–incompatible strains react with T cells (Ts) or T cell–derived factors (TsF) involved in the generation of suppressor activity. The determinants detected by these antibodies are expressed on a small number of peripheral T cells (< 10%) but not on B lymphocytes, thus distinguishing them from I region–encoded class-II antigens. Studies with intra–I region recombinant strains provided further evidence that the gene (*Ia-4*) which controlled determinants expressed on Ts or TsF mapped within the MHC in a new I subregion, designated I-J.

Subsequent studies by numerous investigators have confirmed these observations and have shown that these T-cell determinants, now referred to collectively as "I-J determinants," are expressed on several immuno-regulatory T-cell subsets and T cell–derived factors involved in the generation of suppressor, contrasuppressor, and helper-amplifier activity. Different I-J determinants are selectively expressed on many of these T-cell subsets and factors. Notably, I-J determinants have not been detected on helper T lymphocytes that recognize foreign antigen plus self–class II determinants or on cytolytic T lymphocytes that recognize foreign antigen plus self–class-I determinants. I-J determinants may thus demarcate a small but functionally distinct T-cell lineage. Studies in several systems provide evidence that molecules bearing I-J determinants (I-J molecules) are involved in information trafficking among immunocompetent cells,

---

*Abbreviations: *Ia,* *I* region *associated*; *Igh-V region,* chromosomal segment carrying immunoglobulin heavy chain variable region genes and closely linked genes; *I-J molecules,* molecules bearing I-J determinants; *Ir,* Immune response; *MHC,* major histocompatibility complex; *Th,* helper T lymphocyte; *Ts,* suppressor T lymphocyte; *TsF,* suppressor T cell factor.

0732–0582/87/0410–0405$02.00

most likely serving as self-recognition molecules which channel regulatory factors to their appropriate target cells. The observation that some T-T and T-macrophage interactions are I-J restricted provides additional evidence that I-J molecules may also serve as self-restriction elements. I-J determinants on macrophages are either passively acquired, or, if synthesized by macrophages, are probably distinct from the serologically defined T-cell I-J determinants.

Problems with the original interpretation of the genetic basis for control of I-J determinants arose during molecular genetic analysis of the I region. Crossovers in the strains used to map the *Ia-4* locus and define the I-J subregion were localized to a small segment of DNA within the class-II E_β gene. In addition, no DNA-RNA hybridization was observed with I-J$^+$ somatic T-cell hybrids utilizing DNA probes which overlap the E_β gene and span most of the I region. Thus, although it is clear that polymorphism in I-J determinants is influenced by the I region of the MHC, these studies provide evidence that an I-region gene does not encode I-J molecules.

A logical explanation for this apparent paradox is that I-J molecules are T-cell receptors that recognize self–I-region products and/or that recognize receptors for self–I-region products. Under this model, the gene(s) encoding I-J molecules could map anywhere in the genome. However, expression of and polymorphism in these receptors would be influenced by self–I-region products, presumably by positive selection during ontogeny. Chimera studies showing that I-J phenotype is influenced by the environment in which T cells mature provide strong support for this receptor-ligand model.

One key to resolving the genetic basis for control of I-J determinants is identification of the I-region gene that influences polymorphism in I-J molecules. Recent studies showing that E_α transgenic mice display an altered I-J phenotype provide formal proof that class-II genes are intimately involved in this process. Molecular, functional, and serological analyses suggest it is extremely unlikely that serologically defined T-cell I-J determinants reside on modified or altered class-II molecules in the transgenic mice. Rather, I-J determinants appear to reside on molecules whose expression is influenced by class-II molecules (e.g. T-cell receptors).

Final resolution of this puzzle is dependent on a molecular characterization of I-J molecules and genes and on an understanding of the interplay between these genes and class-II genes. In the end, it seems probable that I-J molecules will prove to be a novel set of self-receptors used by T cells (exclusive of MHC-restricted helper T cells or cytolytic T cells) to traffic regulatory signals among immunocompetent cells. This system may have evolved to allow specific T-cell regulation of immune responses by soluble mediators at long range.

Definition of the I-J Subregion

The I region of the MHC was discovered and defined by McDevitt and his colleagues (1) in the early 1970s. They showed that a gene which regulated high or low responsiveness to foreign protein antigens, designated *Immune response-1* (*Ir-1*), mapped within the H-2 gene complex between the *H-2K* and *Ss* (*C4*) loci. Crossovers on opposite sides of the *Ir-1* locus defined the segment of chromosome known as the I region (see Table 1 for general map of MHC regions). Data with I-region recombinant strains provided evidence for at least three Ir genes (*Ir-1A, Ir-1B, Ir-1C*), each mapping in a distinct I subregion (I-A, I-B, I-C) (1–5). Subsequent study in several laboratories showed that antibodies produced in I region–incompatible strains of mice reacted predominantly with B lymphocytes (>95%) and macrophages. The antigens detected by these antisera were referred to as *I* region *a*ssociated or Ia antigens. Genes *Ia-1* and *Ia-3* controlling Ia antigens were mapped to the I-A and I-C subregions, respectively. At that time, it was not clear whether Ir and Ia loci were the same or distinct. Later study showed that *Ir-1C* and *Ia-3* map in the I-E subregion (3–5).

Today, we know that Ia antigens are the molecular products of Ir genes (6, 7). These antigens reside on noncovalently associated cell surface dimers

Table 1 Evidence that a distinct I region locus, *Ia-4*, mapping in a new subregion, I-J, controls determinants expressed on allotype Ts

Antibody[a]	Strain	I						I-J determinant expressed[c]
		K	A	(J)	E	S	D[b]	
Anti-*Ia-4*^{k}	B10.A	k	k	k	k	d	d	+
	B10	b	b	b	b	b	b	−
	B10.A(4R)	k	k/b	b	b	b	b	−
	B10.A(5R)	b	b/k	k	d	d		+
	B10.A(3R)	b	b	b/k	d	d		−
Anti-*Ia-4*^{s}	B10.S(7R)	s	s	s	s	s	d	+
	B10.A	k	k	k	k	d	d	−
	B10.S(9R)	s	s/k	k	d	d		−
	B10.HTT	s	s	s/k	k	d		+

[a] Anti-*Ia-4*^{k} [(B10.T(6R) × B10.D2]F1 anti-B10.AQR; Anti-*Ia-4*^{s} {A.TL anti-A.TH or [A.TH × B10.A. × A.TLJFL anti-B10.HTT}.

[b] Haplotype origin of region or subregion. K region is marked by the *H-2K* locus, I-A subregion by *Ir-1A* or *Ia-1* (now A_α and A_β), I-J subregion by *Ia-4*, I-E subregion by *Ir-1C* or *Ia-3* (now E_α), S region by *Ss* (now C4), and D region by *H-2D*. The E_β gene spans the I-A and I-E subregions.

[c] Tested by ability of unstimulated spleen and lymph node cells to absorb antibody reactive with allotype Ts. (+) = strain expresses the I-J determinant; (−) = strain does not express the I-J determinant. Adapted from Ref. 27.

($\alpha : \beta$ complexes, M_r 28,000–35,000) which are also referred to as class-II molecules (8). Genes encoding these molecules map in the I-A subregion (A_α and A_β, equivalent of earlier *Ir-1A* or *Ia-1* designations) and the I-E subregion (E_α, equivalent of earlier *Ir-1C* or *Ia-3* designations). The E_β gene spans the I-A and I-E subregions and is a site for a recombination hotspot (9–12). Four other class-II β genes have been identified ($A_{\beta 2}$, $A_{\beta 3}$, $E_{\beta 2}$, $E_{\beta 3}$). $A_{\beta 3}$ and $E_{\beta 3}$ are pseudogenes. Although $A_{\beta 2}$ and $E_{\beta 2}$ mRNA has been detected, it is not clear whether this mRNA is translated. The A_α, A_β, and E_β genes are extremely polymorphic, while limited polymorphism has been observed with E_α. A_α chains (M_r 35,000) pair with A_β chains (M_r 28,000), and E_α chains (M_r 33,000) pair with E_β chains (M_r 30,000). Although it has been shown that an A_β chain can pair with an E_α chain in L cells cotransfected with A_β^d and E_α^k genes, it is not clear whether this occurs in normal cells (13). All mouse strains studied so far express $A_\alpha : A_\beta$ complexes on the cell surface. However, some strains fail to synthesize E_α and/or E_β chains due to partial deletions in the structural genes encoding these molecules or synthesis of nonfunctional mRNA (14). In cases where E_α genes are not expressed, E_β chains are synthesized but remain in the cytoplasm (15). In cases where E_β genes are not expressed, E_α chains are synthesized and are found on the cell surface but in reduced amount (16). Normal expression of E_β (or E_α) chains in the above strains can be obtained by crossing with a strain that carries a functional E_α (or E_β) gene (15, 16) or by transfecting cells (13) or mice (17–19) with functional E_α (or E_β) genes. Transgenic mice of the latter type were recently used to prove formally that class-II molecules are the molecular products of Ir genes (17–19). Class-II molecules function as self-restriction elements which together with foreign antigen are recognized by helper T lymphocytes, and which play an important role in helper-T-cell activation and interaction (20). Why different alleles at class-II loci confer high or low responsiveness to different foreign protein antigens has not been fully resolved (21, 22). Class-II molecules also elicit strong T-cell proliferative responses in allogenic mixed lymphocyte culture. The *Ir-1B* locus and I-B subregion may be artifacts due to epistasis between I-A and I-E subregion class-II genes (23, 24). Although it is clear that no class-II gene maps in the I-C subregion (11), it has been reported that this chromosomal segment controls antigens expressed on T cells (reviewed in 4). For the purposes of this review, we omit further consideration of the I-B and I-C subregions.

In 1976, it was observed that I-region antisera also contained antibody reactive with a small subset ($< 10\%$) of peripheral T cells involved in the generation of idiotype(25)- or allotype(26, 27)-specific suppressor activity (Ts) and with a T cell–derived factor (28) involved in the generation of antigen-specific suppressor activity (TsF). Study in the allotype suppression system showed that antibodies reactive with Ts did not react with B lymphocytes, indicating that the Ts determinant was distinct from previously defined class-II antigens (27). Antibody reactive with allotype Ts also did not react with helper T lymphocytes (Th) (26). Studies in both the allotype(27, 29)- and antigen(28)-specific suppressor systems showed that numerous antisera containing polyvalent antibody reactive with class-II molecules did not react with Ts or TsF. This provides strong evidence that the serologically defined T-cell determinant does not reside on modified or altered class-II molecules. Most important, study in the allotype(27)- and antigen(28)-specific suppressor systems showed that intra-I region recombinant strains, which appeared to express the same class-II molecules on B cells, differentially expressed the Ts determinant (Table 1). This raised the possibility that an I-region locus, *Ia-4*, controlled determinants expressed on Ts or TsF, and that crossovers on opposite sides of this locus in the recombinant strains separated the locus from other I-region loci and defined a new subregion, designated I-J (27, 28). For example, recombinant strains B10.A(3R) (abbrev. 3R) and B10.A(5R) (abbrev. 5R) both typed *Ia-1*^{b} and *Ia-3*^{k}. However, strain 3R typed *Ia-4*^{b}, mapping the *Ia-4* locus to the left (K-end) of *Ia-3*, and strain 5R typed *Ia-4*^{k}, mapping the *Ia-4* locus to the right (D-end) of *Ia-1*. The *Ia-4* locus thus appeared to map between the *Ia-1* and *Ia-3* loci in a segment of chromosome, the I-J subregion, which is defined by the crossovers in strains 3R (I-A^b I-J^b/I-E^k) and 5R (I-A^b/I-J^k I-E^k). Similar data were obtained with a second pair of independently derived intra–I region recombinant strains, B10.HTT (*Ia-4*^{s}, I-A^s I-J^s/I-E^k) and B10.S(9R) (abbrev. 9R) (*Ia-4*^{k}, I-A^s/I-J^k I-E^k) (Table 1). Within a year, this finding was confirmed in several other laboratories (30–38).

Several key assumptions were made in mapping the *Ia-4* locus and defining the I-J subregion. First, it was assumed that the recombinant strains carry and express the same alleles at class-II loci. Evidence supporting this assumption comes from extensive serological and functional but limited biochemical (peptide map analysis, 39; two-dimensional gel electrophoretic analysis, 6, 15; and P. P. Jones, personal communication) and molecular genetic analyses (40, 41). Second, it was assumed that the position and number of loci carried by each H-2 haplotype is invariant. Third, it was assumed that single crossovers occurred in the construction of the recombinant strains. Fourth, it was assumed that unequal or intragenic crossovers did not occur. Fifth, it was assumed that mutations did not occur during or subsequent to the derivation of the recombinant strains. Finally, it was assumed that the recombinant strains were congenic with respect to the MHC and do not differ at unlinked (non-MHC) loci. This latter assumption was particularly important, since strains 3R and 5R were

derived from (B10 × A) (H-2^b × H-2^a) backcross progeny, and strains
B10.HTT and 9R were derived from A and B10 background progenitor
mice (42–45).

General Properties of I-J Determinants and Biology of I-J Molecules

Serologically detected I-J determinants are found on a number of immunocompetent cell types and on T cell–derived antigen-specific soluble regulatory factors (reviewed in 46–51). Based on antibody-plus-complement
mediated cytolysis, less than 10% of peripheral T cells bear I-J determinants (27). Failure to detect I-J determinants on peripheral T cells by
immunofluorescence analysis suggests that a very small proportion
(< 1%?) of T cells are I-J$^+$ (52, 53). In general, these determinants are
expressed on T cells and T cell–derived factors that regulate helper-T-cell
activity. These include inducer, transducer (also referred to as acceptor or
amplifier), and effector T-cell subsets; factors involved in the generation of
suppressor (46–51) and contrasuppressor (47) activity; and T-lymphocyte
subsets which amplify helper-T-cell activity (49) or humoral tumor immunity (54). I-J determinants have generally not been detected on helper
T lymphocytes that recognize foreign antigen together with self–class-II
molecules or on cytolytic T lymphocytes that recognize foreign antigen
together with self–class-I molecules (49). This latter observation is consistent with the small number of T cells which bear I-J determinants, and
it raises the possibility that I-J determinants demarcate a distinct T-cell
lineage. Serologically detected I-J determinants are also found on non-
T: non-B cells (probably macrophages) involved in mitogen responses (55),
the generation of suppressor activity (56, 57), and the generation of helper
activity (58). Passively administered I-J antibody augments immunity in
vivo against foreign antigens (34) and certain tumors (54, 59), presumably
by blocking suppressor activity and/or augmenting contrasuppressor or
helper amplifier activity.

Studies in several laboratories have clearly shown that different I-J
determinants are expressed on some of the above cell types and factors
(60–62). For example, in studies with alloantisera, distinct I-J determinants
were detected on suppressor T cells and on helper amplifier T cells (63),
contrasuppressor T cells (64), or macrophages (65). Whether the later three
cell types express the same or different I-J determinants has not been
resolved. Studies with monoclonal I-J antibodies have shown further that
factors involved in suppressor inducer, transducer, and effector activity
share some I-J determinants but uniquely express others (62, 66). This is
analogous to the public and private determinants detected on class-I and

class-II molecules. These studies raise the possibility that multiple loci encode I-J determinants.

I-J determinants have also been detected by nonserological assays (3, 53). For example, (AKR × 5R)F1 mice but not (AKR × 3R)F1 mice develop good cellular immunity against tumor-specific antigens on an AKR thymoma (32). T-cell responses to H-2D region molecules in vivo (skin graft rejection) (67, 68) and in vitro (proliferation) (69) are influenced by I-J determinants, as is the bactericidal activity of macrophages during a graft-vs-host reaction (70). In the above systems, I-J disparity or incompatibility appears to suppress the response. Although it has been reported that allogeneic I-J determinants can specifically stimulate in mixed lymphocyte culture (71), we routinely obtained a different result. Rather, we find that stimulation in I-J incompatible strains breaks self-tolerance to class-II molecules and leads to the generation of autoreactive class II–specific T cells (72). Others have observed that stimulation in I-J incompatible strains can lead to the production of allogeneic effector factor (AEF) (73). Whether suppressor T cells and other regulatory T-cell types are activated or inactivated by stimulation between I-J incompatible strains remains to be resolved. Finally, some T-T and T-macrophage interactions are I-J restricted (see below).

Limited biochemical studies with suppressor factors provide evidence that T-cell I-J determinants reside on glycoprotein molecules of approximately M_r 25,000–30,000 (46, 50, 51, 61, 74–76). In addition, there are reports that I-J determinants are found on larger (M_r 45,000) (46) and smaller (M_r 15,000) (77) polypeptides. Glycosidase digestion analysis of T-cell I-J determinants suggests that carbohydrates can influence I-J cell surface expression (78). It is interesting that trypsin-digested T cells, but not macrophages, appear to resynthesize cell surface I-J molecules (78). This latter observation raises the possibility that I-J molecules on macrophages may be passively acquired.

T cells and T cell–derived factors bearing I-J determinants are often genetically restricted in their interaction with other cells (46, 60, 61, 74, 75, 79). For example, as initially reported (28), I-J$^+$ TsF could only act on I-J-compatible T cells to effect suppression, i.e. TsF was I-J restricted in its activity. Subsequent studies have shown that some T-macrophage interactions are also I-J restricted and that the activity of I-J$^+$ TsF can be MHC or Igh-V-region restricted. As discussed below, these restrictions are probably due to recognition of determinants controlled by MHC-linked or by immunoglobulin heavy-chain variable-region (Igh-V)-linked genes. The restriction observed varies with the system, function, and pathway examined. The I-J restricting element on macrophages may be passively acquired or may be synthesized by macrophages. If synthesized, this restriction element is probably distinct from the serologically defined T-cell I-J determinants.

The observations that I-J determinants are selectively expressed on functionally distinct cell types and that I-J$^+$ factors produced by T cells are often genetically restricted in their activity provides tentative evidence that I-J molecules play a role in cellular interaction. As originally proposed, I-J molecules may be part of the regulatory language used for communication among responding lymphocytes (26, 28, 29). More recent studies strongly support this hypothesis and provide further evidence that I-J molecules represent a set of self–T-cell receptors which are important in trafficking regulatory signals within the immune system. Below, we consider one of the systems in which this was demonstrated.

Suppressor factors are usually composed of two molecules, one that binds antigen and is I-J$^-$ (antigen binding molecule, or ABM) and another that does not bind antigen and is I-J$^+$ (I-J molecule) (46–51, 60, 75). In some systems, these molecules are covalently associated. Although these molecules remain to be well characterized chemically, they are functionally distinct. Where examined, the I-J molecule usually determines the self-restricted activity of the factor. For example, in the SRBC feedback suppression system (47, 60), ABM and I-J molecules are secreted by different T-cell subsets in both the induction and effector phases of the response. In the induction phase, the suppressor-inducer antigen-binding molecule (ABM$_{si}$) is secreted by an L3T4$^+$, Ly-2$^-$, I-J$^-$ T cell, while the suppressor-inducer I-J molecule (I-J$_{si}$) is secreted by an L3T4$^-$, Ly-2$^-$, I-J$^+$ T cell. Together, both molecules constitute a factor (T-suppressor-inducer factor or TsiF) that activates Ly-1$^+$, 2$^+$ transducer cells in the feedback suppression system. Initial studies (80) showed that TsiF is restricted in its activity by a gene linked to the immunoglobulin heavy-chain variable-region gene complex (Igh-V-region restricted). Subsequent study showed that the I-J molecule, and not the antigen-binding molecule associated with TsiF, determines the Igh-V region restriction (81, 82). Evidence that this restriction is influenced by the environment in which T cells mature can be drawn from some chimera studies (83, 84) but not others (85). This was the first clear demonstration that I-J molecules associated with soluble regulatory factors determine the restricted activity of the factors. These data also raise the possibility that I-J molecules may be self receptors for Igh-V-region controlled determinants.

Similar results were obtained with a second factor, T suppressor effector factor (TseF), which inhibits helper-T-cell activity (47, 60). This factor is also comprised of two molecules secreted by different cells. One cell is L3T4$^-$, Ly-2$^+$, I-J$^-$ and secretes a molecule that binds antigen and determines the specificity of the effector process (suppressor-effector antigen binding molecule or ABM_{se}). The second cell is L3T4$^-$, Ly-2$^-$, I-J$^+$ and secretes a molecule that does not bind antigen and is I-J$^+$ (I-J$_{se}$ molecule). ABM_{se} induces secretion of I-J$_{se}$ in an MHC(I-E region)-restricted manner. Interaction between ABM_{se} and I-J$_{se}$ is MHC (I-J) restricted, while interaction between TseF and helper T cells is Igh-V-region restricted. As observed with TsiF, the I-J$_{se}$ molecule determines the Igh-V-region restricted activity of the factor (86), and, depending on the study, this restriction is (83, 84) or is not (87) influenced by the chimeric environment in which T cells mature. The MHC-restricted interaction between ABM_{se} and I-J$_{se}$ may be due to a self–anti-self interaction, e.g. ABM_{se} may recognize self–class-II molecules (I-E restricted), while I-J$_{se}$ may recognize receptors for self–class-II molecules (I-J restricted). I-J$_{si}$ and I-J$_{se}$ molecules are serologically distinct (66), as are ABM_{si} and ABM_{se} (88). Mixing experiments with I-J and ABM molecules suggests that the biological signal delivered by the factors (induction-activation or suppression-inactivation) is determined by the ABM molecule (89).

The observation that I-J molecules associated with suppressor factors determine the restricted activities of the factors has now been confirmed in several other suppressor systems. In the ABA system, I-J molecules were shown to determine the Igh-V-region restricted activity of an inducer factor (90). In the KLH system, I-J molecules determine the I-J–restricted activity of a transducer factor, and this restriction is influenced by the chimeric environment in which T cells mature (75). Finally, studies with an insulin-specific suppressor factor show that I-J molecules determine the MHC-restricted activity of the factor (91). That a complete correlation exists between the I-J determinant expressed and the I-J restriction observed with F_1-derived hybridoma suppressor factors in the NP system is compatible with the I-J molecule determining the restricted activity of the factor (92).

The above studies provide clear evidence that I-J molecules are intimately involved in channeling information among lymphocytes, probably serving as self-receptors for MHC and/or Igh-V-region controlled determinants. These determinants could be encoded by genes mapping on chromosome 17 (MHC) or chromosome 12 (Igh-V region), or they could reside on receptors for MHC or Igh-V region encoded determinants. Recent studies showing that a suppressor-inducer factor binds to class-II molecules is compatible with recognition of class-II determinants by I-J molecules, antigen-binding molecules, or possibly by both molecules together (93). Idiotypic antibodies produced against I-J antibody may prove useful in identifying the molecule(s) recognized by I-J molecules (94).

Although the majority of I-J$^+$ suppressor factors appear to be composed of two molecules, some of which are covalently associated, there are a few

I-J$^+$ suppressor factors that appear to be a single polypeptide (46, 50, 51). Notably, one factor, specific for GAT and of M$_r$ 25,000, appears to bind antigen, to bear I-J determinants, and to mediate suppressor activity (46). If this molecule proves to be a single polypeptide, this would be the first evidence that at least some I-J molecules (rather than molecules associated with I-J molecules) can function as antigen receptors.

One of the more intriguing observations with regard to functional study of I-J determinants is the finding that monoclonal I-J antibodies block autologous and allogeneic mixed lymphocyte reactions (MLR) (61, 95, 96). Inhibition of the proliferative response was observed at the level of the responder cell and was influenced by both MHC and Igh-V-region linked genes (95). The interpretation given was that I-J determinants may reside on T-cell receptors that recognize MHC and Igh-V-region controlled products. One problem with this interpretation is that T cells that respond in auto- or allo-MLR are predominantly helper T cells that recognize foreign antigen plus self–class-II molecules (20). These cells, also referred to as Th$_{MHC}$, have routinely typed I-J$^-$ (49). In addition, although there is evidence that regulatory helper-T-cell subsets can recognize a portion of Ig molecules, there is no evidence that Th$_{MHC}$ display this recognition capability (97). Finally, it was reported that the response of strain B10.BR (I-J^k) to 5R (I-J^k) but not 3R (I-J^b) was completely inhibited (>90%) by I-J^k monoclonal antibody and that this blocking occurred at the level of the responder cell (95). Since stimulation in these combinations appears to be primarily against class-II determinants, and both stimulator strains appear to express the same class-II determinants, it is difficult to envision how I-J determinants could reside on responding Th. Rather, these data seem more compatible with activation of suppressor cells and/or inactivation of contrasuppressor or helper amplifier cells by I-J antibody.

Problems with the Original Definition of the I-J Subregion

Molecular genetic studies provide evidence against our original postulate that a distinct locus, *Ia-4*, mapping in the I-J subregion, controls determinants found on Ts (9). Based on restriction enzyme polymorphism analysis, crossovers in several key strains used to map the *Ia-4* locus and to define the I-J subregion have been localized to a recombination hotspot within the E_β gene (40, 41). In particular, it has been shown that the crossovers in strains 3R and 5R occurred in a 1-kb segment of DNA situated entirely within the intron between the $\beta1$ and $\beta2$ exons (98). The nucleotide sequence of this DNA segment, plus approximately 1-kb on either side, is identical in the two strains. It is thus clear that no polymorphic gene or coding element maps in the I-J subregion, as defined by the crossovers in strains 3R and 5R. This raises the possibility that the crossovers have nothing to do with determining the different I-J phenotypesin these strains. Whether the same holds for the other two strains used to define the I-J subregion, B10.HTT and 9R, remains to be determined.

A second series of experiments showed that DNA probes that span the E_β gene and most of the I region failed to hybridize with any mRNA in numerous I-J$^+$ somatic T-cell hybrids (9, 99). No DNA rearrangements within or near the E_β gene were observed in the hybrids. Assuming the hybrids contained I-J mRNA, which was not formally proven, these data would argue that I-J molecules are not encoded (even in part) by an I-region gene.

Serological studies provide tentative evidence that a non-H-2 gene can influence the qualitative nature of I-J determinants (51, 78, 100, 101). For example, it was observed that F1 hybrids between two I-J^{k-} strains (e.g. B10 and 3R, both *I-J*^{b}) typed I-J^{k+} (100). One cannot explain this result using the original definition of the I-J subregion. Rather, it would appear that interaction between two genes—at least one of them mapping in the MHC—controls I-J phenotype. Suggestive evidence that the second gene does not map in the MHC comes from the observation that F1 hybrids between 3R and A.BY (*H-2*^{b}, like B10) type I-J^k (51). Presumably, strain 3R provides the appropriate allele at the MHC gene, while strain B10 (but not strain A.BY) provides the appropriate allele at the non-MHC gene. Strain 3R may carry a strain A–derived allele at the non-MHC locus, While the I-J^{k+} strain 5R may carry a strain B10–derived allele at the non-MHC locus. Formal proof that a non-MHC gene is involved must come from segregation and/or molecular analysis.

The observation that interaction between two genes may influence the qualitative nature of I-J determinants should not be confused with the observation that a gene, Jt, *mapping on chromosome 4, controls I-J cell surface expression.* Serological studies have shown that I-J^k monoclonal antibodies do not react with peripheral T lymphocytes in the *H-2*^{k} strain AKR/J, as judged by complement mediated cytotoxicity (100). The ability to lysc AKR/J T cells with I-J^k antibody plus complement is controlled by the chromosome 4 *Jt* gene, as shown with recombinant inbred and congenic strains. However, this does not prove that the *Jt* gene either encodes or influences the qualitative nature of I-J molecules. In fact, two studies (102, 103) have now shown that strain AKR/J does synthesize I-J^k molecules associated with suppressor factors. Possibly, the *Jt* gene controls I-J cell surface expression (transportation to or insertion in the cell membrane, quantitative levels?) or the relative number of I-J$^+$ T cells in the periphery. Further study is required to sort this out.

The I-J Puzzle

Studies with numerous congenic and recombinant strains, and with transgenic mice (see below), provide unequivocal evidence that polymorphism in I-J molecules is influenced by the I region of the MHC. However, molecular genetic studies provide compelling evidence that I-J molecules are not encoded by a gene mapping in the I region, and serological studies provide tentative evidence that a non-MHC gene (exclusive of Jt on chromosome 4) can influence the qualitative nature of I-J determinants. How can one resolve this apparent paradox? Namely, how could an I-region gene influence polymorphism in I-J molecules which are probably encoded by a gene mapping elsewhere? In particular, how do the strains (e.g. 3R and 5R) used to define and map control of I-J determinants differ?

Possible Solutions to the I-J Puzzle

The most logical explanation for control of I-J determinants was initially proposed by Schrader in 1979 (104). He hypothesized that I region–controlled T-cell determinants might reside on receptors for self–class-II molecules, i.e. I-J antigens are anti–class-II idiotypes. These receptors could be encoded by non-MHC genes, but expression of these receptors would be selected for during ontogeny by I region–encoded class-II products. I-J polymorphism would thus reflect class-II polymorphism. The concept that I-J molecules may be non–MHC encoded T-cell receptors that recognize self–class-II molecules and/or that recognize self–class-II receptors has subsequently been espoused by numerous investigators (40, 60, 75, 78, 79, 93, 99, 105–107). Strong support for the receptor nature of I-J molecules can be drawn from chimera experiments, particularly those showing that I-J phenotype is influenced by the environment in which T cells mature (61, 75, 96). For example, bone marrow cells from B6 (I-J^b, H-2^b) mice that had matured in a lethally irradiated adult C3H (I-J^k, H-2^k) host were shown to express serologically detected T-cell I-J^k determinants, and vice versa (108, 109). Whether H-2 or non-H-2 genes influence I-J phenotype in these chimeras was not resolved. These putative receptors are probably distinct from those expressed by MHC-restricted helper or cytolytic T cells, since the majority of I-J^+ somatic T-cell hybrids have deleted the T-cell-receptor β genes contributed by the suppressor-T-cell fusion partner (110). The observation in another system (93) that an I-J^+ suppressor factor binds class-II determinants is consistent with recognition of class-II molecules by I-J molecules and/or antigen binding molecules. Heterogeneity in I-J determinants could readily be accounted for by expression of unique self-receptors by different T-cell clones or by functionally distinct T-cell subsets. Under this model, I-J determinants on macrophages are either passively acquired or possibly represent a restriction element which selects for self–T cell receptors (79).

Other explanations, of course, are possible (40, 51, 74, 99, 107, 111). For example, it has been proposed that I-J determinants reside on modified or altered class-II molecules (40, 99, 100, 112). This would occur if the gene that encodes I-J molecules used at least a part of a class-II gene or if another gene posttranslationally modified (e.g. glycosylated) class-II molecules in T cells. Tentative support for this model was drawn from the observation that a monoclonal class-II antibody reacted with an I-J$^+$ suppressor factor (112). Failure to detect DNA rearrangements in the I region and to detect mRNA that hybridizes with class-II DNA probes in I-J$^+$ somatic-T-cell hybrids (99), coupled with the original observation that polyvalent alloantisera against class-II determinants do not react with I-J molecules (26, 28, 29), all argue strongly against this model. Reactivity of monoclonal class-II antibody with an I-J$^+$ factor may be due to cross-reactivity with an internal image of a class-II determinant, e.g. class-II antibody may cross-react with a receptor that recognizes a receptor for class-II molecules. Another possible explanation for genetic control of I-J determinants is that an I-region DNA regulatory element controls expression of I-J structural genes (40, 99). The major problem with this model is that it is not readily apparent how the regulatory element could influence polymorphism in the I-J structural gene. Also, one must postulate that this regulatory element is transcriptionally silent in established I-J$^+$ somatic-T-cell hybrids. Finally, other possible explanations stem from the observation that the crossovers in the strains utilized to define and map control of I-J determinants occurred in a recombination hotspot within the E_β gene (40, 41, 98). Perhaps a crossover in this area precludes crossovers in other nearby areas, and I-J structural genes map within or near the MHC (41). Alternatively, this recombination hotspot might influence mutations (gene conversion events?) elsewhere in the MHC in neighboring I-J structural genes.

At least three major pieces of the I-J puzzle must be assembled before the puzzle can be solved. First, the MHC gene which influences I-J polymorphism must be identified. Are class-II genes involved, as many have speculated, or, does a novel I region or MHC gene play a role? Formal proof that class-II genes influence I-J polymorphism comes from studies with E_α transgenic mice (see below). Second, the structural gene that encodes I-J molecules must be identified and characterized. Is this gene unlinked to the MHC, and where does it map? Third, what is the nature of the interplay between the genes that determines I-J phenotype. Are I-J molecules a novel set of self–T-cell receptors for I-region products (which seems most likely) or is some other type of gene-gene interaction involved?

Answers to the latter two questions must await future genetic, functional, and molecular study. Below, we consider the evidence that formally proves that MHC class-II genes influence I-J phenotype.

Class-II Genes Influence I-J Phenotype

To determine the influence of class-II genes on I-J phenotype, backcross progeny from an E_α^k transgenic mouse were examined (113). The transgenic mouse was produced by injecting the E_α^k gene into an ($H-2^b \times$ $H-2^s$)F2 embryo (17). A successful transfectant was backcrossed to $H-2^b$ homozygous mice, and a male offspring, who was heterozygous for the MHC and the transfected E_α^k gene, was backcrossed again to $H-2^b$. Mice carrying the $H-2^b$ and $H-2^s$ haplotypes carry nonfunctional E_α genes, and routinely type I-J^{k-}. Although these mice synthesize E_β chains, they are found in the cytoplasm. In contrast, strains carrying a functional E_α gene express $E_\alpha : E_\beta$ complexes on the cell surface. At least some of these latter strains (e.g. 5R, 9R) type I-J^{k+}. Of ten backcross progeny, six carried and expressed the E_α^k gene, while four did not. Five of the six E_α^k-positive progeny synthesized I-J^{k+} molecules associated with a T-suppressor-inducer factor (TsiF). All four E_α^k-negative progeny typed I-J^{k-} (113). Failure of one E_α^k-positive animal to type I-J^{k+} may have been due to environmental variability in TsiF production, the influence of another segregating gene on I-J expression, or possibly to variable expression of the transfected E_α^k gene in different tissues (e.g. the thymus?). Clearly, the E_α^k gene influences I-J phenotype in the transgenic mice. We assume class-II genes influence I-J polymorphism in strains carrying the same MHC type as the transgenic mice as well as other MHC types.

Future Questions and Study

Although the above study with transgenic mice provides a very important piece to the I-J puzzle, it does not resolve how class-II genes influence I-J phenotype. As discussed above, initial serological studies (26, 28, 29) together with more recent molecular studies (99) provide overwhelming evidence that serologically defined T-cell I-J determinants do not reside on modified or altered class-II products. Rather, these results are more compatible with expression of I-J determinants on products of other genes whose expression is influenced by class-II genes. Combined with studies in chimeras which show that I-J phenotype is influenced by the environment in which T-cells mature (108, 109), and studies which show that T cell–receptor β genes are frequently deleted in I-J$^+$ somatic T-cell hybrids (110), these data provide evidence that I-J molecules are a novel set of T-cell receptors which recognize self–class-II molecules or which recognize (or are recognized by) receptors for self–class-II molecules. In the transgenic mice, these receptors could be selected for by E_α^k chains, or, more likely, by cell surface $E_\alpha : E_\beta$ complexes. The concept that at least some regulatory T cells use self-receptors that are distinct from the conventional T-cell-receptor $\alpha : \beta$ complex draws strong support from the observation that the T-cell-receptor β gene is not rearranged in an I-A subregion–restricted suppressor-T-cell hybridoma (114). Formal proof that a distinct gene encodes I-J molecules and that I-J molecules are a novel set of self–T cell receptors must come from molecular analysis of I-J structural genes and products.

It will be particularly interesting to determine how strains like 3R and 5R, which were originally used to define and map control of I-J determinants, differ. Molecular genetic studies show that the segment of chromosome bounded by the crossovers in these strains does not contain a polymorphic coding element (98, 99). This raises the possibility that the crossovers play no role in determining the I-J phenotype in the two strains, at least not in a direct way (see below for other possibilities). Given that class-II molecules influence I-J phenotype, and assuming that another gene encodes I-J molecules, there are three possible ways the strains could differ (107). First, the two strains may express a different class-II–encoded determinant. In the receptor-ligand model, this difference could select for different self–T cell repertoires in the two strains. Evidence against this possibility comes from the observation that both strains appear to carry and express the same class-II alleles. However, it is still possible that an undetected mutation has occurred in one of the class-II genes (A_α, A_β, E_α, or E_β) carried by one of the strains. For example, I-J restriction elements on macrophages in suppressor pathways may reside on a mutated (or altered; see possibility three below) class-II molecule. Such determinants may not be recognized by antibody but could select for the serologically detected T-cell I-J determinants. Second, the two strains may carry different alleles at the genes which encode I-J molecules. Although these genes may not be linked to the MHC, their expression would still be influenced by class-II genes. In the receptor-ligand model, I-J receptor molecules would be selected for by self–class-II molecules during ontogeny. This alternative is most compatible with the observation that F1 hybrids between two I-J^{k-} strains type I-J^{k+} (100). Under this view, I-J determinants on macrophages are probably passively acquired T-cell receptors. Finally, the two strains may carry different alleles at another locus which controls or modifies expression of class-II or I-J structural genes. For example, class-II molecules in the two strains may be posttranslationally modified in different ways. In the receptor-ligand model, although serologically detected I-J determinants may not reside on these products, they could select for I-J receptor molecules. Regardless of the interpretation, the genetic difference between the strains was either derived from one of the two progenitor strains (B10 or A) or resulted from mutation. Similar arguments can be made to account for the genetic difference between strains B10.HTT and 9R, although the two pairs of strains may be polymorphic at different loci, e.g. 3R and 5R may carry different alleles at the loci which encode I-J molecules, while B10.HTT and 9R may differ in the expression of a class-II determinant.

The observation that crossovers in the strains used to define I-J determinants occurred in a recombination hotspot within the E_β gene (40, 41, 98) may have additional implications for control of I-J polymorphism. In addition to generating hybrid E_β genes that derive the $\beta1$ exon from one parent and the $\beta2$ exon from another, such hotspots could have other genetic effects. For example, if strains 3R and 5R differ in the expression of a class-II determinant, this is probably due to mutation. Could the recombination hotspot influence mutations (gene conversion events?) in neighboring class-II genes? Alternatively, if strains 3R and 5R carry different alleles at the I-J structural locus and this locus is not linked to the MHC, why did this non-MHC locus segregate with the MHC in the establishment of the congenic strains? If strains B10.HTT and B10.S(9R) also differ at the same non-MHC locus, it seems unlikely that this is a fortuitous event. Could the MHC recombination event in any way influence the segregation of unlinked genes?

Perhaps the most difficult part of the I-J puzzle will be elucidating the interplay between class-II genes and I-J structural genes which determines I-J phenotype in different strains (106, 107). One must consider the possibility that polymorphism in both sets of genes may influence I-J phenotype, and that different genes within one set may be operable in different strains. For example, which class-II gene determines the I-J phenotype in strains B10 and 5R? Strain B10 expresses $A_\alpha^b : A_\beta^b$ complexes on the cell surface and synthesizes E_β^b chains which are found in the cytoplasm (E_α^b gene is nonfunctional). Strain 5R expresses cell surface $A_\alpha^b : A_\beta^b$ and $E_\alpha^k : E_\beta^b$ complexes. Strain B10 types I-J^{b+}, I-J^{k-}, while strain 5R types I-J^{b-}, I-J^{k+}. Based on studies with E_α^k transgenic mice (113), it would seem reasonable to conclude that cell surface $E_\alpha^k : E_\beta^b$ complexes influence the I-J^k phenotype in strain 5R. The situation with strain B10, however, is not at all clear. If cell surface $A_\alpha^b : A_\beta^b$ complexes determine the I-J^b phenotype in this strain, two additional questions must be addressed. First, how can one produce antibody against I-J^b determinants in strain 5R which also expresses $A_\alpha^b : A_\beta^b$ complexes? Second, why do other strains which carry nonfunctional E_α alleles and express different $A_\alpha : A_\beta$ complexes also type I-J^{b+}, I-J^{k-} [e.g. B10:A(4R) ($A_\alpha^k : A_\beta^b$), B10.GD ($A_\alpha^d : A_\beta^d$)]? Possibly, $A_\alpha : A_\beta$ complexes do not determine the I-J^b phenotype in these strains. For example, other candidates in strain B10 include cryptically expressed E_β^b monomeric or homodimeric cell surface molecules, mixed $A_\alpha^b : E_\beta^b$ heterodimers, or expression of other class-II products (e.g. $A_{\beta 2}$, $E_{\beta 2}$). Of these three possibilities, there is only suggestive evidence for the latter two (6, 11, 13). Resolution of the genetic difference between strains 5R and 3R (which also expresses $A_\alpha^b : A_\beta^b$ and $E_\alpha^k : E_\beta^b$ complexes like 5R, but types I-J^b like B10) should help resolve these questions and may well reveal novel aspects of class-II gene expression.

A molecular characterization of I-J products should also resolve two other pieces of the I-J puzzle. First, what is the molecular basis for the influence of chromosome 12 genes on I-J expression and function? For example, I-J molecules determine the Igh-V region–restricted activity of several suppressor factors (81, 82, 86, 90). Are I-J molecules self-receptors for chromosome 12–controlled determinants, as is suggested by some chimera studies? If so, does a single polypeptide or do tightly associated polypeptides determine the apparent Igh-V region and class-II (or anti–class II) recognition capabilities of I-J molecules? Alternatively, I-J molecules may bear chromosome 12–controlled determinants and may be recognized by other self-receptors. This latter interpretation is consistent with the observation that antibody against the chromosome 12–encoded Meth A antigen reacts with I-J molecules (115). One obvious possibility is that I-J molecules are chromosome-12–encoded receptors which are expressed in T cells and which recognize class-II molecules or anti–class-II receptors. Other possibilities are that I-J molecules are posttranslationally modified by a chromosome 12 product, or, simply that they bear a determinant that is cross-reactive with the Meth-A antigen. Resolution of these questions should provide further insight into more basic immunoregulatory mechanisms.

Second, what is the molecular basis for reactivity of I-J antibody with a glycosylation inhibiting factor (GIF) which inhibits assembly of N-linked oligosaccharides to IgE-binding factors and which appears to be a phosphorylated derivative of lipomodulin, a phospholipase inhibitor (77)? GIF and lipomodulin are synthesized by both T cells and macrophages. GIF has been shown to be part of a factor that is involved in the effector phase of antigen-specific suppression. Possibly, lipomodulin or GIF polypeptides are tightly associated with I-J molecules. Alternatively, I-J antibody may fortuitously cross-react with these molecules. Regardless of the interpretation, it is extremely unlikely that T cell–derived I-J molecules that restrict the activity of suppressor factors, that adaptively acquire specificity in allogeneic chimeras, and that display an altered phenotype in E_α^k transgenic mice are lipomodulin or GIF.

In conclusion, the genetic basis for control of I-J determinants is one of the more intriguing puzzles in immunology today. Although it is clear that MHC class-II genes influence polymorphism in I-J molecules, it remains to be determined how this is accomplished. The most likely explanation is that I-J molecules represent a novel set of T-cell receptors that recognizes self-class-II molecules and/or receptors for class-II molecules. Such a system would allow for the delivery of specific regulatory signals to appropriate target cells at long range. Molecular characterization of I-J molecules and I-J structural genes, coupled with an understanding of the interplay between I-J structural genes and class-II genes, is required before this puzzle can be completely assembled.

Literature Cited

1. McDevitt, H. O., Deak, B. D., Shreffler, D. C., Klein, J., Stimpfling, J. H., Snell, G. D. 1972. Genetic control of the immune response. Mapping of the *Ir-1* locus. *J. Exp. Med.* 135: 1259
2. Benacerraf, B. 1981. Role of MHC gene products in immune regulation. *Science* 212: 1229
3. Murphy, D. B. 1981. Genetic fine structure of the H-2 gene complex. In *The Role of the Major Histocompatibility Complex in Immunobiology*, ed. M. E. Dorf, p. 1. New York: Garland STPM
4. Klein, J., Figuera, F., Nagy, Z. A. 1983. Genetics of the major histocompatibility complex: The final act. *Ann. Rev. Immunol.* 1: 119
5. Sachs, D. H. 1985. The major histocompatibility complex. In *Fundamental Immunology*, ed. W. E. Paul, p. 303. New York: Raven
6. Lerner, E. A., Matis, L. A., Janeway, C. A., Jones, P. P., Schwartz, R. H., Murphy, D. B. 1980. Monoclonal antibody against an Ir gene product? *J. Exp. Med.* 152: 1085
7. Baxevanis, C. N., Wernet, D., Nagy, Z. A., Maurer, P. H., Klein, J. 1980. Genetic control of T-cell proliferative responses to Poly (Glu⁴⁰ Ala⁶⁰) and Poly (Glu⁵¹, Lys³⁴, Tyr⁵): Subregion-specific inhibition of the responses with monoclonal Ia antibodies. *Immunogenetics* 11: 617
8. Klein, J. 1981. The histocompatibility-2 (H-2) complex. In *The Mouse in Biomedical Research*, ed. H. L. Foster, p. 119. New York: Academic Press
9. Hood, L., Steinmetz, M., Malissen, B. 1983. Genes of the major histocompatibility complex of the mouse. *Ann. Rev. Immunol.* 1: 529
10. Mengle-Gaw, L., McDevitt, H. O. 1985. Genetics and expression of mouse Ia antigens. *Ann. Rev. Immunol.* 3: 367
11. Flavell, R. A., Allen, H., Burkly, L. C., Sherman, D. H., Waneck, G. L., Widera, G. 1986. Molecular biology of the H-2 histocompatibility complex. *Science* 233: 437
12. Steinmetz, M., Stephan, D., Fischer-Lindahl, K. 1986. Gene organization and recombinational hotspots in the murine major histocompatibility complex. *Cell* 44: 895
13. Germain, R. N., Malissen, B. 1986. Analysis of the expression and function of class II major histocompatibility complex encoded molecules by DNA-mediated gene transfer. *Ann. Rev. Immunol.* 4: 281
14. Mathis, D., Benoist, C., Williams, V. E., Kanter, M., McDevitt, H. O. 1983. Several mechanisms can account for defective E_α gene expression in different mouse haplotypes. *Proc. Natl. Acad. Sci. USA* 80: 273
15. Jones, P. P., Murphy, D. B., McDevitt, H. O. 1978. Two-gene control of the expression of a murine Ia antigen. *J. Exp. Med.* 148: 925
16. Murphy, D. B., Jones, P. P., Loken, M. R., McDevitt, H. O. 1980. Interaction between I region loci influences the expression of a cell surface antigen. *Proc. Natl. Acad. Sci. USA* 77: 5404
17. Lemeur, M., Gerlinger, P., Benoist, C., Mathis, D. 1985. Correcting an immune-response deficiency by creating E_α gene transgenic mice. *Nature* 316: 38
18. Yamamura, K., Kikutani, H., Folsom, V., Clayton, V., Kimoto, M., Akira, S., Kashiwamura, S., Tonegawa, S., Kishimoto, T. 1985. Functional expression

of a microinjected E_α^d gene in C57BL/6 transgenic mice. *Nature* 316: 67

19. Pinkert, C. A., Widera, G., Cowing, C., Heber-Katz, E., Palmiter, R. D., Flavell, R. A., Brinster, R. L. 1985. Tissue-specific, inducible and functional expression of the E_α^d MHC class II gene in transgenic mice. *EMBO J.* 4: 2225

20. Schwartz, R. H. 1985. T-lymphocyte recognition of antigen in association with gene products of the major histocompatibility complex. *Ann. Rev. Immunol.* 3: 237

21. Klein, J., Nagy, Z. A. 1982. MHC restriction and Ir genes. *Adv. Cancer Res.* 37: 234

22. Paul, W. E. 1985. Immune response genes. See Ref. 5, p. 439

23. Baxevanis, C. N., Nagy, Z. A., Klein, J. 1981. A novel type of T-T cell interaction removes the requirement for I-B region in the H-2 complex. *Proc. Natl. Acad. Sci. USA* 78: 3809

24. Simpson, E. A., Lieberman, R., Ando, I., Sachs, D. H., Paul, W. E., Berzofsky, J. 1986. How many class II immune response genes? A reappraisal of the evidence. *Immunogenetics* 23: 302

25. Hammerling, G. J., Black, S. J., Segal, S., Eichmann, K. 1975. Cellular expression of Ia antigens and their possible role in immune reactions. In *Membrane Determinants Regulating Immune Reactivity*, ed. V. P. Eijsvoogel, D. Roos, W. P. Zeijlemaker, p. 367. New York: Academic Press

26. Okumura, K., Herzenberg, L. A., Murphy, D. B., McDevitt, H. W., Herzenberg, L. A. 1976. Selective expression of H-2 (I region) loci controlling determinants on helper and suppressor T lymphocytes. *J. Exp. Med.* 144: 685

27. Murphy, D. B., Herzenberg, L. A., Okumura, K., Herzenberg, L. A., McDevitt, H. O. 1976. A new I subregion (I-J) marked by a locus (*Ia-4*) controlling surface determinants on suppressor T lymphocytes. *J. Exp. Med.* 144: 699

28. Tada, T., Taniguichi, M., David, C. S. 1976. Properties of the antigen-specific suppressive T-cell factor in the regulation of antibody responses in the mouse. IV: Special subregion assignment of the gene(s) that codes for the suppressive T cell factor in the H-2 histocompatibility complex. *J. Exp. Med.* 144: 713

29. Murphy, D. B., Okumura, K., Herzenberg, L. A., Herzenberg, L. A., McDevitt, H. O. 1977. Selective expression of separate I-region loci

in functionally different lymphocyte subpopulations. *Cold Spring Harbor Symp. Quant. Biol.* 41: 497

30. Frelinger, J. A., Neiderhuber, J. E., Shreffler, D. C. 1976. Effects of anti-Ia sera in mitogenic responses. III. Mapping the genes controlling expression of Ia determinants in concanavalin A reactive cells to the I-J subregion of the H-2 gene complex. *J. Exp. Med.* 144: 1141

31. Greene, M. I., Pierres, A., Dorf, M. E., Benacerraf, B. 1977. The I-J subregion codes for determinants on suppressor factors which limit the contact sensitivity response to picryl chloride. *J. Exp. Med.* 146: 293

32. Meruelo, D., Deak, B., McDevitt, H. O. 1977. Genetic control of cell-mediated responsiveness to an AKR tumor associated antigen. Mapping of the locus involved to the I region of the H-2 complex. *J. Exp. Med.* 146: 1367

33. Okuda, K., David, C. S., Shreffler, D. C. 1977. The role of gene products of the I-J subregion in mixed lymphocyte reactions. *J. Exp. Med.* 146: 1561

34. Pierres, M., Germain, R. N., Dorf, M. E., Benacerraf, B. 1977. Potentiation of a primary in vivo antibody response by alloantisera against gene products of the I region of the H-2 complex. *Proc. Natl. Acad. Sci. USA* 74: 3975

35. Theze, J., Waltenbaugh, C., Dorf, M. E., Benacerraf, B. 1977. Immunosuppressive factor(s) specific for L-glutamic acid^{50}-L-tyrosine^{50} (GT). II. Presence of I-J determinants on the GT-suppressive factor. *J. Exp. Med.* 146: 287

36. Hammerling, G. J., Eichmann, K. 1978. Demonstration of T cell specific Ia antigens coded for by the I-J subregion on idiotype suppressor T cells. In *Ir Genes and Ia Antigens*, ed. H. O. McDevitt, p. 157. New York: Academic Press

37. Neiderhuber, J. E., Mayo, L., Shreffler, D. C. 1978. The requirement for Ia positive macrophages in the primary in vitro humoral response. See Ref. 36, p. 393

38. Zan-bar, I., Murphy, D. B., Strober, S. 1978. Cellular basis of tolerance to serum albumin in adult mice. I. Characterization of T suppressor and T helper cells. *J. Immunol.* 120: 497

39. Cook, R. G., Vitetta, E. S., Uhr, J. W., Capra, J. D. 1979. Structural studies on the murine Ia alloantigens. V. Evidence that the structural gene for the I-E/C beta polypeptide is encoded within the I-A subregion. *J. Exp. Med.* 149: 981

40. Steinmetz, M., Minard, K., Horvath, S., McNicholas, J., Frelinger, J., Wake, C., Long, E., Mach, B., Hood, L. 1982. A molecular map of the immune response region from the major histocompatibility complex of the mouse. *Nature* 300: 25

41. Kobori, J. A., Winoto, A., McNicholas, J., Hood, L. 1984. Molecular characterization of the recombination region of six major histocompatibility complex (MHC) I-region recombinants. *J. Mol. Cell. Immunol.* 1: 125

42. Stimpfling, J. H., Richardson, A. 1965. Recombination within the histocompatibility-2 locus of the mouse. *Genetics* 51: 831

43. Stimpfling, J. H., Reichert, A. E. 1970. Strain C57BL/10ScSn and its congenic resistant sublines. *Transplant. Proc.* 2: 39

44. Meo, T., David, C. S., Nabholz, M., Miggiano, V., Shreffler, D. C. 1973. Demonstration by MLR test of a previously unsuspected intra-H-2 crossover in the B10.Htt strain: Implications concerning location of MLR determinants in the Ir region. *Transplant. Proc.* 5: 1507

45. Klein, J., Hauptfeld, M., Hauptfeld, V. 1974. Evidence for a third, Ir-associated histocompatibility region in the H-2 complex of the mouse. *Immunogenetics* 1: 45

46. Webb, D. R., Kapp, J. A., Pierce, C. W. 1983. The biochemistry of antigen-specific T cell factors. *Ann. Rev. Immunol.* 1: 423

47. Green, D. R., Flood, P. M., Gershon, R. K. 1983. Immunoregulatory T-cell pathways. *Ann. Rev. Immunol.* 1: 439

48. Dorf, M. E., Benacerraf, B. 1984. Suppressor cells and immunoregulation. *Ann. Rev. Immunol.* 2: 127

49. Tada, T. 1985. Help, suppression, and specific factors. See Ref. 5, p. 481

50. Asherson, G. L., Colizzi, V., Zembala, M. 1986. An overview of T-suppressor cell circuits. *Ann. Rev. Immunol.* 4: 37

51. Hayes, C. E. 1986. I-J: A rogue's riddle. *BioEssays* 4: 278

52. Stout, R. D., Murphy, D. B., McDevitt, H. O., Herzenberg, L. A. 1977. The Fc receptor on thymus-derived lymphocytes. IV. Inhibition of binding of antigen-antibody complexes to Fc receptor-positive T cells by anti-Ia sera. *J. Exp. Med.* 145: 187

53. Murphy, D. B. 1978. The I-J subregion of the murine H-2 gene complex. *Springer Semin. Immunopathol.* 1: 111

54. Meruelo, D., Flieger, N., Smith, D., McDevitt, H. O. 1980. *In vivo* or *in vitro* treatments with anti-I-J alloantisera abolish immunity to AKR leukemia. *Proc. Natl. Acad. Sci. USA* 77: 2178

55. Habu, S., Yamauchi, K., Gershon, R. K., Murphy, D. B. 1981. A non-T:non-B cell bears I-A, I-E, I-J, and Tla (Qa-1?) determinants. *Immunogenetics* 13: 215

56. Minami, M., Henji, N., Dorf, M. E. 1982. The mechanism responsible for the induction of I-J restrictions on T_{S3} suppressor cells. *J. Exp. Med.* 156: 1502

57. Lowy, A., Tominaga, A., Drebin, J. A., Takaoki, M., Benacerraf, B. 1983. Identification of an I-J$^+$ antigen-presenting cell required for third order suppressor cell activation. *J. Exp. Med.* 157: 353

58. Neiderhuber, J. E., Allen, P., Mayo, L. 1979. The expression of Ia antigen determinants on macrophages required for the *in vitro* antibody response. *J. Immunol.* 122: 1342

59. Greene, M. I., Dorf, M. E., Pierres, M., Benacerraf, B. 1977. Reduction of syngeneic tumor growth by an anti-I-J alloantiserum specific for suppressor T cells. *Proc. Natl. Acad. Sci. USA* 74: 5118

60. Murphy, D. B., Horowitz, M. C., Homer, R. J., Flood, P. M. 1985. Genetic, serological, and functional analysis of I-J molecules. *Immunol. Rev.* 83: 79

61. Tada, T., Uracz, W., Abe, R., Asano, Y. 1985. I-J as an inducible T cell receptor for self. *Immunol. Rev.* 83: 105

62. Waltenbaugh, C., Lei, H. Y. 1984. Regulation of immune responses by I-J gene products as detected by anti-I-J monoclonal antibodies. *J. Immunol.* 133: 1730

63. Tada, T., Okumura, K., Tokuhisa, T., Nonaka, M., Taniguichi, M. 1978. Heterogeneity of I-J subregion gene products: Their expression on functionally different subsets of T cells. *J. Immunol.* 121: 1607

64. Gershon, R. K., Eardley, D. D., Durum, S., Green, D. R., Shen, F. W., Yamauchi, K., Cantor, H., Murphy, D. B. 1981. Contrasuppression: A novel immunoregulatory activity. *J. Exp. Med.* 153: 1533

65. Murphy, D. B., Yamauchi, K., Habu, S., Eardley, D. D., Gershon, R. K. 1981. T cells in a suppression circuit and non-T:non-B cells bear different I-J determinants. *Immunogenetics* 13: 205

66. Flood, P. M., Waltenbaugh, C. W., Tada, T., Chue, B., Murphy, D. B.

1986. Serologic heterogeneity in I-J determinants associated with functionally distinct T cell regulatory factors. *J. Immunol.* 137: 2237

67. Streilein, J. W., Klein, J. 1979. Neonatal tolerance to K and D region alloantigens of H-2 complex: I-J region requirements. *Transplant. Proc.* 11: 732

68. Holan, V., Hilgert, I., Chutna, J., Hasek, M. 1980. The regulatory role of I-J subregion in neonatal tolerance induction to H-2D alloantigens. *J. Immunogenetics* 7: 221

69. Czitrom, A. A., Sunshine, G. H., Mitchison, N. A. 1980. Suppression of the proliferative response to H-2D by I-J subregion gene products. *Immunogenetics* 11: 97

70. Zingernagel, R. M. 1980. Activation or suppression of bactericidal activity of macrophages during a graft-vs-host reaction against I-A and I-J region differences, respectively. *Immunogenetics* 10: 373

71. Okuda, K., David, C. S., Shreffler, D. C. 1977. The role of gene products of the I-J subregion in mixed lymphocyte reactions. *J. Exp. Med.* 146: 1561

72. Horowitz, M. C., Murphy, D. B. 1983. Induction of autoreactive T cells by stimulation across the I-J subregion. In *Ir Genes, Past, Present, and Future*, ed. C. W. Pierce, S. E. Cullen, J. A. Kapp, B. O. Schwartz, D. C. Shreffler, p. 29. Clifton, NJ: Humana Press

73. Delovitch, T. L., Sohn, V. 1978. In vitro analysis of allogeneic lymphocyte interaction. III. Generation of a helper allogeneic effect factor (AEF) across an I-J subregion disparity. *J. Immunol.* 122: 1528

74. Klein, J., Ikezawa, Z., Nagy, Z. A. 1985. From LDH-B to J: An involuntary trip. *Immunol. Rev.* 83: 61

75. Taniguichi, M., Takayuki, S. 1985. "I-J" as an idiotypic marker on the antigen-specific suppression T cell factor. *Immunol. Rev.* 83: 125

76. Daly, M. J., Nakamura, M., Gefter, M. L. 1986. Functional and biochemical characterization of a secreted I-J⁺ suppressor factor that binds to immunoglobulin. *J. Exp. Med.* 163: 1415

77. Jardieu, P., Akasaki, M., Ishizaka, K. 1986. Association of I-J determinants with lipomodulin/macrocortin. *Proc. Natl. Acad. Sci. USA* 83: 160

78. Hayes, C. E., Klyczek, K. K. 1985. The I-J glycoprotein: Genetic control, biochemistry, and function. *Immunol. Rev.* 83: 41

79. Dorf, M. E., Benacerraf, B. 1985. I-J as a restriction element in the suppressor

T cell system. *Immunol. Rev.* 83: 23

80. Eardley, D. D., Shen, F. W., Canton, H., Gershon, R. K. 1979. Genetic control of immunoregulatory circuits. Genes linked to the Ig locus govern communications between regulatory T cell sets. *J. Exp. Med.* 150: 44

81. Yamauchi, K., Chao, N., Murphy, D. B., Gershon, R. K. 1982. Molecular composition of an antigen-specific, Ly-1 suppressor inducer factor: One molecule binds antigen and is I-J⁻: another is I-J⁺, does not bind antigen, and imparts an Igh-variable region linked restriction. *J. Exp. Med.* 155: 655

82. Flood, P. M., Lowy, A., Tominaga, A., Chue, B., Greene, M. I., Gershon, R. K. 1983. The nature of Igh-V region restricted T cell interactions: Genetic restriction of an antigen-specific suppressor-inducer factor is imparted by an I-J⁺ antigen-nonspecific molecule. *J. Exp. Med.* 158: 1938

83. Flood, P. M., Yamauchi, K., Singer, A., Gershon, R. K. 1982. Homologies between cell interaction molecules controlled by MHC and Igh-V linked genes that T cells use for communication: Tandem "adaptive differentiation" of producer and acceptor cells. *J. Exp. Med.* 156: 1390

84. Yamauchi, K., Flood, P. M., Singer, A., Gershon, R. K. 1983. Homologies between cell interaction molecules controlled by MHC and Igh-V linked genes that T cells use for communication: Both molecules undergo "adaptive differentiation" in the thymus. *Eur. J. Immunol.* 13: 285

85. Noguchi, M., Ogasawara, M., Iwabuchi, K., Ogasawara, K., Ishihara, T., Good, R. A., Morikawa, K., Onoe, K. 1985. Recipient micro-environment does not dictate the Igh-V restriction specificity of T cell suppressor inducer factor (TsiF) from allogeneic bone marrow chimera mice. *J. Immunol.* 135: 2557

86. Flood, P. M., Louie, D. C. 1984. Mechanisms of Ly-2 suppressor cell activity: Activation of an Ly-1 I-J⁺ cell is required to transduce the suppressive signal. *J. Exp. Med.* 159: 1413

87. Noguchi, M., Onoe, K., Ogasawara, M., Iwabuchi, K., Geng, L., Ogasawara, K., Good, R. A., Morikawa, K. 1985. H-2-incompatible bone marrow chimeras produce donor-H-2-restricted Ly-2 suppressor T-cell factor(s). *Proc. Natl. Acad. Sci. USA* 82: 7063

88. Ferguson, T. A., Iverson, G. M. 1986. Isolation and characterization of an antigen-specific suppressor inducer mol-

ecule from serum of hyperimmune mice by using a monoclonal antibody. *J. Immunol.* 136: 2896

89. Flood, P. M., Chue, B., Whitaker, B. 1985. Information transfer between T cell subsets is directed by I-J⁺ antigen nonspecific molecules. *J. Immunol.* 135: 933

90. Lowy, A., Flood, P. M., Tominaga, A., Drebin, J. A., Dambrauskas, J., Gershon, R. K., Greene, M. I. 1984. Analysis of hapten-specific T suppressor factors: Genetic restrictions of TsF_1 activity analyzed with synthetic hybrid suppressor molecules. *J. Immunol.* 132: 640

91. Jensen, P. E., Kapp, J. A. 1986. Insulin-specific suppressor T cell factors. *J. Immunol.* 136: 1309

92. Minami, M., Aoki, I., Honji, N., Waltenbaugh, C. R., Dorf, M. E. 1983. The role of I-J and Igh determinants in F_1 derived suppressor factor in controlling restriction specificity. *J. Exp. Med.* 158: 1428

93. Waltenbaugh, C. W., Sun, L., Lei, H.-Y. 1986. Regulation of immune responses by I-J gene products. VI: Recognition of I-E molecules by I-J-bearing suppressor factors. *J. Exp. Med.* 163: 797

94. Zupko, K., Waltenbaugh, C., Diamond, B. 1985. Use of anti-idiotypic antibodies to identify a receptor for the T-cell I-J determinant. *Proc. Natl. Acad. Sci. USA* 82: 7399

95. Uracz, W., Abe, R., Tada, T. 1985. Involvement of I-J epitopes in the self and allorecognition sites of T cells: Blocking of syngeneic and allogeneic mixed lymphocyte reaction cells by monoclonal anti-I-J antibodies. *Proc. Natl. Acad. Sci. USA* 82: 2905

96. Tada, T., Asano, Y. 1986. Somatic generation of a genetic polymorphism: Towards the solution of the I-J enigma. *BioEssays* 4: 283

97. Bottomly, K. 1984. All idiotypes are equal, but some are more equal than others. *Immunol. Rev.* 79: 45

98. Kobori, J. A., Strauss, E., Minard, K., Hood, L. 1986. Molecular analysis of the hotspot of recombination in the murine major histocompatibility complex. *Science* 234: 173

99. Kronenberg, M., Steinmetz, M., Kobori, J., Kraig, E., Kapp, J. A., Pierce, C. W., Sorensen, C. M., Suzuki, G., Tada, T., Hood, L. 1983. RNA transcripts for I-J polypeptides are apparently not encoded between the I-A and I-E subregions of the murine major histocompatibility complex. *Proc.*

Natl. Acad. Sci. USA 79: 5704

100. Hayes, C. E., Klyczek, K. K., Krum, D. P., Whitcomb, R. M., Hullett, D. A., Cantor, H. 1984. Chromosome 4 *Jt* gene controls murine T cell surface I-J expression. *Science* 223: 559

101. Klyczek, K. K., Cantor, Hl., Hayes, C. E. 1984. T cell surface I-J glycoprotein, concerted action of chromosome-4 and -17 genes forms an epitope dependent on α-D-mannosyl residues. *J. Exp. Med.* 159: 1604

102. Flood, P. M., Murphy, D. B. 1985. The putative $I-J^{k-}$ strain AKR/J synthesizes $I-J^{k+}$ molecules: Implications for Jt gene control of I-J expression. *J. Mol. Cell. Immunol.* 2: 95

103. Waltenbaugh, C., Sun, L., Lei, H.-Y. 1985. I-J expression is not associated with murine chromosome 4. *Eur. J. Immunol.* 15: 922

104. Schrader, J. W. 1979. Nature of the T-cell receptor. *Scand. J. Immunol.* 10: 387

105. Tada, T. 1984. Points of contact between network and circuit. In *Progress in Immunology*, ed. Y. Yamamura, T. Tada, p. 595. Tokyo: Academic Press

106. Murphy, D. B. 1984. Genetic control of I-J determinants—A commentary. *J. Mol. Cell. Immunol.* 1: 133

107. Murphy, D. B. 1985. Commentary on the genetic basis for control of I-J determinants. *J. Immunol.* 135: 1543

108. Sumida, T., Sado, T., Kojima, M., Oneo, K., Kamisaku, H., Taniguichi, M. 1985. I-J as an idiotype of the recognition component of antigen-specific suppressor T-cell factor. *Nature* 316: 738

109. Uracz, W., Asano, Y., Abe, R., Tada, T. 1985. I-J epitopes are adaptively acquired by T cells differentiated in the chimeric condition. *Nature* 316: 741

110. Kronenberg, M., Siu, G., Hood, L. E., Shastri, N. 1986. The molecular genetics of the T cell antigen receptor and T cell antigen recognition. *Ann. Rev. Immunol.* 4: 529

111. Hansen, T. H., Spinella, D. G., Lee, D. R., Shreffler, D. C. 1984. The immunogenetics of the mouse major histocompatibility gene complex. *Ann. Rev. Genet.* 18: 99

112. Ikezawa, Z., Baxevanis, C. N., Arden, B., Tada, T., Waltenbaugh, C. R., Nagy, Z. A., Klein, J. 1983. Evidence for two suppressor factors secreted by a single cell suggests a solution to the J-locus paradox. *Proc. Natl. Acad. Sci. USA* 80: 6637

113. Flood, P. M., Benoist, C., Mathis, D., Murphy, D. B. 1986. Altered I-J pheno-

type in E_α transgenic mice. *Proc. Natl. Acad. Sci. USA.* 83: 8308

114. Blanckmeister, C. A., Yamamoto, K., Davis, M. M., Hammerling, G. J. 1985. Antigen-specific, I-A restricted suppressor hybridomas with spontaneous cytolytic activity. *J. Exp. Med.* 162: 851

115. Flood, P. M., DeLeo, A. B., Old, L. J., Gershon, R. K. 1985. Investigations into the nature of Igh-V region restricted T cell interactions using antibodies to antigens on methylcholanthrene-induced sarcomas: I. Analysis of an Ly-1 I-J$^+$ suppressor inducer factor. *J. Immunol.* 134: 1665

Ann. Rev. Immunol. 1987. 5 : 429–59

B-CELL STIMULATORY FACTOR-1/INTERLEUKIN 4[1]

William E. Paul and Junichi Ohara

Laboratory of Immunology, National Institute of Allergy and Infectious Diseases, National Institutes of Health, Bethesda, Maryland 20892

Introduction

B-cell stimulatory factor-1 (BSF-1) is a 20,000-M_r protein produced by some activated T cells. First described as a costimulant of the entry into S-phase of resting mouse B cells incubated with anti-immunoglobulin (Ig) antibodies (1), BSF-1 has now been shown to act on resting B cells to increase their expression of class-II major histocompatibility complex (MHC) molecules (2, 3) and to increase markedly the production of IgG1 (4–6) and IgE (7) by B-cell populations stimulated with lipopolysaccharide (LPS). Although initially believed to be principally active on B cells, BSF-1 has been shown to increase the viability and to stimulate growth of normal T cells (8, 9) and of certain T-cell lines (10–13), to act as a costimulant for growth of some mast cell lines (10, 14), and to act on several other hematopoietic lineage cells (15, 16), including granulocyte, megakaryocyte, and erythroid precursors and macrophages.

It has been proposed that BSF-1 be designated interleukin-4 (IL-4) (17). Because of the likelihood that there are a large number of distinct lymphocyte products for which an interleukin designation may soon be sought, the nonspecialist may find it increasingly difficult to associate an interleukin name with the functions of a particular molecule. For that reason, we prefer to use the historic or "trivial" name of the lymphokine, but in association with an interleukin designation, to avoid any possible confusion. Thus, at the beginning of any discussion of a paper in which BSF-1 is used, we refer to it as BSF-1/IL-4 and thereafter, as BSF-1.

Purification and Chemical Characterization of BSF-1

BSF-1 was initially identified in the supernatant fluids of cells of the thymoma line EL-4 which had been stimulated with phorbol myristate

---

[1] The US Government has the right to retain a nonexclusive, royalty-free license in and to any copyright covering this paper.

429

acetate (PMA) (1). Our own group has purified it from this source (18–20). BSF-1 is also produced by many long-term T-cell lines when stimulated by antigen and antigen-presenting cells or by mitogens such as concanavalin A (Con A); others have purified it from such sources (10, 12, 21, 22).

BSF-1 was first partially purified for functional studies by chromatography on phenyl Sepharose, using a gradient of ethanediol to elute the activity (18). This purification was then superseded by the use of trimethylsilyl-controlled pore glass bead adsorption and elution with acetonitrile followed by reverse phase high pressure liquid chromatography (RP-HPLC) (19). Homogeneous BSF-1 has now been obtained by affinity chromatography using a Sepharose column to which a monoclonal rat anti–BSF-1 antibody (23) had been conjugated, followed by RP-HPLC (20) and, independently, by RP-HPLC combined with two rounds of ion-exchange chromatography (12). The material obtained from these purification schemes behaves as a single protein on sodium dodecyl sulfate–polyacrylamide gel electrophoresis (SDS-PAGE). Its apparent molecular weight is 20,000. N-terminal amino acid sequences of 20 and 24 residues have been reported, respectively, by Grabstein et al (12) and by our own group (20) (Figure 1). These sequences are in agreement and, as described below, are in agreement with the amino acid sequence inferred from the nucleotide sequence of the cDNA clones derived by Noma et al (17) and by Lee et al (24).

Digestion of BSF-1 with endoglycosidase F reduces its M_r to 15,000–16,000 (20), indicating that the molecule has complex glycosidic chains. The endoglycosidase F-digested material has full biologic activity, demonstrating that the complex sugar side chains are not required for function. Digestion with endoglycosidase H does not change the M_r of BSF-1, and this suggests that it has no high mannose side chains. Reduction of BSF-1 causes a loss of biologic activity, and the reduced, but not the unreduced, material binds ^{14}C-iodoacetamide, indicating the existence of disulfide bonds in the intact molecule.

The monoclonal anti–BSF-1 antibody 11B11 immunoprecipitates ^{125}I-labeled BSF-1 and binds to BSF-1 on Western blotting (23). This antibody also binds to another molecule (M_r 14,000) in partially purified preparations of BSF-1. The 14,000-M_r molecule was purified both by elution from gel slices and by affinity chromatography followed by RP-HPLC. A detailed analysis suggests that the 14,000-M_r molecule is not BSF-1 or a BSF-1–related peptide. Thus, purified 14,000-M_r material displayed little or no BSF-1 activity and failed to inhibit the binding of ^{125}I-BSF-1 to its receptor. A rabbit antibody to a C-terminal BSF-1 peptide (residues 100–113; see Figure 1) binds to BSF-1 on Western blotting but not to the

```
-20
Met Gly Leu Asn Pro Gln Leu Val Val Ile
-10
Leu Leu Phe Phe Leu Glu Cys Thr Arg Ser
 1                                     10
HIS ILE HIS GLY  ?  ASP LYS ASN HIS LEU
His Ile His Gly Cys Asp Lys Asn His Leu
                                       20
ARG GLU ILE ILE GLY ILE LEU ASN GLU VAL
Arg Glu Ile Ile Gly Ile Leu Asn Glu Val

THR GLY GLU GLY                        30
Thr Gly Glu Gly Thr Pro Cys Thr Glu Met
                                       40
Asp Val Pro Asn Val Leu Thr Ala Thr Lys
                                       50
Asn Thr Thr Glu Ser Glu Leu Val Cys Arg
                                       60
Ala Ser Lys Val Leu Arg Ile Phe Tyr Leu
                                       70
Lys His Gly Lys Thr Pro Cys Leu Lys Lys
                                       80
Asn Ser Ser Val Leu Met Glu Leu  Gln Arg
                                       90
Leu Phe Arg Ala Phe Arg Cys Leu Asp Ser
                                      100
Ser Ile Ser Cys Thr Met Asn Glu Ser Lys
                                      110
Ser Thr Ser Leu Lys Asp Phe Leu Glu Ser
                                      120
Leu Lys Ser Ile Met Gln Met Asp Tyr Ser
```

Figure 1 Mouse BSF-1/IL-4 protein and DNA sequences. The amino acid sequence of mouse BSF-1 inferred from the nucleotide sequences of the BSF-1 cDNA clones (17, 24) is shown in upper and lower case letters (i.e. His), and the amino acid sequences determined by automated sequence analyses (12, 20) are shown in upper case letters (i.e. HIS). Sites of potential N-glycosylation are indicated in bold letters as are cysteines. The underlined sequence (100–113) was synthesized and an anti-peptide antibody prepared. The amino acid numbering commences with the first amino acid detected by N-terminal sequencing.

14,000-M_r molecule. The amino acid composition of the 14,000-M_r molecule is strikingly different from BSF-1. Furthermore, the 14,000-M_r molecule does not change its mobility on SDS-PAGE when digested with endoglycosidase F. From this we conclude that this molecule is not a fragment of BSF-1 but is probably a contaminant with a fortuitous cross-reaction with 11B11.

BSF-1 has generally been measured in terms of a unit (U) based on its activity in the anti-IgM-stimulatory assay. The concentration of BSF-1 required to half-maximally stimulate ^{3}H-thymidine uptake by 5×10^4 resting splenic B cells is defined as U/ml (19). It should be noted that the reproducibility of this unit from laboratory to laboratory has not been determined; we suggest that in the future a unit may best be determined based on inhibition of binding of ^{125}I-BSF-1 to its receptor. We have estimated that a unit of BSF-1 weighs 12–19 pg (20), while Grabstein et al (12) have estimated it to be ~ 3 pg. This difference may reflect differences in the biologic assay used for measuring a unit.

Derivation of cDNA Clones for BSF-1

Complementary DNA clones for BSF-1 were obtained essentially simultaneously in Kyoto and Palo Alto. Noma et al (17) used the production of RNA under the influence of the SP6 promoter by inserts into a vector designated pSP6-K2. The RNA was injected into *Xenopus* oocytes, and activity was monitored by the ability of the oocyte supernatant to cause LPS-stimulated cells to secrete IgG1. To construct this library, mRNA was obtained from an alloreactive mouse T-cell line designated 2.19, derived from a B6.C-H-2^{bm12} donor. This line secretes IgG1-inducing factor (BSF-1) upon stimulation with C57BL/6 spleen cells plus interleukin-2 (IL-2) or with concanavalin A (Con A). The RNA was initially size fractionated on sucrose density gradients. BSF-1 mRNA has an apparent S value of 8–9 (25). Such mRNA was used to prepare cDNA, which was then cloned into the pSP6K system. The frequency of BSF-1 cDNA clones in this library was approximately 1/1000.

Lee et al (24) cloned BSF-1 from a cDNA library constructed from the T-cell line Cl.Ly1$^+$2$^-$/9 which had been stimulated with Con A. They used expression in COS-7 cells to obtain a product of their cDNA. Initially, two assays to detect the BSF-1 gene were employed. They were the mast cell growth factor-2 (MCGF-2) and the T-cell growth factor-2 (TCGF-2) assays. As we discuss later in this review, both MCGF-2 (10, 14) and TCGF-2 (10) are activities of BSF-1. Lee et al (24) demonstrated that the supernatant from COS-7 cells transiently expressing the BSF-1 cDNA clone had the capacity to induce class-II MHC molecules on B cells and to enhance IgG$_1$ and IgE secretion by LPS-stimulated B cells. We tested a sample of this COS-7 supernatant and demonstrated that it had BSF-1 activity in the anti-IgM costimulation assay and that monoclonal anti-BSF-1 antibody blocked its action (J. Hu-Li et al, unpublished observations). Lee et al isolated their BSF-1 cDNA clone from a population of ~ 5000 cDNA clones prepared from unfractionated RNA obtained from Cl.Ly1$^+$2$^-$/9 cells.

The high frequencies of BSF-1 cDNA in the libraries constructed by Noma et al (17) and Lee et al (24) might suggest that BSF-1 mRNA is generally a high frequency message. Our own experience was that only one BSF-1 cDNA clone was found in a cDNA library of 30,000 members, which had been constructed from size fractionated mRNA obtained from

PMA-stimulated EL-4 cells (26). Indeed, EL-4 cells, when analyzed by Northern blotting, express much lower concentrations of BSF-1 mRNA than do several T-cell lines.

The amino acid sequence inferred from the nucleotide sequence of the BSF-1 cDNA clones is presented in Figure 1. It contains an open reading frame sufficient to code for 140 amino acids. Based on the N-terminal amino acid sequence, the secreted protein begins with the histidine which is 21 amino acids from the initial methionine. This histidine is designated as residue 1 of the protein. The first 20 amino acids (-20 to -1) are highly hydrophobic, presumably comprising the leader sequence. The BSF-1 sequence expresses some homology in limited regions with GM-CSF and interferon gamma (IFN_γ) (17).

The open reading frame of the BSF-1 nucleotide sequence, taking into account that the first amino acid of the protein is histidine, is sufficient to code for a protein of 120 amino acids. There are three sites for the potential attachment of N-linked sugars (Figure 1). As discussed above, treatment of BSF-1 with endoglycosidase F reduces its M_r from 20,000 to 15,000–16,000, indicating that one or more of these sites is glycosylated. There are six cysteines; as already noted, the intact molecule does not label with ^{14}C-iodoacetamide, although the reduced molecule does. This would indicate that all the cysteines are involved in disulfide bonds in the native protein. Furthermore, since BSF-1 migrates with an M_r of 20,000 both in the presence and absence of reducing agents, we conclude that each cysteine is involved in an intrachain disulfide bond.

We have prepared a synthetic peptide consisting of residues 100–113 (20). Rabbit antibodies to this peptide immunoprecipitate ^{125}I-BSF-1 and bind to BSF-1 by Western blotting. This provides further evidence to support the position that the BSF-1 purified from EL-4 supernatant fluids by RP-HPLC and affinity chromatography is the same molecule as that produced by recombinant DNA techniques. Furthermore, Southern blotting suggests that BSF-1 is a single copy gene (27).

Very recently, Yokota et al (28) have obtained a clone from a human cDNA library, prepared from Con A–stimulated cells of the human T-cell line 2F1, which codes for a molecule that has $\sim 50\%$ homology with mouse BSF-1 at the level of inferred amino acid sequences (Figure 2). This molecule stimulates thymidine uptake by some human T-cell lines and displays costimulatory action with anti-IgM antibodies in the initiation of DNA synthesis by human peripheral blood B cells. These results strongly indicate that this cDNA encodes human BSF-1/IL-4, which is particularly interesting since human BSF-1 had not been identified by conventional bioassays (29). However, some human T-cell products which have growth promoting activity on human T cells, but are distinct from IL-2, will

```
                        10                        20
Mouse   M G L N P Q L V V I L L F F L E C T R S
Human   _____ T S ___ L P P _ F _ L _ A _ A G N


                                30                        40
Mouse   H I H G - - C D K N H L R E I I G I L N E V
Human   F V ___ H K ___ - I T _ Q _____ K T ___ S L


                                50                        60
Mouse   T G E - G T P C T E M D V P N V L T A T K
Human   _ - _ Q K _ L _____ L T _ T D I F A _ S _


                                70                        80
Mouse   N T T E S E L V C R A S K V L R I F Y L
Human   _____ K _ T F _____ A T _____ Q ___ S


                                90                        100
Mouse   K H G K - T P C L K K N S S V L M E L - Q R
Human   H _ E _ D _ R ___ G A T A Q Q F H R H K _ -


                                110
Mouse   L F R A F - - H C L D S - - - - - - -
Human   _ I _ - _ L K R - ___ R N L W G L A G


                                120                        130
Mouse   S I S C T M - - - N E S K S T S L K D F L E S
Human   L N ___ P V K E A _ Q _ - - _ - _ E N _____ R


                                140
Mouse   L K S I M Q M D Y S
Human   ___ T ___ R E K ___ K C S S
```

Figure 2 Comparison of inferred amino acid sequences of mouse and human BSF-1/IL-4. Inferred amino acid sequences from cDNA clones for mouse (17, 24) and human (28) BSF-1 are compared using the one-letter code for amino acids. The comparison is derived from that of Yokota et al (28). Those positions at which the human sequence is the same as the mouse sequence are shown by __, whereas the positions where a gap has been introduced to maximize homology are indicated by a –.

probably prove to be BSF-1. Human BSF-1 has no activity on mouse cells, and mouse BSF-1 fails to stimulate human cells (30). As is described in more detail below, mouse BSF-1 does not bind to receptors on human B or T cells.

High Affinity Receptors for BSF-1

The nature of the cellular receptor for BSF-1 has become a matter of considerable interest. The capacity of certain monoclonal antibodies to mimic some of the functions of BSF-1 has prompted the suggestion that the surface molecules recognized by these antibodies might function as the BSF-1 receptor. Subbarao & Mosier (31) demonstrated that a monoclonal anti-Lyb2.1 antibody, which is specific for the B-cell alloantigen Lyb2.1, caused B-cell DNA synthesis when present at high concentrations. Yakura et al (32) then showed that a different anti-Lyb2.1 monoclonal antibody, used at lower concentration, synergized with anti-IgM antibodies in the stimulation of B-cell DNA synthesis. This anti-Lyb2.1 antibody did not display synergy with the supernatant of Con A–stimulated FS6-14.13 T-lymphocyte hybridoma cells, which contains BSF-1 or a BSF-1–like activity. The capacity of anti-Lyb2.1 to synergize with anti-IgM but not with BSF-1 suggested that it exerted a BSF-1–like function. This stimulation of B-cell DNA synthesis was only observed with the cells of mouse strains that expressed the Lyb2.1 alloantigen, strongly supporting the conclusion that the activity was due to the binding of anti-Lyb2.1 antibodies to Lyb2 molecules on the cell surface. To determine whether the action of anti-Lyb2.1 might represent interaction with the cellular receptor for BSF-1, Yakura et al (32) examined the capacity of anti-Lyb2.1 to block the absorption of BSF-1 activity by cells that expressed the Lyb2.1 alloantigen (B6-Lyb-2.1 cells). Their results indicated that the alloantibody did, indeed, block absorption by these cells but failed to block absorption by cells from congenic C57BL/6 mice which are Lyb 2.1$^-$. This led them to suggest that Lyb2 was the physiologic receptor for BSF-1.

On the other hand, it should be noted that the action of anti-Lyb2.1 differs from that of BSF-1 in some potentially important ways. BSF-1, even at very high concentrations, does not cause DNA synthesis by resting B cells unless a costimulant such as anti-IgM or LPS is also used (1); by contrast, anti-Lyb2.1, by itself, may cause substantial ^{3}H-thymidine uptake by B cells (31). Indeed, Subbarao & Mosier have reported that F(ab) fragments of anti-Lyb2.1 will display such activity (33). A second important difference between anti-Lyb2.1-function and that of BSF-1 is that the action of anti-Lyb2.1 appears to be limited to cells bearing the alloantigen (i.e. B cells). BSF-1 is now known to act on a variety of Lyb2$^-$ hematopoietic lineage cells including T cells and mast cells; these cells have recently been shown by binding studies to express cellular receptors for BSF-1.

Mishra et al (34) reported that G48, a monoclonal antibody to the LFA-1 α chain, displays BSF-1–like activities on resting B cells. They have shown that G48 will synergize with anti-IgM in the stimulation of B-cell DNA synthesis and will induce the expression of class-II MHC molecules on resting B cells. Furthermore, G48 antibodies acting with LPS will induce IgG1 secretion, a property of BSF-1. The latter action of G48 was only observed when the monoclonal antibody was conjugated to Sepharose beads. It has not yet been determined whether G48 mimics other biologic effects of BSF-1, such as effects on T cells and mast cells. However, Glimcher (35) has recently observed that BSF-1 induces the expression of class-II MHC molecules in cells of a pre–B cell line which is LFA-1$^-$, strongly suggesting that LFA-1 is not required for the action of BSF-1 on this cell type.

Direct measurement of binding of BSF-1 to receptors on cell surfaces has been made possible by the purification of BSF-1 and the radioiodination of the purified molecule. ^{125}I-BSF-1 binds to spleen cells in a saturable manner; virtually all the binding can be inhibited with a large excess of nonradioactive BSF-1 (36) (Figure 3). Scatchard analysis of the binding data indicates that DBA/2 spleen cells possess 450 ± 38 receptors per cell, and these bind BSF-1 with an equilibrium constant of $3.3 \pm 0.7 \times 10^{10}$ M^{-1}. A plot of bound BSF-1 versus the ratio of bound to free BSF-1 gives no evidence of binding heterogeneity, suggesting that only a single class of receptors binds BSF-1 under the conditions employed. However, since the total concentration of BSF-1 used in these experiments is still relatively low, we cannot exclude the existence of a second class of receptors with a substantially lower binding constant for BSF-1.

The binding of BSF-1 to its receptor is inhibited by the monoclonal anti–BSF-1 antibody 11B11 which is also a potent inhibitor of the biologic function of BSF-1 (23). On the other hand, antibody to the BSF-1 peptide (100–113) fails to inhibit binding of BSF-1 to its receptor or to block the biologic action of BSF-1. Neither interleukin-2 (IL-2), nor IFN$_\gamma$ in large excess, blocks the binding of ^{125}I-BSF-1. The latter finding is important

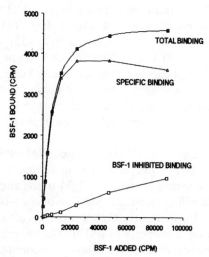

Figure 3   Binding of ^{125}I-BSF-1 to spleen cells. ^{125}I-BSF-1 was incubated with spleen cells in medium only or in the presence of a 100-fold molar excess of non-radioactive BSF-1. The amount of radioactivity bound to the cells is indicated by "Total Binding" and "BSF-1 Inhibited Binding," respectively. Specific Binding is the difference between Total and Inhibited Binding.

because IFN$_\gamma$ inhibits many of the actions of BSF-1 on resting B cells (37, 38). Furthermore, neither anti-Lyb2.1 nor several anti–LFA-1 antibodies, including antibodies to the LFA-1 α chain, block the binding of ^{125}I-BSF-1 to its receptor. This further argues against the possibility that either Lyb2 or LFA-1 is the ligand binding chain of the BSF-1 receptor.

The binding of BSF-1 to cells is reversible, and treatment with dilute acid elutes BSF-1 from membrane receptors. BSF-1 bound to receptors is quite rapidly endocytosed and is released principally, if not totally, in the form of trichloroacetic acid–soluble fragments. Whether the BSF-1 receptor is also degraded upon endocytosis or is recycled to the cell membrane has not yet been established.

BSF-1 bound to receptors on spleen cell surfaces can be cross-linked with bifunctional agents such as disuccinimidyl suberate (DSS) or dithiobis(succinimidyl proprionate) (DSP). Under these conditions, a portion of the ^{125}I-BSF-1 becomes associated with a complex with a mobility of $\sim$80,000 M_r by SDS-PAGE (36). This complex is not formed if excess nonradioactive BSF-1 is present, if the cross-linking agent is not added, or if cells lacking BSF-1 receptors are used. These results indicate that the BSF-1 which was cross-linked had been bound to the cell through its receptor, and they also suggest that the BSF-1 receptor or a component of that receptor is included in the 80,000 M_r complex. Since the M_r of BSF-1 is 20,000, the membrane component present in this complex presumably has an M_r of $\sim$60,000.

The BSF-1 receptor is found on several types of cells of hematopoietic derivation (Table 1). Small, dense splenic B cells and small, dense lymph node T cells express 250–300 receptors per cell. B lymphoma cells have slightly larger numbers of receptors (400–800/cell), and the BSF-1–sensitive T-cell–line HT-2 possesses substantially more receptors (2000/cell). BSF-1 receptors are also found in considerable numbers on Abelson

Table 1 BSF-1 receptor expression on cells of hematopoietic lineage

Type	Cell Designation	Receptors/Cell
B cell	Resting	310
	BAL 17 (B lymphoma)	420
	R8/205 (AbMulV-pre B cell)	810
T cell	Resting	280
	HT-2 (IL-2 dependent on T cell line)	2080
Macrophage	P388D1	1100
Mast cell	ABFTL-2	2170
Myeloid	DA-1	670
Human	Various lymphoid cells	Undectable

murine leukemia virus–transformed mast cell lines. Equilibrium binding studies indicate that the receptor on the mast cell line ABFTL-2 has a binding affinity similar to that of the spleen cell receptor (36); DSS-cross-linking studies show that the binding of ^{125}I-BSF-1 to the ABFTL-2 receptor creates a complex of the same size as found for the complex of BSF-1 with the spleen cell receptor. These results indicate that the mast cell receptor is very likely to be the same molecular entity as the spleen cell receptor.

In addition to B cells, T cells, and mast cell lines, we have observed binding of ^{125}I-BSF-1 to the macrophage line P388D1 and to the myeloid line DA-1. BSF-1 fails to bind both to normal human B and T cells and to Raji cells, as well as to the human T-cell line Jurkat. This result is in keeping with the failure of BSF-1 to act on human cells in functional assays (30).

Resting B cells stimulated with LPS show increased receptor expression within 24 hr, reaching a maximum of 1000–1500 receptors/cell. Receptor up-regulation in B cells is also achieved with anti-IgM antibodies and, perhaps most interestingly, with BSF-1 itself. Indeed, up-regulation in spleen cells in response to BSF-1 is detectable within 6 hr and is maximal by 18 hr (39) (Figure 4). The capacity of BSF-1 to up-regulate its own receptor on B cells may explain some of the actions of BSF-1 on resting cells. As will be discussed subsequently, BSF-1 acts on resting B cells to make them more responsive to subsequent stimulation with anti-IgM antibodies or with anti-IgM plus additional BSF-1 (40–42). At least

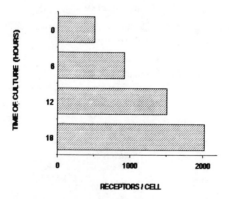

Figure 4 BSF-1 causes spleen cells to increase the expression of the BSF-1 receptor. Small spleen cells, prepared by density gradient centrifugation, were cultured for various periods of time with BSF-1 (100 U/ml) and the number of receptors per cell measured at the end of the incubation period.

some of this preparative effect may be due to or associated with receptor upregulation. Initial work with resting T cells indicates that receptor number is strikingly enhanced as a result of culturing with BSF-1 or concanavalin A.

Based on the evidence accumulated thus far, we consider the nature of the BSF-1 receptor to be still unresolved. Neither Lyb2.1 or LFA-1, by itself, appears likely to be the ligand-binding component of the receptor. However, it would be premature to conclude that these membrane molecules play no role in the action of BSF-1 on B cells. Lyb2.1 or LFA-1 might be part of a receptor complex or may have some role in regulating receptor function.

Action of BSF-1 on Resting B Cells

BSF-1 was initially described on the basis of its activity as a costimulant of B-cell DNA synthesis. Resting B cells cultured with low concentrations of anti-IgM antibodies (1–5 μg/ml) enter the G1 phase of the cell cycle but synthesize little if any DNA (43). The addition of supernatant fluids obtained from cells of the thymoma line EL-4 that had been stimulated with PMA allowed anti-IgM-activated B cells to enter S-phase (1). The molecule responsible for this B-cell growth-promoting activity in EL-4 supernatants could be separated from interleukin-2 by gel filtration, and T-cell lines could absorb IL-2 from EL-4 supernatant fluids without diminishing this B cell–stimulating activity. In addition, the colony stimulating activity in EL-4 supernatant was well separated from the B-cell growth promoting activity and EL-4 supernatants contained no IFN$_\gamma$. This led to the conclusion that the B-cell growth promoting activity in EL-4 supernatants was likely to represent a previously unrecognized entity. The material was initially designated B-cell growth factor (BCGF); later, this was modified by some authors to BCGF-1 (44). However, an international group met in Kyoto in 1983 and proposed that the factor be renamed B-cell stimulatory factor-1 (45), and still more recently the designation interleukin-4 has been suggested (17). As already noted, we will refer to this entity as BSF-1/IL-4.

Recently, Stein et al (46) have demonstrated that BSF-1 acts as a costimulant with antigens, as well as with anti-IgM, in proliferative responses of antigen-specific B cells. They prepared trinitrophenyl (TNP)-specific B cells and stimulated these cells in vitro with various TNP-conjugates. TNP-Ficoll, the prototypic type 2 antigen (47), is a semi-synthetic polymer possessing multiple TNP groups on each molecule of Ficoll. It stimulated DNA synthesis by B cells in the presence of BSF-1, but not in its absence. This response was enhanced by interleukin-1 (IL-

1), just as the response of resting B cells to BSF-1 plus anti-IgM is enhanced by IL-1, particularly when cells are cultured at low density (48). By contrast, TNP-specific B cells did not respond to TNP-conjugated to ovalbumin even in the presence of BSF-1; such B cells were capable of responding to the type-1 antigen TNP-LPS without the need for the addition of BSF-1 (46). These results indicate that BSF-1 plays a principal role in the response of B cells to type-2 antigens, but itself does not replace the action of T cells in responses to thymus-dependent antigens. It is not required for responses to type-1 antigens. Since TNP-Ficoll should be an efficient cross-linker of membrane Ig, these results are in keeping with the important role of BSF-1 in B-cell responses mediated by cellular signaling induced by cross-linkage of membrane Ig.

Since anti-IgM antibodies are known to act on resting B cells to increase inositol phospholipid metabolism (49–52), to elevate intracellular free calcium concentration ($[Ca^{++}]_i$) (53), to initiate protein phosphorylation (54), and to stimulate cells to enter the G1 phase of the cell cycle (43), it seemed reasonable to postulate that BSF-1 would function in late G1 phase to stimulate the activated cell to synthesize DNA. In terms of the paradigm of regulation of fibroblast growth (55), BSF-1 was expected to be a "progression factor." However, this was soon shown to be an inaccurate description of BSF-1 action on B cells. The first evidence to challenge this idea was the demonstration that a delay in the addition of BSF-1 to cultures of B cells stimulated with anti-IgM diminished the degree of DNA synthesis that occurred at the time when the first cohort of cells normally entered S-phase (56). Even clearer evidence that BSF-1 could not be regarded as a simple progression factor was provided by experiments which demonstrated that BSF-1 could act on resting B cells in the absence of any recognized costimulant.

Resting B cells cultured with BSF-1 demonstrated a small but significant increase in cell volume and an improved viability over a 24–48 hr period (40, 57). Alderson et al (58) have provided convincing evidence that these effects of BSF-1 are mediated by direct action of the lymphokine on B cells. They cultured individual B cells in Terasaki culture dishes and demonstrated that BSF-1, by itself, caused an increase in cell diameter and substantially increased the survival of B cells in these single cell cultures.

Resting B cells treated with BSF-1 demonstrated striking increases in the density of class-II MHC molecules on their surface (2, 3). This increase can be observed within 6 hr and reaches a magnitude of sixfold or more within 24 hr. It is, perhaps, the most sensitive indicator of BSF-1 action (9); as little as 0.1 U/ml of BSF-1 will cause detectable increases in expression of class-II MHC molecules.

The mechanism by which BSF-1 stimulates enhancement of class-II

MHC molecule expression is not clear. This induction can be achieved in the presence of 1 mM EGTA/Mg (52, 59), a concentration sufficient to chelate virtually all extracellular calcium. This indicates that calcium influx is not critical for induction of class-II molecules by BSF-1. By contrast, the increase in B-cell expression of class-II molecules caused by treatment with anti-IgM antibodies (60) is completely inhibited by the depletion of extracellular calcium (52, 59). Similarly, induction of class-II molecules by BSF-1 is resistant to cyclosporin A, whereas induction in response to anti-IgM is blocked by this agent (61).

Polla et al have demonstrated that BSF-1 causes induction of expression of class-II molecules in the R8.205 pre–B cell line, which normally does not express class-II molecules (62). The induction of expression of class-II molecules in this cell line depends on new transcription of class-II mRNA, as demonstrated by nuclear run-off experiments (63).

A second striking effect of BSF-1 on resting B cells is the preparation of these cells to respond more promptly to subsequent stimulation with anti-IgM plus additional BSF-1 or to high concentrations of anti-IgM alone (40–42). Resting B cells treated with anti-IgM antibodies and BSF-1 do not begin to enter S-phase until 30 hr after the initiation of culture; some cells do not enter S-phase for the first time until 45 hr of culture (42, 43). Pretreatment of resting B cells in BSF-1 for a period of 24 hr speeds their subsequent response to anti-IgM plus BSF-1 by about 12 hr. BSF-1–precultured cells begin to enter S-phase at 18 hr of the "secondary" culture. This effect of BSF-1 on resting B cells can be obtained in serum-free medium, strongly suggesting that it does not depend upon costimulation by others agents. As noted previously, treatment of resting B cells with BSF-1 causes increased expression of BSF-1 receptors on B cells. This may partially explain the enhanced reactivity of precultured cells to anti-IgM and BSF-1. However, receptor up-regulation is probably not the only effect of BSF-1 that is important in the accelerated responsiveness of resting B cells, since preculture with BSF-1 prepares the cells to respond more promptly to high concentrations of anti-IgM without BSF-1. In these experiments, BSF-1 was excluded from the secondary culture by washing after the preculture and by the addition of monoclonal anti-BSF-1 antibody to the secondary culture to inhibit the action of any "carried over" BSF-1 or of any BSF-1 that might have been produced during the secondary culture (42).

BSF-1 action in preparing resting B cells for more rapid responses to subsequent stimuli requires the prolonged presence of the lymphokine. Approximately 12 hr of preculture is required for half-maximal preparation; this is shown either by delaying the addition of BSF-1 during the preculture or by adding monoclonal anti–BSF-1 antibody at various times to terminate the action of BSF-1 (42). Thus, BSF-1 appears to cause the production of some type of cellular signal that is required over an extended period of time and that is only produced during the time that the lymphokine is bound to receptors on the cell surface.

The mechanism through which BSF-1 prepares B cells to respond more promptly to subsequent stimuli has not been established. BSF-1 does not cause a detectable increase in inositol phospholipid metabolism in resting B cells nor does it cause increased $[Ca^{++}]_i$ or translocation of cytosolic protein kinase C to the membrane (52, 64). Furthermore, B cells prepared to respond more promptly as a result of BSF-1 pretreatment do not display enhanced accumulation of inositol phosphate or greater elevation of $[Ca^{++}]_i$ in response to anti-IgM (52). Justement et al (64) have recently reported that BSF-1 causes phosphorylation of a 44,000-M_r protein in isolated membrane preparations—the only early consequence of BSF-1 treatment that has been observed.

Treatment of resting B cells with BSF-1 does not cause them to increase RNA content sufficiently to be scored as having entered the G_{1a} phase of the cell cycle (41). Such cells do show small but significant increases in cell volume (40, 57).

These results indicate that BSF-1 exerts important actions on resting B cells and thus would place it in the category of a "competence" factor, or more properly a "cocompetence" factor, since preparation to enter S-phase requires the joint action of BSF-1 and anti-IgM on the resting B cells. Nonetheless, BSF-1 action is probably more complicated. B-cell blasts prepared by culture with anti-IgM and BSF-1 can be stimulated to enter S-phase by the addition of high concentrations of BSF-1 without a costimulant (42, 57). In these experiments, the possible action of residual anti-IgM in the "secondary" culture was inhibited both by washing the cells and by the addition of an IgM myeloma protein to neutralize any remaining anti-IgM. In general, the late-G_1 action of BSF-1 requires more than 10 U/ml of BSF-1, whereas some of the actions of BSF-1 on resting B cells can be obtained with less than 1 U/ml; this strongly suggests that the major locus of BSF-1 action may prove to be the resting B cell.

The concept that anti-IgM and BSF-1 jointly act in G_0 and in early G_1 raises the question of the nature of the physiologic signal for the entry into S-phase of these stimulated B cells. As already noted, high concentrations of BSF-1 will cause DNA synthesis by in vitro–generated B-cell blasts. It is possible both that such concentrations are achieved in the context of the intimate association of lymphokine-secreting T cells and responsive B cells which occurs in cognate interactions and also that BSF-1 does act as a progression factor in this model of B-cell activation. On the other hand, DNA synthesis responses of B cells to high concentrations of anti-IgM or to LPS are independent of a requirement for added BSF-1 and cannot be inhibited by the addition of monoclonal anti-BSF-1 antibody (42). Thus, these ligands may directly act as progression factors or may cause the autocrine production of a B-cell growth factor. Recently, human B cells and B lymphomas have been reported to secrete such a B-cell growth factor, one that acts on late-G_1–B cells (65, 66). In the mouse, a cDNA clone for the late-G1 acting factor BCGF-2 has recently been obtained (67). This lymphokine will be amenable to careful examination for its role as a progression factor in both cognate and receptor cross-linkage–dependent B-cell stimulation.

IFN$_\gamma$ is a potent inhibitor of the action of BSF-1 on resting B cells (37, 38). If IFN$_\gamma$ is added with BSF-1 to resting cells, it blocks the preparation of these cells to respond more promptly to subsequent stimulation (37), and it inhibits the induction of class-II MHC molecules normally caused by BSF-1 (38). In addition, it blocks the BSF-1–stimulated production of IgG$_1$ and IgE by B cells (68). However, IFN$_\gamma$ does not appear to be a specific inhibitor of BSF-1 activity. It does not inhibit the costimulatory activity of BSF-1 on B cells which have been prepared by preculture with BSF-1 (37). Moreover, IFN$_\gamma$ does not block the action of BSF-1 on T cells, and IFN$_\gamma$ does not compete with ^{125}I-BSF-1 for binding of BSF-1 receptors on B cells (36).

BSF-1 Promotes the Expression of IgG$_1$ and IgE

B cells treated with LPS will secrete IgM and IgG$_3$, but little or no IgG$_1$. The addition of a factor derived from the T-cell–line PK 7.1 to LPS-stimulated B cells causes the secretion of IgG$_1$ and results in partial suppression of IgG$_3$ production (4). The factor mediating this activity was designated B-cell differentiation factor for IgG$_1$ (BCDF$_\gamma$) by Vitetta, Krammer, and their colleagues. Sideras et al described a similar activity in supernatants of several stimulated T-cell lines; they designated this material IgG$_1$-inducing factor (5). The biochemical characterization of IgG$_1$-inducing factor by Sideras et al (21) indicated that it was similar to BSF-1 in both molecular weight and isoelectric point. Direct evidence that BSF-1 and BCDF$_\gamma$ were the same molecule came from experiments in which highly purified EL-4–derived BSF-1 was shown to have the capacity to induce IgG$_1$ production of LPS-stimulated cells and in which monoclonal anti–BSF-1 antibody was demonstrated to inhibit the IgG$_1$-inducing activity of PK 7.1 supernatants (6). Indeed, the biologic screening procedure through which Noma et al isolated a BSF-1 cDNA clone was the induction of IgG$_1$ secretion in LPS-stimulated cell populations (17).

BSF-1 was then shown to promote the secretion of IgE in LPS-stimulated cultures of B cells. Normally, IgE content in LPS-stimulated cultures is below the sensitivity of the detection method (i.e. < 1–2 ng/ml). Cultures stimulated with LPS and BSF-1 express up to 4 μg of IgE per ml (7; C. Snapper, W. E. Paul, manuscript in preparation). In general, the concentration of BSF-1 required to obtain IgE expression in LPS-stimulated B cells is two- to fivefold higher than that required for IgG$_1$ production (69). The IgE-promoting activity of BSF-1 has also been verified with BSF-1 produced by recombinant DNA technology (24).

BSF-1 has now been demonstrated to be involved in IgE production in vivo (70). Mice infected with the larvae of the helminth *Nippostrongylus brasiliensis* display an approximately 100-fold increase in serum IgE concentrations within two weeks. The administration of ascitic fluid containing monoclonal anti–BSF-1 antibody at the same time that larvae are injected into the mouse diminishes the increase in serum IgE by a factor of 5 to 10, whereas a control ascitic fluid has little or no effect (Table 2).

Anti-BSF-1 antibody has also been shown to inhibit increases in serum IgE in a second in vivo model. Administration of anti-IgD antibody to mice leads to an initial wave of B-cell activation (71), presumably due to receptor cross-linkage by the anti-IgD antibody. Four days after the infusion of the anti-IgD antibody, there is a striking lymphocyte proliferation which is T-cell dependent and, in association, an ~100-fold increase in serum IgG$_1$ concentration. This effect appears to represent the appearance of T cells that recognize antigenic determinants on the anti-

Table 2 Anti-BSF-1 antibody blocks in vivo IgE production to *N. brasiliensis* (Nb) and to anti-IgD[a]

Treatment	Number	Serum IgE (μg/ml)
Infection with Nb		
None	6	<0.1
Nb	6	14.2
Nb + anti-BSF-1	6	1.3
Nb + control ascites	6	6.4
Injection of anti-IgD		
None	3	0.4
Anti-IgD	3	13.6
Anti-IgD + pur. anti-BSF-1	3	2.6
Anti-IgD + rat IgG1	3	18.5

[a] BALB/c mice which were infected with *N. brasiliensis* larvae received 1 ml of ascites containing monoclonal anti-BSF-1 or a control antibody on the day of infection and 7 days later. Serum IgE concentrations were measured 13–14 days after infection. BALB/c injected intravenously with goat anti-mouse IgD (800 μg) received 20 mg of purified anti-BSF-1, a monoclonal rat IgG$_1$ antibody, or of rat IgG$_1$ purified from normal serum. Serum IgE concentrations were measured 7 days later.

IgD antibody and which presumably establish a cognate interaction with B cells to which anti-IgD has bound. Finkelman et al (72) have recently demonstrated that T cells harvested 4–8 days after injection of anti-IgD antibody actively produce BSF-1. This led to the prediction that mice receiving anti-IgD antibody would display increases in serum IgE level as well as in serum IgG_1. Measurement of serum IgE concentrations 6–8 days after injection of anti-IgD revealed that these concentrations had risen 30- to 100-fold (70, 73). Furthermore, the increase in serum IgE could be inhibited by injection of purified monoclonal anti-BSF-1 at the time of injection of the anti-IgD (70). A control rat IgG_1 had no inhibitory activity (Table 2).

The amounts of anti–BSF-1 used in these experiments were quite high. Inhibition of increases in serum IgE in response to *N. brasiliensis* could be obtained with 0.2 ml/mouse of ascitic fluid containing the monoclonal rat anti–BSF-1 antibody 11B11, but dramatic inhibition required 1 ml/ mouse. Experiments involving purified antibody utilized 20 mg/mouse, the amount of 11B11 in 1 ml of ascitic fluid. The need for such large amounts of antibody is not fully understood. It is not due to rapid loss of antibody from the injected animals, because rat IgG is found in the serum of the mice at the termination of the experiment. It is known that a large molar excess of 11B11 is required to block the binding of ^{125}I-BSF-1 to its receptor (36), presumably because of the high affinity of the lymphokine-receptor interaction. In addition, it is quite possible that BSF-1 is produced in the context of a cognate T cell–B cell interaction in which the T cell and B cell are physically linked and in which the lymphokine may be directionally secreted from the surface of the T cell that is bound to the B cell (74). Under such circumstances, antibody may be at a distinct disadvantage in competing with lymphokine for binding to receptor.

Equally interesting is the apparent specificity of the inhibition. There was no detectable diminution in the serum concentration of IgG_1 in anti-BSF-1 treated mice, and the number of class-II molecules on the surface of the B cells were only modestly diminished as a result of the admin-istration of 11B11 to mice which received anti-IgD or were infected with *N. brasiliensis* larvae. This may reflect the fact that the higher con-centrations of BSF-1 are required to induce IgE production than are required for either IgG_1 stimulation or for enhancement of expression of class-II MHC molecules. Thus, the IgE response could be blocked if anti-BSF-1 antibody reduced BSF-1 concentrations to levels that were still adequate for the IgG_1 response and for the enhancement of expression of class-II MHC molecules, but which were below the threshold for IgE production. Alternatively, the IgE response may depend exclusively on a differentiation pathway that involves the action of BSF-1, whereas the

IgG$_1$ response and the enhanced expression of class-II MHC molecules may utilize multiple stimulatory mechanisms.

This in vivo experiment establishes that BSF-1 plays a critical role in the IgE response of mice to helminthic infections and suggests that BSF-1 is important in the IgE increases observed in parasitic infections in general. The role of BSF-1 in allergy is obviously a crucial issue which requires examination. In addition, these experiments demonstrate that BSF-1 enhances IgE responses not only in B cells stimulated with LPS. Activation by other mechanisms also renders B cells sensitive to the IgE-stimulating effects of BSF-1.

Vitetta and colleagues have examined several aspects of the process through which the B cells stimulated with LPS and BSF-1 develop into IgG$_1$-secreting cells. They have shown that cells lacking IgG$_1$ on their surface can be stimulated by LPS and BSF-1 to secrete IgG$_1$; this strongly suggests that the combination of agents causes Ig class "switching" (75). Elegant support for the concept that BSF-1 promotes class switching has recently been obtained by Severinson & Stavnezer (76) who showed that the I.29 cell line—known to switch to the expression of IgA, IgG2a, and IgE upon stimulation with LPS—would also switch to IgG$_1$ expression if cultured with LPS, anti-idiotype antibody, and BSF-1.

Vitetta and her colleagues have provided evidence that the B cells that switched under the influence of BSF-1 and LPS either already had the capacity to switch to IgG$_1$ or acquired this capacity as a result of culture with LPS (77). These conclusions were drawn from limiting dilution experiments in which B cells were cultured for three days with LPS, which stimulated cell division. Individual culture wells then split. Replicas were cultured with LPS and BSF-1. It was reasoned that if cells had already acquired the possibility to switch to IgG1 production by the completion of the initial culture in LPS, then there should be a statistically significant concordance in IgG1 production in replica cultures. Indeed, that result was observed.

Snapper et al (69) have recently shown that the effect of BSF-1 on the production of IgG$_1$ does not require the simultaneous presence of LPS. Resting B cells, depleted of membrane IgG–bearing cells by electronic cell sorting, could be treated with BSF-1 for 48 hr in the absence of LPS or any other costimulant. At the end of that time, the cells were washed and LPS was added. The preculture with BSF-1 led to a very significant enhancement in IgG$_1$ production, providing strong evidence that even the IgG$_1$-inducing activity of BSF-1 can be exerted on resting B cells. Indeed, Snapper et al (69) further showed that BSF-1 could be added for a 48-hr period prior to LPS, at the initiation of culture with LPS or after 48 hr culture with LPS. In each case, BSF-1 markedly enhanced the IgG$_1$

production after 6 days of culture with LPS. This leads us to propose that LPS and BSF-1 act independently and cause distinct intracellular changes, both of which are necessary for the expression of IgG_1. This also leads us to suggest that the mode of BSF-1 action in switching to IgG_1-production may not be fundamentally different from its other actions on resting B cells, such as receptor up-regulation, induction of class-II MHC molecules, and preparation of the cells to respond more promptly to subsequent growth stimuli.

BSF-1 Action on T-Cell Lines and Normal T Cells

BSF-1 was initially shown to act on cells other than B cells by Mosmann et al (10). They had observed that the supernatant fluid of stimulated T cells of the line MB2-1 contained an activity which caused both survival and proliferation of cells of the IL-2–dependent line HT-2. The activity of this factor was not blocked by a monoclonal antibody to IL-2 (S4B6), strongly suggesting that it was not IL-2. They designated this material T-cell growth factor-2 (TCGF-2). TCGF-2 copurified with the activity in MB2-1 supernatant which induced expression of class-II MHC molecules, suggesting that TCGF-2 might be an activity of BSF-1. This was established by the demonstration that purified EL-4–derived BSF-1 had TCGF-2 activity and that the TCGF-2 activity of both MB2-1 supernatant fluid and of EL-4–derived BSF-1 could be inhibited by the 11B11 monoclonal anti–BSF-1 antibody but not by S4B6. The action of authentic IL-2 on HT-2 cells was blocked by S4B6 but not by 11B11. BSF-1 produced by recombinant DNA technology also proved to have TCGF-2 activity, unequivocally establishing that TCGF-2 activity was a property of BSF-1 (24).

Other T-cell lines have been shown to be sensitive to BSF-1. CTLL cells respond to BSF-1 by incorporating ^{3}H-thymidine, although these cells are in general less sensitive to BSF-1 than HT-2 cells (11). Indeed, both HT-2 and CTLL cells are substantially less sensitive to BSF-1 than to IL-2. The maximal degree of ^{3}H-thymidine incorporation achieved in HT-2 and in CTLL cells in response to BSF-1 is always less than the maximal response to IL-2; for CTLL this difference may be fivefold or more. Furthermore, responses of HT-2 and CTLL cells generally require considerably more (~ 30-fold) BSF-1 than do responses of normal B cells.

Whereas HT-2 and CTLL cells respond to BSF-1 alone, the responsiveness of cells of the conalbumin-specific T-cell line D10 is both more complex and more interesting. Kaye et al (78) demonstrated that D10 cells could be stimulated to proliferate by interleukin-1 (IL-1) and concanavalin A (Con A) or by IL-1 and antibodies to clonotypic determinants on the receptor of D10 cells. D10 cells also respond to IL-2 alone. An apparent paradox, noted in these studies of D10 cells, is that although the cells responded to IL-2 added exogenously, they were unresponsive to the T-cell growth factor which they themselves produced (79). Thus, supernatant fluids from stimulated D10 cells contained an activity which stimulated HT-2 cells but which failed, *by itself*, to stimulate D10 cells. The TCGF activity of D10 supernatants was then shown to be due to BSF-1, and authentic BSF-1 differed from IL-2 in its action on D10 cells in that BSF-1 did not cause ^{3}H-thymidine uptake without costimulation by IL-1 (13).

BSF-1 appears to play an important role in the growth of D10 cells. Monoclonal anti–BSF-1 antibody blocks the proliferation of D10 cells to IL-1 and Con A or to conalbumin and antigen-presenting cells, while antibodies to the IL-2 receptor cause no inhibition of these responses (13). The production of BSF-1 by D10 cells is stimulated by Con A alone, and BSF-1 can replace Con A in synergizing with IL-1 to cause ^{3}H-thymidine uptake by D10 cells.

These results indicate that BSF-1 acts on D10 cells as a growth-promoting factor in association with IL-1, although it remains to be understood why D10 cells are more sensitive to IL-2 than to BSF-1 if BSF-1 is the "normal" growth stimulant of D10 cells.

Normal T cells are also responsive to BSF-1. Survival rates of highly purified small, dense T cells over a period of 5–7 days are about fivefold higher in the presence of BSF-1 than in its absence, although BSF-1, by itself, causes no ^{3}H-thymidine uptake by resting T cells (8). Resting T cells show striking proliferative responses to PMA plus BSF-1 (8). Virtually all resting T cells respond to this combination of stimulants by enlarging in volume, and more than 50% will enter S-phase within 48 hr of stimulation. The response of resting T cells to PMA plus BSF-1 is density-independent and can be observed at cell numbers as low as 700/culture well. Both L3T4$^+$ and Lyt2$^+$ T cells respond, although the degree of ^{3}H-thymidine uptake by Lyt2$^+$ cells is somewhat greater than that of L3T4$^+$ cells. Resting T cells also respond to BSF-1 plus Con A or phytohemagglutinin (8, 9), although these responses are cell density dependent.

BSF-1 must be added within 4 hr of the time of addition of PMA or the response at 48 hr will be diminished, suggesting that BSF-1 is required very early in the response of T cells to PMA plus BSF-1 (8). However, BSF-1 also is needed late in this response. T cells cultured for 24 hr in BSF-1 plus PMA and then washed will not enter S-phase at 48 hr unless BSF-1 is re-added. T cells prepared by culture with PMA plus BSF-1 for 24 hr are also prepared to respond to IL-2. Such cells display IL-2 receptors in considerably higher density than do cells cultured with PMA alone or BSF-1 alone.

The finding that resting T cells precultured with BSF-1 and PMA express

IL-2 receptors and become responsive to IL-2 raises the question of whether BSF-1 acts on T cells to induce growth by causing IL-2 to be produced or whether it acts as an authentic growth factor independent of the action of IL-2.

Several lines of evidence lead to the conclusion that BSF-1 action on T cells is not dependent on the production of IL-2 (80). The synthesis of DNA by resting T cells in response to PMA plus BSF-1 is not inhibited by a mixture of monoclonal antibodies to the IL-2 receptor and to IL-2 (Table 3). This is true under conditions in which any one of the reagents used alone would block the response of resting T cells to Con A. Secondly, no IL-2 can be demonstrated in culture supernatants of resting T cells stimulated with PMA plus BSF-1, and in situ hybridization with ^{35}S-labeled RNA probes complementary to IL-2 mRNA fails to reveal the presence of IL-2 mRNA in T cells stimulated with PMA plus BSF-1. Finally, the response of T cells to PMA plus BSF-1 is not inhibited by cyclosporin A (CsA), which has been shown to inhibit transcription of IL-2 mRNA (81). Thus, it seems very unlikely that BSF-1 operates by stimulating IL-2 production.

In addition, resting T cells activated for 24 hr with Con A in the presence of accessory cells will enter S-phase when stimulated with either BSF-1 or with IL-2, with generally similar kinetics (8, 9). These cells thus respond to BSF-1 in a manner resembling that of HT-2 or CTLL cells. The response of Con A–activated T cells to BSF-1 is not inhibited by CsA and is not blocked by a mixture of anti–IL-2 and anti–IL-2 receptor antibodies (80).

These results lead us to postulate that BSF-1 acts as both a "cocompetence factor" and as a "progression factor" in the response of resting T cells to PMA plus BSF-1. It is required for resting cells to become sensitive

Table 3 Anti-IL-2 and anti-IL-2 receptor antibodies do not block T-cell responses to BSF-1 plus PMA[a]

	^{3}H-thymidine uptake (CPM)	
Stimulant	Control antibody	Anti-IL-2 + anti-IL-2R
None	200	140
Con A + acc. cells	16,920	1,940
BSF-1 + PMA	13,090	12,110

[a] BALB/c lymph node T cells ($2.5 \times 10^4/0.2$ ml/well) were cultured with medium only, Con A plus irradiated T-depleted spleen cells (acc. cells) or BSF-1 (100 U/ml) +PMA (1 ng/ml) in the presence of the control rat IgG monoclonal antibody 50C1 or a mixture of the monoclonal rat IgG anti-IL-2 antibody S4B6.1 and the anti-IL-2 receptor (IL-2R) antibody PC61. All antibodies were in the form of 1 : 300 dilutions of ascitic fluid. Cells were cultured for 48 hr and ^{3}H-thymidine was then added for a 16-hr period. Responses are recorded as mean counts per minute per culture well.

to the growth promoting activities of BSF-1 itself or of IL-2, and it can act on prepared cells in late G1 to cause them to enter S-phase. Its actions thus have many similarities to those of IL-2. Indeed, IL-2 can be used in place of BSF-1 with PMA in the stimulation of resting T cells.

Cells cultured with PMA and IL-2 enter S-phase with kinetics very similar to that of cells cultured with PMA and BSF-1. IL-2 is required early in the culture but must also be present in the latter part of G1. One interpretation of the response of resting T cells to IL-2 plus PMA is that PMA causes the appearance of a small number of IL-2 receptors but does not stimulate IL-2 production. IL-2 is known to cause up-regulation of IL-2 receptors (82–84) and, thus, is likely to enhance receptor expression on these cells. When the cells have attained an appropriate density of IL-2 receptors and have undergone certain other activation events, they become sensitive to the growth-promoting or progression factor activity of IL-2 and enter the S-phase of the cell cycle.

This model does not easily apply to the response of resting T cells to PMA plus BSF-1. Resting T cells express BSF-1 receptors and increase the number of such receptors in response to BSF-1 alone. No greater increase in receptor number is observed in cells cultured with PMA plus BSF-1. Thus, the costimulatory activity of PMA must be due to some activity other than regulation of BSF-1 receptor production or BSF-1 secretion.

BSF-1 also acts upon immature thymocytes. Palacios et al (85) have shown that BSF-1 plus PMA causes growth of thymocytes from mouse fetuses of 14–15 days of gestation. Maintenance of such cells in culture for 14–21 days with BSF-1 gave rise to some cells which expressed low levels of Lyt-2 and developed cytotoxic activity for syngeneic and allogeneic spleen cells which had been activated with LPS. This cytotoxic activity was markedly enhanced by the addition of PHA to the effector-target cell interaction.

BSF-1 Production

BSF-1 was originally detected in the supernatants of PMA-stimulated cells of the EL-4 line. Studies of mRNA content by translation in *Xenopus* oocytes or by Northern blot analysis indicate that resting EL-4 cells have little or no BSF-1 mRNA. mRNA for BSF-1 can be detected within 6 hr of stimulation of the cells with PMA. It reaches a maximum at 24 hr and then declines (25). Studies on the D10 cell line indicate that production of BSF-1 in response to Con A is inhibited by CsA (86), suggesting that this agent blocks transcription of the BSF-1 gene as well as the IL-2 gene. BSF-1 has been shown to be synthesized by some T-cell lines upon stimulation with IL-2, suggesting the possibility of a growth factor cascade (87).

Several T-cell lines have been described that make BSF-1 upon stimulation with Con A or with antigen and antigen-presenting cells. Cells of the T-cell lines 2.19 and Cl.Ly1$^+$2$^-$/9, from which the cDNA libraries used in cloning of the BSF-1 gene were constructed, express considerably more BSF-1 mRNA than do EL-4 cells. Whether this relationship in amounts of BSF-1 message will be observed when a wider range of normal T cells are sampled is not known.

Mosmann and colleagues examined a large panel of L3T4$^+$ T-cell clones and found that they could be divided into two broad categories, Th1 cells and Th2 cells (88). The Th1 cells produced IL-2 and IFN$_\gamma$ upon stimulation but failed to produce BSF-1; by contrast, Th2 cells produced BSF-1 but failed to produce IL-2 or IFN$_\gamma$. Killar et al examined a separately derived panel of T-cell lines and found that the production of BSF-1 and IFN$_\gamma$ tended to be mutually exclusive and that the BSF-1–producing cells, but not the IFN$_\gamma$-producing cells, expressed helper function for anti-phosphoryl choline antibody responses (89).

A characteristic of the BSF-1 producing lines is that they required IL-2 supplementation in order to grow, whereas the IL-2–producing lines could be maintained by periodic stimulation with antigen and antigen-presenting cells without any requirement for exogenous IL-2. These results, together with other data already described, indicate that BSF-1 is not as efficient as a growth-promoting agent for T cells as is IL-2. The physiologic role of BSF-1 in T-cell growth is uncertain; it appears likely that its action on B cells, possibly in the context of cognate T cell–B cell interactions, may prove to be its major role in in vivo immune responses.

The Role of BSF-1 in Cognate T Cell–B Cell Interactions

BSF-1 has properties which suggest that it may play an important role in cognate T cell–B cell interactions. In such interactions, B cells bind antigen to their receptors, process that antigen, and reexpress it on their surface in association with class-II MHC molecules. T cells recognize the resultant antigen–class-II molecular complex and are stimulated as a result of this interaction. Such stimulation appears also to require some other action of antigen-presenting cells; small B cells have been shown to be capable of mediating such stimulatory activity (90).

As a result of this interaction, T cells potentially capable of producing BSF-1 would be stimulated to secrete this lymphokine. The production of BSF-1 in the microenvironment of the T cell–B cell interaction might lead to several outcomes, each potentially important in the activation process. Among these would be an increase in the expression of class-II MHC molecules by resting B cells. This would enhance the capacity of such cells to act as antigen-presenting cells, since the number of class-II molecules on the surface of an antigen-presenting cell has been shown to increase its ability to stimulate a specific T cell, presumably by increasing the likelihood that antigen–class-II molecular complexes will form (91).

Furthermore, BSF-1 could prepare the B cell for responses to other stimulants, could act as a costimulant, and if present in sufficiently high concentrations, might also act as a B-cell growth factor. Indeed, BSF-1 made locally would also aid in the switching of the B cells to the secretion of IgG$_1$ and IgE. Locally produced BSF-1 might enhance the responsiveness of the T cell engaged in the cognate T cell–B cell interaction and might aid in the recruitment of new T cells to the ongoing response. Although relatively little direct information has been obtained on the role of BSF-1 in cognate interactions, DuBois et al (92) have shown that a cloned T-cell line which does not secrete BSF-1 can stimulate B-cell entry into S-phase only if exogenous BSF-1 is provided. In these experiments, it was demonstrated that BSF-1 did not replace the need for T cells that recognize the antigen–class-II molecular complex on the B cell, indicating that BSF-1 is not the only entity required for cognate interactions. Indeed, if the concept that class-II molecules are capable of transducing activation signals into B cells proves to be correct, then the role of BSF-1 in increasing the expression of class-II molecules on the cell surface would be even more important in the B-cell activation process.

It should also be pointed out that the cognate T cell–B cell interaction microenvironment might prove to be an important immunoregulatory site. If a given B cell were to interact with both a Th1 and a Th2 cell at the same time, the production of both IFN$_\gamma$ and BSF-1 might lead to an inhibition of the BSF-1 effect. That would lead to the apparently paradoxical outcome that a Th1 cell, normally regarded as a helper-inducer cell, could also function as a suppressor cell, particularly in the stimulation of resting B cells.

Action of BSF-1 on Cells of Other Hematopoietic Lineages

The capacity of BSF-1 to stimulate cells other than lymphocytes was first described by Smith & Rennick (22) in their studies of the biologic activities of Cl.Ly1$^+$2$^-$/9 supernatants and by Mosmann et al (10) in their examination of the lymphokines found in the supernatant fluids of Th2 cell lines. As described above, these supernatants contained a TCGF activity (TCGF-2) distinct from IL-2 which proved to be due to BSF-1. Copurifying with TCGF-2 was an activity which could be detected on an interleukin-3 (IL-3)–dependent mast cell line. This activity, designated mast cell growth factor-2 (MCGF-2), was detected by the enhancement of growth of cells receiving an optimal concentration of IL-3. By itself, MCGF-2 did not appear to cause growth of the mast cells.

MCGF-2 was shown to be BSF-1 by the same set of experiments used to prove the TCGF-2 was BSF-1. Independently, Ruede and colleagues (14) had observed that a factor with the properties of BSF-1 enhanced the responsiveness to IL-3 of newly prepared mast cell lines.

The action of BSF-1 on IL-3–dependent mast cell lines is in keeping with the finding that Abelson virus–transformed, factor-independent mast cell lines express 1500–3000 BSF-1 receptors per cell (36). More surprising are the results of Northern analyses, which indicate that 6 of 11 Abelson-virus–transformed mast lines express BSF-1 mRNA (93). Furthermore, supernatant fluid from the lines that express BSF-1 mRNA contain BSF-1–like activity, as shown by the capacity of these supernatant fluids to synergize with anti-IgM in the stimulation of DNA synthesis by resting B cells. The BSF-1–like activity of the mast cell line supernatants is inhibited by monoclonal anti–BSF-1 antibody, indicating that it is BSF-1. M. Brown in our laboratory, in collaboration with J. Pierce, has shown that long-term IL-3–dependent mast cell lines also express BSF-1 mRNA (93). These results suggest that the expression of BSF-1 mRNA by mast cell lines growing under the influence of IL-3 may be a relatively common event. This expression could reflect BSF-1 production by normal mast cells, or it may be a prelude to malignant transformation of these cells. The possibility that such events occur in vivo requires careful consideration.

The action of BSF-1 can also be observed on other cell types. Meltzer et al (94) have observed that BSF-1 can stimulate macrophage activation. Macrophages treated with BSF-1 acquire increased cytotoxic capacity and display enhanced expression of class-II MHC molecules (16, 94).

Peschell et al (15) have examined the effect of BSF-1 on the appearance of bone marrow–derived soft agar colonies. To exclude the possible indirect action of BSF-1 in that system, T cells were eliminated from the bone marrow cell populations by treatment with anti–Thy 1 antibody and complement. In selected experiments, B cells and pre–B cells were also removed by treatment with monoclonal anti-B220 antibodies. Such bone marrow cells developed few colonies in response to BSF-1 alone, but BSF-1 synergized with other "colony stimulating factors" to stimulate colonies of several types. For example, BSF-1 synergized with G-CSF to cause the appearance of granulocytic colonies; it synergized with recombinant erythropoietin (rEpo) to cause colony formation by immature erythroid precursors (BFU-e) (Table 4), and it enhanced the capacity of rEpo to stimulate colony formation by mature erythroid precursors (CFUe). BSF-1 also synergized in the formation of megakaryocytic colonies. For this cell type, the costimulants are IL-1, rEpo, and a factor [megakaryocyte promoting activity (Mk-PA)] found in the supernatant fluid of stimulated T cells of the hybridoma line FS7-20.6.18 (95). Thus, BSF-1 plus either

Table 4 BSF-1 costimulates colony formation by immature erythroid precursors (BFU-e)[a]

Stimulants	BFU-e (colonies/40,000 cells)
None	< 1
BSF-1 (30 U/ml)	< 1
IL-3 (50 U/ml)	< 1
rEpo (1 U/ml)	2.0
BSF-1 + rEpo	22.3
IL-3 + rEpo	26.8

[a] BALB/c bone marrow cells (40,000/dish) were depleted of mature T cells and seeded into soft agar with medium only, BSF-1, IL-1, rEpo, BSF-1 + rEpo or IL-3 + rEpo. BFU-e were enumerated by counting erythroid colonies 7 days later.

IL-1, rEpo, or Mk-PA will cause the formation of megakaryocytic colonies. This activity of BSF-1 has some similarities to that of IL-3, which shows striking enhancement by IL-1, rEpo, or Mk-PA in its stimulation of megakaryocyte precursors. Furthermore, BSF-1 and IL-3 do not synergize with one another in the stimulation of megakaryocyte colonies, further suggesting they have somewhat similar effects. However, IL-3 by itself does cause the appearance of some megakaryocytic colonies, whereas BSF-1 has little or no independent capacity to cause formation of such colonies.

BSF-1 has also been shown to stimulate DNA synthesis by cells of the myeloid line DA-1 and to cause differentiation of certain monocyte-macrophage cell lines.

These results indicate that BSF-1 can act on essentially all cells of hematopoietic origin. The mode of action of BSF-1 on nonlymphoid cells has not yet been examined in detail, but it seems logical to assume that it will prove to exert both activation and growth-promoting activities.

Conclusions

It is now clear that BSF-1 is a lymphokine with a broad range of biologic activities. It has been shown to act on normal B and T lymphocytes, on mast cell lines, on immature and mature erythroid precursors, on granulocyte precursors, on megakaryocyte precursors, and on cells of the macrophage lineage. It affects the activation, growth, and program of gene expression in these cells. Thus, BSF-1 is a T-cell product which may have a broad range of regulatory effects on the immune and hematopoietic systems.

Efforts to understand the molecular basis of its action will be critical both to furthering our general knowledge of lymphocyte growth regulation and in the development of pharmacologic approaches to the control of

BSF-1 functions. A striking first step in this direction is the demonstration that a monoclonal anti–BSF-1 antibody markedly inhibits the increase in serum IgE in mice infected with the helminthic parasite *N. brasiliensis*. This establishes a role for BSF-1 in the in vivo regulation of IgE production. However, since "switching" to IgE secretion requires relatively large amounts of BSF-1 in vitro, it is possible that many other in vivo functions of BSF-1 have not been observed because anti–BSF-1 antibody treatment cannot lower local BSF-1 concentrations sufficiently to block these activities. A detailed analysis of the in vivo properties of BSF-1 and of its potential immunotherapeutic role is now a major research priority. It is only through such studies that an appreciation of the physiologic functions of BSF-1 will be obtained.

ACKNOWLEDGMENTS

We wish to thank our scientific colleagues who aided in the preparation of this review by providing preprints and reprints of their manuscripts and, in many cases, allowing us to quote from their unpublished work. We also wish to thank Abul Abbas, Melissa Brown, Robert Coffman, John Coligan, Fred Finkelman, Ronald Germain, Laurie Glimcher, Ira Green, Andrew Lichtman, W. Lee Maloy, James Mond, Tim Mosmann, Junichiro Mizuguchi, David Parker, Jacalyn Pierce, Christian Peschell, Evelyn Rabin, Ethan Shevach, Clifford Snapper, Ellen Vitetta, Jane Hu-Li, Cynthia Watson, and Charles Hoes for their collaboration in the studies described here. The expert editorial assistance of Shirley Starnes is gratefully acknowledged.

Literature Cited

1. Howard, M., Farrar, J., Hilfiker, M., Johnson, B., Takatsu, K., Hamaoka, T., Paul, W. E. 1982. Identification of a T cell–derived B cell growth factor distinct from interleukin 2. *J. Exp. Med.* 155: 914
2. Noelle, R., Krammer, P. H., Ohara, J., Uhr, J. W., Vitetta, E. S. 1984. Increased expression of Ia antigens on resting B cells: An additional role for B-cell growth factor. *Proc. Natl. Acad. Sci. USA* 81: 6149
3. Roehm, N. W., Liebson, J., Zlotnick, A., Kappler, J., Marrack, P., Cambier, J. C. 1984. Interleukin-induced increase in Ia expression by normal mouse B cells. *J. Exp. Med.* 160: 679
4. Vitetta, E. S., Brooks, K., Chen, Y. W., Isakson, P., Jones, S., Layton, J., Mishra, G. C., Pure, E., Weiss, E., Word, C., Yuan, D., Tucker, P., Uhr, J.

W., Krammer, P. H. 1984. T cell-derived lymphokines that induce IgM and IgG secretion in activated murine B cells. *Immunol. Rev.* 78: 137
5. Sideras, P., Bergstedt-Lindqvist, S., MacDonald, H. R., Severinson, E. 1985. Secretion of IgG1 induction factor by T cell clones and hybridomas. *Eur. J. Immunol.* 15: 586
6. Vitetta, E. S., Ohara, J., Myers, C., Layton, J., Krammer, P. H., Paul, W. E. 1985. Serological, biochemical, and functional identity of B cell-stimulatory factor 1 and B cell differentiation factor for IgG1. *J. Exp. Med.* 162: 1726
7. Coffman, R. L., Ohara, J., Bond, M. W., Carty, J., Zlotnick, E., Paul, W. E. 1986. B cell stimulatory factor-1 enhances the IgE response of lipopolysaccharide-activated B cells. *J. Immunol.* 136: 4538
8. Hu-Li, J., Shevach, E. M., Mizuguchi,

J., Ohara, J., Mosmann, T., Paul, W. E. 1987. B cell stimulatory factor-1 (interleukin 4) is a potent costimulant for normal resting T lymphocytes. *J. Exp. Med.* 165: 157

9. Severinson, E., Naito, T., Tokumoto, H., Hama, K., Honjo, T. 1986. Interleukin 4: (IgG1 induction factor) A multifunctional lymphokine acting also on T cells. Submitted for publication

10. Mosmann, T. R., Bond, M. W., Coffman, R. L., Ohara, J., Paul, W. E. 1986. T cell and mast cell lines respond to B cell stimulatory factor-1. *Proc. Natl. Acad. Sci. USA* 83: 5654

11. Fernandez-Botran, R., Krammer, P. H., Diamenstein, T., Uhr, J. W., Vitetta, E. S. 1986. B cell stimulatory factor 1 (BSF-1) promotes growth of helper T cell lines. *J. Exp. Med.* 164: 580

12. Grabstein, K., Eiseman, J., Mochizuki, D., Shanebeck, K., Conlon, P., Hopp, T., March, C., Gillis, S. 1986. Purification to homogeneity of B cell stimulatory factor. A molecule that stimulates proliferation of multiple lymphokine-dependent cell lines. *J. Exp. Med.* 163: 1405

13. Lichtman, A., Kurt-Jones, E. A., Abbas, A. K. 1987. B cell stimulatory factor-1 and not interleukin-2 is the autocrine growth factor for some helper T lymphocytes. *Proc. Natl. Acad. Sci. USA.* In press

14. Schmitt, E., Fassbender, B., Spaeth, E., Schwarzkopf, R., Reude, E. 1986. Characterization of a T cell-derived lymphokine acting synergistically with IL-3 on the growth of murine mast cells and inducing suboptimal proliferation of IL-2 dependent T cells. Submitted for publication

15. Peschell, C., Paul, W. E., Ohara, J., Green, I. 1986. Effects of B-cell stimulatory factor-1/interleukin-4 on hematopoietic progenitor cells. Submitted for publication

16. Zlotnick, A., Daine, B., Ransom, J., Zipori, D. 1986. Effects of recombinant B cell growth factor 1 on a macrophage cell line. *J. Leukocyte Biology* 40: 314

17. Noma, Y., Sideras, T., Naito, T., Bergstedt-Lindqvist, S., Azuma, C., Severinson, E., Tanabe, T., Kinashi, T., Matsuda, F., Yaoita, Y., Honjo, T. 1986. Cloning of cDNA encoding the murine IgG1 induction factor by a novel strategy using SP6 promoter. *Nature* 319: 640

18. Farrar, J. J., Howard, M., Fuller-Farrar, J., Paul, W. E. 1983. Biochemical and physicochemical characterization of mouse B-cell growth factor: A lympho-

kine distinct from interleukin-2. *J. Immunol.* 131: 1838

19. Ohara, J., Lahet, S., Inman, J., Paul, W. E. 1985. Partial purification of BSF-1 by HPLC. *J. Immunol.* 135: 2518

20. Ohara, J., Coligan, J., Zoon, K., Maloy, W. L., Paul, W. E. 1986. Rapid purification, N-terminal sequencing, and chemical characterization of mouse B cell stimulatory factor-1/interleukin-4. Submitted for publication

21. Sideras, P., Bergstedt-Lindqvist, S., Severinson, E. 1985. Partial biochemical characterization of IgG$_1$-inducing factor. *Eur. J. Immunol.* 15: 593

22. Smith, C. A., Rennick, D. M. 1986. Characterization of a murine lymphokine distinct from interleukin 2 and interleukin 3 (IL-3) possessing a T-cell growth factor activity and a mast-cell growth factor activity that synergizes with IL-3. *Proc. Natl. Acad. Sci. USA* 83: 1857

23. Ohara, J., Paul, W. E. 1985. B cell stimulatory factor BSF-1: Production of a monoclonal antibody and molecular characterization. *Nature* 315: 333

24. Lee, F., Yokota, T., Otsuka, T., Meyerson, P., Villaret, D., Coffman, R., Mosmann, T., Rennick, D., Roehm, N., Smith, C., Zlotnick, A., Arai, K. 1986. Isolation and characterization of a mouse interleukin cDNA clone that expresses B-cell stimulatory factor 1 activities and T-cell- and mast-cell-stimulating activities. *Proc. Natl. Acad. Sci. USA* 83: 2061

25. Brown, M. A., Watson, C., Ohara, J., Paul, W. E. 1986. *In vitro* translation of B-cell stimulatory factor-1 in *Xenopus laevis* oocytes. *Cell. Immunol.* 98: 538

26. Brown, M. A. Unpublished observation

27. Otsuka, T., Villaret, D., Yokota, T., Takebe, Y., Lee, F., Arai, N., Arai, K. 1986. Structural analysis of the mouse chromosomal gene encoding interleukin 4 which expresses B cell, T cell and mast cell stimulating activities. *Nucleic Acids Res.* In press

28. Yokota, T., Otsuka, T., Mosmann, T., Banchereau, J., DeFrance, T., Blanchard, D., DeVries, J., Lee, F., Arai, K. 1986. Isolation and characterization of a human interleukin cDNA clone, homologous to mouse BSF-1, which expresses B cell and T cell stimulating activities. *Proc. Natl. Acad. Sci. USA* 83: 5894

29. Kishimoto, T. 1985. Factors affecting B-cell growth and differentiation. *Ann. Rev. Immunol.* 3: 133

30. Mosmann, T. R., Yokota, T., Kastelein, R., Zurawski, S. M., Arai, N., Takebe,

Y. 1986. Species specificity of T cell stimulating activities of IL-2 and BSF-1 (IL-4): Comparison of normal and recombinant, mouse and human IL-2 and BSF-1 (IL-4). Submitted for publication

31. Subbarao, B., Mosier, D. E. 1983. Induction of B lymphocyte proliferation by monoclonal anti-Lyb2 antibody. *J. Immunol.* 130: 2033

32. Yakura, H., Kawabata, I., Ashida, T., Shen, F. W., Katagiri, M. 1986. A role of Lyb-2 in B cell activation mediated by a B cell stimulatory factor. *J. Immunol.* 137: 1475

33. Sabbarao, B., Mosier, D. E. 1984. Activation of B lymphocytes by monovalent anti-Lyb 2 antibodies. *J. Exp. Med.* 159: 1796

34. Mishra, G. C., Berton, M. T., Oliver, K. G., Krammer, P. H., Uhr, J. W., Vitetta, E. S. 1986. A monoclonal anti-mouse LFA-1α antibody mimics the biological effects of B cell stimulatory factor-1 (BSF-1). *J. Immunol.* 137: 1590

35. Glimcher, L. 1986. Personal communication

36. Ohara, J., Paul, W. E. 1987. High affinity receptors for B cell stimulatory factor-1 (interleukin-4) expressed on lymphocytes and other cells of hematopoietic lineage. *Nature.* In press

37. Rabin, E. M., Mond, J. J., Ohara, J., Paul, W. E. 1986. Interferon-γ inhibits the action of B cell stimulatory factor (BSF)-1 on resting B cells. *J. Immunol.* 137: 1573

38. Mond, J. J., Carman, J., Sarma, C., Ohara, J., Finkelman, F. D. 1986. IFN_γ suppresses B cell-stimulatory factor 1 induction of class II MHC determinants on B cells. *J. Immunol.* 137: 3534

39. Ohara, J., Paul, W. E. 1986. Unpublished observations

40. Rabin, E. M., Ohara, J., Paul, W. E. 1985. B cell stimulatory factor (BSF)-1 activates resting B cells. *Proc. Natl. Acad. Sci. USA* 82: 2435

41. Oliver, K., Noelle, R. J., Uhr, J. W., Krammer, P. H., Vitetta, E. S. 1985. B-cell growth factor (B-cell growth factor 1 or B-cell stimulating factor, provisional 1) is a differentiation factor for resting B cells and may not induce growth. *Proc. Natl. Acad. Sci. USA* 82: 2465

42. Rabin, E. M., Mond, J. J., Ohara, J., Paul, W. E. 1986. B cell stimulatory factor (BSF)-1 prepares resting B cells to enter S phase in response to anti-IgM and to lipopolysaccharide. *J. Exp. Med.* 164: 517

43. DeFranco, A., Raveche, E., Asofsky, R., Paul, W. E. 1982. Frequency of B lymphocytes responsive to anti-immunoglobulin. *J. Exp. Med.* 155: 1523

44. Swain, S. L., Howard, M., Kappler, J., Marrack, P., Watson, J., Booth, R., Wetzel, G. D., Dutton, R. W. 1983. Evidence for two distinct classes of murine B cell growth factors with activities in different functional assays. *J. Exp. Med.* 158: 822

45. Paul, W. E. 1984. Nomenclature of lymphokines which regulate B-lymphocytes. *Molec. Immunol.* 21: 343

46. Stein, P., DuBois, P., Greenblatt, D., Howard, M. 1986. Induction of antigen-specific proliferation in affinity-purified small B lymphocytes: Requirement for BSF-1 by type 2 but not type 1 thymus-independent antigens. *J. Immunol.* 136: 2080

47. Mosier, D. E., Zitron, I. M., Mond, J. J., Ahmed, A., Scher, I., Paul, W. E. 1977. Surface immunoglobulin D as a functional receptor for a subclass of B lymphocytes. *Immunol. Rev.* 37: 89

48. Howard, M., Mizel, S., Lachman, L., Ansel, J., Johnson, B., Paul, W. E. 1983. Role of interleukin 1 in anti-immunoglobulin-induced B cell proliferation. *J. Exp. Med.* 257: 1529

49. Coggeshall, K. M., Cambier, J. 1984. B cell activation. VIII. Membrane immunoglobulins transduce signals via activation of phosphatidyl inositol hydrolysis. *J. Immunol.* 133: 3382

50. Grupp, S. A., Harmony, J. A. K. 1985. Increased phosphatidyl inositol metabolism is an important but not obligatory early event in B lymphocyte activation. *J. Immunol.* 134: 4087

51. Bijsterbosch, M., Meade, C. J., Turner, G. A., Klaus, G. G. B. 1985. B lymphocyte receptors and phosphoinositide degradation. *Cell* 41: 999

52. Mizuguchi, J., Beaven, M. A., Ohara, J., Paul, W. E. 1986. BSF-1 action on resting B cells does not require elevation of inositol phospholipid metabolism or increased $[Ca^{2+}]_i$. *J. Immunol.* 137: 2215

53. Pozzan, T., Arslan, P., Tsien, R. Y., Rink, J. J. 1982. Anti-immunoglobulin, cytoplasmic free calcium and capping in B lymphocytes. *J. Cell Biol.* 94: 335

54. Hornbeck, P., Paul, W. E. 1986. Anti-immunoglobulin and phorbol ester induce phosphorylation of proteins associated with the plasma membrane and cytoskeleton in B lymphocytes. *J. Biol. Chem.* 261: 14817

55. Singh, J. P., Chaikin, M. A., Pledger, W. J., Scher, C. D., Stiles, C. D. 1983.

Persistence of the mitogenic response to platelet-derived growth factor (competence) does not reflect a long-term interaction between the growth factor and the target cell. *J. Cell Biol.* 96 : 1497

56. Howard, M., Paul, W. E. 1983. Regulation of B-cell growth and differentiations by soluble factors. *Ann. Rev. Immunol.* 1 : 307

57. Thompson, C. B., Schaefer, M. E., Finkelman, F. D., Scher, I., Farrar, J., Mond, J. J. 1985. T cell-derived B cell growth factor(s) can induce stimulation of both resting and activated B cells. *J. Immunol.* 134 : 369

58. Alderson, M. R., Pike, B. L., Nossal, G. J. V. 1987. Single cell studies on the role of B-cell stimulatory Factor 1 in B-cell activation. *Proc. Natl. Acad. Sci. USA.* In press

59. Dennis, G. J., Mizuguchi, J., Finkelman, F. D., Ohara, J., Mond, J. J. 1986. Calcium dependence of the induction of B cell class II molecules and of B cell proliferation stimulated by B cell mitogens and purified growth factors. Submitted for publication

60. Mond, J. J., Seghal, E., Kung, J., Finkelman, F. D. 1981. Increased expression of I-region-associated antigen (Ia) on B cells after cross-linking of surface immunoglobulin. *J. Immunol.* 127 : 881

61. O'Garra, A., Warren, D. J., Holman, M., Popham, A. M., Sanderson, C. J., Klaus, G. G. B. 1986. Effects of cyclosporin on responses of murine B cells to T cell-derived lymphokines. *J. Immunol.* 137 : 2220

62. Polla, B. S., Poljak, A., Geier, S., Nathenson, S. G., Ohara, J., Paul, W. E., Glimcher, L. H. 1986. Three distinct signals can induce class II gene expression in a murine pre-B cell line. *Proc. Natl. Acad. Sci. USA* 83 : 4878

63. Polla, B. S., Poljak, A., Ohara, J., Paul, W. E., Glimcher, L. H. 1986. Regulation of class II gene expression : Analysis in B cell stimulatory factor 1-inducible murine pre B cell lines. *J. Immunol.* 137 : 3332

64. Justement, L., Chen, Z., Harris, L., Ransom, J., Sandoval, V., Smith, C., Rennick, D., Roehm, N., Cambier, J. 1986. BSF1 induces membrane protein phosphorylation but not phosphoinositide metabolism, Ca^{2+} mobilization, protein kinase C translocation, or membrane depolarization in resting murine B lymphocytes. *J. Immunol.* 137 : 3664

65. Gordon, J., Ley, S. C., Melamed, M. D., English, L. C., Hughes-Jones, N. C. 1984. Immortalized B lymphocytes pro-

duce B cell growth factor. *Nature* 310 : 145

66. Jurgenson, C. H., Ambrus, J. L., Fauci, A. S. 1986. Production of B cell growth factor by normal human B cells. *J. Immunol.* 136 : 4542

67. Kinashi, T., Harada, N., Severinson, E., Tanabe, T., Sideras, P., Kinishi, M., Azuma, C., Tominaga, A., Bergstedt-Lindqvist, S., Takahashi, M., Matsuda, F., Yaoita, Y., Takatsu, K., Honjo, T. 1986. *Nature* 334 : 70

68. Coffman, R. L., Carty, J. 1986. A T cell activity that enhances polyclonal IgE production and its inhibition by interferon-γ. *J. Immunol.* 136 : 949

69. Snapper, C., Ohara, J., Paul, W. E. 1986. BSF-1 action as a "switch-factor" can be mediated on resting B cells. Submitted for publication

70. Finkelman, F. D., Katona, I., Urban, J. F. Jr., Snapper, C. M., Ohara, J., Paul, W. E. 1986. Suppression of *in vivo* polyclonal IgE responses by monoclonal antibody to the lymphokine BSF-1. *Proc. Natl. Acad. Sci. USA.* In press

71. Finkelman, F. D., Scher, I., Mond, J. J., Kessler, S., Kung, J. T., Metcalfe, E. S. 1982. Polyclonal activation of the murine immune system by an antibody to IgD. II. Generation of polyclonal antibody production and cells with surface IgG. *J. Immunol.* 129 : 638

72. Finkelman, F. D., Ohara, J., Goroff, D. K., Smith, J., Villacreses, N., Mond, J. J., Paul, W. E. 1986. Production of BSF-1 during an *in vivo*, T-dependent immune response. *J. Immunol.* 137 : 2878

73. Finkelman, F. D., Snapper, C. M., Mountz, J. D., Katona, I. M. 1987. Polyclonal activation of the murine immune system by a goat antibody to mouse IgD. IX. Induction of a polyclonal IgE response. *J. Immunol.* In press

74. Kupfer, A., Swain, S. L., Janeway, C. A. Jr., Singer, S. J. 1986. The specific direct interaction of helper T cells and antigen-presenting B cells. *Proc. Natl. Acad. Sci. USA* 83 : 6080

75. Isakson, P. C., Puré, E., Vitetta, E. S., Krammer, P. H. 1982. T cell-derived B cell differentiation factor(s). Effect on the isotype switch of murine B cells. *J. Exp. Med.* 155 : 734

76. Stavenzer, J., Severinson, E. 1986. The I.29 B cell lymphoma as a model for memory IgM^+ cells precommitted to an IgA, IgE, or IgG2a switch. In *Mucosal Immunity and Infection of Mucosal Surfaces,* ed. W. Strober, M. Lamon, J. McGhee. Oxford : Oxford Univ. Press. In press

77. Layton, J. E., Vitetta, E. S., Uhr, J. W.,

Krammer, P. H. 1984. Clonal analysis of B cells induced to secrete IgG by T cell-derived lymphokine(s). *J. Exp. Med.* 160: 1850

78. Kaye, J., Gillis, S., Mizel, S. B., Shevach, E. M., Malek, T. R., Dinarello, C. A., Lachman, L. B., Janeway, C. A. 1984. Growth of a cloned helper T cell line induced by a monoclonal antibody specific for the antigen receptor: Interleukin 1 is required for the expression of receptors for interleukin 2. *J. Immunol.* 133: 1339

79. Horowitz, J. B., Kaye, J., Conrad, P. J., Katz, M. E., Janeway, C. A. Jr. 1986. Autocrine growth inhibition of a cloned line of helper T cells. *Proc. Natl. Acad. Sci. USA* 83: 1886

80. Brown, M. A., Hu-Li, J., Shevach, E. M., Chused, T., Mosmann, T., Paul, W. E. 1986. The activity of BSF-1 as a T cell growth stimulant is not dependent upon IL-2 production. Manuscript in preparation

81. Elliot, J. F., Lin, Y., Mizel, S. B., Bleackley, R. C., Harnish, D. G., Paetkau, V. 1984. Induction of interleukin-2 messenger RNA inhibited by cyclosporin A. *Science* 226: 1439

82. Malek, T. R., Ashwell, J. D. 1985. Interleukin-2 upregulates expression of its receptor on a T-cell clone. *J. Exp. Med.* 161: 1575

83. Depper, J. M., Leonard, W. J., Drogula, C., Krönke, M., Waldmann, T. A., Greene, W. C. 1985. Interleukin-2 (IL-2) augments transcription of the interleukin-2 receptor gene. *Proc. Natl. Acad. Sci. USA* 82: 4230

84. Smith, K., Cantrell, D. 1985. Interleukin-2 regulates its own receptors. *Proc. Natl. Acad. Sci. USA* 82: 864

85. Palacios, R., Sideras, P., von Boehmer, H. 1986. Recombinant interleukin-4/BSF-1 promotes growth and differentiation of intrathymic T cell precursors from fetal mice *in vitro*. Submitted for publication

86. Lichtman, A. H., Williams, M. E., Ohara, J., Paul, W. E., Faller, D. V., Abbas, A. K. 1986. Altered growth factor responses in a Kirsten murine sarcoma virus infected T cell clone. Submitted for publication

87. Howard, M., Matis, L., Malek, T., Shevach, E., Kell, W., Cohen, D., Nakanishi, K., Johnson, B., Paul, W. E. 1983. Interleukin-2 induces antigen-reactive T cell lines to secrete BCGF-1. *J. Exp. Med.* 158: 2024

88. Mosmann, T. R., Cherwinski, H., Bond, M. W., Giedlin, M. A., Coffman, R. L. 1986. Two types of murine helper T cell clones. I. Definition according to profiles of lymphokine activities and secreted proteins. *J. Immunol.* 136: 2348

89. Killar, L., MacDonald, J., West, J., Woods, A., Bottomly, K. 1986. Cloned Ia-restricted T cells that do not produce interleukin-4/B cell stimulatory factor-1 fail to help antigen-specific B cells. Submitted for publication

90. Ashwell, J. D., DeFranco, A. L., Paul, W. E., Schwartz, R. H. 1984. Antigen presentation by resting B cells: Radiosensitivity of the antigen-presentation function and two distinct pathways of T cell activation. *J. Exp. Med.* 159: 881

91. Matis, L. A., Glimcher, L. H., Paul, W. E., Schwartz, R. H. 1983. The magnitude of response of histocompatibility-restricted T cell clones is a function of the product of the concentrations of antigen and Ia molecules. *Proc. Natl. Acad. Sci. USA* 80: 6019

92. DuBois, P., Stein, P., Ennist, D., Greenblatt, D., Mosmann, T., Howard, M. 1986. Requirement for BSF-1 in the induction of antigen-specific B cell proliferation by a thymus dependent antigen and a carrier-reactive T cell line. *J. Immunol.* In press

93. Brown, M. A., Pierce, J., Watson, C. J., Falco, J., Ihle, J. N., Paul, W. E. 1986. B cell stimulatory factor-1 is produced constitutively by Abelson MuLV-transformed mast cell lines. Submitted for publication

94. Meltzer, M. S., Crawford, R. M., Finbloom, D. S., Ohara, J., Paul, W. E. 1986. BSF-1: a macrophage activation factor. Submitted for publication

95. Long, M. W., Shapiro, D. N. 1985. Immune regulation of in vitro megakaryocyte development. Role of T lymphocytes and Ia antigen expression. *J. Exp. Med.* 162: 2053

Ann. Rev. Immunol. 1987. 5 : 461–75

BIOPHYSICAL ASPECTS OF ANTIGEN RECOGNITION BY T CELLS[1]

Tania H. Watts[2] and Harden M. McConnell

Stauffer Laboratory for Physical Chemistry, Stanford University, Stanford, California 94305

INTRODUCTION

Biological cells can communicate with one another through the release and binding of water soluble substances and also through direct contact between their plasma membranes. A particularly challenging problem in physical chemistry is to determine the molecular events that take place at the interface between the plasma membranes of two cells that specifically "recognize" one another. One approach to this problem is to replace one cell of a pair of interacting cells by a reconstituted membrane that contains just those molecules thought to be responsible for cellular recognition. Interacting pairs of immunological cells provide excellent opportunities for such studies. In fact there are now a number of experiments in which reconstituted membranes are used to mimic class-II MHC-restricted peptide antigen presentation to T-helper cells (1–10) and to elicit cytotoxic T-cells by alloantigens (11–17). This review emphasizes the use of planar membranes on solid supports for studies of the nature of the ligand of the T-cell receptor.

---

[1] Abbreviations: MHC, major histocompatibility complex; R_T, the antigen-specific MHC-restricted T-cell receptor; T_c, cytotoxic cell; T_H, helper cell; Ag, foreign antigen; M, class-I and class-II MHC proteins; M_I, class-I MHC protein; M_{II}, class-II MHC protein; APC, antigen presenting cell, the ligand of T_H cells; APT, antigen presenting target, the target of T_c cells; Ia, I-associated antigens (I-A or I-E); IL-2, interleukin 2; FCS, fetal calf serum; and e, enhancement in Texas red fluorescence due to energy transfer.

[2] Present address: Department of Immunology, University of Toronto, Toronto, Ontario, Canada M5S 1A8.

461

0732–0582/87/0410–0461$02.00

The receptors (R_T) on cytotoxic T cells (T_c cells) and on helper T cells (T_H cells) are proteins encoded by members of the immunoglobulin supergene family (18). They consist of α and β chains, each with constant and variable region domains analogous to those of immunoglobulins. Both T_c and T_H appear to use the same pool of variable region genes. Comparison of the predicted protein sequences of a number of T-cell-receptor genes with different specificities has not led to an assignment of specific subregions responsible for antigen or MHC recognition (19–20). For a comprehensive review of T-cell-receptor gene structure, the reader is referred to Kronenberg et al (19). The receptor considered here is the $\alpha\beta$ heterodimer. Recently it has been demonstrated that an alternate population of T-cells express a functional γ-chain and lack the $\alpha\beta$ heterodimer (21).

Much evidence now suggests that the receptors R_T on cytotoxic T cells and on helper T cells have ligands each of which involves both specific antigen (Ag) and MHC molecules (M molecules). For a review see Schwartz (22). In the case of T_c cells, the ligand of R_T is in most cases an association of Ag and class-I M molecules (M_I molecules) on the surface membranes of the target cells. In the case of T_H cells, the ligand of R_T is thought to be an association of Ag and class-II M molecules (M_{II} molecules) on the surface membranes of antigen presenting cells. In other words a good working hypothesis is that similar ternary complexes of R_T, Ag, and M are involved in the triggering of responses by these two types of T cells and that these ternary complexes are present at the interfaces between the membranes of the relevant pairs of cells.

For T_H-cell recognition of some protein antigens there is a requirement for antigen processing (23–25). Numerous examples now show that the processed antigen can be replaced by proteolytic fragments or specific synthetic peptides, thereby allowing mapping of the regions recognized by T_H cells (for review see 26).

In a similar way, compelling evidence has recently been obtained that T_c cells can recognize peptides derived from influenza nucleoprotein on the surfaces of target cells (27). Thus, in these cases, the ternary complexes responsible for T_H-cell triggering involve R_T, M_{II}, and specific peptide Ag, whereas the ternary complexes responsible for T_c-cell triggering involve R_T, M_I, and specific peptide Ag. For simplicity, the following discussion is limited to peptide antigens. For a discussion of the peptides recognized by T cells the reader is referred to the accompanying review by Livingstone & Fathman (26). The evidence that a single ternary complex is involved in the R_T-ligand interaction is indirect but compelling. Some of this evidence is summarized below. We use the notation *APC* for cells that present antigen to T_H cells and *APT* for cells that are targets for T_c cells.

THE TERNARY COMPLEX BETWEEN T-CELL RECEPTOR, MHC PROTEIN, AND PEPTIDE ANTIGEN

A summary follows of some of the evidence for the involvement in T-cell recognition of antigen of a single ternary complex composed of R_T, M_I or M_{II} molecules, and peptide antigen:

1. Kappler et al were able to fuse two T_H hybridomas that had different specificities (each specific for a combination of Ag and M_{II}) and thus showed that the specificities do not mix (28). More recently, Dembic et al transferred the α- and β-chain R_T genes from a class-I MHC-restricted T_c-cell clone to a second T_c clone with a different specificity. The transfected α and β genes endowed the second cell with the specificity of the first (29). *Thus, a single receptor (an $\alpha\beta$ heterodimer) is involved in MHC-restricted recognition of antigen.* The γ chain is not required for the function of most of the T cells studied to date (30, 31).

2. Numerous studies show that the specificity of a given T_H hybridoma or clone is jointly sensitive to the amino acid sequence of both the peptide antigen and the M_{II} molecules (for review see 22 and 26). These results show that MHC-restricted peptide antigen recognition is specific for all three molecules, and the simplest picture is a ternary complex between R_T, peptide, and M_{II}. However, such data do not provide information on whether the complex is $1:1:1$, nor in fact do they exclude the (unlikely) possibility that peptide binds to one R_T and M_{II} to a second but otherwise identical R_T.

3. Babbitt et al (32) have obtained evidence that a relatively weak but selective binding takes place between fluorescently modified peptide from hen egg lysozyme (NBD-HEL-46-61) and detergent solubilized I-A^k (and not with I-A^d). This peptide is presented by I-A^k expressing APC to the T_H hybridoma 3A9. The reported dissociation constant is $k = 2 \times 10^{-6}$ mole/l. Further studies by Babbitt et al (9) show that other peptides that inhibit the binding of NBD-HEL-(46–61) to detergent solubilized I-A^k also inhibit the presentation of HEL(46–61) to the T_H hybridoma (3A9) by both planar membranes containing I-A^k and by intact APC. These peptide–I-A binding results leave little doubt that molecular contacts between peptide and Ia are involved in peptide antigen presentation.

   Buus et al (33) have carried out a similar study using the chicken ovalbumin peptide OVA(323-339) and I-A^d purified from the membranes of APC. Again, weak but selective binding was observed. No binding to I-A^k or I-E^k was observed. The peptide OVA(323-339) is presented to the T_H hybridoma 3DO-54.8 in the context of I-A^d (34).

Phillips et al have demonstrated an association between "processed" insulin and I-A on APC using photoaffinity cross-linking reagents (35). In contrast to the results of Babbitt et al and of Buus et al, this group's results showed labeling of the I-A of all mouse haplotypes tested, although the intensity of the labeling differed from strain to strain.

4. Watts et al (7) have observed Förster energy transfer between a fluorescein-labeled peptide antigen from chicken ovalbumin (ov323-339) and Texas red–labeled I-A^d, mediated by the presence of the specific T_H hybridoma 3DO-54.8. This experiment is also most easily interpreted in terms of the formation of a ternary complex, in which the distance between donor fluorophore and acceptor fluorophore is equal to or less than 40 Å.

5. Extensive studies on the response of T_H cells to varying concentrations of cytochrome c peptides and varying I-A densities have been used by Schwartz and colleagues as evidence for the formation of R_T complexes with equimolar ratios of Ia and peptide molecules (10, 36).

6. Allogeneic responses to a single molecule, alloantigen, by T_c and T_H cells have always been a strong hint that the normal ligand for R_T is a single macromolecular structure, M_I plus peptide, or M_{II} plus peptide. A great deal of evidence favors this point of view. Isolated counter examples (6, 37, 38), especially those involving multivalent antigens in the absence of MHC, certainly do not detract from the view that a single ternary complex is the usual trigger signal for the T-cell response.

T-CELL RECOGNITION OF ANTIGEN IN RECONSTITUTED SUPPORTED PLANAR MEMBRANES

Reconstituted planar membranes offer an interesting and possibly unique approach for investigating the physical chemistry of antigen recognition by T lymphocytes. This is because such membranes can be used to study T-cell responses to reconstituted membranes of defined composition that mimic APC or APT. The same membranes can also be used to study the physical chemistry of membrane-membrane recognition, for example, using fluorescently labeled antigens together with fluorescence spectroscopy. For details concerning the preparation and properties of supported planar membranes, the reader is referred to a recent general review (39). For the purposes of the present review, the reader need only note that detergent dialyzed lipid vesicles containing M_I or M_{II} molecules will fuse with one another on a clean glass slide or on beads, producing a continuous supported bilayer in which the M molecules are functional.

The lipids in the upper leaflet facing the aqueous solution diffuse freely, but the M molecules are essentially immobilized. A similar procedure can be employed using alkylated glass slides or glass beads, in which case the outer lipid monolayer is membrane-like and the opposing alkyl chains are anchored to the glass.

Cytotoxic T Cells

Brian & McConnell (16) used supported planar membranes containing an M_I molecule H-$2K^k$ to elicit allogeneic responses by pre–T_c cells. These membranes are far more efficient in eliciting allogeneic cytotoxic T cells than are comparable concentrations of alloantigen in phospholipid vesicles. This may be due simply to a more efficient contact of cells resting on the supported membranes. Supported membranes can also be used for cell-to-membrane binding assays, as has been shown by the specific binding of clonal cytotoxic T cells to target-supported membranes containing phospholipid and purified alloantigen (40).

Helper T Cells

A number of laboratories have shown that supported planar membranes containing purified I-A molecules can serve to present peptide antigen to T_H cells, resulting in the production of IL-2. Watts et al (2, 3) showed that membranes containing affinity purified I-A^d together with soluble peptide antigen from chicken ovalbumin are recognized by the I-A^d–restricted T-cell hybridoma 3DO-54.8. (I-A^d in micron-sized unilamellar vesicles is completely ineffective.) The planar membranes containing I-A^d trigger a peptide-antigen–specific T_H response at peptide concentrations 50 times lower than are required to get a response from planar membranes prepared from A20-1.11 membranes (3). In these experiments I-A density was 15 times lower in the plasma membranes than in the reconstituted membranes.

It has been argued that antigen processing is not essential for antigen recognition by T cells. The argument asserts that blocking antigen presentation by chloroquine treatment or glutaraldehyde fixation of APC may have multiple effects. Supported planar membranes containing only I-A and lipid do not present native ovalbumin to 3DO-54.8 cells (2). This result shows that antigen processing is necessary at least for presentation of some antigens to some T_H cells.

In other laboratories: Babbitt et al (9) have reported that planar membranes containing I-A^k are able to present antigen to $T_H(3A9)$; Gay et al (8) have also used planar and bead supported membranes to study I-A–dependent antigen presentation to T_H cells.

Fox et al (10) have carried out a quantitative analysis of the dose-response of the T_H hybridoma 2B4.11 to planar membranes containing purified I-E^k together with peptide antigen. The planar membranes were from liposomes fused with glass beads. An object of this work was to use dose-response data to infer the stoichiometry of antigen and I-E molecules in the putative ternary molecular complex. These investigators first made log plots of the two-dimensional concentrations of I-E^k, and of the three-dimensional concentrations of cytochrome c peptide, together required to produce a given activity of IL-2. Such plots yield slopes of -2, which might be interpreted as a signal complex composed of one molecule of peptide and two molecules of I-E. However, when these plots are normalized to a given fraction of the maximum response for a given two-dimensional I-E concentration, then such plots yield slopes of -1, which have been taken to indicate a $1:1$ stoichiometry of Ia and peptide antigen in the signal complex (10).

To discuss these interesting data of Fox et al, it is useful to consider the following equilibria:

$$p + Ia \rightleftharpoons pIa \qquad\qquad 1.$$

$$[pIa] = k[p][Ia]. \qquad\qquad 2.$$

Here $[p]$ is the concentration of peptide in three-dimensional solution, assumed to be in large excess, and $[Ia]$ is the two-dimensional concentration of Ia on the supported membrane. Thus:

$$[pIa] = k[p][Ia]_0/(1 + k[p]). \qquad\qquad 3.$$

We further assume that the next step in the reaction, the binding of R_T to pIa, can also be described by a law of mass action, with binding constant K. That is,

$$R + pIa \rightleftharpoons RpIa. \qquad\qquad 4.$$

Two limiting cases may be considered: The total available number of R_T receptors $[R]_0$ greatly exceeds the number of available Ia molecules $[Ia]_0$, or vice versa. For these two cases,

$$[R]_0 \gg [Ia]_0:$$

$$[RpIa] = Kk[p][R]_0[Ia]_0/(1 + Kk[p][R]_0); \qquad\qquad 5.$$

$$[Ia]_0 \gg [R]_0$$

$$[RpIa] = Kk[p][R]_0[Ia]_0(1 + Kk[p][Ia]_0. \qquad\qquad 6.$$

We note that if the normalized production of IL-2 is some monotonic function of the fractional occupation of the T-cell receptor, $[RpIa]/[R]_0$, then when $[Ia]_0$ is in excess, one expects a slope of -1 in plots of $\ln[p]$ vs $\ln[Ia]_0$, for a fixed normalized response. While such assumptions of mass action may ultimately prove to be gross oversimplifications, they are none-theless useful guides in the design of experiments.

Previously unpublished data by Watts & McConnell on the response of 3DO-54.8 cells to planar membranes containing varying densities of I-A^d are shown in Figure 1. These data (Figure 1, lower), like those of Fox et al (10), show that the concentration of Ia molecules in the supported membrane can be controlled and measured quantitatively. In these experiments the response of the 3DO-54.8 cells to increasing I-A concentration $[I\text{-}A^d]_0$ appears to saturate at large values of $[I\text{-}A^d]_0$. This saturation is

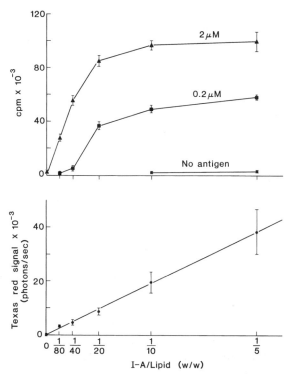

Figure 1 Effect of I-A density on the response of 3DO-54.8 cells to planar membranes and antigen: (*a*) IL-2 release by 3DO-54.8 cells was measured using the CTLL indicator line as described in (2). I-A was reconstituted with varying ratios of protein to lipid. The concentration of antigen (a synthetic peptide presenting residue 323-339 of ovalbumin) is given beside each curve. It can be seen that at higher I-A density, less peptide is required to get a comparable response. (*b*) In order to ensure that the I-A density in the final planar membrane could be predicted from the ratio of I-A/lipid used in the reconstitution of Texas red, I-A was used and the fluorescence spectra of the planar membranes obtained using fluorescence microspectro described in (7). A ratio of 1:10 by wt I-A:lipid equals about 800 lipid molecules/I-A and 2500 I-A/μ^2.

certainly not described by Equation 5 but might be described by Equation 6, which would imply that the response saturates because of saturation of R_T by ligand.

The interpretation of IL-2 production in some experiments in terms of equilibrium constants of ternary signal complexes implies that these constants under the described conditions (T_H hybridomas, peptide antigens) are not very large. That is, the effective binding constant of peptide antigen to a complex of I-A and/or R_T cannot be very large. In fact, data such as those in Figure 1 suggest that the binding constant of the peptide in the signal complex is the same order of magnitude as the binding constants for peptide to I-A in detergents, described by Babbitt et al (32), and the approximate binding of ovalbumin (323-329) to I-A^d reported by Buus et al (33). However, see later discussion.

Fluorescence Spectroscopy

Supported membranes on glass slides are ideal for high sensitivity fluorescence microscopic studies of membrane-membrane recognition, when one or more molecules involved in this recognition event can be made fluorescent. The data in Figure 1 provide a simple example, where the fluorescence emission from Texas red-labeled I-A^d is used to quantitate the concentration of this molecule in the planar membrane.

Most of the experiments in this laboratory (7) have employed a two-dimensional I-A^d concentration equal to 2×10^{11} molecules/cm^2, the protein molecule bearing about five covalently linked Texas red fluorophores. In our set-up, this produces about 20,000 photons/sec due to the weak, direct excitation of Texas red by the argon ion line at 488 nm. Experiments employing fluorescein-N-terminal–labeled peptides at a concentration of 20 μM, and using an 800-A-deep evanescent wave field for fluorescence excitation, detect the emission from about 10^{11} molecules/cm^2. This produces about 240,000 photons/sec at the emission peak due to direct excitation of the fluorescein peptide by the 488 nm argon ion laser line.

We now make an order-of-magnitude estimate of the sensitivity of this system to detect complex formation between Texas red–labeled I-A^d and fluorescein-labeled peptide, based on Förster energy transfer from fluorescein to Texas red.

The enhancement e in the Texas-red fluorescence is defined as the ratio of the intensity of the Texas-red fluorescence including energy transfer stimulation to the Texas-red fluorescence in the absence of energy transfer. For a complex between labeled I-A^d and peptide, this specific transfer is of the order of magnitude:

$$e(\text{specific}) = \frac{k[p]\tau}{1+k[p]} \cdot \frac{\varepsilon_F(488)}{\varepsilon_{TR}(488)}, \qquad 7a.$$

where [p] is the free peptide concentration; k is the stability constant defined above in Equation (2); $\varepsilon_F(488)$ and $\varepsilon_{TR}(488)$ are the molar extinction coefficients for fluorescein and Texas red at the argon ion laser line (488 nm); and τ is probability of Förster resonance energy transfer from fluorescein to Texas red in the complex. When $e = 1$, there is no enhancement, and when $e = 2$ the enhancement is 100%, i.e. the Texas-red emission intensity is doubled. For a transfer efficiency between $\tau = 0.1$ and 1.0 and for $\varepsilon_F(488)/\varepsilon_{TR}(488) = 24$, an enhancement of the order of 1–10 is expected, provided $k[p] \gg 1$. As a positive control for transfer, it was found that fluorescinated MKD6 bound to Texas red–labeled I-A^d showed an enhancement $e = 1.90$ when the antibody was labeled with 8 fluoresceins per antibody, and $e = 1.25$ when there was 1 fluorescein per antibody.

We can make an order-of-magnitude estimate of a collision probability that a fluorescein donor be within 40 Å of a Texas-red acceptor, for a peptide concentration of p(molar). This is:

$$e(\text{collision}) = 160\,p\tau. \qquad\qquad 7b.$$

For $p \simeq 10^{-5}$ molar, or less, enhancement due to nonspecific collisions is negligible. Under some circumstances nonspecific Förster transfer between donor peptide and acceptor Texas red is observed on the planar membrane (T. Watts, H. Gaub, H. M. McConnell, unpublished). In the presence of cells, strong transfer is always observed when the buffer solution does not contain fetal calf serum. In some experiments nonspecific transfer was observed even in the presence of fetal calf serum, presumably related to some imperfection in the membrane preparation. Specific transfer is, by our definition, transfer that can be reduced or eliminated with excess nonfluorescent specific peptide.

Watts et al (7) have reported an enhancement of $e = 1.2$–1.7 in energy transfer from a fluorescinated 17-residue peptide to Texas red–labeled I-A^d in the presence of the peptide specific T-cell hybridoma. In our experiments, in the absence of cells and in the presence of fetal calf serum, we find the enhancement to be less than 1.1, when $[p] = 20 \times 10^{-6}$ M. Thus, in our system, either the stability constant k is much less than that observed by others (32, 33), or the transfer efficiency is lower than expected, or binding of peptide to I-A^d is blocked by components in the serum. As discussed later, we consider the latter possibility to be the more likely. It is entirely plausible that energy transfer is more efficient in the ternary complex than in the binary complex. There is certainly no reason to regard the energy transfer results as being inconsistent with the stability constants reported by Buus et al (33) and Babbitt et al (32). In our experiments it was found that a tenfold excess of unlabeled peptide eliminated the transfer, evidently by competitive binding. The transfer was also blocked by anti–

I-A^d antibody, which binds to I-A^d. It is also worth noting that the energy transfer experiments differ from the binding studies described (32, 33); in (7) the energy transfer experiments, the membranes are washed with fetal calf serum before adding peptide.

The energy transfer results are consistent with a model in which peptide, I-A^d, and the T-cell receptor form a ternary complex. Also in this model, the fluorescein fluorophore on the peptide and one or more Texas-red molecules on the I-A^d molecule are within 40 Å of one another. Note that for a transfer enhancement of $e = 1.25$–1.90, a significant fraction of the Texas red–labeled I-A^d molecules in the planar membrane must be involved in the ternary complex formation. If only 1% were involved, the enhancement of Texas-red emission could hardly be much more than $e = 0.01$–0.1 and would probably not be detectable under the experimental conditions employed by Watts et al (7). In these experiments the supported planar membrane was covered to 90% with a "lawn" of 3DO-54.8 cells, in contact or near contact with one another. For a cell 10 μ in diameter, there should be $\pi r^2 \times 1.6 \times 10^{11} = 0.9 \times 10^5$ I-A^d per T_H cell. There are presumably also T_H receptors involved of the order of 10^4–10^5. In recent work we have found that energy transfer between fluorescein-labeled peptide and Texas red–labeled I-A^d can be observed in the absence of T_H cells, provided serum and competing peptides are also absent. See later discussion.

The 3DO-54.8 cells are not strongly adherent to supported planar membranes containing I-A^d, either with or without peptide. Nonetheless, if 10^5 receptors of the T cell engaged in transmembrane signaling, one can estimate that of the order of at least a few percent of the T-cell surface would have to be in contact with the supported membrane. In contrast, the 3DO-54.8 cells form strong conjugates with antigen presenting cells, suggesting that accessory molecules on the presenting cells as well as on the T_H cells may be involved in strengthening the interaction between the two membranes (A. Brian, T. Watts, unpublished).

It is also worth noting that to detect energy transfer Watts et al had to use peptide concentrations that saturate with respect to IL-2 production. Thus, fewer signal complexes are required for T-cell activation than are necessary to detect energy transfer.

Molecular Events in Antigen Recognition

Since termolecular collisions are extremely improbable on purely statistical grounds, the termolecular complexes must be considered to arise from one or some combination of three distinct pathways:

$$M + Ag \rightleftharpoons MAg \tag{8a}$$

$$R_T + MAg \rightleftharpoons MAgR_T \tag{8b}$$

$$R_T + Ag \rightleftharpoons R_TAg \qquad\qquad 9a.$$

$$R_TAg + M \rightleftharpoons R_TAgM \qquad\qquad 9b.$$

$$R_T + M \rightleftharpoons R_TM \qquad\qquad 10a.$$

$$R_TM + Ag \rightleftharpoons R_TAgM. \qquad\qquad 10b.$$

Consider only APC and T_H cells. By definition, reactions 8a and 9a are presumed to take place in the absence of cell-cell contact, when the two cells are far apart. In reaction 10b antigen only binds to the R_T complex when the APC and T_H cells are in contact, and (perhaps transient) R_T complexes are formed. There is very little evidence for the binding of monovalent antigen to T_H cells, and thus reaction-pathway 9a is unlikely. There are a few cases where T-cell-antigen binding and activation have been demonstrated in the absence of M molecules (6, 37, 31, 41). In general these have involved multivalent antigens (with the exception of the studies in 37). Providing reaction 9a has some small but finite stability constant, such results are not surprising, because cross-linking of the R_T complex by multivalent anti-R_T or anti-T3 antibodies leads to T-cell activation in the absence of M molecules (reviewed in 42, 43). Walden et al (6) have demonstrated that when antigen is covalently linked to liposomes it triggers M-restricted IL-2 production at low antigen densities, but at high antigen densities, the M molecule restriction is eliminated.

A particularly interesting study relevant to the quantitative aspects of antigen presentation has been carried out by Lanzavecchia (44). In this work Epstein-Barr virus–transformed human B-cell clones with surface Ig receptors specific for tetanus toxoid were found to present antigen in an MHC-restricted manner to T-cell clones. The specific antibody-dependent protein-antigen uptake was found to be 10^4 times more efficient than is nonspecific uptake. In this and other studies (45), protein antigen concentrations on the order of 10^{-12} M can be detected on APC by T cells. Clearly, the peptide antigens derived from the toxoid protein antigen must somehow be tightly associated with the APC membrane. A tight association of peptide antigen with APC is also indicated by antigen pulsing experiments (for example see Ref. 46). In this laboratory we have recently found that when vesicles consisting of pure I-A^d and lipid are mixed with peptide antigen and then dialyzed exhaustively, these prepulsed vesicles retain enough antigen to stimulate IL-2 release from 3DO-54.8 cells when the liposomes are converted to planar membranes (48). The ability to retain fluorescent peptide antigen on the liposome surface was likewise dependent on the presence of I-A in the liposomes. The affinity of peptide for I-A in our experiments appears to be orders of magnitude higher than the affinity of peptide for I-A in detergent micells (32, 33).

In conclusion, we speculate briefly on the structure of the ternary recognition complex. Since peptides with lengths of the order of magnitude of 10–20 amino acids need not have well-defined secondary and tertiary structures in aqueous solutions, it is plausible to assume that the three-dimensional configuration of any given peptide in the ternary recognition complex will be determined largely by the structures of the complex of M and peptide. Current evidence is that antibodies and their protein antigens undergo no large conformation changes on binding to one another, so we assume that no large conformation changes are involved in step 8b, the binding of R_T to the MAg complex. On the other hand, since peptide antigens have no well-defined tertiary structures and the M molecules must be able to bind a variety of peptide antigens, it is possible that substantial structural changes are involved in the formation of the AgM complexes. The self–non-self discrimination does not take place at the level of the M-molecule, since an autologous lysozyme peptide can compete for peptide antigen binding to I-A^k and can inhibit presentation of this antigen to peptide-reactive T_H cells (9). It is possible that antigen presentation by pathway (8) is kinetically controlled. That is, the peptides that are presented are those that bind tightly and that are also present at a high (perhaps transient) concentration relative to competing peptides. The tight binding of the 17–amino acid peptide from chicken ovalbumin to I-A^d in lipid bilayers can be competed with nonspecific peptides from a digest of hen egg lysozyme when specific and nonspecific peptides are added to the I-A^d preparations at the same time, but not when the specific peptide is added first (48).

The evident low specificity but high affinity of M molecules for peptides (in the absence of serum) suggests that a generic structural principle is involved. Amphiphilic peptides may bind to complementary hydrophobic/hydrophilic regions on the M molecules. It is possible that serum albumin reduces peptide binding to IA, either by binding peptide, by providing amphiphilic substances (fatty acids) that compete with peptide, or by binding directly to IA. This could account for an apparent blocking of peptide binding to I-A^d by serum (48).

In this review we have described a number of unsolved problems and made some speculations. We believe some of these interesting problems can be solved using supported reconstituted membranes and techniques of physical chemistry. Of particular immediate interest are kinetic measurements of the binding of peptides to M molecules (including competitive binding), and elucidation of the pathways whereby protein antigens are degraded to peptides, followed by the formation of peptide MAg complexes.

ACKNOWLEDGMENT

This work was supported by NIH Grant 5R01 AI 13587 and DOD Equipment Grant N00014-84-G-0210. T. H. Watts was supported by a fellowship from the Medical Research Council of Canada and a research allowance from the Alberta Heritage Fund for Medical Research.

Literature Cited

1. Curman, B., Ostberg, L., Peterson, P. A. 1978. Incorporation of murine MHC antigens into liposomes and their effect in secondary mixed lymphocyte reaction. *Nature* 272: 545–47
2. Watts, T. H., Brian, A. A., Kappler, J. W., Marrack, P., McConnell, H. M. 1984. Antigen presentation by supported planar membranes containing affinity purified I-A^d. *Proc. Natl. Acad. Sci. USA* 81: 7564–68
3. Watts, T. H., Gariepy, J., Schoolnik, G., McConnell, H. M. 1985. T cell activation by peptide antigen: Effect of peptide sequence and method of antigen presentation. *Proc. Natl. Acad. Sci. USA* 82: 5480–84
4. Walden, P., Nagy, Z. A., Klein, J. 1985. Induction of regulatory T-lymphocyte responses by liposomes carrying major histocompatibility complex molecules and foreign antigen. *Nature* 315: 327–29
5. Walden, P., Nagy, Z. A., Klein, J. 1986. Antigen presentation by liposomes: Inhibition by antibody. *Eur. J. Immunol.* 16: 717–20
6. Walden, P., Nagy, Z. A., Klein, J. 1986. Major histocompatibility complex-restricted and unrestricted activation of helper T cell lines by liposome-bound antigen. *J. Mol. Cell. Immunol.* 2: 191–97
7. Watts, T. H., Gaub, H. E., McConnell, H. M. 1986. T cell-mediated association of peptide antigen and major histocompatibility complex protein detected by energy transfer in an evanescent wavefield. *Nature* 320: 176–79
8. Gay, D., Coeshott, C., Golde, W., Kappler, J., Marrack, P. 1986. The major histocompatibility complex-restricted antigen receptor on T cells. IX. Role of accessory molecules in recognition of antigen plus isolated I-A. *J. Immunol.* 136: 2026–32
9. Babbitt, B. P., Matsueda, G., Haber, E., Unanue, E. R., Allen, P. M. 1986. Antigen competition at the level of peptide-Ia binding. *Proc. Natl. Acad. Sci. USA* 83: 4509–13
10. Fox, B. S., Quill, H., Carlson, L., Schwartz, H. 1986. Quantitative analysis of the T cell response to antigen and planar membranes containing purified Ia molecules. *J. Immunol.* Submitted
11. Lawman, M. J. P., Naylor, P. T., Huang, L., Courtney, R. J. 1981. Cell-mediated immunity to herpes simplex virus: Induction of cytotoxic T lymphocyte responses by viral antigens incorporated into liposomes. *J. Immunol.* 126: 304–8
12. Hermann, S. H., Mescher, M. F. 1981. Secondary cytolytic T lymphocyte stimulation by purified H-2K^k in liposomes. *Proc. Natl. Acad. Sci. USA* 78: 2488–92
13. Hermann, H. S., Weinberger, O., Burakoff, S. J., Mescher, M. J. 1982. Analysis of the two-signal requirement for precursor T lymphocyte activation using H-2K^k in liposomes. *J. Immunol.* 128: 1968–74
14. Cartwright, G. S., Smith, L. M., Heinzelmann, E. W., Ruebush, M. J., Parce, J. W., McConnell, H. M. 1982. H-2K^k and VSV-G proteins are not extensively associated in reconstituted membranes recognized by T cells. *Proc. Natl. Acad. Sci. USA* 79: 1506–10
15. Albert, F., Boyer, C., Lesserman, L. D., Schmitt-Verhulst, A. M. 1983. Immunopurification and insertion into liposomes of native and mutant H-2K^b: Quantification by solid phase radioimmunoscopy. *Mol. Immunol.* 20: 655–62
16. Brian, A. A., McConnell, H. M. 1984. Allogeneic stimulation of cytotoxic T cells by supported planar membranes. *Proc. Natl. Acad. Sci. USA* 81: 6159–63
17. Goldstein, S. A. N., Mescher, M. F. 1985. Carbohydrate molecules of major histocompatibility complex class I alloantigens are not required for their recognition by T lymphocytes. *J. Exp. Med.* 162: 1381–86
18. Hood, L., Kronenberg, M., Hunka-

piller, T. 1985. T cell antigen receptors and the immunoglobulin supergene family. *Cell* 40 : 225–29

19. Kronenberg, M., Jin, G., Hood, L. E., Shastri, N. 1986. The molecular genetics of the T-cell antigen receptor and T-cell antigen recognition. *Ann. Rev. Immunol.* 4 : 529–91

20. Fink, P. J., Matis, L. A., McElliot, D. L., Bookman, M., Hedrick, S. M. 1986. Correlations between T-cell specificity and the structure of the antigen receptor. *Nature* 321 : 219–26

21. Brenner, M. B., McLean, J., Dialynas, D. P., Strominger, J. L., Smith, J. A., Owen, F. L., Seidman, J. G., Ip, S., Rosen, F., Krangel, M. S. 1986. Identification of a putative second T-cell receptor. *Nature* 322 : 145–49

22. Schwartz, R. S. 1985. T lymphocyte recognition of antigen in association with gene products of the major histocompatibility complex. *Ann. Rev. Immunol.* 3 : 237–61

23. Chesnut, K. S., Colon, S., Grey, H. 1982. Requirements for the processing of antigen by antigen-presenting B cells. I. Functional comparison of B cell tumors and macrophages. *J. Immunol.* 129 : 2382–88

24. Ziegler, K., Unanue, E. 1982. Decrease in macrophage antigen catabolism caused by ammonia and chloroquine is associated with inhibition of antigen presentation to T cells. *Proc. Natl. Acad. Sci. USA* 79 : 175–78

25. Shimonkevitz, R., Kappler, J., Marrack, P., Grey, H. 1983. Antigen recognition by H-2-restricted T cells. I. Cell free antigen processing. *J. Exp. Med.* 158 : 303–16

26. Livingstone, A. M., Fathman, C. G. 1987. Structure of antigeneic epitopes recognized by T-cells. *Rev. Immunol.* 5 : In press

27. Townsend, A. R. M., Rothbard, J., Gotch, F. M., Bahadur, G., Wraith, D., McMichael, A. J. 1986. The epitopes of influenza nucleoprotein recognized by cytotoxic T lymphocytes can be defined with short synthetic peptides. *Cell* 44 : 959–68

28. Kappler, J. W., Skidmore, B., White, J., Marrack, P. 1981. Antigen-inducible, H-2-restricted, interleukin-2-producing T cell hybridomas. Lack of independent antigen and H-2 recognition. *J. Exp. Med.* 153 : 1198–1214

29. Dembic, Z., Haas, W., Weiss, S., McCubrey, J., Kiefer, H., von Beohmer, H., Steinmetz, M. 1986. Transfer of specificity by murine and T-cell receptor genes. *Nature* 320 : 232–38

30. Rupp, F., Frech G., Hengartner, H., Zinkernagel, R. M., John, R. 1986. No functional γ-chain transcripts detected in an alloreactive cytotoxic T-cell clone. *Nature* 321 : 876–78

31. Reilly, E. B., Krantz, D. M., Tonegawa, S., Eisen, H. N. 1986. A functional γ-chain formed from known γ-gene segments is not necessary for antigen-specific responses of murine cytotoxic T lymphocytes. *Nature* 321 : 878–80

32. Babbitt, B. P., Allen, P. M., Matsueda, G., Haber, E., Unanue, E. R. 1985. Binding of immunogenic peptides to Ia histocompatibility molecules. *Nature* 317 : 359–61

33. Buus, S., Colon, S., Smith, C., Freed, J. H., Miles, C., Grey, H. M. 1986. Interaction between a processed ovalbumin peptide and Ia molecules. *Proc. Natl. Acad. Sci. USA* 83 : 4509–13

34. Shimonkevitz, R., Colon, S., Kappler, J. W., Marrack, P., Grey, H. M. 1984. Antigen recognition by H-2 restricted T cells II. A tryptic ovalbumin peptide that substitutes for processed antigen. *J. Immunol.* 133 : 2067–74

35. Phillips, M. L., Yip, C. C., Shevach, E. M., Delovitch, T. L. 1986. Photoaffinity labeling demonstrates binding between Ia molecules and nominal antigen on antigen-presenting cells. *Proc. Natl. Acad. Sci. USA* 83 : 5634–38

36. Ashwell, J. D., Fox, B. S., Schwartz, R. H. 1986. Functional analysis of the interaction of the antigen-specific T cell receptor with its ligands. *J. Immunol.* 136 : 757–68

37. Carel, S., Bron, C., Corradin, G. 1983. T cell hybridoma specific for cytochrome c peptide : Specific peptide binding and interleukin 2 production. *Proc. Natl. Acad. Sci. USA* 80 : 4832–36

38. Silicano, R. F., Keegan, A. D., Dintzis, R. Z., Dintzis, A. M., Shin, H. Y. 1985. The interaction of nominal antigen with T cell antigen receptors. I. Specific binding of multivalent nominal antigen to cytolytic T cell clones. *J. Immunol.* 135 : 906–14

39. McConnell, H. M., Watts, T. H., Weis, R. M., Brian, A. A. 1986. Supported planar membranes in studies of cell-cell recognition in the immune system. *Biochem. Biophys. Acta* 864 : 95–106

40. Nakanishi, M., Brian, A. A., McConnell, H. M. 1973. Binding of cytotoxic T-lymphocytes to supported lipid monolayers containing trypsinized H-2K^k. *Molec. Immunol.* 20 : 1227–31

41. Rao, A., Ko, W. W.-P., Fass, S. J., Cantor, H. 1984. Binding of antigen in the absence of histocompatibility pro-

teins by arsonate reactive T-cell clones. *Cell* 36: 879–86

42. Haskins, K., Kappler, J., Marrack, P. 1984. The major histocompatibility complex-restricted antigen receptor on T cells. *Ann. Rev. Immunol.* 2: 51–66

43. Weiss, A., Imboden, J., Hardy, K., Manger, B., Terhost, C., Stobo, J. 1986. The role of the T3/antigen receptor complex in T cell activation. *Ann. Rev. Immunol.* 4: 593–619

44. Lanzavecchia, A. 1985. Antigen-specific interaction between T and B cells. *Nature* 314: 537–39

45. Abbas, A. K., Haber, S., Rock, K. L. 1985. Antigen presentation by hapten-specific B lymphocytes. II. Specificity and properties of antigen-presenting B lymphocytes, and function of immunoglobulin receptors. *J. Immunol.* 135: 1661–67

46. Rock, K. L., Benacerraf, B. 1983. Inhibition of antigen-specific T lymphocyte activation by structurally related Ir gene-controlled polymers. *J. Exp. Med.* 157: 1618–34

47. Amit, A. G., Mariuzza, R. A., Phillips, S. E. V., Poljak, R. J. 1986. Three-dimensional structure of an antigen-antibody complex at 6 Å resolution. *Nature* 313: 156–58; Three-dimensional structure of an antigen-antibody complex at 2.8 Å resolution. *Science* 233: 747–53

48. Watts, T. H., McConnell, H. M. 1986. *Proc. Natl. Acad. Sci. USA.* In press

Ann. Rev. Immunol. 1987. 5 : 477–501

THE STRUCTURE OF T-CELL EPITOPES

Alexandra M. Livingstone and C. Garrison Fathman

Division of Immunology, Department of Medicine,
Stanford University Medical Center, Stanford, California 94305

INTRODUCTION

The humoral and cellular arms of the adaptive immune response, mediated by B and T lymphocytes respectively, differ fundamentally in the way in which they recognize antigen. The B-cell receptor for antigen (immunoglobulin) can bind to soluble antigen in a manner similar to that of many well-characterized receptor-ligand systems. In contrast, the T-cell antigen receptor can generally recognize antigen only in association with cell surface molecules encoded by the major histocompatibility complex (MHC). This requirement, known as MHC restriction, ensures that T-cell activation or effector function occurs only in an appropriate cellular context. Antigen-specific T-cell activation thus results from the formation of a ternary complex involving the T-cell receptor (TcR), nominal antigen, and class-I or class-II MHC molecules (1, 2).

Characterization of the antigen-specific receptors on T-helper and cytotoxic cells unfortunately did not suggest any structural basis for the MHC-restricted recognition of antigen. On the contrary, the genes encoding the α and β chains of the TcR heterodimer showed marked homology with immunoglobulin genes, and similar mechanisms of DNA rearrangement appeared to be involved in the generation of both T- and B-cell repertoires (3–5). Experiments to determine the mechanism underlying MHC restriction have therefore increasingly focused on the way in which nominal antigen might interact with MHC molecules to form a single antigenic structure recognized by the TcR.

This review is concerned with the characterization of antigenic determinants seen by T cells. Almost all the studies in this area have concentrated on the response of MHC class II (Ia)–restricted T cells to soluble

0732–0582/87/0410–0477$02.00

protein antigens, for several reasons. In the first place, many of these antigens, such as lysozyme, cytochrome *c*, and myoglobin, belong to sets of homologous proteins with known primary amino acid sequence and tertiary structure. Residues that were important for T-cell stimulation could be identified by analyzing the response to several members of a protein family and comparing their sequences. Secondly, T-cell-helper/proliferative responses to many protein antigens were shown to be specific for sequential determinants. This meant that the epitopes seen by these T cells could be identified relatively easily using proteolytic fragments or synthetic peptides. In addition, the response to many of these antigens was under immune response (*Ir*) gene control. The determinant selection theory argued that Ir gene effects were determined by the ability of antigen to form an appropriate immunogenic association with Ia molecules on the antigen-presenting cell surface (6-8); many groups have therefore sought to identify T-cell epitopes on these proteins, in order to analyze the specificity of antigen-Ia interactions.

The recognition of antigens such as viruses, or minor histocompatibility antigens by class I–restricted T cells (predominantly of cytotoxic phenotype) was until recently thought to be qualitatively different from the way in which class II–restricted T cells recognized soluble protein antigens. It was assumed that these antigens were integral membrane proteins associated in some way on the cell surface with the appropriate MHC molecule. The principal reason for thinking this was that cytotoxic T cells could kill targets not known to have any antigen-processing capability. Several problems arose with this model; perhaps the most important was the almost universal failure to block cytotoxic responses with antisera specific for the nominal antigen, although antibodies specific for the class I–restricting molecules could block lysis very effectively (9). Townsend and colleagues have recently shown, however, that class I–restricted cytotoxic T cells can recognize fragments of molecules that are not integral membrane proteins (10). They found that an influenza-specific, class I–restricted cytotoxic-T-cell clone was specific for the viral nucleoprotein. Most importantly, this clone could be activated by a short synthetic peptide of nucleoprotein, presented on the surface of uninfected cells expressing the appropriate MHC class-I molecules. These experiments suggest that the recognition of viral proteins by class I–restricted cytotoxic T cells may in all essentials be similar to the recognition of soluble protein antigens by class II–restricted T-helper cells (11).

In this review, we address in detail the structure of the epitopes recognized by helper T cells. The observed differences between such epitopes and those recognized by B cells are examined, and a hypothesis to explain the utility of such distinct recognition systems is presented.

T-CELL EPITOPES CAN BE DEFINED BY
PRIMARY AMINO ACID SEQUENCE

The first evidence that T-cell recognition of antigen was qualitatively different from that of B cells came from experiments by Gell & Benacerraf (12), at a time when lymphocytes had not yet been divided into T- and B-cell subsets. They compared humoral and cellular responses in guinea pigs to a series of protein antigens and found that while the antibody response was specific for the form of antigen used for immunization (native versus denatured), delayed type hypersensitivity (DTH) responses (later shown to be T-cell dependent) could be elicited equally well using either native or denatured antigen. They proposed that while the antigenic structures seen by antibodies depended on the tertiary configuration of the immunizing protein, the cells effecting the DTH response might recognize epitopes sufficiently defined by primary amino acid sequence. Sela (13) had named such antigenic determinants "sequential epitopes" to distinguish them from "conformational epitopes," dependent on tertiary protein structure.

Subsequent experiments with a wide range of protein antigens supported the contention that T cells of helper/proliferative phenotype recognized denatured antigens. Thompson et al (14) showed that although there was essentially no serological cross-reaction between reduced, carboxymethylated hen egg lysozyme (RCM-HEL) and native HEL, splenic lymphocytes from guinea pigs immunized with either native HEL or RCM-HEL proliferated strongly to both forms of this antigen. Ishizaka et al (15) found that denatured ragweed antigen E could induce a specific T-helper response, even though the denaturation had destroyed the IgE-binding properties of this antigen. Similarly, Schirrmacher & Wigzell (16) showed that, while bovine serum albumin (BSA) denatured by methylation no longer bound BSA-specific antibodies, T cells from mice immunized with BSA could provide help when challenged with methylated BSA. Direct evidence that T cells could recognize sequential determinants came from experiments in which cleavage fragments of proteins such as staphylococcal nuclease, cytochrome c, and lysozyme were shown to stimulate T cells from animals immunized with the native protein (17–19). Finally, an important experiment by Chesnut et al (20) demonstrated that native and denatured forms of ovalbumin primed the same population of T-helper cells.

These results all suggested that protein antigens were denatured and perhaps enzymatically cleaved, prior to recognition by T cells. Studies by Rosenthal and colleagues (6, 7) had shown that the T-cell response to insulin was critically dependent upon an interaction between antigen and macrophages of the appropriate strain. It was suggested that this macro-phage-dependent step was in fact a requirement for antigen dena-turation/degradation ("antigen processing") by the macrophage. This was confirmed in a series of experiments (20, 22, 23), reviewed recently by Unanue (21), where lysosomotrophic agents such as chloroquine and ammonium chloride (which raise lysosomal pH and thus inhibit proteolytic activity), or fixation of antigen-presenting cells with glutaraldehyde or paraformaldehyde, were used to block antigen processing. Shimonkevitz et al (23) showed that ovalbumin-specific T-cell hybridomas were unable to respond to native or to denatured ovalbumin presented on glutaraldehyde-fixed macrophages but could respond to proteolytic digests or cyanogen bromide fragments of ovalbumin presented by these cells—a result sug-gesting that cleavage of this antigen was essential for T-cell recognition. Other experiments indicated that cleavage was not invariably necessary for the effective presentation of all protein antigens. Allen & Unanue (24) analyzed the antigen processing requirements of two T-cell hybridomas, both specific for hen egg lysozyme (HEL). The first clone behaved like the ovalbumin-specific clones described above : Fixed antigen presenting cells, or cells incubated with chloroquine, could only stimulate with enzymic digests of HEL, and not with native HEL or HEL denatured by reduction and carboxymethylation (CM-HEL). The second clone, however, could be stimulated by CM-HEL (but not by native HEL), presented either by fixed cells or in the presence of chloroquine. This suggested that unfolding of the molecule was sufficient to expose the epitope seen by this particular T-cell hybridoma. Similar results were obtained by Streicher et al (25), who showed that a T-cell clone specific for sperm whale myoglobin could respond to denatured but intact myoglobin presented by cells treated with a variety of lysosomal inhibitors. Streicher et al argued that the function of antigen processing was to expose hydrophobic residues hidden inside the native protein, so that they could interact with Ia on the antigen-presenting cell surface. They suggested that proteolytic cleavage was an effective, if rather unsubtle, method of unfolding proteins to expose hydro-phobic surfaces.

The objection has been raised that treatment with lysosomal inhibitors such as chloroquine, or fixation with aldehydes, may have important effects on cellular metabolism. Evidence suggests that lysosomal drugs can inhibit both the recycling of intracellular vesicles (26) and endosome fusion (27). The fact remains, however, that T cells can respond to antigen "processed" to some extent (either by denaturation or by cleavage) in circumstances where they are unable to respond to native antigen. Walden et al (28) have reported that proteins such as insulin do not necessarily need to be processed before they can be recognized by T cells. They found that native antigen incorporated into Ia-bearing lipid vesicles was able to stimulate antigen-specific T cells. They could not, however, rule out the possibility that the antigen had been denatured during incorporation into the vesicles. The clearest demonstration of a requirement for antigen processing comes from the experiments of Watts et al (29), who showed that an ovalbumin-specific hybridoma could respond to a peptide digest of ovalbumin but not to native ovalbumin when both were presented on planar membranes containing only lipid and purified Ia.

While T cells of helper/proliferative phenotype generally see processed antigen, not all T cell epitopes can be defined solely by primary amino acid sequence. Barcinski & Rosenthal (31) showed that the T cell epitope involving the α loop of the insulin A chain was destroyed if the disulfide bond holding this loop together was reduced. More recently, Naquet et al (30) have shown that insulin-specific T cell clones recognize an epitope consisting of the intact α loop of the A chain, disulfide bonded to a portion of the B chain. There was no significant stimulation of these T-cell clones by oxidized insulin A chain or B chain. Thus, the T-cell epitope(s) seen by these clones involved two separate disulfide-bonded peptides, one of which included an intrachain disulfide-bonded loop. This observation appears, however, to be the exception rather than the rule. Hen egg lysozyme, for instance, has four internal disulfide bonds, yet short peptides based on primary amino acid sequence have been able to stimulate T-cell pro-liferation in all cases examined so far.

STRATEGIES FOR IDENTIFYING T-CELL EPITOPES

T-cell epitopes on protein antigens have generally been identified in one of two ways. The first method works only for homologous series of proteins, such as lysozymes, insulins, myoglobins, and cytochromes, where the amino acid sequences from several members of the protein family are known. Residues important for T-cell activation can be identified by correlating the ability to stimulate a T-cell response with amino acid sequences. Alternatively, proteins can be cleaved either enzymatically or with reagents such as cyanogen bromide, and the fragments tested to see whether they stimulate a T-cell response. Epitope assignments made by either of these methods can be confirmed using synthetic peptide analogues of the sequence in question.

A number of antigenic sites were identified in this way on proteins such as insulin (31, 32), staphylococcal nuclease (17), cytochrome c (18, 33), myoglobin (34–36), and ovalbumin (37). Several important points could be made from these experiments. In the first place, the T-cell response was often exquisitely specific for the immunizing antigen. In some instances, homologous proteins or peptides differing from the immunizing protein at only a single position were unable to stimulate a T-cell response (18, 31). Second, it was found that a protein could have distinct antigenic sites which stimulated different functional T-cell subsets. An analysis of the response to hen egg lysozyme in a "nonresponder" strain demonstrated that T-suppressor cells were directed against a site involving the N- and C-terminal portions of the molecule, while T-helper cells were specific for determinants in the middle and C-terminal cyanogen bromide cleavage fragments (38, 39). In addition, MHC-disparate strains were found to respond preferentially to particular antigenic regions (31, 39–43).

Over the last few years, a number of important questions about the nature of the interaction between the T-cell receptor, antigen and the restricting MHC molecule have been answered using synthetic peptides to analyze the requirements for T-cell activation. The minimum peptide length necessary for T-cell stimulation has been determined using over-lapping sets of synthetic peptides, and the importance of particular residues has been examined using peptides substituted at various positions. These studies, together with analyses of the effect of peptide length on antigenic potency, have led to the suggestion that the ability of peptides to adopt appropriate secondary structure may be of crucial importance for anti-genicity. In addition, comparison of known epitopes has allowed the formulation of several hypotheses about what makes a particular sequence antigenic for T cells. The repeated observation that one immunodominant region can stimulate T-cell responses that are extremely heterogeneous in fine specificity emphasizes the importance of using T-cell clones, rather than uncloned T-cell lines, for these studies. Very recently, interactions between antigenic peptides and MHC molecules have been demonstrated directly using a number of biophysical techniques. These studies are dis-cussed in more detail in the remainder of this review.

SIZE OF T-CELL EPITOPES

T-cell epitopes on protein antigens or polypeptides seem in general to involve at least seven amino acids. The first evidence for this came from experiments with the synthetic polypeptide poly-L-lysine. Schlossman (46) analyzed the proliferative response of lymphocytes from guinea pigs immu-nized with poly-L-lysine and found that oligomers with seven contiguous L-lysine residues, such as [(L-lysine)$_7$,D-lysine,L-lysine (L$_7$DL)], were sti-mulatory, while oligomers of the same length and amino acid composition, but with fewer than seven adjacent L-lysines (e.g. L$_4$DL$_4$), were unable to stimulate proliferation. Further experiments in this system (47) demonstrated that T cells were responsible for the proliferative response and confirmed that peptides shorter than seven amino acids were unable to stimulate proliferation.

Similar results have since been obtained for T-cell epitopes on native proteins. The same general approach described below was used in all these experiments. Sets of overlapping peptides synthesized by serially removing one amino acid from either end were generated and tested for their ability to stimulate the T-cell clone or hybridoma in question. This strategy has been successfully applied to the fine specificity mapping of epitopes on a number of globular protein antigens (48–53). The consensus is that these T-cell epitopes contain approximately seven amino acids. It should be pointed out that since these peptides are synthesized from the C-terminus, it is very much easier to make sets of peptides where the C-terminus is constant and the N-terminus varied than it is to make peptides where the C-terminus is varied. For this reason, the N-terminal limits of these epitopes have frequently been more precisely defined than the C-termini.

Experiments in this laboratory highlight some of the complications involved in this kind of analysis. We have used sets of peptides shortened one residue at a time from both the N- and C-termini to map an epitope on sperm whale myoglobin to the seven amino acid sequence 112–118 (53). Clone 9.4 (made by A. J. Infante) responded well to peptide 110–121, progressively less well to peptides 111–121 and 112–121, and not at all to peptide 113–121. We therefore concluded that residue 112 was essential for stimulation. To test whether peptide 113–121 failed to stimulate simply because it was too short, this experiment was repeated with a longer set of peptides, where the C-terminus was kept constant at residue 124 instead of 121. The results were identical: Peptide 112–124 was stimulatory, while peptide 113–124 was not. Clone 9.4 was then assayed on a set of peptides where the N-terminus was kept constant at position 102. It responded strongly to peptide 102–120, less well to peptide 102–118, and not to peptide 102–117, indicating that residue 118 was essential for stimulation. The epitope seen by this clone could therefore be defined by the seven amino acid sequence 112–118. Further analysis, however, suggested that the definition of an epitope could be rather more complicated. While the experiments described above showed that residues outside the sequence 112–118 were not essential for stimulation, peptide 112–118 failed to stimulate any proliferation, even at high concentrations (50 μM). The slightly longer peptide 111–118 did stimulate at the highest concentration, and this suggested that a minimum peptide length, as well as the appropriate amino acid sequence, was required for stimulation. The epitope seen by clone 9.4 could thus be defined either as residues 112–118 (as determined with the overlapping sets of peptides) or as residues 111–118 (the shortest peptide to stimulate proliferation).

These experiments highlight one problem that continually arises in this kind of study: The designation of any peptide as stimulatory or non-stimulatory is an arbitrary one. In our experiments, any peptide that failed to stimulate proliferation at 50 μM was considered to be nonantigenic. However, a peptide that was nonantigenic at this concentration might perhaps stimulate, albeit weakly, at still higher concentrations. At what concentration does one stop? One example of this dilemma comes from experiments by Schwartz et al (50). The cytochrome c peptide 97–103 was found to be nonstimulatory at 10 μM for a particular T-cell clone, even though it spanned the two residues (99 and 103) known to be essential for stimulation. Theoretical considerations (discussed in the next section) suggested that this peptide should be able to adopt a conformation thought to be important for antigenicity. It was therefore tested on a second clone, apparently of the same specificity as the first, that responded to much lower doses of antigen. This second clone did respond to peptide 97–103 at the very high concentration of 100 μM; the epitope seen by this clone could thus be identified as the seven amino acid sequence 97–103. The definition of an epitope thus depends both on the antigen sensitivity of the T-cell clone in question and on arbitrarily imposed limits on the maximum concentration of peptide.

THE EFFECT OF PEPTIDE LENGTH ON ANTIGENIC POTENCY

The relationship between peptide length, secondary structure, and anti-genicity has been analyzed in some detail by Schwartz and colleagues, working with Ia-restricted T-cell clones and hybridomas specific for an immunodominant epitope in the C-terminal cyanogen bromide fragment of cytochrome c. Experiments with synthetic peptide analogues of this fragment had suggested that residues 99 and 103 of moth cytochrome c were essential for T-cell stimulation (54, 55). However, peptide 97–103, which included both these residues, was unable to stimulate proliferation. The shortest peptide that could stimulate a T-cell response corresponded to the moth sequence 94–103, and maximum stimulation was only achieved with the much longer peptide 88–103. These results suggested that residues 88–98 were somehow important for the antigenic potency of these peptides.

One possibility was that these residues stabilized a particular secondary structure. Pincus et al (56) hypothesized that the lysine residue at position 99 had to adopt a particular orientation and that this was readily achieved only when the peptide was in an α helical conformation. They further suggested that residues 88–98 might be important because they stabilized this conformation. They therefore used a computer program to determine the preferred minimum energy conformations of various regions of this cytochrome c fragment. Residues 99–103 alone showed no obvious preference for any particular secondary structure, but the addition of residues 94–98 resulted in an α helical minimum energy conformation in a nonpolar environment. Schwartz et al (50) tested this hypothesis further by measuring the α helical content of stimulatory and nonstimulatory peptides by circular dichroism. In aqueous solution, none of the peptides had any detectable α helical content; in a relatively nonpolar solvent, however, many of the peptides adopted an α helical conformation. Moreover, there was a quantitative correlation between the tendency to form helical structures in solution and the ability to stimulate a T-cell response. From these results, Schwartz et al concluded that the antigenic potency of these cytochrome c peptides depended upon their tendency to adopt an α helical conformation in a nonpolar environment.

Watts et al (48) have also speculated that the antigenic potency of peptides might depend upon their ability to adope an α helical structure. Building on the idea that the antigen-Ia binding site might involve hydrophobic interactions (25, 41), they showed that while the linear sequence of the ovalbumin peptide 323–339 did not have a particularly obvious hydrophobic region, a hydrophobic surface could be generated by modeling the peptide as an α helix. DeLisi & Berzofsky (58) have suggested that secondary structures such as α helices are important for antigenicity precisely because they form stable amphipathic structures (see below).

The effects of increasing length on peptide antigenicity have now been demonstrated in a number of different systems. The idea that the addition of residues outside the epitope increases antigenic potency by stabilizing appropriate secondary structure is clearly seductive. The results are, however, compatible with the alternative hypothesis that while these residues are not essential for stimulation, they do in fact contact the T-cell receptor and/or the restricting Ia molecule and contribute directly to the affinity of the interaction between nominal antigen, Ia, and the T-cell receptor. The resolution of this question will probably have to wait until the structure of this ternary complex is elucidated.

MODELS FOR PREDICTING T-CELL EPITOPES

Studies on the specificity of B-cell responses to protein antigens have suggested that antibodies are directed predominantly against sites with certain characteristics, such as hydrophilicity and a high degree of mobility. In other words, these sites tend to be in highly exposed areas such as loops and turns (57). It would obviously be useful to have similar guidelines for identifying sequences likely to act as T-cell epitopes. Two models have recently been proposed for the prediction of T-cell antigenic sites on proteins on the basis of primary sequence alone. Both have fulfilled the obvious criterion for a good model—the ability to make correct predictions. They differ completely, however, in the premises upon which these predictions are founded.

DeLisi & Berzofsky (58) have proposed that T-cell epitopes are likely to involve those sequences of a protein which can adopt stable amphipathic conformations. They suggest that such conformations might be induced or stabilized by hydrophobic interactions with structures such as Ia molecules on the antigen-presenting cell surface. Antigen processing experiments (24, 25) had shown that protein cleavage was not always necessary for effective presentation of antigen to T cells; that denaturation was sometimes sufficient. From this, Streicher and colleagues (25) argued that the function of antigen processing was to expose hydrophobic residues hidden inside the native protein, so that they could interact with Ia on the antigen presenting cell surface. They suggested that proteolytic cleavage was simply a rather drastic, but effective, method of unfolding proteins. DeLisi & Berzofsky used theoretical calculations to estimate the probability that a particular sequence would adopt an amphipathic conformation. They determined the periodicity of hydrophobic residues in the primary amino acid sequences of protein antigens and looked for segments with periodicity corresponding to ordered secondary structures such as α helices, or β-pleated sheets. They then asked whether these regions contained any of the known T-cell antigenic sites. At this time, 12 antigenic sites on 6 proteins (cytochrome c, insulin, lysozyme, myoglobin, ovalbumin, and influenza hemagglutinin) had been identified. Of these sites 10 fell within sequences which, on the basis of these calculations, could be modeled as amphipathic α helices.

This paper (58) made a prediction which, unknown to the authors, had already been verified in our laboratory. An analysis of the primary sequence of sperm whale myoglobin identified three blocks of sequence where the periodicity of hydrophobicity corresponded to that of an α helix. Two of these regions included known T-cell epitopes (51, 52). It was suggested that if there were another immunodominant site on this protein, then it might well fall within the third block of sequence (residues 69–81). We had in fact already identified an epitope within this region, contained within the 9-residue sequence 70–78 (53).

DeLisi & Berzofsky pointed out two constraints that applied to the interpretation of their model. In the first place, even though a protein might have several regions of amphipathic structure known to include T-cell epitopes, a single inbred strain should not be expected to respond to all these epitopes. Any one site might fail to stimulate a T-cell response in a given strain either because it could not associate with Ia, as proposed by the determinant selection model, or because of other constraints on the T-cell response. Secondly, while they argued strongly that T-cell epitopes were likely to have amphipathic properties, this did not mean that all sequences with amphipathic properties would be T-cell antigenic sites; it simply suggested that T-cell epitopes would most probably be located within such sequences.

Rothbard et al (59) have expressed reservations about this model, based principally on the fact that the number of epitopes analyzed in this study was small, while proteins with α helical structure in the native molecule were overrepresented. They adopted an empirical, rather than theoretical, approach to the identification of T-cell epitopes. They analyzed the primary amino acid sequences of the 28 T-cell antigenic sites known at the time, to see whether a pattern common to all these sequences could be found. Each residue was assigned to one of four categories: (a) hydrophobic; (b) charged; (c) polar; and (d) glycine and/or proline. Of these 28 antigenic sites, 27 were found to include one or more regions of sequence where a charged residue, or glycine, was followed by two hydrophobic residues. Since this pattern had been identified using a database of known T-cell epitopes, it clearly had to be tested for the ability to predict T-cell antigenic sites before it could be considered as characteristic of T-cell epitopes in general. Townsend et al (10) had demonstrated that short synthetic peptide analogues of influenza nucleoprotein could stimulate influenza-specific cytotoxic T cells. Rothbard and colleagues found that the cytotoxic response against influenza in one particular mouse strain was directed principally against the first 87 residues of this protein. This sequence was found to include three sites where a charged residue or glycine was followed by two hydrophobic residues. A peptide spanning one of these sites was synthesized and was shown to substitute in cytotoxic assays for the intact virus, the intact nucleoprotein, and for a deleted form of the nucleoprotein containing the first 87 residues (59).

The test for these models must surely lie in their ability to make correct predictions. Both models have correctly predicted a previously unknown T-cell antigenic site. One important difference between these models, however, is that while DeLisi & Berzofsky suggested an explanation for the importance of amphipathic structures, the three-residue template used by Rothbard et al was chosen on an empirical, rather than theoretical basis. The generality of these models can only be confirmed by the identification of new T-cell epitopes on a rather more diverse selection of proteins than have until now been studied.

SPECIFICITY OF THE INTERACTION BETWEEN Ia AND ANTIGEN

Although it is generally agreed that activation of the T cells studied in these systems requires the associative recognition of antigen and Ia by the T-cell receptor, not all the models for ternary complex formation require the physical association of antigen with the Ia molecule (reviewed by Schwartz: 2). Such an interaction is, however, central to the determinant selection hypothesis of immune responsiveness. In this section, we discuss experiments concerned with two questions: (a) Does nominal antigen bind to Ia on the antigen presenting cell surface; and (b) do such associations show specificity that can be correlated with Ir gene effects?

The first experiments to address these questions looked at specific competition of T-cell activation by molecules that could bind to the restricting Ia molecule but were not stimulatory for the T-cell population in question. Werdelin (60) looked at the T-cell response in guinea pigs to two synthetic polymers, DNP-poly-L-lysine (DNP-PLL) and poly-$(Glu^{20}Lys^{80})_n$ (GL). Both DNP-PLL and GL elicited T-cell-proliferative responses in strain-2 guinea pigs, but DNP-PLL-specific T cells were not stimulated by GL. This T-cell response could, however, be blocked if the antigen-presenting cells were pulsed with nonstimulatory GL before being exposed to DNP-PLL. This result was compatible with the idea that both antigens were binding to the same site on the Ia molecules. There was, moreover, some evidence that this interaction was of relatively high affinity. Antigen presenting cells could be pulsed with GL, then washed, yet after 2 hr the ability of these cells to present DNP-PLL was still inhibited. This competition was apparently specific: GL did not block T-cell responses to ovalbumin, while another synthetic polymer, poly-L-arginine, did not block the response to DNP-PLL.

Rock & Benacerraf (61) demonstrated that this type of antigen competition was specific and that it operated at the level of the antigen presenting cell. The synthetic copolymer GAT could be recognized in association either with I-A^d or with I-A^b by appropriately restricted T-cell-hybridoma clones. Another synthetic polymer, GT, blocked the response of I-A^d restricted clones to GAT, while the response of I-A^b-restricted clones to GAT was unaffected. Rock & Benacerraf showed that (b × d)F_1 antigen presenting cells cultured with GAT and GT were unable to stimulate I-A^d-restricted clones but could still stimulate I-A^b-restricted clones very effectively. They also showed that prepulsing the I-A^d-restricted T-cells with GT had no inhibitory effect on the ability to respond to GAT. These experiments argued against toxicity and nonspecific blocking mechanisms and supported the hypothesis that GAT and GT bound specifically to the same site on the I-A^d molecule.

The experiments of Werdelin and of Rock & Benacerraf were certainly compatible with a physical association between the synthetic polymers and Ia, but they did not demonstrate this directly. A number of groups have now used biophysical or biochemical techniques to look at interactions between antigen and Ia. The biophysical aspects of such interactions have been discussed in detail in the review by Watts & McConnell (62) in this volume, but we shall outline these experiments briefly again, in order to discuss the implications of the results for the specificity of T-cell activation.

Equilibrium dialysis has been used by two groups to show binding of antigenic peptides to the appropriate Ia molecules. Babbit et al (63) demonstrated that a fluorescently labeled hen egg lysozyme peptide HEL(46–61), which spans an immunodominant epitope on HEL seen by I-A^k-restricted T-cell hybridomas (41), was bound by detergent-solubilized I-A^k. The binding was weak but specific; the peptide did not bind to I-A^d. Similar experiments by Buus et al (64) showed that the chicken ovalbumin peptide OVA(323–339), which defines an ovalbumin epitope seen in association with I-A^d (37) bound weakly to purified I-A^d, but not to I-A^k or I-E^k.

In a subsequent paper, Babbit et al (65) showed that peptides which could competitively inhibit the response of a T-cell hybridoma to the HEL peptide 46–61 could also inhibit the binding of this peptide to purified Ia in equilibrium dialysis experiments. Allen et al (49) had previously analyzed the response of a T-cell hybridoma specific for HEL(52–61) to a series of substituted or truncated peptides spanning various parts of the HEL sequence 46–61; they had identified a series of these peptides that failed to stimulate proliferation. Babbit et al showed that some of these nonstimulatory peptides could block the cell response to HEL(46–61) and that these same nonstimulatory peptides could inhibit the binding of fluoresceinated HEL(46–61) to detergent-solubilized I-A^k in equilibrium dialysis experiments. The ability to block the T-cell response to HEL(46–61) correlated with the ability to inhibit the binding of fluoresceinated HEL(46–61) to I-A^k. This meant that the binding of HEL(46–61) to I-A^k in equilibrium dialysis experiments reflected a physiological interaction necessary for T-cell stimulation. It was particularly interesting that peptide Gln^{49} Phe^{56} 49–61, the sequence of which was identical with that of mouse lysozyme, inhibited the binding of HEL(46–61) to I-A^k in solution very effectively. Peptides that failed to block the T-cell response to HEL(46–61) also failed to inhibit the binding of HEL(46–61) to I-A^k. Fibrinopeptide B, which can stimulate a T-cell proliferative response restricted by I-A^k (66), was also unable to inhibit this interaction, a result that could suggest two discreet binding sites for nominal antigen on the I-A^k molecule. Alternatively, HEL(46–61) and fibrinopeptide B might bind to the same site on I-A^k, but the interaction between fibrinopeptide B and I-A^k could be weak relative to that between HEL(46–61); if so, it should be possible to block the response of a fibrinopeptide B-specific, I-A^k-restricted T-cell clone with HEL(46–61). The evidence that Ia molecules have multiple binding sites for antigen is discussed in more detail in the next section.

Phillips et al (70) used another approach to demonstrate an interaction between antigen and Ia. They incubated radioiodinated, photoreactive beef insulin with antigen-presenting cells. After varying periods of incubation, the cells were exposed to light, which triggered the photoreactive group, allowing it to bind covalently to adjacent molecules. When the cell membranes were solubilized and run out on gels, the radiolabel was found to be associated with two bands running at the appropriate size for Ia α and β chains. Moreover, the polypeptides running in these positions could be immunoprecipitated with haplotype-specific Ia alloantisera. These experiments showed that the photolabeled insulin bound specifically to Ia and not to other molecules on the antigen-presenting cell surface. However, this binding was not haplotype specific. The antigen bound not only to I-A^b and I-A^d, which are known to present insulin to T-cell hybridomas, but also to Ia molecules on cells from a panel of nonresponder haplotypes. In addition, it was pointed out that while photolabeled insulin coupled to cellular components could stimulate appropriate T-cell hybridomas, there was no evidence that the antigen responsible for this T-cell activation was still in the form seen in the gels. It was possible that these T cells recognized the antigen only after it had been cleaved from the Ia molecule.

Using Ia-containing planar membranes, Watts et al (71) were able to observe energy transfer between fluorescein-labeled OVA(323–339) and Texas red-labeled I-A^d, indicating that the distance between donor and acceptor fluorophores was 40 Å or less. In contrast to the equilibrium dialysis experiments (63, 64), the peptide-Ia association responsible for this energy transfer was originally observed only in the presence of the appropriate T cell, specific for OVA(323–339) plus I-A^d. Further experiments by Watts & McConnell (72), however, showed that energy transfer between OVA(323–339) and I-A^d could be seen in the absence of T cells, if Ia and peptide were allowed to interact in serum-free conditions. First, they incubated either intact antigen presenting cells or a lysate of these cells with OVA(323–339) for 3 hr at 37°C and showed that membrane preparations from the lysed cells could stimulate T-cell activation almost as effectively as membranes prepared from the intact cells, even after exhaustive dialysis. This meant that the ability of pulsed cells to stimulate in these experiments could not be explained by the continual release of antigen from intracellular pools but must instead be the result of a strong association between antigen and cell membrane components. They then pulsed phospholipid vesicles containing only lipid and I-A^d with fluoresceinated OVA(323–339), dialysed them, and showed that they too could stimulate the appropriate T-cell hybridoma. Stimulation was reduced if a proteolytic digest of hen egg lysozyme (pHEL) was added to OVA(323–339) at the time of the antigen pulse, but there was no inhibition if pHEL was added after the pulse, during the incubation of pulsed vesicles with the T cells. In addition, pHEL could not block the T-cell response to unpulsed vesicles in the presence of exogenous OVA(323–339). The association between OVA(323–339) and I-A^d was shown to be specific; this peptide did not bind to vesicles containing glycophorin instead of Ia.

These results suggested that the association between peptide and Ia was susceptible to competition by other protein fragments but was extremely stable once formed. They also demonstrated that this association could be stabilized by antigen-specific T cells, so that it was no longer vulnerable to competition. The energy transfer experiments were therefore repeated in serum-free conditions. Fluoresceinated OVA(323–339) was added in buffer with 1% fetal calf serum, as in the previously reported experiments (71), or in buffer without added serum, to planar membranes containing Texas-red-labeled I-A^d. In the absence of serum, an increase in Texas-red fluorescence was observed. The presence of 1% fetal calf serum or of unlabeled peptide reduced the fluorescence to background levels. While the initial association between peptide and Ia required quite high concentrations of peptide, the actual concentration of antigen required to stimulate T cells appeared to be extremely low. It was estimated that one T-cell could be activated by about 10^3 peptide molecules bound to Ia.

These results may explain why it has been so difficult to demonstrate inhibition of Ia-restricted T-cell responses to antigenic peptides by competition with nonstimulatory peptides known to bind to the Ia molecule in question (2). Watts & McConnell demonstrated, both in energy transfer and in antigen pulsing experiments, that T cells could stabilize antigen-Ia interactions in the presence of competing peptides or serum (72). In the experiments by Babbit et al (65), where T-cell responses were successfully blocked by a series of nonstimulatory peptides, the competing peptides were incubated with Ia for 30 min before the antigenic peptide and T cells were added.

The experiments discussed here show convincingly that antigen can bind specifically to Ia and not to other cell surface molecules and that this interaction between antigen and Ia reflects the associative recognition of these molecules by antigen-specific T cells. It is quite possible, however, that Ia-antigen binding sufficiently strong to be demonstrated in the absence of T cells will be the exception rather than the rule. The HEL and OVA peptides used in these studies both span "immunodominant" epitopes. These epitopes are arguably immunodominant precisely because they form such a strong association with Ia. Schwartz (2) has suggested that the experiments of Werdelin (60) and Rock & Benacerraf (61) also belong in a special category; that the highly charged DNP-PLL, GAT, and GT peptides might bind nonspecifically to the cell surface and then bind specifically, but with relatively low binding affinities, to Ia.

It is too early to say whether these kinds of experiment support the determinant selection theory of immune responsiveness. They certainly confirm a central assumption of this theory, that T cells recognize a physical association between antigen and Ia. It seems reasonable to extrapolate from this that if an antigenic fragment is unable to associate with Ia, then it will not trigger an immune response. However, the converse is not necessarily true; peptides from protein antigens may associate with Ia, yet fail to stimulate an immune response because this would break self tolerance. There is clearly at least some specificity at the level of peptide-Ia binding. The hen egg lysozyme peptide HEL(46–61) bound to I-A^k but not to I-A^d (63), while the ovalbumin peptide OVA(323–339) bound to I-A^d but not to I-A^k or I-E^k (64). In both cases, these results reflected the specificity of the observed immunodominant T-cell response in the appropriate strains. The generality of these observations, however, has still to be tested. This should not be too difficult, since a reasonable number of epitopes seen preferentially in association with particular Ia molecules has now been defined. Whether there are multiple, distinct binding sites for antigen on any one Ia molecule has yet to be determined. Work from our laboratory has clearly demonstrated that there exist more than one "restriction site" on Ia molecules (67, 68); the possible location of such sites has been reviewed elsewhere (69). These results are not, however, incompatible with the idea that antigenic peptides bind to one particular region of the Ia molecule.

IMMUNODOMINANT REGIONS ON PROTEIN ANTIGENS STIMULATE HETEROGENEOUS T-CELL RESPONSES

A number of investigators have shown that the T-cell response to certain well-defined epitopes can be quite heterogeneous in fine specificity. Allen et al (49) showed that two I-A^k-restricted T-cell hybridomas, which both responded to a peptide spanning residues 52–61 of hen egg lysozyme, differed strikingly in their response to a panel of longer peptides. The response of clone 3A9 increased as the peptides became longer; clone

2A11, in contrast, responded strongly to the shortest stimulatory peptide (52–61), but progressively less well to the longer peptides. Another antigenic region of hen egg lysozyme has also been found to stimulate a heterogeneous T-cell response (44, 73). T-cell clones derived from H-2^b mice immunized with the middle cyanogen bromide fragment (residues 13–105) of hen egg lysozyme were all specific for epitopes within the tryptic peptide T11 (residues 74–96), but the T-cell clones showed three different patterns of reactivity to a panel of gallinaceous lysozymes. The clones were tested on shorter peptides and on peptides spanning residues 74–96, substituted to mimic the species variant lysozymes. All the clones responded to peptide 81–96 but differed in their responses to the substituted peptides. The differences in specificity could be attributed to sequence differences within the T11 region itself. To see whether this result was an isolated instance, B10.A mice were immunized with the HEL T11 peptide. The clones fell into two distinct groups. All the I-A^k-restricted clones responded to peptide 74–86, but not to peptide 81–96, while all the I-E^k-restricted clones responded to peptide 85–96. Again, clones within these two groups differed in fine specificity, as determined by their response to the panel of substituted peptides. Thus, within the immunodominant region 74–96, there are three distinct antigenic regions, spanning residues 74–86, 85–96, and 81–96, seen by T cells in association with I-A^k, I-E^k, and I-A^b, respectively, each one of which can stimulate a heterogeneous T-cell response.

Our analysis of the T-cell response to sperm whale myoglobin in DBA/2 mice has revealed similar levels of complexity. We have focused on the response to two overlapping antigenic sites in the middle cyanogen bromide fragment (residues 56–131). The first site involves residues 111–118, and was identified by use of overlapping sets of peptides spanning residues 100–121 (53). The same set of peptides were used to characterize clones specific for the second site, an epitope first described by Berkower et al (35) centering on residue 109.

An analysis of clones recognizing the first antigenic site (residues 111–118) showed that sequences approaching the minimum size for T-cell activation (7–8 amino acids) could be recognized by T cells of at least three different specificities. We isolated a large number of clones reactive with residues 112–118, presented in association with I-E^d. These clones had indistinguishable patterns of response to the panel of peptides spanning residues 100–121. Most of them also responded equally well to sperm whale and horse myoglobins, which have lysine and arginine residues respectively at position 118. One clone, however, responded very poorly to horse myoglobin compared with sperm whale myoglobin, and this suggested that the 118-lysine to 118-arginine substitution had a deleterious effect for the stimulation of this particular clone. In addition, this clone showed a pattern of cross-reaction on allogeneic stimulators, absent from all the other 112–118 reactive clones. A third clone, also I-E^d-restricted, recognized an epitope defined by residues 111–117. Thus the 8-residue sequence 111–118 was shown to include at least three distinct T-cell epitopes, all seen in association with I-E^d.

All the clones specific for the "109" epitope were I-A^d restricted, as observed for another H-2^d mouse strain by Berkower et al (43). We could, however, distinguish at least three reactivity patterns. One set of clones responded well to all the available peptides spanning residue 109; the longest peptides were the most stimulatory. These clones were stimulated by peptides 108–118 and 102–117, indicating that the epitope lay within residues 108–117. The second group of clones also recognized an epitope within residues 108–117 but showed a completely different response to the synthetic peptides. They responded very strongly to peptide 102–118, but there was little or no response to the slightly longer peptides 102–120 and 100–121. A clone representing a third specificity was unable to respond either to peptide 102–117 or 108–118; it did, however, respond to peptides 106–121 and 102–118, indicating that the epitope seen by this clone lay somewhere within the 13-residue sequence 106–118. Similar results, using "109"-specific T-cell clones from B10.D2 mice, have been reported by Cease et al (52).

Three increasingly parsimonious models can be proposed to account for these results: (a) an antigenic peptide might be able to bind to multiple sites on any one Ia molecule; (b) the peptide might bind to only one site on Ia, but would be able to adopt multiple conformations; and (c) the peptide might bind in one conformation to one site on Ia. This third model is the one favored by Cease et al (52). They argue persuasively that their data, and all the other reports of heterogeneous T-cell responses to immunodominant antigenic sites, can be satisfactorily explained by a model where T-cells of multiple specificities are activated by antigen bound in a single conformation to one site on the Ia molecule. If this model is correct, then heterogeneity in the system must be caused entirely by the T-cell repertoire. It should be possible to resolve this question by antigenic competition studies using peptides that distinguish between clones of differing fine specificities.

WHY ARE THERE DIFFERENCES BETWEEN B-CELL AND T-CELL EPITOPES?

As discussed earlier in this review, the first indication that B cells and T cells might be recognizing different epitopes on the same antigen came from the experiments of Gell & Benacerraf (12). The recognition that T cells immunized either with native or with denatured antigen could recognize the alternative form of the antigen, whereas B cells recognized only the form of antigen used for immunization together suggested a fundamental difference between B-cell and T-cell epitopes. The original hypothesis of Gell & Benacerraf was that antigenic structures seen by antibodies depended on tertiary configuration of the immunizing protein, whereas T cells (recognized by the authors as the DTH response) might recognize epitopes defined by primary amino acid sequence. Many other groups of investigators mentioned earlier in this review presented additional evidence that there was a fundamental difference between B- and T-cell epitopes. In addition to these experiments, Senyck et al (74) originally suggested that T- and B-cell epitopes occupied nonoverlapping areas on the immunizing molecule. These studies of the guinea pig responses to glucagon showed that the antibody response was directed primarily against the N-terminal portion of the molecule (residues 1–17), while all detectable T-cell reactivity appeared to be specific for the C-terminal section (residues 18–29). More recent experiments in the lysozyme model suggested that humoral and cellular responses were similarly directed against distinct portions of the molecule. Antibodies bound to an epitope involving the N-terminal residues, while the proliferative T-cell response was focused on the middle and C-terminal cyanogen bromide fragments (38, 39). In fact, the original idea of hapten and carrier determinants suggested that T cells and B cells recognized different epitopes. Haptens were defined as antigenic and were recognized by B cells, but they were not immunogenic. Carriers were portions of molecules required to allow the recognition and production of antihapten antibodies. These results suggest the possibility of a unifying theme for immune response, which was previously suggested by Abbas et al (75). This hypothesis suggests that B cells serve to present antigen to T cells, and this has been clearly documented (76–86). In addition, such antigen presentation should be more efficient if the antigen is recognized by the B cell, and this has also been demonstrated (75, 87, 88). The hypothesis we wish to examine may explain the fundamental difference between epitopes recognized by T cells and B cells. First, antigen-binding B cells recognize antigenic epitopes on the basis of tertiary configuration of the antigen via surface immunoglobulin. The antigen is then internalized and processed before presentation to T cells. The requirement for processing was demonstrated by metabolically inactivating B cells by fixation (75). Such hapten-binding B cells failed to present haptenated antigen if it was added subsequent to fixation or if such hapten-binding cells were fixed after a short (less than 4-hr) pulse with antigen. However, if sufficient time was allowed for antigen recognition and processing (18 hr), such cells could subsequently be fixed and could function to present antigen to the antigen-specific T cells or hybridomas. Once internalized and processed, the tertiary configuration of the protein is disrupted and the carrier determinants, T-cell epitopes as defined in this review, can then be expressed on the surface of the B cell. They form the portion of the ternary complex required for T-cell activation in association with endogenous MHC class-II products present on the B-cell surface. An unanswered question is where the association of antigen and MHC class-II products occurs. Clearly, the results of Watts & McConnell presented earlier in this review demonstrate that it is possible to get appropriate presentation of T-cell epitopes by isolated I-region molecules and to pulse fixed antigen-presenting cells with appropriate peptides for presentation. That such associations between processed antigen and I-region molecules can be formed with surface MHC class-II products does not necessarily mean that this is the usual route. It is more likely that I-region association with T-cell epitopes occurs intracellularly. This could, in fact, occur in the endocytic vesicle, which has been demonstrated to contain not only surface immunoglobulin, but also MHC class-II products from activated B cells (89, 90). This would assume that the MHC class-II product in some way escapes proteolytic degradation, whereas the nominal antigen undergoes appropriate cleavage and/or unfolding, a difficult scenario to accept. Alternatively, freshly synthesized MHC class-II products might join such endocytic vesicles following antigen degradation processes, and the association might be made prior to transport to the cell surface. In any event, this model clearly allows an antigen-specific B cell to present multiple T-cell epitopes to the immune system and thus to enhance its ability to be triggered in a specific manner. Such a model fulfills the known requirements for hapten carrier interaction, T-cell/B-cell MHC restrictions, and hapten-specific carrier-induced antibody responses. Further, it supports the notion that B-cell epitopes should, in most instances, be distinct from T-cell epitopes for the system to function efficiently.

SUMMARY

We have reviewed here studies using synthetic peptides to analyze some of the properties of T-cell epitopes. Several general conclusions can be drawn. First, T-cell epitopes can usually be defined by linear sequences of about seven amino acids. However, the observation that increasing peptide length often results in increased antigenic potency has suggested that antigenicity may crucially depend upon the ability of peptides to adopt appropriate secondary structures. Two models for the prediction of T-cell epitopes on the basis of primary sequence data alone were discussed.

Biophysical studies on the association of peptides with Ia molecules have shown that antigenic peptides bind directly to Ia; the evidence suggests that a binary association between Ia and peptide occurs in the absence of specific T-cells. Finally, a hypothesis to explain the observation that B-cells and T-cells generally recognize distinct epitopes on multideterminant antigens has been examined.

ACKNOWLEDGMENTS

The authors would like to thank Robyn Kizer for her excellent secretarial assistance. This manuscript was supported by National Institutes of Health grants AI 18716, AI 18705, and AI 19512.

Literature Cited

1. Fathman, C. G., Frelinger, J. G. 1983. T-lymphocyte clones. *Ann. Rev. Immunol.* 1: 633
2. Schwartz, R. H. 1985. T lymphocyte recognition of antigen in association with products of the major histocompatibility complex. *Ann. Rev. Immunol.* 3: 237
3. Kronenberg, M., Siu, G., Hood, L. E., Shastri, N. 1986. The molecular genetics of the T-cell antigen receptor and T-cell antigen recognition. *Ann. Rev. Immunol.* 4: 529
4. Fink, P. J., Matis, L. A., McElliot, D. L., Bookman, M., Hedrick, S. M. 1986. Correlations between T-cell specificity and the structure of the antigen receptor. *Nature* 321: 219
5. Robertson, M. 1985. T-cell receptor. The present state of recognition. *Nature* 317: 768
6. Rosenthal, A. S., Barcinski, M. A., Blake, J. T. 1977. Determinant selection is a macrophage-dependent immune response gene function. *Nature* 267: 156
7. Rosenthal, A. S. 1978. Determinant selection and macrophage function in genetic control of the immune response. *Immunol. Rev.* 40: 136
8. Benacerraf, B. 1978. A hypothesis to relate the specificity of T lymphocytes and the activity of I region-specific Ir genes in macrophages. *J. Immunol.* 120: 1809
9. Sherman, L. A., Vitiello, A., Klinman, N. R. 1983. T cell and B cell responses to viral antigens at the clonal level. *Ann. Rev. Immunol.* 1: 63
10. Townsend, A. R. M., Rothbard, J. B.,

Gotch, F. M., Bahadur, G., Wraith, D., McMichael, A. J. 1986. The epitopes of influenza nucleoprotein recognized by cytotoxic T lymphocytes can be defined with short synthetic peptides. *Cell* 44: 959
11. Germain, R. N. 1986. The ins and outs of antigen processing and presentation. *Nature* 322: 687
12. Gell, P. G. H., Benacerraf, B. 1959. Studies on hypersensitivity. II. Delayed hypersensitivity to denatured proteins in guinea pigs. *Immunology* 2: 64
13. Sela, M. 1969. Antigenicity: Some molecular aspects. *Science* 166: 1365
14. Thompson, K., Harris, M., Benjamini, E., Mitchell, G., Noble, M. 1972. Cellular and humoral immunity: A distinction in antigenic recognition. *Nature New Biol.* 238: 20
15. Ishizaka, K., Kishimoto, T., Delepasse, G., King, T. P. 1974. Immunogenic properties of modified antigen E. I. Response of specific determinants for T cells in denatured antigen and polypeptide chains. *J. Immunol.* 113: 70
16. Schirrmacher, V., Wigzell, H. 1974. Immune responses against native and chemically modified albumins in mice. II. Effect of alteration of electric charge and conformation on the humoral antibody response and helper T cell responses. *J. Immunol.* 113: 1635
17. Berzofsky, J. A., Schechter, A. N., Shearer, G. M., Sachs, D. H. 1977. Genetic control of the immune response to staphylococcal nuclease. III. Time-course and correlation between the response to native nuclease and the

498   LIVINGSTONE & FATHMAN

response to its polypeptide fragments. *J. Exp. Med.* 145 : 111

18. Corradin, G., Chiller, J. M. 1979. Lymphocyte specificity to protein antigens. II. Fine specificity of T-cell activation with cytochrome *c* and derived peptides as antigenic probes. *J. Exp. Med.* 149 : 436

19. Maizels, R. A., Clarke, J. A., Harvey, M. A., Miller, A., Sercarz, E. E. 1980. Epitope specificity of the T cell proliferative response to lysozyme : Proliferative T cells react predominantly to different determinants from those recognized by B cells. *Eur. J. Immunol.* 10 : 509

20. Chesnut, K. S., Colon, S., Grey, H. 1982. Requirements for the processing of antigen by antigen-presenting B cells. I. Functional comparison of B cell tumors and macrophages. *J. Immunol.* 129 : 2382

21. Unanue, E. R. 1984. Antigen-presenting function of the macrophage. *Ann. Rev. Immunol.* 2 : 395

22. Ziegler, K., Unanue, E. 1982. Decrease in macrophage antigen catabolism caused by ammonia and chloroquine is associated with inhibition of antigen presentation to T cells. *Proc. Natl. Acad. Sci. USA* 79 : 175

23. Shimonkevitz, R., Kappler, J., Marrack, P., Grey, H. 1983. Antigen recognition by H-2-restricted T cells. I. Cell free antigen processing. *J. Exp. Med.* 158 : 303

24. Allen, P. M., Unanue, E. R. 1984. Differential requirements for antigen processing by macrophages for lysozyme-specific T cell hybridomas. *J. Immunol.* 132 : 1077

25. Streicher, H. Z., Berkower, I. J., Busch, M., Gurd, F. R. N., Berzofsky, J. A. 1984. Antigen conformation determines processing requirements for T-cell activation. *Proc. Natl. Acad. Sci. USA* 81 : 6831

26. Tieze, C., Schlesinger, P., Stahl, P. 1980. Chloroquine and ammonia ion inhibit receptor-mediated endocytosis of mannose-glycoconjugates by macrophages : Apparent inhibition of receptor cycling. *Biochem. Biophys. Res. Commun.* 93 : 1

27. Gordon, A. H., D'Arcy Hart, P., Young, M. R. 1980. Ammonia inhibits phagosome-lysosome fusion in macrophages. *Nature* 286 : 79

28. Walden, P., Nagy, Z. A., Klein, J. 1985. Induction of regulatory T-lymphocyte responses by liposomes carrying major histocompatibility complex molecules and foreign antigen. *Nature* 315 : 327

29. Watts, T. H., Brian, A. A., Kappler, J. W., Marrack, P., McConnell, H. M. 1984. Antigen presentation by supported planar membranes containing affinity purified I-A^d. *Proc. Natl. Acad. Sci. USA* 81 : 7564

30. Naquet, P., Phillips, M. L., Ellis, J., Hodges, R., Singh, B., Delovitch, T. L. 1986. Structural requirements for T cell antigenic sites on insulin. In *Immunogenicity of Protein Antigens : Repertoire and Regulation,* ed. E. Sercarz and J. Berzofsky, pp. CRC Press. In press

31. Barcinski, M. A., Rosenthal, A. S. 1977. Immune response gene control of determinant selection. I. Intramolecular mapping of the immunogenic sites on determinant recognized by guinea pig T and B cells. *J. Exp. Med.* 145 : 726

32. Thomas, D. W., Danho, W., Bullesbach, E., Fohles, J., Rosenthal, A. S. 1981. Immune response gene control of determinant selection. III. Polypeptide fragments of insulin are differentially recognized by T but not by B cells in insulin-immune guinea pigs. *J. Immunol.* 126 : 1095

33. Solinger, A. M., Ultee, M. E., Margoliash, E., Schwartz, R. H. 1979. T lymphocyte response to cytochrome *c*. I. Demonstration of a T cell heteroclitic proliferative response and identification of a topographic antigenic determinant on pigeon cytochrome *c* whose immune recognition requires two complementary major histocompatibility-linked genes. *J. Exp. Med.* 150 : 830

34. Infante, A. J., Atassi, M. Z., Fathman, C. G. 1981. T cell clones reactive with sperm whale myoglobin. Isolation of clones with specificity for individual determinants on myoglobin. *J. Exp. Med.* 154 : 1342

35. Berkower, I., Buckenmeyer, G., Gurd, F. R. N., Berzofsky, J. A. 1982. A possible immunodominant epitope recognized by murine T lymphocytes immune to different myoglobins. *Proc. Natl. Acad. Sci. USA* 79 : 4723

36. Berkower, I., Matis, L. A., Buckenmeyer, G. K., Gurd, F. R. N., Longo, D. L., Berzofsky, J. A. 1984. Identification of distinct predominant epitopes recognized by myoglobin-specific T cells under the control of different *Ir* genes and characterization of representative T cell clones. *J. Immunol.* 132 : 1370

37. Shimonkevitz, R., Colon, S., Kappler, J. W., Marrack, P., Grey, H. M. 1984. Antigen recognition by H-2 restricted T cells. II. A tryptic ovalbumin peptide that substitutes for processed antigen. *J. Immunol.* 133 : 2067

38. Sercarz, E. E., Yowell, R. L., Turkin, D., Miller, A., Araneo, B. A., Adorini, L. 1978. Differential functional specifi-

city repertoires for suppressor and helper T cells. *Immunol. Rev.* 39 : 108

39. Adorini, L., Harvey, M. A., Miller, A., Sercarz, E. E. 1979. Fine specificity of regulatory T cells. II. Suppressor and helper T cells are induced by different regions of hen egg lysozyme in a genetically non-responder mouse strain. *J. Exp. Med.* 150 : 293

40. Katz, M. E., Maizels, R. M., Wicker, L., Miller, A., Sercarz, E. E. 1982. Immunological focusing by the mouse major histocompatibility complex : Mouse strains confronted with distantly related lysozymes confine their attention to very few epitopes. *Eur. J. Immunol.* 12 : 535

41. Allen, P. M., Strydom, D. J., Unanue, E. R. 1984. Processing of lysozyme by macrophages ; Identification of the determinant recognized by two T cell hybridomas. *Proc. Natl. Acad. Sci. USA* 81 : 2489

42. Allen, P. M., McKean, D. J., Beck, B. N., Sheffield, J., Glimcher, L. H. 1985. Direct evidence that a class II molecule and a simple globular protein generate multiple determinants. *J. Exp. Med.* 162 : 1264

43. Berkower, I., Kawamura, H., Matis, L. A., Berzofsky, J. A. 1985. T cell clones to two major T cell epitopes of myoglobin : Effect of I-A/I-E restriction on epitope dominance. *J. Immunol.* 135 : 2628

44. Manca, F., Clarke, J. A., Miller, A., Sercarz, E. E., Shastri, N. 1984. A limited region within hen egg-white lysozyme serves as the focus for a diversity of T cell clones. *J. Immunol.* 133 : 2075

45. Heber-Katz, E., Hansburg, D., Schwartz, R. H. 1983. The Ia molecule of the antigen presenting cell plays a critical role in immune response gene regulation of T cell activation. *J. Mol. Cell. Immunol.* 1 : 3

46. Schlossman, S. 1972. Antigen recognition : The specificity of T cells involved in the cellular immune response. *Immunol. Rev.* 10 : 97

47. Stashenko, P. P., Schlossman, S. 1977. Antigen recognition : The specificity of an isolated T lymphocyte population. *J. Immunol.* 118 : 544

48. Watts, T. H., Gariepy, J., Schoolnik, G., McConnell, H. M. 1985. T cell activation by peptide antigen : Effect of peptide sequence and method of antigen presentation. *Proc. Natl. Acad. Sci. USA* 82 : 5480

49. Allen, P. M., Matsueda, G. R., Haber, E., Unanue, E. R. 1985. Specificity of the T receptor : Two different determinants are generated by the same peptide and the I-A^k molecules. *J. Immunol.* 135 : 368

50. Schwartz, R. H., Fox, B. S., Fraga, E., Chen, C., Singh, B. 1985. The T lymphocyte response to cytochrome *c*. V. Determination of the minimal peptide size required for stimulation of T cell clones and assessment of the contribution of each residue beyond this size to antigenic potency. *J. Immunol.* 135 : 2598

51. Berkower, I., Buckenmeyer, G. K., Berzofsky, J. A. 1986. Molecular mapping of a histocompatibility-restricted immunodominant T cell epitope with synthetic and natural peptides : Implications for T cell antigenic structure. *J. Immunol.* 136 : 2498

52. Cease, K. B., Berkower, I., York-Jolley, J., Berzofsky, J. A. 1986. T cell clones specific for an amphipathic alpha helical region of sperm whale myoglobin show differing fine specificities for synthetic peptides : A multiview/single structure interpretation of immunodominance. *J. Exp. Med.* 164 : 1779

53. Livingstone, A. M., Levy, H., Rothbard, J. B., Fathman, C. G. 1986. Fine specificity of two epitopes on sperm whale myoglobin. Manuscript in preparation

54. Hansburg, D., Fairwell, T., Schwartz, R. H., Appella, E. 1983. The T lymphocyte response to cytochrome *c*. IV. Distinguishable sites on a peptide antigen which affect antigenic strength and memory. *J. Immunol.* 131 : 319

55. Hansburg, D., Heber-Katz, E., Fairwell, T., Appella, E. 1983. Major histocompatibility complex-controlled antigen presenting cell expressed-specificity of T cell antigen recognition. *J. Exp. Med.* 158 : 25

56. Pincus, M. R., Gerewitz, F., Schwartz, R. H., Scheraga, H. A. 1983. Correlation between the conformation of cytochrome *c* peptides and their stimulatory activity in a T lymphocyte proliferation assay. *Proc. Natl. Acad. Sci. USA* 80 : 3297

57. Tanier, J. A., Getzoff, E. D., Paterson, Y., Olson, A. J., Lerner, R. A. 1985. The atomic mobility component of protein antigenicity. *Ann. Rev. Immunol.* 3 : 501

58. DeLisi, C., Berzofsky, J. A. 1985. T-cell antigenic sites tend to be amphipathic structures. *Proc. Natl. Acad. Sci. USA* 82 : 7048

59. Rothbard, J. B., Townsend, A., Edwards, M., Taylor, W. 1986. Pattern recognition among T cell epitopes. *Modern Trends in Human Leukemia VII.* In press

60. Werdelin, O. 1982. Chemically related antigens compete for presentation by

accessory cells to T cell. *J. Immunol.* 129 : 1883

61. Rock, K. L., Benacerraf, B. 1983. Inhibition of antigen-specific T lymphocyte activation by structurally related Ir gene-controlled polymers. Evidence of specific competition for accessory cell antigen presentation. *J. Exp. Med.* 157 : 1618

62. Watts, T. H., McConnell, H. M. 1987. Biophysical aspects of antigen recognition by T cells. *Ann. Rev. Immunol.* In press

63. Babbit, B. P., Allen, P. M., Matsueda, G., Haber, E., Unanue, E. R. 1985. Binding of immunogenic peptides to Ia histocompatibility molecules. *Nature* 317 : 359

64. Buus, S., Colon, S., Smith, C., Freed, J. H., Miles, C., Grey, H. M. 1986. Interaction between a processed ovalbumin peptide and Ia molecules. *Proc. Natl. Acad. Sci. USA* 83 : 4509

65. Babbit, B. P., Matsueda, G., Haber, E., Unanue, E. R., Allen, P. M. 1986. Antigen competition at the level of peptide-Ia binding. *Proc. Natl. Acad. Sci. USA* 83 : 4509

66. Peterson, L. B., Wilner, G. D., Thomas, D. W. 1983. Functional differentiation in the genetic control of murine T lymphocyte responses to human fibrinopeptide B. *J. Immunol.* 130 : 637

67. Beck, B. N., Nelson, P. A., Fathman, C. G. 1983. The I-A^b mutant B6.C-H-2^{bm12} allows definition of multiple T cell epitopes on I-A molecules. *J. Exp. Med.* 157 : 1396

68. Frelinger, J. G., Shigeta, M., Infante, A. J., Nelson, P. A., Pierres, M., Fathman, C. G. 1984. Multiple functional sites on a single Ia molecule defined using T cell clones and antibodies with chain-determined specificity. *J. Exp. Med.* 159 : 704

69. Mengle-Gaw, L., McDevitt, H. O. 1985. Genetics and expression of mouse Ia antigens. *Ann. Rev. Immunol.* 3 : 367

70. Phillips, M. L., Yip, C. C., Shevach, E., Delovitch, T. L. 1986. Photoaffinity labelling demonstrates binding between Ia molecules and nominal antigen on antigen-presenting cells. *Proc. Natl. Acad. Sci. USA.* In press

71. Watts, T. H., Gaub, H. E., McConnell, H. M. 1986. T cell-mediated association of peptide antigen and major histocompatibility complex protein detected by energy transfer in an evanescent wavefield. *Nature* 320 : 176

72. Watts, T. H., McConnell, H. M. 1986. High affinity fluorescent peptide binding to I-A^d in lipid membranes. *Proc. Natl. Acad. Sci. USA.* In press

73. Shastri, N., Oki, A., Miller, A., Sercarz, E. E. 1985. Distinct recognition phenotypes exist for T cell clones specific for small peptide regions of proteins. Implications for the mechanisms underlying major histocompatibility complex-restricted antigen recognition and clonal deletion models of immune response gene defects. *J. Exp. Med.* 162 : 332

74. Senyck, G., Williams, E. B., Nitecki, D., Goodman, J. W. 1971. The functional dissociation of an antigen molecule : Specificity of humoral and cellular immune responses to glucagon. *J. Exp. Med.* 133 : 1295

75. Abbas, A., Haber, E., Rock, K. L. 1985. Antigen presentation by hapten-specific B lymphocytes. II. Specificity and properties of antigen presenting B lymphocytes, and function of immunoglobulin receptor. *J. Immunol.* 135 : 1661

76. Glimcher, L., Kim, K. J., Green, I., Paul, W. E. 1982. Ia antigen-bearing B cell tumor lines can present protein antigen and alloantigen in a major histocompatibility complex-restricted fashion to antigen-reactive T cells. *J. Exp. Med.* 155 : 445

77. McKean, D. J., Infante, A. J., Nilson, A., Kimoto, M., Fathman, C. G., Walker, E., Warner, N. 1981. Major histocompatibility complex-restricted antigen presentation to antigen-reactive T cells by lymphocyte tumor cells. *J. Exp. Med.* 154 : 1419

78. Walker, E., Warner, N. L., Chesnut, R., Kappler, J., Marrack, P. 1982. Antigen-specific, I region-restricted interactions *in vitro* between cell lines and T cell hybridomas. *J. Immunol.* 128 : 2164

79. Chesnut, R. W., Colon, S. M., Grey, H. M. 1982. Antigen presentation by normal B cells, B cell tumors, and macrophages: Functional and biochemical comparison. *J. Immunol.* 128 : 1764

80. Issekutz, T., Chu, E., Geha, R. S. 1982. Antigen presentation by human B cells: T cell proliferation induced by Epstein Barr virus B lymphoblastoid cells. *J. Immunol.* 129 : 1446

81. Chesnut, R. W., Grey, H. M. 1981. Studies on the capacity of B cells to serve as antigen-presenting cells. *J. Immunol.* 126 : 1075

82. Kakiuchi, T., Chesnut, R. W., Grey, H. M. 1983. B cells as antigen-presenting cells: The requirement for B cell activation. *J. Immunol.* 131 : 109

83. Cowing, C., Chapdelaine, J. M. 1983. T cells discriminate between Ia antigens expressed on allogeneic accessory cells and B cells: A potential function for

carbohydrate side chains on Ia. *Proc. Natl. Acad. Sci. USA* 80: 6000

84. Ashwell, J. D., DeFranco, A. L., Paul, W. E., Schwartz, R. H. 1984. Antigen presentation by resting B cells: Radiosensitivity of the antigen-presentation function and two distinct pathways of T cell activation. *J. Exp. Med.* 159: 881

85. Frohman, M., Cowing, C. 1985. Presentation of antigen by B cells: Function dependence on radiation dose, interleukins, cellular activation, and differential glycosylation. *J. Immunol.* 134: 2269

86. Tony, H.-P., Parker, D. C. 1985. MHC-restricted polyclonal B cell responses resulting from helper T cell recognition of anti-immunoglobulin presented by small B lymphocytes. *J. Exp. Med.* 161: 223

87. Lanzavecchia, A. 1985. Antigen-specific interaction between T and B cells. *Nature* 314: 537

88. Malynn, B. A., Wortis, H. H. 1984. Role of antigen-specific B cells in the induction of SRBC-specific T cell proliferation. *J. Immunol.* 132: 2253

89. Pletscher, M., Pernis, B. 1983. Internalized membrane immunoglobulin meets intracytoplasmic DR antigen in human B lymphoblastoid cells. *Eur. J. Immunol.* 13: 581

90. Cresswell, P. 1985. Intracellular class II HLA antigens are accessible to transferrin-neuraminidase conjugates internalized by receptor-mediated endocytosis. *Proc. Natl. Acad. Sci. USA* 82: 8188

Ann. Rev. Immunol. 1987. 5 : 503–40

STRUCTURE, FUNCTION, AND SEROLOGY OF THE T-CELL ANTIGEN RECEPTOR COMPLEX

James P. Allison

Cancer Research Laboratory, Department of Microbiology and Immunology, University of California, Berkeley, California 94720

Lewis L. Lanier

Becton Dickinson Monoclonal Center, Inc., Mountain View, California 94043

INTRODUCTION

The vertebrate immune response involves a complex interplay between different cell types, mediated in part by interacting cell surface structures. Of the cells involved in the immune response, both the B and T lymphocytes have the capacity for specific recognition of a tremendous variety of antigens. Despite this superficial similarity, the processes by which T cells and B cells acquire and carry out their recognition and effector functions differ in several important respects. T cells recognize antigen only on the surface of other cells, and antigens are recognized only in the context of products of the major histocompatibility complex (MHC). The antibody produced by B cells is capable of recognizing antigen both alone and in soluble form. The immunoglobulin molecule serves both recognition and effector functions for the B cell. T cells can be divided into subclasses according to effector function, including cells capable of cytotoxic (Tc) or helper (Th) activity. Each functional class has been demonstrated to employ a distinct set of cellular products in the effector function which can be distinguished from the structure used for antigen recognition. The T-cell antigen receptor, unlike B-cell surface immunoglobulin, is intimately associated with an ensemble of auxiliary proteins that may function in signal transduction.

503

0732–0582/87/0410–0503$02.00

In addition to the receptor complex, T cells express a set of accessory molecules that correlate with the function, or more properly, with the class of MHC antigens used in restricting antigen recognition. B cells do not express similar accessory molecules. Finally, the mechanism by which phenotypically and functionally distinct T cells arise from separate sub-lineages within the thymus has no apparent parallel in the B-cell system.

While there has been a wealth of information gained on the functional properties of T cells, insight into the molecular and cell biology of T cells has until recently been clouded by a lack of knowledge of the molecular nature of the T-cell antigen receptor. In the four years since the identi-fication and initial characterization of the T-cell antigen-receptor protein, considerable progress has been made in unraveling many of the mysteries of the T cell. The genes encoding the α and β chains of the receptor heterodimer have been isolated, and the molecular genetics of the system are already known in considerable detail. This article seeks to summarize our current knowledge of the structure of the T-cell antigen-receptor $\alpha\beta$ heterodimer and its associated proteins and their roles in T-cell function. We also discuss the serology of the $\alpha\beta$ heterodimer in the context of the molecular biology of the system and review recent findings suggesting the possibility of an alternative receptor system involving the product of a third antigen receptor–like gene, γ. Finally, we consider the implications of the data for current theories addressing the mechanisms of T-cell differ-entiation, antigen recognition, and cellular activation.

IDENTIFICATION OF THE T-CELL ANTIGEN RECEPTOR

The strategy that led to the definition of the T-cell antigen receptor was based on the assumption that mature T cells should express on their surface unique epitopes related to their specific recognition structure. To this end, a number of laboratories produced clone-specific monoclonal antibodies to T-cell lymphomas (1, 2), hybridomas, and clones (3–8). Several prop-erties of the antibodies suggested that the target structure might be the antigen-specific receptor. First, the antibodies were clone-specific (1, 3, 4). Consistent with the basic assumption, the antibodies are practically all specific for the cells against which they were raised. The general rule, as well as the exceptions to it, has implications for the structure and function of the antigen receptor—to be discussed later. Second, the antibodies specifically precipitated similar structures. In each case, clonotypic anti-bodies reacted with a glycoprotein with the same architecture—an 80–90 kd disulfide-bonded heterodimer (1, 3, 9). Third, the 90-kd disulfide-bonded heterodimer is expressed as a major component of the cell surface only by T cells, while the major disulfide-bonded multimer of the B-cell surface is the immunoglobulin molecule (1, 10). The reciprocal nature of the expression of the disulfide-bonded multimers in itself suggested a functional role for the molecule. Fourth, as expected for a receptor molecule with clonally elaborated specificity, the subunits of the heterodimer were found to have shared as well as clone-specific peptides (9, 11–14). Fifth, the antibodies blocked (3–6) or stimulated (6, 7) T cells in a clone-specific manner. Finally, clonotypic antibodies to the heterodimer could be used to predict the specificity of T-cell hybridomas (15). Together, these results provide compelling evidence that the heterodimer is the T-cell antigen receptor.

STRUCTURE OF THE T-CELL ANTIGEN RECEPTOR

Architecture of the αβ Heterodimer

The antigen receptor isolated from a variety of T cells by immuno-precipitation with clone-specific monoclonal antibodies has been shown to be composed of an acidic α chain of 39–46 kd and a more basic 40–44 kd β chain (1, 3, 4, 6, 11, 16, 17). That the molecule is a heterodimer composed of a single α chain linked via a disulfide bond to a single β chain was established by two dimensional electrophoresis, employing separation by charge under nonreducing and reducing conditions (13, 18). Electrophoresis of the receptor in gradients of 2-mercaptoethanol indicated that in addition to the interchain disulfide bond, the molecule has intrachain disulfide bonds within both the α and β chains (18). This observation suggested that the subunits had similar domain structures. Comparison of peptide maps of the individual chains isolated from different clones revealed that the α and β chains are very different in primary structure and that the α and β chains both contain common as well as clone-specific peptides (9, 11–14). These results indicated that the α and β chains are encoded by different genes and that the genetic mechanisms involved in generation of the T-cell repertoire are operative on both.

Both subunits of the antigen-receptor heterodimer display charge and size microheterogeneity typical of cell surface proteins subjected to glycosylation. The fact that the antigen receptor binds to, and can be specifically eluted from, lectins indicates that the molecule is a glycoprotein (1, 19, 20). Endoglycosidase cleavage and tunicamycin sensitivity analysis indicate that both subunits are glycosylated (19–22). Both chains carry as many as three N-linked glycans which may be of the complex or high mannose type (19, 20). The maximum size of the deglycosylated cell surface forms of the α and β subunits are 27 kd and 32 kd, respectively. Studies of the biosynthesis of the antigen receptor indicate that the molecule is assembled into a dimer and partially glycosylated within minutes, followed by processing with time to yield the mature forms of the subunits seen on the cell surface (19). Glycosylation of the receptor is of special interest to the biology of T cells for at least two reasons. First, lectins that bind to high mannose oligosaccharides induce T cell–mediated lysis of inappropriate target cells. Indeed, the fact that the antigen receptor carries high mannose oligosaccharides and the observation that a clonotypic antireceptor antibody inhibits lectin-dependent cytotoxicity together suggest that cytolytic activity is induced by binding of the lectin to the antigen receptor (20). Second, glycosylation affects the serological properties of the receptor. Pulse-chase analysis of the biosynthesis of the receptor indicates that in at least one system the appearance of epitopes reactive with a clonotypic antibody requires additional glycosylation beyond that necessary for the expression of epitopes reactive with an antiserum to framework determinants (19). Similarly, a clonotypic antibody to a human T cell detects a subset of the α/β heterodimer species reactive with an antibody to β-chain framework epitopes (M. Brenner, unpublished observation). To the extent that the clonotypic epitopes reflect the conformation of the combining site, it is possible that structural microheterogeneity resulting from differential glycosylation may influence the specificity of the receptor.

T-Cell Receptor Genes

The molecular genetics of the T-cell antigen receptor began with the isolation, via the technique of subtractive hybridization, of genes that are specifically expressed in T cells, that exhibit a high degree of structural homology to immunoglobulin genes, and that detect gene segments which rearrange (25–29). The relationship of the genes to the β and α chains of the receptor was established by comparison of the predicted amino acid sequence with that of isolated antigen-receptor subunits (30, 31). A third gene with structural similarity to the α and β genes, commonly referred to as the γ gene, has also been isolated (32). The product and function of the γ gene, only now coming to light, are discussed below.

The α, β, and γ genes encode proteins with similar domain organization, reminiscent of immunoglobulin. Each has a hydrophobic leader sequence of 18–29 amino acids, a variable region segment of 102–119 amino acids, a constant region segment of 87–113 amino acids, a small connecting peptide, a transmembrane region of 20–24 residues, and a small cytoplasmic region of 5–12 amino acids. The variable and constant regions each contain cysteine residues at positions consistent with the presence of a centrally located disulfide loop of 63–69 amino acids, and the genes each contain a cysteine proximal to the transmembrane region that might be involved in the formation of an interchain disulfide bond. The hydrophobic transmembrane regions of the chains have an unusual feature: they contain positively charged amino acids that may play a role in interaction with components of the T3 complex. The α and β genes each contain at least three sites for N-linked glycosylation.

The structural features of the α/β heterodimer are illustrated in Figure 1. The details are inferred from analyses of the protein and from the gene structure.

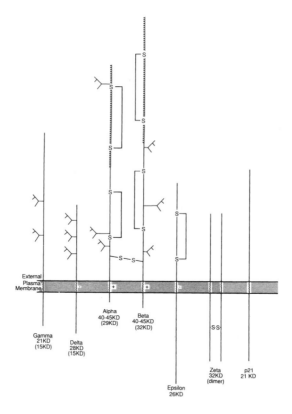

Figure 1   Structure of the murine αβ antigen receptor heterodimer and associated structures of the CD3 complex. The structural features are inferred from studies of the protein and, where possible, from details predicted by the gene sequence. Values in parentheses are the size of the deglycosylated proteins. The existence of intrachain disulfide bonds in the CD3-γ and CD3-δ is possible but has not been directly demonstrated.

MOLECULAR GENETICS OF THE T-CELL ANTIGEN RECEPTOR

As is the case for immunoglobulins, T-cell receptor genes consist of separate germline segments that encode the variable and constant regions of the chains. (For a recent extensive review of T-cell receptor genes, see 50.) The variable region genes are assembled from two or three separate germline segments that somatically rearrange during differentiation to form functional variable region genes. The genomic organizations of the α and β loci in man and mouse are very similar. The variable region of the β chain gene is assembled from three segments, variable (Vβ), diversity (Dβ), and joining (Jβ) (33, 34). There is a relatively small number of Vβ genes, probably less than 40 (35–37), and in some mouse strains, half of these may be deleted with no apparent pathological effect (38). In germline configuration, the β-chain locus has two closely linked constant genes (Cβ1 and Cβ2), each of which is associated with an upstream cluster of functional Jβ gene segments and a single Dβ segment (33, 34, 39–43). The α-chain locus is comprised of a fairly large number of Vα genes (approximately 100) and at least 20 Jα segments spread over 50-kb upstream from a single Cα gene (44, 46–49, 88, 89). Diversity in the variable regions of the α and β chains results from germline diversity of V, D (β only), and J sequences, combinatorial joining of the segments, junctional diversity leading to codon changes at the end of gene segments, N-region diversity arising from addition of template-independent nucleotides at junctions, and by multiple translational reading frames in the Dβ segment (50).

The structure of the γ locus differs substantially from that of α and β. In the mouse, there are three to four constant region genes, each with an upstream J segment (23, 45), and only six Vγ genes (45, 51, 52). In humans, there are at least twelve Vγ genes, two virtually identical Jγ segments, and two Cγ genes (53–57). Diversity of the gamma gene is limited by the small degree of germline Vγ and Jγ diversity and by the fact that the rearranged genes seem to be assembled from only a few of the Vγ and Cγ genes available (51, 52). This is partially offset by a high degree of junctional diversity (51, 58). The γ gene is expressed at high levels in immature thymocytes and at much lower levels in mature T cells (59, 60), and it appears that there is a higher variability of expression of Vγ genes in immature murine T cells (52). There is some indication that the Vγ genes used in the thymus are different from those in peripheral T cells (52). Sterile γ transcripts arising from aberrant rearrangements are regularly found in both MHC class I (61, 62)– and class II (24)–restricted T cells. Thus, it appears that the γ gene product is not essential for the function of mature, MHC-restricted T cells. While the functional significance of the γ gene product has not been completely elucidated, recent data discussed below suggest that it might serve, along with an as yet incompletely characterized partner, as an alternative for the $\alpha\beta$ heterodimer in a receptor/T3 complex on a subset of T cells.

SEROLOGY OF THE $\alpha\beta$ HETERODIMER

Antibodies to the T-cell antigen receptor can be divided into three classes: clonotypic antibodies that react only with the immunizing T cell and with rare T cells of similar specificity, antibodies that react with subsets of T cells, and antibodies that react with the majority of mature T cells. The epitopes defined by these antibodies can in many cases be related to structural features of the receptor, and differential expression of the epitopes on T-cell subsets has important implications as to the origin of the T-cell antigenic repertoire.

Clonotypic Antibodies

Reactivity with the immunizing T-cell lymphoma, clone, or hybridoma, and lack of reactivity with related T cells was initially the sina qua non of antireceptor antibodies (1, 3, 4, 6, 17, 63). MAb 124-40, an alloantibody directed to the antigen receptor of the murine T lymphoma C6VL is illustrative of the extremely restricted distribution of the epitopes defined by many clonotypic antibodies. MAb 124-40 does not react with a detectable fraction of normal T cells or with any other T lymphoma, and the antigen defined is presented, in effect, as a tumor-specific antigen (1). Clonotypic antibodies of extremely restricted reactivity have also been obtained to the receptors of human-T-cell tumors (2, 64) and clones (65, 66). However, the reactivity of clonotypic antibodies is not necessarily restricted to a single clone. In at least one case, the serologically defined clonotype can be clearly associated with the specificity of the T cells. KJ1-26, an antibody, was raised against a clonotypic epitope of DO-11.10, a murine T-T hybridoma that reacts with chicken ovalbumin in the context of Ia^d or Ia^b, a 17 amino acid tryptic peptide of ovalbumin plus Ia^d. On subsequent analysis, KJ1-26 was found to react with Ia^b (3), also with two additional BALB/c-derived, ovalbumin-specific hybridomas obtained from independent fusions (15, 67). The reactive hybridomas alone among several hundred ovalbumin-specific, BALB/c-derived hybridomas, exhibit the same fine specificity as DO-11.10. In addition, the three hybridomas show the same α and β gene rearrangements (67). These observations suggest that both the serologically defined clonotype and the specificity of T cells are inherent in the particular $V\alpha J\alpha/V\beta J\beta$ combination used. Thus, the clonotypic epitope may be a combinatorial idiotype and represent a distinct structural feature of a specific antigen/MHC recognition structure.

Antibodies Reactive with T-Cell Subpopulations

On close scrutiny a number of apparently clonotypic antibodies, particularly heteroantibodies to the human-T-cell receptor, have been found to react with subpopulations of peripheral T cells. An antibody to the receptor heterodimer of a Sézary tumor cell reacts with 1–2% of normal human peripheral T cells (2). In two extensive serological analyses of the human-T-cell leukemia HPB-ALL, antibodies to both private and public epitopes of the antigen receptor were obtained (64, 68). The serologic complexity of the receptor is illustrated by the fact that the 3–5% of normal peripheral blood T cells that react with three antibodies to the HPB-ALL receptor can be further divided into four subsets on the basis of reactivity with two additional antibodies (68).

In some cases subclass reactivity can be correlated with a structural feature of the antigen receptor. Three apparently clonotypic antibodies directed to the receptor of the human T-cell line REX (66) react with about 2% of peripheral T cells and with several of a panel of T-cell clones (65). One of the antibodies defines an epitope of the β subunit of the antigen receptor and reacts only with T-cell clones expressing the same $V\beta$ gene as REX (65). Structural analysis of isolated subunits indicates that the cross-reactive clones can employ different $D\beta$, $J\beta$, and perhaps $C\beta$ gene segments, as well as different α genes in the construction of the receptors (65). Similarly, another antibody that reacts with 5% of human peripheral T cells detects a subset of peripheral T cells that express a particular $V\alpha$ segment (69).

Antireceptor antibodies defining subpopulations of T cells have also been described in the mouse. A rat antibody, KJ16-133, raised against the antigen receptor isolated from a T-cell hybridoma reacts with about 20% of peripheral cells in most but not all mouse strains (70, 71). Immunization of C57L/J mice with BALB.B T cells resulted in an antibody, F23.1, that stains essentially the same fraction of peripheral cells with the same strain distribution as KJ16 (72). The strain-specific distribution of the epitopes defined by KJ16-133 and F23.1 raised the possibility that the murine T-cell antigen receptor expresses allotypic determinants. However, genetic analysis indicates that mouse strains with KJ16/F23.1-reactive T cells share the presence of a family of homologous $V\beta$ genes ($V\beta8$) which is lacking in strains unreactive with KJ16 or F23.1 (38, 73). Thus, it appears that the antibodies define a polyclonal $V\beta$-encoded determinant and not an allelic polymorphism. No genetic or serological data presently exist that demonstrate the existence of antigen receptor allotypes, although the possibility cannot be excluded.

Antibodies to Monomorphic Epitopes

Antibodies to clonotypic and V gene–encoded epitopes have been tremendously valuable tools in unraveling the structure and function of the T-cell antigen receptor, but the limited distribution of the reactive epitopes has restricted the usefulness of the reagents in studies of the ontogeny and differentiation of T cells. A considerable effort in many laboratories has therefore been directed towards the production of antibodies to monomorphic epitopes of the receptor. Rabbit antisera have been produced to murine (11) and human (74, 75) antigen receptors isolated by immunoabsorption with clonotypic antibodies. Chimeric proteins engineered to use immunoglobulin light chains as carriers for constant-region domains of the T-cell antigen receptor have been used to produce similar antisera (76). The antisera precipitate the antigen-receptor heterodimer from most mature T cells, indicating the presence of monomorphic epitopes most likely associated with constant regions of the α or β chains. However, despite their broad reactivity, antisera raised against purified receptor have not proven especially useful in functional studies of T cells, or in tracing cell lineages by flow microfluorometry, because the reactive epitopes are not accessible at the surface of intact cells (11, 74, 75). The results of cross-linking experiments suggest that the inaccessibility of the constant region epitopes might be a result of masking by the associated chains of the CD3 complex, since a clonotypic antibody, but not a rabbit antiserum to monomorphic epitopes of the murine receptor, immunoprecipitates the heterodimer after chemical cross-linking to CD3 (77).

Immunization of mice with human thymocytes has yielded an antibody, WT-31, that recognizes an epitope of the human antigen receptor which is accessible on intact cells (78). WT-31 stains >95% of peripheral T cells, is mitogenic for resting T cells, and blocks cytolysis by CTL clones. The antibody precipitates, albeit weakly, the α/β heterodimer from radiolabeled T cells, and the reactive antigen comodulates with CD3. Binding of WT-31 to HPB-ALL cells is inhibited by antibodies to CD3 but not by antibodies to the clonotypic epitope of the HPB-ALL receptor. This suggests that the antigen reactive with WT-31 is distal to the clonotypic epitope but close to the epitopes recognized by anti-CD3 reagents. Immunoprecipitation of the antigen receptor heterodimer by WT-31 is enhanced when performed under conditions which preserve the association with the CD3 complex (M. Brenner, unpublished observation), suggesting that the antibody recognizes a conformational epitope that is stabilized by CD3. This antibody has proven extremely useful in the study of antigen receptor expression and in the identification of T cells that express CD3 in the absence of the α/β heterodimer.

THE $\alpha\beta$ HETERODIMER AND ANTIGEN/MHC RECOGNITION

The $\alpha\beta$ Heterodimer as Receptor for Antigen and MHC

T cells have a dual specificity for antigen and polymorphic determinants of class-I or class-II MHC molecules (79–81), the phenomenon that is termed MHC restriction. Several lines of evidence suggest that a single receptor recognizes both antigen and MHC. First, antigen recognition is directly influenced by the MHC haplotype of the antigen presenting cell for both class-I and class-II MHC-restricted T cells (82–84). Second, antigen specificity and MHC restriction do not segregate independently in somatic cell hybrids constructed by fusion of two T-cell hybridomas of different antigen/MHC specificity (85). Third, as previously discussed, a clonotypic antibody to the $\alpha\beta$ heterodimer identified independent clones with identical antigen/MHC specificity and alloreactivity (15). The loss of cell surface $\alpha\beta$ heterodimer expression by antigen-specific T-cell hybridomas results in the loss of antigen/MHC-induced proliferation and IL-2 secretion (67). Finally, transfection of functional α and β genes from a murine CTL reactive with the hapten fluorescein in the context of H-2D^d into a CTL specific for the hapten 3-(p-sulphophenyldiazo)-4-hydroxy-phenylacetic acid in the context of H-2K^k produced transfectants that respond to hapten only in the context of the original H-2 haplotype (86). Collectively these studies provide compelling evidence that a single receptor, the $\alpha\beta$ heterodimer, is responsible for both MHC and antigen recognition.

Structural Origins of Functional Diversity of the $\alpha\beta$ Heterodimer

The functional and phenotypic heterogeneity of T cells raises the possibility that the antigen receptor might display function-correlated structural features. One manner by which this could be accomplished is exemplified by the immunoglobulin heavy-chain locus, which uses a set of multiple nonallelic constant-region genes, or isotypes, to produce antibodies with different effector functions (87). However, T cells do not employ isotypes in the production of functionally different antigen receptors. The single constant region gene that has been identified for the α chain (88–90) is used by both cytotoxic and helper T cells (28, 29). The β-chain locus contains two isotypic constant region genes that encode proteins that differ by only four (murine) or six (human) amino acid residues (39, 40, 91–94).

However, expression of the $C\beta$ genes does not correlate with function (95, 96) or specificity (96).

An alternative possibility for the production of functionally distinct receptors is the use of nonoverlapping pools of $V\alpha$ or $V\beta$ gene pools. This is clearly not the case. An antibody apparently directed to epitopes of the human T-cell line REX $V\beta$ gene product reacts with both helper and cytotoxic T cells, and with both class-I– and class-II–reactive cells (65). Similarly, antibodies to epitopes of the murine $V\beta8$ family react with both helper and cytotoxic cells and show no correlation with specificity or MHC restriction (70, 72, 73, 97). These results are consistent with a number of studies documenting a lack of any simple and general correlation between usage of any $V\beta$ or $V\alpha$ gene and phenotype, function, specificity, or MHC restriction (98–100).

Despite the lack of simple rules governing gene usage, in a single system, specific gene segments may be preferentially employed. Thus, three independently derived chicken-ovalbumin–specific, Ia^d-restricted T-cell hybridomas reactive with clonotypic antibody KJ1-26 shared a common fine specificity as well as a particular configuration of $V\alpha/J\alpha$ and $V\beta/J\beta$ (67). Similarly, all of 16 B10.A-derived cytochrome c–specific T-cell clones were found to use members of the same $V\alpha$ family, and most used one of two $V\beta$ genes (101). Similarly, in a panel of 42 H-2K^b-restricted, 2,4,6-trinitrophenyl-specific cytotoxic-T-cell clones, nearly half used the same $V\beta$, $J\beta$, and $D\beta$ elements (102). The limited germline diversity observed in these studies indicates that particular segments of both the α and β genes may be selected by particular antigen/MHC configurations and that both chains contribute to the T-cell response.

An extensive analysis of phenotypically related T-cell clones provides some insight into the origin of receptor specificity for antigen and MHC (101). $V\beta$ gene usage was found to correlate with MHC specificity in a panel of 16 B10.A-derived cytochrome c–specific T-cell clones using essentially the same α-chain gene. While this implicates combinatorial association of particular $V\beta$ genes with MHC restriction, the results do not necessarily indicate the physical association of one chain with antigen or MHC molecules. The receptors of two clones with similar specificity but differing in reactivity with I-A^s were found to be composed of segments identical except for $J\alpha$. The receptor genes of two clones differing in dose response to antigen and in fine specificity were assembled from identical segments, but they differed in junctional sequences in the β chain. These results suggest that the α and β chains each contribute to both antigen specificity and MHC restriction and that the combining site has a specificity for antigen and MHC that is not predictable from the sequence of either chain alone. It is also evident that combinatorial association of α and β

chains, combinatorial joining of gene segments, and junctional diversity can each affect the specificity of T cells and therefore may play a role in development of the T-cell repertoire.

Structure of the αβ Heterodimer Combining Site

In addition to similarities between the T-cell antigen receptor and immunoglobulin chains in domain organization, the presence of 60–70 residue disulfide-bonded intrachain loops, and a fair degree of homology in primary structure, there are a number of observations that suggest a high degree of similarity in the secondary, tertiary, and quaternary structure. First, sequence comparisons indicate that several residues that are highly conserved in immunoglobulins, including those involved in domain-domain interactions, are also conserved in T-cell receptor chains (103, 104). For example, of five positions involved in intra- and interchain immunoglobulin V region interactions, three are conserved in Vα genes and four in Vβ genes (35, 36, 103). Second, algorithms that assess hydropathicity, hydrophobicity, and potential for formation of α helix and β-pleated sheets do suggest that the T-cell receptor chains fold into very similar tertiary configurations (35, 36, 104). This includes the presence of the "antibody fold," a series of multistranded antiparallel β-sheet bilayers characteristic of immunoglobulin domains, in the V and C regions of the α and β chains (104). Thus, the structural features that provide the substructure for antibody combining sites are also present in the T-cell antigen receptor. The combining site of the antibody molecule is known to be formed by hypervariable regions that determine the fine structure superimposed on this scaffolding. The close similarity in the profiles of sequence variability in immunoglobulin V genes and the Vα and Vβ genes (35–37, 47, 48) suggests that the T-cell antigen receptor combining site is fundamentally the same as that of immunoglobulins.

Differences in Antigen Recognition by B and T Cells

The similarity in the structure of the antigen binding sites of B and T cells has important implications. Since the upper limit of the size of the combining site is determined by invariant features, antibodies and the T-cell αβ receptor should accommodate antigens of approximately the same size (104). Thus, the T-cell antigen receptor is inherently no more likely than immunoglobulin to have independent binding sites for antigen and MHC. In fact, antibodies have been produced that exhibit MHC-restricted recognition of antigen (105). It therefore seems likely that there is no fundamental difference in the manner in which B- and T-cell receptors recognize antigen. The features that distinguish the responses of B and T cells to antigens must reside elsewhere.

One fundamental way in which T and B cells differ may be the nature of the antigen recognized. The epitope recognized by the combining site of immunoglobulins consists entirely of antigen. The epitope recognized by the receptor of T cells may be derived in part from antigen and in part from MHC. There is accumulating evidence that antigen and MHC molecules physically interact. An antigenic peptide of lysozyme binds to class-II molecules derived from responders but does not bind to class-II molecules from nonresponders (106). Analysis of the relative avidity of a T-cell clone to an antigenic peptide presented in the context of two different class-II molecules suggests a physical interaction between the peptide and MHC (107). Studies employing measurements of energy transfer to determine spatial relationships indicate that T cells bring about an intimate association between an antigenic peptide and class-II molecules in a planar membrane (108). Taken together, these results strongly suggest that antigen recognition by T cells may involve a ternary complex of the antigen receptor, the antigen, and MHC.

Another way in which T and B cells differ fundamentally is in the expression of cell surface molecules that play a role in activation but that are unrelated to antigen-specific recognition. These include auxiliary proteins involved directly in cellular activation, as well as accessory molecules involved in conjugate formation. The net effect of these structures, which will be discussed in some detail, is to focus the T-cell antigen/MHC recognition structure on events occurring at cell membranes. Thus, it appears that in T cells, but not in B cells, binding of antigen to the receptor is necessary but in itself not sufficient to lead to activation.

THE CD3 COMPLEX

Evidence for Association with the T-Cell Antigen Receptor

Among the myriad of lineage-specific molecules defined on the surface of human T cells, several early observations implicated those of the CD3 complex as having a critical role in T-cell function. CD3 was observed to arise relatively late in intrathymic differentiation and to be expressed on all peripheral cells (109). Under appropriate conditions, antibodies to CD3 were found to block T-cell function or to be mitogenic for T cells (110). While these observations do strongly suggest a linkage to the antigen specific receptor, the invariant nature of the CD3 complex rules it out as the discriminative surface recognition structure. However, several lines of evidence indicate that the CD3 complex is intimately associated with the $\alpha\beta$ heterodimer on the T-cell surface. First, CD3 and $\alpha\beta$ comodulate on T cells and clones (4). Under appropriate conditions, Ti may be observed to coprecipitate with CD3 (9, 22, 112). Treatment of human and murine T cells with bifunctional reagents results in cross-linking of αβ and CD3 (77, 111, 113). Finally, independent selection of mutagenized T cells with antibodies against either CD3 or αβ resulted in loss of both structures (114). Together these data suggest that the functional T-cell-antigen receptor on MHC-restricted T cells is a multisubunit structure composed of the αβ heterodimer and the CD3 complex.

Structure of the CD3 Complex

The human CD3 defined with monoclonal antibodies is composed of at least three distinct polypeptides that have been designated γ, δ, and ε (112, 115). CD3-γ is a 25-kd glycoprotein with a 16-kd peptide backbone (22). CD3-δ and CD3-ε are both 20-kd proteins but have distinctive polypeptide chains. CD3-ε is not glycosylated, while CD3-δ carries N-linked oligosaccharide side chains (115). The bifunctional reagent DSS was found to cross-link the antigen receptor β chain to CD3-γ, suggesting that interactions between these subunits play a role in the αβ/CD3 association (113). The lack of observed cross-linking of other components of the complex may reflect an inaccessibility of reactive amino groups rather than the absence of an intimate spatial relationship, since the α and β chains of the heterodimer also failed to cross-link. Thus, it cannot be concluded that interactions between additional components do not also exist.

Identification and structural studies of the murine CD3 complex have until recently been hampered by the lack of antibodies. The existence of CD3-like structures on murine T cells was initially demonstrated in nearest neighbor analysis of the murine α/β receptor heterodimer using bifunctional cross-linking reagents and clonotypic antibodies (5, 13, 77). More recently, mild detergents (Triton X-100, digitonin) have been used to solubilize cells under conditions that preserve the noncovalent association of the components of the αβ/CD3 complex (116, 117). Under appropriate conditions, immunoprecipitates obtained from murine T cells with clonotypic antireceptor antibodies contain at least four polypeptides in addition to the αβ heterodimer. Three of these are apparent homologues of the human CD3 complex. The murine CD3-γ chain is a 21-kd glycoprotein with a polypeptide core of 16 kd. The murine CD3-δ chain is a 28-kd glycoprotein with a polypeptide core of 16 kd. The murine CD3-ε chain is a nonglycosylated protein of 26 kd which contains an intrachain disulfide bond. Recently, an antiserum to murine CD3 which precipitates the γ, δ, and ε chains has been obtained by immunization of hamsters with TCR/CD3 complex isolated from T lymphoma cells with clonotypic antireceptor antibody (J. P. Allison, unpublished observation). An antiserum of similar reactivity has been produced by immunization of rabbits with a synthetic peptide corresponding to a sequence of the δ chain (118).

Recently a hamster monoclonal antibody has been produced with specificity for the murine CD3-ε chain (J. Bluestone, personal communication).

In addition to γ, δ, and ε chains, the murine antigen receptor complex contains at least two additional components that have not as yet been identified on human T cells. Murine CD3-ZETA is a nonglycosylated, disulfide-bonded homodimer with 16-kd subunits. The murine T-cell receptor complex also contains a fifth chain, p21, which is an endoglycosidase F-resistant, 21-kd protein that is part of a disulfide-linked molecule (119).

Recently, cDNA clones encoding human and mouse CD3-δ (120, 121) and human CD3-ε and CD-γ have been isolated (122, 123). The human CD3-δ gene consists of four domains, a 21-residue leader peptide, an extracellular domain of 79 residues, a transmembrane segment with a central aspartic acid, and a 44-residue intracellular domain, consistent with a primary translation product of 19 kd (120). Two consensus sites for N-linked glycosylation are found in the extracellular domain. The human CD-γ cDNA sequence predicts a 182-residue amino acid sequence with a typical signal peptide, an 89-residue hydrophilic domain with two potential sites for N-linked glycosylation, a hydrophobic domain with a central glutamic acid residue, and a 44-residue cytoplasmic domain (123). Human CD-γ and CD3-δ are highly homologous at the nucleotide and amino acid level, particularly in the intracellular and the membrane-proximal half of the extracellular domain. The high degree of homology suggests that the CD-γ and CD3-δ genes arose by duplication and divergence. The human and murine CD3-δ genes are highly conserved at the amino acid level, especially in the transmembrane and cytoplasmic domains ($>70\%$) (121). The murine CD3-δ gene contains a third site for N-linked glycosylation, consistent with the higher molecular weight observed for the murine CD3-δ protein. The human CD3-ε gene also consists of four domains, a 22-residue leader, a 104-residue extracellular domain, a transmembrane domain with a central aspartic acid residue, and a relatively long (81-residue) cytoplasmic domain (122). Three features of the genes are of special interest. First, none displays significant homology to immunoglobulins, indicating that the CD3 proteins are not members of the immunoglobulin gene superfamily. Second, CD-γ CD3-δ, and CD3-ε all contain a negatively charged residue centrally located in the transmembrane domain. This residue might play a role in stabilizing the TCR/CD3 complex by forming salt bridges with the positively charged lysine residues similarly located in the transmembrane region of the heterodimer α and β chains (25, 28). Finally, CD3-δ and especially CD3-ε contain cytoplasmic domains that are considerably larger than those of the heterodimer α and β chains and which could play a role in interactions with cytoplasmic components involved in signal transduction.

The possible role of the CD3 complex in signal transduction is supported by recent studies demonstrating that components of the complex are phosphorylated upon activation of human or murine T cells (116, 117, 124, 125). Activation of murine T cells by antigen, conconavalin A, or phorbol ester results in phosphorylation of a serine residue on the 21-kd, CD3-γ chain (116, 117, 124). The CD3-ε chain is also phosphorylated when T cells are treated with phorbol ester (116). Antigen activation of murine T cells also results in tyrosine phosphorylation of p21 (119). These results implicate two protein kinase systems in T-cell activation. It is of interest that p21 is phosphorylated on a tyrosine residue, since tyrosine phosphorylation is a relatively rare event usually associated with viral transformation or mitogenesis.

The components of the murine antigen receptor/CD3 complex are presented schematically in Figure 1. The details of the individual components are based on biochemical analysis or are inferred from the gene structure. It should be noted that the γ and δ chain genes contain cysteine residues which could form interchain disulfide loops, but the presence of intrachain disulfide bonds has not yet been demonstrated by analysis of the protein. The stoichiometry of the CD3 components is unknown, as is the nature of the associations. It is also not known whether the complete ensemble is represented in each $\alpha\beta$ complex. However, considerable evidence, discussed below, suggests that the complex may be functionally divided into two domains—the $\alpha\beta$ heterodimer responsible for recognition of antigen and MHC, and CD3 complex responsible for signal transduction.

THE $\alpha\beta$/CD3 COMPLEX AND T-CELL ACTIVATION

The role of the $\alpha\beta$/CD3 complex in cellular activation and in antigen recognition has been established by several lines of experimentation.

Agonist or Antagonist Effects of Antibodies to $\alpha\beta$ and CD3

Antibodies against either clonotypic or common epitopes of the $\alpha\beta$ heterodimer or invariant determinants of CD3 have been shown either to inhibit antigen-specific immune responses or to act as agonists resulting in cellular activation, depending on the experimental conditions. Anti-CD3 and anti-Ti antibodies affect cellular proliferation, cytotoxicity, and lymphokine secretion.

Anti-CD3 antibodies, as well as anti-$\alpha\beta$ antibodies, can inhibit cytotoxicity mediated by antigen-specific Tc against specific target cells (4, 72, 126, 127). In this situation, antibody may sterically inhibit antigen recognition by the cytotoxic cell. In contrast, coculture of antigen-specific Tc with irrelevant targets in the presence of anti-CD3 antibodies results in nonspecific lysis of the target (128–132). Anti-CD3 induced cytotoxicity is mediated by signal transduction resulting from interaction of the Fab portion of the anti-CD3 antibody with the $\alpha\beta$/CD3 complex on the Tc and conjugate formation via the Fc portion of the anti-CD3 antibody binding to Fc receptors present on the target cells (130, 131). This latter phenomenon has been exploited by covalently cross-linking anti-CD3 or anti-$\alpha\beta$ allotype antibodies with other antibodies to achieve antibody-directed targeting of the Tc (133–136).

Under certain conditions, anti-CD3 and anti-$\alpha\beta$ antibodies can inhibit both antigen-specific and mitogen-induced proliferation (17, 137). However, anti-CD3 and anti-$\alpha\beta$ antibodies can also induce proliferation (7, 64, 65, 72, 138–142). This differential response is concentration dependent and biphasic; in general, low concentrations of antibody induce proliferation, whereas higher concentrations of antibody inhibit proliferation induced by other stimuli (e.g. mitogen or antigen). Antibody-induced proliferation requires cross-linking of the anti-CD3 antibodies on the T-cell surface. Monocytes can accomplish antibody cross-linking via their Fc receptors, depending on the isotype of the anti-CD3 or anti-$\alpha\beta$ antibody (143). Alternatively, antibodies can be attached to a solid matrix, such as Sepharose, to accomplish the multimeric binding required for the proliferative signal (72, 138). The mechanisms that result in either inhibition or induction of proliferative signals as a consequence of anti-CD3 or anti-$\alpha\beta$ binding have not been delineated.

Comparable to their effects on CTL activity, anti-CD3 and anti-$\alpha\beta$ antibodies can either trigger or inhibit lymphokine secretion by helper-T-cell lines (3, 138, 144–148). Studies of a human T-cell leukemia line of helper phenotype, JURKAT, indicate that interleukin-2 secretion and gamma interferon synthesis involve a two-signal process (146–148). One signal requires soluble ligand binding to CD3 or to the $\alpha\beta$ heterodimer. The second signal is transmitted via an unidentified receptor that can be activated by phorbol myristate acetate (PMA), an activator of protein kinase C. Immobilization of anti-CD3 or anti-$\alpha\beta$ and the presence of interleukin-1 in the culture overcomes the necessity for PMA in the induction of IL-2 secretion by the JURKAT cell line.

$\alpha\beta$/CD3 Loss Mutants

As previously discussed, several lines of evidence indicate that CD3 and $\alpha\beta$ are intimately associated in the plasma membrane. Additional studies suggest an obligatory requirement for coexpression of CD3 and of products of antigen-receptor genes for surface expression of any components of the complex (114). CD3/$\alpha\beta$ negative mutants of the JURKAT cell line were produced by mutagenesis and antigenic selection using either anti-

CD3 or anticlonotypic antibodies. Independent selection against either CD3 or $\alpha\beta$ resulted in concomitant loss of both structures, indicating that both CD3 and $\alpha\beta$ are necessary for expression of the antigen-receptor complex on the cell surface. In parallel with loss of CD3/$\alpha\beta$, mutant JURKAT cells lost the ability to synthesize IL-2 in response to either anti-CD3 or anti-$\alpha\beta$ antibodies or to the mitogenic lectin PHA (114). It was subsequently established (149) that the CD3/$\alpha\beta$ negative mutants of JURKAT lack functional TCR-β mRNA and β-chain proteins. The expression of cell surface CD3/$\alpha\beta$ complex and the ability to synthesize IL-2 upon stimulation with PHA or anti-CD3 antibodies were restored by transfection of these mutants with competent, cloned TCR-β genes (114). These experiments unequivocally demonstrated the obligate coexpression of CD3 and $\alpha\beta$ heterodimer for cell surface expression of the antigen-receptor complex and the role of the complex in T-cell activation.

Mechanism of Signal Transduction via $\alpha\beta$/CD3

The biochemical process through which ligand binding to CD3/$\alpha\beta$ results in cellular activation is a complex pathway apparently involving multiple signals. Using antibodies against CD3 or $\alpha\beta$ as agonists, several investigators have demonstrated that MAb binding to CD3 or $\alpha\beta$ results in a rapid increase in intracellular $[Ca^{2+}]$ in T leukemia cell lines, as well as antigen-specific T-cell clones (64, 148, 150–156). Perturbation of the $\alpha\beta$/CD3 complex results in activation of a phosphodiesterase that hydrolyzes phospholipid phosphatidylinositol bisphosphate (PIP_2) to generate inositol triphosphate (IP_3) and diacylglycerol (148, 153, 154). Subsequently, IP_3 increases free cytoplasmic $[Ca^{2+}]$ and is converted into inositol bisphosphate (IP_2), inositol phosphate (IP_1), and inositol. Diacylglycerol has been previously shown to activate protein kinase C, which may act as a cofactor in the activation process. In addition to ligand, anti-CD3 or anti-$\alpha\beta$ antibodies, other cofactors (e.g. IL-1) may be required to initiate the activation process, depending upon the metabolic state of the T lymphocyte. Activation of T cells via the $\alpha\beta$/CD3 complex has recently been reviewed (148).

ACCESSORY MOLECULES INVOLVED IN T-CELL FUNCTION

Although the $\alpha\beta$/CD3 complex serves as the trigger for initiation of an antigen-specific immune response, the antigen-receptor complex apparently does *not* initiate primary conjugate formation between T lymphocytes and antigen presented on the plasma membrane of antigen presenting cells or targets of CTL. Several accessory molecules on the T-cell surface have been shown to participate in conjugate formation and activation. As depicted in Figure 2, the LFA-1, CD2 (T11/Leu 5/LFA-2) CD4 (T4/Leu 3) and L3T4, and CD8 (T8/Leu 2) and Ly2 molecules are involved in antigen-specific activation. Despite the functional involvement of these molecules in antigen recognition and cellular activation, it should be noted that none of these structures has been shown to be structurally associated with the $\alpha\beta$/CD3 complex. Rather, these structures may play a role in cell-cell adhesion, may provide alternative pathways of immune activation, or may serve as fail-safe signals to prevent inadvertent activation via the $\alpha\beta$/CD3 pathway.

LFA-1

In both mouse and man, LFA-1 is a glycoprotein complex, composed of an ≈ 177-kd subunit noncovalently associated with an ≈ 95-kd subunit, which is expressed on essentially all hematopoietic cells (157, 158). Essentially all immune functions requiring conjugate formation between hematopoietic cells are inhibited by antibodies against LFA-1. For example, anti–LFA-1 antibodies can inhibit CTL and NK cell–mediated cytotoxicity at the stage of effector-target conjugate formation (158–166), T-cell proliferation (160–165), and self-adhesion of B lymphoblastoid cells (167). Thus, LFA-1 is apparently involved in cell-cell adhesion. The ligand for LFA-1 binding has not been identified, but LFA-1/LFA-1 homologous interactions between cytotoxic effectors and targets has been excluded.

CD2

Like anti-CD3 and anti-$\alpha\beta$, antibodies against CD2 can either inhibit or induce T-cell functions, including proliferation, lymphokine secretion, and cytotoxicity (162–164, 168, 169). It has been proposed that CD2 serves as an alternative pathway for T-cell activation (168). CD2 is expressed in the differentiation pathway of T lymphocytes prior to acquisition of the antigen-receptor complex (170, 171). Antibodies against the $\alpha\beta$/CD3 complex can inhibit activation via CD2, suggesting that the two pathways are functionally related (172). Furthermore, anti-CD2 antibodies have been shown to affect CTL-mediated cytotoxicity (162, 163) and to inhibit the binding of sheep erythrocytes to T lymphocytes (173, 174). Structural

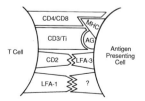

Figure 2 Structures involved in interactions between T cells and antigen presenting cells.

studies have demonstrated that CD2 is an $\approx$ 50-kd glycoprotein (173) that, unlike the $\alpha\beta$ heterodimer, apparently does not possess peptide variability. Recently, the LFA-3 antigen, a 55–70-kd glycoprotein expressed on essentially all hematopoietic cells as well as other nonhematopoietic cells (162, 163), has been implicated as the specific cell-surface receptor for CD2 (175). Additionally, certain human helper T-cell lines have been reported to secrete a soluble lymphokine (designated IL-4A) that apparently binds to CD2 and results in antigen-independent activation of T lymphocytes (176). The relationship between LFA-3 and IL-4A has not been delineated.

CD4 and CD8

CD4 (murine L3T4) and CD8 (murine Ly2) are membrane glycoproteins that have been implicated in recognition of MHC class-II (HLA-DR, DQ, DP and I-A/I-E) and class-I (HLA-A, B, C and H-2) antigens, respectively (177–184). CD4 (L3T4) and CD8 (Ly2) are expressed on reciprocal subsets of mature peripheral T lymphocytes. However, in the thymus the majority of immature thymocytes coexpress both CD4 (L3T4) and CD8 (Ly2) (170, 179).

Both antibody inhibition experiments and functional analysis of MHC restriction of T-cell clones support the concept that these molecules are primarily involved in MHC interactions (177–184). These accessory molecules may be involved in immune activation by enhancing the avidity of T-cell antigen–specific recognition, perhaps by binding monomorphic epitopes of MHC (179, 180, 185). Antibodies against CD8 have been shown to inhibit conjugate formation and subsequent cytotoxicity mediated by CTL (127, 166, 177, 183, 186–189). However, as yet there is no direct evidence that CD4 and CD8 actually bind class-II or class-I molecules. More recently, several investigators have suggested that CD4 and CD8 may be involved in cellular activation, conveying either "on" or "off" signals, depending on the experimental conditions (190, 191). However, these molecules may not be absolutely essential to T-cell function, since it has been demonstrated that cells that do not express CD4 or CD8 can nonetheless be activated (180).

The genes encoding CD4 (L3T4) and CD8 (Ly2) have been cloned and demonstrate sequence homology to the immunoglobulin/T-cell antigen receptor gene family (192–194). It is particularly interesting that the molecules may carry recognition sites similar to the antibody combining site. CD4 and murine L3T4 encode $\approx$ 55-kd glycoproteins that are expressed as monomers on the cell surface of T lymphocytes and monocytes (179, 195). Human CD8 encodes an $\approx$ 32-kd glycoprotein that is expressed on peripheral T lymphocytes as a disulfide-linked homodimer, whereas in thymus some CD8 glycoproteins are assembled with a MHC class I–like molecule designated CD1 (T6/Leu 6) (195, 196). In mouse, Ly2 molecules are expressed on the membrane of T lymphocytes as a disulfide-linked heterodimer composed of Ly2 and Ly3 subunits (197, 198).

T-CELL ACTIVATION BY ANTIGEN-PRESENTING CELLS

The cell surface molecules implicated in the interaction of T cells with antigen-presenting cells are illustrated in Figure 2. A reasonable scenario for T-cell activation involves a sequential series of events leading from interaction of the cells to effector function. The initial event in the interaction involves conjugate formation between the T lymphocyte and an antigen-presenting cell. The initial interaction may not be antigen specific, and it occurs at least in part via molecules involved in cell-cell adhesion. LFA-1 on the T-cell plasma membrane is apparently the predominant structure involved in conjugate formation, although other molecules such as CD4 and CD8 may also participate. Following antigen-independent conjugate formation, the $\alpha\beta$ heterodimer may engage antigen and class-I or class-II MHC molecules, if the appropriate combination is present, in a ternary complex. This binding may be facilitated or stabilized by interactions of CD8 or CD4 on the T-cell surface with appropriate MHC structures on the surface of the antigen presenting cell. Following the specific recognition of antigen/MHC by the $\alpha\beta$ heterodimer, the CD3 complex is triggered and generates a signal that leads to activation of the T cell and, ultimately, to effector function. This scenario is obviously speculative but is consistent with most current data. Elucidation of the details of the molecular events involved in the cellular interactions and the cytoplasmic events leading to activation remains a major task.

OTHER CD3-ASSOCIATED MOLECULES— THE γ GENE PRODUCT?

The third T-cell antigen receptor–like gene, γ, has been enigmatic since it was identified (32). The gene rearranges early in ontogeny (59, 60) and appears to be rearranged in most, if not all, T cells. However, in most functional T-cell clones the transcripts seem to be nonproductive (24, 57, 61, 62). Until recently, no gene product has been associated with the γ gene. However, several recent findings suggest that a protein related to the γ gene may be expressed on T cells.

Several investigators have recently described human-T-cell leukemias (199) and IL-2–dependent T-cell lines derived from peripheral blood (200, 201) and thymus (202) that express CD3 on the cell surface but do not react with an antibody (WT-31) directed against a framework determinant of the $\alpha\beta$ heterodimer (78). Immunoprecipitation with anti-CD3 antibodies has revealed several unique CD3-associated structures on these cell lines. Four distinct CD3-associated structures have been observed.

1. Certain IL-2–dependent T-cell lines established from peripheral blood T lymphocytes (201) and thymocytes (202) fail to react with WT-31 and an anti-β (framework) chain antibody (201). These cells do not produce full-length α or β chain mRNA but do express potentially functional γ-chain mRNA. Immunoprecipitation with anti-CD3 demonstrated the presence of a non-disulfide-linked heterodimer composed of 55–62-kd and 40–44-kd subunits (201, 202). In the case of the peripheral cells, the 55-kd subunit was found to react with an anti-γ peptide antiserum (201). Based on these results, it has been proposed the 40-kd subunit represents the gene product of an unidentified fourth T-cell antigen receptor gene, tentatively designated *TCR-δ*.

2. A human T-cell leukemia, PEER, expresses a 55–60-kd glycoprotein that coprecipitates with CD3 (199). PEER cells lack reactivity with anti-Ti (WT-31) MAb and fail to transcribe α-chain genes, but they synthesize competent γ-chain transcripts (199; and A. Weiss, personal communication). Enzymatic removal of N-linked carbohydrates reveals a polypeptide of 29 kd, consistent with the predicted size of a γ-gene product (199).

3. A clonotypic antibody to an IL-2–dependent T-cell clone established from fetal blood precipitates a disulfide-linked homodimer (44-kd, reduced) associated with CD3 immunoprecipitates (202b). Like the IL-2–dependent T-cell lines described above (201), these fetal T cells failed to react with WT-31 and did not synthesize competent α-chain mRNA, but they did transcribe apparently full-length γ-chain and β-chain mRNA. The fetal T-cell clones killed "NK sensitive" tumor cell targets without deliberate immunization or MHC-restriction, and cytotoxicity was inhibited by the anticlonotypic antibody (200; Th. Hercend, personal communication). This result suggests that the CD3-associated structure is involved in antigenic recognition and cytotoxic function.

4. It has recently been demonstrated that human peripheral blood contains a subset of CD3+ T lymphocytes that express neither CD4 nor CD8 on the cell surface (203). IL-2-dependent cell lines of this phenotype established from several donors were found to mediate non-MHC-restricted cytotoxicity against several tumor cell targets. The cell lines failed to react with WT-31 and did not transcribe functional β- and α-chain mRNA, but they did transcribe γ-chain mRNA (L. L. Lanier, unpublished observation). Unlike the cells described above, these

CD4−, 8− cells express a CD3-associated heterodimer composed of disulfide-linked subunits.

A common feature of each of the CD3+ T-cell lines that failed to react with anti-$\alpha\beta$ or anti-β (framework) antibodies is the presence of an apparently full-length γ transcript, and the absence of α-chain and in some cases also full-length β, mRNA. While there is as yet no direct demonstration that the CD3-associated proteins are products of the γ gene, the data are highly suggestive. The biochemical features of the CD3-associated glycoprotein on the PEER cell line are consistent with the possibility that this protein is a γ-encoded protein (199). One subunit of the CD3-associated, non-disulfide linked heterodimer of the CD3+,$\alpha\beta$−,CD4,8− T cells reacted with an antiserum to a γ synthetic peptide. There is little doubt that firm structural data will be forthcoming.

The diversity of the CD3-associated structures expressed by CD3+,$\alpha\beta$− T cells is intriguing. Non-disulfide-linked heterodimers, disulfide-linked heterodimers, and disulfide-linked homodimers have all been described. This complexity is not necessarily surprising, since only one of the two human Cγ genes whose sequence is known has a cysteine that might be used for interchain disulfide formation (55, 56). Another exciting observation is that in each case save one, clear evidence exists that the structure detected is composed of two dissimilar subunits. This is consistent with structural analyses suggesting that $\gamma\gamma$ homodimers would be unstable due to electrostatic effects (104). This raises the exciting prospect that the γ chain is expressed with a partner, δ, that might be encoded by yet a fourth rearranging gene.

A second common feature shared by the cells that express a putative γ chain is surface phenotype—CD3+,$\alpha\beta$−. A recent study examined the distribution and relative frequency of CD3+,$\alpha\beta$(WT-31)− in normal lymphoid tissues (204). In adult peripheral blood, $\approx 2\%$ of T lymphocytes (range <0.2%–10%) are CD3+,WT-31−, whereas the remaining T lymphocytes are CD3+,WT-31+ (204). In thymus, $\approx 70\%$ of thymocytes are CD3+,WT-31+, and $\approx 0.5\%$ of thymocytes (range <0.2%–$\approx 1\%$) are CD3+,WT-31−. An examination of murine thymus and peripheral lymphoid tissues yielded comparable results (J. Bluestone, personal communication). In both thymus and blood, the CD3+,WT-31− T cells are unique in that these cells express neither CD4 nor CD8, whereas most CD3+,WT-31+ thymocytes express both CD4 and CD8 (204). The CD4−,CD8− phenotype is usually associated with immature cells in the thymic cortex (170, 205) that are precursors of functional T cells (206). CD3+,$\alpha\beta$− T cells may represent a developmental stage of CD3+,$\alpha\beta$+ T lymphocytes, or, alternatively, constitute an independent lineage of T lymphocytes. Insight might come from analysis if similar cells in the mouse, since in the mouse different $V\gamma$ genes are expressed in thymus and in the periphery (52).

ANTIGEN-RECEPTOR EXPRESSION DURING INTRATHYMIC DIFFERENTIATION

The evolution and diversification of the T-cell repertoire is influenced to a large degree by the thymus (207–209). A number of critical yet poorly understood selective events occur during the residency of T-cell precursors in the thymus, resulting in the death of the large majority (>95%) of the differentiating cells (210) and the acquisition of self-tolerance, antigen specificity, and MHC restriction by the surviving cells.

Considerable insight into the process has been obtained from the phenotypic and functional analysis of thymocytes (see 205, 224 for review). The adult thymus can be divided into four major compartments of phenotypically and functionally distinct cells. Approximately 3–15% express neither CD4 (L3T4) nor CD8 (Lyt-2) and are functionally incompetent. Cells of this phenotype are the earliest immigrants in the fetal thymus (211, 212). Adoptive transfer experiments indicate that these double-negative cells can give rise to each of the other populations (206). These data suggest that the double-negative cells represent a population of precursor thymocytes at an intermediate developmental stage between marrow-derived stem cells and functional T cells. The great majority (75–85%) of thymocytes express both CD4 and CD8. The role of these double-positive cells in thymic differentiation is controversial. The double-positive cells may represent an intermediate stage (67), but there is no direct data demonstrating differentiative potential of these cells in vivo or in vitro. Finally, the thymus contains T cells that express only CD4 (8–15%) or CD8 (3–7%). These single-positive cells are functionally mature and probably represent the endpoint of intrathymic differentiation.

Studies of fetal thymocytes, adult thymic subpopulations, and hybridomas constructed from immature thymocytes indicate an ordered rearrangement of T-cell antigen receptor genes in ontogeny and differentiation (59, 60, 213–218). The γ gene appears to be the first to rearrange and become expressed, followed quickly by β and finally by α (see 115 for review). The same order is followed during differentiation in the adult. The double-negative compartment contains cells that have high levels of γ mRNA, substantial levels of full-length β mRNA, and little or no α mRNA (59, 214, 218). Moreover, cells with the β gene in germline configuration and cells with rearranged β genes are found in the double-negative compartment (214). Similarly, studies of murine (219) and human (220, 221) T-cell tumors indicate that double-negative cells generally express β, but not α, mRNA. Double positive cells express full-length β mRNA and may or may not express α mRNA (213, 219–221). The mature, single-positive cells have rearranged β genes and express both β and α mRNA (213, 215, 219–221).

The finding that γ RNA expression precedes β and α led to speculation that the γ chain might associate with β or α chains to form a second type of receptor (59, 222). To date there has been no direct demonstration of association between γ and either β or α. T cells reported to express a putative γ protein all fail to express α RNA, and in some cases they also fail to express full-length β RNA (199, 201, 202). The relevance of the γ gene to the differentiation of cells that ultimately express the $\alpha\beta$ heterodimer and function as MHC-restricted cells is uncertain.

Expression of functional α mRNA appears to be the event that allows cell surface expression of the $\alpha\beta$ heterodimer. Expression of mRNA for CD3-δ (214, 221) and CD3-ε (221) occurs in double-negative cells and may precede β-gene rearrangement and expression (221). In the absence of α transcription, CD3-δ and CD3-ε accumulate in the cytoplasm localized with the perinuclear envelope (221). A similar localization of β chain occurs in cortical cells that lack surface expression of the $\alpha\beta$ heterodimer (223). This pattern of localization is consistent with restriction to the endoplasmic reticulum and lack of transport to the golgi for processing. These observations suggest that expression of α chain and assembly of the entire $\alpha\beta$/CD3 complex are required for transport and processing of any of the components.

Although it has previously been proposed that surface expression of the $\alpha\beta$ antigen receptor is an exclusive property of mature T cells (66, 170), considerable evidence indicates that expression occurs in immature T cells (97, 171, 219, 223). First, the frequency of reactivity of thymocytes with an antibody to murine Vβ8 suggested that double-positive cells express receptor (97). Second, staining of thymus sections revealed surface expression of receptor on cells in the cortex (223). Third, the majority of double-positive cells in a panel of phenotypically characterized primary murine T cell tumors expressed surface heterodimer that immunoprecipitated with an antibody to common epitopes (219). Finally, three-color flow cytometric analysis of human T cells indicates that approximately 50–75% of CD4+,CD8+ cells stained with antibodies to CD3 and with an antibody to invariant epitopes of the $\alpha\beta$ heterodimer (171, 206). Taken together, these data provide compelling evidence that $\alpha\beta$ antigen receptor is expressed on the majority of double-positive thymocytes.

Two observations suggest that expression of heterodimer by immature T cells may be relevant to selective events during thymic differentiation. First, the frequency of heterodimer expression by double-positive cells is probably higher than expected by chance. Since rearrangement is imprecise, only one in three rearrangements can be expected to yield in-frame junctions. The possibility of multiple rearrangements, addition of non-germline nucleotides, and the availability of multiple reading frames preclude an accurate estimate of the frequency of productive rearrangements, but it seems likely that the majority of transcripts would be sterile (50). The fact that fully 75% of double-positive thymocytes express receptor suggests that these cells have been selected for expression of heterodimer protein. Second, surface expression of heterodimer protein by many lymphocytes in the thymic cortex is restricted to the pole adjacent to an epithelial cell (223). This suggests an interaction that may have significance in selection.

A model indicating the possible relationships of the major subpopulations in the human thymus is presented in Figure 3. The available data for the mouse indicate an essentially identical pattern. It should be pointed out that, with the exception of the demonstration in the mouse that the CD4−,CD8−, or double-negative cells, can give rise to each of the other populations (206), no lineal relationships have been directly established. The availability of an increasing number of probes, in the form of monoclonal antibodies and cloned genes for the antigen-receptor

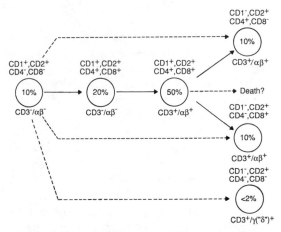

Figure 3   A model of human thymic differentiation. The existence of thymocytes expressing the indicated antigenic phenotypes has been demonstrated by correlated three and four color immunofluorescence and flow cytometric analysis. The frequencies indicated are approximate. The connecting arrows indicate possible relationships between the cell types.

chains and for the relevant accessory molecules, will facilitate the dissection of the events in the evolution of the T-cell response.

SUMMARY AND PERSPECTIVE

It is clear that the receptor for MHC-restricted recognition of antigens by the majority of T cells is the $\alpha\beta$ heterodimer. The specificity of the receptor is determined by events occurring during thymic residence of precursor cells. The $\alpha\beta$ receptor is expressed as part of an ensemble with proteins of the CD3 complex. The function of the $\alpha\beta$ heterodimer is to recognize and bind antigen in a ternary complex with MHC on the surface of antigen presenting cells. Antigen/MHC binding by the $\alpha\beta$ receptor is not fundamentally different from antigen binding by immunoglobulin. The manner in which a T cell interacts with an antigen-presenting cell is influenced by a set of accessory molecules, including two, CD4 and CD8, that correlate with the MHC restriction phenotype of the T cell. Binding of antigen/MHC to the $\alpha\beta$ receptor results in the transmission of a signal, probably via components of the CD3 complex, which results in activation of the cell. While the basic features of the system and some of the rules by which it operates are understood, the processes involved in generation of the repertoire, interaction between the T cell and its target, and signal transduction are only partially understood at present.

In addition to cells expressing the $\alpha\beta$ receptor, both the thymus and the peripheral circulation contain cells that express an alternative structure, probably the product of the γ gene, in association with the CD3 complex. These cells lack expression of CD4 and CD8 and may be precursors that ultimately give rise to cells that employ $\alpha\beta$/CD3 as an antigen receptor. Alternatively, these cells may represent a previously unrecognized class of T cells that share with conventional T cells the common feature of employing the CD3 complex for signal transduction but use a receptor made up in part of the γ chain for antigen recognition. The rules governing the expression of γ are poorly understood, as is the functional potential and biological role of the $\alpha\beta-$,CD3+,CD4−,CD8− cells. The possibility exists that there may be a fourth gene that contributes to antigen-specific recognition by a new class of T cells.

ACKNOWLEDGMENTS

The author thank Art Weiss, Michael Brenner, Larry Samelson, and Jeff Bluestone for sharing manuscripts and data prior to publication, and Nancy Caplan for assistance in preparing the manuscript.

530   ALLISON & LANIER

Literature Cited

1. Allison, J. P., McIntyre, B. W., Bloch, D. 1982. Tumor-specific antigen of murine T lymphoma defined with monoclonal antibody. *J. Immunol.* 129 : 2293–2300
2. Bigler, R. D., Fisher, D. E., Wang, C. Y., Rinnooykan, E. A., Kunkel, H. G. 1983. Idiotype-like molecules on cells of a human T cell leukemia. *J. Exp. Med.* 158 : 1000–5
3. Haskins, K., Kubo, R., White, J., Pigeon, M., Kappler, J., Marrack, P. 1983. The major histocompatibility complex-restricted antigen receptor on T cells. I. Isolation with a monoclonal antibody. *J. Exp. Med.* 157 : 1149–69
4. Meuer, S., Fitzgerald, K., Hussey, R., Hodgdon, J., Schlossman, S., Reinherz, E. 1983. Clonotypic structures involved in antigen-specific human T cell function. Relationship to the T3 molecular complex. *J. Exp. Med.* 157 : 705–19
5. Samelson, L. E., Schwartz, R. H. 1984. Characterization of the antigen-specific T cell receptor from a pigeon cytochrome c-specific T cell hybrid. *Immunol. Rev.* 81 : 131–44
6. Staerz, U. D., Pasternack, M. S., Klein, J. R., Benedetto, J. D., Bevan, M. J. 1984. Monoclonal antibodies specific for a murine cytotoxic T-lymphocyte clone. *Proc. Natl. Acad. Sci. USA* 81 : 1799–1803
7. Kaye, J., Procelli, S., Tite, J., Jones, B., Janeway, C. Jr. 1983. Both a monoclonal antibody and antisera specific for determinants unique to individual cloned helper T cell lines can substitute for antigen and antigen-presenting cells in the activation of T cells. *J. Exp. Med.* 158 : 836–56
8. Lancki, D. W., Lorber, M. I., Loken, M. R., Fitch, F. W. 1983. A clone-specific antibody that inhibits cytolysis of a cytolytic T-cell clone. *J. Exp. Med.* 157 : 921–35
9. Reinherz, E., Meuer, S., Fitzgerald, K., Hussey, R., Hodgdon, J., Acuto, O., Schlossman, S. 1983. Comparison of T3-associated 49- and 42-kilodalton cell surface molecules on individual human T-cell clones: Evidence for peptide variability in T-cell receptor structures. *Proc. Natl. Acad. Sci. USA* 80 : 4104–8
10. Goding, J. W., Harris, A. W. 1981. Subunit structure of cell surface proteins: Disulfide bonding in antigen receptors, Ly-2/3 antigens, and transferrin receptors of murine T and B lymphocytes. *Proc. Natl. Acad. Sci. USA* 78 : 4530–34
11. McIntyre, B. W., Allison, J. P. 1983. The mouse T cell receptor: Structural heterogeneity of molecules of normal T cells defined by xenoantiserum. *Cell* 34 : 739–46
12. Kappler, J., Kubo, R., Haskins, K., Hannum, C., Marrack, P., Pigeon, M., McIntyre, B., Allison, J., Trowbridge, I. 1983. The major histocompatibility complex-restricted antigen receptor on T cells in mouse and man: Identification of constant and variable peptides. *Cell* 35 : 295–302
13. Allison, J. P., Ridge, L., Lund, J., Gross-Pelose, J., Lanier, L. L., McIntyre, B. W. 1984. The murine T cell antigen receptor and associated structures. *Immunol. Rev.* 81 : 145–60
14. Acuto, O., Meuer, S. C., Hodgdon, J. C., Schlossman, S. F., Reinherz, E. L. 1983. Peptide variability exists within the α and β subunits of the T cell receptor for antigen. *J. Exp. Med.* 158 : 1368
15. Marrack, P., Shimonkevitz, R., Hannum, C., Haskins, K., Kappler, J. 1983. The major histocompatibility complex-restricted antigen receptor on T cells. IV. An anti-idiotypic antibody predicts both antigen and I-specificity. *J. Exp. Med.* 158 : 1635–46
16. Samelson, L. E., Germain, R. N., Schwartz, R. H. 1983. Monoclonal antibodies against the antigen receptor on a cloned T-cell hybrid. *Proc. Natl. Acad. Sci. USA* 80 : 6972–76
17. Kaye, J., Janeway, C. A. Jr. 1984. The Fab fragment of a directly activating monoclonal antibody that precipitates a disulfide linked heterodimer from a helper T cell clone blocks activation by either allogeneic-Ia or antigen and self-Ig. *J. Exp. Med.* 159 : 1397
18. Samelson, L. E. 1985. An analysis of the structure of the antigen receptor on a pigeon cytochrome c-specific T cell hybrid. *J. Immunol.* 134 : 2529–35
19. McIntyre, B. W., Allison, J. P. 1984. Biosynthesis and processing of murine T cell antigen receptor. *Cell* 38 : 659–65
20. Hubbard, S. C., Kranz, D. M., Longmore, G. D., Sitkovsky, M. V., Eisen, H. N. 1986. Glycosylation of the T-cell antigen-specific receptor and its potential role in lectin-mediated cytotoxicity. *Proc. Natl. Acad. Sci. USA* 83 : 1852–56
21. Kaye, J., Janeway, C. A. 1984. The alpha and beta subunits of a murine T

cell antigen/Ia receptor have a molecular weight of 31,000 in the absence of N-linked glycosylation. *J. Immunol.* 133 : 2291–93

22. Borst, J., Alexander, S., Elder, J., TerHorst, C. 1983. The T3 complex on human T lymphocytes involves four structurally distinct glycoproteins. *J. Biol. Chem.* 258 : 5135–43

23. Iwamoto, A., Rupp, F., Ohashi, P. S., Walker, C. L., Pircher, H., Joho, R., Hengartner, H., Mak, T. W. 1986. T cell-specific γ genes in C57BL/10 mice : Sequence and expression of new constant and variable region genes. *J. Exp. Med.* 163 : 1203–12

24. Heilig, J. S., Glimcher, L. H., Kranz, D. M., Clayton, L. K., Greenstein, J. L., Saito, H., Maxam, A. M., Burakoff, S. J., Eisen, H. N., Tonegawa, S. 1985. Expression of the T-cell-specific gamma gene is unnecessary in T cell recognizing class II MHC determinants. *Nature* 317 : 68–70

25. Hedrick, S. M., Nielsen, E. A., Kavaler, J., Cohen, D. I., Davis, M. M. 1984. Sequence relationships between putative T-cell receptor polypeptides and immunoglobulins. *Nature* 308 : 153–58

26. Hedrick, S. M., Cohen, D. I., Nielsen, E. A., Davis, M. M. 1984. Isolation of cDNA clones encoding T cell-specific membrane associated proteins. *Nature* 308 : 149–53

27. Yanagi, Y., Yoshikai, Y., Legget, K., Clark, S. P., Aleksander, I., Mak, T. W. 1984. A human T cell-specific cDNA clone encodes a protein having extensive homology to immunoglobulin chains. *Nature* 308 : 145–49

28. Chien, Y., Becker, D., Lindsten, T., Okamura, M., Cohen, D., Davis, M. 1984. A third type of murine T-cell receptor gene. *Nature* 312 : 31–35

29. Saito, H., Kranz, D., Takagaki, Y., Hayday, A., Eisen, H., Tonegawa, S. 1984. A third rearranged and expressed gene in a clone of cytotoxic T lymphocytes. *Nature* 312 : 36–40

30. Acuto, O., Fabbi, M., Smart, J., Poole, C. B., Protentis, J., Royer, H. D., Schlossman, S., Reinherz, E. L. 1984. Purification and NH2-terminal amino acid sequencing of the β subunit of a human T-cell antigen receptor. *Proc. Natl. Acad. Sci. USA* 81 : 3851–55

31. Hannum, C. H., Kappler, J. W., Trowbridge, I. S., Marrack, P., Freed, J. H. 1984. Immunoglobulin-like nature of the α-chain of a human T-cell antigen/MHC receptor. *Nature* 312 : 65–67

32. Saito, H., Kranz, D. M., Takagaki, Y., Hayday, A. C., Eisen, H. N., Tonegawa, S. 1984. Complete primary structure of a heterodimeric T-cell receptor deduced from cDNA sequences. *Nature* 309 : 757–62

33. Chien, Y.-H., Gascoigne, N. R. J., Kavaler, J., Lee, N. E., Davis, M. M. 1984. Somatic recombination in a murine T-cell receptor gene. *Nature* 309 : 322–26

34. Siu, G., Clark, S. P., Yoshikai, Y., Malissen, M., Yanagi, Y., Strauss, E., Mak, T. W., Hood, L. 1984. The human T cell antigen receptor is encoded by variable, diversity, and joining gene segments that rearrange to generate a complete V gene. *Cell* 37 : 393–401

35. Patten, P., Yokota, T., Rothbard, J., Chien, Y.-H., Arai, K.-I., Davis, M. M. 1984. Structure, expression and divergence of T-cell receptor β-chain variable regions. *Nature* 312 : 40–46

36. Barth, R., Kim, B., Lan, N., Hunkapiller, T., Sobieck, N., Winoto, A., Gershenfeld, H., Okada, C., Hansburg, D., Weissman, I., Hood, L. 1985. The murine T-cell receptor employs a limited repertoire of expressed V_β gene segments. *Nature* 316 : 517–23

37. Behlke, M. A., Spinella, D. G., Chou, H., Sha, W., Hartle, D. L., Loh, D. Y. 1985. T-cell receptor β chain expression : Dependence on relatively few variable region genes. *Science* 229 : 566–70

38. Behlke, M. A., Chou, H. S., Huppi, K., Loh, D. Y. 1986. Murine T-cell receptor mutants with deletions of betachain variable region genes. *Proc. Natl. Acad. Sci. USA* 83 : 767–71

39. Malissen, M., Minard, K., Mjolsness, S., Kronenberg, M., Goverman, J., Hunkapiller, T., Prystowsky, M. B., Yoshikai, Y., Fitch, F., Mak, T. W., Hood, L. 1984. Mouse T-cell antigen receptor : Structure and organization of constant and joining gene segments encoding the β polypeptide. *Cell* 37 : 1101–10

40. Gascoigne, N. R. J., Chien, Y.-H., Becker, D. M., Kavaler, J., Davis, M. M. 1984. Genomic organization and sequence of T-cell receptor β-chain constant- and joining-region genes. *Nature* 310 : 387–91

41. Kavaler, J., Davis, M. M., Chien, Y.-H. 1984. Localization of a T-cell receptor diversity region element. *Nature* 310 : 421–23

42. Clark, S. P., Yoshikai, Y., Taylor, S., Siu, G., Hood, L., Mak, T. W. 1984.

Identification of a diversity segment of human T-cell receptor beta-chain, and comparison with the analogous murine element. *Nature* 311: 387–89

43. Siu, G., Kronenberg, M., Strauss, E., Haars, R., Mak, T. W., Hood, L. 1984. The structure, rearrangement and expression of D beta gene segments of the murine T-cell antigen receptor. *Nature* 311: 344–50

44. Winoto, A., Mjolsness, S., Hood, L. 1985. Genomic organization of the genes encoding mouse T-cell receptor alpha-chain. *Nature* 316: 832–36

45. Hayday, A. C., Saito, H., Gillies, S. D., Kranz, D. M., Tanigawa, G., Eisen, H. N., Tonegawa, S. 1985. Structure, organization and somatic rearrangement of T-cell gamma genes. *Cell* 40: 259–69

46. Yoshikai, Y., Clark, S. P., Taylor, S., Sohn, U., Wilson, B., Minden, M., Mak, T. W. 1985. Organization and sequences of the variable, joining and constant region genes of the human T-cell receptor α chain. *Nature* 316: 837–40

47. Arden, B., Klotz, J., Siu, G., Hood, L. 1985. Diversity and structure of genes of the α family of mouse T-cell antigen receptor genes. *Nature* 316: 783–87

48. Becker, D., Patten, P., Chien, Y.-H., Yokota, T., Eshhar, Z., Giedlin, M., Gascoigne, N. R. J., Goodenow, C., Wolf, R., Arai, K.-I., Davis, M. M. 1985. Variability and repertoire size in T-cell receptor V_α gene segments. *Nature* 317: 430

49. Yoshikai, Y., Kimura, N., Toyonaga, B., Mak, T. W. 1986. Sequences and repertoire of human T cell receptor alpha chain variable region genes in mature T lymphocytes. *J. Exp. Med.* 164: 90–103

50. Kronenberg, M., Siu, G., Hood, L. E., Shastri, N. 1986. The molecular genetics of the T-cell antigen receptor and T-cell recognition. *Ann. Rev. Immunol.* 4: 529–91

51. Kranz, D. M., Saito, H., Heller, M., Takagaki, Y., Haas, W., Eisen, H. N., Tonegawa, S. 1985. Limited diversity of the rearranged T-cell γ gene. *Nature* 313: 752–55

52. Garman, R. D., Doherty, P. J., Raulet, D. H. 1986. Diversity, rearrangement, and expression of murine T cell gamma genes. *Cell* 45: 733–42

53. Lefranc, M. P., Forster, A., Rabbitts, T. H. 1986. Rearrangement of two distinct T-cell gamma-chain variable-region genes in human DNA. *Nature* 319: 420–22

54. Lefranc, M. P., Forster, A., Baer, R., Stinson, M. A., Rabbitts, T. H. 1986. Diversity and rearrangement of the human T cell rearranging gamma genes: Nine germ-line variable genes belonging to two subgroups. *Cell* 45: 237–46

55. Quertermous, T., Murre, C., Dialynas, D., Duby, A. D., Strominger, J. L., Waldman, T. A., Seidman, J. G. 1986. Human T-cell γ chain genes: Organization, diversity, and rearrangement. *Science* 231: 252–55

56. Lefranc, M. P., Rabbitts, T. H. 1985. Two tandemly organized human genes encoding the T-cell γ constant-region sequences show multiple rearrangement in different T-cell types. *Nature* 316: 464–66

57. Dialynas, D. P., Murre, C., Quertermous, T., Boss, J. M., Leiden, J. M., Seidman, J. G., Strominger, J. L. 1986. Cloning and sequence analysis of complementary DNA encoding an aberrantly rearranged human T-cell gamma chain. *Proc. Natl. Acad. Sci. USA* 83: 2619–23

58. Quertermous, T., Strauss, W., Murre, C., Dialynas, D. P., Strominger, J. L., Seidman, J. G. 1986. Human T-cell gamma genes contain N segments and have marked junctional variability. *Nature* 322: 184–87

59. Raulet, D. H., Garman, R. D., Saito, H., Tonegawa, S. 1985. Developmental regulation of T-cell receptor gene expression. *Nature* 314: 103–7

60. Snodgrass, H. R., Dembic, Z., Steinmetz, M., von Boehmer, H. 1985. Expression of T-cell antigen receptor genes during fetal development in the thymus. *Nature* 315: 232–33

61. Reilly, E. B., Kranz, D. M., Tonegawa, S., Eisen, H. N. 1986. A functional gamma gene formed from known gamma-gene segments is not necessary for antigen-specific responses of murine cytotoxic T lymphocytes. *Nature* 321: 878–80

62. Rupp, F., Frech, G., Hengartner, H., Zinkernagel, R. M., Joho, R. 1986. No functional gamma-chain transcripts detected in an alloreactive cytotoxic T-cell clone. *Nature* 321: 876–78

63. Kranz, D. M., Sherman, D. H., Sitovsky, M. V., Pasternack, M. S., Eisen, H. N. 1984. Immunoprecipitation of cell surface structures of cloned cytotoxic T lymphocytes by clone-specific antisera. *Proc. Natl. Acad. Sci. USA* 81: 573–77

64. Lanier, L. L., Ruitenberg, J. J., Allison, J. P., Weiss, A. 1986. Distinct epitopes

on the T cell antigen receptor of HPB-ALL tumor cells identified by monoclonal antibodies. *J. Immunol.* 137: 2286

65. Acuto, O., Campen, T. J., Royer, H. D., Hussey, R. E., Poole, C. B., Reinherz, E. L. 1985. Molecular analysis of T cell receptor (Ti) variable region (V) gene expression. Evidence that a single Tiβ V gene family can be used in formation of V domains on phenotypically and functionally diverse T cell populations. *J. Exp. Med.* 161: 1326–43

66. Acuto, O., Hussey, R. E., Fitzgerald, K. A., Protentis, J. P., Meuer, S. C., Schlossman, S. F., Reinherz, E. L. 1983. The human T cell receptor: Appearance in ontogeny and biochemical relationship of α and β subunits on IL-2 dependent clones and T cell tumors. *Cell* 34: 717–26

67. Yague, J., White, J., Coleclough, C., Kappler, J., Palmer, E., Marrack, P. 1985. The T cell receptor: The α and β chains define idiotype, and antigen and MHC specificity. *Cell* 42: 81–87

68. Boylston, A. W., Borst, J., Yssel, H., Blanchard, D., Spits, H., DeVries, J. E. 1986. Properties of a panel of monoclonal antibodies which react with the human T cell antigen receptor on the leukemic line HPB-ALL and a subset of normal peripheral blood T lymphocytes. *J. Immunol.* 137: 741–44

69. Brenner, M. B., Mclean, J., Strominger, J. L. 1986. Human T cell receptor structure and subsets defined by framework monoclonal antibodies. *Fed. Proc.* 45: 377 (Abstr.)

70. Haskins, K., Hannum, C., White, J., Roehm, N., Kubo, R., Kappler, J., Marrack, P. 1984. The major histocompatibility complex-restricted antigen receptor on T cells. VI. An antibody to a receptor allotype. *J. Exp. Med.* 160: 452–71

71. Roehm, N. W., Carbone, C., Kushnir, E., Taylor, B. A., Riblet, R. J., Marrack, P., Kappler, J. W. 1985. The major histocompatibility complex-restricted antigen receptor on T cells: The genetics of expression of an allotype. *J. Immunol.* 135: 2176–82

72. Staerz, U. D., Rammensee, H. G., Benedetto, J. D., Bevan, M. J. 1985. Characterization of a murine monoclonal antibody specific for an allotypic determinant on T cell antigen receptor. *J. Immunol.* 134: 3994–4000

73. Sim, G.-K., Augustin, A. 1985. Vβ gene polymorphism and a major polyclonal T-cell receptor idiotype. *Cell* 42: 89–92

74. Brenner, M. B., Trowbridge, I. S.,

Mclean, J., Strominger, J. L. 1984. Identification of shared antigenic determinants of the putative human T lymphocyte antigen receptor. *J. Exp. Med.* 160: 541–51

75. Fabbi, M., Acuto, O., Bensussan, A., Poole, C. B., Reinherz, E. L. 1985. Production and characterization of antibody probes directed at constant regions of the alpha and beta subunit of the human T cell receptor. *Eur. J. Immunol.* 15: 821–27

76. Traunecker, A., Dolder, B., Karjalainen, K. 1986. A novel approach for preparing anti-T cell receptor constant region antibodies. *Eur. J. Immunol.* 16: 851–54

77. Allison, J. P., Lanier, L. L. 1985. Identification of antigen receptor associated structures on murine T cells. *Nature* 314: 107–8

78. Spits, H., Borst, J., Tax, W., Capel, P. J. A., TerHorst, C., deVries, J. E. 1985. Characteristics of a monoclonal antibody (WT-31) that recognizes a common epitope on the human T cell receptor for antigen. *J. Immunol.* 135: 1922–28

79. Kindred, B., Shreffler, D. C. 1972. H-2 dependence of cooperation between T and B cells in vivo. *J. Immunol.* 109: 940–43

80. Katz, D. H., Hammaoka, T., Benacerraf, B. 1973. Cell interactions between histoincompatible T and B lymphocytes. II. Failure of physiologic cooperative interactions between T and B lymphocytes from allogeneic donor strains in humoral response to hapten protein conjugates. *J. Exp. Med.* 137: 1405–18

81. Zinkernagel, R. M., Doherty, P. C. 1974. Restriction of in vitro T-cell mediated cytotoxicity in lymphocytic choriomeningitis within a syngeneic or semi-allogeneic system. *Nature* 248: 701–2

82. Hunig, T. R., Bevan, M. J. 1982. Antigen recognition by cloned cytotoxic T-lymphocytes follows rules predicted by the altered-self hypothesis. *J. Exp. Med.* 155: 111–25

83. Heber-Katz, E., Schwartz, R. H., Matis, L. A., Hannum, C., Fairwell, T., Appella, E., Hansburg, D. 1982. Contribution of antigen-presenting cell major histocompatibility complex gene products to the specificity of antigen-induced T-cell activation. *J. Exp. Med.* 155: 1086–99

84. Hedrick, S. M., Matis, L. A., Hecht, T. T., Samelson, L. E., Longo, D. L., Heber-Katz, E., Schwartz, R. H. 1982.

The fine specificity of antigen and Ia determinant recognition by T cell hybridoma clones specific for cytochrome c. *Cell* 30: 141–52

85. Kappler, J. W., Skidmore, B., White, J., Marrack, P. 1981. Antigen-inducible, H-2-restricted interleukin-2-producing T cell hybridomas. Lack of independent antigen and H-2 recognition. *J. Exp. Med.* 153: 1198–1214

86. Dembic, Z., Haas, W., Weiss, S., McCubrey, J., Kiefer, H., von Boehmer, H., Steinmetz, M. 1986. Transfer of specificity by murine α and β T-cell receptor genes. *Nature* 320: 232–38

87. Honjo, T. 1983. Immunoglobulin genes. *Ann. Rev. Immunol.* 1: 499–528

88. Hayday, A., Diamond, D., Tanigawa, G., Heilig, J., Folsom, V., Saito, H., Tonegawa, S. 1985. Unusual features of the organization and diversity of T-cell receptor α chain genes. *Nature* 316: 828–32

89. Winoto, A., Mjolsness, S., Hood, L. 1985. Genomic organization of the genes encoding the mouse T-cell receptor α chain: 18 Jα gene segments map over 60 kilobases of DNA. *Nature* 316: 832–36

90. Sim, G. K., Yague, J., Nelson, J., Marrack, P., Palmer, E., Augustin, A., Kappler, J. 1984. Primary structure of human T-cell receptor α-chain. *Nature* 312: 771–75

91. Sims, J. E., Tunnacliffe, A., Smith, W. G., Rabbitts, T. H. 1984. Complexity of human T-cell antigen receptor beta-chain constant- and variable-region genes. *Nature* 312: 541–45

92. Yoshikai, Y., Anatoniou, D., Clark, S. P., Yanagi, Y., Sangster, R., van den Elsen, P., TerHorst, C., Mak, T. W. 1984. Sequence and expression of transcripts of the human T-cell receptor β-chain genes. *Nature* 312: 521–24

93. Jones, N., Leiden, J., Dialynas, D., Fraser, J., Clabby, M., Kishimoto, T., Strominger, J. L., Andrews, D., Lane, W., Woody, J. 1985. Partial primary structure of the alpha and beta chains of human tumor T-cell receptors. *Science* 227: 311–14

94. Tunnacliffe, A., Lefford, R., Milstein, C., Forster, A., Rabbitts, T. H. 1985. Sequence and evolution of the human T-cell receptor β-chain genes. *Proc. Natl. Acad. Sci. USA* 82: 5068–72

95. Kronenberg, M., Goverman, J., Haars, R., Malissen, M., Kraig, E., Phillips, L., Delovitch, T., Suciu-Foca, N., Hood, L. 1985. Rearrangement and transcription of the beta-chain genes of the T-cell antigen receptor in different

types of murine lymphocytes. *Nature* 313: 647–53

96. Royer, H. D., Bensussan, A., Acuto, O., Reinherz, E. L. 1984. Functional isotypes are not encoded by the constant region genes of the β subunit of the T-cell receptor for antigen/major histocompatibility complex. *J. Exp. Med.* 160: 947–52

97. Roehm, N., Herron, L., Cambier, J., DiGuisto, D., Haskins, K., Kappler, J., Marrack, P. 1984. The major histocompatibility complex-restricted antigen receptor on T cells: Distribution on thymus and peripheral T cells. *Cell* 38: 577–84

98. Goverman, J., Minard, K., Shastri, N., Hunkapiller, T., Hansburg, D., Sercarz, E., Hood, L. 1985. Rearranged beta T cell receptor genes in a helper T cell clone specific for lysozyme: No correlation between V beta and MHC restriction. *Cell* 40: 859–67

99. Garman, R. D., Ko, J. L., Vulpe, C. D., Raulet, D. H. 1986. T-cell receptor variable region gene usage in T-cell populations. *Proc. Natl. Acad. Sci. USA* 83: 3987–91

100. Hedrick, S. M., Germain, R. N., Bevan, M. J., Dorf, M., Engel, I., Fink, P., Gascoigne, N., Heber-Katz, E., Kapp, J., Kaufmann, Y., Kay, J., Melchers, F., Pierce, C., Schwartz, R. H., Sørenson, C., Taniguchi, M., Davis, M. 1985. Rearrangement and transcription of a T-cell receptor beta-chain gene in different T-cell subsets. *Proc. Natl. Acad. Sci. USA* 82: 531–35

101. Fink, P. J., Matis, L. A., McElligott, D. L., Bookman, M., Hedrick, S. M. 1986. Correlations between T-cell specificity and the structure of the antigen receptor. *Nature* 321: 219–26

102. Hochgeschwender, U., Weltzien, H. U., Eichmann, K., Wallace, R. B., Epplen, J. 1986. Preferential expression of a defined T-cell receptor β-chain gene in hapten-specific cytotoxic T-cell clones. *Nature* 322: 376–78

103. Kronenberg, M., Siu, G., Hood, L. E., Shastri, N. 1986. The molecular genetics of the T-cell antigen receptor and T-cell antigen recognition. *Ann. Rev. Immunol.* 4: 529–91

104. Novotnoy, J., Tonegawa, S., Saito, H., Kranz, D. M., Eisen, H. N. 1986. Secondary, tertiary, and quaternary structure of T-cell-specific immunoglobulin-like polypeptide chains. *Proc. Natl. Acad. Sci. USA* 83: 742–46

105. Wylie, D. E., Sherman, L. A., Klinman, N. R. 1982. Participation of the major histocompatibility complex in anti-

body recognition of viral antigens expressed on infected cells. *J. Exp. Med.* 155: 403–14

106. Babbitt, B., Allen, P. M., Matsueda, G., Haber, E., Unanue, E. R. 1985. Binding of immunogenic peptides to Ia histocompatibility molecules. *Nature* 317: 359–61

107. Ashwell, J. D., Schwartz, R. H. 1986. T-cell recognition of antigen and the Ia molecule as a ternary complex. *Nature* 320: 176–79

108. Watts, T. H., Gaub, H. E., McConnell, H. M. 1986. T-cell-mediated association of peptide antigen and major histocompatibility complex protein detected by energy transfer in an evanescent wave-field. *Nature* 320: 179–81

109. Reinherz, E. L., Kung, P. C., Goldstein, G., Schlossman, S. F. 1979. A monoclonal antibody with selective reactivity with functionally mature human thymocytes and all peripheral human T cells. *J. Immunol.* 123: 1312–18

110. Reinherz, E. L., Hussey, R. E., Schlossman, S. F. 1980. A monoclonal antibody blocking human T cell function. *Eur. J. Immunol.* 10: 758–62

111. Allison, J. P., McIntyre, B. W., Ridge, L. L., Gross-Pelose, J., Lanier, L. L. 1985. Molecular characterization of the murine T cell antigen receptor and associated structures. *Fed. Proc.* 44: 2870–73

112. Oettgen, H. C., Kappler, J., Tax, W. J., TerHorst, C. 1984. Characterization of the two heavy chains of the T3 complex on the surface of human T lymphocytes. *J. Biol. Chem.* 259: 12039–48

113. Brenner, M. B., Trowbridge, I. S., Strominger, J. L. 1985. Cross-linking of human T cell receptor proteins: Association between the T cell idiotype beta subunit and the T3 glycoprotein heavy subunit. *Cell* 40: 183–90

114. Weiss, A., Stobo, J. D. 1984. Requirement for the coexpression of T3 and the T cell antigen receptor on a malignant human T cell line. *J. Exp. Med.* 160: 1284–99

115. Borst, J., Coligan, J. E., Oettgen, H., Pessano, S., Malin, R., TerHorst, C. 1984. The delta- and epsilon-chains of the human T3/T-cell receptor complex are distinct polypeptides. *Nature* 312: 455–58

116. Samelson, L. E., Harford, J. B., Klausner, R. D. 1985. Identification of the components of the murine T cell antigen receptor complex. *Cell* 43: 223–31

117. Oettgen, H. C., Pettey, C. L., Maloy, W. L., TerHorst, C. 1986. A T3-like protein complex associated with the antigen receptor on murine T cells. *Nature* 320: 272–75

118. Samelson, L. E., Weissman, A. M., Robey, F. A., Berkower, I., Klausner, R. D. 1986. Characterization of antipeptide antibody that recognizes the murine analogue of the human T cell antigen receptor-T3 δ chain. *J. Immunol.* In press

119. Samelson, L. E., Patel, M. D., Weissman, A. M., Harford, J. B., Klausner, R. D. 1986. Antigen activation of murine T cells induces tyrosine phosphorylation of a polypeptide associated with the T cell antigen receptor. *Cell* 46: 1083

120. van den Elsen, P., Shepley, B. A., Borst, J., Coligan, J. E., Markham, A. F., Orkin, S., TerHorst, C. 1984. Isolation of cDNA clones encoding the 20K T3 glycoprotein of human T-cell receptor complex. *Nature* 312: 413–18

121. van den Elsen, P., Shepley, B. A., Cho, M., TerHorst, C. 1985. Isolation and characterization of a cDNA clone encoding the murine homologue of the human 20K T3/T-cell receptor glycoprotein. *Nature* 314: 542–44

122. Gold, D. P., Puck, J. M., Pettey, C. L., Cho, M., Coligan, J., Woody, J. N., TerHorst, C. 1986. Isolation of cDNA clones encoding the 20K non-glycosylated polypeptide chain of the human T-cell receptor/T3 complex. *Nature* 321: 431–34

123. Krissansen, G. W., Owen, M. J., Verbi, W., Crumpton, M. J. 1986. Primary structure of the T3 γ subunit of the T3/T cell antigen receptor complex deduced from cDNA sequences: Evolution of the T3 γ and δ subunits. *EMBO J.* 5: 1799–1808

124. Samelson, L. E., Harford, J., Schwartz, R. H., Klausner, R. D. 1985. A 20-kDa protein associated with the murine T-cell antigen receptor is phosphorylated in response to activation by antigen or concanavalin A. *Proc. Natl. Acad. Sci. USA* 82: 1969–73

125. Cantrell, D. A., Davies, A. A., Crumpton, M. J. 1985. Activators of protein kinase C down-regulate and phosphorylate the T3/T-cell antigen receptor complex of human T lymphocytes. *Proc. Natl. Acad. Sci. USA* 82: 8158–62

126. Meuer, S. C., Acuto, O., Hussey, R. E., Hodgdon, J. C., Fitzgerald, K. A., Schlossman, S. F., Reinherz, E. L. 1983. Evidence for the T3-associated 90K heterodimer as the T-cell antigen receptor. *Nature* 303: 808–10

127. Landegren, U., Ramstedt, U., Axberg, I., Ullberg, M., Jondal, M., Wigzell, H. 1982. Selective inhibition of human T cell cytotoxicity at levels of target recognition or initiation of lysis by monoclonal OKT3 and Leu 2a antibodies. *J. Exp. Med.* 155: 1579–84

128. Spits, H., Yssel, H., Leeuwenberg, J., de Vries, J. E. 1985. Antigen-specific cytotoxic T cell and antigen-specific proliferating T cell clones can be induced to cytolytic activity by monoclonal antibodies against T3. *Eur. J. Immunol.* 15: 88–91

129. Leeuwenberg, J. F. M., Spits, H., Tax, W. J. M., Capel, P. J. A. 1985. Induction of nonspecific cytotoxicity by monoclonal anti-T3 antibodies. *J. Immunol.* 134: 3770–75

130. Mentzer, S. J., Barbosa, J. A., Burakoff, S. J. 1985. T3 monoclonal antibody activation of nonspecific cytolysis: A mechanism of CTL inhibition. *J. Immunol.* 135: 34–38

131. Staerz, U. D., Bevan, M. J. 1985. Cytotoxic T lymphocyte-mediated lysis via the Fc receptor of target cells. *Eur. J. Immunol.* 15: 1172–77

132. Phillips, J. H., Lanier, L. L. 1986. Lectin-dependent and anti-CD3 induced cytotoxicity are preferentially mediated by peripheral blood cytotoxic T lymphocytes expressing Leu-7 antigen. *J. Immunol.* 136: 1579–85

133. Perez, P., Hoffman, R. W., Shaw, S., Bluestone, J. A., Segal, D. M. 1985. Specific targeting of cytotoxic T cells by anti-T3 linked to anti-target cell antibody. *Nature* 316: 354–56

134. Perez, P., Hoffman, R. W., Titus, J. A., Segal, D. M. 1986. Specific targeting of human peripheral blood T cells by heteroaggregates containing anti-T3 crosslinked to anti-target cell antibodies. *J. Exp. Med.* 163: 166–78

135. Staerz, U. D., Kanagawa, O., Bevan, M. J. 1985. Hybrid antibodies can target sites for attack by T cells. *Nature* 314: 628–31

136. Staerz, U. D., Bevan, M. J. 1986. Hybrid hybridoma producing a bispecific monoclonal antibody that can focus effector T-cell activity. *Proc. Natl. Acad. Sci. USA* 83: 1453–57

137. Reinherz, E. L., Hussey, R. E., Schlossman, S. F. 1980. A monoclonal antibody blocking human T cell function. *Eur. J. Immunol.* 10: 758–62

138. Meuer, S. C., Hodgdon, J. C., Hussey, R. E., Protentis, J. P., Schlossman, S. F., Reinherz, E. L. 1983. Antigen-like effects of monoclonal antibodies directed at receptors on human T cell

clones. *J. Exp. Med.* 158: 988–93

139. Bigler, R. D., Posnett, D. N., Chiorazzi, N. 1985. Stimulation of a subset of normal resting T lymphocytes by a monoclonal antibody to a crossreactive determinant of the human T cell antigen receptor. *J. Exp. Med.* 161: 1450–63

140. Boylston, A. W., Cosford, P. 1985. Growth of normal human T lymphocytes induced by monoclonal antibody to the T cell antigen receptor. *Eur. J. Immunol.* 15: 738–42

141. Moretta, A., Pantaleo, G., Lopez-Botet, M., Mingari, M. C., Moretta, L. 1985. Anticlonotypic monoclonal antibodies induce proliferation of clonotype-positive T cells in peripheral blood human T lymphocytes: Evidence for a phenotypic (T4/T8) heterogeneity of the clonotype-positive proliferating cells. *J. Exp. Med.* 162: 1393–98

142. Van Wauwe, J. P., De Mey, J. R., Goossens, J. G. 1980. OKT3: A monoclonal anti-human T lymphocyte antibody with potent mitogenic properties. *J. Immunol.* 124: 2708–13

143. Tax, W., Willems, H., Reekers, P., Capel, J., Koene, R. 1983. Polymorphism in mitogenic effect of IgG1 monoclonal antibodies against T3 antigen on human T cells. *Nature* 304: 445–47

144. Chang, T., Testa, D., Kung, P., Perry, L., Dreskin, H., Goldstein, G. 1982. Cellular origin and interactions involved in γ-interferon production induced by OKT3 monoclonal antibody. *J. Immunol.* 128: 585–89

145. von Wussow, P., Platsoucas, C. D., Wiranowska-Stewart, M., Stewart, W. E. 1981. Human γ interferon production by leukocytes induced with monoclonal antibodies recognizing T cells. *J. Immunol.* 127: 1197–1200

146. Weiss, A., Wiskocil, R. L., Stobo, J. D. 1984. The role of T3 surface molecules in the activation of human T cells: A two-stimulus requirement for IL 2 production reflects events occurring at a pre-translational level. *J. Immunol.* 133: 123–28

147. Wiskocil, R., Weiss, A., Imboden, J., Kamin-Lewis, R., Stobo, J. 1985. Activation of a human T cell line: A two-stimulus requirement in the pretranslational events involved in the coordinate expression of interleukin 2 and γ-interferon genes. *J. Immunol.* 134: 1599–1603

148. Weiss, A., Imboden, J., Hardy, K., Manger, B., TerHorst, C., Stobo, J. 1986. The role of the T3/antigen recep-

tor complex in T-cell activation. *Ann. Rev. Immunol.* 4: 593–619

149. Ohashi, P. S., Mak, T. W., van den Elsen, P., Yanagi, Y., Yoshikai, Y., Calman, A. F., TerHorst, C., Stobo, J. D., Weiss, A. 1985. Reconstitution of an active surface T3/T-cell antigen receptor by DNA transfer. *Nature* 316: 606–9

150. Manger, B., Weiss, A., Weyand, C., Goronzy, J., Stobo, J. D. 1985. T cell activation: Differences in the signals required for IL 2 production by nonactivated and activated T cells. *J. Immunol.* 135: 3669–73

151. O'Flynn, K., Zanders, E., Lamb, J., Beverley, P., Wallace, D., Tatham, P., Tax, W., Linch, D. 1985. Investigation of early T cell activation: Analysis of the effect of specific antigen, interleukin 2 and monoclonal antibodies on intracellular free calcium concentration. *Eur. J. Immunol.* 15: 7–11

152. Oettgen, H. C., TerHorst, C., Cantley, L. C., Rosoff, P. M. 1985. Stimulation of the T3-T cell receptor complex induces a membrane-potential-sensitive calcium influx. *Cell* 40: 583–90

153. Imboden, J. B., Weiss, A., Stobo, J. D. 1985. Transmembrane signalling by the T3-antigen receptor complex. *Immunol. Today* 6: 328–31

154. Imboden, J. B., Stobo, J. D. 1985. Transmembrane signalling by the T cell antigen receptor. Perturbation of the T3-antigen receptor complex generates inositol phosphates and releases calcium ions from intracellular stores. *J. Exp. Med.* 161: 446–56

155. Weiss, A., Imboden, J., Shoback, D., Stobo, J. 1984. Role of T3 surface molecules in human T-cell activation: T3-dependent activation results in an increase in cytoplasmic free calcium. *Proc. Natl. Acad. Sci. USA* 81: 4169–73

156. Weiss, M. J., Daley, J. F., Hodgdon, J. C., Reinherz, E. L. 1984. Calcium dependency of antigen-specific (T3-Ti) and alternative (T11) pathways of human T-cell activation. *Proc. Natl. Acad. Sci. USA* 81: 6836–40

157. Kurzinger, K., Reynolds, T., Germain, R. N., Davignon, D., Martz, E., Springer, T. A. 1981. A novel lymphocyte function-associated antigen (LFA-1): Cellular distribution, quantitative expression, and structure. *J. Immunol.* 127: 596–602

158. Sanchez-Madrid, F., Krensky, A. M., Ware, C. F., Robbins, E., Strominger, J. L., Burakoff, S. J., Springer, T. A. 1982. Three distinct antigens associated with human T lymphocyte-mediated cytolysis: LFA-1, LFA-2, and LFA-3. *Proc. Natl. Acad. Sci. USA* 79: 7489–93

159. Sanchez-Madrid, F., Simon, P., Thompson, S., Springer, T. A. 1983. Mapping of antigenic and functional epitopes on the α- and β-subunits of two related mouse glycoproteins involved in cell interactions, LFA-1 and MAC-1. *J. Exp. Med.* 158: 586–602

160. Springer, T. A., Davignon, D., Ho, M.-K., Kurzinger, K., Martz, E., Sanchez-Madrid, F. 1982. LFA-1 and Lyt-2,3 molecules associated with T lymphocyte-mediated killing; and Mac-1, and LFA-1 homologue associated with complement function. *Immunol. Rev.* 68: 171–95

161. Davignon, D., Martz, E., Reynolds, T., Kurzinger, K., Springer, T. A. 1981. Monoclonal antibody to a novel lymphocyte function-associated antigen (LFA-1): Mechanism of blockade of T lymphocyte-mediated killing and effects on other T and B lymphocyte functions. *J. Immunol.* 127: 590–95

162. Gronkowski, S. H., Krensky, A. M., Martz, E., Burakoff, S. J. 1985. Functional distinctions between the LFA-1, LFA-2 and LFA-3 membrane proteins on human CTL are revealed with trypsin-pretreated target cells. *J. Immunol.* 134: 244–49

163. Krensky, A. M., Sanchez-Madrid, F., Robbins, E., Nagy, J. A., Springer, T. A., Burakoff, S. J. 1983. The functional significance, distribution, and structure of LFA-1, LFA-2 and LFA-3: cell surface antigens associated with CTL-target interaction. *J. Immunol.* 131: 611–16

164. Krensky, A. M., Robbins, E., Springer, T. A., Burakoff, S. J. 1984. LFA-1, LFA-2 and LFA-3 antigens are involved in CTL-target cell conjugation. *J. Immunol.* 132: 2180–82

165. Dongworth, D. W., Gotch, F. M., Hildreth, J. E. K., Morris, A., McMichael, A. J. 1985. Effects of monoclonal antibodies to the α and β chains of the human lymphocyte function-associated (H-LFA-1) antigen on T lymphocyte function. *Eur. J. Immunol.* 15: 888–92

166. Spits, H., van Schooten, W., Keizer, H., van Seventer, G., Van de Rijn, M., TerHorst, C., de Vries, J. E. 1986. Alloantigen recognition is preceded by nonspecific adhesion of cytotoxic T cells and target cells. *Science* 232: 403–5

167. Mentzer, S. J., Gromkowski, S. H., Krensky, A. M., Burakoff, S. J., Martz, E. 1985. LFA-1 membrane molecule in the recognition of homotypic adhesions of human B lymphocytes. *J. Immunol.* 135: 9–11

168. Meuer, S. C., Hussey, R. E., Fabbi, M., Fox, D., Acuto, O., Fitzgerald, K. A., Hodgdon, J. C., Protentis, J. P., Schlossman, S. F., Reinherz, E. L. 1984. An alternative pathway of T-cell activation: A functional role for the 50 KD T11 sheep erythrocyte receptor protein. *Cell* 36: 897–906

169. Siliciano, R. F., Pratt, J. C., Schmidt, R. E., Ritz, J., Reinherz, E. L. 1985. Activation of cytolytic T lymphocyte and natural killer cell function through the T11 sheep erythrocyte binding protein. *Nature* 317: 428–30

170. Reinherz, E. L., Schlossman, S. F. 1980. The differentiation and function of human T lymphocytes. *Cell* 19: 821–27

171. Lanier, L. L., Allison, J. P., Phillips, J. H. 1986. Correlation of cell surface antigen expression on human thymocytes by multi-color flow cytometric analysis: Implications for differentiation. *J. Immunol.* 137: 2501

172. Fox, D. A., Schlossman, S. F., Reinherz, E. L. 1986. Regulation of the alternative pathway of T cell activation by anti-T3 monoclonal antibody. *J. Immunol.* 136: 1945–50

173. Howard, F. D., Ledbetter, J. A., Wong, J., Bieber, C. P., Stinson, E. B., Herzenberg, L. A. 1981. A human T lymphocyte differentiation marker defined by monoclonal antibodies that block E rosette formation. *J. Immunol.* 126: 2117–22

174. Kamoun, M., Kadin, M. E., Martin, P. J., Nettleton, J., Hansen, J. A. 1981. A novel human T cell antigen preferentially expressed on mature T cells and shared by both well and poorly differentiated B cell leukemias and lymphomas. *J. Immunol.* 127: 987–91

175. Shaw, S., Luce, G. E. G., Quinones, R., Gress, R. E., Springer, T. A., Sanders, M. E. 1986. Two antigen-independent adhesion pathways used by human cytotoxic T cell clones. *Nature* In press

176. Milanese, C., Richardson, N. E., Reinherz, E. L. 1986. Identification of a T helper cell-derived lymphokine that activates resting T lymphocytes. *Science* 231: 1118–22

177. Swain, S. L. 1981. Significance of Lyt phenotypes: Lyt2 antibodies block activities of T cells that recognize class 1 major histocompatibility complex antigens regardless of their function. *Proc. Natl. Acad. Sci. USA* 78: 1101–5

178. Engleman, E. G., Benike, C., Glickman, E., Evans, R. L. 1981. Antibodies to membrane structures that distinguish suppressor/cytotoxic and helper T lymphocyte subpopulations block the mixed lymphocyte reaction in man. *J. Exp. Med.* 154: 193–98

179. Dialynas, D. P., Quan, Z. S., Wall, K. A., Pierres, A., Quintans, J., Loken, M. R., Pierres, M., Fitch, F. W. 1983. Characterization of the murine T cell surface molecule, designated L3T4, identified by monoclonal antibody GK1.5: Similarity of L3T4 to the human Leu3/T4 molecule. *J. Immunol.* 131: 2445–51

180. Marrack, P., Endres, R., Shimonkevitz, R., Zlotnik, A., Dialynas, D., Fitch, F., Kappler, J. 1983. The major histocompatibility complex-restricted antigen receptor on T cells. II. Role of the L3T4 product. *J. Exp. Med.* 158: 1077–91

181. Biddison, W. E., Rao, P. E., Talle, M. A., Goldstein, G., Shaw, S. 1984. Possible involvement of the T4 molecule in T-cell recognition of class II HLA antigens: Evidence from studies of CTL-target cell binding. *J. Exp. Med.* 159: 793–97

182. Meuer, S. C., Schlossman, S. F., Reinherz, E. L. 1982. Clonal analysis of human cytotoxic T lymphocytes: T4 and T8 effector T cells recognize products of different major histocompatibility antigens. *Proc. Natl. Acad. Sci. USA* 79: 4395–99

183. Evans, R. L., Wall, D. W., Platsoucas, C. D., Siegal, F. 1981. Thymus-dependent membrane antigens in man: Inhibition of cell-mediated lympholysis by monoclonal antibodies to TH2 antigen. *Proc. Natl. Acad. Sci. USA* 78: 544–48

184. Wilde, D. B., Marrack, P., Kappler, J., Dialynas, D. P., Fitch, F. W. 1983. Evidence implicating L3T4 in class II MHC antigen reactivity: Monoclonal antibody GK1.5 (anti-L3T4a) blocks class II MHC antigen specific proliferation, release of lymphokines, and binding by cloned murine helper T lymphocyte lines. *J. Immunol.* 131: 2178–83

185. Shaw, S., Goldstein, G., Springer, T. A., Biddison, W. E. 1985. Susceptibility of cytotoxic T lymphocyte (CTL) clones to inhibition by anti-T3 and anti-T4 (but not anti-LFA-1) monoclonal antibodies varies with the "avid-

ity" of CTL-target interaction. *J. Immunol.* 134: 3019–26

186. Shinohara, N., Sachs, D. H. 1979. Mouse alloantibodies capable of blocking cytotoxic T-cell function. I. Relationship between the antigen reactive with blocking antibodies and the Lyt-2 locus. *J. Exp. Med.* 150: 432–44

187. Nakayama, E., Shiku, H., Stockert, E., Oettgen, H. F., Old, L. J. 1979. Cytotoxic T cells: Lyt phenotype and blocking of killing activity by Lyt antisera. *Proc. Natl. Acad. Sci. USA* 76: 1977–81

188. Fan, J., Ahmed, A., Bonavida, B. 1980. Studies on the induction and expression of T cell-mediated immunity. X. Inhibition by Lyt2,3 antisera of cytotoxic T lymphocyte-mediated antigen-specific and -nonspecific cytotoxicity: Evidence for the blocking of the binding between T lymphocytes and target cells and not the post-binding cytolytic steps. *J. Immunol.* 125: 2444–53

189. Platsoucas, C. D. 1984. Human T cell antigens involved in cytotoxicity against allogeneic or autologous chemically modified targets. Association of the Leu 2a/T8 antigen with effector-target cell binding and of the T3/Leu 4 antigen with triggering. *Eur. J. Immunol.* 14: 566–77

190. Bank, I., Chess, L. 1985. Perturbation of the T4 molecule transmits a negative signal to T cells. *J. Exp. Med.* 162: 1294–1303

191. Welte, K., Platzer, E., Wang, C. Y., Kan, E. A. R., Moore, M. A. S., Mertelsman, R. 1983. OKT8 antibody inhibits OKT3 inducted IL-2 production and proliferation in OKT8+ cells. *J. Immunol.* 131: 2356–61

192. Maddon, P. J., Littman, D. R., Godfrey, M., Maddon, D. E., Chess, L., Axel, R. 1985. The isolation and nucleotide sequence of a cDNA encoding the T-cell surface protein T4: A new member of the immunoglobulin gene family. *Cell* 42: 93–104

193. Littman, D. R., Thomas, Y., Maddon, P. J., Chess, L., Axel, R. 1985. The isolation and sequence of the gene encoding T8: A molecule defining functional classes of T lymphocytes. *Cell* 40: 237–46

194. Sukhatme, V. P., Sizer, K. C., Vollmer, A. C., Hunkapiller, T., Parnes, J. R. 1985. The T cell differentiation antigen Leu-2/T8 is homologous to immunoglobulin and T cell receptor variable regions. *Cell* 40: 591–97

195. Ledbetter, J. A., Evans, R. L., Lipinski, M., Cunningham-Rundles, C., Good, R. A., Herzenberg, L. A. 1981. Evolutionary conservation of surface molecules that distinguish T lymphocyte helper/inducer and cytotoxic/suppressor subpopulations in mouse and man. *J. Exp. Med.* 153: 310–23

196. Ledbetter, J. A., Tsu, T. T., Clark, E. A. 1985. Covalent association between human thymus leukemia-like antigens and CD8 (Tp32) molecules. *J. Immunol.* 134: 4250–54

197. Ledbetter, J. A., Herzenberg, L. A. 1979. Xenogeneic monoclonal antibodies to mouse lymphoid differentiation antigens. *Immunol. Rev.* 47: 63–90

198. Ledbetter, J. A., Seaman, W. E., Tsu, T. T., Herzenberg, L. A. 1981. Lyt-2 and Lyt-3 antigens are on two different polypeptide sub-units linked by disulfide bonds. Relationship of subunits to T cell cytolytic activity. *J. Exp. Med.* 153: 1503–16

199. Weiss, A., Newton, M., Crommie, D. 1986. Expression of T3 in association with a molecule distinct from the T cell antigen receptor heterodimer. *Proc. Natl. Acad. Sci. USA* 83: 6998

200. Nowill, A., Moingeon, P., Ythier, A., Graziani, M., Faure, F., Delmon, L., Rainaut, M., Forrestier, F., Bohuon, C., Hercend, T. 1986. Natural killer clones derived from fetal (25 wk) blood. Probing the human T cell receptor with WT31 monoclonal antibody. *J. Exp. Med.* 163: 1601–6

201. Brenner, M. B., Mclean, J., Dialynas, D. P., Strominger, J. L., Smith, J. A., Owen, F. L., Seidman, J. G., Ip, S., Rosen, F., Krangel, M. S. 1986. Identification of a putative second T-cell receptor. *Nature* 322: 145–49

202. Bank, I., DePinho, R. A., Brenner, M. B., Cassimeris, J., Alt, F. W., Chess, L. 1986. A functional T3 molecule associated with a novel heterodimer on the surface of immature human thymocytes. *Nature* 322: 179–81

202b. Moingeon, P., Ythier, A., Goubin, G., Faure, F., Nowill, A., Delmon, L., Rainaut, M., Forrestier, F., Daffos, F., Bohuon, C., Hercend, T. 1986. A unique T-cell receptor complex expressed on human fetal lymphocytes displaying natural killer-like activity. *Nature* 323: 638

203. Lanier, L. L., Ruitenberg, J. J., Phillips, J. H. 1986. Human CD3+ T lymphocytes that express neither CD4 nor CD8 antigens. *J. Exp. Med.* 164: 339–44

204. Lanier, L. L., Weiss, A. 1986. Presence of Ti (WT31) negative T lymphocytes in normal blood and thymus. *Nature.* In press

205. Scollay, R., Shortman, K. 1983. Thymocyte subpopulations: An experimental review including flow cytometric cross-correlations between the major urine thymocyte markers. *Thymus* 5: 245–60

206. Fowlkes, B. J., Edison, L., Mathieson, B. J., Chused, T. M. 1985. Early T lymphocytes. Differentiation in vivo of adult intrathymic precursor cells. *J. Exp. Med.* 162: 802–22

207. Zinkernagel, R. M. 1978. Thymus and lymphohemapoietic cells: Their role in T cell maturation in selection of T cell H-2 restriction specificity and in H-2 linked Ir gene control. *Immunol. Rev.* 42: 224–36

208. Fink, P. J., Bevan, M. J. 1978. H-2 antigens of the thymus determine lymphocyte specificity. *J. Exp. Med.* 148: 766–73

209. von Boehmer, H., Haas, W., Jerne, N. K. 1978. Major histocompatibility complex linked immune responsiveness is acquired by lymphocytes of low-responder mice differentiating in thymus of high-responder mice. *Proc. Natl. Acad. Sci. USA* 75: 2439–43

210. McPhee, D., Pye, J., Shortman, K. 1979. The differentiation of T lymphocytes. V. Evidence for intrathymic death of most thymocytes. *Thymus* 1: 151–62

211. van Ewijk, W., Jenkinson, E. J., Owen, J. J. T. 1982. Detection of Thy-1, T-200, Lyt-1, and Lyt-2 bearing cells in the developing lymphoid organs of the mouse embryo *in vivo* and *in vitro*. *Eur. J. Immunol.* 12: 262–71

212. Ceredig, R., MacDonald, H. R., Jenkinson, E. J. 1983. Flow microfluorometric analysis of mouse thymus development *in vivo* and *in vitro*. *Eur. J. Immunol.* 13: 185–90

213. Snodgrass, H. R., Kisielow, P., Kiefer, M., Steinmetz, M., von Boehmer, H. 1985. Ontogeny of the T-cell antigen receptor within the thymus. *Nature* 313: 592–95

214. Samelson, L. E., Lindsten, T., Fowlkes, B. J., van den Elsen, P., TerHorst, C., Davis, M. M., Germain, R. N., Schwartz, R. H. 1985. Expression of genes of the T-cell antigen receptor complex in precursor thymocytes. *Nature* 315: 765–68

215. Royer, H. D., Acuto, O., Fabbi, M., Tizard, R., Ramachandran, K., Smart, J. E., Reinherz, E. L. 1984. Genes enco-

ding the Tiβ subunit of the antigen/MHC receptor undergo rearrangement during intrathymic ontogeny prior to surface T3-Ti expression. *Cell* 39: 261–66

216. Born, W., Yague, J., Palmer, E., Kappler, J., Marrack, P. 1985. Rearrangement of T-cell receptor β-chain genes during T-cell development. *Proc. Natl. Acad. Sci. USA* 82: 2925–29

217. Haars, R., Kronenberg, M., Gallatin, W. M., Weissman, I. L., Owen, F. L., Hood, L. 1986. Rearrangement and expression of T cell antigen receptor and gamma genes during thymic development. *J. Exp. Med.* 164: 1–24

218. Trowbridge, I. S., Lesley, J., Trotter, J., Hyman, R. 1985. Thymocyte subpopulation enriched for progenitors with an unrearranged T-cell receptor β-chain gene. *Nature* 315: 666–69

219. Richie, E. R., McEntire, B., Guyden, J., Lanier, L., Allison, J. P. 1987. Expression of T-cell antigen receptor α and β chain genes in phenotypically characterized murine T lymphomas: Implications for thymocyte differentiation. *J. Immunol.* In press

220. Sangster, R. N., Minowada, J., Suciu-Foca, N., Minden, M., Mak, T. W. 1986. Rearrangement and expression of the alpha, beta, and gamma chain T cell receptor genes in human thymic leukemia cells and functional T cells. *J. Exp. Med.* 163: 1491–1508

221. Furley, A. J., Mizutani, S., Weilbaecher, K., Dhaliwal, H. S., Ford, A. M., Chan, L. C., Molgaard, H. V., Toyonaga, B., Mak, T., van den Elsen, P., Gold, D., TerHorst, C., Greaves, M. F. 1986. Developmentally regulated rearrangement and expression of genes encoding the T cell receptor-T3 complex. *Cell* 46: 75–87

222. Pernis, B., Axel, R. 1985. A one and a half receptor model for MHC-restricted antigen recognition by T lymphocytes. *Cell* 41: 13–16

223. Farr, A. G., Adamson, S. K., Marrack, P., Kappler, J. 1985. Expression of antigen-specific, major histocompatibility complex-restricted receptors by cortical and medullary thymocytes in situ. *Cell* 42: 543–50

224. Mathieson, B. J., Fowlkes, B. J. 1984. Cell surface antigen expression on thymocytes: Development and phenotypic differentiation of intrathymic subsets. *Immunol. Rev.* 82: 141–73

Ann. Rev. Immunol. 1987. 5 : 541–59

RECENT ADVANCES IN THE MOLECULAR BIOLOGY OF HTLV-1: *TRANS*-ACTIVATION OF VIRAL AND CELLULAR GENES

Mitsuaki Yoshida and Motoharu Seiki

Department of Viral Oncology, Cancer Institute Kami-Ikebukuro, Toshima-ku, Tokyo 170, Japan

INTRODUCTION

Human T-cell leukemia virus type 1 (HTLV-1) is the first human retrovirus to be well established (1, 2) and is associated with a unique T-cell malignancy—adult T-cell leukemia (ATL) (3). Since the ATL and HTLV-1 show the same geographic distribution (4, 5, 6) and patients with ATL are always infected with HTLV-1, it has been expected that HTLV-1 would provide a direct clue to understanding the mechanism of tumorigenesis in humans. It was also expected that studies on HTLV-1 would provide strategies for diagnosis, treatment, and prevention of ATL. Studies on HTLV-1 and the related viruses have progressed rapidly, largely as the result of technologies developed in research on animal retroviruses. New aspects of the retroviruses have been discovered, including a novel mechanism of leukemogenesis. This chapter reviews recent advances in the molecular biology of HTLV-1, focusing on the activation of viral and cellular gene expression.

ATL

Adult T-cell leukemia (ATL) was first described by Takatsuki and his colleagues in 1977 (3). ATL affects mostly adults and is frequently associated with hypercalcemia and skin lesions. The malignant cells in most cases are helper T cells (7); they show the highly lobulated nucleus and

0732–0582/87/0410–0541$02.00

express the receptor for interleukin-2 (IL-2R) on their surface (8, 9, 10). However, the IL-2 receptor does not seem to be required for tumor cell growth (10). ATL is endemic in southwestern Japan (4), the West Indies (5) and central Africa (6).

HTLV-1

The HTLV-1 was first isolated from American patients with T-cell lymphoma (1, 11) and subsequently from Japanese patients with ATL (2). The Japanese isolate was initially described as ATLV (2), but later HTLV-1 and ATLV were shown to be the same (12, 13). Many other isolates were then reported from other areas where the disease is endemic and also from those where it is not. The etiological association of HTLV-1 with ATL was shown by the following observations: (a) The geographical distribution of HTLV-1 is the same as that of ATL (4, 5, 14, 15); (b) With only few exceptions, patients with ATL are infected with HTLV-1 (2, 4, 5, 14, 15); (c) Tumor cells of patients with ATL are always infected with HTLV-1 (as evidenced by presence of the integrated proviral genome), whereas nonleukemic cells in the patients are mostly not infected (16, 17, 18, 19); (d) In vitro transmission of HTLV-1 frequently immortalizes the recipient T cells (9, 20, 21).

Viral infection shows familial aggregation, and the virus is thought to be transmitted from mother to child (15) (probably through milk; 22), from husband to wife (15), and also by means of blood transfusion (23). All these transmissions seem to require the transfer of living infected cells from virus carriers to recipients. Seropositive people are mostly over 20 years old, and the positive rates increase with age (15). Only a small proportion of the infected population develops leukemia, and these only after long latency.

The HTLV-1 genome contains open reading frames for *gag*, *pol*, and *env* as replication competent retroviruses (24). However, the unique feature of this virus is the presence of an unusual sequence now termed "pX" between the *env* gene and the 3′ long terminal repeat (LTR) (24). This sequence is not a typical oncogene derived from cellular genome (24), and its presence has established HTLV-1 as a new retrovirus group. It is now a common marker for members of the so-called HTLV family, which includes HTLV-2 (25, 26), STLV-1 (27) and BLV (28, 29), in addition to HTLV-1. The biological significance of the expression of this gene is a main subject of this chapter.

HTLV-2

HTLV-2 was isolated from a patient with hairy T-cell leukemia (25); the virus is structurally related to, but distinct from, HTLV-1 (30). On average, sequence homology between HTLV-1 and -2 is about 60% but is much higher in the pX region (26, 31). Since only two cancer patients carrying HTLV-2 have so far been found, the pathogenicity of HTLV-2 remains unclear.

VIRAL INVOLVEMENT IN LEUKEMOGENESIS: PROPOSAL FOR *TRANS*-ACTING VIRAL FUNCTION

Association of HTLV-1 with ATL was clearly demonstrated by extensive seroepidemiology which showed the same geographical distributions of HTLV-1 and ATL and infection with HTLV-1 in all patients with ATL. Analysis of provirus integration in primary leukemic cells provide an insight into the mode of viral involvement in the development of ATL.

Causative Role of HTLV-1

A crucial question in studies of HTLV-1 was whether the viral function is involved in the mechanism of leukemogenesis in ATL directly or indirectly. This was tested by analysis of the mode of provirus integration in primary tumor cells from ATL patients (16, 17, 18, 19), because the proviral genome is integrated into the host chromosomal genome for the viral gene expression. Leukemic or lymphoma cells from all ATL cases so far tested were found to contain the proviral genome. Furthermore, analysis of the viral-cellular junction showed that the primary tumor cells of all cases were monoclonal (16, 17, 18). No exception was found among ATL patients in the Japanese endemic area (16). The HTLV-1 infection into the ATL tumor cell itself and the monoclonality of tumor cells with respect to the provirus integration site both strongly suggest that HTLV-1 infects the target cells directly, and these then become malignant; that is, HTLV-1 infection played a causative role in the development of ATL at the single cell level.

In respect to HTLV-1 involvement in ATL, Shimoyama et al (32) recently reported a few exceptions. Six patients diagnosed as ATL were negative for antibodies against HTLV-1, and the tumor cells also contained no detectable provirus genome. Criteria for the diagnosis may still be problematic in certain cases, but it is apparent that ATL-like diseases can be induced without HTLV-1 infection. Nevertheless, the great majority of cases of ATL are etiologically associated with HTLV-1 infection.

Absence of Common Site for Provirus Integration

Two mechanisms are known in tumorigenesis induced by retroviruses: One is through the actions of viral oncogenes themselves; the other is through *cis*-activation of an adjacent cellular proto-oncogene by the integrated proviral LTR. Since HTLV-1 has no typical oncogene in its genome (24), a viral promoter insertion mechanism was suspected in the development of ATL. However, this mechanism was shown to be unlikely by the absence of a common integration site of HTLV-1 provirus genome in primary leukemic cells (33). A common site for proviral integration was sought by testing cellular DNA rearrangements induced by provirus integration, using cellular DNA probes flanked to integrated proviruses, but no common site was detected. That the provirus was integrated into different chromosomes in ATL cells from patients with ATL also supported this conclusion (33). These findings suggested that *cis*-acting viral promoter can not explain ATL development induced by HTLV-1. Eventually, a *trans*-acting viral function was expected to be involved in leukemogenesis, which can function without regard to the site of provirus integration.

A possible role for *trans*-acting viral function was also supported by in vitro immortalization of infected T-cells: HTLV-1 transmission by cocultivation of primary lymphocytes with X-irradiated cells that produce HTLV-1 can easily immortalize the recipient T cells of human (9, 20, 21), monkey (34), rabbit (35), rat (36), and cat (37). Similar effects were also observed with HTLV-2 (38) and STLV-1 (39). Cell-free preparations of HTLV-1 were also able to immortalize infected T cells (40), clearly eliminating the possibility that some cellular agents are transferred by the cocultivation technique. Frequent immortalization or transformation in vitro has never been reported with chronic leukemia virus, which has no oncogene. Therefore, a unique viral function, which can act in *trans*, was suspected to induce in vitro immortalization. The proposed *trans*-acting viral function for leukemogenesis and in vitro immortalization might be a novel mechanism not known in other retroviral carcinogenesis. These unique properties of HTLV-1 prompted us to study the unique sequence "pX."

IDENTIFICATION OF pX PROTEINS

Gene Products of the pX Sequence

The unique sequence pX was first identified in sequence analysis of HTLV-1 by Seiki et al (24); it has open reading frames (ORF) possibly coding for four proteins (Figure 1). This unique sequence is the region most highly homologous to HTLV-2 (26, 31), and this suggests its importance for viral replication.

A product encoded by the gene p40^x (also reported as a 42-kd protein), was first identified independently in four laboratories by means of different

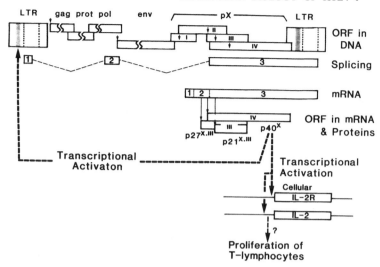

Figure 1 Schematic illustration of pX gene arrangements in HTLV-1 genome, mechanism of expression of pX genes, and their function. In ORFs in DNA sequence, the first initiation codon ATG in each ORF is indicated with small vertical arrow (↓) and initiation codons AUG are naturally at the first of ORFs in a mRNA. Two AUGs for p40^x and p27$^{x\text{-}III}$ located in exon 2 thus originated from the very 3′-region of *pol* gene. An AUG for p21$^{x\text{-}III}$ is derived from ORF III in DNA sequence. Shaded regions in two LTRs represent an enhancer that consisted of three direct repeats of 21 nucleotides which respond to the p40^x mediated *trans*-activation.

strategies (41–44). Kiyokawa et al (41) synthesized decapeptides coded by the 3′ terminal regions of ORFs I–IV in the pX sequence of HTLV-1. Using antisera against these peptides, they identified a 40-kd protein encoded by ORF IV. Lee et al (42) analyzed amino acid sequence of tryptic peptides of HTLV-1 42-kd protein. This protein was immunoprecipitated with patient serum and identified as a pX protein coded by ORF IV. The identification was based on a previously published nucleotide sequence (24). They used a term "x- lor" (long open reading frame) for ORF IV. Slamon et al (43) and Miwa et al (44) treated cell extract infected by HTLV-1 or HTLV-2 with anti-peptide sera and identified 40-kd protein for HTLV-1 and 38-kd for HTLV-2. The apparent molecular weight of the protein coded by ORF IV was reported as 40 kd, 41 kd, and 42 kd. However, the cDNA sequence of the mRNA coding for this protein predicted a molecular weight of 39,842 daltons for the product (see later section) (45). Furthermore, there is no evidence for possible modification of the product. Therefore, we will use the term "p40^x" for this protein throughout this chapter.

Nuclear association of the p40^x with the HTLV-1–producing cell line was discovered by subcellular fractionation (46), although a significant amount of p40^x was detected in the cytoplasmic fraction. More directly, the great majority of p40^x was found in nuclei, by cytochemical staining of the cells with antibodies (47, 48). A short half-life of p40^x was also described (47), suggesting some regulatory functions of the p40^x.

For a while, the p40^x was thought to be the only product of the pX sequence of HTLV-1. Thus the functions of pX were thought to be conferred by p40^x. Later, Kiyokawa et al (49) discovered a second and third pX protein. They found two novel proteins, 27 kd and 21 kd, by immunoblot assay with rabbit antibodies against synthetic peptide; these were predicted from ORF III in the pX sequence of HTLV-1. These two proteins were detected in all HTLV-1–producing cell lines but not in uninfected human T-cell lines; thus, these researchers concluded that these two proteins were encoded by ORF III in pX sequence of HTLV-1. They are provisionally named p27^{x-III} and p21^{x-III}. The p27^{x-III} and p21^{x-III} are phosphoproteins; p27^{x-III} is localized in nuclei, but p21^{x-III} is found in cytoplasm (49).

Mechanism and Regulation of pX Gene Expression

The three pX gene products were discovered to be encoded by overlapping ORFs III and IV of HTLV-1 (49). However, neither ORFs has an ATG codon in the 5′ region for initiating translation at the 5′ region (Figure 1). Seiki et al (45) analyzed a cDNA clone derived from a 2.1 kb pX mRNA, which codes for p40^x, and unequivocally demonstrated that the mRNA was formed by a double splicing. The first exon was derived from the 5′ half of the R sequence in the LTR, the second from the 3′ region of *pol*, and the third from the pX sequence. As a result of the second splicing between *pol* and pX, the ATG codon used for *env* gene translation is joined to ORF IV and is used for initiation of p40^x translation. Thus, only the first methionine is fused onto p40^x from the *env* domain. The molecular weight predicted from the cDNA sequence is 39,842 daltons, which exactly corresponds to the 40 kd estimated by gel electrophoresis (45). A nuclease S1 protection assay on mRNA was also carried out for the second splicing (56, 57, 58) and the same splicing was proposed.

The key sequences for these splicings for expression of p40^x in HTLV-1 are also found at similar positions in HTLV-2, BLV, and STLV-1 (45), but not in HIV. This suggests that the unusual splicing mechanism for the pX mRNA is common to all four members of the HTLV family. In fact, the predicted splicing was shown on HTLV-2. The conservation of these splicing signals strongly suggests that expression of the pX gene is biologically important.

The ORF III which encodes p27^{x-III} and p21^{x-III} overlaps mostly ORF

IV which codes for p40^x after double splicing (45). However, no evidence for alternatively spliced mRNA other than 2.1-kb pX mRNA was found by blotting or by a nuclease S1 protection assay. The nucleotide sequence of the cDNA clone corresponding to the 2.1-kb pX mRNA (45) gives ORF III starting with ATG, which is located in the second exon; thus, it came from the 3'-region of the *pol* gene (Figure 1). It was therefore speculated that this 2.1-kb pX mRNA also codes for p27^{x-III} and p21^{x-III}. This was directly demonstrated by Nagashima et al (50) who detected both the p40^x and p27^{x-III} in cells transfected with cDNA clone. Furthermore, they prepared pure mRNA using the SP6 vector and polymerase system and translated it in vitro. They showed that three pX proteins, p40^x, p27^{x-III} and p21^{x-III} were synthesized from a pure mRNA (50), and by inducing mutagenesis at ATG codons, they unequivocally demonstrated independent translation of these pX proteins from three ATG codons in two ORFs III and IV: the first ATG codon in mRNA for p27^{x-III}, the second for p40^x, and the fourth for p21^{x-III}. The first ATG codon for p27^{x-III} is located in exon 2; it thus originated from the 3'-region of the *pol* sequence. The N-terminal 19 amino acids of p27^{x-III} are encoded by the *pol* sequence but with a different ORF. Of these 19 amino acids, 9 are basic, suggesting this protein has affinity for nucleic acid (50). This may be consistent with the nuclear localization of this protein (49), but not p21^{x-III}. Furthermore, the basic signal is highly conserved among the HTLV family, which thus suggests the biological importance in viral replication (51). Equivalent proteins in HTLV-2 and BLV have not been reported yet. The p27^{x-III} and p21^{x-III} have the identical amino acid sequence except for a simple deletion of the N-terminal 78 amino acids in p21^{x-III}. They show an unusually high content of proline (21%) and serine (15%) which suggests unique protein structures (50). The biological significance of three proteins translated from a single mRNA, and thus under coordinate control, is not understood; one may speculate that these proteins function in the same or in a closely associated process.

How is expression of pX regulated? With respect to the expression of viral genes, it is known that primary tumor cells do not express detectable amounts of the viral antigens, *gag* and *env*, but the expression can be induced in vitro by cultivation for a few days (53). Specific analysis of expression of pX (49) showed also that the pX proteins, p40^x, p27^{x-III}, and p21^{x-III}, were not expressed in freshly isolated leukemic cells; however, these cells can express an anomalous amount of p40^x and p27^{x-III} after being cultured for 2 days. Synthesis of the pX proteins seems to be regulated in a way similar to that of other viral proteins. The in vivo suppression of viral gene expression seems to be partly but not entirely exerted by antibodies against viral proteins (52); it could not be explained by methylation of the proviral genomes (54, 55). Although leukemic cells or most of the infected cells do not express viral antigens in vivo, the viral proteins must be synthesized in certain infected cells, because high titers of antibodies are persistently present in infected people. In fact, almost all patients and carriers have antibodies against the viral *gag* and/or *env* proteins. With respect to the pX proteins, one third of those infected with HTLV-1 have detectable antibodies against p40^x (49); however, no HTLV-1 carriers or patients had significant levels of antibodies against p27^{x-III} and/or p21^{x-III} when 50 independent sera were tested (49). The reason for this difference is not known.

IDENTIFICATION OF *TRANS*-ACTING FACTOR

Trans-*Activation of the LTR Function*

Although some retroviruses such as avian acute leukemia virus MC29 can induce many types of tumors, many viruses exhibit target specificity for the pathogenicity. The LTR sequences of mouse retroviruses are known to exert this tissue specificity (59). For example, the tissue specificities of two retroviruses can be interchanged by interchange of their LTRs (60). ATL associated with HTLV-1 infection is a malignancy of helper type T cells almost exclusively, although the virus can infect many types of cells, such as B cells, fibroblasts and epithelial cells (61, 62). Therefore, it was initially speculated that the LTR of HTLV-1 showed T-cell specificity; however, this turned out not to be the case. Sodroski et al (63) first tested this possibility using a plasmid pLTR-CAT containing a gene for CAT (chloramphenicol acetyltransferase) under the control of the LTR of HTLV-1 and -2. After transfection of the pLTR-CAT into various cell lines, the CATase activity was measured by acetylation of chloramphenicol. The activity of the LTR-CAT was found to be almost equal in epithelial, fibroblast, B-, and T-cell lines, clearly demonstrating no significant T-cell specificity of the LTRs.

More surprisingly, Sodroski et al (63) and also Fujisawa et al (64) found that the LTR function in HTLV-1-infected cell lines was stimulated more than 100-fold relative to the SV40 promoter. On the other hand, such stimulation was not observed in uninfected human T-cell lines or in cells that did not express the viral antigens even when infected with HTLV-1. Therefore, the activation was presumptively mediated by a viral factor which acts in *trans* because the assay was carried out with unintegrated plasmids. Sodroski et al (63) observed virus-specific activation of the LTR of HTLV-1 and -2: Both LTRs of HTLV-1 and -2 were activated in HTLV-1–infected cell lines, whereas only HTLV-2 LTR was activated in cells infected with HTLV-2. This virus specificity supports the idea that the factor is a viral rather than a cellular protein.

The pX protein p40^x was proposed to be a *trans*-acting factor activating the LTR (63, 64), since activation of the LTR was also detected in cells—human cell line C81-66-45 and rat cell line TARL-2—which were thought to express only p40^x. Today, however, TARL-2 cells are known to express other pX proteins also, i.e. p27^{x-III} and p21^{x-III} (49). Direct evidence that pX proteins are in fact *trans*-activators of the transcription from the LTR was reported independently from several laboratories. Sodroski et al (58) and Felber et al (65) constructed plasmids in which the pX sequence was directed by HTLV-1 LTR or metallothioneine promoter, and the product could be expressed after splicing of the transcript. When the pLTR-CAT was cotransfected with these pX-expression plasmids, high expression of CATase was detected. Mutations in the pX sequence or antisense insertion abolished the activity. Thus, the expression of the pX gene appeared to be essential for the activation. These *trans*-activations of LTR in cotrans-fection assays were observed in many types of cells; this indicates that pX expression is directly involved in the activation. As described in the previous section, the pX can code for three proteins. To identify which protein is required for *trans*-activation, Seiki et al (66, 67) constructed plasmids consisting of metallothioneine promoter and a genomic or cDNA form of the pX sequence of HTLV-1. They then introduced frame-specific mutations which generated a termination codon in one ORF but did not induce gross amino acid alteration in the other ORFs. In quantitative analysis of the activities of these mutants, they clearly showed that p40^x alone is sufficient for *trans*-activation of the LTR function, and that neither p27^{x-III} nor p21^{x-III} is required.

Similar studies were also carried out on the pX sequence of HTLV-2 (58), expressing the p38^x by fusion of the p38^x-coding sequence to the 5'-most sequence of the CAT gene coding sequence driven by the LTR. Thus, the construction encodes p38^x but may not synthesize one of the two other possible pX proteins encoded by ORFs Xb or Xc. (Which one is not clearly identified yet for HTLV-2.) Since this plasmid could stimulate the HTLV-2 LTR function, p38^x alone might be enough for the *trans*-activation.

The pX function of HTLV-2 was also demonstrated to be essential for the efficient transcription of the integrated proviral genome. This was demonstrated by the following observations: an integrated proviral ge-nome with a deletion in pX coding sequence was not transcribed efficiently, but the transcription of the defective proviral genome was induced by super infection of HTLV-2 or transfection of the wild type HTLV-2 DNA (68). However, it is not yet clear which pX product is responsible for this activation. For the ORFs coding for p40^x in HTLV-1 and for p38^x in

HTLV-2, the terms "*tat*-1 and *tat*-2" were proposed by Sodroski et al (58). However, they also use *tat* for the *trans*-activating gene of HIV, which operates at a posttranscriptional level (69). It might be better to use an alternative terminology for the different functions.

The Sequence Responsible for p40^x Mediated Trans-*activation*

The LTR was specifically activated by p40^x of HTLV-1 or p38^x of HTLV-2 (58, 63, 64, 65), although p38^x of HTLV-2 activates some adenovirus promoter less efficiently (70). This specificity suggests the presence of sequences responsible for p40^x mediated *trans*-activation. Fujisawa et al (71) studied HTLV-1 LTR and identified an enhancer-like element which is responsible for p40^x-mediated activation by mutagenesis at the region upstream of the site for initiation of transcription. Serial deletion mutations in this region suggested the presence of two or three responsible elements which corresponded to three direct repeats of 21 nucleotides. Conservation of these direct repeats in HTLV-1 and -2 (26, 31) also suggested the functional importance of these repeats. In fact, insertion of two repeats of the 21b into an enhancerless SV40 promoter made this plasmid responsible for the activation by the p40^x. Furthermore, the site and orientation of the insertion did not affect the activity (71). Therefore, the unit of direct repeat was concluded to be an enhancer of the LTR and was thought to be responsible for the *trans*-activation mediated by p40^x (Figure 1). Pavlakis and his colleagues (personal communication) also showed that synthetic oligonucleotides of 21b behave as an enhancer responding to p40^x stimulation when it was inserted into a promoter sequence as oligomers.

On the other hand, Rosen et al (72) mapped the sequence responsible for the *trans*-activation at the region downstream of the enhancer. At present, the reason for this apparent difference between the conclusions of Fujisawa et al (71) and Rosen et al (72) is not understood. It could be due to sequence variations in different isolates of the LTR or to different cell lines used for the assays. The whole activity of the enhancement could not be explained by the enhancer alone. Deletion in the U5 and R regions in the LTR also reduced the CAT gene expression but did not affect the *trans*-activation by p40^x (71). Nevertheless, these elements are required for the maximum expression of the LTR function.

In case of the HTLV-2 LTR, Rosen et al (73) proposed that an enhancer in the LTR is responsible for the activation by p38^x in a way that differs from their own results on HTLV-1 (72). But this conclusion concerning HTLV-2 is the same as that proposed for HTLV-1 by Fujisawa et al (71).

In the case of both HTLV-1 and HTLV-2, it is not known whether the pX proteins interact with the enhancers directly or indirectly through cellular factors.

Significance of the Trans-*Activation in Viral Replication*

As was discussed in the previous section, expression of $p40^x$ of HTLV-1 or $p38^x$ of HTLV-2 activates the LTR function and enhances transcription from the LTR in the plasmids, thus eventually enhancing the CAT gene expression. This activation also works on the integrated proviral genome of HTLV-1 (M. Yoshida et al, unpublished) and HTLV-2 (68), thereby inducing expression of the viral genes. It is likely, therefore, that the LTR in the integrated proviral genome induces low levels of transcription after integration into the host cell genome, and this results in low levels of the pX mRNA. The low levels of $p40^x$ or $p38^x$ would activate the LTR function, in turn producing high levels of pX proteins, which further activate the LTR to give greater levels of the viral mRNA (Figure 1). The low activity or specificity of the LTR is thus amplified by expression of pX proteins. Therefore, the *trans*-acting function of the pX proteins is indispensable for the efficient viral gene expression and thus for viral replication. In this respect, the pX genes are functionally analogous to *trans*-acting transcriptional regulatory genes in some DNA tumor viruses such as E1A of adenoviruses and the gene for T antigen of SV40. In fact, the functional similarities between HTLV-2 pX and E1A have been demonstrated (74): The pX gene product of HTLV-2 can activate transcription from adenovirus E1A-dependent early promoters and furthermore, E1A protein can activate the HTLV-2 LTR, though at a much lower level.

Trans-*Activation of Cellular Genes*

Immediately after the discovery of *trans*-activation at the transcriptional level, such a mechanism was suspected to activate certain cellular gene expression, which may be associated with proliferation of T-cells. Primary tumor cells of patients with ATL were shown to express the IL-2R on their surfaces. The number of the IL-2R increased after in vitro cultivation of the tumor cells, concomitant with expression of viral proteins (8, 9, 10), including pX gene products, and over-expression of IL-2R was toxic to the cells in the presence of IL-2 (75). Even a B-cell line infected with HTLV-1 expresses the IL-2R (76). Thus, a viral function was suspected to enhance the IL-2R expression.

Greene et al (77) recently reported that pX expression of HTLV-2 induced expression of the genes for interleukin 2 (IL-2) and its receptor (IL-2R) gene expressions. They inserted the pX sequence of HTLV-2 into a retroviral vector, and the rescued virus was used to infect Jurkat cells, a human leukemia T-cell line with helper-type phenotype. Four independent cell clones were found to produce IL-2 constitutively and to express IL-2R on their cell surfaces, although the expression was rather weak. It was concluded that the production of IL-2 and IL-2R resulted from activation at the transcriptional level because their mRNA were elevated. Three cell clones infected with virus carrying the pX sequence in an antisense orientation did not express the IL-2 and IL-2R. Furthermore, neither IL-2 nor IL-2R was expressed in Raji cells. Their conclusion was that expression of the *tat* of HTLV-2 in some but not all cells induced the expression of IL-2 and IL-2R (Figure 1). Since these two proteins are required for T-cell proliferation, this induction by an HTLV-2 gene may induce preferential growth of infected T-cells in vivo.

The activation of expression of the genes for IL-2 and IL-2R was shown by Inoue et al (78) to be an early cellular response to $p40^x$ expression (see Figure 1). Transient expression of pX genes in human T-cell lines such as Jurkat or HSB-2 induced IL-2 and IL-2R within 20 hr almost to maximum levels and their mRNA accumulated. Therefore, the activation of the IL-2R gene is an early event after pX gene expression. The induction was related to the fate of the plasmids. ORF-specific mutagenesis in the pX sequence clearly demonstrated that expression of $p40^x$ alone was enough to induce the IL-2 and IL-2R and that neither $p27^{x-III}$ nor $p21^{x-III}$ is required. This requirement for the activation was the same as in activation of the unintegrated LTR, which suggests that transcription of these cellular genes is by *trans*-activation (Figure 1). An enhancer of 21b repeats in the LTR was shown to be responsible for $p40^x$-mediated activation (71, 73); however, the available sequences for IL-2 and IL-2R do not show any significant homology to the 21b.

Transient induction of IL-2R showed a unique cell type specificity (78). The expression of $p40^x$ induces IL-2 and IL-2R in Jurkat and HSB-2 cells but not in human T-cell lines CEM and Molt 3 or in human B-cell lines such as Raji and Ball-1. In the first two cell lines, the genes for IL-2 and IL-2R were induced by $p40^x$, or by phytohemaglutinin (PHA) and 12-O-tetradecanoylphorbol-13-acetate (TPA) (79). Inoue et al therefore concluded that some cellular factor or state of cellular differentiation keeps the gene available for activation by $p40^x$. Thus, it is conceivable that only certain T cells infected with HTLV-1 may initiate altered proliferation without stimulation by antigen—possibly by an autocrine mechanism (Figure 1). This unique specificity may also account in part for the long latency for the leukemogenesis after HTLV-1 infection.

POSSIBLE MECHANISM OF EARLY STAGE ATL DEVELOPMENT

As we said in the Introduction, studies on primary tumor cells had proposed a *trans*-acting viral function associated with leukemogenesis in ATL. The independent findings that a pX protein of HTLV-1 activates in *trans* not only viral genes but also some cellular genes fits with this theory of *trans*-acting viral function. Since the genes activated by $p40^x$ are IL-2 and IL-2R, which are required for T-cell growth, the $p40^x$ is an attractive candidate which can account for an early step in leukemogenesis (Figure 2). Infection of HTLV-1 into T-cells leads to expression of $p40^x$ at low levels. This $p40^x$ stimulates the viral promoter and induces high expression of viral antigens, including $p40^x$. The accumulated $p40^x$ promptly activates transcription of the IL-2 and IL-2R in particular T-cells that were differentiated into a certain stage of T-cell maturation. These would then be induced to undergo uncontrolled growth—possibly by an autocrine mechanism—without any stimulation by antigen. This model predicts multiclonal expansion of the subpopulation of infected T-cells at the initial stage of development of ATL. A similar pattern of proliferation of the infected cells in vitro was described (40, 80). In vitro infected cells at first proliferated multi- or oligoclonally. This was followed by clonal selection of certain infected cells. In vivo random provirus integration was found among some patients with smoldering ATL (M. Yoshida, unpublished), an early clinical stage of ATL development (81). This multiclonal expansion

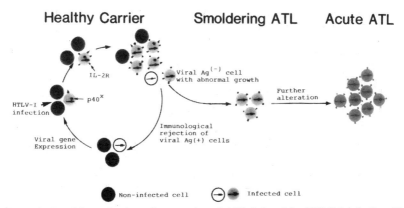

Figure 2 Possible mechanism of progression of ATL induced by HTLV-1 infection. ■, $p40^x$ encoded by pX sequence of HTLV-1; ▲, the envelope glycoprotein of HTLV-1; ●, cellular coded IL-2R.

of certain infected T cells may increase the target size for malignant transformation. This hypothetical mechanism might be able to explain the pathogenic specificity of HTLV-1 for helper type T cells.

Primary ATL cells express the IL-2R constitutively (8, 9, 10); however, no significant expression of pX proteins has been generally found in fresh tumor cells from ATL patients when either antigen (49, 54, 55) or mRNA was assessed (54, 55; and M. Yoshida, unpublished). There seem to be three possible explanations of these results. The first is that $p40^x$ is no longer required for the constitutive expression of the IL-2R on ATL cells in vivo but may still be required at the early stage of leukemogenesis, as discussed above. This is a reasonable explanation in light of the following considerations (see Figure 2): T-cells activated through the endogenous $p40^x$ would express viral antigens including the envelope glycoproteins which are exposed on the cell surface. These glycoproteins are targets of host immune surveillance, as is evidenced by the cytotoxic effects of anti-envelope antibodies or patient sera (82). Eventually all cells expressing the viral antigens, that is, all cells driven by the $p40^x$ would be rejected by the host. Only those cells that did not express the viral antigens would survive. Later, these antigen-negative infected cells would begin again to express viral antigens, including $p40^x$, thus entering into the second cycle of cell propagation. These cycles would be repeated in so-called healthy virus carriers for 20 or 30 years or longer. During these repeating cycles, a second event may take place in a cell driven by $p40^x$ activation, resulting in $p40^x$-independent growth and expression of the IL-2R. Thus, this cell can escape from the host immune system, thereby establishing the smoldering ATL, which is the first clinical stage of ATL progression (81). Patients with smoldering ATL show only a small proportion of peripheral blood lymphocytes, that are morphologically abnormal but these are infected with HTLV-1 and already monoclonally expanded (M. Yoshida, unpublished). The putative second event for malignant transformation, which may or may not be associated with viral function, is likely a rare event, since less than 1% of HTLV-1 carriers ultimately develop ATL (Figure 2). Cultures of lymphocytes from patients with ATL start with growth IL-2 dependent cells; subsequently one of these becomes IL-2 independent, but without virus infection this conversion seems to be infrequent. Thus, the putative second event might be associated with viral function. Alternatively, some chromosomal alteration could be the second event, because ATL leukemic cells show chromosomal abnormalities with extraordinary frequency (83, 84). No abnormality was found to be universal among ATL patients. Thus, an additional assumption is required for this explanation.

A second possible explanation for the absence of $p40^x$ expression in fresh ATL cells is that leukemic cells in peripheral blood could all be in the resting state and not proliferating; thus, absence of p40^x expression may not necessarily exclude the possibility of the function of p40^x expression in maintenance of constitutive expression of the IL-2R or of the malignant state. However, this explanation seems to be unlikely, since enlarged lymphnodes where malignant cells are thought to proliferate failed to score positively for p40^x expression in immuno-staining or RNA blotting assays (M. Yoshida, unpublished). The third possible explanation for the absence of p40^x in fresh ATL is that trace levels of p40^x which could not be detected with the usual techniques are still enough for maintenance of constitutive expression of IL-2R or of the malignant state.

All these discussions are based on the findings that p40^x of HTLV-1 and p38^x of HTLV-2 can activate T-cell growth factor IL-2 and its receptor IL-2R in Jurkat and HSB cell lines. However, no similar effects of pX proteins in primary T-cells could be demonstrated experimentally, yet, because of their fragile properties. Of course, pX gene products may modify expression of many other genes, such as those for the major histocompatibility class-II antigens (9), other lymphokines, or ATL-derived factor (ADF) (85). An answer to a rather simple question—whether expression of pX proteins in T-cells is enough to transform primary cells—would provide direct evidence for the hypothesis discussed above and thus should be a critical subject for research in the future.

Literature Cited

1. Poiesz, B. J., Ruscetti, F. W., Gazdar, A. F., Bunn, P. A., Minna, J. D., Gallo, R. C. 1980. Detection and isolation of type C retrovirus particles from fresh and cultured lymphocytes of a patient with cutaneous T cell lymphoma. *Proc. Natl. Acad. Sci. USA* 77: 7415–19

2. Yoshida, M., Miyoshi, I., Hinuma, Y. 1982. Isolation and characterization of retrovirus from cell lines of human adult T cell leukemia and its implication in the disease. *Proc. Natl. Acad. Sci. USA* 79: 2031–35

3. Uchiyama, T., Yodoi, J., Sagawa, K., Takatsuki, K., Uchino, H. 1977. Adult T cell leukemia: Clinical and hematological features of 16 cases. *Blood* 50: 481–91

4. Hinuma, Y., Nagata, K., Misoka, M., Nakai, M., Matsumoto, T., Kinoshita, K., Shirakawa, S., Miyoshi, I. 1981. Adult T cell leukemia: Antigen in an ATL cell line and detection of antibodies to the antigen in human sera. *Proc. Natl. Acad. Sci. USA* 78: 6476–80

5. Blattner, W. A., Kalyanaraman, V. S., Robert-Guroff, M., Lister, A., Galton, D. A. G., Sarin, P. S., Crawford, M. H., Catovski, D., Greaves, M., Gallo, R. C. 1982. The human type C retrovirus, HTLV, in blacks from the Caribbean region, and relationship to adult T cell leukemia/lymphoma. *Int. J. Cancer* 30: 257–64

6. Hunsmann, G., Schneider, J., Schmitt, J., Yamamoto, N. 1983. Detection of serum antibodies to adult T-cell leukemia virus in non-human primates and in people from Africa. *Int. J. Cancer* 32: 329–32

7. Hattori, T., Uchiyama, T., Toibana, K., Takatsuki, K., Uchino, H. 1981. Surface phenotype of Japanese adult T-cell leukemia cells characterized by monoclonal antibodies. *Blood* 58: 645–47

8. Depper, J. M., Leonard, W. J., Kronke, M., Waldmann, T. A., Greene, W. C. 1984. Augmented T-cell growth factor receptor expression in HTLV-1-infected human leukemic T-cells. *J. Immunol.* 133: 1691–95

9. Popovic, M., Lange-Wantzin, G., Sarin,

P. S., Mann, D., Gallo, R. C. 1983. Transformation of human umbilical cord blood T cells by human T cell leukemia/lymphoma virus. *Proc. Natl. Acad. Sci. USA* 80: 5402–6

10. Yodoi, J., Uchiyama, T., Maeda, M. 1983. T-cell growth factor receptor in adult T-cell leukemia. *Blood* 62: 509–11

11. Poiesz, B. J., Ruscetti, F. W., Reitz, M. S., Kalyanaraman, V. S., Gallo, R. C. 1981. Isolation of a new type C retrovirus (HTLV) in primary uncultured cells of a patient with Sezary T-cell leukemia. *Nature* 294: 268–71

12. Watanabe, T., Seiki, M., Yoshida, M. 1983. Retrovirus terminology. *Science* 222: 1178

13. Watanabe, T. 1984. HTLV type I (U.S. isolate) and ATLV (Japanese isolate) are the same species of human retrovirus. *Virology* 133: 238–41

14. Kalyanaraman, V. S., Sarngadharan, M. G., Nakao, Y., Ito, Y., Aoki, T., Gallo, R. C. 1982. Natural antibodies to the structural core protein (p24) of the human T cell leukemia (lymphoma) retrovirus found in sera of leukemia patients in Japan. *Proc. Natl. Acad. Sci. USA* 79: 1653–57

15. Tajima, K., Tominaga, S., Suchi, T., Kawagoe, T., Komoda, H., Hinuma, Y., Oda, T., Fujita, K. 1982. Epidemiological analysis of the distribution of antibody to adult T-cell leukemia virus. *Gann* 73: 893–901

16. Yoshida, M., Seiki, M., Yamaguchi, K., Takatsuki, K. 1984. Monoclonal integration of HTLV in all primary tumors of adult T-cell leukemia suggests causative role of HTLV in the disease. *Proc. Natl. Acad. Sci. USA* 81: 2534–37

17. Yoshida, M., Seiki, M., Hattori, S., Watanabe, T. 1984. Genome structure of human T-cell leukemia virus and its involvement in the development of adult T-cell leukemia. In *Human T-cell Leukemia/Lymphoma Virus,* ed. R. C. Gallo, M. Essex, L. Gross, pp. 141–48. Cold Spring Harbor, NY: Cold Spring Harbor Lab.

18. Yoshida, M., Hattori, S., Seiki, M. 1985. Molecular biology of human T-cell leukemia virus associated with adult T-cell leukemia. *Curr. Top. Microbiol. Immunol.* 115: 157–75

19. Wong-Staal, F., Hahn, H., Manzari, V., Colonbini, S., Franchini, G., Gelman, E. P., Gallo, R. C. 1983. A survey of human leukemia for sequences of a human retrovirus. *Nature* 302: 626–28

20. Miyoshi, I., Kubonishi, I., Yoshimoto, S., Akagi, T., Ohtsuki, Y., Shiraishi, Y., Nagata, K., Hinuma, Y. 1981. Type C

virus particles in a cord T cell line derived by cocultivating normal human cord leukocytes and human leukemic T cells. *Nature* 294: 770–71

21. Yamamoto, N., Okada, M., Koyanagi, Y., Kannagi, M., Hinuma, Y. 1982. Transformation of human leukocytes by cocultivation with an adult T cell leukemia virus producer cell line. *Science* 217: 737–39

22. Kinoshita, K., Hino, S., Amagasaki, T., Ikeda, S., Yamada, Y., Suzuyama, J., Momita, S., Toriya, K., Kamihira, S., Ichimaru, M. 1984. Demonstration of adult T-cell leukemia virus antigen in milk from three sero-positive mothers. *Gann* 75: 103–5

23. Okochi, K., Sato, H., Hinuma, Y. 1983. A retrospective study on transmission of adult T cell leukemia virus by blood transfusion; sero-conversion in recipients. *Vox Sang* 46: 245–53

24. Seiki, M., Hattori, S., Hirayama, Y., Yoshida, M. 1983. Human adult T cell leukemia virus: Complete nucleotide sequence of the provirus genome integrated in leukemia cell DNA. *Proc. Natl. Acad. Sci. USA* 80: 3618–22

25. Kalyanaraman, V. S., Sarngadharan, M. G., Robert-Guroff, M., Miyoshi, I., Blayney, D., Golde, D., Gallo, R. C. 1982. A new subtype of human T-cell leukemia virus (HTLV-II) associated with a T-cell variant of hairy cell leukemia. *Science* 218: 571–73

26. Shimotohno, K., Wachsman, W., Takahashi, Y., Golde, D. W., Miwa, M., Sugimura, T., Chen, I. S. Y. 1984. Nucleotide sequence of the 3' region of an infectious human T-cell leukemia virus type II genome. *Proc. Natl. Acad. Sci. USA* 81: 6657–61

27. Watanabe, T., Seiki, M., Tsujimoto, H., Miyoshi, I., Hayami, M., Yoshida, M. 1985. Sequence homology of the simian retrovirus (STLV) genome with human T-cell leukemia virus type I (HTLV-I). *Virology* 144: 59–65

28. Rice, N. R., Stephens, R. M., Couez, D., Deschamp, J., Kettmann, R., Burny, A., Gilden, R. 1984. The nucleotide sequence of the *env* and post-*env* region of bovine leukemia virus. *Virology* 138: 82–93

29. Sagata, N., Yasunaga, T., Tsuzuku-Kawamura, J., Ohishi, K., Ogawa, Y., Ikawa, Y. 1985. Complete nucleotide sequence of the genome of bovine leukemia virus: Its evolutionary relationship to other retroviruses. *Proc. Natl. Acad. Sci. USA* 82: 677–81

30. Chen, I. S. Y., McLaughlin, J., Gasson, J. C., Clark, S. C., Golde, D. W. 1983.

Molecular characterization of genome of a novel human T-cell leukemia virus. *Nature* 305: 502–5

31. Haseltine, W. A., Sodroski, J., Patarca, R., Briggs, D., Perkins, D., Wong-Staal, F. 1984. Structure of 3′ terminal region of type II human T lymphotropic virus: Evidence for new coding region. *Science* 225: 419–21

32. Shimoyama, M., Kagami, Y., Shimotohno, K., Miwa, M., Minato, K., Tobinai, K., Suemasu, K., Sugimura, T. 1986. Adult T-cell leukemia/lymphoma not associated with human T-cell leukemia virus type I. *Proc. Natl. Acad. Sci. USA* 83: 4524–28

33. Seiki, M., Eddy, R., Shows, T. B., Yoshida, M. 1984. Nonspecific integration of the HTLV provirus genome into adult T-cell leukemia cells. *Nature* 309: 640–42

34. Miyoshi, I., Tauchi, H., Fujishita, M., Yoshimoto, S., Kubonishi, I., Ohtsuki, Y., Shiraishi, Y., Akagi, T. 1982. Transformation of monkey lymphocytes with adult T-cell leukemia virus. *Lancet* 1: 1016

35. Miyoshi, I., Yoshimoto, S., Taguchi, H., Kubonishi, I., Fujishita, M., Ohtsuki, Y., Shiraishi, Y., Adagi, T. 1983. Transformation of rabbit lymphocytes with T-cell leukemia virus. *Gann* 74: 1–4

36. Tateno, M., Kondo, N., Itoh, T., Chubachi, T., Togashi, T., Yoshiki, T. 1984. Rat lymphoid lines with human T-cell leukemia virus production. I. Biological and serological characterization. *J. Exp. Med.* 159: 1105–16

37. Clapham, P., Nagy, P., Weiss, R. A. 1984. Pseudotypes of human T-cell leukemia virus types 1 and 2: Neutralization of patients' sera. *Proc. Natl. Acad. Sci. USA* 81: 3083–86

38. Chen, I. S. Y., Quan, S. G., Golde, D. W. 1983. Human T-cell leukemia virus type II transforms normal human lymphocytes. *Proc. Natl. Acad. Sci. USA* 80: 7006–9

39. Tsujimoto, H., Komuro, A., Iijima, K., Miyamoto, J., Ishikawa, K., Hayami, M. 1985. Isolation of simian retroviruses closely related to human T-cell leukemia virus by establishment of lymphoid cell lines from various non-human primates. *Int. J. Cancer* 35: 377–84

40. de Rossi, A., Aldovini, A., Franchini, G., Mann, D., Gallo, R. C., Wong-Staal, F. 1985. Clonal selection of T-lymphocytes infected by cell-free human T-cell leukemia/lymphoma virus type 1: Parameters of virus integration and expression. *Virology* 143: 640–45

41. Kiyokawa, T., Seiki, M., Imagawa, K., Shimizu, F., Yoshida, M. 1984. Identification of a protein (p40^x) encoded by a unique sequence pX of human T-cell leukemia virus type I. *Gann* 75: 747–51

42. Lee, T. H., Coligan, J. E., Sodroski, J. G., Haseltine, W. A., Salahuddin, S. Z., Wong-Staal, F., Gallo, R. C., Essex, M. 1984. Antigens encoded by the 3′-terminal region of human T-cell leukemia virus: Evidence for a functional gene. *Science* 226: 57–61

43. Slamon, D. J., Shimotohno, K., Cline, M. J., Golde, D. W., Chen, I. S. Y. 1984. Identification of the putative transforming protein of the human T-cell leukemia viruses HTLV-I and -II. *Science* 226: 61–65

44. Miwa, M., Shimotohno, K., Hoshino, H., Fujino, M., Sugimura, T. 1984. Detection of pX proteins in human T-cell leukemia virus (HTLV)-infected cells by using antibody against peptide deduced from sequences of X-IV DNA of HTLV-I and X-C DNA of HTLV-II proviruses. *Gann* 75: 752–55

45. Seiki, M., Hikikoshi, A., Taniguchi, T., Yoshida, M. 1985. Expression of the pX gene of HTLV-I: General splicing mechanism in the HTLV family. *Science* 228: 1532–34

46. Goh, H. G., Sodroski, J., Rosen, C., Essex, M., Haseltine, W. A. 1985. Subcellular localization of the product of the long open reading frame of human T-cell leukemia virus type I. *Science* 227: 1227–29

47. Slamon, D. J., Press, M. F., Souza, L. M., Murdock, D. C., Cline, M. J., Golde, D. W., Gasson, J. C., Chen, I. S. Y. 1985. Studies of the putative transforming protein of the type I human T-cell leukemia virus. *Science* 228: 1427–30

48. Kiyokawa, T., Kawaguchi, T., Seiki, M., Yoshida, M. 1985. Association with nucleus of pX gene product of human T-cell leukemia virus type I. *Virology* 147: 462–65

49. Kiyokawa, T., Seiki, M., Iwashita, S., Imagawa, K., Shimizu, F., Yoshida, M. 1985. p27^{x-III} and p21^{x-III}, proteins encoded by the pX sequence of human T-cell leukemia virus type I. *Proc. Natl. Acad. Sci. USA* 82: 8359–63

50. Nagashima, K., Yoshida, M., Seiki, M. 1986. A single species of pX mRNA of HTLV-1 encodes *trans*-activator p40^x and two other phosphoproteins. *J. Virology*. In press

51. Sagata, N., Yasunaga, T., Ikawa, Y. 1985. Two distinct polypeptides may be

translated from a single spliced mRNA of the X genes of human T-cell leukemia and bovine leukemia virus. *FEBS Lett.* 192: 37–42

52. Tochikura, T., Iwahashi, M., Matsumoto, T., Koyanagi, Y., Hinuma, Y., Yamamoto, N. 1985. Effect of human serum anti-HTLV antibodies on viral antigen induction in vitro cultured peripheral lymphocytes from adult T-cell leukemia patients and healthy virus carriers. *Int. J. Cancer* 36: 1–7

53. Hinuma, Y., Gotoh, Y., Sugamura, K., Natgata, K., Goto, T., Nakai, M., Kamada, N., Natsumoto, T., Kinoshita, K. 1982. A retrovirus associated with human adult T-cell leukemia: *in vitro* activation. *Gann* 73: 341–44

54. Kitamura, T., Takano, M., Hoshino, H., Shimotohno, K., Shimoyama, M., Miwa, M., Takaku, F., Sugimura, T. 1985. Methylation pattern of human T-cell leukemia virus *in vivo* and *in vitro*: pX and LTR regions are hypomethylated *in vivo*. *Int. J. Cancer* 35: 629–35

55. Clarke, M. F., Trainor, C. D., Mann, D. L., Gallo, R. C., Reitz, M. S. 1984. Methylation of human T-cell leukemia virus proviral DNA and viral RNA expression in short- and long-term cultures of infected cells. *Virology* 135: 97–104

56. Wachsman, W., Shimotohno, K., Clark, S. C., Golde, D. W., Chen, I. S. Y. 1984. Expression of the 3′ terminal region of human T-cell leukemia viruses. *Science* 226: 177–79

57. Wachsman, W., Golde, D. W., Temple, P. A., Orr, E. C., Clark, S. C., Chen, I. S. Y. 1985. HTLV x-gene product: Requirement for the *env* methionine initiation codon. *Science* 228: 1534–36

58. Sodroski, J., Rosen, C., Goh, W. C., Haseltine, W. 1985. A transcriptional activator protein encoded by the x-lor region of the human T-cell leukemia virus. *Science* 228: 1430–34

59. Chatis, P. A., Holland, C. A., Hartley, J. W., Rowe, W. P., Hopkins, N. 1983. Role for the 3′ end of the genome in determining disease specificity of friend and moloney murine leukemia viruses. *Proc. Natl. Acad. Sci. USA* 80: 4408–11

60. Celander, D., Haseltine, W. A. 1984. Tissue-specific transcription preference as a determinant of cell tropism and leukemogenic potential of murine retroviruses. *Nature* 312: 159–62

61. Clapham, P., Nagy, K., Cheingsong-Popov, R., Weiss, R. A. 1983. Productive infection and cell free transmission of human T-cell leukemia virus in a non-

lymphoid cell line. *Science* 222: 1125–27

62. Yoshikura, H., Nishida, J., Yoshida, M., Kitamura, Y., Takaku, F., Ikeda, S. 1984. Isolation of HTLV derived from Japanese adult T cell leukemia patients in human diploid fibroblast strain MIR90 and the biological characters of the infected cells. *Int. J. Cancer* 33: 745–49

63. Sodroski, J. G., Rosen, C. A., Haseltine, W. A., 1984. Trans-acting transcriptional activation of the long terminal repeat of human T lymphotropic viruses in infected cells. *Science* 225: 381–85

64. Fujisawa, J., Seiki, M., Kiyokawa, T., Yoshida, M. 1985. Functional activation of long terminal repeat of human T-cell leukemia virus type I by *trans*-acting factor. *Proc. Natl. Acad. Sci. USA* 82: 2277–81

65. Febler, B. K., Paskalis, H., Kleinman-Ewing, C., Wong-Staal, F., Pavlakis, G. N. 1985. The pX protein of HTLV-I is a transcriptional activator of its long terminal repeats. *Science* 229: 675–79

66. Seiki, M., Inoue, J., Takeda, T., Hikikoshi, A., Sato, M., Yoshida, M. 1985. The p40ˣ of human T-cell leukemia virus type I is a trans-acting activator of viral gene transcription. *Gann* 76: 1127–31

67. Seiki, M., Inoue, J., Takeda, T., Yoshida, M. 1986. Direct evidence that p40ˣ of human T-cell leukemia virus type I is a trans-acting transcriptional activator. *EMBO J.* 5: 561–65

68. Chen, I. S. Y., Slamon, D. J., Rosenblatt, J. D., Shah, N. P., Quan, S. G., Wachsman, W. 1985. The x gene is essential for HTLV replication. *Science* 229: 54–58

69. Rosen, C. A., Sodroski, J. G., Goh, W. C., Dayton, A. I., Lippke, J., Haseltine, W. A. 1986. Post-transcriptional regulation accounts for the *trans*-activation of the human T-lymphotropic virus type III. *Nature* 319: 555–59

70. Chen, I. S. Y., Cann, A. J., Shah, N. P., Gaynor, R. B. 1985. Functional relation between HTLV-2 *x* and adenovirus E1A proteins in transcriptional activation. *Science* 230: 570–73

71. Fujisawa, J., Seiki, M., Sato, M., Yoshida, M. 1986. A transcriptional enhancer sequence of HTLV-I is responsible for *trans*-activation mediated by *p40* of HTLV-I. *EMBO J.* 5: 713–18

72. Rosen, C. A., Sodroski, J. G., Haseltine, W. A. 1985. Location of *cis*-acting regulatory sequences in the human T-cell leukemia virus type 1 long terminal repeat. *Proc. Natl. Acad. Sci. USA* 82: 6502–6

73. Rosen, C. A., Sodroski, J. G., Kettman,

R., Haseltine, W. A. 1986. Activation of enhancer sequences in type II human T-cell leukemia virus and bovine leukemia virus long terminal repeats by virus-associated *trans*-acting regulatory factors. *J. Viol.* 57: 738–44

74. Chen, I. S. Y., Cann, A. J., Shah, N. P., Gaynor, R. B. 1985. Functional relation between HTLV-2 *x* and adenovirus E1A protein in transcriptional activation. *Science* 230: 570–73

75. Sugamura, K., Nakai, S., Fujii, M., Hinuma, Y. 1985. Interleukin 2 inhibits in vitro growth of human T-cell lines carrying retrovirus. *J. Exp. Med.* 161: 1243–48

76. Sugamura, K., Fujii, M., Kobayashi, N., Sakitani, M., Hatanaka, M., Hinuma, Y. 1984. Retrovirus-induced expression of interleukin 2 receptors on cells of human β-cell lineage. *Proc. Natl. Acad. Sci. USA* 81: 7441–45

77. Greene, W. C., Leonard, W. J., Wano, Y., Svetlik, P. B., Peffer, N. J., Sodroski, J. G., Rosen, C. A., Goh, W. C., Haseltine, W. A. 1986. *Trans*-activator gene of HTLV-II induces IL-2 receptor and IL-2 cellular gene expression. *Science* 232: 877–80

78. Inoue, J., Seiki, M., Taniguchi, T., Tsuru, S., Yoshida, M. 1986. Induction of interleukin 2 receptor gene expression by p40^x encoded by human T-cell leukemia virus type 1. *EMBO J.* In press

79. Greene, W. C., Robb, R. J., Depper, J. M., Leonard, W. J., Drogula, C., Svetlik, P. B., Wong-Staal, F., Gallo, R. C., Waldmann, T. 1984. *J. Immunol.* 133: 1042–47

80. Hahn, B., Gallo, R. C., Franchini, G., Popovic, M., Aoki, T., Salahuddin, S. Z., Markham, P. D., Wong-Staal, F. 1984. Clonal selection of human T-cell leukemia virus-infected cells *in vivo* and *in vitro*. *Mol. Biol. Med.* 2: 29–36

81. Yamaguchi, K., Nishimura, H., Kohrogi, H., Jono, M., Miyamoto, Y., Takatsuki, K. 1983. A proposal for smoldering adult T-cell leukemia (smoldering ATL): A clinico-pathologic study of 5 cases. *Blood* 62: 758–66

82. Kiyokawa, T., Yoshikura, H., Hattori, S., Seiki, M., Yoshida, M. 1984. Envelope proteins of human T cell leukemia virus: Expression in E. coli and its application to studies of env gene functions. *Proc. Natl. Acad. Sci. USA* 81: 6202–7

83. Miyamoto, K., Tomita, N., Ishii, A., Nonaka, H., Kondo, T., Tanaka, T., Kitajima, K.-I. 1984. Chromosome abnormalities of leukemia cells in adult patients with T-cell leukemia. *J. Natl. Cancer Inst.* 73: 353–62

84. Fukuhara, S., Hinuma, Y., Gotoh, Y. I., Uchino, H. 1983. Chromosome aberrations in T lymphocytes carrying adult T-cell leukemia-associated antigens (ATLA) from healthy adults. *Blood* 61: 205–7

85. Teshigawara, K., Maeda, K., Nishino, K., Nikaido, T., Uchiyama, T., Tsudo, M., Wano, M., Yodoi, J. 1985. Adult T-cell leukemia cells produce a lymphokine that augments interleukin 2 receptor expression. *J. Mol. Cell. Immunol.* 2: 17–26

Ann. Rev. Immunol. 1987. 5 : 561–84

THE STRUCTURE OF THE CD4 AND CD8 GENES

Dan R. Littman

Department of Microbiology and Immunology, University of California, San Francisco, California 94143

INTRODUCTION

The CD4 and CD8 glycoproteins are nonpolymorphic members of the immunoglobulin gene superfamily which are expressed on the surface of functionally distinct populations of T lymphocytes. These molecules, also known respectively as T4/Leu3 and T8/Leu2 in human, and as L3T4 and Lyt2/Lyt3 in mouse, appear to be essential for effective cell-cell interactions, resulting either in target cell lysis or in activation of T lymphocytes. Expression of the CD4 and CD8 molecules correlates with the specificity of T cells for either class-II or class-I major histocompatibility complex (MHC) molecules on target cells (1–7). Thus, CD4 and CD8 may serve to increase the avidity of cell-cell interactions by directly binding to monomorphic determinants of the appropriate MHC molecule on target cells (2, 8, 9).

The view that the CD4 and CD8 molecules act solely as avidity enhancers has been challenged by recent findings which suggest that these molecules have a role in transmembrane signal transduction (10–14a). However, no experiments to date have directly tested the ability of CD4 or CD8 to bind ligand or to transduce physiologically relevant signals. It is also not known whether CD4 and CD8 perform analogous functions on separate populations of T cells. The basis for intrathymic selection of CD4$^+$ and CD8$^+$ cells that express appropriate MHC-restricted receptors likewise remains a mystery. The elucidation of these problems awaits functional experiments utilizing the newly isolated genes that encode the CD4 and CD8 glycoproteins (15–24). The emphasis of this review hence is on the isolation and structure of these genes and on the glimpses that the predicted structures of the glycoproteins give us into their potential

0732–0582/87/0410–0561$02.00

functions. In addition, recent evidence that the human CD4 molecule serves as the receptor for human immunodeficiency virus (HIV), the etiologic agent in acquired immunodeficiency syndrome (AIDS), is discussed.

STUDIES ON THE FUNCTION OF THE CD4 AND CD8 GLYCOPROTEINS

Early interest in the CD4 and CD8 glycoproteins centered on the role of these molecules as markers of T-cell differentiation (25, 26). Expression of these molecules could be correlated with discrete stages in maturation of thymocytes and T cells. A small fraction of thymocytes express neither CD4 nor CD8; these immature "double negative" cells are thought to be precursors to thymocytes that express these antigens. The majority of thymocytes are double-positive cells expressing both CD4 and CD8; and the more mature thymocytes, as well as peripheral T cells, exclusively express either CD4 or CD8. The heterodimeric T-cell receptor (TcR) is absent from double-negative cells but is found on the surface of about 50% of double-positive cells and of all mature single-positive thymocytes (27). It is not known whether the mature thymocytes are derived from the TcR$^+$ double-positive cells or from the immature double-negative precursors. The former possibility would present an interesting problem in the regulation of gene expression, since a previously expressed gene or its product would have to be shut off during development.

Recent experiments have revealed a small population of double-negative cells among peripheral T cells (28). These cells are different from the majority of double-negative cells in the thymus and can be distinguished by their CD3$^+$ phenotype in the absence of the surface $\alpha\beta$ receptor heterodimer (28). Instead, these cells appear to express another T-cell-receptor (TcR) chain, Tγ, in association with the CD3 complex (29–31). The elucidation of the function and specificity of these cells in the absence of expression of CD4 and CD8 may shed some light on the role of these glycoproteins during intrathymic differentiation.

Initial studies on the functions of CD4 and CD8 in peripheral T cells indicated that antibodies against these glycoproteins block a wide variety of T-cell functions. Antibodies against the 55-kilodalton (kd) CD4 glycoprotein block most helper-T-cell functions, the mixed lymphocyte reaction, and the induction of helper activity (6, 26, 32, 33); and antibodies against the 34-kd human CD8 glycoprotein and the heterodimeric murine homologue block most cytotoxic-T-cell (CTL) activity (25, 26). It was thus initially thought that CD8$^+$ cells comprise the majority of CTLs and that CD4$^+$ cells make up the helper cell population. This concept required reevaluation after Swain (1) demonstrated that antisera against murine

CD8 blocks both helper and cytotoxic cell functions specific for class-I MHC antigens, but not for class-II antigens. Several other laboratories subsequently showed that CD4$^+$ cytotoxic-T-cell clones have reactivity with class-II MHC molecules on target cells while CD8$^+$ cells react with class-I molecules (3–7). The view that emerged from these findings is that a more stringent functional association exists between expression of CD4 or CD8 and expression of T-cell-receptors restricted to interact with either class-II or class-I MHC molecules.

Few experiments have addressed the potential role of CD4 and CD8 in thymocyte development. However, the segregation of expression of these molecules with MHC specificity of the T-cell receptor makes a compelling argument in favor of involvement of these molecules in the selection of appropriate receptors. Results of studies that may have some bearing on this issue are difficult to interpret. Neonatal administration of antibodies against class-II MHC molecules blocks the development of class-II restricted CD4$^+$ cells, suggesting that there is interaction between cells that become CD4$^+$ and class-II MHC molecules during thymic maturation (34). Other in vivo studies have demonstrated that anti-CD4 antibodies, administered in conjunction with immunizing antigens, lead to sustained tolerance to the antigen (35, 36); however, it is not yet known which cells are the targets of these antibodies. This approach may be useful for antigen-specific abrogation of immune responses and has been successfully used to reverse autoimmune experimental allergic encephalomyelitis in mice (37).

Based on the dichotomy of MHC recognition by mature T cells, it may be speculated that the expression of either CD4 or CD8 determines the MHC specificity of the T-cell-receptor $\alpha\beta$ heterodimer which is selected during thymic ontogeny. Alternatively, the specificity of the TcR may trigger a particular program of gene expression, resulting in surface expression of either CD4 or CD8.

Several experiments have supported the view that CD4 and CD8 are accessory molecules that are required when the avidity of the TcR-antigen interaction is not sufficient to result in T-cell activation (8, 38, 39, 40). It has been shown that anti-CD4 and anti-CD8 antibodies act by preventing the formation of cell : cell conjugates (6, 41, 42). Inhibition of CTL : target cell conjugate formation with "cold targets" correlates with the ability to block cytotoxicity with anti-CD4 antibody (43); thus, CTL clones with high affinity interactions are relatively resistant to blocking with anti-CD4. Experiments performed with antigen-specific helper-T-cell hybridomas have demonstrated an inverse correlation between the amount of anti-CD4 antibody required to block T-cell activation and the amount of antigen required to obtain a particular level of activation (8). Moreover, antibody against the class-II molecule on the antigen-presenting cell results in increased sensitivity to anti-CD4 antibody in clones that have relatively "low avidity" for antigen. Similar experiments with murine CD8 also appear to demonstrate that CD8 is essential in low avidity T cell : target cell interactions but may be dispensable in secondary CTLs, which presumably have higher affinity receptors (40).

Several studies have attempted to confirm the hypothesis that CD4 and CD8 molecules interact with MHC gene products. Experiments on the activation of a rare CD4$^+$ T-cell hybridoma reactive with the class-I alloantigen, H-2D^d, have provided some support for this hypothesis (9, 44). Anti-CD4 antibody was found to have no effect on activation when the alloantigen was presented on Ia$^-$ cells; however, if the antigen-presenting cell expressed surface class-II molecules, there was a decrease in response following anti-CD4 treatment. It is anticipated that direct protein-protein interactions can be more readily studied in reconstituted systems; in one such system, in which T-cell hybridomas can be stimulated by a combination of peptide antigen and purified MHC class-II molecules inserted into lipid bilayers, the response was still susceptible to anti-CD4 antibody treatment (45). In these systems, inhibition with anti-CD4 cannot be solely due to steric interactions with other T-cell surface molecules, since variant T-cell hybrids that express vastly lower levels of surface CD4 exhibit a dramatic decrease in their reactivity to antigen (8).

While these results favor direct interaction between CD4 and CD8 and MHC molecules, several recent studies suggest that anti-CD4 antibodies may also impair function at other levels and that CD4 may not solely be involved in MHC recognition but may have an active role in transmembrane signaling. Activation of T cells with either anti-CD3 or concanavalin A, in the absence of stimulatory cells, can be blocked by prior incubation with anti-CD4 antibodies (10–13). Previous blocking studies thus need to be reassessed, since the results may be due to a negative signal mediated by CD4-antibody interaction. This interpretation may also apply to the finding that T cells reactive with a class-II $\beta 1$ domain within a transfected class I–class II chimeric molecule continue to be susceptible to blocking with anti-CD4 antibody (46). This result suggests that the presumptive monomorphic region of class-II molecules that interacts with CD4 is located in the relatively polymorphic $\beta 1$ domain; however, the inhibitory effect of the antibody may be independent of the cell-cell contact and thus may have no bearing on CD4–class II MHC interactions.

Other recent studies have demonstrated that human CD4 is internalized following treatment with phorbol esters (14, 47). Internalization is accompanied by phosphorylation of a cytoplasmic serine residue (48a). The CD4 glycoprotein thus resembles the epidermal growth factor receptor and other receptors in which internalization follows protein kinase C-mediated phosphorylation of cytoplasmic serine or threonine residues (48).

In antigen-specific T cell–B cell interactions, the CD4 glycoprotein is concentrated at the interface between the two cells (A. Kupfers, personal communication). In the formation of nonspecific conjugates, however, there is no evidence of redistribution of the CD4 molecule. Taken together, these recent results, combined with the finding that the cytoplasmic tail of the CD4 molecule is strongly conserved in evolution (21), argue that the CD4 glycoprotein is a dynamic entity which may mediate signals from extracellular sources.

ISOLATION OF THE CD4 AND CD8 GENES

The CD4 and CD8 molecules are expressed at very low levels on the cell surface. Purification of sufficient protein for sequence analysis is hence a difficult task, which has, nonetheless, been accomplished by several groups (23, 49). The approach that was successfully used in the initial cloning of the CD4 and CD8 genes (15, 16, 18) circumvented isolation of protein and, instead, relied on the ability of murine Ltk$^-$ cells to express the human-T-cell antigens following cotransfection with pTK and total human T-cell DNA (Figure 1). Cotransfection of L cells results in the chromosomal integration of approximately 1000 kilobases (kb) of high-molecular-weight donor DNA in addition to the selectable gene (50); since the human genome consists of approximately 3×10^6 kb, it is predicted that 1/1000 to 1/10,000 cotransfected colonies will take up a given fragment of donor human DNA. This approach has a number of constraints, including the size of the transcription unit, the ability of the promoter to function in L cells, and the requirement that a single gene be sufficient to permit surface expression of the desired antigen. It has nevertheless proven useful for the isolation of a growing number of genes.

Upon cotransfection with human-T-cell DNA, we found that approximately 1/5000 colonies stably expressed the CD8 glycoprotein, but only one in more than 10^5 colonies expressed CD4 (16). Moreover, this frequency for CD8 was maintained in secondary transfections, but we were unable to obtain any secondary CD4 transformants. This difference between CD4 and CD8 transfections is most likely due to the large size of the CD4 gene (see below). Cotransfected colonies expressing CD4 and CD8 were detected and easily isolated using an in situ erythrocyte rosetting assay (Figure 1), which employed monoclonal anti-CD4 and anti-CD8 antibodies followed by human erythrocytes bearing coupled rabbit-anti-mouse IgG. A variation of this approach permitted isolation of clones expressing CD8 following sequential sorts on the fluorescence activated

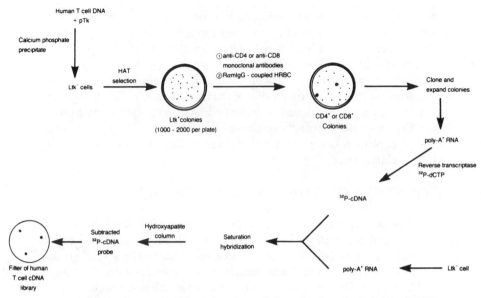

Figure 1   Scheme for gene isolation using gene transfer and subtractive hybridization. The in situ rosetting assay is performed with rabbit antimouse IgG-coupled human erythrocytes (RαmIgG-HRBC).

cell sorter (51). This general method has also allowed transfer of genes encoding HLA class-I molecules (16, 51), T1 (Leu 1) (16, 51), transferrin receptor (T9) (16, 52), nerve growth factor (NGF) receptor (53), and the CD2 (T11 or Leu 5) glycoprotein (16). The transferrin and NGF receptor genes were cloned following isolation of human-specific "Alu" sequences from secondary L-cell transfectants (52, 53).

Since the human CD8 gene lacked closely-linked "Alu" sequences, isolation of the gene required preparation of probes enriched in CD8 sequences by subtractive hybridization (15, 16). Briefly, ^{32}P-labeled cDNA was prepared from the transfected L cells and was hybridized to excess RNA from untransfected L cells. Nonhybridizing sequences were isolated by hydroxyapatite chromatography, and the resulting probe was used to screen cDNA libraries prepared from human T cells or from other transfected L cells. CD8 cDNA clones were readily obtained by this subtractive method, since the level of CD8 expression in the transfected L cells was very high. (By utilizing cell sorting, it has been possible to select L cells that have amplified the cotransfected pTK and CD8 genes on double minute chromosomes (54).) This method was found to be sufficiently sensitive to allow detection of cDNA clones represented by 5–

10 copies of RNA per transfected cell (D. Littman, R. Axel, unpublished results). We therefore screened the library with probe prepared from the single CD4 transfectant, which expressed a very low level of CD4 mRNA, and isolated a cDNA corresponding to the CD4 transcript (18).

The murine homologues of CD4 and CD8 (also known as L3T4 and Lyt2) were subsequently isolated by screening mouse-T-cell cDNA libraries at low stringency with the human cDNAs (19–21a). The identities of the mouse and human clones were confirmed by demonstrating that Ltk⁻ cells that were cotransfected with genomic clones of CD8 or with expression vectors containing full-length cDNAs of CD4 or CD8 all displayed the appropriate antigen on the cell surface. The human CD8 gene was found within a 10-kb HindIII-Bam fragment, which could transfer the phenotype to L cells (16; D. Littman, unpublished results); and the murine Lyt2 gene is encoded by a 5.2-kb HindIII genomic fragment which directs the synthesis of both the 34- and 38-kd α' and α forms in cotransfected L cells (19, 20).

The rat cDNA corresponding to the Lyt3 chain of the murine CD8 complex has also been recently isolated using conventional means of protein purification and sequencing, followed by screening a cDNA library with synthetic oligonucleotide probes (23). The rat 32-kd CD8 (Lyt2 equivalent) and CD4 cDNAs have been isolated using the murine Lyt2 and human CD4 cDNAs to probe rat-T-cell cDNA libraries (22, 24).

STRUCTURE OF THE CD8 GENES AND GLYCOPROTEINS

The subunit composition of the mouse and human CD8 molecules has been the subject of considerable interest, partly because some investigators believed CD8 to be part of the T-cell-antigen receptor (55, 56). While this possibility was effectively ruled out by the demonstration that CD8 is not polymorphic (57), only the recent isolation of the genes encoding mouse and human CD8 (15, 16, 19, 20) and rat Lyt3 (23) has allowed a fuller understanding of the structure of the molecules.

Unlike CD4, which is found as a 55-kd monomer in both mouse and human (58, 59), the CD8 molecule is surprisingly different in the two species. The 34-kd human CD8 molecule forms disulfide-linked homodimers and homomultimers on the surface of peripheral T cells (60); in the thymus, however, this chain forms complexes with the 45-kd CD1 (T6) glycoprotein, a presumptive class-I MHC molecule similar to murine TL antigen (61). In contrast, the murine CD8 homologue consists of two chains, the 34-kd or 38-kd Lyt2 chain (also designated α' and α) and the 30-kd Lyt3 (or β) glycoprotein, which form disulfide-linked $\alpha\beta$ and $\alpha'\beta$

heterodimers and higher multimers (62–64). There is no difference in the basic composition of the murine CD8 complex expressed in thymus and peripheral T cells, although there appears to be a lower level of the 34-kd subunit relative to the 38-kd subunit in the periphery (65).

The hypotheses that the Lyt2 antigenic determinants should be assigned to the α and α' polypeptides and the Lyt3 determinants to the β polypeptide (63, 64, 66) have been confirmed by recent studies using the cloned murine and rat CD8 genes (19, 20, 23; see Table 1). Following transfection of the murine CD8 (Lyt2) gene into mouse L cells, there is surface expression of dimers and multimers reactive with anti-Lyt2 antibodies and consisting of both 34- and 38-kd α' and α subunits (19); like the human CD8 molecule, these chains can also be expressed in the absence of the Lyt3 (β) subunit. Whether both α and α' are required to form these dimers, or whether they can function normally in the absence of the β chain, is still unknown. Transfection and expression of the second chain of rodent CD8 (the 30-kd β chain in mouse and the 37-kd chain in rat) have not yet been reported; however, there is good agreement between the cDNA and peptide sequences of the 37-kd chain of rat CD8 and the limited peptide sequence of mouse Lyt3 (23). The murine homologue will have to be isolated and expressed in transfected cells to confirm that it encodes the Lyt3 chain.

The sequence of the CD8 cDNAs obtained from human, mouse, and rat has in large part clarified the structure of the various chains, although the nature of the association between human CD8 and CD1 remains to be elucidated, as does the significance of the difference in CD8 structure between the species. The human CD8 molecule has been shown to correspond to the 38-kd α chain of murine CD8 and to the 32-kd subunit of rat CD8. The 34-kd α' chain, found only in mice, is a truncated form of the α chain, resulting from alternate splicing of exons encoding the cytoplasmic tail (20, 67; see below). The rodent-specific Lyt3 (β) chain (37-

Table 1 CD8 subunits in different species

	Human	Mouse	Rat
Antigen	T8/Leu2	Lyt-2/Lyt3	OX-8
"Lyt-2"	34 kd (0*)	38 kd (α) (3)	32 kd (1)
	—	34 kd (α') (3)	—
"Lyt-3"	—	30 kd (β) (1)	37 kd (3)
Membrane complex	homodimer	αβ heterodimer α'β heterodimer	32/37 kd heterodimer

* Number in parentheses denotes the sites for asn-linked glycosylation in each subunit.

kd CD8 subunit in rat) is another member of the immunoglobulin gene superfamily and has a structure similar to that of Lyt2 (23; Table 1). Whether an Lyt3 homologue exists and is expressed in human cells is not yet known.

Sequence Comparison of the CD8 Polypeptides

The membrane-bound human CD8 protein and its murine and rat homologues (the α chain of Lyt2 and the 32-kd rat CD8 subunit) contain each an amino-terminal hydrophobic signal sequence followed by an immunoglobulin-like domain, a short extracellular hinge region, a hydrophobic transmembrane segment, and a highly charged cytoplasmic domain (Figure 2; Refs. 16, 17, 19, 20, 22). The cleavage sites for the signal peptides (21 and 27 amino acid residues in human and mouse, respectively) have been determined by amino terminal sequencing of the mature CD8 glycoproteins (49, 66). The external domain of the mature protein displays striking homology with the numerous members of the immunoglobulin gene superfamily. The greatest homology is with variable regions of immunoglobulin light chains (28% homology for human CD8, 30% for Lyt2). The rat 32-kd glycoprotein, however, has roughly equal homology to V_κ, V_λ, and Thy 1 (22). Alignment of the homologous sequences indicates that the amino acid residues required for the formation of the immunoglobulin fold, such as the two cysteines involved in intrachain disulfide bond formation and the invariant tryptophan, are located in the expected positions. In addition, the predicted β-pleated sheet structure of the CD8 V-like domains corresponds to that of immunoglobulin V regions, suggesting that the basic folding pattern of CD8 is similar to that of the immunoglobulins (17, 19).

The Ig-like domain is followed by a short region, rich in prolines, threonines, and serines, having significant homology to immunoglobulin hinge sequences (17). In its amino acid composition, this domain resembles the similarly placed, highly glycosylated membrane-proximal domain of the LDL receptor (68). Like the LDL receptor domain, this hinge region of CD8 is also likely to contain sites for O-linked glycosylation (60). While the human CD8 glycoprotein has no sites for N-linked oligosaccharides (the *asn-pro-ser* site in the V-like domain is not an appropriate signal), the rat and murine homologues have one and three sites, respectively, in good agreement with the biochemical data.

The external domains of CD8 are anchored by a hydrophobic transmembrane region, which is absent in an alternately processed secreted form of the human molecule (see below). Both the human and mouse molecules have 28 amino acid cytoplasmic tails consisting of numerous basic residues. This cytoplasmic region is absent in the α' form of murine CD8 as result of alternative splicing (see below).

In addition to the cysteine residues which form the Ig fold, several other cysteines are conserved between human and murine CD8 and are likely to be involved in the formation of disulfide-linked dimers and multimers. The V-like domain contains one additional residue; the hinge, transmembrane, and cytoplasmic segments each contain two conserved cysteine residues. The presence of CD8 multimers indicates that more than one of these other cysteines can form interchain disulfide bonds.

Homology between the human and murine CD8 molecules is uneven throughout the length of the polypeptides. The greatest conservation is in the transmembrane and cytoplasmic domains (79% and 55%, respectively), while the least is in the external V-like domain (42%). Surprisingly, the mouse and rat CD8 molecules differ significantly, with the greatest divergence also in the V-like segment (60% identity). It has been suggested that the divergence of the V-like region reflects coevolution of human and murine class-I MHC molecules, the proposed ligands for the CD8 molecules (19). However, this sequence in CD8 appears to be diverging much more rapidly than that of other Ig family members. Since there is 70% identity of the class-I MHC external domain residues between mouse and human, the proposed coevolution of CD8 and class-I molecules would have had to tolerate much more change in the CD8 molecule. Conservation of the C-terminal domains may be due to requirements for interaction of this region of CD8 with other T-cell proteins; it suggests that CD8 may have some as yet undetected role in transmembrane signaling.

The cDNA encoding the rat equivalent of murine Lyt3 has recently been isolated and sequenced (23). In rat, this gene encodes the larger of the CD8 subunits, a 37-kd glycoprotein. The sequence of the rat 37-kd cDNA predicts a protein with an overall structure very similar to that of the other chain of CD8. A leader sequence of 21 residues is followed by a V-like region, which has homology with numerous members of the Ig family, including V_H, V_λ, V_κ, and T-cell-receptor V_α and V_β domains. The level of homology varies from 24% to 29%. The homology to the N-terminal domain of the 32-kd rat CD8 (corresponding to human CD8 and murine Lyt2) is only 21%. Most surprising is the finding that the sequence which follows the 37-kd protein V domain is almost identical to J sequences of immunoglobulins and T-cell receptors. Unlike these receptor genes, whose V regions are separated from the J sequences in the germ line and undergo rearrangement during differentiation, the gene encoding the rat 37-kd CD8 glycoprotein has no intervening sequence between the V-like and J-like domains. The 37-kd rat CD8 molecule also contains a hinge-like membrane-proximal domain, rich in prolines and threonines, a hydrophobic transmembrane domain, and a basic cytoplasmic tail of 19 amino acids. It contains three sites for N-linked glycosylation, compared to a single site for the murine 30-kd equivalent Lyt3 glycoprotein.

The structures of the two chains of CD8 in rat and, presumably, in mouse, suggest that these molecules and the two chains of the T-cell receptor evolved from a common ancestor (23). In the process, the CD8 chains would have lost the C-region domains, while the T-cell receptors would have acquired intervening sequences between V and J, eventually resulting in the capacity to rearrange multiple V regions during differentiation (23). The similarity in structure of the CD8 and TcR heterodimers suggests that during evolution these two sets of molecules acquired the capacity to recognize MHC molecules. The T-cell-receptor V regions would have evolved to interact with polymorphic domains of class-I and class-II MHC molecules, while the CD8 molecule would have evolved to recognize only monomorphic determinants of class-I molecules.

Alternative Splicing of the CD8 Gene Product

Both the human and mouse CD8 glycoproteins are expressed in multiple forms as a consequence of alternative splicing of the primary transcripts. The major protein domains described above are encoded by separate exons. In human CD8, the signal sequence, V-like domain, membrane-proximal hinge, transmembrane domain, and cytoplasmic domain are encoded each by separate exons (16). In mouse Lyt2, the leader and V-domains are encoded by a single exon, probably due to intron loss during evolution (19, 67, Figure 2).

The cytoplasmic domains of CD8 are encoded by two separate exons (19, 20, 67). The 34-kd α' form of murine Lyt2 is a product of alternate splicing of the penultimate cytoplasmic exon (20). Removal of this 31-

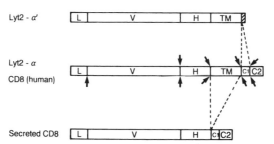

Figure 2 Representation of the alternate forms of CD8 RNA in mouse and human. Only the coding regions of the transcripts are included. The positions of the arrows indicate locations of introns in the murine (top) and human (bottom) CD8 genes. (L: leader sequence; V: Variable-like domain; H: Hinge domain; TM: transmembrane domain; C1, C2: cytoplasmic exons.)

nucleotide exon results both in the loss of 11 amino acid residues and in a frame shift leading to a termination codon 6 nucleotides downstream. Consequently, the α' glycoprotein contains a cytoplasmic tail of only 3 amino acids, not unlike the short tails of immunoglobulins and T-cell-receptor proteins. Some evidence suggests that the α' product is favored in the thymus and that this splicing process is regulated at various stages of T-cell development (20, 65). In addition, transfection of L cells with a 5.2-kb HindIII fragment containing the Lyt2 gene results in equimolar expression of both mRNAs and of both α and α' in the cell surface CD8 oligomers (19, 20). The predicted α' sequence, based on the sequence of a cDNA clone and on S1 nuclease protection experiments, is in good agreement with biochemical data which suggested that the cytoplasmic portion of the α' chain of Lyt2 is less than 1000 daltons, compared to 2000–5000 daltons for the cytoplasmic tails of the α and β chains (69).

A similar splicing product is either absent or barely detectable in human T cells (20; A. Norment, D. Littman, unpublished results). However, a different RNA processing event results in a secreted form of the human CD8 molecule. Previous studies had demonstrated that supernatants of CD8$^+$ human leukemic cell lines as well as serum from patients with CD8$^+$ T-cell leukemias contain a sizable amount of immunoreactive soluble CD8 glycoprotein (70, 71). Since a labeled 27-kd form of CD8 was released from T cells following cell surface iodination, it was concluded that the soluble form of CD8 found in supernatants and in serum resulted from a specific cleavage of the membrane-bound molecule (71). However, murine surface CD8 is known to be much more sensitive to trypsin treatment than other T-cell surface molecules, such as Lyt1 (63), suggesting that protease activity in the media selectively cleaves the human CD8 molecule.

During the course of the characterization of the human CD8 cDNA clones, we found one clone that had spliced out the exon encoding the transmembrane domain (D. Littman, R. Axel, unpublished result). Subsequent S1 nuclease studies have demonstrated that this alternate transcript may account for as much as 30% of the CD8 mRNA in leukemic T-cell lines (A. Norment, D. Littman, unpublished results). Moreover, injection into Xenopus oocytes of synthetic RNA, prepared in vitro from the alternate cDNA clone, results in secretion of a 30-kd soluble form of the CD8 molecule (A. Weiss, V. Lingappa, D. Littman, unpublished results); and expression of this cDNA in transfected L cells is accompanied by secretion of immunoreactive CD8 (A. Norment, D. Littman, unpublished results). These studies indicate that alternative splicing of the human CD8 mRNA gives rise to a protein that has no transmembrane domain and can thus be secreted. At least some of the immunoreactive CD8 protein identified in culture supernatants and in leukemic sera can probably be accounted for by this process. Note that similar splicing results in alternate secreted forms of human class-I MHC molecules (72); the transmembrane domains of the HLA molecules are also encoded by discrete exons that can be excised during posttranscriptional processing. The function, if any, of these secreted products has yet to be determined.

STRUCTURE OF THE CD4 GENE AND ITS PRODUCTS

The CD4 polypeptides are approximately twice the size of the CD8 molecules and consist of 5 external domains, a stretch of hydrophobic transmembrane residues, and a cytoplasmic tail of 40 residues (18, 21; Figure 3). The mature CD4 mRNA encodes a protein with an amino-terminal hydrophobic signal sequence which, like CD8, is followed by a domain having extensive homology to immunoglobulin light-chain variable regions. The homology in both amino acid sequence (28% for murine CD4, 35% for human CD4) and predicted secondary structure suggests that this domain bears a close resemblance to variable regions in its folding pattern. The homology between the V-like domains of CD4 and CD8 is very limited and supports the notion that these molecules interact with different ligands.

Unlike CD8, which has a short hinge region separating its V-like domain from the membrane anchor, CD4 contains another 270 external amino acid residues that have no strong homology to other proteins in the data banks. It has been argued that this long sequence can be divided into three domains which also have some structural homology to Ig family domains, particularly regions of the poly-Ig receptor (24). Indeed, the first and third of these regions each has two cysteines which are conserved between the rodent and human CD4 molecules and may be involved in intradomain disulfide bonds. In addition, each of these regions is encoded on a separate

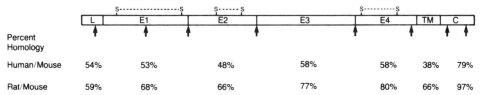

Percent Homology	L	E1	E2	E3	E4	TM	C
Human/Mouse	54%	53%	48%	58%	58%	38%	79%
Rat/Mouse	59%	68%	66%	77%	80%	66%	97%

Figure 3 Interspecies homologies of the individual domains of CD4. The arrows indicate the location of introns within the coding region of the CD4 gene; there is at least one more intron, located within the 5′ untranslated region, which is not included here. The putative disulfide loops within three of the extracellular domains are also indicated. The individual domains are the leader (L), extracellular domains 1–4, including the V-like region E1, the transmembrane domain (TM), and the cytoplasmic domain (C).

exon, suggesting that they do indeed form discrete domains (E2-E4) within the glycoprotein (S. Gettner, P. Estess, D. Littman, unpublished results; Figure 3). Surprisingly, the V-like domain (E1) is encoded on two separate exons, unlike other Ig family domains (21; see below). The large intron separating these exons is likely to have been inserted during evolution and may contain sequences required for regulation of CD4 gene expression in different tissues.

The human CD4 gene has been mapped to the short arm of chromosome 12 and thus is not linked to any known immunoglobulin gene family loci (73). This region of chromosome 12 shows striking homology to the distal segment of mouse chromosome 6. Although the murine CD4 gene has not yet been mapped, it is likely to be located on this segment of chromosome 6 and would thus be linked to the Lyt2, Lyt3, and Ig-κ loci (73).

Sequence Comparison of the CD4 Polypeptides

The most striking finding in a comparison of the amino acid sequence of the murine, rat, and human CD4 molecule is the evolutionary conservation of the 40 amino-acid-long cytoplasmic tail (Figure 3). Of the first 32 residues, 29 are identical in human and mouse; there is a single amino acid difference between rat and mouse CD4 cytoplasmic domains. There is much greater divergence in the other domains, with overall homologies of 55% between human and mouse and 74% between rat and mouse. The two sites for N-linked glycosylation in human CD4 are conserved in the murine homologue, and the six extracellular cysteines, located in pairs in three of the external domains, are also conserved between the species (Figure 3).

The finding of strong conservation in the cytoplasmic domain of CD4 argues that this sequence is essential for the function of the molecule and that the CD4 glycoprotein transmits signals across the plasma membrane. Both the human and rodent CD4 molecules contain five cytoplasmic serines and threonines. One of the serines in human CD4 is phosphorylated following activation of T cells with phorbol esters or with specific antigen (48a). The mechanism of phorbol ester–induced internalization of CD4 may thus be similar to that for transferrin receptor (74) and EGF receptor (48), which are phosphorylated following activation of protein kinase C.

THE ROLE OF HUMAN CD4 IN HIV INFECTION

The major clinical features of AIDS are a profound immunodeficiency and a severe neurological syndrome (75, 76). The clinical manifestations appear to correlate with infection of mononuclear cells and cells within the central nervous system with human immunodeficiency virus (HIV), the etiologic agent of AIDS. Numerous studies indicate that this tissue tropism of HIV is dependent on the expression of CD4 on target cells. The CD4 glycoprotein thus appears to form at least part of the receptor for HIV on T cells and monocytes/macrophages; and it is also likely to be the receptor on cells within the CNS.

The immune dysfunction in AIDS correlates with the virtual disappearance of CD4$^+$ T lymphocytes from the circulation (77). Loss of CD4$^+$ T cells can be reproduced in vitro by culturing lymphocytes with HIV (78). Moreover, only CD4$^+$ cells have been readily infected with HIV in vitro, and monoclonal antibodies directed against several epitopes of the CD4 glycoprotein can readily block infection (79, 80). Infection with HIV is followed by a significant decrease in the level of surface CD4 and by resistance to superinfection with VSV(HIV) pseudotypes (79). Other retroviruses have also been shown to down-regulate their receptors following infection. These results are therefore consistent with CD4 acting as the receptor for the HIV retrovirus.

One of the monoclonal antibodies, OKT4, which recognizes an epitope absent in 4% of the American Black population (81), has no effect on blocking infection (79, 82, 83). The finding that T cells from these homozygous OKT4$^-$, OKT4A$^+$ individuals are still susceptible to infection with HIV is consistent with the fact that this epitope is not involved in viral binding (82). McDougal and his colleagues exploited this finding and obtained convincing evidence for direct interaction between CD4 and a component of the HIV envelope glycoprotein. In their initial study, they demonstrated that HIV and OKT4A monoclonal antibody reciprocally inhibit each other in binding to CD4$^+$ T cells, while the OKT4 MAb does not inhibit the HIV binding. When virus was bound to ^{125}I-surface-labeled T cells, anti-HIV antibodies immunoprecipitated only the labeled 55-kd CD4 glycoprotein and no other cellular proteins; conversely, when ^{35}S-labeled virus was bound to cells, only the gp110 envelope glycoprotein was coprecipitated with the OKT4 mAb (83). Definitive proof of the role of CD4 in HIV in section has now been obtained by the demonstration that human cell lines expressing transfected human CD4 acquire the ability to bind and internalize HIV (84).

The CD4-envelope glycoprotein interaction appears to result not only in viral binding and internalization but also in fusion of cellular membranes. Infected T cells readily fuse with uninfected CD4$^+$ T cells, but not with CD4$^-$ T-cell variants (85). Transient transfection of the CD4$^+$ T-cell line Jurkat with constructs that express the HIV envelope glycoprotein also leads to syncitium formation, which is readily blocked by OKT4A but not by OKT4 monoclonal antibodies (86). Since only a small fraction of T lymphocytes have been shown to harbor HIV, this CD4-mediated process may provide a means through which a single infected cell can recruit numerous uninfected cells to form syncitia which are then eliminated by the reticuloendothelial system.

The CD4-gp110 interaction therefore appears to be crucial both for the viral life cycle and for subsequent cytopathic effects. While there is extensive polymorphism in the sequence of the envelope glycoprotein of different viral isolates (87), it is expected that the region of gp110 which interacts with CD4 is conserved in all infectious isolates. Segments of gp110 that are relatively constant have been identified (87), but their role in binding has yet to be determined. The elucidation of the precise topology of CD4-gp110 binding is likely to provide important insights into vaccine strategy and potential treatment of infection with HIV.

Expression of CD4 mRNA in Mouse and Human Brain

The tissue specificity of HIV infection, which includes not only hematopoietic cells but also the central and peripheral nervous system (88–90), suggested that the viral receptor may be present on cells other than T lymphocytes. Indeed, the CD4 glycoprotein is also expressed in cells of the macrophage and neutrophil lineage, as well as in lymphoblastoid B-cell lines, all of which can be infected with HIV (79, 91, 92).

In AIDS patients suffering from neurological symptoms, HIV RNA has been detected by in situ hybridization in brain sections (88). However, the cell type infected has not yet been determined. We have found a substantial level of the 3.3 kb CD4 mRNA in human brain (N. Lonberg, E. Lacy, S. Gettner, D. Littman, manuscript in preparation). Since transcripts for CD3 and Tiβ are not found in RNA preparations from brain, it is likely that the CD4 transcript is expressed in bone marrow–derived macrophage-like cells, which have been shown to be infectable with HIV (94). CD4 may also be expressed in brain parenchymal cells which may be susceptible to HIV infection. Although expression of CD4 glycoprotein in the brain has yet to be convincingly demonstrated, the finding of the mRNA suggests that the CD4 molecule serves as the viral receptor in brain as well as in cells within the immune system.

In mouse, different forms of CD4 mRNA are expressed in T cells and in brain. While a 3.2-kb transcript is found in T cells, only a 2.4-kb mRNA is expressed in brain. cDNA cloning and sequencing, combined with primer extension and RNase protection analyses, indicate that this brain-specific transcript is a truncated form of the T-cell message that initiates within the exon encoding the second membrane-proximal domain (E3) of T cell CD4 (N. Lonberg, E. Lacy, S. Gettner, D. Littman, in preparation). The brain-specific transcriptional unit thus appears to utilize a promoter that spans the preceding intron and part of the exon in which a TATA box is located. It is not yet known if this short transcript is translated into a functional brain-specific protein.

EVOLUTION OF THE CD4 AND CD8 GENES

The evolution of the immunoglobulin gene superfamily has been the subject of several recent reviews (95, 96). Members of this family encode cell surface proteins that have strategically conserved amino acid residues, notably cysteines forming an intrachain disulfide bond, which are involved in generating an immunoglobulin fold. These immunoglobulin segments are encoded by single exons of approximately 300 nucleotides. The studies described in this review clearly place the CD4 and CD8 genes within this superfamily and, in addition, provide insight into means by which members of the family may diverge. The Lyt3 gene provides the first example of V- and J-like segments linked within an exon in a gene that does not undergo rearrangement (23); and the CD4 gene is, to this date, the unique member of the superfamily in which the sequence encoding the Ig domain is interrupted by an intron (21).

The structure of the CD8 molecules suggests that the CD8 genes evolved from immunoglobulin V regions that have lost the capacity to rearrange. This view is supported by the finding that the Lyt2 and Lyt3 genes are closely linked to the Ig κ locus on mouse chromosome 6 (97, 98) and that the human CD8 gene is linked to the κ locus on human chromosome 2 (99). In the case of the Lyt2 gene, the sequence characteristic of rearrangement signals is absent from the intron located 3′ to the V-like domain (67). The Lyt3 gene, however, clearly has a conserved J-like sequence immediately 3′ to the V-like region, and both segments are encoded on a single exon (23). The Lyt3 gene may thus represent a progenitor gene, into which an intervening sequence bearing rearrangement signals was introduced during evolution, giving rise to rearranging immunoglobulin and T-cell-receptor genes. The alternative evolutionary model favored by Johnson & Williams (23) holds that the Lyt2/Lyt3 genes and the T-cell-receptor genes evolved from a common precursor, having an intron between V and J; the former set of genes would have lost the intron as well as the linked C region; while the latter set would have acquired sequences within the intron enabling the V and J regions to undergo rearrangement. A final possibility is that the Lyt3 gene is the product of a germline V-J rearrangement.

Studies of numerous genes have given rise to the general concept that individual structural domains of proteins are encoded by separate exons. These findings have led to the hypothesis that exons migrate in the course of evolution and are utilized when juxtaposed with appropriate regulatory regions and other exons (100, 101). Introns are thus essentially considered to be remnants, which are often lost during evolution. The intron-exon structure of the CD4 gene conforms in large part to this view since gene segments encoding each of the four external domains, the transmembrane region, and the cytoplasmic tail are separated by introns (Figure 3; S. Gettner, P. Estess, D. Littman, unpublished results). The surprising finding is that both the mouse and human genes contain an additional long intron (at least 18-kb in human CD4) that interrupts the coding region for the Ig-like domain (21). This finding suggests that this intron was inserted in the germ line prior to divergence of humans from rodents (Figure 4). Alternatively, the CD4 V-like region may represent the ancestral Ig domain which, following duplication, lost the intron (Figure 4). This is very unlikely given the considerable homology between CD4 and immunoglobulin variable regions, from which CD4 is thought to have diverged quite recently. The location of this intron thus argues against the hypothesis that introns evolve solely from the shuffling of exons during evolution.

Aside from framework residues which they share with other V region domains, the Ig segments of CD4 and CD8 display very little homology with each other. In addition, the CD4 glycoprotein has three additional external domains, while both chains of CD8 have short hinge regions. The difference in basic structure between CD4 and CD8 casts some doubt on the generally held belief that these two molecules perform essentially identical functions on different populations of T cells. It is much more likely that CD4 and CD8 may have some analogous functions, such as binding to a monomorphic domain of MHC molecules, but that they differ in effector functions. For example, the transmembrane and cytoplasmic

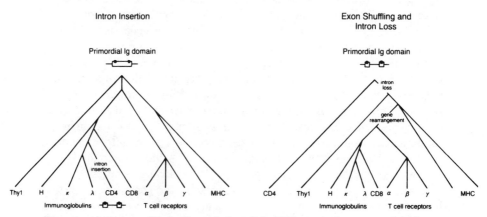

Figure 4   Alternate models for the evolution of the CD4 gene in the context of the immunoglobulin gene superfamily.

domains of these molecules may interact with different sets of membrane-bound or cytoplasmic molecules to transmit distinct signals; in addition, the CD4 molecule, which in its response to phorbol esters behaves like a growth factor receptor, may have additional ligand binding sites that could be utilized not only by T lymphocytes but also by cells within the central nervous system.

CONCLUSIONS AND FUTURE DIRECTIONS

Much progress has recently been made in understanding the structure both of the CD4 and CD8 genes and of their products. Both glycoproteins are members of the immunoglobulin gene superfamily and are likely to be cell surface recognition structures which interact with other members of this family, the class-I and class-II MHC molecules. Despite these advances, little has yet been learned about the functions of these molecules and the regulation of their expression. The ability to manipulate the genes that encode these glycoproteins will facilitate studies that directly address the issue of ligand specificity of CD4 and CD8. In addition, the functions of these molecules in the selection of appropriate MHC-restricted receptors during thymic ontogeny may now be approached through the use of transgenic mice. Expression of CD4 and CD8 in inappropriate thymocyte compartments and of class-I or class-II restricted $\alpha\beta$ chains of the TcR during differentiation may shed some light on the mechanism of intra-thymic selection. Such studies await the elucidation of the regulatory sequences of the CD4 and CD8 genes.

The CD4 molecule is of special interest because it appears to mediate signals and is also the portal of entry into the cell for HIV. However, very little effort has been made to understand the biochemistry of the CD4 glycoprotein; although this molecule has been shown to be present as a monomer in T-cell membranes, no experiments have rigorously tested the possibility that CD4 interacts with other membrane proteins. The highly conserved cytoplasmic domain suggests that this segment of the molecule interacts with other cellular components, which may be involved in activation of the T cell. Future studies will concentrate on pinpointing the segment(s) of CD4 involved in binding to the HIV envelope glycoprotein and on determining if the same or different regions of CD4 are involved in the normal function of this molecule. Such studies may allow development of peptides that interfere with binding of HIV to T cells and to other cells which express CD4. In addition, the role of the cytoplasmic segment, both in signal transduction and in internalization of HIV, will be amenable to studies using the cloned CD4 gene.

ACKNOWLEDGMENTS

I am grateful to Arthur Weiss for his comments on the manuscript. This work was supported by NIH grant AI23513 and by a Searle Scholar Award from the Chicago Community Trust.

Literature Cited

1. Swain, S. L. 1981. Significance of Lyt phenotypes: Lyt 2 antibodies block activities of T cells that recognize class I major histocompatibility complex antigens regardless of their function. *Proc. Natl. Acad. Sci. USA* 78: 7101–5
2. Swain, S. L. 1983. T cell subsets and the recognition of MHC class. *Immunol. Rev.* 74: 129–42
3. Krensky, A. M., Reiss, C. S., Mier, J. W., Strominger, J. L., Burakoff, S. J. 1982. Long-term human cytolytic T-cell lines allospecific for HLA-DR6 antigen are OKT4⁺. *Proc. Natl. Acad. Sci. USA* 79: 2365–69
4. Spits, H., Borst, J., Terhorst, C., de Vries, J. E. 1982. The role of T cell differentiation markers in antigen-specific and lectin-dependent cellular cytotoxicity mediated by T8⁺ and T4⁺ human cytotoxic T cell clones directed at class I and class II MHC antigens. *J. Immunol.* 129: 1563–69
5. Meuer, S. C., Schlossman, S. F., Reinherz, E. 1982. Clonal analysis of human cytotoxic T lymphocytes: T4⁺ and T8⁺ effector T cells recognize products of different major histocompatibility complex regions. *Proc. Natl. Acad. Sci. USA* 79: 4395–99
6. Wilde, D. B., Marrack, P., Kappler, J., Dialynas, D. P., Fitch, F. W. 1983. Evidence implicating L3T4 in class II MHC antigen reactivity; monoclonal antibody GK1.5 (anti-L3T4a) blocks class II MHC antigen-specific proliferation, release of lymphokines, and binding by cloned murine helper T lymphocyte lines. *J. Immunol.* 131: 2178–83
7. Swain, S. L., Dialynas, D. P., Fitch, F. W., English, M. 1984. Monoclonal antibody to L3T4 blocks the function of T cells specific for class II major histocompatibility complex antigens. *J. Immunol.* 132: 1118–23
8. Marrack, P., Endres, R., Schimonkevitz, R., Zlotnik, A., Dialynas, D., Fitch, F., Kappler, J. 1983. The major histocompatibility complex-restricted antigen receptor on T cells. II. Role of the L3T4 product. *J. Exp. Med.* 158:

1077–91
9. Greenstein, J. L., Kappler, J., Marrack, P., Burakoff, S. J. 1984. The role of L3T4 in recognition of Ia by a cytotoxic, H-2Dᵈ-specific T cell hybridoma. *J. Exp. Med.* 159: 1213–24
10. Wassmer, P., Chan, C., Logdberg, L., Shevach, E. M. 1985. Role of the L3T4-antigen in T cell activation: II. Inhibition of T cell activation by monoclonal anti-L3T4 antibodies in the absence of accessory cells. *J. Immunol.* 135: 2237–42
11. Bank, I., Chess, L. 1985. Perturbation of the T4 molecule transmits a negative signal to T cells. *J. Exp. Med.* 162: 1294–1303
12. Tite, J. P., Sloan, A., Janeway, C. A. Jr. 1986. The role of L3T4 in T cell activation: L3T4 may be both an Ia-binding protein and a receptor that transduces a negative signal. *J. Mol. Cell. Immunol.* 2: 179–90
13. Bekoff, M., Kubo, R., Grey, H. M. 1986. Activation requirements for normal T cells: Accessory cell-dependent and -independent stimulation by anti-receptor antibodies. *J. Immunol.* 137: 1411–19
14. Hoxie, J. A., Matthews, D. M., Callahan, K. J., Cassel, D. L., Cooper, R. A. 1986. Transient modulation and internalization of T4 antigen induced by phorbol esters. *J. Immunol.* 137: 1194–1201
14a. Emmrich, F., Strittmatter, U., Eichmann, K. 1986. Synergism in the activation of human CD8 T cells by crosslinking the T-cell receptor complex with the CD8 differentiation antigen. *Proc. Natl. Acad. Sci. USA* 83: 8298–8302
15. Kavathas, P., Sukhatme, V. P., Herzenberg, L. A., Parnes, J. R. 1984. Isolation of the gene coding for the human T lymphocyte differentiation antigen Leu-2 (T8) by gene transfer and cDNA subtraction. *Proc. Natl. Acad. Sci. USA* 81: 7688–92
16. Littman, D. R., Thomas, Y., Maddon, P. J., Chess, L., Axel, R. 1985. The isolation and sequence of the gene en-

coding T8: A molecule defining functional classes of T lymphocytes. *Cell* 40: 237–46

17. Sukhatme, V. P., Sizer, K. C., Vollmer, A. C., Hunkapiller, T., Parnes, J. R. 1985. The T cell differentiation antigen Leu-2/T8 is homologous to immunoglobulin and T cell receptor variable regions. *Cell* 40: 591–97

18. Maddon, P. J., Littman, D. R., Godfrey, M., Maddon, D. E., Chess, L., Axel, R. 1985. The isolation and nucleotide sequence of a cDNA encoding the T cell surface protein T4: A new member of the immunoglobulin gene family. *Cell* 42: 93–104

19. Nakauchi, H., Nolan, G. P., Hsu, C., Huang, H. S., Kavathas, P., Herzenberg, L. A. 1985. Molecular cloning of Lyt-2, a membrane glycoprotein marking a subset of mouse T lymphocytes: Molecular homology to its human counterpart, Leu-2/T8, and to immunoglobulin variable regions. *Proc. Natl. Acad. Sci. USA* 82: 5126–30

20. Zamoyska, R., Vollmer, A. C., Sizer, K. C., Liaw, C. W., Parnes, J. R. 1985. Two Lyt-2 polypeptides arise from a single gene by alternative splicing patterns of mRNA. *Cell* 43: 153–63

21. Littman, D. R., Gettner, S. N. 1987. Unusual intron in the Ig-like domain of the newly isolated mouse CD4 (L3T4) gene. *Nature* 325: 453–55

21a. Tourvieille, B., Gorman, S. D., Field, E. H., Hunkapiller, T., Parnes, J. R. 1986. Isolation and sequence of L3T4 complementary DNA clones: Expression in T cells and brain. *Science* 234: 610–14

22. Johnson, P., Gagnon, J., Barclay, A. N., Williams, A. F. 1985. Purification, chain separation and sequence of the MRC OX-8 antigen, a marker of rat cytotoxic T lymphocytes. *EMBO J.* 4: 2539–45

23. Johnson, P., Williams, A. F. 1986. Striking similarities between antigen receptor J pieces and sequence in the second chain of the murine CD8 antigen. *Nature* 323: 74–76

24. Clark, S. J., Jefferies, W. A., Barclay, A. N., Brown, W., Gagnon, J., Williams, A. F. 1987. Peptide and nucleotide sequences of rat CD4 (W3/25) antigen: Derivation from a structure with four Ig-related domains. *Proc. Natl. Acad. Sci. USA.* In press

25. Cantor, H., Boyse, E. A. 1977. Lymphocytes as models for the study of mammalian cellular differentiation. *Immunol. Rev.* 33: 105–24

26. Reinherz, E. L., Schlossman, S. F. 1980. The function and differentiation of human T cells. *Cell* 19: 821–27

27. Lanier, L. L., Allison, J., Phillips, J. 1986. Correlation of cell surface antigen expression on human thymocytes by multicolor flow cytometric analysis: Implications for differentiation. *J. Immunol.* 137: 2501–7

28. Lanier, L., Weiss, A. 1986. Presence of Ti (WT31) negative T lymphocytes in normal blood and thymus. *Nature.* 324: 268–70

29. Weiss, A., Newton, M., Crommie, D. 1986. Expression of T3 in association with a molecule distinct from the T cell antigen receptor heterodimer. *Proc. Natl. Acad. Sci. USA* 83: 6998–7002

30. Brenner, M. B., Mclean, J., Dialynas, D. P., Strominger, J. L., Smith, J. A., Owen, F. L., Seidman, J. G., Ip, S., Rosen, F., Krangel, M. S. 1986. Identification of a putative second T-cell receptor. *Nature* 332: 145–49

31. Bank, I., DePinho, R. A., Brenner, M. B., Cassimeris, J., Alt, F. W., Chess, L. 1986. A functional T3 molecule associated with a novel heterodimer on the surface of immature human thymocytes. *Nature* 322: 179–81

32. Webb, M., Mason, D. W., Williams, A. F. 1979. Inhibition of the mixed lymphocyte response with a monoclonal antibody specific for a rat T lymphocyte subset. *Nature* 282: 841–43

33. Rogozinski, L., Bass, Y., Glickman, E., Talle, M. A., Goldstein, G., Wang, J., Chess, L., Thomas, Y. 1984. The T4 surface antigen is involved in the induction of helper function. *J. Immunol.* 132: 735–39

34. Kruisbeek, A. M., Mond, J. J., Fowlkes, B. J., Carmen, J. A., Bridges, S., Longo, D. L. 1985. Absence of the Lyt-2$^-$, L3T4$^+$ lineage of T cells in mice treated neonatally with anti-I-A correlates with absence of intrathymic I-A-bearing antigen-presenting cell function. *J. Exp. Med.* 161: 1029–47

35. Benjamin, R. J., Waldmann, H. 1986. Induction of tolerance by monoclonal antibody therapy. *Nature* 320: 449–51

36. Gutstein, N. L., Seaman, W. E., Scott, J. H., Wofsy, D. 1986. Induction of immune tolerance by administration of monoclonal antibody to L3T4. *J. Immunol.* 137: 1127–32

37. Waldor, M. K., Sriram, S., Hardy, R., Herzenberg, L. A., Lanier, L. L., Lim, M., Steinman, L. 1985. Reversal of experimental allergic encephalomyelitis with monoclonal antibody to a T-cell subset marker. *Science* 227: 415–17

38. Lakey, E. K., Margoliash, E., Fitch, F. W., Pierce, S. K. 1986. Role of L3T4 and Ia in the heteroclitic response of T cells to cytochrome c. *J. Immunol.* 136: 3933–38

39. MacDonald, H. R., Glasebrook, A. L., Bron, C., Kelso, A., Cerottini, J.-C. 1982. Clonal heterogeneity in the functional requirement for Lyt-2,3 molecules on cytolytic T lymphocytes (CTL): Possible implications for the affinity of CTL antigen receptors. *Immunol. Rev.* 68: 88–115

40. Shimonkevitz, R., Luescher, B., Cerottini, J. C., MacDonald, H. R. 1985. Clonal analysis of cytolytic T lymphocyte-mediated lysis of target cells with inducible antigen expression: Correlation between antigen density and requirement for Lyt-2/3 function. *J. Immunol.* 135: 892–99

41. Landegren, U., Ramstedt, U., Axberg, I., Ullberg, M., Jondal, M., Wigzell, H. 1982. Selective inhibition of human T cell cytotoxicity at levels of target recognition or initiation of lysis of monoclonal OKT3 and Leu-2a antibodies. *J. Exp. Med.* 155: 1579–84

42. Spits, H., van Schooten, W., Keizer, H., van Seventer, G., van de Rijn, M., Terhorst, C., deVries, J. E. 1986. Alloantigen recognition is preceded by nonspecific adhesion of cytotoxic T cells and target cells. *Science* 232: 403–5

43. Biddison, W. E., Rao, P. E., Talle, M. A., Goldstein, G., Shaw, S. 1984. Possible involvement of the T4 molecules in T cell recognition of class II HLA antigens: Evidence from studies of CTL-target cell binding. *J. Exp. Med.* 159: 783–97

44. Greenstein, J. L., Malissen, B., Burakoff, S. J. 1985. Role of L3T4 in antigen-driven activation of a class I-specific T cell hybridoma. *J. Exp. Med.* 162: 369–74

45. Watts, T. H., Brian, A. B., Kappler, J. W., Marrack, P., McConnell, H. M. 1984. Antigen presentation by supported planar membranes containing affinity purified I-A^d. *Proc. Natl. Acad. Sci. USA* 81: 7564–68

46. Golding, H., McCluskey, J., Munitz, T. I., Germain, R. N., Margulies, D. H., Singer, A. 1985. T-cell recognition of a chimaeric class II/class I MHC molecule and the role of L3T4. *Nature* 317: 425–27

47. Solbach, W. 1982. Tumor-promoting phorbol esters selectively abrogate the expression of the T4 differentiation antigen expressed on normal and malignant (Sezary) T helper lympho-

cytes. *J. Exp. Med.* 156: 1250–55

48. Davis, R. J., Czech, M. P. 1984. Tumor-promoting phorbol diesters mediate phosphorylation of the epidermal growth factor receptor. *J. Biol. Chem.* 259: 8545–49

48a. Acres, R. B., Conlon, P. J., Mochizuki, D. Y., Gallis, B. 1986. Rapid phosphorylation and modulation of the T4 antigen on cloned helper T cells induced by phorbol myristate acetate or antigen. *J. Biol. Chem.* In press

49. Snow, P. M., Keizer, G., Coligan, J. E., Terhorst, C. 1984. Purification and N-terminal amino acid sequence of the human T cell surface antigen T8. *J. Immunol.* 133: 2058–66

50. Wigler, M., Sweet, R., Sim, G. K., Wold, B., Pellicer, A., Lacy, E., Maniatis, T., Silverstein, S., Axel, R. 1979. Transformation of mammalian cells with genes from procaryotes and eucaryotes. *Cell* 16: 777–85

51. Kavathas, P., Herzenberg, L. A. 1983. Stable transformation of mouse L cells for human membrane T-cell differentiation antigens, HLA and β2-microglobulin: Selection by fluorescence-activated cell sorting. *Proc. Natl. Acad. Sci. USA* 80: 524–28

52. Kuhn, L. C., McClelland, A., Ruddle, F. H. 1984. Gene transfer, expression, and molecular cloning of the human transferrin receptor gene. *Cell* 37: 95–103

53. Chao, M. V., Bothwell, M. A., Ross, A. H., Koprowski, H., Lanahan, A. A. 1986. Gene transfer and molecular cloning of the human NGF receptor. *Science* 232: 518–21

54. Kavathas, P., Herzenberg, L. A. 1983. Amplification of a gene coding for human T-cell differentiation antigen. *Nature* 306: 385–87

55. Nakayama, E., Shiku, H., Stockert, E., Oettgen, H. F., Old, L. J. 1979. Cytotoxic T cells: Lyt phenotype and blocking of killing activity by Lyt antisera. *Proc. Natl. Acad. Sci. USA* 76: 1977–81

56. Hollander, N., Pillemer, E., Weissman, I. L. 1980. Blocking effect of Lyt-2 antibodies on T cell functions. *J. Exp. Med.* 152: 674–87

57. Rothenberg, E., Triglia, D. 1983. Lyt-2 glycoprotein is synthesized as a single molecular species. *J. Exp. Med.* 157: 365–70

58. Dialynas, D. P., Quan, Z. S., Wall, K. A., Pierres, A., Quintans, J., Loken, M. R., Pierres, M., Fitch, F. W. 1983. Characterization of the murine cell sur-

face molecule, designated L3T4, identified by monoclonal antibody GK1.5: Similarity of L3T4 to the human Leu3/T4 molecule and the possible involvement of L3T4 in class II MHC antigen reactivity. *J. Immunol.* 131: 2445–51

59. Terhorst, C., van Agthoven, A., Reinherz, E., Schlossman, S. 1980. Biochemical analysis of human T lymphocyte differentiation antigens T4 and T5. *Science* 209: 520–21

60. Snow, P. M., Terhorst, C. 1983. The T8 antigen is a multimeric complex of two distinct subunits on human thymocytes but consists of homomultimeric forms on peripheral blood T lymphocytes. *J. Biol. Chem.* 258: 14675–81

61. Snow, P. M., Van de Rijn, M., Terhorst, C. 1985. Association between the human thymic differentiation antigens T6 and T8. *Eur. J. Immunol.* 15: 529–32

62. Reilly, E. B., Auditore-Hargreaves, K., Hammerling, U., Gottlieb, P. D. 1980. Lyt-2 and Lyt-3 alloantigens: Precipitation with monoclonal and conventional antibodies and analysis on one- and two-dimensional polyacrylamide gels. *J. Immunol.* 125: 2245–51

63. Ledbetter, J. A., Seaman, W. E., Tsu, T. T., Herzenberg, L. A. 1981. Lyt-2 and Lyt-3 antigens are on two different polypeptide subunits linked by disulfide bonds; relationship of subunits to T cell cytolytic activity. *J. Exp. Med.* 153: 1503–16

64. Ledbetter, J. A., Seaman, W. E. 1982. Lyt 2, Lyt 3 macromolecules: Structural and functional studies. *Immunol. Rev.* 68: 197–218

65. Walker, I. D., Murray, B. J., Hogarth, P. M., Kelso, A., McKenzie, I. F. C. 1984. Comparison of thymic and peripheral T cell Lyt-2/3 antigens. *Eur. J. Immunol.* 14: 906–10

66. Walker, I. D., Hogarth, P. M., Murray, B. J., Lovering, K. E., Classon, B. J., Chambers, G. W., McKenzie, I. F. C. 1984. Ly antigens associated with T cell recognition and effector function. *Immunol. Rev.* 82: 47–77

67. Liaw, C. W., Zamoyska, R., Parnes, J. R. 1986. Structure, sequence, and polymorphism of the Lyt-2 T cell differentiation antigen gene. *J. Immunol.* 137: 1037–43

68. Brown, M. S., Goldstein, J. L. 1986. A receptor-mediated pathway for cholesterol homeostasis. *Science* 232: 34–47

69. Luescher, B., Rousseaux-Schmid, M., Nain, H. Y., MacDonald, H. R., Bron, C. 1985. Biosynthesis and maturation of the Lyt-2/3 molecular complex in mouse thymocytes. *J. Immunol.* 135: 1937–44

70. Fujimoto, J., Levy, S., Levy, R. 1983. Spontaneous release of the Leu-2 (T8) molecule from human T cells. *J. Exp. Med.* 159: 752–66

71. Fujimoto, J., Stewart, S. J., Levy, R. 1984. Immunochemical analysis of the released Leu-2 (T8) molecule. *J. Exp. Med.* 160: 116–24

72. Krangel, M. S. 1986. Secretion of HLA-A and -B antigens via an alternative RNA splicing pathway. *J. Exp. Med.* 163: 1173–90

73. Isobe, M., Huebner, K., Maddon, P. J., Littman, D. R., Axel, R., Croce, C. M. 1986. The gene encoding the T-cell surface protein T4 is located on human chromosome 12. *Proc. Natl. Acad. Sci. USA* 83: 4399–4402

74. May, W. S., Jacobs, S., Cuatrecasas, P. 1984. Association of phorbol ester-induced hyperphosphorylation and reversible regulation of transferrin membrane receptors in HL-60 cells. *Proc. Natl. Acad. Sci. USA* 81: 2016–20

75. Gluckman, J. C., Klatzmann, D., Montagnier, L. 1986. Lymphadenopathy-associated-virus infection and acquired immunodeficiency syndrome. *Ann. Rev. Immunol.* 4: 97–117

76. Petito, C. K., Namia, B. A., Eun-Sook, C. H. O., Jordon, B. D., George, D. C., Price, R. W. 1985. Vacuolar myelopathy pathologically resembling subacute combined degeneration in patients with the acquired immunodeficiency syndrome. *New Engl. J. Med.* 312: 874–79

77. Gottlieb, M. S., Groopman, J. E., Weinstein, W. M., Fahey, J. L., Detels, R. 1983. The acquired immunodeficiency syndrome. *Ann. Int. Med.* 99: 208–20

78. Klatzmann, D., Barre-Sinoussi, F., Nugeyre, M. T., Dauguet, C., Vilmer, E., Griscelli, C., Brun-Vezinet, F., Rouzioux, C., Gluckman, J. I., Chermann, J. C., Montagnier, L. 1984. Selective tropism of lymphadenopathy associated virus (LAV) for helper-inducer T lymphocytes. *Science* 225: 59–63

79. Dalgleish, A. G., Beverley, P. C. L., Clapham, P. R., Crawford, D. H., Greaves, M. F., Weiss, R. A. 1984. The CD4 (T4) antigen is an essential component of the receptor for the AIDS retrovirus. *Nature* 312: 763–66

80. Klatzmann, D., Champagne, E., Chamaret, S., Gruest, J., Guetard, D.,

Hercend, T., Gluckmann, J.-C., Montagnier, L. 1984. T-lymphocyte T4 molecule behaves as the receptor for human retrovirus LAV. *Nature* 312: 767–68

81. Stohl, W., Kunkel, H. G. 1984. Heterogeneity in expression of the T4 epitope in black individuals. *Scand. J. Immunol.* 20: 273–78

82. Hoxie, J. A., Flaherty, L. E., Haggarty, B. S., Rackowski, J. L. 1986. Infection of T4 lymphocytes by HTLV-III does not require expression of the OKT4 epitope. *J. Immunol.* 136: 361–63

83. McDougal, J. S., Kennedy, M. S., Sligh, J. M., Cort, S. P., Mawle, A., Nicholson, J. K. 1986. Binding of HTLV/LAV to T4+ T cells by a complex of the 110K viral protein and the T4 molecule. *Science* 231: 382–85

84. Maddon, T. J., Dalgleish, A. G., McDougal, J. S., Clampham, T. R., Weiss, R. A., Axel, R. 1986. The T4 Gene: encodes the AIDS virus receptor and is expressed in the Immune System and the brain. *Cell* 47: 333–48

85. Lifson, J. D., Reyes, G. R., McGrath, M. S., Stein, B. S., Engleman, E. G. 1986. AIDS retrovirus induced cytopathology: Giant cell formation and involvement of CD4 antigen. *Science* 232: 1123–27

86. Sodroski, J., Goh, W. C., Rosen, C., Campbell, K., Haseltine, W. A. 1986. Role of HTLV-III/LAV envelope in syncytium formation and cytopathicity. *Nature* 322: 470–74

87. Starcich, B. R., Hahn, B. H., Shaw, G. M., McNeely, P. D., Modrow, S., Wolf, H., Parks, E. S., Parks, W. P., Joseph, S. F., Gallo, R. S., Wong-Staal, F. 1986. Identification and characterization of conserved and variable regions in the envelope gene of HTLV-III/LAV, the retrovirus of AIDS. *Cell* 45: 637–48

88. Shaw, G. M., Harper, M. E., Hahn, B. H., Epstein, L. G., Gajdusek, D. C. 1985. HTLV-III infection in brains of children and adults with AIDS encephalopathy. *Science* 227: 177–82

89. Ho, D. D., Rota, T. R., Schooley, R. T., Kaplan, J. C., Allan, J. D., Groopman, J. E., Kesnick, L., Felsenstein, D., Andrews, C. A., Hirsch, M. S. 1985. Isolation of HTLV-III from cerebrospinal fluid and neural tissues of pati-

ents with neurologic syndrome related to the acquired immunodeficiency syndrome. *N. Engl. J. Med.* 313: 1498–1504

90. Levy, J. A., Hollander, H., Shimabukuro, J., Mills, J., Kaminsky, L. 1985. Isolation of AIDS-associated retroviruses from cerebrospinal fluid and brain of patients with neurological symptoms. *Lancet* 2: 586–88

91. Jefferies, W. A., Green, J. R., Williams, A. F. 1985. Authentic T helper CD4 (W3/25) antigen on rat peritoneal macrophages. *J. Exp. Med.* 162: 117–27

92. Levy, J. A., Shimabukuro, J., McHugh, T., Casavant, C., Stites, D., Oshiro, L. 1985. AIDS-associated retroviruses (HIV) can productively infect other cells besides human T-helper cells. *Virology* 147: 441–48

93. Deleted in proof

94. Koenig, S., Gendelman, H. E., Orenstein, J. M., DalCanto, M. C., Pezeshkpour, G. H., Yungbluth, M., Janotta, F., Aksamit, A., Martin, M. A., Fauci, A. S. 1986. Detection of AIDS virus in macrophages in brain tissue from AIDS patients with encephalopathy. *Science* 233: 1089–93

95. Hood, L., Kronenberg, M., Hunkapiller, T. 1985. T cell antigen receptors and the immunoglobulin supergene family. *Cell* 40: 225–29

96. Hunkapiller, T., Hood, L. 1986. The growing immunoglobulin gene superfamily. *Nature* 323: 15–16

97. Gibsen, D. M., Taylor, B. A., Cherry, M. 1978. Evidence for close linkage of a mouse light chain marker with the Ly-2,3 locus. *J. Immunol.* 121: 1585–90

98. Gottlieb, P. D. 1974. Genetic correlation of a mouse light chain variable region marker with a thymocyte surface antigen. *J. Exp. Med.* 140: 1432–37

99. Sukhatme, V. P., Vollmer, A. Cd., Erikson, J., Isobe, M., Croce, C., Parnes, J. R. 1985. Gene for the human T cell differentiation antigen Leu-2/T8 is closely linked to the *k* light chain locus on chromosome 2. *J. Exp. Med.* 161: 429–34

100. Doolittle, R. 1978. Genes in pieces: Were they ever together? *Nature* 272: 581–82

101. Gilbert, W., Marchionni, M., McKnight, G. 1986. On the antiquity of introns. *Cell* 46: 151–54

Ann. Rev. Immunol. 1987. 5 : 585–620

GENES OF THE T-CELL ANTIGEN RECEPTOR IN NORMAL AND MALIGNANT T CELLS

Barry Toyonaga and Tak W. Mak

The Ontario Cancer Institute, and Department of Medical Biophysics, University of Toronto, 500 Sherbourne Street, Toronto, Ontario, Canada M4X 1K9

INTRODUCTION

Continuous surveillance for invasions by foreign antigens (Ag) is the responsibility of the vertebrate immune system. A response is first elicited when foreign antigens are identified by receptor proteins produced by B and T lymphocytes known as the immunoglobulin (Ig) and T-cell-antigen receptors (TcR), respectively. The structures of Ig proteins and genes have been well characterized. The Ig molecule has a variable domain at its amino terminus responsible for the Ag specificity, and this is attached to a more constant domain at the carboxyl terminus (1, 2). Depending on the isotype of the Ig constant domain, various effector functions are possible. Whether bound to the B-cell surface or as a secreted molecule, the Ig recognizes and binds to a specific Ag.

The most striking contrast between B- and T-cell activation is that T cells require dual recognition of Ag and a polymorphic gene product of the major histocompatibility complex (MHC). This phenomenon is known as MHC-restricted recognition (3). A further constraint is that the MHC product recognized must be that of the host organism (i.e. self MHC). The "education" of T cells is believed to be in the thymus at the level of the surface membrane bound TcR.

The TcR was first identified as a heterodimeric protein composed of an acidic (α) and basic (β) chain, each a glycosylated polypeptide (4–6). Taking different approaches, two laboratory groups isolated similar cDNA clones, one from a human (7) and the other from a murine source (8). Later, comparison to protein sequence data determined that both cDNAs coded

0732–0582/87/0410–0585$02.00

for the TcR-β chain (9). Application of similar techniques resulted in the isolation of cDNAs coding for the TcR-α chain (10–13). Molecular studies revealed that the TcR genes are independent from Ig genes but possess Ig-like structural features, variable and constant domains (9–13). Also, as do Ig genes, noncontiguous V, D, and J germline TcR gene segments are capable of undergoing somatic rearrangements leading to the creation of unique V regions for each maturing T cell (14–21). The data collected so far support hypothesis of the relation of TcR and Ig genes through a common ancestral gene. While searching for the TcR-α-chain genes, Saito et al also isolated a T-cell-specific gene with an Ig-like structure, the T-cell-γ gene (Tcγ), capable of undergoing somatic rearrangement (22).

With the discovery of these three genes (Figure 1), the techniques of molecular biology have been applied to studies of the structural relationship between the TcR- and MHC-restricted antigen recognition, the thymic selection of T cells, and the role of TcR genes as analytical and diagnostic probes for the evaluation of proliferative disorders and autoimmune diseases. In this article, we summarize the current molecular biology of the TcR genes. In many examples, especially those of the TcR-α– and TcR-β–chains, the structures of the human genes are analyzed in detail. Several comprehensive reviews with emphasis on the molecular genetics of the murine TcR have been published (23–25). The use of the TcR genes in the analysis of T-cell malignancies is presented from the point of view of defining clonality and lineage. In addition, the possible activation of cellular genes involved in these proliferative disorders through chromosomal translocation is reviewed.

MOLECULAR CHARACTERISTICS

Chromosomal Location

The chromosomal locations of the human and mouse TcR-α– (26–33) and -β– (34–37) chain genes as well as that of the Tcγ (28, 29) gene

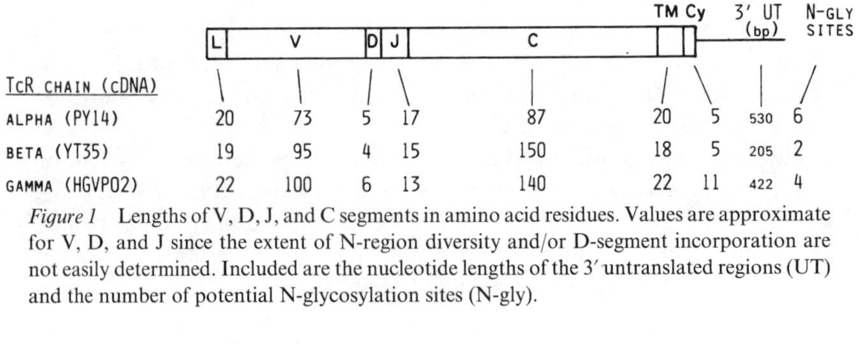

TcR CHAIN (cDNA)	L	V	D	J	C	TM	Cy	3′ UT (bp)	N-GLY SITES
ALPHA (PY14)	20	73	5	17	87	20	5	530	6
BETA (YT35)	19	95	4	15	150	18	5	205	2
GAMMA (HGVP02)	22	100	6	13	140	22	11	422	4

Figure 1 Lengths of V, D, J, and C segments in amino acid residues. Values are approximate for V, D, and J since the extent of N-region diversity and/or D-segment incorporation are not easily determined. Included are the nucleotide lengths of the 3′ untranslated regions (UT) and the number of potential N-glycosylation sites (N-gly).

Table 1 Chromosomal locations of related genes in the human and mouse

Gene	Human	Mouse
TcR-α	14q11	14C–D
TcR-β	7q32–35	6B
Tcγ	7p15	13A2–3

are shown in Table 1. These data provide conclusive evidence that the immunorecognition genes of B and T cells are distinct from one another. However, the comparison of structural details of Ig and TcR genes supports the hypothesis that their evolutionary development is from a common ancestor thought to be the progenitor of the Ig gene superfamily (38).

Germline Organization

Using the originally isolated cDNA clones as molecular probes, researchers were able to characterize the TcR-α–, TcR-β–, and Tcγ–chain germline loci within the human genome. The results of the parallel study of the murine genome are presented in several comprehensive reviews (23–25). As the similarities between the human and murine data are numerous, only those results that differ are presented.

In general, the TcR-α, TcR-β and Tcγ chains are products of somatic rearrangement processes that bring together noncontiguous germline gene segments in an apparently random fashion (Figure 2). The TcR gene

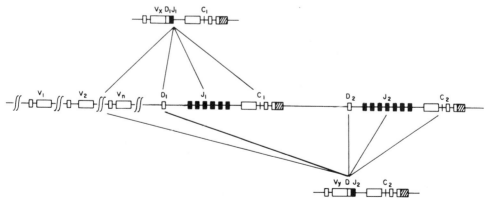

Figure 2 Combinatorial possibilities leading to the formation of active/rearranged TcR-β–chain genes using noncontiguous germline gene segments. Similar processes are thought to be involved in the rearrangement of TcR-α– and Tcγ-chain genes.

segments are reminiscent of Ig variable (V), diversity (D), joining (J) and constant (C) gene segments and as such are referred to in the same manner (14–21). The generation of a diverse Ag/MHC recognition repertoire by this assembly process is discussed in a later section.

The germline genomic organizations of the human TcR-α–, TcR-β–, and Tcγ–gene loci are shown in Figure 3. Ig-like recombinational recognition sequences flank the human TcR-α–, TcR-β–, and Tcγ–chain gene segments as they do those of the mouse. The "12/23" recombination rule established for the somatic rearrangement of Ig genes also holds for TcR genes.

In the TcR-α locus, only one constant region gene has been found, with numerous J segments upstream of the C_α gene (19–21). The 5′ most J_α gene segment has yet to be found. At present, overlapping clones covering a distance of over 65 kb 5′ to the single C_α gene locus have been isolated. Based upon a statistical analysis of usage in TcR α-chain cDNAs examined, an estimate of approximately 30–60 unique J_α gene segments can be made (39).

The TcR-β locus is probably the most well-defined of the three TcR gene loci shown. A comparison of the human and murine loci show conservation of the coding regions that is remarkable when one considers that the two species diverged some 70 million years ago (Figures 4a–d).

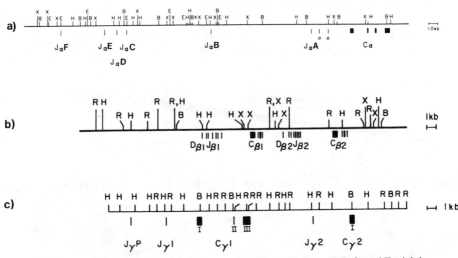

Figure 3 Germline genomic restriction maps of the human TcR-α–, TcR-β– and Tcγ joining (J), diversity (D), and constant (C) gene segment loci. Only coding regions that have been sequenced are indicated by filled boxes below each map (i.e. many unsequenced J_α gene segments exist over the region 5′ to the C_α). Pseudo genes (ϕ), and the restriction sites HindIII (H), BamH1 (B), EcoR1 (E and R) and Xbal (X) are indicated.

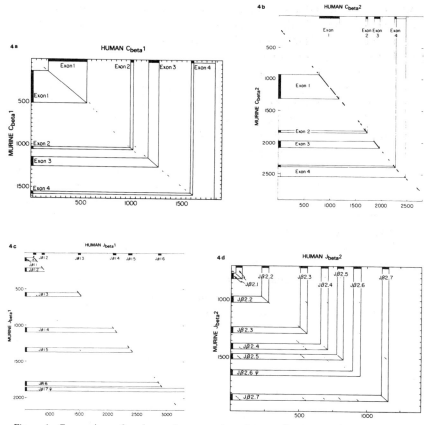

Figure 4   Comparison of analogous human and murine germline TcR-β–chain gene loci by dot matrix analysis (131). Coding positions are indicated by filled boxes along the axes. Two murine pseudo J_β gene segments are marked (ψ).

a. $C_\beta 1$ DNA sequence comparison (span = 25, homology = 65%)
b. $C_\beta 2$ DNA sequence comparison (span = 25, homology = 65%)
c. $J_\beta 1$ DNA sequence comparison (span = 25, homology = 75%)
d. $J_\beta 2$ DNA sequence comparison (span = 25, homology = 75%).

Several interesting features of the human and mouse TcR-β loci become evident upon closer examination. First, the human loci contains six $J_\beta 1$ and seven $J_\beta 2$ gene segments, all potentially functional (17). In contrast, each murine J_β cluster contains six functional segments and one pseudo-gene segment (16–17, 40). Second, the human and murine $D_\beta 1.1$ gene segments are identical, whereas the $D_\beta 2.1$ gene segments are different in both sequence and length (16–18, 40). Third, upon comparison of $C_\beta 1$ with

$C_\beta 2$ genomic loci, an unexplained homology spanning approximately 95 nucleotides is found extending 3' of the first exon/intron boundary (17). No other introns of either human or murine C_β genes display this degree of conservation. It is interesting that this homologous intron stretch forms part of a larger 220-bp fragment that straddles the exon/intron boundary and contains an alternate open translational reading frame of 73 amino acids complete with RNA donor/acceptor splice signals at either end (17).

Finally, a new murine TcR-β–chain constant region–like exon, $C_\beta 0$, has been reported by Behlke & Loh (41). Sequence analysis of murine TcR-β–chain cDNAs revealed the presence of a conserved 72-bp insertion between J_β and C_β gene segments (41). This insertional element was shown to correspond to an exon located in the murine germline genome, between the $J_\beta 1$ cluster and the $C_\beta 1$ gene. Apparently $C_\beta 0$ can be inserted via an alternative splicing pathway available to $C_\beta 1$ containing TcR-β messages. $C_\beta 0$ sequences are observed in 1% to 18% of the murine TcR-β–chain cDNAs isolated, and the proportion seems to depend on the source of isolation. No report of a similar exon in the human TcR-β–chain cDNAs has been made to date (42–45).

Neither of the C_β regions described above has been linked to corresponding V-gene segments in the human germline genome. In the case of the murine TcR-β–chain locus, however, a V_β segment ($V_\beta 14$) was found in an inverted orientation 10 kb 3' to the $C_\beta 2$ gene (46).

Little more than the partial restriction map and sequences of the human Tcγ J and C genes are known. Only two highly homologous C_γ region genes have been found so far (45–50). The germline map of this locus (Figure 3) summarizes the available data. Note that only the first exon of $C_\gamma 2$ has been mapped and sequenced in the germline state. However, nine germline V_γ gene segments have been sequenced and mapped onto two unlinked regions spanning a total of 54 kb (50). It is thought that this represents most if not all of the human V_γ germline repertoire (49–51).

The murine γ-chain locus is more extensively characterized (Figure 5). It is now known that at least four constant region genes are present in the mouse genome. One is part of a pseudogene VJC complex (52) and is absent in some strains of mice (53). Another constant region gene ($C_\gamma 4$) is only partially homologous to the original C_γ genes ($C_\gamma 1$ and $C_\gamma 2$) (53) and may exist as an isotype-like constant region (53). Its functional importance in T cells is unknown. The $C_\gamma 4$ gene and associated V ($V_\gamma 1$) and J gene segments are linked together on a 10.8-kb EcoR1 fragment, in an orientation inverse from the original $V_\gamma 2$, and $C_\gamma 2$ genes (53). This design allows each of the genes ($C_\gamma 4$ and $C_\gamma 2$) to rearrange and be transcribed with its respective V gene segments ($V_\gamma 1$ and $V_\gamma 2$) on the same allele without affecting the other. It is interesting that $C_\gamma 4$ does not cross-hybridize with

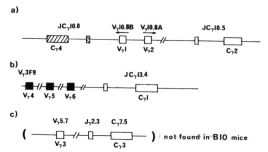

Figure 5   Schematic diagram of germline genomic organization of Tcγ genes. In the BALB/c mice genome, all of a, b, and c are present. Genomes of B10 mice contain only a and b as shown, but are missing c. Introns are not shown.

any of the other three C_γ genes, raising a question as to whether or not other TcR or Tcγ constant regions (isotypes) exist and whether they may perform functionally important roles in T cells. Several V_γ gene segments ($V_\gamma 4$, $V_\gamma 5$ and $V_\gamma 6$) are now known to be associated with $C_\gamma 1$ gene (53–55). They are expressed mainly in early thymic ontogeny (54).

Evolution of the Immunoglobulin Gene Superfamily

Over the past few years, the immunoglobulin gene superfamily has taken shape (38) with the isolation and characterization of a diverse group of cell surface molecules such as Thy1, Poly Ig receptor, T3, T4, T6, T8, Leu-1, T11, class-I and class-II MHC molecules, and the TcR genes. Although not all of these molecules are known to be involved in interactions between immune recognition cells, they exhibit deduced protein sequence homologies. While most of them show a closer relationship to the variable domain of Ig, the MHC-gene products and the constant domain of the TcR genes are more structurally related to the constant domain of Ig (38). A precise estimate of how these genes have evolved and their relationship to each other is difficult to assess based on their amino acid sequences. Their homologies to each other may have been maintained because of functional constraints upon their structure. On the other hand, these genes may simply have a very low rate of evolutionary divergence. A more precise picture of the relationship among these genes may be obtained by comparison of their sequences at the DNA level. Hypothetical evolutionary relationships among the three related families, MHC genes, the TcR genes, and the Ig heavy- and light-chain genes, can be determined based on comparisons of their constant region–domain nucleotide sequences (S.

Clark, B. Wilson, unpublished data). Figure 6 illustrates that the relationships among genes from the same subfamily frequently are not as close as those of genes from different subfamilies. For example, the TcR-β chain gene is more closely related to MHC, Ig-λ, and Ig heavy-chain genes than to either the α or γ chains of the TcR or the κ chain of Ig. Based on the observed degree of homologies among these genes, it would seem that the IgH and Igλ have branched off from the TcR-β genes, which in turn could have evolved from the MHC genes. Divergence points in this figure are predicted assuming that MHC genes were first to arrive, followed by TcR genes, and then Ig genes. This is consistent with the postulated order of appearance of immune system genes as one ascends the evolutionary ladder.

Thymic Ontogeny

The rearrangement and expression of TcR-α–, TcR-β–, and Tcγ-chain genes during thymic ontogeny have been extensively studied. If one looks at the expression of T-cell-receptor messages in murine thymus during embryogenesis (56–58), Tcγ-chain messages appear first, followed by TcR-β–chain and then TcR-α–chain messages. The day-14 fetal thymus was taken as the earliest stage of thymic development. It contains developing T cells which are mainly Thy-1$^+$ and Lyt2$^-$ L3T4$^-$ with low levels of IL-2-receptor expression. These cells express Tcγ but not TcR-β or TcR-α transcripts. By day 15, the short 1.0-kb TcR-β message is transcribed, and only low levels of the longer 1.3-kb message are detectable. By day 16, TcR-α–chain transcription begins and Tcγ levels decrease significantly. Levels of TcR-β–chain expression, greater than that found in mature T cells, are first seen on day 16 and continue until birth. The expression of

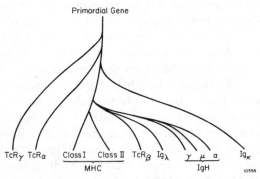

Figure 6 Evolutionary relationships between members of the immunoglobulin gene super-family based upon similarities of constant region–like DNA sequences.

TcR-α chain only reaches significant levels on day 17. The T-cell receptor is first seen on the surface at this time. The level of TcR-β–chain messages is elevated in thymocytes compared to peripheral T cells (59).

The order of TcR gene expression is generally agreed, based on DNA rearrangements found in heterogeneous T-cell populations, T-cell lymphoma lines or thymocyte hybridomas. For example, Born et al have reported $D_\beta 1$-$J_\beta 1$ and $D_\beta 2$-$J_\beta 2$ rearrangements in early thymic development (days 14–17), before complete TcR rearrangement and expression on the cell surface (day 16) (60). Analysis of thymic leukemia cell lines and functional T-cell clones also supports the hypothesis that rearrangement and expression of TcR-β–chain genes occur prior to those of the TcR-α–chain genes (61, 62). However, no compelling data has been reported that suggest an obligatory ordered rearrangement of Tcγ-chain genes before TcR-β–chain genes. Rearranged Tcγ-chain genes in tumors in the presence of germline configurations of TcR-β–chain genes have been observed (63). In contrast, rearranged TcR-β–chain genes in tumor cells with germline configurations of the Tcγ-chain genes have also been seen (62). The search for an ordered sequence of rearrangement of Tcγ before TcR-β is also clouded by the observation of Griesser et al that Tcγ-chain–gene rearrangement can also be found in a high proportion of non-T-cell tumors (63). Examination of thymocyte subpopulations from the human thymus also supports hypothesis of the rearrangement and expression of TcR-β–chain genes prior to those of the TcR-α–chain genes (64, 65).

The Functions of the TcR-αβ Heterodimer

Gene transfer experiments have formally demonstrated the functions of the TcR-αβ heterodimer. Ohashi et al were able to reconstitute a functional TcR/T3 complex by gene transfer confirming that the TcR-αβ heterodimer is indeed a signal transducing molecule and may require the T3 molecules to surface (66). By the transfer of TcR-αβ–chain genes from a T-cell specific for fluorescein (FL) and the mouse class-I product D^d, Dembic et al were able to convert a T-cell hybridoma specific for the hapten (SP) and class-I K^k to a cell that exhibits reactivity against both FL-D^d and SP-K^k (67). This result strongly supports the hypothesis that the TcR-αβ heterodimer may be sufficient for the recognition of both antigen and MHC components. More recent experiments from Germain and coworkers (personal communication) also confirmed this observation. They were able to confer functional specificities by transferring TcR-α– and TcR-β–chain genes from murine helper T cells that had reactivities towards defined antigens and specific Ia molecules. All of these experiments are consistent with the earlier observations of Kappler and Marrack and coworkers who demonstrated that the dual recognition of antigen and MHC molecules is probably mediated by a single receptor (68).

Possible Structural-Functional Relationships of α-β Heterodimers

Attempts to correlate the primary structures of TcR-α– and TcR-β–chain genes with the functional specificities of the T cells from which they were derived have been initiated using functional, helper, and killer T cells. In the analysis of five independent helper-T-cell clones specific for pigeon cytochrome c in the context of several Ia molecules, Hedrick and his coworkers (69) have found an unusually high occurrence of V-gene segments belonging to the same V_α family. It is possible that the V_α-gene segments of this family are preferentially used in the recognition of cytochrome c and Ia molecules. The primary structures of the TcR-α– and TcR-β–chain genes have also been determined in cytotoxic-T-cell (CTL) clones recognizing the hapten AED in the context of different determinants of murine class-I MHC gene products (70). In these studies, no particular V_α or V_β genes were found to be utilized preferentially. Instead, the use of a specific combination of J_α-J_β ($J_\alpha 810$ and $J_\beta 2.6$) was noted in two independent CTLs that had specificities of AED-K^b and AED-D^b (70). Also, a CTL specific for AED-D^b (70) and an alloreactive CTL specific for D^b (71) use the same V_β- and J_β-gene segments. Perhaps the unique combination of V_β and J_β is related to the recognition of the D^b molecule. A direct and simple structural-functional relationship between T-cell specificities and TcR-α– and -β–chain primary structures will apparently be difficult to establish. Further support comes from the observation that cells recognizing both class-I and class-II MHC gene products can utilize the same V or J gene segments (71–73). The most striking example is that a CTL with an alloreactivity to D^b has the same V_α and V_β segments as a helper T cell recognizing chicken red blood cell in the context of I-A^b (73).

Tcγ-Chain Function

Although extensive characterization of the structure and the diversity of Tcγ-chain genes has been carried out, its role remains unknown. Initially, Tcγ-chain transcripts were found mainly in class I–restricted CTLs (74, 75); it was later shown that some helper-T-cell clones also transcribe these genes (76). In fact, high levels were found in many autoreactive L3T4$^+$ helper clones (76). The relation of Tcγ-chain expression to its functional significance in T cells is still not known. In addition, it was found that the vast majority of messages found in these CTLs were nonfunctional (77, 78). Since then, an analysis of 21 human Tcγ-chain transcripts from peripheral blood T lymphocytes and thymocytes showed that all of them contained deletions or frameshift mutations or insertions (51). Very recently, two laboratories have found potentially functional transcripts of Tcγ-chain genes in certain special settings. For example, Jones et al were able to demonstrate that two functional Tcγ-chain transcripts were found in a cDNA library constructed from RNA in a mixed-leukocyte reaction (79). Yoshikai et al reported that while the TcR-α– and TcR-β–chain transcripts from young (8 week) athymic mice were not functional, 4 of 4 Tcγ messages contained full-length open reading frames (80). This data suggests that Tcγ-chain genes may be part of a second immunorecognition structure in either prethymic or extrathymic pathways, perhaps in a lineage where TcR-α– and TcR-β–chain heterodimers are not utilized. Brenner et al (81), Bank et al (82), and Weiss et al (83) recently reported finding human Tcγ-chain polypeptides in cells from patients with immunodeficiencies and in leukemia cell lines lacking either the TcR-α– or TcR-β–chain heterodimer; this is consistent with the hypothesis. A 45-kd polypeptide together with the Tcγ-chain polypeptide was found in some cases (81, 82), while only the 55-kd Tcγ was demonstrated in another (83). The identity of the 45-kd polypeptide is not known. It is possible that a fourth rearranging TcR-like chain is present. Other explanations include the pairing of $C_\gamma 1,2$ with $C_\gamma 4$ (53) or the association of a nonrearranging gene product (e.g. MHC class I) with the Tcγ chain.

TcR V-Region Repertoire

Most, if not all, of the ability of T cells to recognize a potentially infinite array of antigen/MHC combinations is thought to reside in the N-terminal portions of the TcR-α and β chains which have been shown to have clonally unique structures.

The variable regions of TcR-α, TcR-β and Tcγ chains are encoded by genes produced by the somatic rearrangement of noncontiguous variable (V), sometimes diversity (D) and joining (J) germline gene segments (14– 21). The nomenclature has been adopted from that established for Ig genes because of the structural and postulated functional similarities that exist between the Ig and TcR molecules. In addition to the generation of a large recognition repertoire by combinatorial joining of these gene segments which exist in multiple copies within the genome, other Ig-like V-region diversification mechanisms are utilized. Junctional flexibility, the imprecise joinings of the gene segments, and N-region diversification, the addition of nongenomic nucleotides at junction points, all occur during the somatic rearrangement process. Conclusive evidence for the process of somatic hypermutation, common in the generation of Ig molecules, has not yet been presented for TcR or Tcγ genes. The overall T-cell–recognition repertoire is therefore a fundamental function of the number of potentially functional

V, D, and J gene segments in the genome. Two approaches have been taken by a number of groups to estimate the V-gene segment repertoires of human and murine genomes. First, the number of hybridizing bands observed in Southern blots of enzyme restricted germline genomic DNA probed by non-cross-hybridizing V-gene segments can be tabulated. This method is dependent to some extent on the hybridization/washing conditions used. Secondly, more quantitative estimates have been determined by a statistical analysis of V region DNA sequences of TcR cDNAs isolated from various sources. The validity of assumptions made in the treatment of such data have been discussed in detail (39, 42–45, 84). The reliability or significance of the estimates is dependent on the random sampling of cDNAs derived from genes that are all equally utilized. The accuracy, on the other hand, will depend on the algorithms used to analyze the tabulated frequencies of observation. Obviously as the number of unique observations approaches the total number of observations, accuracy decreases regardless.

Estimates reported for V_α and V_β gene segment repertoires are collected in Table 2. Calculations have been performed on data collected from single individuals where possible (HAVP, HAVT, HBVP, and HBVT). As well,

Table 2   Statistical estimates of V-gene-segment repertoires

Source	Number of cDNA[a] containing V_α	Number of unique V_α	Most probable number of V_α (probability)	95% upper bound of V_α repertoire
		TcR-α chain		
HAVP	18	13	24 (0.28)	60
HAVT	10	8	19 (0.37)	112
all[b]	31	23	47 (0.21)	92

Source	Number of cDNA[a] containing V_β	Number of unique V_β	Most probable number of V_β (probability)	95% upper bound of V_β repertoire
		TcR-β chain		
HAVP	10	9	42 (0.43)	881
PL[c]	26	18	32 (0.24)	61
PH	15	12	30 (0.31)	110
HBVT	11	8	15 (0.36)	54
all[b]	76	46	68 (0.15)	90

[a] Each cDNA is made up of a unique combination of V, D, J, C and N-region sequences.
[b] All cDNA/genomic data available.
[c] This data was collected from libraries derived from more than one individual.
Abbreviations and references: See Figures 8 and 9.

estimates using all of the data available are given. In the last case, identical cDNAs (VDJC) are counted only once. The estimates are very approximate since the existence of polymorphisms and somatic hypermutation effects cannot be determined.

The estimated human TcR V_α-gene-segment repertoire is similar to that of the mouse, with approximately 50 members in over a dozen families. The number of human TcR–V_β-gene segments is estimated to be about 70, dispersed throughout 20 families, definitely greater than that found in the mouse (at most 20 segments). This human V_β estimate is probably an underestimation since the data collected in Table 3 and presented in Table 6a support the existence of a subset of V_β families that are used more frequently than the rest. The number of human TcR–V_β-gene segments could be in the hundreds (see HBVP in Table 2).

Results of genomic Southern blots of human germline DNA probed with selected TcR-α– and TcR-β–chain cDNAs are shown in Figures 7a and 7b. V-region repertoire estimates based upon the number of non-constant region hybridizing bands are consistent with those arrived at by statistical calculations. The nomenclature for the family assignments is described below. The results are in agreement with the empirical rule that the cross-hybridization threshold can be approximated by 75% nucleotide sequence homology (see HBVP25 and HBVP50 in Figure 7b).

$V\alpha$-gene segments from cDNA and genomic clones (with introns removed) reported so far have been aligned with one another. Spaces have been added at equivalent positions in the nucleotide and deduced amino acid sequences so that regions of homologous residues were maximized. Figure 8 shows the deduced amino acid alignment. Using the sequences as they have been aligned, the frequency of the most common residue at a given position (vertical columns within Figure 8) was calculated as a percentage of the total number of sequences. These results, consensus sequences, are displayed as a function of frequency from 10% to 100% above the alignment. Pairwise comparisons of the V_α nucleotide sequences resulted in the identification of those sequences which shared more than 75% of homology over their overlapping regions. The nucleotide sequences corresponding to the leader residues were not included in the comparisons.

It has been shown that gene fragments with at least 75% nucleotide sequence homology will cross hybridize. This empirical definition has been used to assign V_α-gene segments as members of noncross-hybridizing families (39). Family assignments are given in Figure 8, maintaining the previously established nomenclature in which unique members of the same family share the first digit and differ in the second (i.e. $V_\alpha 1.1$ and $V_\alpha 1.2$ are two cross-hybridizing members of the $V_\alpha 1$ family). Nomenclature to define these V_α and J_α families has been established by Yoshikai et al (39). This

Table 3 Composition of TcR-β–chain cDNA clones

Entry	V_β	D_β	J_β	C_β	Clone (comments)
1	1.1	1.1/2.1	2.3	2	PL5.6
2	1.1	2.1	2.3	2	PL6.4
3	1.1	1.1/2.1	2.1	2	PL6.1
4	1.1	2.1	2.1	2	PL5.2
5	1.2	1.1	1.5	1	HBVT96
6	1.2	2.1	2.2	2	HBVT73
7	2.1	—	2.1	2	PL2.13
8	2.1	2.1	2.1	2	PL6.21, MOLT4
9	2.2	2.1	2.3	2	MT11
10	2.3	—	1.1	NR	PH34
11	2.3	1.1	1.3	NR	PH7
12	3.1	1.1	2.5	2	PL4.4
13	3.1	1.1	1.5	1	PL4.22
14	3.1	—	1.6	1	PL3.10
15	3.1	2.1	2.5	2	HBVT22
16	3.2	1.1	2.1	2	DT259
17	3.3	2.1	2.7	2	HBVP55 different N region
18	3.3	2.1	2.7	NR	PH21 sequences
19	3.3	—	2.3	2	PL8.1
20	4.1	2.1	2.3	2	PL2.14 1 nucleotide
21	4.1	—	1.1	1	PL5.7 difference in **V**
22	4.1	2.1	2.5	NR	2G2
23	4.2	1.1	1.1	1	DT110
24	4.3	1.1	2.3	2	HBVP48
25	5.1	1.1	1.1	1	PL7.16 different N region
26	5.1	1.1	1.1	1	PL4.16 sequences
27	5.1	2.1	2.3	2	HPVP51
28	5.2	—	2.2	2	PL2.5
29	5.3	2.1	2.5	2	12A1, HPB-ALL, HPBβ2
30	5.4	2.1	2.5	NR	PH24
31	6.1	1.1	2.1	2	PL4.14
32	6.1	2.1	2.5	2	HPB-MLT, 4D1
33	6.1	1.1	1.1	1	HBVP04
34	6.1	2.1	2.3	2	HBVP50
35	6.2	1.1	1.5	1	PL5.10
36	6.3	1.1	1.5	1	ATL122
37	6.3	—	2.7	NR	PH5
38	6.3	2.1	2.1	2	HBVT23
39	6.3	2.1	2.7	2	HBVT10, HBVT41, HBVT65
40	6.4	2.1	2.5	2	HBVP25
41	6.5	—	2.1	2	HBVT116

Table 3 (*continued*)

Entry	V_β	D_β	J_β	C_β	Clone (comments)	
42	6.5	—	2.7	2	HBVT11	
43	6.6	1.1	1.5	1	HBVT45	
44	6.7	2.1	2.7	NR	PH16	
45	6.7	2.1/1.1	2.6	NR	PH79	
46	6.8	—	1.3	NR	PH22	
47	6.9	1.1	1.5	NR	L17	
48	7.1	1.1	1.1	1	PL4.9	
49	7.2	1.1/2.1	2.3	2	PL4.19	
50	8.1	1.1	1.2	1	YT35	
51	8.1	1.1	1.1	NR	PH11	
52	8.2	1.1	1.4	1	PL3.3	
53	8.2	—	—	—	HBV32SP	
54	8.3	NR	NR	NR	HBVP41	
55	8.4	2.1	2.6	NR	PH8	
56	9.1	2.1	2.6	2	PL2.6	
57	9.1	—	2.3	—	CEM2	
58	10.1	1.1	2.1	2	PL3.9	
59	10.2	1.1	1.5	1	ATL121	
60	11.1	—	2.2	2	PL3.12	
61	11.2	—	2.3	NR	PH15	
62	12.1	—	2.5	2	PL4.2	
63	12.2	1.1	1.3	1	HBVP54	
64	12.2	—	2.5	NR	PH27	
65	12.3	—	1.1	1	PL4.24	
66	12.3	1.1	1.2	1	HBVP34	
67	12.4	1.1	1.5	1	PL5.3	
68	12.4	NA	NA	NA	CEM1	
69	15.1	1.1	1.5	1	ATL21	
70	15.1	2.1	2.7	NR	PH32	
71	16.1	1.1	1.2	1	HBVP42	
72	17.1	—	2.1	2	HBVT02	
73	18.1	—	2.1	2	HBVT56	
74	18.2	1.1	1.4	NR	PH29	
75	18.2	2.1	2.1	NR	PH26	
76	19.1	1.1	1.6	1	HBVT72	
77	20.1	—	1.2	NR	HUT	
78	P	P	2.4	2	HBVP22, HBVP68	
79	P	2.1	2.1	2	HBVP15, HBVP37	different N region
80	P	2.1	2.1	2	HBVP31	sequences
81	—	—	2.5	2	HBVP58, HBVP63	
82	P	P	2.1	2	PL3.1	

Abbreviations: NA, not available; NR, not reported; —, unassigned; P, partial rearrangement products (DJC, JC).

(See Figure 9 for references.)

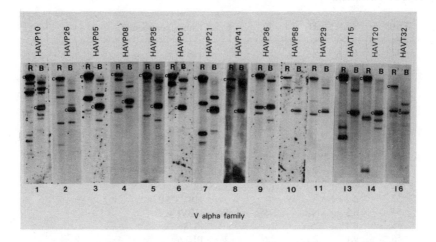

V alpha family

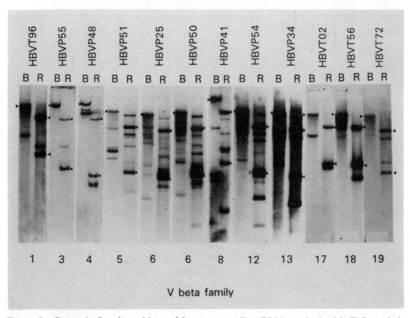

V beta family

Figure 7   Genomic Southern blots of human germline DNA probed with TcR-α–chain cDNAs and TcR-β–chain cDNAs. cDNA clone nomenclature is consistent with that used in Figures 8 and 9. DNA has been digested with either EcoR1 (R) or BamH1 (B). Constant regions of TcR-α and TcR-β chains are indicated by "c" and triangles respectively. cDNAs were isolated from peripheral blood T lymphocyte (HAVP, HBVP) and thymocyte (HAVT, HBVT) libraries.

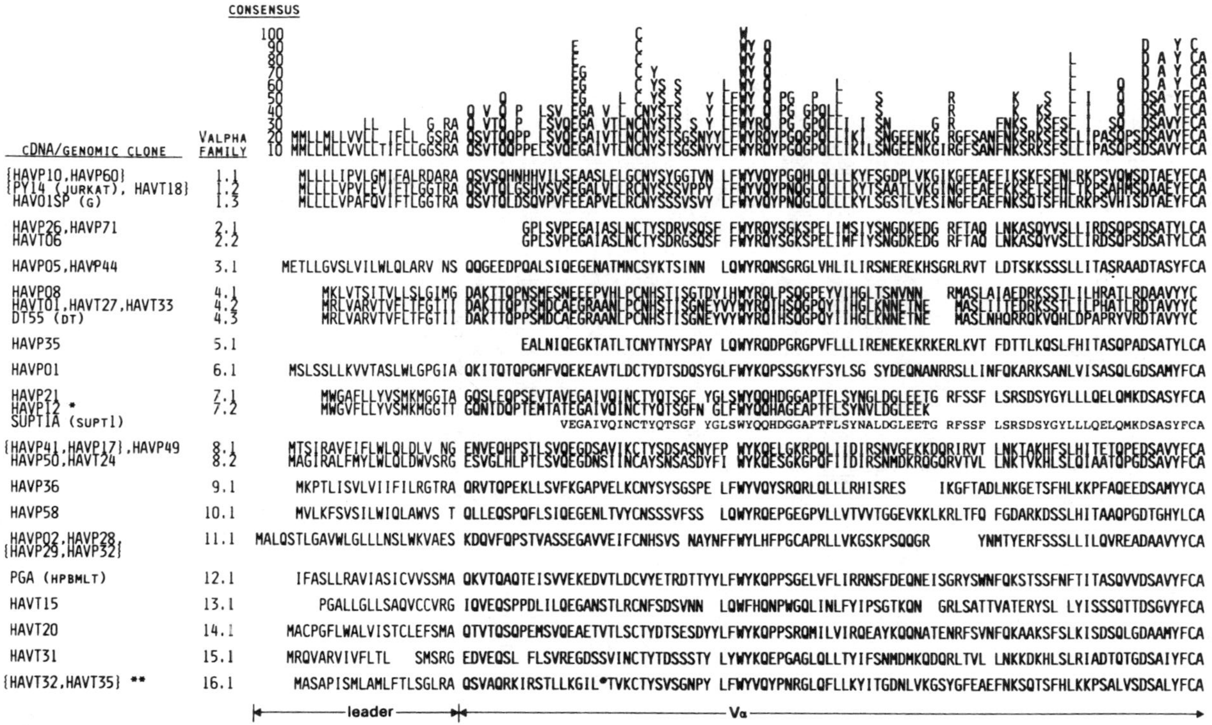

Figure 8 Deduced amino acid sequences of human-TcR-V_α segments. Spaces at related positions in both the nucleotide and deduced amino acid sequences were added to maximize regions of homology. HAVP12(*) has a deletion of the 3' V and the entire J-gene segment. C_α is found directly attached to the partial V gene. **An apparent single nucleotide deletion has led to a translational reading frameshift (●). A genomic V_α sequence (G) has been included along with cDNA sequences from various sources; T-cell clone specific for Diptheria toxoid (DT, F. Triebal, personal communication), SUPT1 (Ref. 116), HPBMLT (PGA, Ref. 12), JURKAT (PY14, Ref. 13), peripheral blood T lymphocytes from a single individual (HAVP, Ref. 39), and thymocytes from a single individual (HAVT, Ref. 43).

was extended to include more sequences from thymocytes V_α sequences (43). The total number of individual V_α segments in humans is comparable to that found in mice (85).

A similar analysis of all known human V_β-gene segments from cDNA and genomic clones (introns removed) has been performed (42–45). The deduced amino acid alignments for the unique V_β chains are displayed in Figure 9. A more complete compilation of TcR-β–chain V, D, J, and C gene

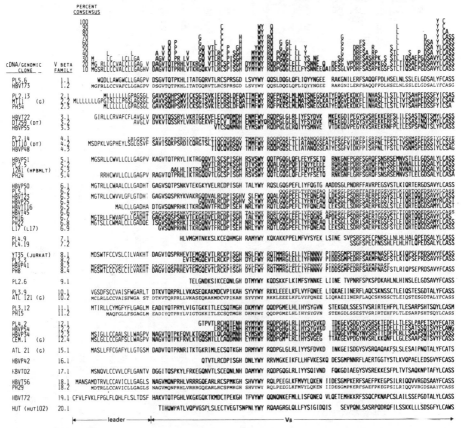

Figure 9 Deduced amino acid sequences of human-TcR-V_β segments. Spaces at related positions in both the nucleotide and deduced amino acid sequences were added to maximize regions of homology. Introns have been removed from rearranged genomic sequences obtained from ATL cell lines (ATL21, ATL121, MT11; see 132) and CEM (CEM.1; see 133). cDNA sequences are from various sources; T-cell clones specific for diphtheria toxoid (DT; see 42), HPBMLT (12A1; see 134), L17 (L17; see 134), JURKAT (YT35; see 7), HUT (HUT 102; see 134), peripheral blood T lymphocytes (PL-different individuals, HBVP-single individual; see 45, 42), thymocytes from a single individual (HBVT; see 43), and tonsil (PH; see 41). Sequences shown in small capital letters were not used in the consensus calculation.

segment usage in cDNA and genomic clones from the current literature is presented in Table 3. Nomenclature of V_β families was adopted from the studies of Kimura et al (42) and Concannon et al (45). In the interim, with the addition of new V_β sequences, new families have been defined and some old ones deleted (43). The elimination of families $V_\beta13$ and $V_\beta14$ has taken place as a result of a report of new V_β-gene segments which share > 75% nucleotide homology with members of two different families. In this case, $V_\beta13$ and $V_\beta14$ members have merged with $V_\beta12$ and $V_\beta3$ members, respectively. To avoid confusion, we have adopted a convention in which members are merged into the lower numbered family and the vacated family designation left alone. A consensus sequence analysis is presented above the alignments. Several sequences (shown in small capitals in Figure 9) have been omitted from the consensus calculation but have been assigned unique family designations for completeness. It is possible that sequences that only differ in the leader portion originated from unique germline sources, or they may be simply polymorphic variants of the same gene ($V_\beta1.2, 4.1, 6.6, 10.2, 11.2, 18.2$). Two others, $V_\beta6.2$ and $V_\beta8.3$, cannot be uniquely assigned because of insufficient sequence data.

Finally in Figure 10, the deduced amino acid alignments of human-T-cell V_γ genes (48–51, 86, 87) are presented with family assignments. Again, several sequences have been excluded from the consensus sequence calculation but have been presented for completeness in the figure for the following reasons. $V_\gamma1.9$ appears to be identical to $V_\gamma1.2$ except for non-V_γ-like sequences at its N terminus. $V_\gamma1.10$ is most likely the unspliced analog of $V_\gamma1.2$. $V_\gamma1.3$ and $V_\gamma1.11$ differ only in the leader sequence. $V_\gamma2.1$ and $V_\gamma2.2$ are identical throughout except for one amino acid residue. Partial sequences which may define other families, $V_\gamma3$ and $V_\gamma4$, are shown. This presentation is, therefore, consistent with the current literature which supports the existence of 12–15 human germline V_γ genes making up four families (48, 49, 86). Of all the human $Tc\gamma$-chain cDNAs isolated from the peripheral blood T lymphocyte and thymus libraries, not one potentially functional $Tc\gamma$ gene was found (51). Each cDNA contained some deleterious mutation caused by insertions, deletions, partial rearrangements, or incomplete mRNA splicing (51).

Notice from Figures 8–10 that well-defined, distinct hypervariable and framework regions found in Ig genes (88) are not observed. Instead, more diffuse regions of variability and conservation are found. Even so, those framework-like amino acid residues highly conserved in all three types of V-gene segments, as indicated by the consensus patterns of V_α, V_β, and V_γ, are very similar to those found in the variable gene segments of Ig molecules.

As mentioned in the discussion of germline organization, the J_α gene

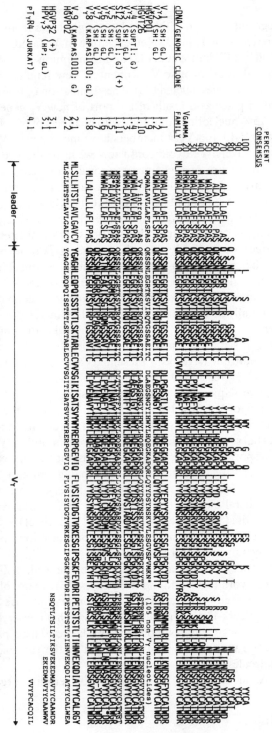

Figure 10 Deduced amino acid sequences of human-Tcγ-V segments. Spaces at related positions in both the nucleotide and deduced amino acid sequences were added to maximize regions of homology. Germline (GL), rearranged genomic (G), and cDNA sequences are shown derived from various sources; human placenta (HP; see 86), B-lymphoblastoid cell line (SH; see 50), transformed cord T-cell line (KARPAS1010; see 50), SUPT1 (see 50), Jurkat (pTγR4; see 50), peripheral blood T lymphocytes from a single individual (HGVP; see 51), and thymocytes from a single individual (HGVT; see 51). Additional clones using the same Vγ-gene segment but are otherwise unique (+); Vγ1.11 is found in HGVP03, HPBMLT (pTγ2, see 87) and HP−Vγ1.1. Vγ3.1 are found in HGVP06, HGVP08, HGVP10, and HGVT25. Sequences shown in small capital letters were not used in the consensus calculation.

segment loci boundaries have not been determined. So far, J_α genes have been localized to a stretch of over 65-kb in the human genome (E. Champagne, unpublished data). In the sequencing of cDNAs from V_α genes, J_α gene sequences were also determined (39, 43) (Figure 11). Each unique J_α (large capital letters in Figure 11) is used in the consensus calculation shown above the alignments. All reported J_α sequences have been included for comparison. A germline V_α sequence has been included to indicate the approximate V_α boundary. The extent of junctional diversity is most evident from this compilation. Since only six human J_α germline gene segments have been sequenced and mapped with respect to the C_α gene (19). (J_αA–J_αF, see Figure 3), each unique cDNA J_α gene has been assigned a letter (J_αG → J_αX) upon sequencing and comparative analysis. Table 4b summarizes the types of human J_α genes and the frequencies at which they have been observed in cDNAs sequenced to date. Using the statistical estimation process and underlying assumptions described above in the V-gene analysis, there are approximately 30–60 J_α gene segments in the human genome. Depending on the data used, the 95% upper bound is statistically estimated to be in the range of 150 to 200 different genes. A similar frequency analysis of J_β genes from cDNA clones is presented in Table 5b. The most probable value for the number of J_β gene segments is in agreement with the 13 known germline segments (Figure 12a). Therefore, based on the available data, there is no evidence for the existence of other J_β genes in the human genome. All known germline J_β segments have been found utilized. The three known J_γ gene segments are presented in Figure 12b.

Table 4   Composition of TcR-α–chain cDNA clones

cDNA library	Number of unique cDNA	Number of V_α	Number of unique V_α	V_α family[a,b] 1	2	3	4	5	6	7	8	9	10	11	12	13	14	15	16
HAVP	21	18	13	1	2	2	1	1	1	2	3	1		1		3			
HAVT	12	10	8	1	1	3				1						1	1	1	1
other	6	4	4	1			1			1					1				

cDNA library	Number of J_α	Number of unique J_α	J_α gene segment A	B	C	D	E	F	G	H	I	J	K	L	M	N	O	P	Q	R	S	T	U	V	W	X
HAVP	20	17	1	1					2	1	1	1	1	1	1	1	2	1	1	1	1	2	1			
HAVT	12	10	1							2	1	1			1							1	1	1	2	1
other	5	3	3	1				1																		

[a] See Figure 8 for references.
[b] Others: PY14, DT55, SUPT1A, PGA.

```
                                                        PERCENT
                                                        CONSENSUS

                                          E  GI         100
                                       EG GI            90
                                       EG GI  L V       80
                              D A Y C  L  FG I  E V P   70
                              D A Y CA K  FG I  E V P   60
                              D A Y CA K FG I  E V P    50
                              DSA YFCA K FG I  E V P    40
                              DSAVYFCA  FGKGT  L  P     30
                    JALPHA    DSAVYFCA  GGSGGKI EFGKGTRLTVP  20
  cDNA/GENOMIC CLONE FAMILY   DSAVYFCA  VGGGSGGKLIEFGKGTRLTVIP  10
```

cDNA/GENOMIC CLONE	JALPHA FAMILY		
GERMLINE	A	/GYSSASKIIFGSGTRLSIRP/	
HAVP35		DSATYLCA LAPSYSSASKIIFGSGTRLSIRP	NIQN
PGA (HPBMLT)		DSAVYFCA LDSSASKIIFGSGTRLSIRP	NIQN
SUPT1B		DTAVYYCA RVRRRYSSASKIIFGSGTRLSIR	
HAVTI8, PY14 (JURKAT)		DAAEYFCAVSDLEPNSSASKIIFGSGTRLSIRP	NIQN
GERMLINE	B	/MDSSYKLIFGSGTRLLVRP/	
HAVP42		/MDSSYKLIFGSGTRLLVRP	HIQN
SUPT1A (SUPT1)		DSASYFCAVPPtg MDSSYKLIFGSGTRLLVA	
GERMLINE	C	/SSGSARQLTFGSGTQLTVLP/	
GERMLINE	D	/TTDSWGKFEFGAGTQVVVTP/	
GERMLINE	E	/EGQGFSFIFGKGTRLLVKP/	
GERMLINE	F	/NSGNTPLVFGKGTRLSVIA/	
DT55 (DT)		DTAVYYC IVRAINSGNTPLVFGKGTRLSVIA	NIQN
HAVP49	G	DSAVYFCA AKRKASSNTGKLIFGGTTLQVKP	DIQN
HAVP29, HAVP32		DAAVYYC GLPSNTGKLIFGQTTLQVKP	DIQN
HAVP36	H	DSAMYYCA LSVYNQGGKLIFGQGTELSVKP	NIQN
HAVT20		DAAMYFCA IIDNQGGKLIFGQGTELSVKP	NIQN
HAJT23		NQGGKLIFGQGTELSVKP	NIQN
HAVP02	I	DAAVYYCA VEVPNTDKLIFGTGTRLQVFP	NIQN
HAVP08	J	DAAVYYCI RANAGGTSYGKLTFGQGTILTVHP	NIQN
HAVT06		DSATYLCA VKPAGGTSYGKLTFGQGTILTVHP	NIQN
HAVP44	K	DTASYFCATPPLSSGGSNYKLTFGGTLLTVNP	NIQN
HAVT33		DTAVYYCI VRGNSGGSNYKLTFGKGTLLTVNP	NIQN
HAVP26	L	DSATYLCA LRDGQKLLFARGTMLKVDL	NIQN
HAVP50	M	DSAVYFCA EIGGEKLVFGQGTRLTINP	NIQN
HAVP10, HAVP60	N	DTAEYFCA VNEYDYKLSFGAGTTVTVRA	NIQN
HAVP41, HAVP17	O	DSAVYFCA ASRKDSGGYQKVTFGTGTKLQVIP	NIQN
HAVP25		GGYQKVTFGTGTKLQVIP	NIQN
HAVP05	P	DTASYFCA TDGNRDDKIIFGKGTRLHILP	NI
HAVP71	Q	DSATYLCA VNYPRGTTLGRLYFGRGTQLTVWP	DIQN
HAVT32, HAVT35		DSALYFCA VRPDRGSTLGRLYFGRGTQLTVWP	DIQN
HAJP28	R	LRARNNARLMFGDGTQLVVKP	NI
HAVP21	S	DSASYFCA VFNQAGTALIFGKGTTLSVSS	NIQN
HAVP58	T	DTGHYLCA GVSSGGSYIPTFGRGTSLIVHP	YIQN
HAVT31		DSAIYFCA ESKTPSRPTFGRGTSLIVHP	YIQN
HAVP51		IPTFGRGTSLIVHP	YIQN
HAVP01	U	DSAMYFCA SREGSGNQFYFGTGTSLTVIP	NI
HAVT24		DSAVYFCA EGPPTGNQFYFGTGTSLTVIP	NIQN
HAVT01	V	DTAVYYCI ALYSGAGSYQLTFGKGTKLSVIP	NIQN
HAVT27	W	DTAVVYCI VRDWVGGGADGLTFGKGTHLIIQP	YIQN
HAVT15		DSGVYFCA ALDLWGGADGLTFGKGTHLIIQP	YIQN
HAJT17	X	ctatgtgaagatcacctagMLNFGKGTELIVSL	DIQN

(GERMLINE VALPHA)   DTAEYFCAVSgacacagtgcctgagactgcaggagagctgaacacaaacc

◄—Vα—►◄—Dα?►◄————Jα———► |◄—Cα—►

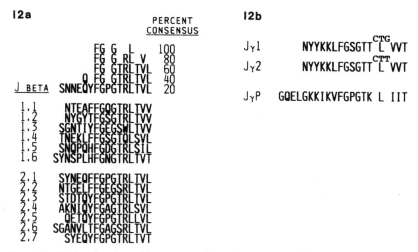

Figure 12 Deduced amino acid sequences of J regions: 12a. TcR-β–chain J segments. 12b. Tcγ-chain J segments. A silent single nucleotide change is the only difference between $J_\gamma 1$ and $J_\gamma 2$ (small capital letters).

Only two diversity gene segments, both used in the TcR-β chain, have been sequenced and mapped onto the human genome (17). Because of the combined effects of junctional flexibility, N-region diversification, the brevity of D_β-gene segments, and the availability of multiple D_β translational reading frames, it is not always possible to identify them in cDNA sequences (42). Some assignments have been made but are not unequivocal (Table 3). Thus, if similar diversity gene segments are used in the creation of TcR-α and Tcγ chains it is unlikely that they will be found by the analysis of cDNA sequences.

In summary, the data support the potential for extensive V region repertoires in both TcR-α and TcR-β chains. Very rough calculations of V-region diversity are summarized in Table 6. On a very simplistic level it appears the combined repertoire of the human TcR-α and TcR-β chains

---

Figure 11 Deduced amino acid sequences of human TcR-Jα-gene segments. Spaces at related positions in both the nucleotide and deduced amino acid sequences were added to maximize regions of homology. Sequences from cDNA/genomic clones used in Figure 6 as well as six cDNAs that do not contain V_γ (HAVP25, HAVP42, HAVP51, HAJP28, HAJT17, HAJT23; Ref. 39, 43) and the six germline J_α segments are included (/ indicate germline boundaries). Sequences shown in small capital letters were not used in the consensus calculation. Small lower case letters denote nucleotide residues. In $J_\alpha X$ an in-frame stop codon is underlined and the heptamer of the putative recombination recognition signal is boxed. A germline V_α sequence is shown at the bottom for the definition of V-J boundaries.

equals if not exceeds that of Ig. Thus, the repertoire is at least sufficient for the recognition of Ag/MHC. Whether all possible TcR-$\alpha\beta$ combinations are allowed has yet to be determined.

Knowledge of the cellular origins of the V_α and V_β region containing cDNAs (Figures 8 and 9) makes it possible to tabulate the frequencies at which the known V and J gene segments have been observed (Tables 4 and 5). Without more V_α sequence data, it is difficult to determine whether there is preferential use of certain V_α families in different tissues (Table 4a). On the other hand the more extensive V_β data appear to be consistent with the more frequent use of several V_β-gene-segment families (Table 5a). Thus, as shown in Table 3, certain V_β families are represented in the cDNAs derived from independent sources (V_β1-6, V_β8). Both J_α and J_β gene segments are used in a random fashion.

It is interesting that similar frequencies of potentially functional TcR-α cDNAs were isolated from both thymus and peripheral blood T lymphocyte libraries. Table 7 also shows that the same is true for potentially functional TcR-β cDNA clones. In contrast, occurrence of Tcγ clones in the same libraries are an order of magnitude lower, and as discussed above, none of those sequenced appears to be capable of producing a functional protein (51).

THE T-CELL ANTIGEN RECEPTOR INVOLVEMENT IN HUMAN DISORDERS

Rearrangement of TcR Genes in Malignancies and Proliferative Disorders

The cloning of the TcR genes provides clonal markers useful in the evaluation of malignant and nonmalignant samples from patients with lymphoproliferative disorders (89). Studies have indicated that the TcR-β–chain gene is a useful marker for the T-cell lineage. The analysis of the structure of these genes has also been useful for establishing clonality of the malignant cells in leukemia or lymphoma patients at the time of presentation or relapse. Table 8 summarizes the results of several studies carried out with the TcR-β–chain genes on a variety of T-cell and non–T-cell disorders (63, 90–102).

The use of the Tcγ genes to evaluate the clonality of the malignant cells in lymphomas and as a marker for the T-cell lineage has also been examined (63). In these studies, the authors have also compared the clonal rearrangement of Tcγ and TcR-β–chain genes to those of the rearrangement of Ig heavy- and light-chain genes in the same samples. Since rearrangements of the Tcγ-chain genes were also found in non–T-cell lymphomas, such as B-cell lymphoma and B-cell lines, it was concluded that rearrangement of

Table 5 Human TcR V_β and J_β use in cDNA clones

cDNA library	Number of unique cDNA	Number of v_β	Number of unique V_β	V_β family[a,b,c] 1	2	3	4	5	6	7	8	9	10	11	12	(13) (deleted)	(14) (deleted)	15	16	17	18	19	20
HBVP	13	10	9			1	1	1	3		1				2				1				
PL	27	26	18	4	2	4	2	3	2	2	1	1	1	1	3								
PH	15	15	12		2	1		1	4		2			1	1			1			2		
HBVT	11	11	8	2		1			5											1	1	1	
Other	15	15	15		2	1	2	1	3		1	1	1		1								1

cDNA library	Number of unique cDNA	$C_\beta1$	$C_\beta2$	$J_\beta1$ 1	2	3	4	5	6	$J_\beta2$ 1	2	3	4	5	6	7
HBVP	13	4	8	1	2	1				2		3	1	1		1
PL	27	10	17	5			1	3	1	7	2	5		2	1	
PH	15	NR	NR	2		2	1			1		1		2	2	4
HBVT	11	3	8					2	1	4	1			1		2
Other	15	5	5	1	2			4		2		1		3		

[a] See Figure 9 for references.
[b] Others: see Figure 9 plus MOLT-4 (Ref. 136), 2G2 (Ref. 134), 4D1 (HPBMTL, Ref. 134), ATL122 (Ref. 132) and CEM.2 (CEM, Ref. 133).
[c] NR—not reported.

Table 6 Estimate of human TcR repertoire

	α	β
V	50	70
D	?	2
J	50	13
C	1	2
Combinations	2.5×10^3	3.6×10^3
N sequences added	2.5×10^5	3.6×10^5
$\alpha\beta$ Combinations	Approximately 10^{11}	

Table 7 Expression of α-, β-, and γ-chain T-cell-receptor messages

	Thymocytes		Mature T lymphocytes	
α	0.02%	$(24/10^5)$	0.02%	$(84/4 \times 10^5)$
β	0.1%	$(110/10^5)$	0.04%	$(168/4 \times 10^5)$
γ	0.003%	$(3/10^5)$	0.003%	$(13/4 \times 10^5)$

the Tcγ gene can be used as clonal marker but not in the identification of cell lineage (63). These studies also indicate that, in at least two tumors, somatic rearrangement of the Tcγ-chain genes may occur before the TcR-β–chain genes (63). Preliminary data using numerous J$_\alpha$-gene segment region probes have been reported (61). These studies are consistent with the hypothesis that the TcR-α–chain genes are rearranged after the TcR-β genes.

The ability to detect clonal population of T cells provides a means of monitoring effect of treatment, detecting of early relapse, and identifying progression of disease. The availability of these probes has also been useful in the study of lymphoproliferative disorders such as T-cell lymphocytosis and angioimmunoblastic lymphadenopathy in which morphologically normal T lymphocytes predominate. In such cases, the TcR-β– or Tcγ-chain genes are useful markers indicating whether a clonal T-cell malignancy or polyclonal expansion of T cells exists.

Involvement of TcR Genes in Chromosomal Translocations

Chromosomal translocations are common among a number of tumors. The most well-studied translocations associated with some malignancies involve chromosomes 9 and 22 in chronic myelogeneous leukemia, and the *c-myc*-Ig translocation in Burkitt's lymphoma (103, 104). In some T-

Table 8 Rearrangement of TcR$_\beta$ chain genes in specimens from patients with leukemia or lymphoma

	Number	Rearranged TcR$_\beta$	Rearranged Ig H C	References
T-cell malignancies				
T-ALL	59	59	9 (8 not done)	91–97, 135
ATL	10	10	0 (6 not done)	92, 95, 96, 100
T-cell lymphoma	43	39	0 (11 not done)	92, 93, 95, 97–99
T-cell prolymphocyte leukemia	14	14	0 (1 not done)	90, 93, 94, 97, 135
Sezary Syndrome/ mycosis fungoides	49	48	1 (39 not done)	92–97, 100, 101
T-CLL	24	24	0 (8 not done)	92–94, 97, 100
T8 lymphocytosis	5	3	0 (4 not done)	95, 96, 100
B-cell and pre–B-cell malignancies				
ALL	46	13	46	90, 97
Lymphoma	19	3	19	97, 99
B-CLL	13	3	13	96, 97
Myeloma	2	0	2	97
Other hematopoietic malignancies				
Lennerts	6	6	0	93, 99
Hodgkins disease	9	4	0 (1 not done)	95, 99
AIL	12	9	4 (also TcR$_\beta$)	95, 99
AML	30	3	1 (also TcR$_\beta$)	90, 102

Abbreviations: T-ALL, T cell precursor acute lymphoblastic leukemia; ATL, adult T cell leukemia; T-CLL, T cell chronic lymphocytic leukemia; ALL, common ALL non-B, non T ALL, cALLa; B-CLL, B cell chronic lymphocytic leukemia; AIL, angioimmunoblastic lymphadenopathy; AML, acute myeloblastic leukemia.

cell malignancies and stimulated T lymphocytes, recurring chromosomal translocations have also been observed. The most common involves the region of 14q11 to 14q13. This locus has since been identified as the TcR-α–chain gene locus (26, 29, 33). Translocations involving the 7q33 and 7p13 loci of the TcR-β– (36, 105) and Tcγ-chain genes, respectively, have also been described. Table 9 summarizes a list of such translocations described by numerous researchers. Most surprising is the finding that although most translocations in B-cell malignancies involve the *c-myc* gene locus of chromosome 8 (103, 104), a variety of loci appear to be involved in translocations to the TcR genes (106–130). Some of these occur in positions on chromosomes that may be related to existing oncogenes. Others, however, appear to be at positions where no oncogenes have yet been described. It is of interest to note that translocations involving TcR and Ig gene loci are detected occasionally in stimulated T cells and in samples from patients with ataxia telangiectasia (128–130).

Table 9   Translocations involving T-cell-receptor genes

Disease	Translocation	Loci involved	Reference
Pre T (T-cell) leukemia/lymphoma			
	t(1:14) (p32;q11)	L-myc?-TcR$_\alpha$ N-ras?-TcR$_\alpha$ src-2?-TcR$_\alpha$	106
	t(6:14) (q21;q11)	c-myb-TcR$_\alpha$	C. Thompson, personal communication
	t(8,14) (q24;q11)	c-myc-TcR$_\alpha$	107–110
	t(10;14) (q23;q11.2)	TdT?-TcR$_\alpha$ onc?-TcR$_\alpha$	111
	t(11;14) (p13;q11)	Wilm's?-TcR$_\alpha$ onc?-TcR$_\alpha$	107, 112–114
	t(12,14) (q24;q11)	onc?-TcR$_\alpha$	112
	inv(14) (q11;q32)	IgH-TcR$_\alpha$	115–117
	t(14;14) (q11;q32)	IgH?-TcR$_\alpha$ akt?-TcR$_\alpha$ onc?-TcR$_\alpha$	112, 118
	t(7;9) (q34;q33)	TcR$_\beta$-abl?	119, 120 J. Sklar, personal communication
Pre B (B-cell) leukemia/lymphoma			
	inv(14) (q11;32)	IgH-TcR$_\alpha$	121
Ataxia-telangiectasia or "stimulated" T-cells			
	inv(7) (p15;q33)	TcR$_\gamma$?-TcR$_\beta$?	122–130
	t(7:14) (p13;q11)	TcR$_\gamma$?-TcR$_\alpha$?	
	t(7;14) (q34;q11)	TcR$_\beta$?-TcR$_\alpha$?	
	inv(14) (q11;q32)	TcR$_\alpha$?-IgH?	
	t(14;14) (q11;q32)	TcR$_\alpha$?-IgH?	

Although it has not been proven that the TcR genes are directly involved in the translocations in each one of these cases, it is unlikely that these translocations into TcR gene regions are purely coincidental. In at least three cases, molecular data exist which support the involvement of the TcR-α–chain gene locus. Two are reciprocal translocations between 11p13 and 14q11 (113, 114) and between 8q24 and 14q11 (108, 110). A third is an inversion between 14q12 and 14q32 (116, 117). It is not known what genes are involved at 11p13, but the genes at 8q24 and 14q32 have been identified. In the two cases involving 8q24, translocations involving the TcR-α–chain genes and the oncogene *c-myc* have been confirmed (108, 110). It is possible that the activation of the *c-myc* genes by the TcR-α–chain genes occurs in the same manner that this oncogene is thought to be activated by Ig genes. The inversion involved in 14q12 and 14q32 leads to a site-specific recombination between an Ig heavy-chain variable-region gene and a J_α-gene segment of the TcR-α gene (116, 117). It is tempting to speculate that such a structure, if expressed on the cell surface, may play a role in the malignant transformation of the T cell.

SUMMARY

In this review, the germline genomic and cDNA structures and the evolution of the human TcR-α–, TcR-β–, and Tcγ-chain genes have been reviewed and discussed. The estimated V-gene segment repertoires established for the human TcR-α–, TcR-β–, and Tcγ-chain genes have also been summarized. The use of TcR and Tcγ genes in the evaluation of hematopoietic malignancies and identification of chromosomal translocations has also been presented. It is hoped that such information will serve as a reference point from which the study of TcR genes can further progress. This will almost certainly include the use of TcR and perhaps Tcγ genes in the study of the dual recognition of antigen and MHC gene products, the role of these genes in thymic education/selection, and the role(s) of these genes in oncogenesis processes and autoimmune settings.

ACKNOWLEDGMENTS

The literature search for this article was completed in October 1986. The authors thank Dr. Mark Minden, Dr. Marciano Reis, and Nicolette Caccia for their help in the preparation of this review. We also thank John Cargill for computational assistance in sequence analyses and both the Art and Photography departments at the Ontario Cancer Institute for their help. The authors are grateful to Diana Quon for excellent manuscript preparation. This work was supported by grants from the Medical Research Council of Canada, the National Cancer Institute of Canada, and the Natural Sciences and Engineering Research Council of Canada, and by a special research fund from the University of Toronto. B. Toyonaga is a Postdoctoral Fellow of the Medical Research Council of Canada.

Literature Cited

1. Davies, D. R., Metzger, H. 1983. Structural basis of antibody function. *Ann. Rev. Immunol.* 1: 63–86
2. Honjo, T. 1985. Immunoglobulin genes. *Ann. Rev. Immunol.* 1: 499–528
3. Zinkernagel, R. M., Doherty, P. C. 1974. Restriction of *in vitro* T cell-mediated cytotoxicity in lymphocytic choriomeningitis within a syngeneic or semiallogeneic system. *Nature* 248: 701–2
4. Allison, J. P., McIntyre, B. W., Bloch, D. 1982. Tumor-specific antigen of murine T lymphoma defined with monoclonal antibody. *J. Immunol.* 129: 2293–2300

5. Haskins, K., Kubo, R., White, J., Pigeon, M., Kappler, J., Marrack, P. 1983. The major histocompatibility complex-restricted antigen receptor on T cells. I. Isolation with a monoclonal antibody. *J. Exp. Med.* 157: 1149–69

6. Meuer, S. C., Fitzgerald, K. A., Hussey, R. E., Hodgdon, J. C., Schlossman, S. F., Reinherz, E. L. 1983. Clonotypic structures involved in antigen-specific human T-cell function. Relationship to the T3 molecular complex. *J. Exp. Med.* 157: 705–19

7. Yanagi, Y., Yoshikai, Y., Leggett, K., Clark, S. P., Aleksander, I., Mak, T. W. 1984. A human T cell-specific cDNA clone encodes a protein having extensive homology to immunoglobulin chains. *Nature* 308: 145–49

8. Hedrick, S. M., Cohen, D. I., Nielsen, E. A., Davis, M. M. 1984. Isolation of cDNA clones encoding T cell-specific membrane-associated proteins. *Nature* 308: 149–53

9. Acuto, O., Fabbi, M., Smart, J., Poole, C., Protentis, J., Royer, H., Schlossman, S., Reinherz, E. L. 1984. Purification and N-terminal sequencing of the β subunit of a human T cell antigen receptor. *Proc. Natl. Acad. Sci. USA* 81: 3851–55

10. Chien, Y., Becker, D., Lindsten, T., Okamura, M., Cohen, D., Davis, M. 1984. A third type of murine T-cell receptor gene. *Nature* 312: 31–35

11. Saito, H., Kranz, D., Takagaki, Y., Hayday, A., Eisen, H., Tonegawa, S. 1984. A third rearranged and expressed gene in a clone of cytotoxic T lymphocytes. *Nature* 312: 36–40

12. Sim, G. K., Yague, J., Nelson, J., Marrack, P., Palmer, E., Augustin, A., Kappler, J. 1984. Primary structure of human T-cell receptor α-chain. *Nature* 312: 771–75

13. Yanagi, Y., Chan, A., Chin, B., Minden, M., Mak, T. W. 1985. Analysis of cDNA clones specific for human T cells and the α and β chain of the T cell receptor hetcrodimer from a human T cell line. *Proc. Natl. Acad. Sci. USA* 82: 3430–34

14. Siu, G., Clark, S., Yoshikai, Y., Malissen, M., Yanagi, Y., Strauss, E., Mak, T. W., Hood, L. 1984. The human T cell antigen receptor is encoded by variable, diversity, and joining gene segments that rearrange to generate a complex V gene. *Cell* 37: 393–401

15. Toyonaga, B., Yanagi, Y., Suciu-Foca, N., Minden, M. D., Mak, T. W. 1984. Rearrangement of the T cell receptor gene YT35 in human DNA from thymic leukemic T cell lines and functional helper, killer and suppressor T cell clones. *Nature* 311: 385–87

16. Malissen, M., Minard, K., Mjolsness, S., Kronenberg, M., Goverman, J., Hunkapiller, T., Prystowsky, M. B., Fitch, F., Yoshikai, Y., Mak, T. W., Hood, L. 1984. Mouse T-cell antigen receptor: Structure and organization of constant and joining gene segments encoding the β polypeptide. *Cell* 37: 1101–10

17. Toyonaga, B., Yoshikai, Y., Vadasz, V., Chin, B., Mak, T. W. 1985. Organization and sequences of the diversity, joining and constant region genes of the human T cell receptor β chain. *Proc. Natl. Acad. Sci. USA* 82: 8624–28

18. Gascoigne, N., Chien, Y., Becker, D., Kavaler, J., Davis, M. 1984. Genomic organization and sequence of T cell receptor β-chain constant and joining region genes. *Nature* 310: 387–91

19. Yoshikai, Y., Clark, S. P., Taylor, S., Sohn, V., Wilson, B., Minden, M., Mak, T. W. 1985. Organization and sequences of the variable, joining, and constant region genes of the human T cell receptor α chain. *Nature* 316: 837–40

20. Winoto, A., Mjolsness, S., Hood, L. 1985. Genomic organization of genes encoding mouse T cell receptor α chain. *Nature* 316: 832–36

21. Hayday, A. C., Diamond, D. J., Tanigawa, G., Heilig, J. S., Folsom, V., Saito, H., Tonegawa, S. 1985. Unusual organization and diversity of T cell receptor α chain genes. *Nature* 316: 828–32

22. Saito, H., Kranz, D. M., Takagaki, Y., Hayday, A., Eisen, H., Tonegawa, S. 1984. Complete primary structure of a heterodimeric T cell receptor deduced from cDNA sequences. *Nature* 309: 757–62

23. Davis, M. M. 1985. Molecular genetics of the T-cell-receptor beta chain. *Ann. Rev. Immunol.* 3: 537–60

24. Kronenberg, M., Siu, G., Hood, L. E., Shastri, N. 1986. The molecular genetics of the T-cell antigen receptor and T-cell antigen recognition. *Ann. Rev. Immunol.* 4: 529–91

25. Marrack, P., Kappler, J. 1986. The antigen-specific, major histocompatibility complex-restricted receptor on T cell. *Adv. Immunol.* 38: 1–30

26. Caccia, N., Bruns, G. A. P., Kirsch, I. R., Hollis, G. F., Bertness, V., Mak, T. W. 1985. T-cell receptor α chain genes

are located on chromosome 14q12 in humans. *J. Exp. Med.* 151: 1255–60

27. Dembic, Z., Bannwarth, W., Taylor, B. A., Steinmetz, M. 1985. The gene encoding the T-cell receptor α-chain maps close to the Np-2 locus on mouse chromosome 14. *Nature* 314: 271–73

28. Kranz, D. M., Saito, H., Disteche, C. M., Swisshelm, K., Pravtcheva, D., Ruddle, F. H., Eisen, H. N., Tonegawa, S. 1985. Chromosomal locations of the murine T-cell receptor alpha-chain gene and the T-cell gamma gene. *Science* 227: 941–45

29. Rabbitts, T. H., Lefranc, M. P., Stinson, M. A., Sims, J. E., Schroeder, J., Steinmetz, M., Spurr, N. L., Solomon, E., Goodfellow, P. N. 1985. The chromosomal location of T-cell receptor genes and a T-cell rearranging gene: Possible correlation with specific translocations in human T-cell leukaemia. *EMBO J.* 4: 1461–65

30. Roehm, N. W., Carbone, C. Kushnir, E., Taylor, B. A., Riblet, R. J., Marrack, P., Kappler, J. W. 1985. The major histocompatibility complex-restricted antigen receptor on T cells: The genetics of expression of an allotype. *J. Immunol.* 135: 2176–82

31. Epstein, R., Roehm, N., Marrack, P., Kappler, J., Davis, M., Hedrick, S., Cohn, M. 1985. Genetic markers of the antigen-specific T-cell receptor locus. *J. Exp. Med.* 161: 1219–24

32. Collins, M. K. L., Goodfellow, P. N., Spurr, N. K., Solomon, E., Tanigawa, G., Tonegawa, S., Owen, M. J. 1985. The human T-cell receptor α-chain gene maps to chromosome 14. *Nature* 314: 273–74

33. Croce, C. M., Isobe, M., Palumbo, A., Puck, J., Ming, J., Tweardy, D., Erikson, J., Davis, M., Rovera, G. 1985. Gene for α-chain of human T-cell receptor: Location on chromosome 14 region involved in T-cell neoplasms. *Science* 227: 1044–47

34. Caccia, N., Kronenberg, M., Saxe, D., Haars, R., Bruns, G., Goverman, J., Malissen, M., Willard, H., Yoshikai, Y., Simon, M., Hood, L., Mak, T. 1984. The T-cell receptor β chain genes are located on chromosome 6 in mice and chromosome 7 in humans. *Cell* 37: 1091–99

35. Barker, P. E., Ruddle, F. H., Royer, H. D., Acuto, O., Reinherz, E. L. 1984. Chromosomal location of human T-cell receptor gene Tiβ. *Science* 226: 348–49

36. Morton, C. C., Duby, A. D., Eddy, R. L., Shows, T. B., Seidman, J. G. 1985.

Genes for β chain of human T-cell antigen receptor map to regions of chromosomal rearrangement in T cells. *Science* 228: 582–85

37. Lee, N. E., D'Eustachio, P., Pravtcheva, D., Ruddle, F. H., Hedrick, S. M., Davis, M. M. 1984. Murine T-cell receptor β chain is encoded on chromosome 6. *J. Exp. Med.* 160: 905–13

38. Barclay, A. N., Johnson, P., McCaughn, G. W., Williams, A. F. 1987. Immunoglobulin-related structures associated with vertebrate cell surfaces. In *The T Cell Receptor*, ed. T. W. Mak. New York: Plenum

39. Yoshikai, Y., Kimura, N., Toyonaga, B., Mak, T. W. 1986. Sequence and repertoire of human T cell receptor α chain variable region genes in mature T lymphocytes. *J. Exp. Med.* 164: 90–103

40. Clark, S. P., Yoshikai, Y., Siu, G., Taylor, S., Hood, L., Mak, T. W. 1984. Identification of a diversity segment of the human T-cell receptor beta chain, and comparison to the analogous murine element. *Nature* 911: 387–89

41. Behlke, M. A., Loh, D. Y. 1986. Alternative splicing of murine T cell receptor β-chain transcripts. *Nature* 322: 379–82

42. Kimura, N., Toyonaga, B., Yoshikai, Y., Minden, M. D., Mak, T. W. 1986. Sequence and diversity of human T cell receptor β chain genes. *J. Exp. Med.* 164: 739–50

43. Kimura, N., Toyonaga, B., Yoshikai, Y., Mak, T. W. 1987. Sequences of human T cell receptor α and β chain variable region genes in thymocytes. *Eur. J. Immunol.* Submitted

44. Behlke, M. A., Spinella, D. G., Chou, H., Sha, W., Hartl, D. L., Loh, D. Y. 1985. T-cell receptor β chain expression: Dependence on relatively few variable region genes. *Science* 229: 566–70

45. Concannon, P., Pickering, L. A., Kung, P., Hood, L. 1986. Diversity and structure of human T-cell receptor β-chain variable region genes. *Proc. Natl. Acad. Sci. USA* 83: 6598–6602

46. Malissen, M., McCoy, C., Blanc, D., Trucy, J., Devaux, C., Schmitt-Verhulst, A., Fitch, F., Hood, L., Malissen, B. 1986. Direct evidence for chromosomal inversion during T-cell receptor β-gene rearrangements. *Nature* 319: 28–33

47. Murre, C., Waldmann, R. A., Morton, C. C., Bongiovanni, K. F., Waldmann, T. A., Shows, T. B., Seidman, J. G.

1986. Human γ-chain genes are rearranged in leukaemic T cells and map to the short arm of chromosome 7. *Nature* 316: 549–52

48. Lefranc, M.-P., Forster, A., Rabbits, T. H. 1986. Rearrangement of two distinct T-cell γ-chain variable-region genes in human DNA. *Nature* 319: 420–22

49. Quertermous, T., Murre, C., Dialynas, D. P., Duby, A. D., Strominger, J. L., Waldmann, T. A., Seidman, J. G. 1985. Human T-cell gamma genes: Organization, diversity and rearrangement. *Science* 231: 252–55

50. Lefranc, M.-P., Forster, A., Baer, R., Stinson, M. A., Rabbitts, T. H. 1986. Diversity and rearrangement of the human T cell rearranging γ genes: Nine germ-line variable genes belonging to two subgroups. *Cell* 45: 237–46

51. Yoshikai, Y., Koga, Y., Kimura, N., Mak, T. W. 1987. Repertoire of the human T cell gamma genes: High frequency of non-functional transcripts in thymus and mature T cells. *Eur. J. Immunol.* In press

52. Hayday, A. C., Saito, H., Gilles, S. D., Kranz, D. M., Tanigawa, G., Eisen, H. N., Tonegawa, S. 1985. Structures, organization and somatic rearrangement of T cell gamma genes. *Cell* 40: 259–69

53. Iwamoto, A., Rupp, F., Ohashi, P. S., Walk, C. C., Pircher, H., Joho, R., Hengartner, H., Mak, T. W. 1986. T cell specific γ genes in C57BL/10 mice: Sequence and expression of new constant and variable regions genes. *J. Exp. Med.* 163: 1203–12

54. Garman, R. D., Doherty, P. J., Raulet, D. H. 1986. Diversity rearrangement and expression of murine T cell gamma genes. *Cell* 45: 733–42

55. Heilig, J. S., Tonegawa, S. 1986. Diversity of murine gamma genes and expression in fetal and adult T lymphocytes. *Nature* 322: 836–40

56. Snodgrass, H. R., Dembic, Z., Steinmetz, M., von Boehmer, H. 1985. Expression of T-cell antigen receptor genes during fetal development in the thymus. *Nature* 313: 232–33

57. Raulet, D. H., Garman, R. D., Saito, H., Tonegawa, S. 1985. Developmental regulation of T-cell receptor gene expression. *Nature* 314: 103–7

58. Samelson, L. E., Lindsten, T., Fowlkes, B. J., van den Elsen, P., Terhorst, C., Davis, M. M., Germain, R. N., Schwartz, R. H. 1985. Expression of genes of the T-cell antigen receptor complex in precursor thymocytes. *Nature* 315: 765–68

59. Yoshikai, Y., Yanagi, Y., Suciu-Foca, N., Mak, T. W. 1984. Presence of T cell receptor mRNA in functionally distinct T cells and elevation during intrathymic differentiation. *Nature* 310: 506–8

60. Born, W., Yague, J., Palmet, E., Kappler, J., Marrack, P. 1985. Rearrangement of T-cell receptor β-chain genes during T cell development. *Proc. Natl. Acad. Sci. USA* 82: 2925–29

61. Sangster, B., Minowada, J., Suciu-Foca, N., Minden, M., Mak, T. W. 1986. Rearrangement and expression of the α, β and γ T cell receptor genes in human leukemias and functional T cells. *J. Exp. Med.* 163: 1491–1507

62. Spolski, R., Risser, R., Mak, T. W. 1987. Control of the rearrangement and expression of T cell receptor genes: Studies using cell lines with immature and natural phenotypes. *J. Exp. Med.* Submitted

63. Griesser, H., Feller, A., Lennert, K., Tweedale, M., Messner, H. A., Zalcberg, J., Minden, M. D., Mak, T. W. 1986. The structure of the T cell gamma chain gene in lymphoproliferative disorders and lymphoma cell lines. *Blood* 68: 592–94

64. Royer, H. D., Ramarli, D., Acuto, O., Campen, T. J., Reinherz, E. L. 1985. Genes encoding the T-cell receptor α and β subunits are transcribed in an ordered manner during intrathymic ontogeny. *Proc. Natl. Acad. Sci. USA* 82: 5510–14

65. Collins, M. K. L., Tanigawa, G., Kissonherghis, A.-M., Ritter, M., Price, K. M., Tonegawa, S., Owen, M. J. 1985. Regulation of T-cell receptor gene expression in human T-cell development. *Proc. Natl. Acad. Sci. USA* 82: 4503–7

66. Ohashi, P., Mak, T. W., Van den Elsen, P., Yanagi, Y., Yoshikai, Y., Calman, A. F., Terhorst, C., Stob, J. D., Weiss, A. 1985. Reconstitution of an active T3/T cell antigen receptor in human T cells by DNA transfer. *Nature* 316: 602–6

67. Dembic, Z., Haas, W., Weiss, S., McCubrey, J., Kiefer, H., von Boehmer, H., Steinmetz, M. 1986. Transfer of specificity by murine α and β T-cell receptor genes. *Nature* 320: 232–38

68. Marrack, P., Shimonkevitz, R., Hannum, C., Haskins, K., Kappler, J. 1983. The major histocompatibility complex-restricted antigen receptor on T cells. IV. An anti-idiotypic antibody

predicts both antigen and I-specificity. *J. Exp. Med.* 158: 1635–46

69. Fink, P. J., Matis, L. A., McElligott, D. L., Bookman, M., Hedrick, S. M. 1986. Correlation between T-cell specificity and the structure of the antigen receptor. *Nature* 321: 219–26

70. Iwamoto, A., Ohashi, P., Pircher, H., Walker, C. L., Michalopoulos, E. E., Rupp, F., Hengartner, H., Mak, T. W. 1987. T cell receptor variable gene usage in a specific cytotoxic T cell response: Primary structure of antigen specific receptors of four hapten-specific cytotoxic T cell clones. Submitted

71. Rupp, F., Acha-Orbea, H., Hengartner, H., Zinkernagel, R., Joho, R. 1985. Identical V$_\beta$ T-cell receptor genes used in alloreactive cytotoxic and antigen plus I-A specific helper T-cells. *Nature* 315: 425–27

72. Goverman, J., Minard, K., Shastri, N., Hunkapiller, T., Hansburg, D., Sercarz, E., Hood, L. 1985. Rearranged β T-cell receptor specific for lysozyme: No correlation between V$_\beta$ and MHC restriction. *Cell* 40: 859–67

73. Rupp, F., Brecher, J., Miedlin, M. A., Mosmann, T., Zinkernagel, R., Hengartner, H., Joho, R. H. 1987. T cell receptors with identical V but different D-J gene segments have distinct specificities yet cross-reactive idiotypes. *Proc. Natl. Acad. Sci. USA.* In press

74. Kranz, D. M., Saito, H., Heller, M., Takagaki, Y., Haas, W., Eisen, H., Tonegawa, S. 1985. Limited diversity of the rearranged T cell γ gene. *Nature* 313: 752–55

75. Heilig, J. S., Glimcher, L. H., Kranz, D. M., Clayton, L. K., Greenstein, J. L., Saito, H., Maxam, A. M., Burakoff, S. J., Eisen, H. N., Tonegawa, S. 1985. Expression of the T cell specific γ gene is unnecessary in T cells recognizing class II MHC determinants. *Nature* 317: 68–70

76. Zauderer, M., Iwamoto, A., Mak, T. W. 1986. γ gene rearrangement and expression in autoreactive T cells. *J. Exp. Med.* 163: 1314–18

77. Rupp, F., Frech, G., Hentgartner, H., Zinkernagel, R. M., Joho, R. 1986. No functional γ-chain transcripts detected in an alloreactive cytotoxic T cell clone. *Nature* 321: 876–78

78. Reilly, E. B., Kranz, D. M., Tonegawa, S., Eisen, H. N. 1986. A functional γ gene formed from known γ gene segments is not necessary for antigen-specific responses of murine cytotoxic T lymphocytes. *Nature* 321: 878–80

79. Jones, B., Mjolsness, S., Janeway, C., Hayday, A. 1986. Transcripts of functionally rearranged gamma genes in T cells of adult immunocompetent mice. *Nature* 323: 635–38

80. Yoshikai, Y., Reis, M., Mak, T. W. 1986. Athymic mice express high level of functional γ chain but very low level of α and β chain T cell receptor messages. *Nature* 324: 482–85

81. Brenner, M. B., Mclean, J., Dialynas, D. P., Strominger, J. L., Smith, J. A., Owen, F. L., Seidman, J. G., Ip, S., Rosen, F., Krangel, M. S. 1986. Identification of a putative second T cell receptor. *Nature* 322: 145–49

82. Bank, I., DePinho, R. A., Brenner, M. B., Cassimeris, J., Alt, F. W., Chess, L. 1986. A functional T3 molecule associated with a novel heterodimer on the surface of immature human thymocytes. *Nature* 322: 179–81

83. Weiss, A., Newton, M., Crommie, D. 1986. Expression of T3 in association with a molecule distinct from the T cell antigen receptor heterodimer. *Proc. Natl. Acad. Sci. USA* 83: 6998–7002

84. Barth, R., Kim, B., Lan, N., Hunkapiller, T., Sobieck, N., Winoto, A., Gershenfeld, H., Okada, C., Hansburg, D., Weissman, I., Hood, L. 1985. The murine T-cell receptor employs a limited repertoire of expressed V$_\alpha$ gene segments. *Nature* 316: 517–23

85. Arden, B., Klotz, J., Siu, G., Hood, L. 1985. Diversity and structure of genes of the α family of mouse T-cell antigen receptor. *Nature* 316: 783–87

86. Quertermous, T., Strauss, W., Murre, C., Dialynas, D. P., Strominger, J. G. 1986. Human T cell γ genes contain N segments and have marked junctional variability. *Nature* 322: 184–87

87. Dialynas, D. P., Murre, C., Quertermous, T., Boss, J. M., Leiden, J. M., Geidman, J. G., Strominger, J. L. 1986. Cloning and sequence analysis of complimentary DNA encoding an aberrantly rearranged human T cell γ chain. *Proc. Natl. Acad. Sci. USA* 83: 2619–23

88. Wu, T. T., Kabat, E. A. 1970. An analysis of the sequences of the variable regions of Bence Jones proteins and myeloma light chains and their implications for antibody complimentarity. *J. Exp. Med.* 132: 211–50

89. Minden, M. D., Mak, T. W. 1986. The structure of the T cell antigen receptor genes in normal and malignant T cells. *Blood* 68: 327–36

90. Minden, M. D., Toyonaga, B., Ha, K., Yanagi, Y., Chin, B., Gelfand, E., Mak, T. 1985. Somatic rearrangement

of T-cell antigen receptor gene in human T-cell malignancies. *Proc. Natl. Acad. Sci. USA* 82: 1224–27

91. Tawa, A., Hozumi, N., Minden, M., Mak, T. W., Gelfand, E. W. 1985. Rearrangement of the T cell receptor β chain gene in non-T, non-B acute lymphoblastic leukemia of childhood. *N. Engl. J. Med.* 312: 1033–37

92. Flug, F., Pelicci, P.-G., Bonetti, F., Knowles, D. M. II, Dalla-Favera, R. 1985. T-cell receptor gene rearrangements as markers of lineage and clonality in T-cell neoplasms. *Proc. Natl. Acad. Sci. USA* 82: 3460–64

93. O'Connor, N. T. J., Weatherall, D. J., Feller, A. C., Jones, D., Pallesen, G., Stein, H., Wainscoat, J. S., Gatter, K. C., Isaacson, P., Lennert, K., Ramsey, A., Wright, D. H., Mason, D. Y. 1985. Rearrangement of the T-cell receptor beta chain gene in the diagnosis of lymphoproliferative disorders. *Lancet* i: 1295–97

94. Rabbitts, T. H., Stinson, A., Forster, A., Foroni, L., Luzzatto, L., Catovsky, D., Hammarstrom, L., Smith, C. I. E., Jones, D., Karpas, A., Minowada, J., Taylor, A. M. R. 1985. Heterogeneity of T-cell receptor beta chain rearrangements in human leukemias and lymphomas. *EMBO J.* 4: 2217–24

95. Bertness, V., Kirsch, I., Hollis, G., Johnson, B., Bunn, P. A. Jr. 1985. T-cell receptor gene rearrangements as clinical markers of human T-cell lymphomas. *N. Engl. J. Med.* 313: 534–38

96. Waldmann, T. W., Davis, M. M., Bongiovanni, K. F., Korsmeyer, S. J. 1985. Rearrangements of gene for the antigen receptor on T cells as markers of lineage and clonality in human lymphoid neoplasms. *N. Engl. J. Med.* 313: 776–83

97. Pelicci, P.-G., Knowles, D. M. II, Dalla-Favera, R. 1985. Lymphoid tumors displaying rearrangements of both immunoglobulin and T cell receptor genes. *J. Exp. Med.* 162: 1015–24

98. Isaacson, P. G., Spencer, J., Connolly, C. E., Pollock, D. J., Stein, H., O'Connor, N. T. J., Bevan, D. H., Kirkham, H., Wainscoat, J. S., Mason, D. Y. 1985. Malignant histiocytosis of the intestine: A T-cell lymphoma. *Lancet* ii: 688–91

99. Griesser, H., Feller, A., Lennert, K., Minden, M. D., Mak, T. W. 1986. Rearrangement of the β chain of the T cell antigen receptor and immunoglobulin genes in lymphoproliferative disorders. *J. Clin. Invest.* 78: 1179–84

100. Aisenberg, A. C., Krontiris, T. G., Mak, T. W., Wilkes, B. M. 1985.

101. Rearrangement of the gene for the beta chain of the T-cell receptor in T-cell chronic lymphocytic leukemia and related disorders. *N. Engl. J. Med.* 313: 529

101. Weiss, L. M., Hu, E., Wood, G. S., Moulds, C., Cleary, M. L., Warnke, R., Sklar, J. 1985. Clonal rearrangements of T-cell receptor genes in mycosis fungoides and dermatopathic lymphadenopathy. *N. Engl. J. Med.* 313: 539–44

102. Cheng, G., Minden, M. D., Mak, T. W., McCulloch, E. A. 1986. T cell receptor and immunoglobulin gene rearrangements in acute myeloblastic leukemia: Evidence for lineage infidelity. *J. Exp. Med.* 163: 414–24

103. Klein, G., Klein, E. 1985. Evolution of tumours and the impact of molecular oncology. *Nature* 315: 190–95

104. Klein, G. 1983. Specific chromosomal translocations and the genesis of B-cell-derived tumors in mice and men. *Cell* 32: 311–15

105. Lebeau, M. M., Diaz, M. O., Rowley, J. D., Mak, T. W. 1985. Chromosomal localization of the human T-cell receptor β-chain genes. *Cell* 41: 335

106. Kurtzberg, J., Gibner, S. H., Hershfield, M. S. 1985. Establishment of the DU.528 human lymphohemopoietic stem cell line. *J. Exp. Med.* 162: 1561–78

107. Williams, D. L., Look, A. T., Melvin, S. L., Roberson, P. K., Dahl, G., Flake, T., Stass, S. 1984. New chromosomal translocations correlate with specific immunophenotypes of childhood acute lymphoblastic leukemia. *Cell* 36: 101–9

108. Mathieu-Mahul, D., Caubet, J. F., Bernheim, A., Mauchauffe, M., Palmer, E., Berger, R., Larsen, C.-J. 1985. Molecular cloning of a DNA fragment from human chromosome 14 (14q11) involved in T cell malignancies. *EMBO J.* 4: 3427–33

109. Erikson, J., Finger, L., Sun, L., ar-Rushdi, A., Nishikura, K., Minowada, J., Finan, J., Emanuel, B. S., Nowell, P. C., Croce, C. M. 1986. Deregulation of c-myc by translocation of the α-locus of the T cell receptor in T cell leukemias. *Science* 232: 884–86

110. Shima, E., LeBeau, M. M., McKeithan, T. W., Minowada, J., Showe, L. C., Mak, T. W., Minden, M. D., Rowley, J. D., Diaz, M. O. 1986. Gene encoding the α chain of the T cell receptor is moved immediately downstream of c-myc in a chromosomal 8;14 translocation in a cell line from a

human T cell leukemia. *Proc. Natl. Acad. Sci. USA* 83: 3439–43

111. Smith, S. D., Morgan, R., Link, M. P., McFall, P., Hecht, F. 1986. Cytogenetic and immunophenotypic analysis of cell lines established from patients with T-cell leukemia/lymphoma. *Blood* 67: 650–56

112. Sadamori, N., Kusano, M., Nishino, K., Tagawa, M., Yao, E.-I., Yamada, Y., Amagasaki, T., Kinoshita, K.-I., Ichimaru, M. 1985. Abnormalities of chromosome 14 at band 14q11 in Japanese patients with adult T-cell leukemia. *Cancer Genet.* 17: 279–82

113. Erikson, J., Williams, D. L., Finan, J., Nowell, P. C., Croce, C. M. 1985. Locus of the α-chain of the T cell receptor is split by chromosome translocation in T cell leukemias. *Science* 229: 784–86

114. Lewis, W. H., Michaelopoulos, E. E., Williams, D. L., Minden, M. D., Mak, T. W. 1985. Breakpoints in the human T-cell antigen receptor α-chain locus in two T-cell leukemia patients with chromosomal translocations. *Nature* 317: 544–46

115. Baer, R., Chen, K.-C., Smith, S. D., Rabbitts, T. H. 1985. Fusion of an immunoglobulin variable gene and a T cell receptor constant gene in the chromosome 14 inversion associated with T cell tumors. *Cell* 43: 705–13

116. Denny, C. T., Yoshikai, Y., Mak, T. W., Smith, S. D., Hollis, G. H., Kirsch, I. R. 1986. A chromosome 14 inversion in a T cell lymphoma is caused by site-specific recombination between immunoglobulin and T cell receptor loci. *Nature* 320: 549–51

117. Zech, L., Gahrton, G., Hammarstrom, L., Juliusson, G., Mellstedt, H., Robert, K. H., Smith, C. I. E. 1984. Inversion of chromosome 14 marks human T-cell chronic lymphocytic leukaemia. *Nature* 308: 858–60

118. Shah-Reddy, I., Mayeda, K., Mirchandani, I., Koppitch, F. C. 1982. Sezary syndrome with a 14:14 (q12:q31) translocation. *Cancer* 49: 75–79

119. Hecht, F., Morgan, R., Gemmill, R. M., Hecht, B. K.-M., Smith, S. D. 1985. Translocations in T-cell leukemia and lymphoma. *N. Engl. J. Med.* 313: 758–59

120. Hecht, F., Morgan, R., Hecht, B. K.-M., Smith, S. D. 1984. Common region on chromosome 14 in T-cell leukemia and lymphoma. *Science* 226: 1445–47

121. Smith, S. D., Kirsch, I. R. 1987. A common mechanism of chromosomal inversion in B and T-cell tumors:

Implications for lymphocyte development. *Science*. Submitted

122. Fukuhara, S., Hinuma, Y., Gotoh, Y.-I., Uchino, H. 1983. Chromosome aberrations in T lymphocytes carrying adult T cell leukemia-associated antigens (ATLA) from healthy adults. *Blood* 61: 205–7

123. Hecht, F., Kaiser-McCaw, B., Peakman, D., Robinson, A. 1975. Non-random occurrence of 7–14 translocations in human lymphocyte cultures. *Nature* 255: 243–44

124. Aurias, A., Dutrillaux, B., Buriot, D., Lejeune, J. 1980. High frequencies of inversions and translocations of chromosomes 7 and 14 in ataxia telangiectasia. *Mutation Res.* 69: 369–74

125. Taylor, A. M. R. 1982. Ataxia-telangiectasia. A cellular and molecular link between cancer, neuropathology, and immune deficiency, ed. B. A. Bridges, D. G. Harnden, pp. 53–81. New York: Wiley

126. Welch, J. P., Lee, C. L. Y. 1975. Non-random occurrence of 7–14 translocations in human lymphocyte cultures. *Nature* 255: 241–42

127. Battey, J., Moulding, C., Taub, R., Murphy, W., Stewart, T., Potter, H., Lenoir, G., Leder, P. 1983. The human c-myc oncogene: Structural consequences of translocation into the IgH locus in Burkitt lymphoma. *Cell* 34: 779–87

128. Scheres, J. M. J. C., Hustinx, T. W. J., Weemaes, C. M. R. 1980. Chromosome 7 in ataxia-telangiectasia. *J. Pediatr.* 97: 440–41

129. Kaiser-McCaw, B., Hecht, F., Harnden, D. G., Teplitz, R. L. 1975. Somatic rearrangement of chromosome 14 in human lymphocytes. *Proc. Natl. Acad. Sci. USA* 72: 2071–75

130. Kohn, P. H., Whang-Peng, J., Levis, W. R. 1982. Chromosomal instability in ataxia telangiectasia. *Cancer Genet. Cytogenet.* 6: 289–302

131. Staden, R. 1982. An interactive graphics program for comparing and aligning nucleic acid and amino acid sequences. *Nucleic Acids Res.* 10: 2951–61

132. Ikuta, K., Ogura, T., Shimiza, A., Honjo, T. 1985. Low frequency of somatic mutation in β-chain variable region genes of human T cell receptors. *Proc. Natl. Acad. Sci. USA* 82: 7701–5

133. Duby, A. D., Seidman, J. G. 1986. Abnormal recombination products result from aberrant DNA rearrangement of the human T cell antigen receptor β-

chain gene. *Proc. Natl. Acad. Sci. USA* 83: 4890–94

134. Leiden, J. M., Strominger, J. L. 1986. Generation of diversity of the β chain of the human T-lymphocyte receptor for antigen. *Proc. Natl. Acad. Sci. USA* 83: 4456–60

135. Baer, R., Chen, K.-C., Smith, S. D., Rabbitts, T. H. 1984. Fusion of an immunoglobulin variable gene and a T cell receptor constant gene in the chromosome 14 inversion associated with T cell tumors. *Cell* 43: 705–13

136. Tunnacliffe, A., Keford, R., Milstein, C., Forster, A., Rabbitts, T. H. 1985. Sequence and evolution of the human T cell antigen receptor β chain genes. *Proc. Natl. Acad. Sci. USA* 82: 5068–72

Ann. Rev. Immunol. 1987. 5 : 621–69

LYMPHOCYTE HORMONE RECEPTORS[1]

Marshall Plaut

Division of Clinical Immunology, Department of Medicine,
Johns Hopkins University School of Medicine
at The Good Samaritan Hospital, Baltimore, Maryland 21239

INTRODUCTION

This review describes the functional effects on lymphocytes of receptors for a wide variety of hormones and autacoids. Lymphocytes include T cells and B cells. The lineage of NK and K cells is unclear, but in this review we consider them lymphocytes. T cells are divided into subsets, including regulatory cells (helper T cells, suppressor T cells, contrasuppressor T cells) and effector T cells (delayed hypersensitivity T cells and cytotoxic T cells—CTL). Cell surface glycoproteins separate mouse and human T cells into two broad subsets. Most mouse helper and delayed hypersensitivity T cells are L3T4$^+$ Lyt 1$^+$ Lyt 2$^-$ Lyt 3$^-$. Most mouse suppressor T cells and CTL are L3T4$^-$ Lyt 1$^-$ Lyt 2$^+$ Lyt 3$^+$. As defined by the nomenclature

---

[1] Abbreviations: ADA, adenosine deaminase; ADCC, antibody-dependent cellular cytotoxicity; CTL, cytotoxic T lymphocyte; E-rosette, rosettes between sheep erythrocytes and human T cells; GIF, glycosylation inhibitory factor; H1 and H2, histamine-type 1 and -type 2 receptor; HCSF, histamine cell stimulatory factor (stimulates histamine synthesis); HSF, histamine-induced suppressor factor; 5-HT, 5-hydroxytryptamine; IBMX, iso-butylmethylxanthine, an antagonist of adenosine R$_i$ and R$_a$ receptors; IP$_3$, inositol-1.4.5-triphosphate; K, killer cells (kill via ADCC mechanism); LPS, lipopolysaccharide; LTB$_4$, leukotriene B$_4$; MLC, mixed lymphocyte culture; NC, natural cytotoxic cells; NK, natural killer cells; NSAIA, nonsteroidal anti-inflammatory agents (e.g. indomethacin); MLC, mixed lymphocyte culture; P-site, intracellular adenosine binding site, PBL, peripheral blood lymphocyte; PBMC, peripheral blood mononuclear cell; PFC, plaque forming cell; PGE$_2$, prostaglandin E$_2$; POMC, pro-opio melanocortin; PNA, peanut agglutinin; R, receptor; R$_a$, R$_i$, adenosine receptors ("R" refers to specificity for the ribose moiety of adenosine), which activate (a) and inhibit (i) adenylate cyclase; SEA, staphylococcus enterotoxin A; SP, substance P; VIP, vasoactive intestinal peptide.

621

0732–0582/87/0410–0621$02.00

of "cluster of differentiation" (CD) antigens, human helper and delayed hypersensitivity T cells are CD4$^+$ CD8$^-$ (using the old nomenclature, with commercial antibodies, either OKT4$^+$ OKT8$^-$ or Leu 3$^+$ Leu 2$^-$), while human suppressor T cells and CTL are CD4$^-$ CD8$^+$ (OKT4$^-$ OKT8$^+$ or Leu 3$^-$ Leu 2$^+$) (1). Some monoclonal antibodies define these subsets even further. Thus, Leu 2$^+$ Leu 9.3$^+$ cells are CTL, and Leu 2$^+$ Leu 9.3$^-$ are suppressor T cells (2).

Lymphocytes bear receptors for a variety of agonists that deliver both positive and negative signals. These agonists include lymphokines like IL-2 and monokines like IL-1, complement products, antibodies (via Fc receptors), and helper and suppressor factors of antigenic, idiotypic, and isotypic specificity. An additional series of agonists that interact with specific lymphocyte receptors are hormones and "autacoids." Hormones are defined as chemical substances formed in one part of the body, which circulate and interact with organs distinct from the site of origin of the substance. The term autacoid has been used to mean a substance which is formed locally and acts locally, such as histamine and PGE$_2$ (3). Some of the apparent distinctions between hormones and autacoids are artificial, in that some hormones are formed not only by distant endocrine tissue but also locally by inflammatory cells or lymphocytes.

Lymphocyte hormone receptors have been of interest to investigators for several reasons: (a) The existence of such receptors implies that the immune system interacts with the endocrine and neural systems. (b) Many of the components of the hormone receptor signals are phylogenetically primitive means of intercellular communication (4). (c) Hormones and autacoids are "natural" substances that might modulate the immune system yet be relatively nontoxic. (d) Hormone receptor agonists and antagonists are well characterized, so that accurate pharmacologic characterization of lymphocyte receptors can be performed. (e) Lymphocyte hormone receptors are selectively distributed among lymphocyte subsets (3), and hormone receptor numbers and/or responsiveness to a hormone may be altered following lymphocyte activation (5, 6). (f) Receptor numbers may be altered in specific disease states (7). Indeed peripheral blood mononuclear cells (PBMC) or peripheral blood lymphocytes (PBL) are convenient cell types to study receptor numbers. Even in nonimmunologic diseases, changes of lymphocyte receptor numbers and/or responsiveness may reflect similar changes in other organs. (g) Inhibitors of each hormone-receptor system (e.g. nonsteroidal anti-inflammatory agents (NSAIA), which inhibit prostanoid synthesis, or cimetidine, which antagonizes histamine H2 receptors) profoundly affect some in vitro and even some in vivo models and may have particularly strong effects in specific disease states; this suggests that endogenous hormones and autacoids may play a role in normal homeostasis and have an exaggerated role in the patho-physiology of some diseases (8, 9).

In this review I consider the following, in the same sequence, for each of the hormones discussed:

A. Definition of receptor, by radiolabeled ligand studies, and/or by effects of agonists and antagonists on binding and lymphocyte function.

B. Mechanisms of signal transmission (e.g. activation of adenylate cyclase). Certain agonists appear to deliver "compartmentalized" signals (10). Two adenylate cyclase-stimulating agonists may trigger different cAMP-dependent protein kinases and thereby induce distinct functional effects (11).

C. Biologic effects of hormones on lymphocyte function:

    1. Modulation of CTL, natural killer (NK), and antibody dependent cellular cytotoxicity (ADCC) effector function. These activities appear to be mediated by direct lymphocyte-hormone interaction in a short-term assay, and accessory cells do not modulate these functions.

    2. Modulation of lymphocyte migration, E-rosetting of T cells [i.e. rosetting of human T cells and sheep erythrocytes, now known to occur via interactions of CD2 (T11) glycoproteins on the T cell with the sheep erythrocyte membrane], terminal secretion of antibody from mature B cells (in a 1-hr assay).

    3. Proliferative responses to T- and B-cell mitogens, alloantigens, and soluble antigens. These are more complex assays than those in 1, in that at least three steps can be modulated (12, 13):

        ● Secretion of IL-2 (and other soluble factors). (This step depends both on effects on antigen presentation and IL-1 secretion by antigen-presenting cells and on direct effects on the IL-2-secreting cell.)

        ● Expression of IL-2 receptors.

        ● Biochemical pathways of IL-2-dependent proliferation distal to the IL-2 receptor.

    These three steps are dependent on IL-1 and other factors. Activated suppressor cells often inhibit IL-2 secretion and may also inhibit IL-2-dependent lymphocyte proliferation.

    4. Modulation of primary or secondary CTL generation.

    5. Modulation of primary or secondary antibody responses.

For both items 4 and 5, modulation can be of any of three events analogous to those described above that modulate proliferation (12, 13): (a) secretion of IL-2 and other growth and differentiation factors; (b) expression (on CTL precursors or B-cell precursors) of receptors

for IL-2 and other factors; and (c) biochemical pathways, distal to the receptor for factors, of factor-dependent growth and differentiation.
6. Induction or down-regulation of suppressor cells.
7. In vivo effects of hormone.
D. Key observations concerning the mechanism of action.
E. Heterogeneity of responsiveness to hormone among different lymphocyte subsets, and the correlation between this responsiveness and the overall functional effects of hormone.
F. Heterogeneity of responsiveness during the immune response—e.g. insulin receptors are absent on resting lymphocytes but expressed following lymphocyte activation (6).
G. Heterogeneity of responsiveness in different disease states.
H. Physiologic role of hormone-lymphocyte interaction; mechanisms of endogenous production of hormone.
I. Miscellaneous remarks: Is the hormone agonist or antagonist useful clinically?

Because of space limitations, and because several lymphocyte hormone receptor systems have been reviewed elegantly in recent years, this review is not exhaustive but emphasizes recent developments and insights into the importance of each lymphocyte hormone receptor.

Hormones and autacoids are discussed in an arbitrary order. I have chosen not to discuss certain hormones such as thymic hormones like thymopoietin and thymulin, and nonglucocorticoid steroids such as estrogen and testosterone.

Moreover, I have not discussed receptors for "immunologic" autacoids, i.e. monokines and lymphokines such as IL-1, IL-2, IL-3, etc. To discuss these receptors adequately would require a separate review.

GENERAL COMMENTS

Receptors

In general, the actions of the autacoids and hormones are initiated by interaction with specific receptors (7). The receptor specificity of agonist effects has not been demonstrated in all studies. Indeed, many of the actions on lymphocytes of some agonists such as β-endorphin appear to be not via their classical receptor (17). High concentrations of many agents, including receptor agonists and antagonists, may stimulate cell function via cellular structures other than receptors. To define receptors, it is necessary to utilize a set of criteria, with both radioligand binding assays and pharmacologic assays (18). For binding assays, the preferable ligand is an antagonist. In well-studied systems, such as β-adrenergic receptors, agonists convert the high affinity state of receptor to the low affinity state. The affinity of the ligand should be sufficiently high to measure in a standard binding assay. If the dissociation constant (K_D) is $> 10^{-7}$ M, the rate at which ligand dissociates (off-rate) is generally so rapid that it is technically difficult, if not impossible, to obtain accurate binding data. A series of receptor agonists should compete with the ligands in a predictable order of potency and induce functional effects with the same order of potency. This order of potency should be the same as that in other tissues. Antagonists should specifically and competitively inhibit radioligand binding. The K_D for the antagonist should be the same for inhibiting binding and for inhibiting the biological effects of agonist. The K_D should be similar to that in other tissues. The K_D should be constant for inhibiting the effects of several different agonists.

With these criteria, receptors can be defined by quantitative pharmacologic studies of their biological effects even in the absence of a radioligand binding assay. Conversely, in the absence of competitive antagonists, agonist studies are useful but do not completely define the system. If lymphocytes bear a unique receptor subtype, the relative potencies of antagonists will be different from their potencies on the classic receptor. Furthermore, the initial studies of β-adrenergic receptors used the agonist ^{3}H-epinephrine (19)—a ligand that binds to the receptor but with a much lower affinity than it does to catecholamine metabolizing enzymes. The differences between the binding of ^{3}H-epinephrine and true receptor binding were not realized until a series of antagonists were tested against the labeled ligand. Binding studies with labeled β-adrenergic antagonists do identify the relevant receptor (7).

Signal Transmission Following Receptor Activation

Lymphocyte hormone receptors described in this review are all located on the cell surface, except for glucocorticoid receptors, which are cytoplasmic. The binding of hormones to cell surface receptors results in activation of intracellular enzyme systems. The autacoids have been divided into agonists which activate adenylate cyclase and those which have "opposite" effects. Adenylate cyclase activation results in the formation of cAMP, an intracellular "second messenger" which generally mimics the effect of the stimulus (within the constraints of the compartmentalization concept described above). Agonists that have opposite effects may either inhibit adenylate cyclase and/or stimulate the production of an alternative second messenger. Candidates for the role of alternative second messengers include cyclic GMP (20) and products of phospholipase C (which include inositol 1, 4, 5-triphosphate (IP_3) and diacylglycerol (21, 22). Calcium, which is mobilized by IP_3, is another candidate (21).

The best studied consequences of receptor activation are in studies of β-adrenergic receptor linkage to adenylate cyclase; these have been reviewed extensively elsewhere (7). In this system, the interaction of hormone and receptor results in low affinity H-R binding. The H-R complex is linked to a regulatory protein (N). The HRN complex represents a high affinity binding state of hormone. The guanine nucleotide GTP converts HRN to a GTP-N complex and H-R. The GTP-N complex interacts with the catalytic unit (C) of adenylate cyclase to form the active form (N-GTP-C) of the catalytic site of adenylate cyclase. This activated enzyme converts ATP to cAMP. A GTPase destroys the N-GTP-C complex. cAMP typically stimulates metabolic pathways by interacting with cAMP-dependent protein kinases to phosphorylate specific substrate proteins. Several mechanisms modulate this system. One of these is desensitization, which appears to involve "uncoupling" of the HR-N linkage, through rapid alteration in receptor recycling or cAMP-dependent phsophorylation of the receptor; another mechanism is long-term alterations in receptor synthesis and metabolism. Other mechanisms probably exist to alter receptors, coupling, and cAMP-dependent protein kinases or other post-cAMP biochemical events.

The receptors that transmit inhibitory signals to adenylate cyclase, such as adenosine R_i receptors (23), parallel the receptors that transmit activating signals, although they are linked to an inhibitory protein (N_i) which associates with the catalytic site of adenylate cyclase in such a way as to inhibit rather than to activate it.

Other receptors are linked to other putative second messengers. These other second messengers are derived either from linkage of receptor to guanylate cyclase, an enzyme that converts GTP to cyclic GMP, or by linkage to phospholipase C, an enzyme that stimulates phosphatidylinositol (PI) turnover (and the formation of IP_3) and formation of diacylglycerol. IP_3 stimulates calcium mobilization, and diacylglycerol stimulates protein kinase C activation (21, 22). The relation of the guanylate cyclase to the phospholipase C systems is unknown.

Functional Effects of Receptor Activation

It is generally true that hormones that activate adenylate cyclase inhibit lymphocyte function. However, since an elevation of cAMP early in an immune response can activate B cells and can also inactivate suppressor T cells, cAMP may under certain circumstances enhance lymphocyte function, especially antibody responses. Agonists that act in opposite ways, including agents that inhibit adenylate cyclase or agents that apparently activate guanylate cyclase and/or phospholipase C, are generally stimulatory. Glucocorticoids are generally inhibitory. Many hormones activate two receptors, with opposite effects. Thus, adenosine at low concentrations activates R_i receptors, but at somewhat higher concentrations activates R_a receptors, and at very high concentrations acts on intracellular P sites (24). Histamine at low concentrations activates H1 receptors, and at high concentrations activates H2 receptors. Epinephrine acts on α-adrenergic and β-adrenergic receptors. Insulin at low concentrations acts on insulin receptors and at high concentrations also acts on somatomedin C receptors (25).

Regulation of Hormone Receptor Systems

Individual differences in hormone receptors or in hormone responsiveness depend in part on genetic factors, e.g. H2-linked mouse genes determine both basal and stimulated cAMP levels (14, 15). Perhaps comparable genes define the linkage between HLA-B12 expression and decreased responsiveness of human lymphocytes to inhibition by histamine and prostaglandin (16). Besides these inherited differences, there are differences induced by immunization and/or lymphocyte activation, and abnormalities associated with specific acquired disease states.

EFFECTS OF cAMP ON THE IMMUNE RESPONSE

Many of the experiments to evaluate the role of cAMP on the immune response have dealt not only with cAMP or cAMP analogues such as dibutryl cAMP, but also with autacoids that act on adenylate cyclase–linked receptors, such as isoproterenol (a β-adrenergic agonist), PGE_2, and histamine. The concept that autacoids that stimulated increases in cAMP modulate immune responses was novel when it was first introduced approximately 15 years ago, mainly by Lichtenstein and Braun, but it is now accepted. Fifteen years ago, it was known that agents that increased cAMP levels enhanced the secretory activity of many cell types. On the other hand, Lichtenstein showed that cAMP agonists inhibited basophil histamine release (26), and Braun showed that poly A : U enhanced the formation of plaque forming cells (PFC) at low concentrations and inhibited at high concentrations; theophylline at high concentrations in vitro, acting as a phosphodiesterase inhibitor, enhanced the effects of poly A : U. Some of these effects were reproduced in vivo. Braun assumed that these agents acted through cAMP levels (27, 28). Many other experiments followed. Studies on lymphocytes demonstrated that cAMP agonists do inhibit the effector function of mouse CTL generated in vivo (29), but inhibit only poorly CTL generated in vitro (30). They inhibit the cytotoxic effector function of murine and human killer (K) and NK cells (31, 33), E-rosetting, and anti-IgG-induced translational motility of mouse splenic

B lymphocytes (34). They also inhibit secretion of antibody by terminally differentiated B cells (35). Such agonists are potent inhibitors of proliferation of both T and B cells. The capacity of cAMP agonists to inhibit proliferation is reversed by cGMP in some murine systems (36, 37). Agonists that increase cAMP levels appear to act both by inhibiting the production of IL-2 and by inhibiting IL-2-dependent proliferation (38). The latter effect presumably results from blocking of a biochemical step(s) during proliferation, but inhibition of IL-2-receptor display and/or transformation of IL-2 receptors to a low affinity state (39) have not been ruled out. The inhibitory actions of cAMP agonists on T cells are more profound than on B cells, in that, for example, the inhibition of LPS-induced mouse B-cell proliferation is minimal (40). Antigen- and mitogen-induced lymphokine secretion is also inhibited by cAMP in many (41) but not all (42) experimental models.

Agonists that increase intracellular cAMP levels have complex, dual effects on in vitro CTL and antibody responses. Both CTL generation and PFC responses are inhibited by dibutyryl cyclic AMP when the agent is present throughout the culture. However, dibutyryl cyclic AMP when present during only the first 12 hours of culture enhances CTL generation and PFC responses (43, 47). Agents that are active for a long time, such as dibutyryl cyclic AMP, are more effective at enhancing than agents, such as the β-adrenergic agonist isoproterenol, that induce only transient increases in cAMP (46). When cAMP is present in culture for more than 24 hr, it inhibits CTL generation and PFC responses. Antibody-forming cell responses may be divided into two stages: a 24-hr culture of rabbit B cells with anti-immunoglobulin (first signal) and a 6-day culture of activated B cells with T cell–derived soluble factors. When this is done, both anti-immunoglobulin and T-cell factors stimulate biochemical changes, including increases in the amount of, and the phosphorylation of, nonhistone proteins. Dibutyryl cAMP enhances PFC responses and the amount and the phosphorylation of nonhistone proteins when it is present only in the first stage; it inhibits PFC responses and the amount and phosphorylation of nonhistone proteins when it is present only in the second stage (48, 49).

Two distinct mechanisms account for the enhancement of PFC responses by cAMP when it is present early: inactivation of suppressor T cells (50) and direct stimulation of B-cell differentiation (46, 49). The blockade of suppression is presumably via the antiproliferative activity of cAMP agonists (51), although it seems to contradict other models in which the same agonists, added to resting cells, activate suppressor cells (see reviews 8, 9). The direct effect on B cells is best illustrated in a model (52) where spleen cells from nude (nu/nu) mice are induced to form antibody by the combination of IL-1 and dibutyryl cAMP. Neither IL-1 nor dibutyryl cAMP alone are sufficient. T cells substantially enhance the response of nu/nu spleen cells to IL-1, but dibutyryl cAMP markedly inhibits the response of nu/nu spleen cells to T cells and IL-1. This observation suggests that dibutyryl cAMP directly enhances B-cell function. Inhibition works at least in part by blocking secretion of lymphokines: Dibutyryl cAMP inhibits IL-2 and T-cell replacing factor (TRF) secretion but, unexpectedly, enhances B-cell growth factor (BCGF) secretion (52).

The inhibitory effect of cAMP agonists when present late probably is both by inhibition of secretion of growth factors made late in culture and by an anti-proliferative action on B and T cells. Even though cAMP is a weak inhibitor of lipopolysacharride (LPS)-mediated proliferation in mouse B cells, it is capable of inhibiting B cells late in the immune response.

The results of in vivo experiments are even more complex. Several agonists that increase cAMP levels, including cholera toxin and PGE_2 analogues, inhibit cell-mediated immune functions (16, 29). Cholera toxin (a potent stimulator of adenylate cyclase) enhances the immune response both in vivo and in vitro under certain conditions. However, in adrenal-ectomized mice, cholera toxin under the same conditions suppresses the PFC response (53). Presumably cholera toxin not only inhibits the antibody response but also stimulates the adrenal gland to produce glu-cocorticoids. Glucocorticoids may enhance because the dose and timing of the release of glucocorticoids selectively inhibit suppressor T cells.

SPECIFIC HORMONE RECEPTORS

Glucocorticoids

Glucocorticoids have been used as therapeutic agents because of their potent anti-inflammatory effects in vivo. Although they appear to have immunosuppressive properties, some older in vitro studies demonstrated inhibitory effects only at extremely high concentrations. Under other experimental conditions, mitogen-induced proliferation of human PBMC is inhibited by low concentrations of glucocorticoids (53a). The most convincing studies with doses comparable to therapeutic ones concerned killing of subpopulations of mouse thymocytes in vivo (54, 55). Similar lytic events occur in vitro (56). Such in vivo studies have defined sensitive species like mouse, whose thymocytes are killed by glucocorticoids, vs resistant species such as guinea pig and man, whose thymocytes are not killed. In vivo antibody responses are decreased in sensitive species but not in resistant species. In resistant species, glucocorticoid administration results in marked decreases in circulating monocytes and lymphocytes, but these changes result from altered recirculation patterns rather than death.

Possibly these cells now "home" to the bone marrow (57). New in vitro model systems have defined potent effects of glucocorticoids on T cells of both sensitive and resistant species. Several recent reviews have examined glucocorticoid-lymphocyte interaction (57, 58).

Lymphocytes possess glucocorticoid receptors. These receptors have been defined both by binding of labeled glucocorticoids and by the relative potency of a series of glucocorticoid agonists to modulate cell function (59–61). Glucocorticoid receptor antagonists have only recently been defined (62) and have not been used in these types of experiment. In some lymphocyte systems but not all, progesterone acts as a receptor antagonist (63). Lymphocytes possess receptors with K_D of 5.5×10^{-9} M for dexamethasone; activation of T cells by mitogen induces a two- to three-fold increase in receptors in at least some lymphocyte populations (60, 64, 65).

The prevailing view is that glucocorticoids interact with cytoplasmic receptors, and that the glucocorticoid-receptor complexes are translocated to the nucleus where they stimulate synthesis of new messenger RNA and thus of new proteins. Of the proteins stimulated by glucocorticoids, the one that appears to have most relevance to their anti-inflammatory role is lipocortin, a potent inhibitor of phospholipases including phospholipase A2 (59, 66). This inhibitory action prevents production of both cyclooxygenase and lipoxygenase products of arachidonic acid metabolism, because it prevents arachidonic acid release from cell membrane phospholipids; it also prevents release of platelet activating factor from membrane phospholipids. Lipocortin is important not only because of this action but also because of the effects on the immune response of "glycosylation inhibitory factor" (GIF) (discussed below). Lipocortin has been shown to reproduce the effects of glucocorticoids in some systems (including systems to study GIF production). However, lipocortin has not been evaluated in most of the in vitro assays of glucocorticoids on proliferation and antibody production.

Glucocorticoids kill lymphocytes, especially mouse thymocytes, by lysing them; this results in both membrane and DNA damage. Endogenous endonucleases in the cell are activated (67). These endonuclease(s) are also activated during killing by CTL but not killing by complement. However, many of the effects of glucocorticoids do not occur through their lytic effects.

Splenic CTL from in vivo–immune mice are inhibited by concentrations of dexamethasone as low as 10^{-9} M if the glucocorticoid is preincubated with such CTL for 12–24 hr. The relative inhibitory potency of glucocorticoid agonists corresponds to their potencies on classical glucocorticoid receptors (61). Glucocorticoids do not inhibit the effector function of long-term CTL clones (68). However, cloned CTL are long-term in vitro cells; such cells are biochemically distinct from CTL generated in vivo (30), and they require IL-2 for growth. Glucocorticoids inhibit T cells primarily by inhibiting IL-2 synthesis (58, 68, 69), so the lack of effect of glucocorticoids on CTL clones, which are provided with IL-2, is not unexpected. Glucocorticoids modestly inhibit the cytotoxicity mediated by antibody dependent and PHA-dependent killers in human peripheral blood (see review 57). Glucocorticoids inhibit human NK activity, but only if present during preincubation and during the lytic assay. Surprisingly, 10 μM prednisolone inhibits NK activity while 0.1 μM enhances (70). In contrast to these confusing effects of glucocorticoids themselves, lipocortin inhibits the effector function of human K and NK cells (71). This apparent discrepancy between the actions of glucocorticoids and lipocortin is not resolved.

Glucocorticoids inhibit anti-immunoglobulin mediated capping (72). Glucocorticoids also inhibit colony formation by T cells (73).

Glucocorticoids inhibit lymphocyte proliferation in response to T-cell activating signals such as mitogens, alloantigens, and other antigens (see review 57, 58, 69). Proliferation in the autologous mixed lymphocyte culture (MLC) is especially sensitive to inhibition by low concentrations of glucocorticoids (74). Glucocorticoids have variable effects on antibody responses, including inhibition, no effect, and enhancement. The most impressive effects are enhancement of immunoglobulin synthesis in 12-day cultures of PWM-stimulated human PBMC (57, 75–77).

Glucocorticoids in vitro decrease suppressor T cells and have variable effects on suppressor cell generation (see review 57). However, one type of suppressor factor is induced by glucocorticoids: GIF, the phosphorylated form of lipocortin. This molecule is a potent inhibitor of the N-glycosylation of IgE binding factor (and probably other biologically important molecules); it thereby favors the production of IgE suppressor factor over IgE potentiating factor and thus inhibits IgE antibody production (78–79). The clinical relevance of glucocorticoid inhibition of IgE antibody production is not proven.

Two key observations have resulted in a much clearer understanding of the mechanisms of action of glucocorticoids: First, they act by inducing synthesis of new proteins, so that many glucocorticoid effects require preincubation for hours. Second, they interfere with lymphocyte activation by inhibiting IL-1 and IL-2 synthesis, and many inhibitory glucocorticoid effects are entirely reversed by adding IL-1 and IL-2 (58, 61, 68, 69). Glucocorticoids inhibit IL-2 secretion both through inhibition of secretion by macrophages and monocytes of IL-1 and through direct inhibitory effects on T cells themselves (58).

Mature in vivo mouse CTL are inhibited by low concentrations (10^{-9}

M–10^{-8} M) of dexamethasone if and only if CTL are preincubated for 24 hr. These effects are reversed by IL-2 but not IL-1. High concentrations (10^{-6} M) of dexamethasone inhibit CTL activity, and such effects are not reversed by IL-2; they represent a direct (lytic) effect on the CTL (61). Further analysis of the level of messenger RNA (mRNA) confirms the importance of IL-2 inhibition. Glucocorticoids in optimal doses inhibit formation of mRNA for IL-2 receptors by 50% but inhibit the mRNA for IL-2 by 100%. The 50% decrease in mRNA for IL-2 receptors results in virtually no reduction in lymphocyte responses to IL-2, while the complete inhibition of mRNA for IL-2 is associated with a lack of IL-2 production (80).

Lipocortin appears to be an important immunomodulatory protein induced by glucocorticoids. Lipocortin inhibits arachidonic acid release, and two studies suggests that glucocorticoids inhibit lymphocyte function simply by blocking production of one critical arachidonic acid-derived product. One study has demonstrated that the inhibitory effect of glucocorticoids on proliferation and IL-2 production is entirely reversed by 5×10^{-9} M LTB$_4$. The action of LTB$_4$ is specific, in that LTC$_4$ and LTD$_4$ do not reverse its action (81). Moreover, LTB$_4$ does not reverse the inhibition of proliferation induced by histamine, PGE$_2$, or interferon-γ. Glucocorticoids stimulate immunoglobulin production in a 12–14 day in vitro culture of otherwise unstimulated human PBMC; and 3 doses, 24 hr apart, of 10^{-10} M LTB$_4$ reverse the stimulatory effects of glucocorticoids on immunoglobulin production (77). Thus, if these studies can be confirmed, LTB$_4$ may be an essential intermediate in T-cell production of IL-2.

Although all lymphocytes possess glucocorticoid receptors, T-cell subsets are differentially sensitive to glucocorticoid effects on function. Studies with mouse peripheral T cells and mature thymocytes indicate that glucocorticoids added prior to activation inhibit the generation of all T-cell functions (help, suppression, proliferation, and CTL). With activation (e.g. by antigen), macrophages or IL-1 reverse the inhibitory effects of glucocorticoids on "type 1" helper T cells (unlinked recognition of antigen), but "type 2" helper T cells (linked recognition), suppressor T cells, and CTL are still sensitive to glucocorticoid inhibition (82–84). IL-1 reverses the inhibitory effect of glucocorticoids on proliferation of PBL but not of thymocytes (85). Inhibition by glucocorticoids of CTL generation is reversed by IL-2 (61). These results suggest that in at least some helper and proliferating T-cell subsets, activation by antigen and IL-1 makes these cells resistant to glucocorticoids. Such resistance might be due to their capacity to make endogenous IL-2, although experiments to address this hypothesis have not been reported. CTL precursors are resistant only when IL-2 is added.

In addition to subset differences, genetic factors regulate glucocorticoid sensitivity. H-2^a mice are more sensitive to glucocorticoid-induced lysis of thymocytes than are H-2^b mice (86).

These differences in responsiveness of subsets to glucocorticoids are not correlated with glucocorticoid receptor numbers. Antigen induces a 50% increase in glucocorticoid receptors in adrenalectomized mice, and mitogen induces a 2–3-fold increase in receptors (57, 64, 87). The relatively modest changes in receptor numbers that occur after mitogen or antigenic stimulation of T lymphocytes do not appear to cause significant alteration in responsiveness to glucocorticoids. Furthermore, in vivo glucocorticoids markedly decrease the number of $Fc_\mu R^+$ T cells and also modestly decrease numbers of $Fc_\gamma R^+$ T cells and monocytes. The remaining cells, although enriched in putative suppressor ($Fc_\gamma R^+$) T cells, are decreased in T-suppressor function. The $Fc_\mu R^+$ T cells and the $Fc_\gamma R^+$ T cells have equal numbers of receptors per cell (88, 89).

Virtually all endogenous glucocorticoids are presumably derived from the adrenal glands and are not locally produced. The levels of glucocorticoids that modulate lymphocyte function are attainable physiologically especially under conditions of stress. Lymphokines are reported to increase glucocorticoid levels (90), perhaps via stimulating adrenocorticotrophic hormone ACTH (91). It is noteworthy that lymphocytes synthesize ACTH-like molecules which can, theoretically, increase glucocorticoid output.

There are numerous clinical situations in which glucocorticoids are useful, such as during exacerbations of autoimmune diseases and as treatment following transplantation of allografts. Although the capacity of glucocorticoids to prevent allograft rejection is in part through anti-inflammatory effects, immunomodulatory effects may also be important. There are no immunologic diseases known to be caused by abnormalities of lymphocyte glucocorticoid receptors. It is of interest that glucocorticoids stimulated Ig synthesis by blocking suppressor T-cell function in vitro in a small number of patients with common variable hypogammaglobulinemia, and in one patient they also stimulated Ig synthesis in vivo (92). However, this effect is not confined to glucocorticoids, in that cimetidine has been shown to inhibit excess suppressor cell function in this disease also (93). Paradoxically, in at least one disease—chronic active hepatitis—glucocorticoids restore deficient suppressor function (94).

Neuropeptide Hormones

The effects of neuropeptide hormones on the immune system have been reviewed extensively in recent years (91, 95, see also *J. Immunol.* 135:

739s–863s, 1985). I will describe only a few salient features of several hormones.

ACTH   There are receptors for ACTH on thymocytes and lymphocytes, defined by a binding assay. There are two K_Ds of ^{125}I-ACTH, 1×10^{-10} M and 5×10^{-9} M, corresponding to those for receptors on adrenal cells (96). ACTH suppresses antibody production, but the mechanism of this suppression is unknown, although ACTH stimulates human B-cell growth but not T-cell proliferation (see review 91). ACTH also suppresses interferon production in one model (91).

This agonist is of particular interest because ACTH and β-endorphin are both products of pro-opiomelanocortin (POMC). Certain viral infections of lymphocytes and corticotropin releasing factors both appear to induce parallel production of ACTH and β-endorphin by lymphocytes in vitro and in vivo while glucocorticoids inhibit such production, presumably by modulating transcription of the gene for POMC (97, 98, 98a). Thus, there seems to be reciprocal regulation of neuropeptide production during the immune response. IL-1 stimulates in vitro ACTH production from pituitary tumor cells (99). It is not known whether IL-1 can stimulate ACTH production by lymphocytes.

α- AND β-ENDORPHIN, MET-ENKEPHALIN, AND LEU-ENKEPHALIN   Receptors for these agents, which are classical opiate receptors, are defined via inhibition by opiate antagonists such as naloxone and by direct binding studies. Receptors are present on peripheral blood lymphocytes, cultured lymphocytes, and the lymphoblastoid line RPMI 8237; the K_D for β-endorphin is 3×10^{-9} M, but the binding is not blocked by naloxone (17, 91, 95, 100, 101). Thus, the lymphocyte receptor is not the classical central nervous system opiate receptor, which recognizes the amino terminal end of β-endorphin, but rather a different receptor that recognizes the carboxy terminal end. Despite these results with direct binding studies, functional studies demonstrate that some actions of these agonists are via classical (blocked by naloxone) opiate receptors: α-endorphin and the enkephalins suppress PFC formation, via opiate receptors. In contrast, β-endorphin stimulates rat lymphocyte and inhibits human lymphocyte proliferation, although these effects are apparently not via classic opiate receptors. β-endorphin enhances NK activity and interferon production by NK cells. The enkephalins enhance NK activity but do not affect K-cell activity. β-endorphin stimulates NK activity via opiate receptors, but the increase in interferon production is only partially related to activation of opiate receptor (91, 95). β-endorphin apparently is produced not only in neural tissue, but also by lymphocytes, in parallel with ACTH (see above) (97, 98).

ARGININE VASOPRESSIN (AVP), OXYTOCIN   These agents have been studied in a model system in which Staphylococcus enterotoxin A (SEA)–stimulated Lyt2$^+$ spleen cells produce interferon-γ upon stimulation with IL-2. If Il-2 is omitted, several agonists, including AVP and oxytocin, replace IL-2. Studies with several derivatives related to AVP and oxytoxin suggest that the effect is via a specific receptor (see reviews 102, 103).

THYROTROPIN (TSH)   TSH is reported to enhance antibody responses. SEA-stimulated T cells will make TSH but not POMC (see review 91).

GROWTH HORMONE   Growth hormone is reported to act as an auxiliary factor in CTL generation. (The requirement for growth hormone is detected when assays are performed in serum-free medium.) Super-physiologic levels of growth hormone enhance T-cell growth (104; see review 105).

SOMATOSTATIN   Somatostatin receptors have been described by direct binding of the hormone to human peripheral blood monocytes and lymphocytes, but the results are of uncertain significance: there being only 250 receptors per lymphocyte, and these having a very low affinity (K_D of 5×10^{-7} M) (see reviews 95, 106). This peptide has been reported to act by inhibiting adenylate cyclase (107). Somatostatin inhibits the proliferative response of mouse and human T lymphocytes to T-cell mitogens and inhibits the con A–induced secretion of IgA from cultured mouse lymphocytes (108). However, IgG and IgM secretion are not significantly changed. It is interesting that the RBL (rat basophil tumor) cell line contains somatostatin-like peptides, suggesting that the agonist may be produced locally by inflammatory cells (109).

SUBSTANCE P (SP)   Direct binding assays with human PBL and SP reveal a very low affinity receptor (K_D of approximately 2×10^{-7} M). Compelling evidence for the receptor specificity of binding is that D-pro^2-D-phe^7-D-trp^9-SP, a SP antagonist, inhibits binding. The T lymphoblastoid cell line IM-9 binds labeled SP much more avidly ($K_D = 7 \times 10^{-10}$M) (106, 110) than do PBL. The discrepancy in K_Ds is not explained (but see below).

SP directly stimulates proliferation and also augments PHA-induced stimulation, perhaps by acting on both macrophages and T cells. The enhanced proliferation is inhibited by D-pro^2-D-phe^7-D-trp^9-SP (111). SP also increases immunoglobulin secretion by con A–stimulated mouse lymphoid tissue (108). These stimulatory effects are detectable at concentrations as low as 10^{-9} M, suggesting that the true K_D is much less than 2×10^{-7} M.

There appear to be subsets of lymphocytes with receptors for SP, in that 21% of resting T cells (18% of Leu-3a$^+$ and 9% of Leu-2b$^+$), 35% of

PHA-stimulated T cells, and 0% of B cells bear receptors (112). The preferential binding to helper rather than suppressor T cells has not yet been correlated with the stimulatory effects of SP.

VASOACTIVE INTESTINAL PEPTIDE (VIP)   VIP binds to receptors on human and mouse T cells but not to B cells. Direct binding studies indicate that mouse T lymphocytes bear 5000 receptors per cell with a K_D of 3×10^{-10} M and that human lymphocytes have a K_D of 5×10^{-10} M (106, 113, 114). Stimulation of VIP receptors activates adenylate cyclase, resulting in increases in intracellular cAMP levels (113, 114).

VIP inhibits NK cell function, but preincubation with 10^{-10} M–10^{-8} M VIP for 30–60 min, followed by washing, results in augmented NK function. This augmentation is via augmenting effector-target conjugate formation (115). VIP inhibits the proliferative responses of mouse lymphocytes to T-cell mitogens. Its effects on Ig synthesis induced by these mitogens depend on both the Ig isotype and the source of lymphocytes. IgA synthesis is increased in lymph node and spleen but inhibited in Peyer's patches; IgM synthesis is unaffected in lymph node and spleen but is increased in Peyer's patches (108).

Not only Ig synthesis but also lymphocyte migration is influenced by VIP. In vitro exposure to VIP results in a 50% decrease in VIP receptors. Such "desensitized" T cells when tested by in vivo transfer have decreased migration to Peyer's patches and mesenteric lymph nodes (sites where interaction with high endothelial venules determines migration) but have normal migration to spleen and intestine (116). This suggests that VIP-VIP receptor interactions are important for the interaction of T cells with high endothelial venule binding sites.

The concentration of VIP in the blood is reported to be 10^{-11} M, with up to 10^{-9} M concentrations obtained at nerve endings. VIP appears to be released from inflammatory cells including neutrophils and rat peritoneal mast cells (115).

Insulin

Insulin receptors have been defined by direct binding studies on T lymphocytes and the human T lymphoblastoid cell line IM-9. The K_D for insulin receptors on rat and human T cells is 5×10^{-8} M–1×10^{-9} M (117, 118).

The intracellular message activated by insulin is not known. Various theories suggest that insulin inhibits adenylate cyclase, activates guanylate cyclase, or acts in another way (119). Two distinct effects have been reported, at low and high concentrations of insulin. Whether this reflects interactions of insulin with insulin receptors alone or with both insulin and somatomedin C receptors (25, 119) is not known.

Insulin augments the activity of rat splenic CTL, but only under very specific experimental conditions: when the CTL are obtained early after in vivo alloimmunization and when insulin is mixed with CTL 4–6 minutes before targets are added (117, 120). Insulin also augments the activity of splenic CTL from mice early after immunization (32). Direct binding studies demonstrate that viral-specific CTL clones do not bear insulin receptors (121). Functional studies of insulin on these clones were not done. These results differ from data indicating that Lyt 2^+, 3^+ cells (presumably CTL) from short-term MLC cultures bear insulin receptors (122).

Insulin inhibits P388D$_1$ killing of antibody-coated target cells (123). Insulin has no effect on killing by mouse NK cells (32).

Insulin potentiates proliferation in MLC (105). (This effect is detectable only when assays are done in serum-free medium.) Growth hormone and transferrin also potentiate. Insulin appears to be essential for proliferative responses during MLC but is less important for the generation of CTL during such MLC. Insulin also potentiates LPS-induced proliferation of B cells and primary in vitro antibody responses to LPS.

In contrast to the stimulatory effects of insulin on growth, high concentrations of insulin inhibit IL-2-induced proliferation of mouse splenic T and B lymphocytes, and the formation of PFC in response to sheep red blood cells. This action of insulin is apparently not via true insulin receptors but is rather via somatomedin C (insulin growth factor-1) receptors activated by physiologic concentrations of somatomedin C and by superphysiologic concentrations of insulin (120).

Perhaps the most interesting observation concerning insulin receptors on lymphocytes is that resting T lymphocytes lack insulin receptors, but activated (e.g. in a MLC) T and B cells bear insulin receptors (118, 124, 125). Noncytolytic T-cell clones generate insulin receptors after antigenic stimulation (121). During a MLC, insulin receptors increase on about day 3, i.e. one day before CTL are detected. The insulin receptors are mainly only on the Lyt 2^+, 3^+ cells (122). The mechanism of generation of insulin receptors is unknown, but activated T cells are required to cooperate with B cells in order to stimulate the formation of insulin receptors on B cells (125). This suggests either that there is direct contact between T cells and B cells, or that a soluble factor from T cells mediates the stimulation of insulin receptors.

Insulin receptors on lymphocytes are useful markers of events in at least one disease state, diabetes mellitus. When resting T cells are activated, they express insulin receptors. If resting cells are exposed to high levels of insulin, they are desensitized and express fewer receptors after activation. Patients with type-II diabetes mellitus have high levels of circulating insulin, and their activated lymphocytes express decreased numbers of insulin receptors. The mechanism by which resting T cells are desensitized by insulin is unknown (126). Perhaps T cells have low numbers of receptors (sufficient to receive a signal, but too low to detect by standard binding assays), or perhaps monocytes in culture (which do bear receptors) transmit the information to the lymphocytes.

5-hydroxytryptamine (5-HT)

This agent, also referred to as serotonin, is a vasoactive amine. In mice, 5-HT is a more potent constrictor of endothelial cells of postcapillary venules than is histamine, whereas in humans the reverse is true. Receptors have been defined on the basis of the effects of competitive antagonists on their biologic activity. It has been suggested that 5-HT activates 5-HT_2 receptors via stimulating phospholipase C (127). Many of the biologic effects of 5-HT require 10^{-3} M concentrations and are not defined in terms of receptor specificity. 5-HT acts synergistically with histamine in inducing normal human PBL to generate a lymphokine (a chemotactic factor for monocytes) (128). It inhibits human lymphocyte PHA-induced proliferation, via decreasing IL-2 receptor expression and also inhibits IgG and IgM synthesis in vivo and in vitro (129). In contrast to these effects with high concentrations of 5-HT, 5-HT at concentrations as low as 10^{-8} M decreases interferon-γ-induced Ia expression on macrophages. This action is via specific, 5-HT_2 receptors (130). It is not known whether 5-HT_2 receptors are present on lymphocytes.

Cholinergic

Muscarinic cholinergic receptors have been defined by the binding of the labeled antagonist, 3-quinuclidinyl benzilate, to rat, mouse, and human lymphocytes, with a K_D of 3×10^{-9} M (131). These results were confirmed by measurement of K_D values for this and several other muscarinic cholinergic antagonists, both for competing for binding and for blocking agonist-induced augmentation of the cytotoxicity of rat CTL.

Muscarinic agonists may activate cells via inhibition of adenylate cyclase, activation of guanylate cyclase, or another pathway. None of these possibilities has been proven definitively.

Cholinergic agonists (10^{-11} M carbamyl choline or acetyl choline) augment the cytotoxicity of rat CTL, and these effects, as mentioned above, are blocked by muscarinic but not nicotinic cholinergic antagonists (132). These agonists augment only when two conditions are met: they are used with CTL obtained early in the immune response, i.e. day 6 following in vivo grafting, at a time when the receptors were increased on T cells; and they are preincubated with CTL for 5 min before target cells are added. (They do not augment day 9–11 CTL, when the number of receptors on

T cells has decreased below the basal levels and they do not augment if they are not preincubated.) The augmenting effect of agonists reportedly has been reproduced with day-7 immune-mouse-splenic CTL according to one report (120), but extensive experiments by another investigator failed to demonstrate any effect of carbamyl choline on mouse splenic CTL (133). The reason for these differences in results is unclear. Observations that the receptor appears to be relatively "fragile" [it is destroyed by NH_4Cl (131)] may offer a partial explanation.

In contrast to the results cited above, cholinergic agonists are reported to augment human and mouse NK activity, but only at very high concentrations ($10^{-3}–10^{-5}$ M carbamyl choline), and 10^{-6} M atropine blocks the enhancement of human NK activity (33). These agents are reported to increase human ADCC activity modestly. Acetylcholine also induces lymphokine (interferon-γ) production from SEA-stimulated Lyt 2^+ spleen cells (102).

Muscarinic cholinergic receptors are present on T lymphocytes but not B lymphocytes. Their density increases, then decreases after in vitro activation of T lymphocytes. Changes in receptor number seem to correlate well with changes in the biologic activity of the agonists (131).

Adenosine

Considerable interest has focused on the physiologic role of adenosine in modulating T cells because of the clinical observation that adenosine deaminase (ADA) deficiency is the basic defect of a subset of patients with severe combined immunodeficiency. One of several explanations of the effects of ADA deficiency is that excess endogenous adenosine is markedly immunosuppressive.

Adenosine receptor binding studies on neutrophils have been reported, with a K_D of 2×10^{-7} M for 5'-N-ethyl-carboxamide adenosine (NECA) and 2-chloroadenosine, although the amount of specific binding observed was very low (134).

Adenosine receptors on lymphocytes have been defined not by binding studies but by the functional effects of agonists and antagonists. The agonists include adenosine itself, N^6-phenylisopropyl adenosine (PIA) stereoisomers, (−) R-PIA and (+) S-PIA, and NECA; the antagonists include methylxanthines like theophylline and isobutylmethylxanthine (IBMX). The three defined sites of cell activation by adenosine are R_a and R_i receptors (adenosine receptors) ("R" refers to specificity for the ribose moiety of adenosine) which activate (a) or inhibit (i) adenylate cyclase and intracellular P sites (intracellular adenosine binding site) ("P" refers to the specificity for the purine moiety of adenosine). All three of these sites appear to involve either inhibition or activation of adenylate cyclase (23,

24). Human PBMC bear all three sites. R_i receptors are stimulated by low $(10^{-7}$ M) concentrations of adenosine and $(-)$ R-PIA $> (+)$ S-PIA $>$ NECA and mediate decreases in cAMP levels; R_a receptors are stimulated by intermediate $(10^{-5}$ M) concentrations of adenosine and NECA $> (-)$ R-PIA $> (+)$ S-PIA and mediate increases in cAMP levels; intracellular P sites are stimulated by high $(10^{-3}$ M) concentrations of adenosine and mediate decreases in cAMP levels (135; G. Marone, R. Petracca, S. Vigorita, M. Plaut, in preparation). The functional correlates of all three of these actions of adenosine are not yet known. However, R_a receptors appear to mediate inhibition of lymphocyte function. Adenosine inhibits mouse CTL activity, and this effect is blocked by methylxanthines (136; M. Plaut, unpublished). Adenosine inhibits E-rosetting, and theophylline blocks this effect. Unexpectedly, theophylline alone has the same effect as adenosine on E-rosetting, suggesting either that the adenosine receptor involved is atypical or that occupancy of the receptor (rather than activation by an agonist) is sufficient to inhibit rosetting (137). These possibilities have not been explored further. The subset of T cells whose rosetting is inhibited by theophylline (i.e. the "sensitive" subset) represents a population enriched in suppressor cell activity, whereas the resistant subset is enriched in helper cells. Preincubation of T cells for 30 min at 37° with 10^{-5} M adenosine activates human suppressor cells. This concentration of adenosine also increases cAMP levels. Precursor T cells in human peripheral blood ($Fc_\gamma R^-$, T4$^+$, T8$^-$) not only become radioresistant suppressor cells that suppress PHA-induced and PWM-induced proliferation as well as pokeweed mitogen–induced immunoglobulin production of human PBMC; these cells also develop the phenotype of mature suppressor cells ($Fc_\gamma R^+$, T8$^+$) (138, 139). However, the T5 marker is not increased (139), a peculiar observation since T5 and T8 are supposed to be markers of the same cell surface glycoprotein. Also, despite the apparent preferential effect of adenosine on T4$^+$ cells, adenosine induces equal increases in cAMP levels (via R_a receptors) in T4$^+$ and T8$^+$ cells (G. Marone, R. Petracca, S. Vigorita, M. Plaut, in preparation).

Some of the effects of adenosine on lymphocytes (and human mast cells) do not appear to be mediated by the receptors so far identified (140, 141). The precise mechanism is not clear.

The role of endogenous adenosine in influencing human lymphocyte function is unknown, although its possible importance in ADA deficiency has been mentioned briefly above. Patients with systemic lupus appear to have a defect in adenosine-induced suppression (142, 143). Some investigators have postulated that some of the therapeutic actions of theophylline occur via inhibition of endogenous adenosine produced at sites of inflammation. Endogenous adenosine is produced by human neutrophils

(143a) and stimulated rat peritoneal mast cells (144), so it might be made by other inflammatory cells.

Beta-Adrenergic

The β-adrenergic receptor has been extremely well characterized because of the availability of many high affinity labeled antagonists. Both by ligand binding studies and by pharmacologic measurement of inhibition of agonist-induced cAMP increases, β-adrenergic receptors have been characterized on human and mouse lymphocytes, with a K_D of the antagonist, ^{125}I-cyanopindolol, of 2×10^{-11} M, and a K_D of epinephrine of 5×10^{-5} M (7).

Activation of β-adrenergic receptors stimulates adenylate cyclase, which results in increased levels of intracellular cAMP.

β-adrenergic agonists inhibit mouse CTL (29) and human-NK (33) mediated killing. These inhibitory effects are receptor specific but weaker than those of other agonists that increase cAMP levels, probably because β-adrenergic agonists induce only transient increases in cAMP. β-adrenergic agonists also inhibit T-cell-mediated proliferation. They have variable effects on antibody responses, in that they enhance when present early and inhibit when present late (48).

Heterogeneity of β-adrenergic responsiveness has been demonstrated in a number of experimental and clinical models. In several models, there is a spontaneous loss of β-adrenergic responsiveness in culture. Among T lymphocyte subsets, mouse thymocytes have a much larger cAMP response to isoproterenol than splenic T cells (145). Surprisingly, mouse thymocytes have a much lower density of receptors than spleen cells, so the high responsiveness of thymocytes is probably due to efficient receptor-adenylate cyclase coupling (146). β-adrenergic receptor numbers are known to be stimulated by exogenous factors, such as those found in extracts of embryos, but the nature of these factors is unknown (147). In human T cell subpopulations, Leu 3^+ (helper) have the fewest receptors; Leu 2^+ Leu 9.3^+ (CTL) next and Leu 2^+ Leu 9.3^- (suppressor) the most. The cAMP response of these subpopulations to agonists parallels these receptor numbers (148). However, the functional significance of a fourfold greater receptor number on suppressors than helpers is unknown.

The most interesting clinical model of abnormalities in receptors is of β-adrenergic (mainly β_2-adrenergic) hyporesponsiveness in asthmatics, an abnormality found in several organ systems including lymphocytes (7). Hyporesponsiveness is apparently reproduced in lymphocytes of mild asthmatics, 24 hr after antigen challenge (149). The mechanism of hyporesponsiveness appears to reflect reduced numbers of receptors. Lymphocytes from asthmatics are reported to have decreased β-adrenergic receptor density, but the study design was not optimal. Hyporesponsive lymphocytes obtained after antigen challenge appear to have both decreased β-adrenergic receptor numbers and decreased coupling. In one model of experimental asthma in animals, β-adrenergic receptors in mixed lung cells were decreased, and α_1 adrenergic receptors were increased (150). The significance of β-adrenergic hyporesponsiveness in asthmatics is controversial. Some data suggest that cells from asthmatics have an intrinsic hyporesponsiveness, whereas other results suggest that their cells are not hyporesponsive except after therapy with β-adrenergic agonists, which induce desensitization (reviewed in 7). For example, neutrophils of asthmatic patients are not hyporesponsive: the β-adrenergic responsiveness of, and the numbers of β-adrenergic receptors on, neutrophils are the same in asthmatics and normals. However, in asthmatic patients who receive treatment with β-adrenergic agonists, the β-adrenergic receptors on neutrophils are down-regulated (151). Antibodies to β_2-adrenergic receptors have been found in asthmatics (152). These antibodies might mediate an intrinsic β-adrenergic hyporesponsiveness. However, the incidence of antibody to these receptors is similar in asthmatics and normals, whereas the incidence of β-adrenergic hyporesponsiveness is much higher in asthmatics than normals, so the significance of the antibody is unknown. The results of antigen challenge, if confirmed, are particularly interesting because they do not involve therapy with β-adrenergic agonists (although they might reflect release of endogenous catecholamines). The results suggest that the immunologically mediated inflammatory response regulates lymphocyte receptor display.

Other disease states are associated with altered β-adrenergic receptor status in several cell types (reviewed in 7). In hyperthyroidism, β-adrenergic receptor numbers and affinity increase in most tissues, but lymphocyte β-adrenergic receptors are reported both to increase and to remain constant. In hypothyroidism, β-adrenergic receptors decrease and receptor-cyclase coupling is normal or decreased, depending on the tissues. In endogenous depression and psychomotor agitation, isoproterenol-stimulated lymphocyte cAMP levels are reduced, although the numbers of receptors have not been analyzed (153). In patients treated with glucocorticoids, β-adrenergic responsiveness increases, apparently due to increases both in receptor numbers and in receptor-cyclase coupling (153a).

Leukotriene B_4 (LTB_4)

The effects of LTB_4 have been extensively reviewed (154). LTB_4 is a major product of the lipoxygenase pathway of arachidonic acid metabolism. LTB_4 receptors have been defined on the basis of both the relative potency of a series of potential agonists (155) and the binding of labeled agonists

(156, 157). There are no receptor antagonists available to confirm the specificity, but the specificity of biologic effects has been evaluated with other lipoxygenase products. 15-HETE reproduces some of the effects of LTB_4. The leukotrienes, LTC_4, LTD_4, and LTE_4 have biologic properties distinct from those of LTB_4.

The mechanism of action of LTB_4 is still under investigation. It has been shown to induce increased levels of cGMP in peanut agglutinin $(PNA)^+$ thymocytes (158), but this effect has been thought to be less important than its apparent capacity to stimulate phospholipase C (159), resulting in PI turnover, the production of diacylglycerol and IP_3 and activation of protein kinase C.

LTB_4 increases the cytotoxic activity of human NK and natural cytotoxic (NC) cells (see review 154). It enhances random migration of mouse $Lyt1^+$ and 2^+ T cells (160). It inhibits proliferation, variably, to mitogens and alloantigens (154, 161). LTD_4 and LTE_4 also inhibit mouse splenic T-cell proliferation but do not modulate human T-cell responses. LTB_4 inhibits immunoglobulin synthesis and variably inhibits lymphokine (leukocyte inhibitory factor) production stimulated by PHA (154). Many of the apparently variable effects of LTB_4 in different systems can be explained by the following observations: LTB_4 acts on both $T4^+$ and $T8^+$ cells, but the functional consequences are opposite: LTB_4 inhibits proliferation of $T4^+$ cells while it enhances proliferation of $T8^+$ cells (157, 162). LTB_4 at concentrations as low as 10^{-12} M acts on both $T4^+$ and $T8^+$ populations to induce differentiation into suppressor cells, which are potent inhibitors of immunoglobulin production stimulated by pokeweed mitogen. This differentiation step is accompanied by the expression of suppressor phenotype cell surface antigens (i.e. $T8^-$ cells express T8 antigens) (163, 164).

The induction of suppressor cells by LTB_4 requires monocytes and is blocked by indomethacin (164). Indeed, in the presence of indomethacin, LTB_4 stimulates T4 cells to produce IL-2 (164). Indomethacin presumably acts via inhibition of the production of PGE_2 or other cyclo-oxygenase products, but other possible roles of indomethacin cannot be ruled out. LTB_4 stimulates monocytes to produce both IL-1 and prostaglandins, and it enhances LPS-stimulated IL-1 production (165). In one murine model (already described for the effects of AVP and oxytocin), SEA stimulates spleen cells to produce interferon-γ but fails to stimulate Lyt 2^+ enriched spleen cells unless IL-2 is added. LTB_4 (and LTC_4) can replace IL-2 in stimulating the production of interferon-γ, but they cannot replace the requirement for IL-2 for stimulating the proliferation and growth of these cells. In addition, LTB_4 enhances lymphokine (interferon-γ) production stimulated by IL-2 (102, 166).

LTB_4 receptors are present on subsets of both helper and suppressor T cells (157). Vitamin D_3 induces receptors on the hematopoietic cell line HL-60 (167). It seems likely that receptors on lymphocytes are also inducible. The relationship between receptor numbers and biologic effects is not known.

LTB$_4$ is produced from arachidonic acid via the lipoxygenase pathway. Although several early reports claim that lymphocytes do not make LTB$_4$ (168), other reports indicate that lymphocytes do synthesize LTB$_4$ (81). Resolution of these conflicting data will help explain the remarkable observation of Goodwin's group, that secreted LTB$_4$ (or perhaps intracellular LTB$_4$) may provide an essential signal in lymphocyte activation. Goodwin's data suggest that in glucocorticoid-treated lymphocytes, LTB$_4$ entirely reverses the glucocorticoid-induced inhibition of lymphocyte activation (77, 81). This pathway is apparently separate from the monocyte dependent pathway that leads to activation of suppressor cells by LTB$_4$.

Prostaglandin E_2 (PGE$_2$)

PGE$_2$ is a major cyclo-oxygenase product of arachidonic acid metabolism of activated cells, notably macrophages and monocytes. The effects of PGE$_2$ on lymphocytes have been extensively reviewed (8, 169). PGE$_2$ is the one hormone or autacoid most clearly implicated as an endogenous regulator of lymphocyte function, based on in vitro studies indicating that NSAIA, which are presumed to act mainly via inhibiting prostaglandin synthetase, markedly alter lymphocyte responses. Because of the complexities of data with NSAIA, we have dealt with these data separately at the end of this section.

Several studies have utilized ^{3}H-PGE$_2$ to define PGE receptors on lymphocytes (170, 172). Unfortunately, no pharmacologic studies are available to confirm the specificity of the binding assay because there are no receptor antagonists available. These studies demonstrate a K_D of about 2×10^{-9} M, but only a few hundred receptors per cell. The number of counts bound is very low and barely above the background counts per minute. Other evidence of receptor specificity is based on the relative potency of effects on lymphocyte function of PGEs, PGAs, and PGFs, but this type of specificity evaluation has been done to a limited extent with the binding studies (170). Note that PGE$_2$, the major endogenous product, and its isomer, PGE$_1$, have virtually identical biologic effects, so that we often mention PGE, PGE$_1$, and PGE$_2$ interchangeably.

PGE acts by stimulating adenylate cyclase and increasing intracellular cAMP levels (see review 8).

PGE$_2$ at concentrations of approximately 10^{-8} M–10^{-7} M inhibits a variety of lymphocyte functions including the effector function of CTL and NK cells, E-rosetting, and proliferation to the mitogens PHA and con

A, but it does not inhibit proliferation to pokeweed mitogen (see review 8). PGE_2 also inhibits antigen and mitogen-induced lymphokine (IL-2) production (38). It inhibits IL-2-dependent proliferation, presumably by inhibiting a biochemical event distal to the IL-2 receptor (173); the effects of PGE_2 on IL-2 receptor expression are not known. The effects of prostaglandin on immunoglobulin production are complex. In some mouse systems, PGE_2 inhibits the antibody-forming cell response. In other systems, notably those described by Goodwin, in humans, NSAIA such as indomethacin decrease the production of rheumatoid factor, total IgG, and total IgM; PGE_2 reverses this effect (i.e. PGE_2 is immunostimulatory) (173). Perhaps the paradoxical enhancing effect of PGE_2 reflects a minimal effect on inhibition of T-cell proliferation to mitogens like pokeweed mitogen combined with an enhancing effect on B-cell function.

Preincubation with PGE_2 enhances the effector activity of human NK cells. Preincubation of lymphoid cells with PGE_2 stimulates suppressor cells in several models. (The capacity of PGE_2 to inhibit proliferation may be either direct or by means of activation of suppressor cells.) The PGE_2-induced suppressor cell is a monocyte in man (174, 175) and is apparently both a monocyte and a T cell in the mouse (176). There are two major suppressor molecules produced by monocytes, PGE_2 and H_2O_2 (177). The PGE_2 system is the one more commonly operative in in vitro systems. This may be an artifact of in vitro systems, since such systems fail to recirculate or metabolize PGE_2 as efficiently as in vivo systems, but this remains to be determined.

Indomethacin and other inhibitors of the cyclo-oxygenase pathway of arachidonic acid metabolism block the activation of suppressors in several systems. Much work has been done on the system originally described by Webb, of prostaglandin-induced T suppressor (PITS), a suppressor cell which is a potent inhibitor of primary but not secondary PFC responses in vitro (178). Indomethacin enhances antibody responses in vivo and in vitro. One particularly interesting in vitro system has recently been described by Kato & Askenase (179): A T suppressor factor (TsF)–producing cell produces an inactive TsF when the cells have been incubated with indomethacin. Such inactivated TsF can be reconstituted to an active TsF by incubating with PGE_2 or PGE_1, suggesting that the prostaglandin molecule is itself incorporated into the TsF molecule. However, PGE_2 does not activate suppressors in all systems. Thus, in guinea pigs, pre-incubation with histamine produces "HSF" while preincubation with PGE_2 does not (180).

Responsiveness to prostaglandin varies between different cell populations and in certain disease states. Whether this altered responsiveness is due to altered receptor numbers or to events distal to the receptor remains to be determined, because, as noted above, the assay for PGE_2 receptors is of uncertain reliability. Prostaglandin responsiveness (in the macrophage-like cell line, U937) is decreased by T-cell factors and by vitamin D_3 (181). Cultured T cells have decreased responsiveness to prostaglandin, but three apparently distinct mechanisms account for this. First, there is an apparent decrease in receptors in human PBMC (170, 182). Second, there is a biochemical event distal to the receptor and even distal to cAMP elevation in cultured mouse splenic CTL, in that such cells show a normal increase of cAMP in response to PGE_2, but inhibition of their cytotoxic activity by PGE_2 is markedly decreased (183). Third, in contrast to the results of the first alternative, there is an altered distribution of T-cell populations in cultured human PBMC. PGE inhibits the proliferation of most cells but enhances the responsiveness of low density T cells. The relative proportion of low density cells is increased in cultured human peripheral blood mononuclear cells (184). Furthermore, prostaglandin acts selectively on subsets of T cells; it inhibits alloantigen-induced proliferation of allogeneic Lyt 1^+ clones but not Lyt 2^+ clones (185). It inhibits the mitogen-induced proliferation of human T cells, but under certain circumstances (preincubation) it enhances the proliferation of $T8^+$ T cells (186). Finally, prostaglandin enhances the proliferative response of some populations of immature thymocytes (187).

Lymphocytes from many patient populations have altered prostaglandin responsiveness. PBMC from HLA-B12 patients have, relative to the remainder of the population, decreased responsiveness to PGE-mediated inhibition of T-cell proliferation; PBMC from these individuals also have a comparable decreased responsiveness to histamine (16). PBMC of newborns (188) and patients with juvenile periodontitis (189) are also hyporesponsive to PGE_2. In contrast, PBMC obtained from patients undergoing a stress (e.g. post-surgery) (190), or with rheumatoid arthritis (191), or of age > 70 years (192, 193), are hyperresponsive to PGE_2 (see review 8). PGE_2 is reported to induce suppressor cells in nonatopic patients but not atopic patients (171). It is surprising that PGE_2 induces suppressor cells capable of inhibiting DNA synthesis equally in nonatopic and atopic patients but induces suppressor cells capable of inhibiting protein synthesis only in nonatopic patients. With helper T cell–enriched populations, PGE_2 suppresses protein synthesis more in nonatopics than in atopics. With suppressor T-cell-enriched populations, PGE_2 enhances protein synthesis in nonatopic patients. These results are consistent with observations that PGE has distinct effects on $T4^+$ and the $T8^+$ cells, but they also indicate that some atopic patients have an abnormality, although a very subtle one, that can be detected only by a very specific assay, of inhibition of protein synthesis. Lymphocyte receptors for PGE_2 are reported to be decreased in atopic patients (171). Thus, a decrease in receptors is apparently associated with this subtle abnormality, but with a normal capacity to induce suppression of DNA synthesis.

Prostaglandins are synthesized endogenously by many cells that are activated so that arachidonic acid may be released and metabolized by the cyclo-oxygenase pathway. Macrophages are the major source of PGE_2 in immune reactions, and macrophage populations differ in their capacity to produce PGE_2 upon stimulation by activating agents (194). Macrophage populations that migrate into the peritoneal cavity in response to certain irritants like thioglycolate have a markedly reduced capacity to produce PGE_2, compared to resident macrophages (195). In one disease state, Hodgkin's disease, PBMC are unresponsive to several T-cell mitogens (196). It appears that the peripheral monocytes of these patients produce enhanced levels of PGE_2. Indomethacin is capable of restoring the responsiveness of lymphocytes to mitogens in vitro, suggesting that PBMC from these patients may be unresponsive because they are producing increased amounts of PGE.

Many in vitro systems have demonstrated that NSAIA alter lymphocyte function and have suggested that PGE_2 is an important endogenous mediator. However, as pointed out by Goodwin & Ceuppens (8), many patients take NSAIA but have few if any alterations in their lymphocyte function, so the word "important" may be incorrect. Several lines of evidence suggest that PGE is an endogenous mediator. PGE_2 is implicated especially as a feedback inhibitor of IL-1 production (197) in that activated macrophages produce not only IL-1 but also a negative signal, i.e. PGE_2. Several stimuli are known to be potent inhibitors of PGE production; PGE inhibits a variety of immune functions and stimulates suppressor activity; NSAIA like indomethacin block PGE production and have effects on the immune response opposite to that of PGE (8).

PGE_2 has been implicated not only as an endogenous mediator in certain experimental situations but also as an essential intermediate in transmitting inhibitory signals to lymphocytes. This possibility has been raised by the intriguing observation (198) that the inhibition of proliferation of human PBMC by PGE_2, histamine, isoproterenol, and hydrocortisone all were quantitatively correlated in different individuals, and inhibition by all four of these agonists was reversed partially by indomethacin. Interferon-γ mediated inhibition of proliferation was not reversed by indomethacin. However, the PGE_2 production by PBMC was not enhanced by any of these agonists. Indeed, PGE_2 production was slightly inhibited by histamine and isoproterenol and markedly inhibited by hydrocortisone and interferon-γ (198). Indeed, glucocorticoids inhibit all cyclo-oxygenase pathways (59). Another intriguing observation is that indomethacin reverses the inhibition of histamine release induced by several cAMP agonists including PGE_2 (199). In these models indomethacin must be acting by mechanisms distinct from inhibition of the synthesis of PGE_2 and other cyclo-oxygenase products. Indomethacin is known to inhibit cAMP-dependent protein kinase (200) and phosphodiesterase (201), and it may "shunt" arachidonic acid metabolism from the cyclooxygenase to the lipoxygenase pathway (201a), but the precise mechanism of action of indomethacin is unknown.

Several in vivo effects of PGE_2 and of indomethacin have been described, which are for the most part consistent with their in vitro effects. Several studies with PGE derivatives with a long half-life (PGE_1 and PGE_2 themselves have very short half-lives in vivo) on allograft rejection in mice have yielded inconsistent results (reviewed in 202), but in a rat allograft model these agents enhance graft survival and significantly reduce the activity of CTL (202, 203). The reason for the discrepant results among these studies is unclear.

Despite the short half-life of PGEs, PGE_1 has been shown to inhibit autoimmune disease in NZB/W mice. The mechanism by which PGE_1 works is not clear. NZB/W mice have abnormal Tdt^+ thymocytes. PGE_1 may act on these cells. PGE_1 is known to stimulate steroidogenesis, but this is not the mechanism of action of PGE_1, since PGE_1 is effective in inhibiting autoimmune disease even when the mice are adrenalectomized (204).

In vivo, indomethacin stimulates a significant increase in the primary PFC response to T-dependent antigens (in vitro and in vivo) (178). It shortens allograft survival (reversed by PGE_2) (205). It also potentiates the antitumor effects of BCG (206). However, the effects of these agents on tumors are complex since indomethacin may slow tumor growth in certain models by blocking PGE-stimulated bone resorption (possibly important in determining the extent of bone metastases) (207), whereas in others it may potentiate the growth of tumor cells (whose growth is inhibited by PGE_2) (208) (see review 8).

Indomethacin reverses the anergy in some patients with common variable hypogammaglobulinemia (209) and with rheumatoid arthritis (see 8). These particular patients presumably have a hyperactive prostaglandin system, either due to increased production, increased lymphocyte responsiveness, or both. However, as discussed above, the precise mechanism of action of indomethacin is unknown.

Histamine

Histamine is the decarboxylated form of histidine and is widely distributed among mammalian species. It is released from mast cells and basophils during immediate hypersensitivity reactions and induces increases in vascular permeability by contracting the endothelial cells in post-capillary venules. It is only recently that histamine has been shown to act via two distinct receptors, designated H1 and H2. Histamine also interacts with cells other than endothelial cells, including lymphocytes, and most of its effects on the latter are via H2 receptors. The effects of histamine on lymphocytes have been extensively reviewed (9, 210).

Several reports have attempted to define H1 and H2 receptors on lymphocytes by direct binding assays with ^{3}H-histamine (211), and putative receptors have been described. However, the "binding affinities" of these putative receptors is very low ($K_D = 10^{-6}$ M)—probably too low for reliable binding assays—and the binding assays include long wash times which would be expected to dissociate ligand from these low affinity receptors. ^{3}H-histamine binding sites do not appear to be true receptors but are probably sites of uptake, whose specificity matches neither H1 nor H2 receptors (212, 213). Recent studies with the labeled antagonist tiotidine have identified H2 receptors on certain tissue types (213, 214) but so far have not described lymphocyte H2 receptors. H1 receptors have been putatively identified on human T and B lymphocytes and monocytes, but the specificity has not been defined. The numbers of receptors on some cells is unexpectedly high (35,000 on T8$^+$ cells), and the receptor affinity unexpectedly changes in culture (215).

The signals transduced by stimulating H1 receptors are unknown, although stimulation of guanylate cyclase has been noted in some models, and stimulation of adenylate cyclase in another (see below). Stimulation of H2 receptors activates adenylate cyclase and results in increases in cAMP levels (see review 9).

The biologic activities of histamine on lymphocytes include two that appear to be via H1 receptors: stimulation of natural suppressor activity in mouse spleen (216), and stimulation of the production of lymphocyte migration inhibitory factor (Ly MIF) by human PBMC (217). One action of histamine, the induction of contrasuppressor cells, is via a putative H1 receptor (218). There are several other actions of histamine that are of uncertain specificity (see review 9).

Histamine increases cAMP levels, with the greatest increase after 5–10 minutes in mouse spleen cells and human peripheral blood leukocytes; 10^{-5} M histamine stimulates the maximal response, via H2 receptors (183). Unexpectedly, the maximal cAMP increase occurs after 1 min in cloned natural suppressor cells. 10^{-3} M histamine is more potent than 10^{-5} M in stimulating cAMP increases in these cells. Several new derivatives of histamine increase cAMP levels in cloned natural suppressor cells, but these derivatives do not activate myocardial adenylate cyclase, an observation suggesting that the derivatives stimulate H1 receptors and that somehow H1 receptors (as well as H2 receptors) increase cAMP levels in these particular cells (217–219).

Several reports indicate that cloned mouse cell lines (natural suppressors, helper T cells, and cytotoxic T cells) and several T-cell lines bear receptors that mediate histamine-stimulated adenylate cyclase activation (216, 220).

Histamine inhibits the effector activity of mouse splenic CTL, via H2 receptors (221). Histamine has been reported to have multiple effects on NK cells, including inhibition of the effector activity of mouse and human NK cells (receptor specificity undefined) (32, 33), inhibition of NK activity minimally and without receptor specificity (222), no effect (223), and enhancement of NK-cell activity but only if monocytes are present (this effect is via H2 receptors and is not mediated through interferon) (224). The reasons for these discrepant effects on NK cells are not known, but the magnitude of the effects of histamine that are reported are small.

Histamine inhibits the proliferative response to mitogens, alloantigens, and soluble antigens in mouse, guinea pig, and human, presumably via H2 receptors (see review 9). The receptor specificity of this effect has been questioned, primarily because nordimaprit, an inactive congener of the H2 agonist dimaprit, is a potent inhibitor of proliferation. However, the concentrations of nordimaprit that inhibit are toxic and are not reversed by H2 antagonists. High concentrations (10^{-3} M) of dimaprit also are toxic. Histamine and low concentrations of dimaprit do inhibit lymphocyte functions, and these actions are reversed by H2 antagonists (225–228).

The histamine-mediated inhibition of proliferation is associated with inhibition of IL-2 secretion, and reportedly does not affect IL-2 receptor display or metabolic events distal to the IL-2 receptor (216, 229). If it is preincubated with cells before the addition of antigen, histamine increases IL-2 secretion (216). Histamine inhibits the terminal antibody secretory event of mature B-cells (35), but the receptor specificity of this effect is not known. Histamine induces suppressor cells and suppressor factors, which inhibit IgG secretion and PFC responses (discussed below) (210).

When histamine is incubated with human lymphocytes for 4 hr, two types of lymphotactic factors are secreted: a stimulatory factor of molecular weight 56,000 (230, 231), and two inhibitory factors with molecular weights 70,000–80,000 and 30,000–40,000 (217), which have been called lymphocyte chemoattractant factor (LCF) and Ly MIF, respectively. The LCF appears to be stimulated by histamine through H2 receptors on T8[+] cells, and the Ly MIF via stimulation of H1 receptors on T4[+] cells (217, 231). The cells producing these factors are absorbed on histamine-solid phase conjugates.

Several investigators have analyzed the effects of preincubation of histamine with guinea pig, mouse, and human lymphocytes prior to adding an activating stimulus. Four apparently contradictory effects are reported: suppressor cell inactivation; suppressor cell activation; no effect; and concomitant activation of suppressor and contrasuppressor cells. All of these effects are real, but the majority of studies have described activation of suppressor cells.

Suppressor cell inactivation has been defined in an experimental model in which preincubation with histamine and other cAMP agonists prevents the antigen-induced immunosuppression of PFC responses (51, 232). The receptor specificity has not been defined.

Suppressor cell activation has been well documented: The suppressor cells suppress the proliferative response to mitogens and antigens (8, 210). Histamine acts directly on a cell population in mouse spleen, a natural suppressor cell, to increase cAMP levels and to activate the cell. The activation of these suppressors is via H1 receptors while the increase in cAMP is apparently via both H1 and H2 receptors (216, 219). On the other hand, the more "typical" suppressor T cells require activation. Of these typical suppressors, those in mouse spleen which suppress proliferation, and those in human PBL which inhibit proliferation to PHA, appear to be H2-receptor specific. For example, starting with putative suppressor precursors in human peripheral blood (Leu 2^+ Leu 9.3^-), a 30-sec exposure to histamine followed by a 12-hr-incubation results in mature suppressors (2). In addition to these effects of histamine, impromidine, a specific H2 agonist, when preincubated with human PBMC for 30 min, induces suppressors and also induces an increase in Fc_γ receptors and $T8^+$ cells. Both of these effects are blocked by cimetidine (138, 139). The induction of suppressors in human PBMC is blocked by high concentrations of a variety of metabolic inhibitors including azide, 2-deoxyglucose, actinomycin D, puromycin, and cycloheximide and also by X-irradiation (but not mitomycin C) (233). These results suggest that histamine induces human suppressor T-cells (via an H2 receptor) accompanied by phenotypic changes in some cell populations.

Several investigators find only suppressor phenotype cells (i.e. Leu 2^+, or more precisely Leu 2^+ Leu 9.3^-) T cells become suppressor cells, but others claim that the H2 agonist, impromidine, also induces helper phenotype cells ($T4^+$ $T8^-$) cells to mature rapidly to a new phenotype, $T4^-$ $T8^+$ (although the possibility that these cells are $T4^+$ $T8^+$ has not been ruled out) (138, 139). This change in surface phenotype has been reported with suppressors induced not only with impromidine, but also with adenosine and LTB_4 (163). However, the T5 marker is not increased (138–139), a peculiar observation since T5 and T8 are supposed to be markers of the same cell surface glycoprotein.

Another, possibly similar, histamine-induced suppressor cell is reported to secrete a soluble factor which, after prolonged incubation, inhibits NK activity, and the inhibition is reversed by interferon and IL-2 (234). In one suppressor model in the guinea pig, histamine activates suppressors to produce a histamine-induced suppressor factor (HSF) (180, 210); this effect is presumably via an H2 receptor but, unexpectedly, other agonists that increase cAMP levels, and dibutyryl cAMP itself, do not stimulate the formation of HSF. These results suggest that this H2 receptor is atypical in not stimulating increases in cAMP, or that the cAMP signal provided by histamine is unique. HSF is a product not only of guinea pig cells but also of human cells (210). However, the quantitative relationship between induction of suppressor cells from human blood and induction of HSF has not been established.

While some investigators have obtained suppression with histamine, others have used the H2 agonists dimaprit or impromidine to obtain these effects (138, 139, 218), perhaps because histamine induces suppressors and contrasuppressors, via H2 receptors and H1 receptors, respectively. The evidence to support this concept is based on a system in which mouse spleen cells from TNBS-treated mice regulate the primary anti-TNP-self CTL response. Histamine has no consistent effect on this system. However, a 30–60 min incubation with 10^{-3} M dimaprit induces suppressors, and 10^{-4} M 2-pyridylethylamine (an H1 agonist) reverses the effect of dimaprit, presumably by inducing contrasuppressors (218). The suppressor and contrasuppressor data are convincing, but the concentrations of dimaprit and 2-pyridylethylamine are very high. In particular, 10^{-3} M dimaprit induces toxic effects, unrelated to H2 receptor activation (225–228). The relationship of these findings to the action of histamine itself is not convincing.

Human suppressor cells work directly or via the soluble HSF (described above) of molecular weight 25,000–40,000 (235), although suppressor effects have not always been correlated with HSF. HSF is effective at inhibiting proliferation only in high concentrations but is effective at dilutions of 1/1000 to inhibit MLC-induced plaque-forming cell responses in man (236). The reason for this high potency of HSF in inhibiting PFC responses has not been found.

The effects of histamine on human T cells (i.e. apparent preferential stimulation of suppressors) appear to correspond to cAMP responses of T-cell subsets (237). Suppressor T cells (i.e. Leu 2^+, Leu 9.3^-) are more responsive to histamine than CTL (Leu 2^+, Leu 9.3^+) and helper T cells (Leu 3^+, Leu 2^-), and the cAMP response is blocked by H2 antagonists. After lymphocytes are stimulated with PHA for 48 hr, the cAMP response to histamine is markedly increased in CTL and helper T cells but is approximately the same as the original responsiveness in suppressor T cells. The functional significance of this observation is unknown. Also, subsets of Leu 3^+ cells, Leu 3^+ Leu 8^- (helper) and Leu 3^+ Leu 8^+ (containing inducers of suppressors) respond to histamine, although the Leu 8^+ subset responds somewhat better (238).

The responsiveness to histamine varies among mouse T-cell populations. Immature thymocytes do not respond to histamine with any increase in cAMP, whereas cortisone-resistant thymocytes respond with a twofold increase and splenic T cells with a fourfold increase (145). Functional T-cell populations differ in histamine responsiveness. The activity of splenic CTL early in the immune response (days 8–10) is not inhibited by histamine, but the activity of those obtained late in immunization (days 14) is inhibited (221). Furthermore, the capacity of guinea pig lymphocytes to make HSF peaks two weeks after immunization with antigen and the level achieved is considerably higher than that by nonimmune guinea pig lymphocytes (180). CTL in the peritoneum and CTL from in vitro cultures are not inhibited by histamine. When CTL are transferred to syngeneic nonimmune mice, they assume a level of responsiveness that corresponds to the environment to which they home. Thus, after transfer of 10-day immune cells, the activity of spleen-homing CTL is inhibited 25% by histamine, whereas that of peritoneal-homing CTL is inhibited 10%; this suggests that H2 receptors are not an intrinsic property of a lymphocyte but are regulated by factors in the local environment (228; M. Plaut, L. Nordin, A. R. Jacques, M. C. Liu, submitted).

Histamine conjugates of albumin bound to solid phase gels such as Sepharose are reported to bind subpopulations of cells. Several experimental results suggest that suppressors bind preferentially. Such gels deplete HSF-secreting cells, and cells adherent to the gels (presumably stimulated by the histamine conjugate) make HSF, suggesting that the binding was histamine specific (180, 210). Other reports suggest that all cells bind histamine coupled to gels via a non-H2 receptor mechanism. The histamine conjugates themselves appear to have weak H1 activity but no H2 activity (239–240). Thus, the utility of these conjugates is uncertain.

Responsiveness to histamine varies among individuals, particularly in association with disease states. H2 receptors are reported to be decreased in patients with histiocytosis X (241), but this result is based on a receptor binding assay of uncertain reliability. Histamine responsiveness of PBMC is decreased among HLA-B12 individuals (16). Recent studies have emphasized abnormalities in allergic diseases and asthma. In experimental asthma of sheep, putative H2 receptors with airways are decreased in sheep airways (242). In atopic humans, histamine-mediated induction of suppressor cells is markedly deficient, presumably due to a decrease in H2 receptors on suppressor cells (243), although this receptor assay is of uncertain reliability. Antigen challenge of allergic asthmatics results in a decreased histamine responsiveness of the lymphocytes obtained 24 hr later (244). Although these results seem consistent, other reports indicate that lymphocytes of atopics are equally as or more responsive to histamine or H2 receptor agonists than are nonatopics, usually when histamine is added simultaneously with mitogen in a proliferation assay (171, 245, 249). The reasons for these discrepant results are not known.

If H2 receptor variation reflects a physiologically important role for histamine, then endogenous histamine may be an important lymphocyte modulator, and H2 blockers such as cimetidine should modulate the immune response. In one mouse model, cimetidine enhanced the survival of male skin grafts on females (250), apparently via inhibiting immune function. Cimetidine also has been shown to have no effect (e.g. on allograft survival) (251). Most of the effects of cimetidine are immunostimulatory. Cimetidine enhances delayed hypersensitivity in patients with ulcer disease (252), but not patients with allergic diseases (253) (see review 8). Cimetidine has been found to inhibit tumor growth in several models (254), but not all, to increase schistosome granuloma size (255), and to enhance modestly antibody formation (256, 257) (see review 210). The H2 specificity of these effects has been evaluated by the use of other H2 antagonists. In one report only cimetidine, and not ranitidine (another H2 antagonist), increased delayed hypersensitivity to sheep red cells (258), indicating that H2 receptors are not involved. However, in another study cimetidine and ranitidine both decreased the formation of suppressor T cells, and the potency of the two H2 receptor antagonists paralleled the relative potency of these antagonists on H2 receptors (259). The targets in these in vivo studies of the H2 antagonists (lymphocytes or other cells) are not known. If all the published data are accurate, then cimetidine has both H2-specific and nonspecific immunostimulatory effects.

Two interesting clinical studies have demonstrated immunostimulatory effects in patients with specific disease states. In four patients with chronic mucocutaneous candidiasis and cutaneous anergy to Candida, administration of cimetidine reversed the anergy, although with no clinical benefits (260). In another study (93), of patients with common variable hypogammaglobulinemia and excess suppressor cells, cimetidine treatment reduced the activity of suppressor T cells, reduced the number of T8$^+$ cells, and (in one patient) increased the immunoglobulins synthesized in vitro. In those patients with common variable hypogammaglobulinemia who did not have excess suppressor cells, cimetidine had no effect. The simplest interpretation is that these disease states have reduced immune responses

because of excess endogenous histamine production and/or excess responsiveness to histamine and that cimetidine reverses this. Alternatively, cimetidine might act nonspecifically to enhance immune response, by acting on
lymphocyte or nonlymphocyte target cells in vivo. It has not been possible
to demonstrate comparable in vitro effects of cimetidine on lymphocytes
of these patients (261), suggesting that the in vivo target of cimetidine may
not be the lymphocyte.

Other experiments have attempted to evaluate circumstances under
which endogenous histamine is produced or secreted. The sources of
endogenous histamine could be histamine released from granules in basophils and mast cells, or newly synthesized histamine, perhaps synthesized
by mast cell precursors in lymphoid organs. Two distinct factors, IL-3 and
histamine cell stimulatory factor (HCSF), are T cell–derived lymphokines
that stimulate histidine decarboxylase activity and histamine synthesis
(262–264). Lymphocytes from immune mice make much greater quantities
of histamine synthesizing factors than do lymphocytes from nonimmune
mice. Immunization also increases the levels of histamine synthesized in
response to factors, perhaps because of an increase in mast cell precursors.
Because of such results, the effect of histidine decarboxylase inhibitors on
the immune response has been evaluated. Several old studies show that
these inhibitors reduced rat immune responses (265–267), but the inhibitors
were not specific. Recent studies with α-fluoro-methyl-histidine (FMH), a
potent and specific histidine decarboxylase inhibitor (268), demonstrate
modest immunostimulatory effects in vivo (269). FMH increases the level
of CTL activity twofold, but only on about day 20 after immunization
(M. Plaut, unpublished). Similar effects have been observed in CTL
responses in mast cell–deficient, W/W^v mice (M. Plaut, unpublished).
These results suggest that histamine is an endogenous mediator of immune
responses but that its relative importance as an endogenous mediator is
demonstrable only at very specific stages of immune response, such as day
20 after in vivo alloimmunization. Since the effects of FMH are difficult
to reproduce in vitro, the mechanism of action of FMH is uncertain.

CONCLUDING REMARKS

Lymphocytes bear receptors for a large variety of hormones and autacoids.
These receptors are not randomly distributed but are present on distinct
subsets of cells and may have altered distribution in certain disease states.
The functional effects of hormones depend on the subsets of cells that are
activated, the state of cell activation at the time that hormones interact,
and the nature of the signal provided by hormone-receptor interaction.

This review is biased in favor of my view that some or all of lymphocyte receptor–hormone interactions may be physiologically or pathologically relevant. I have pointed out weaknesses in some data defining receptor-specific actions of hormones, and I have discussed a number of inconsistent results and confusing issues. I close by reviewing some important issues that still need to be explored.

One major area that needs to be explained concerns the activation of human suppressor cells. Thus, agonists that increase cAMP levels, when added for brief periods to stimulated cells, can inactivate suppressor-T-cell function; yet when preincubated with resting cells these agonists can activate suppressor-T-cell function. Moreover, PGE_2, adenosine, and histamine apparently activate suppressor cells via adenylate cyclase stimulation, but LTB_4, which is believed to act by another mechanism, has the same actions as these other agents. PGE_2 and LTB_4 are also parallel in their capacity to interact with helper and suppressor phenotype human T cells: LTB_4 and perhaps PGE_2 appear to inhibit the proliferative response of $T4^+$ cells while enhancing the proliferative response of $T8^+$ cells (157, 162, 186). The activation of suppressor cells can involve an alteration in surface phenotype ($T4^+$, $T8^-$ cells become $T8^+$), yet the capacity to stimulate $T4^+$ cells to become suppressors is inconsistent in different laboratories (2, 138, 139) despite similar experimental designs.

A second major question concerns apparent opposite effects of hormone in what seem like minor changes in experimental situations. Thus, LTB_4 is a potent activator of suppressor cells, yet in the presence of indomethacin it stimulates proliferation (81, 164).

A third, related question concerns unusual effects of hormones in special models. In at least three experimental situations, all characterized by deficiency of IL-2 (and perhaps of other T-cell factors), "inhibitory" hormones become lymphocyte activators. In glucocorticoid-inhibited T cells, LTB_4 stimulates proliferation (81). In SEA-stimulated Lyt 2^+ spleen cells, LTB_4 and LTC_4 (as well as acetylcholine, AVP, and oxytocin) stimulate lymphokine (interferon-γ) production (102, 166). In nu/nu spleen cells stimulated by antigen and IL-1, dibutyryl cAMP substitutes for T-cell factors in inducing PFC responses (52).

A fourth question concerns the mechanism by which NSAIA such as indomethacin, and cimetidine, are immunostimulatory. Indomethacin has effects ranging from immunostimulation to blockade of the capacity of LTB_4 to induce suppressors, to partial blockade of the capacity of histamine, isoproterenol, and hydrocortisone to inhibit T-cell proliferation. These observations suggest that indomethacin modulates one or more biochemical pathways involved in interaction of many hormones with lymphocytes. In some models, indomethacin acts specifically via inhibiting PGE_2 synthesis, as PGE_2 reverses the effect of indomethacin (173, 179,

205). In most other models, either PGE_2 has not been tested, or PGE_2 does not reverse the indomethacin effect (199; cf 198). Similarly, cimetidine appears to be immunostimulatory in certain models via blockade of the effects of endogenous histamine (259), but in other models it appears to act independently of histamine (258).

A fifth question concerns the relationship between changes in hormone responsiveness and changes in hormone receptor numbers (vs changes in coupling to adenylate cyclase or other enzyme, or in other events distal to the receptor), since all possible combinations have been described. The changes (with days after immunization) of cholinergic responsiveness of rat CTL activity are exactly parallel to changes in receptor number (131). The decrease in β-adrenergic responsiveness of PBL of patients after antigen challenge is due to decreases both in receptors and in coupling (149). The alterations of glucocorticoid responsiveness of lymphocytes appear to be unrelated to receptor numbers (57). The enhanced β-adrenergic responsiveness of thymocytes vs splenocytes is accompanied by the opposite changes in receptor numbers (146). (Receptors are decreased on thymocytes.) Thus, the mechanisms that modulate post-receptor biochemical events must be understood better since they may be critical to regulation of lymphocyte function.

ACKNOWLEDGMENTS

The author wishes to thank Mrs. Karen Nelson for expert secretarial assistance; Drs. Philip Askenase, Edward Goetzl, James Goodwin, Kenneth Melmon, Ross Rocklin, and Esther Sternberg for access to preprints of their manuscripts; and Drs. Anne Kagey-Sobotka and Robert P. Schleimer for critical reviews of this manuscript. This work was supported in part by NIH grant HL 26489. This is publication No. 664 of the O'Neill Research Laboratories, The Good Samaritan Hospital.

Literature Cited

1. Bernard, A., Bernstein, I., Boumsell, L., Dausset, J., Evans, R., Hansen, J., Haynes, B., Kersey, J., Knapp, W., McMichael, A., Milstein, C., Reinherz, E., Ritts, R. D., Schlossman, S. F. 1984. Differentiation human leukocyte antigens: A proposed nomenclature. *Immunol. Today* 5: 158–59

2. Sansoni, P., Silverman, E. D., Khan, M. M., Melmon, K. L., Engleman, E. G. 1985. Immunoregulatory T cells in man. Histamine-induced suppressor T cells are derived from a Leu 2 + (T8 +) subpopulation distinct from that which gives rise to cytotoxic T cells. *J. Clin. Invest.* 75: 650–56

3. Khan, M. M., Melmon, K. L. 1985. Are autacoids more than theoretic modulators of immunity? *Clin. Immunol. Rev.* 4: 1–30

4. Roth, J., LeRoith, D., Collier, E. S., Weaver, N. R., Watkinson, A., Cleland, C. F., Glick, S. M. 1985. Evolutionary origins of neuropeptides, hormones and receptors: Possible applications to immunology. *J. Immunol.* 135: 816–19S

5. Plaut, M. 1981. Increased responsive-

658 PLAUT

ness to cyclic AMP-active agents during the immune response *in vivo*. *Immunopharmacology* 3: 107–16
6. Helderman, J. H., Reynolds, T. C., Strom, T. B. 1978. The insulin receptor as a universal marker of activated lymphocytes. *Eur. J. Immunol.* 8: 589–95
7. Stiles, G. L., Caron, M. G., Lefkowitz, R. J. 1984. β-adrenergic receptors: Biochemical mechanisms of physiological regulation. *Physiol. Rev.* 64: 661–743
8. Goodwin, J. S., Ceuppens, J. 1983. Regulation of the immune response by prostaglandins. *J. Clin. Immunol.* 3: 295–315
9. Plaut, M., Lichtenstein, L. M. 1982. Histamine and immune responses. In *Pharmacology of Histamine Receptors*, eds. C. R. Ganellin, M. E. Parsons, pp. 392–435. London: Wright
10. Earp, H. S., Steiner, A. L. 1978. Compartmentalization of cyclic nucleotide-mediated hormone action. *Ann. Rev. Pharmacol. Toxicol.* 18: 431–59
11. Byus, C. V., Klimpel, G. R., Lucas, D. O., Russell, D. H. 1977. Type I and type II cyclic AMP-dependent protein kinase as opposite effectors of lymphocyte mitogenesis. *Nature* 268: 63–64
12. Wagner, H., Hardt, C., Heeg, J., Pfizenmaier, K., Solbach, W., Bartlett, R., Stockinger, H., Röllinghoff, M. 1980. T-T cell interactions during cytotoxic T lymphocytes (CTL) responses: T cell derived helper factor (interleukin 2) as a probe to analyze CTL responsiveness and thymic maturation of CTL progenitors. *Immunol. Rev.* 51: 215–55
13. Farrar, W. L., Cleveland, J. L., Beckner, S. K., Bonvini, E., Evans, S. W. 1986. Biochemical and molecular events associated with interleukin 2 regulation of lymphocyte proliferation. *Immunol. Rev.* 92: 49–65
14. Lafuse, W., Meruelo, D., Edidin, M. 1979. The genetic control of liver cAMP levels in mice. *Immunogenetics* 9: 57–65
15. Lafuse, W., Edidin, M. 1980. Influence of the mouse major histocompatibility complex, H-2, on liver adenylate cyclase activity and on glucagon binding to liver cell membranes. *Biochemistry* 19: 49–54
16. Staszak, C. S., Goodwin, J. S., Troup, G. M., Pathak, D. R., Williams, R. C. Jr. 1980. Decreased sensitivity to prostaglandin and histamine in lymphocytes from normal HLA-B12 individuals: A possible role in autoimmunity. *J. Immunol.* 125: 181–88

17. Puppo, F., Corsini, G., Mangini, P., Bottaro, L., Barreca, T. 1985. Influence of β-endorphin on phytohemagglutinin-induced lymphocyte proliferation and on the expression of mononuclear cell surface antigens *in vitro*. *Immunopharmacology* 10: 119–25
18. Furchgott, R. F. 1970. Pharmacological characteristics of adrenergic receptors. *Fed. Proc.* 29: 1352–61
19. Lefkowitz, R. J. 1975. Heterogeneity of adenylate cyclase-coupled β-adrenergic receptors. *Biochem. Pharmacol.* 24: 583–90
20. Hadden, J. W., Coffey, R. G. 1982. Cyclic nucleotides in mitogen-induced lymphocyte proliferation. *Immunol. Today* 3: 299–305
21. Nishizuka, Y. 1986. Studies and perspectives of protein kinase C. *Science* 233: 305–12
22. Berridge, M. J. 1984. Inositol triphosphate and diacylglycerol as second messengers. *Biochem.* 220: 345–60
23. Londos, C., Cooper, D. M. F., Wolff, J. 1980. Subclasses of external adenosine receptors. *Proc. Natl. Acad. Sci. USA* 77: 2551–54
24. Daly, J. W. 1982. Adenosine-receptors: Targets for drugs. *J. Med. Chem.* 25: 197–207
25. Hunt, P., Eardley, D. D. 1986. Suppressive effects of insulin and insulin-like growth factor-1 (IgF1) on immune responses. *J. Immunol.* 136: 3994–99
26. Lichtenstein, L. M., Margolis, S. 1968. Histamine release in vitro. Inhibition by catecholamines and methylxanthine. *Science* 161: 902–3
27. Ishizuka, M., Braun, W., Matsumoto, T. 1971. Cyclic AMP in immune responses. I. Influences of poly (A:U) and cyclic AMP on antibody formation *in vitro*. *J. Immunol.* 107: 1027–35
28. Braun, W., Ishizuka, M. 1971. Cyclic AMP in immune responses. II. Phosphodiesterase inhibitors as potentiators of polynucleotide effects on antibody formation. *J. Immunol.* 107: 1036–42
29. Henney, C. S. 1973. On the mechanism of T-cell mediated cytolysis. *Transplant. Rev.* 17: 37–70
30. Plaut, M. 1979. The role of cyclic AMP in modulating cytotoxic T. lymphocytes. I. *In vivo*-generated cytotoxic lymphocytes, but not *in vitro*-generated cytotoxic lymphocytes, are inhibited by cyclic AMP-active agents. *J. Immunol.* 123: 692–701
31. Garovoy, M. R., Strom, T. B., Kaliner, M., Carpenter, C. B. 1975. Antibody-dependent lymphocyte mediated

cytotoxicity mechanism and modulation by cyclic nucleotides. *Cell. Immunol.* 20: 197–204

32. Roder, J. C., Klein, M. 1979. Target-effector interaction in the natural killer cell system. IV. Modulation by cyclic nucleotides. *J. Immunol.* 123: 2785–90

33. Katz, P., Zaytoun, A. M., Fauci, A. S. 1982. Mechanisms of human cell-mediated cytotoxicity. I. Modulation of natural killer cell activity by cyclic nucleotides. *J. Immunol.* 129: 287–96

34. Shreiner, G. F., Unanue, E. R. 1975. The modulation of spontaneous and anti-Ig-stimulated motility of lymphocytes by cyclic nucleotides and adrenergic and cholinergic agents. *J. Immunol.* 114: 802–8

35. Melmon, K. L., Bourne, H. R., Weinstein, Y., Shearer, G. M., Kram, J., Bauminger, S. 1974. Hemolytic plaque formation by leukocytes *in vitro*: Control by vasoactive hormones. *J. Clin. Invest.* 53: 13–21

36. Diamantstein, T., Ulmer, A. 1975. The antagonistic action of cyclic GMP and cyclic AMP on proliferation by B and T lymphocytes. *Immunology* 28: 113–19

37. Weinstein, Y., Segal, S., Melmon, K. L. 1975. Specific mitogenic activity of 8-Br-guanosine 3′,5′-monophosphate (Br-cyclic GMP) on B lymphocytes. *J. Immunol.* 115: 112–17

38. Baker, P. E., Fahey, J. V., Munck, A. 1981. Prostaglandin inhibition of T-cell proliferation is mediated at two levels. *Cell. Immunol.* 61: 52–57

39. Birchenall-Sparks, M. C., Farrar, W. L., Rennick, D., Kilian, P. L., Ruscetti, F. W. 1986. Regulation of expression of the interleukin-2 receptor on hematopoietic cells by interleukin-3. *Science* 233: 455–58

40. Vischer, T. L. 1976. The differential effect of cyclic AMP on lymphocyte stimulation by T- or B-cell mitogens. *Immunology* 30: 735–39

41. Pick, E. 1974. Soluble lymphocytic mediators. I. Inhibition of macrophage migration inhibitory factor production by drugs. *Immunology* 26: 649–58

42. Henney, C. S., Gaffney, J., Bloom, B. R. 1974. On the relation of products of activated lymphocytes to cell-mediated cytolysis. *J. Exp. Med.* 140: 837–52

43. Kamat, R., Henney, C. S. 1976. Studies on T cell clonal expansion. II. The *in vitro* differentiation of pre-killer and memory T cells. *J. Immunol.* 116: 1490–95

44. Cook, R. G., Stavitsky, A. B., Schoenberg, M. D. 1975. Regulation of the *in vitro* early anamnestic antibody response by exogenous cholera enterotoxin and cyclic AMP. *J. Immunol.* 114: 426–34

45. Watson, J., Epstein, R., Cohn, M. 1973. Cyclic nucleotides as intracellular mediators of the expression of antigen-sensitive cells. *Nature* 246: 405–9

46. Burchiel, S. W., Melmon, K. L. 1979. Augmentation of the *in vitro* humoral immune response by pharmacologic agents. I. An explanation for the differential enhancement of humoral immunity via agents that elevate cAMP. *J. Immunopharmacol.* 1: 137–50

47. Teh, H.-S., Paetkau, V. 1976. Regulation of immune responses. I. Effects of cyclic AMP and cyclic GMP on immune induction. *Cell. Immunol.* 24: 209–19

48. Kishimoto, T., Ishizaka, K. 1976. Regulation of the antibody response *in vitro*. X. Biphasic effect of cyclic AMP on the secondary anti-hapten antibody response to antiimmunoglobulin and enhancing soluble factor. *J. Immunol.* 116: 534–41

49. Kishimoto, T., Nishizawa, Y., Kikutani, H., Yamamura, Y. 1977. Biphasic effect of cyclic AMP on IgG production and on the changes of non-histone nuclear proteins induced with anti-immunoglobulin and enhancing soluble factor. *J. Immunol.* 118: 2027–33

50. Teh, H.-S., Paetkau, V. 1976. Regulation of immune responses. II. The cellular basis of cyclic AMP effects on humoral immunity. *Cell. Immunol.* 24: 220–29

51. Mozes, E., Weinstein, Y., Bourne, H. R., Melmon, K. L., Shearer, G. M. 1974. *In vitro* correction of antigen-induced immune suppression. Effects of histamine, dibutyryl cyclic AMP and cholera enterotoxin. *Cell. Immunol.* 11: 57–63

52. Gilbert, K. M., Hoffman, M. K. 1985. cAMP is an essential signal in the induction of antibody production by B cells but inhibits helper function of T cells. *J. Immunol.* 135: 2084–89

53. Lyons, S. F., Friedman, H. 1978. Cellular mechanisms of cholera toxin-mediated modulation of *in vitro* hemolysin formation by mouse immunocytes. *J. Immunol.* 120: 452–58

53a. Nowell, P. C. 1961. Inhibition of human leukocyte mitosis by prednisolone *in vitro*. *Cancer Res.* 21: 1518–21

54. Claman, H. N. 1972. Corticosteroids and lymphoid cells. *N. Engl. J. Med.* 287: 388–97

55. Claman, H. N., Moorhead, J. W., Benner, W. H. 1971. Corticosteroids and lymphoid cells *in vitro*. I. Hydrocortisone lysis of human, guinea pig, and mouse thymus cells. *J. Lab. Clin. Med.* 78: 499–507

56. Galili, N., Galili, U., Klein, E., Rosenthal, L., Nordenskjold, B. 1980. Human T lymphocytes become glucocorticoid-sensitive upon immune activation. *Cell. Immunol.* 50: 440–44

57. Cupps, T. R., Fauci, A. S. 1982. Corticosteroid-mediated immunoregulation in man. *Immunol. Rev.* 65: 133–55

58. Smith, K. A. 1980. T-cell growth factor. *Immunol. Rev.* 51: 337–57

59. Schleimer, R. P. 1985. The mechanisms of antiinflammatory steroid action in allergic diseases. *Ann. Rev. Pharmacol. Toxicol.* 25: 381–412

60. Lippman, M., Barr, R. 1977. Glucocorticoid receptors in purified subpopulations of human peripheral blood lymphocytes. *J. Immunol.* 118: 1977–81

61. Schleimer, R. P., Jacques, A., Shin, H. S., Lichtenstein, L. M., Plaut, M. 1984. Inhibition of T cell-mediated cytotoxicity by anti-inflammatory steroids. *J. Immunol.* 132: 266–71

62. Gaillard, R. C., Paffet, D., Riondel, A. M., Saurat, J-H. 1985. RU 486 inhibits peripheral effects of glucocorticoids in humans. *J. Clin. Endocrinol. Metabol.* 61: 1009–11

63. Distelhorst, C. W., Rogers, J. C. 1979. Glucocorticoids inhibit trypsin-induced DNA release from phytohemagglutinin-stimulated blood lymphocytes. *J. Immunol.* 123: 487–95

64. Crabtree, G. R., Munck, A., Smith, K. A. 1980. Glucocorticoids and lymphocytes. II. Cell cycle-dependent changes in glucocorticoid receptor content. *J. Immunol.* 125: 13–17

65. Katz, P., Zaytoun, A. M., Lee, J. H. Jr. 1985. Characterization of corticosteroid receptors in natural killer cells: Comparison with circulating lymphoid and myeloid cells. *Cell. Immunol.* 94: 347–52

66. Hirata, F., Notsu, Y., Iwata, M., Parente, L., DiRosa, M., Flower, R. J. 1982. Identification of several species of phospholipase inhibitory protein(s) by radioimmunoassay for lipomodulin. *Biochem. Biophys. Res. Comm.* 109: 223–30

67. Cohen, J. J., Duke, R. C. 1984. Glucocorticoid activation of a calcium-dependent endonuclease in thymocyte nuclei leads to cell death. *J. Immunol.* 132: 38–42

68. Gillis, S., Crabtree, G. R., Smith, K. A. 1979. Glucocorticoid-induced inhibition of T cell growth factor production. II. The effect on the *in vitro* generation of cytolytic T cells. *J. Immunol.* 123: 1632–38

69. Gillis, S., Crabtree, G. R., Smith, K. A. 1979. Glucocorticoid-induced inhibition of T cell growth factor production. I. The effect on mitogen-induced lymphocyte proliferation. *J. Immunol.* 123: 1624–31

70. Hoffman, T., Hirata, F., Bougnoux, P., Fraser, B. A., Goldfarb, R. H., Herberman, R. B., Axelrod, J. 1981. Phospholipid methylation and phospholipase A_2 activation in cytotoxicity by human natural killer cells. *Proc. Natl. Acad. Sci. USA* 78: 3839–43

71. Hattori, T., Hirata, F., Hoffman, T., Hizuta, A., Herberman, R. B. 1983. Inhibition of human natural killer (NK) activity and antibody dependent cellular cytotoxicity (ADCC) by lipomodulin, a phospholipase inhibitory protein. *J. Immunol.* 131: 662–65

72. Ashman, R. F., Karlan, B. R. Y. 1981. Inhibition of antigen-induced and anti-immunoglobulin-induced capping by hydrocortisone and propranolol. *Immunopharmacology* 3: 41–43

73. Claesson, M. H., Ropke, C. 1983. Colony formation by subpopulations of T lymphocytes. IV. Inhibitory effect of hydrocortisone on human and murine T cell subsets. *Clin. Exp. Immunol.* 54: 554–60

74. Ilfeld, D. N., Krakauer, R. S., Blaese, R. M. 1977. Suppression of the human autologous mixed lymphocyte reaction by physiologic concentrations of hydrocortisone. *J. Immunol.* 119: 428–34

75. Cooper, D. A., Duckett, M., Petts, V., Penny, R. 1979. Corticosteroid enhancement of immunoglobulin synthesis by pokeweed mitogen-stimulated human lymphocytes. *Clin. Exp. Immunol.* 37: 145–51

76. Grayson, J., Dooley, N. J., Koski, I. R., Blaese, R. M. 1981. Immunoglobulin production induced *in vitro* by glucocorticoid hormones. T cell-dependent stimulation of immunoglobulin production without B cell proliferation in cultures of human peripheral blood. *J. Clin. Invest.* 68: 1539–47

77. Goodwin, J. S., Atluru, D. 1986. Mechanism of action of glucocorticoid-induced immunoglobulin production: Role of lipoxygenase metabolites of arachidonic acid. *J. Immunol.* 135: 3455–60

78. Ishizaka, K. 1984. Regulation of IgE synthesis. *Ann. Rev. Immunol.* 2: 159–82

79. Akasaki, M., Jardieu, P., Ishizaka, K. 1986. Immunosuppressive effects of glycosylation inhibiting factors on the IgE and IgG antibody response. *J. Immunol.* 136: 3172–79

80. Reed, J. C., Abidi, A. H., Alpers, J. D., Hoover, R. G., Robb, R. J., Nowell, P. C. 1986. Effect of cyclosporin A and dexamethasone on interleukin 2 receptor gene expression. *J. Immunol.* 137: 150–59

81. Goodwin, J. S., Atluru, D., Sierakowski, S., Lianos, E. A. 1986. Mechanism of action of glucocorticosteroids. Inhibition of T cell proliferation and interleukin-2 production by hydrocortisone is reversed by leukotriene B_4. *J. Clin. Invest.* 77: 1244–50

82. Bradley, L. M., Mishell, R. I. 1981. Differential effects of glucocorticosteroids on the functions of helper and suppressor T lymphocytes. *Proc. Natl. Acad. Sci. USA* 78: 3155–59

83. Bradley, L. M., Mishell, R. I. 1982. Selective protection of murine thymic helper T cells from glucocorticoid inhibition by macrophage-derived mediators. *Cell. Immunol.* 73: 115–127

84. Bradley, L. M., Mishell, R. I. 1982. Differential effects of glucocorticosteroids on the functions of subpopulations of helper T lymphocytes. *Eur. J. Immunol.* 12: 91–94

85. Ranelletti, F. O., Musiani, P., Maggiano, N., Lauriola, L., Piantelli, M. 1983. Modulation of glucocorticoid inhibitory action on human lymphocyte mitogenesis: Dependence on mitogen concentration and T cell maturity. *Cell. Immunol.* 76: 22–28

86. Tyan, M. L. 1979. Genetic control of hydrocortisone-induced thymus atrophy. *Immunogenetics* 8: 177–81

87. Crabtree, G. R., Munck, A., Smith, K. A. 1980. Glucocorticoids and lymphocytes. I. Increased glucocorticoid receptor levels in antigen-stimulated lymphocytes. *J. Immunol.* 124: 2430–35

88. Saxon, A., Stevens, R. H., Ramer, S. J., Clements, P. J., Yu, D. T. Y. 1978. Glucocorticoids administered *in vivo* inhibit human suppressor T lymphocyte function and diminish B lymphocyte responsiveness in *in vitro* immunoglobulin synthesis. *J. Clin. Invest.* 61: 922–30

89. Fauci, A. S., Murakami, T., Brandon, D. D., Loriaux, D. L., Lipsett, M. B. 1980. Mechanisms of corticosteroid action on lymphocyte subpopulations. VI. Lack of correlation between glucocorticoid receptors and the differential effects of glucocorticoids on T cell subpopulations. *Cell. Immunol.* 49: 43–50

90. Besedovsky, H. O., del Rey, A. E., Sorkin, E. 1985. Immune-neuroendocrine interactions. *J. Immunol.* 135: 750s–54

91. Blalock, J. E. 1985. Relationships between neuroendocrine hormones and lymphokines. *Lymphokines* 9: 1–13

92. Waldmann, T. A., Blaese, R. M., Broder, S., Krakauer, R. S. 1978. Disorders of suppressor immunoregulatory cells in the pathogenesis of immunodeficiency and autoimmunity. *Ann. Intern. Med.* 88: 226–38

93. White, W. B., Ballow, M. 1985. Modulation of suppressor-cell activity by cimetidine in patients with common variable hypogammaglobulinemia. *N. Engl. J. Med.* 312: 198–202

94. Nouri-Aria, K. T., Hegarty, J. E., Alexander, G. J. M., Eddleston, A. L. W. F., Williams, R. 1982. Effect of corticosteroids on suppressor-cell activity in "autoimmune" and viral chronic active hepatitis. *N. Engl. J. Med.* 307: 1301–4

95. Payan, D. G., McGillis, J. P., Goetzl, E. J. 1986. Neuroimmunology. *Advances in Immunology.* 39: 299–323

96. Johnson, H. M., Smith, E. M., Torres, B. A., Blalock, J. E. 1982. Neuroendocrine hormone regulation of *in vitro* antibody production. *Proc. Natl. Acad. Sci. USA* 79: 4171–74

97. Smith, E. M., Harbour-McMenamin, D., Blalock, J. E. 1985. Lymphocyte production of endorphins and endorphin-mediated immunoregulatory activity. *J. Immunol.* 135: 779s–82s

98. Smith, E. M., Meyer, W. J., Blalock, J. E. 1982. Virus-induced corticosterone in hypophysectomized mice: A possible lymphoid-adrenal axis. *Science* 218: 1311–12

98a. Smith, E. M., Morrill, A. C., Meyer, W. J. III, Blalock, J. E. 1986. Corticotropin releasing factor induction of leukocyte-derived immunoreactive ACTH and endorphins. *Nature* 321: 881–82

99. Woloski, B. M. R. N. J., Smith, E. M., Meyer, W. J. III, Fuller, G. M., Blalock, J. E. 1985. Corticotropin-releasing activity of monokines. *Science* 230: 1035–37

100. Hazum, E., Chang, K. J., Cuatrecasas, P. 1979. Specific opiate receptors for beta endorphin. *Science* 205: 1033–35

101. Gilman, S. C., Schwartz, J. M., Milner, R. J., Bloom, F. E., Feldman, J. D.

1982. β-endorphin enhances lymphocyte proliferative responses. *Proc. Natl. Acad. Sci. USA* 79: 4226–30

102. Johnson, H. M. 1985. Mechanism of interferon γ production and assessment of immunoregulatory properties. *Lymphokines* 2: 33–46

103. Johnson, H. M., Torres, B. A. 1985. Regulation of lymphokine production by arginine vasopression and oxytocin: Modulation of lymphocyte function by neurohypophyseal hormones. *J. Immunol.* 135: 773s–75s

104. Snow, E. C., Feldbush, T. L., Oaks, J. A. 1981. The effect of growth hormone and insulin upon MLC responses and the generation of cytotoxic lymphocytes. *J. Immunol.* 126: 161–64

105. Snow, E. C. 1985. Insulin and growth hormone function as minor growth factors that potentiate lymphocyte activation. *J. Immunol.* 135: 776s–78s

106. Payan, D. G., Goetzl, E. J. 1985. Modulation of lymphocyte function by sensory neuropeptides. *J. Immunol.* 135: 783s–86s

107. Bourne, H. R. 1986. GTP-binding proteins: One molecular machine can transduce diverse signals. *Nature* 321: 814–16

108. Stanisz, A. M., Befus, D., Bienenstock, J. 1986. Differential effects of vasoactive intestinal peptide, substance P, and somatostatin on immunoglobulin synthesis and proliferation by lymphocytes from Peyer's patches, mesenteric lymph nodes, and spleen. *J. Immunol.* 136: 152–56

109. Shanahan, F., Denbury, J. A., Fox, J., Bienenstock, J., Befus, D. 1985. Mast cell heterogeneity: Effects of neuroenteric peptides on histamine release. *J. Immunol.* 135: 1331–37

110. Payan, D. G., Brewster, D. R., Goetzl, E. J. 1984. Stereospecific receptors for substance P on cultured IM-9 lymphoblasts. *J. Immunol.* 133: 3260–65

111. Payan, D. G., Goetzl, E. J. 1983. Specific suppression of human T lymphocyte function by leukotriene B₄. *J. Immunol.* 131: 551–53

112. Payan, D. G., Brewster, D. R., Missirian-Bastian, A., Goetzl, E. J. 1984. Substance P recognition by a subset of human T lymphocytes. *J. Clin. Invest.* 74: 1532–39

113. O'Dorisio, M. S., Wood, C. L., O'Dorisio, T. M. 1985. Vasoactive intestinal peptide and neuropeptide modulation of the immune response. *J. Immunol.* 135: 792s–96s

114. Danek, A., O'Dorisio, M. S., O'Dorisio, T. M., George, J. M. 1983. Specific binding sites for vasoactive intestinal polypeptide on nonadherent peripheral blood lymphocytes. *J. Immunol.* 131: 1173–77

115. Rola-Pleszczynski, M., Bolduc, D., St-Pierre, S. 1985. The effects of vasoactive intestinal peptide on human natural killer cell function. *J. Immunol.* 135: 2569–73

116. Ottaway, C. A. 1984. *In vitro* activation of receptors for vasoactive intestinal peptide changes the *in vivo* localization of mouse T cells. *J. Exp. Med.* 160: 1054–69

117. Helderman, J. H., Strom, T. B. 1977. The emergency of insulin receptors upon alloimmune cells in the rat. *J. Clin. Invest.* 59: 338–44

118. Helderman, J. H., Strom, T. B. 1978. The specific insulin binding site on T and B lymphocytes: A marker of cell activation. *Nature* 274: 62–63

119. Cheng, K., Larner, J. 1985. Intracellular mediators of insulin action. *Ann. Rev. Physiol.* 47: 405–24

120. Strom, T. B., Bear, R. A., Carpenter, C. B. 1975. Insulin-induced augmentation of lymphocyte-mediated cytotoxicity. *Science* 187: 1206–08

121. Braciale, V. L., Gavin, J. R., Braciale, T. J. 1982. Inducible expression of insulin receptors on T lymphocyte clones. *J. Exp. Med.* 156: 664–69

122. Helderman, J. H., Strom, T. B., Dupuy-D'Angeac, A. 1979. A close relationship between cytotoxic T lymphocytes generated in the mixed lymphocyte culture and insulin receptor-bearing lymphocytes: Enrichment by density gradient centrifugation. *Cell. Immunol.* 46: 247–58

123. Bar, R. S., Kahn, C. R., Koren, H. S. 1977. Insulin inhibition of antibody-dependent cytotoxicity and insulin receptors in macrophages. *Nature* 265: 632–35

124. Strom, T. B., Bangs, J. D. 1982. Human serum-free mixed lymphocyte response: The stereospecific effect of insulin and its potentiation by transferrin. *J. Immunol.* 128: 1555–59

125. Helderman, J. H. 1983. T cell cooperation for the genesis of B-cell insulin receptors. *J. Immunol.* 131: 644–50

126. Helderman, J. H. 1984. Acute regulation of human lymphocyte insulin receptors. Analysis by the glucose clamp. *J. Clin. Invest.* 74: 1428–35

127. Sternberg, E. 1986. Modulation of interferon-gamma induction of Ia expression: Implications for mechanisms of interferon-gamma signal trans-

duction. In: *The Year in Immunology,* *1985–1986.* Vol. 2, ed. J. M. Cruse, R. E. Lewis, pp. 152–56. Basel: Karger

128. Foon, K. A., Wahl, S. M., Oppenheim, J. J., Rosenstreich, D. L. 1976. Serotonin-induced production of a monocyte chemotactic factor by human peripheral blood leukocytes. *J. Immunol.* 117: 1545–52

129. Slauson, D. O., Walker, C., Kristensen, F., Wang, Y., DeWeck, A. L. 1984. Mechanisms of serotonin-induced lymphocyte proliferation inhibition. *Cell. Immunol.* 84: 240–52

130. Sternberg, E. M., Trial, J., Parker, C. W. 1986. Effect of serotonin on murine macrophages: Suppression of Ia expression by serotonin and its reversal by 5-HT$_2$ serotonergic receptor antagonists. *J. Immunol.* 137: 276–82

131. Strom, T. B., Lane, M. A., George, K. 1981. The parallel, time-dependent, bimodal change in lymphocyte cholinergic binding activity and cholinergic influence upon lymphocyte-mediated cytotoxicity after lymphocyte activation. *J. Immunol.* 127: 705–10

132. Strom, T. B., Sytkowski, A. J., Carpenter, C. B., Merrill, J. P. 1974. Cholinergic augmentation of lymphocyte-mediated cytotoxicity. A study of the cholinergic receptor of cytotoxic T lymphocytes. *Proc. Natl. Acad. Sci. USA* 71: 1330–33

133. Henney, C. S. 1974. Relationship between the cytolytic activity of thymus-derived lymphocytes and cellular cyclic nucleotide concentrations. In: *Cyclic AMP, Cell Growth and the Immune Response,* ed. W. Braun, L. M. Lichtenstein, C. Parker, pp. 195–208. New York: Springer-Verlag

134. Cronstein, B. N., Rosenstein, E. D., Kramer, S. B., Weissman, G., Hirschhorn, R. 1985. Adenosine: A physiologic modulator of superoxide anion generation by human neutrophils. Adenosine acts via an A2 receptor on human neutrophils. *J. Immunol.* 135: 1366–76

134a. Fredholm, B. B., Sandberg, G. 1983. Inhibition by xanthine derivatives of adenosine receptor-stimulated cyclic adenosine 3′,5′-monophosphate accumulation in rat and guinea-pig thymocytes. *Br. J. Pharmacol.* 80: 639–644

135. Marone, G., Plaut, M., Lichtenstein, L. M. 1978. Characterization of a specific adenosine receptor on human lymphocytes. *J. Immunol.* 121: 2153–59

136. Wolberg, G., Zimmerman, T. P., Hiemstra, K., Winston, M., Chu, L. C. 1975.

Adenosine inhibition of lymphocyte-mediated cytolysis: Possible role of cyclic adenosine monophosphate. *Science* 187: 957–59

137. Moroz, C., Bessler, H., Djaldetti, M., Stevens, R. H. 1981. Human adenosine receptor bearing lymphocytes: Enumeration, characterization, and distribution in peripheral blood lymphocytes. *Clin. Immunol. Immunopathol.* 18: 47–53

138. Birch, R. E., Polmar, S. H. 1982. Pharmacologic modification of immunoregulatory T lymphocytes. I. Effect of adenosine, H1 and H2 histamine agonists upon T lymphocyte regulation of B lymphocyte differentiation *in vitro. Clin. Exp. Immunol.* 48: 218–30

139. Birch, R. E., Polmar, S. H. 1982. Pharmacologic modification of immunoregulatory T lymphocytes. II. Modulation of T lymphocyte cell surface characteristics. *Clin. Exp. Immunol.* 48: 231–38

140. Samet, M. K. 1986. Evidence against functional adenosine receptors on murine lymphocytes. *Int. J. Immunopharmacol.* 8: 179–88

141. Peachell, P. T., Columbo, M., Kagey-Sobotka, A., Lichtenstein, L. M., Marone, G. 1986. Differential activity of adenosine on mediator release from basophils and human lung mast cells. *Fed. Proc.* 45: 1105 (Abst.)

142. Mandler, R., Birch, R. E., Polmar, S. H., Kammer, G. M., Rudolph, S. A. 1982. Abnormal adenosine-induced immunosuppression and cAMP metabolism in T lymphocytes of patients with systemic lupus erythematosus. *Proc. Natl. Acad. Sci. USA* 79: 7542–46

143. Kammer, G. M., Birch, R. E., Polmar, S. H. 1983. Impaired immunoregulation in systemic lupus erythematosus: Defective adenosine-induced suppressor T lymphocyte generation. *J. Immunol.* 130: 1706–12

143a. Cronstein, B. N., Kramer, S. B., Weissmann, G., Hirschhorn, R. 1983. Adenosine: A physiological modulator of superoxide anion generation by human neutrophils. *J. Exp. Med.* 158: 1160–77

144. Marquardt, D. L., Gruber, H. E., Wasserman, S. I. 1984. Adenosine release from stimulated mast cells. *J. Allergy Clin. Immunol.* 44: 1534 (Abst.)

145. Roszkowski, W., Plaut, M., Lichtenstein, L. M. 1977. Selective display of histamine receptors on lymphocytes. *Science* 195: 683–85

146. Pochet, R., Delespesse, G. 1983. β-adrenoreceptors display different efficiency on lymphocyte subpopulations. *Biochem. Pharmacol.* 32: 1651–55

147. Lipschultz, S., Shanfeld, J., Chacko, S. 1981. Emergency of β-adrenergic sensitivity in the developing chicken heart. *Proc. Natl. Acad. Sci. USA* 78: 288–92

148. Khan, M. M., Sansoni, P., Silverman, E. D., Engleman, E. G., Melmon, K. L. 1986. Beta-adrenergic receptors on human suppressor, helper, and cytolytic lymphocytes. *Biochem. Pharmacol.* 35: 1137–42

149. Meurs, H., Koeter, G. H., Vries, K., Kauffman, H. F. 1984. Dynamics of the lymphocyte beta-adrenergic receptor system in patients with allergic bronchial asthma. *Eur. J. Respir. Dis.* 65: 47–61

150. Barnes, D. B., Dollery, C. T., MacDermot, J. 1980. Increased pulmonary α-adrenergic receptors in experimental asthma. *Nature* 285: 569–71

151. Galant, S. P., Duriseti, L., Underwood, S., Alfred, S., Insel, P. A. 1980. Beta-adrenergic receptors of polymorphonuclear particulates in bronchial asthma. *J. Clin. Invest.* 65: 577–85

152. Fraser, C. M., Venter, J. C., Kaliner, M. 1981. Autonomic abnormalities and autoantibodies to beta-adrenergic receptors. *N. Engl. J. Med.* 305: 1165–70

153. Mann, J. J., Brown, R. P., Halper, J. P., Sweeney, J. A., Kocsis, J. H., Stokes, P. E., Bilezikian, J. P. 1985. Reduced sensitivity of lymphocyte beta-adrenergic receptors in patients with endogenous depression and psychomotor agitation. *N. Engl. J. Med.* 313: 715–20

153a. Davies, A. O., Lefkowitz, R. J. 1984. Regulation of β-adrenergic receptors by steroid hormones. *Ann. Rev. Physiol.* 46: 119–30

154. Rola-Pleszczynski, M. 1985. Immunoregulation by leukotrienes and other lipoxygenase metabolites. *Immunol. Today* 6: 302–07

155. Goetzl, E. J., Pickett, W. C. 1981. Novel structural determinants of the human neutrophil chemotactic activity of leukotriene B. *J. Exp. Med.* 153: 482–87

156. Goldman, D. W., Goetzl, E. J. 1982. Specific binding of leukotriene B4 to receptors on human polymorphonuclear leukocytes. *J. Immunol.* 129: 1600–4

157. Payan, D. G., Missirian-Bastian, A., Goetzl, E. J. 1984. Human T lympho-

cyte subset specificity of the regulatory effects of leukotriene B_4. *Proc. Natl. Acad. Sci. USA* 81: 3501–5

158. Mexmain, S., Cook, J., Aldigier, J-C., Gualde, N., Rigand, N. 1985. Thymocyte cyclic AMP and cyclic GMP response to treatment with metabolites issued from the lipoxygenase pathway. *J. Immunol.* 135: 1361–65

159. Farrar, W. L., Anderson, W. B. 1985. Interleukin 2 stimulates association of protein kinase C with plasma membrane. *Nature* 315: 233–35

160. Jordan, M. L., Hoffman, R. A., Debe, E. F., Simmons, R. L. 1986. *In vitro* locomotion of allosensitized T lymphocyte clones in response to metabolites of arachidonic acid is subset specific. *J. Immunol.* 137: 661–68

161. Payan, D. G., Goetzl, E. J. 1983. Specific suppression of human T lymphocyte function by leukotriene B_4. *J. Immunol.* 131: 551–60

162. Gaulde, N., Atluru, D., Goodwin, J. S. 1985. Effects of lipoxygenase metabolites of arachidonic acid on proliferation of human T cells and T cell subsets. *J. Immunol.* 134: 1125–29

163. Atluru, D., Goodwin, J. S. 1984. Control of polyclonal immunoglobulin production from human lymphocytes by leukotrienes; leukotriene B_4 induces an OKT8(+), radiosensitive suppressor cell from resting, human OKT8(−) T cells. *J. Clin. Invest.* 74: 1444–50

164. Rola-Pleszczynski, M. 1985. Differential effects of leukotriene B_4 on T4+ and T8+ lymphocyte phenotype and immunoregulatory functions. *J. Immunol.* 135: 1357–60

165. Rola-Pleszczynski, M., Lemaire, I. 1985. Leukotrienes augment interleukin 1 production by human monocytes. *J. Immunol.* 135: 3958–67

166. Farrar, W. L., Humes, J. L. 1985. The role of arachidonic acid metabolism in the activities of interleukin 1 and 2. *J. Immunol.* 135: 1153–59

167. Goldman, D. W., Olson, D. M., Payan, D. G., Gifford, L. A., Goetzl, E. J. 1986. Development of receptors for leukotriene B_4 on HL-60 cells induced to differentiate by 1α,25-dihydroxyvitamin D_3. *J. Immunol.* 136: 4631–36

168. Goldyne, M. E., Burrish, G. F., Poubelle, P., Borgeat, P. 1984. Arachidonic acid metabolism among human mononuclear leukocytes: lipoxygenase-related pathways. *J. Biol. Chem.* 259: 8815–19

169. Goodwin, J. S., Webb, D. R. 1980. Regulation of the immune response

by prostaglandins. *Clin. Immunol. Immunopathol.* 15: 106–22

170. Goodwin, J. S., Wiik, A., Lewis, M., Bankhurst, A. D., Williams, R. C. 1979. High affinity binding sites for prostaglandin E on human lymphocytes. *Cell. Immunol.* 43: 150–59

171. Rocklin, R. E., Thistle, L., Audera, C. 1985. Decreased sensitivity of atopic mononuclear cells to prostaglandin E_2 (PGE2) and prostaglandin D_2 (PGD2). *J. Immunol.* 135: 2033–39

172. Eriksen, E. F., Richelsen, B., Beck-Nielsen, H., Melsen, F., Nielsen, H. K., Mosekilde, L. 1985. Prostaglandin E_2 receptors on human peripheral blood monocytes. *Scand. J. Immunol.* 21: 167–72

173. Ceuppens, J., Goodwin, J. S. 1982. Endogenous prostaglandin E enhances polyclonal immunoglobulin production by tonically inhibiting T suppressor cell activity. *Cell. Immunol.* 70: 41–54

174. Kennedy, M. S., Stobo, J. D., Goldyne, M. E. 1980. *In vitro* synthesis of prostaglandins and related lipids by populations of human peripheral blood mononuclear cells. *Prostaglandins* 20: 135–41

175. Kurland, J. I., Bockman, R. 1978. Prostaglandin E production by human blood monocytes and mouse peritoneal macrophages. *J. Exp. Med.* 147: 952–60

176. Webb, D. R., Nowowiejski, I. 1978. Mitogen-induced changes in lymphocyte prostaglandin levels: A signal for the induction of suppressor cell activity. *Cell. Immunol.* 41: 72–78

177. Metzger, Z., Hoffeld, J. T., Oppenheim, J. J. 1980. Macrophage mediated suppression. I. Evidence for participation of both hydrogen peroxide and prostaglandins in suppression of murine lymphocyte proliferation. *J. Immunol.* 124: 983–88

178. Webb, D. R., Wieder, K. J., Nowowiejski, I. 1982. Studies on the nonspecific regulation of immunocompetent cell function by prostaglandins. In: *Immunopharmacology and the Regulation of Lymphocyte Function*, ed. D. R. Webb, pp. 143–58. New York: Marcel Dekker

179. Kato, K., Askenase, P. W. 1984. Reconstitution of an inactive antigen-specific T cell suppressor factor by incubation of the factor with prostaglandins. *J. Immunol.* 133: 2025–31

180. Rocklin, R. E., Greineder, D. K., Melmon, K. L. 1979. Histamine-induced suppressor factor (HSF): Further studies on the nature of the stimulus

and the cell which produces it. *Cell. Immunol.* 44: 404–15

181. Goldring, S. R., Amento, E. P., Roelke, M. S., Krane, S. M. 1986. The adenosine 3′:5′-cyclic monophosphate response to prostaglandin E_2 is altered in U937 cells in association with maturational events induced by activated T lymphocytes and 1,25-dihydroxyvitamin D_3. *J. Immunol.* 136: 3461–66

182. Goodwin, J. S., Messner, R. P., Penke, J. T. 1978. Prostaglandin suppression of mitogen-induced lymphocytes *in vitro*. Changes with mitogen dose and preincubation. *J. Clin. Invest.* 62: 753–60

183. Plaut, M., Marone, G., Gillespie, E. 1983. The role of cyclic AMP in modulating cytotoxic T lymphocytes. II. Sequential changes during culture in responsiveness of cytotoxic lymphocytes to cAMP-active agents. *J. Immunol.* 131: 2945–52

184. Stobo, J. D., Kennedy, M. S., Goldyne, M. E. 1979. Prostaglandin E modulation of the mitogenic response of human T cells. *J. Clin. Invest.* 64: 1188–92

185. Deleted in proof.

186. Gualde, N., Goodwin, J. S. 1982. Effects of prostaglandin E_2 and preincubation on lectin-stimulated proliferation of human T cell subsets. *Cell. Immunol.* 70: 373–78

187. Bach, M. A., Fournier, C., Bach, J. F. 1975. Regulation of theta-antigen expression by agents altering cyclic AMP level and by thymic factor. *Ann. N.Y. Acad. Sci.* 249: 316–20

188. Johnsen, S., Olding, L. B., Westberg, N. G., Wilhelmsson, L. 1982. Strong suppression by mononuclear leukocytes from human newborns on maternal leukocytes: Mediation by prostaglandins. *Clin. Immunol. Immunopathol.* 23: 606–15

189. Page, R. C., Clagett, J. A., Engel, L. D., Wilde, G., Sims, T. 1978. Effects of prostaglandin on the antigen- and mitogen-driven responses of peripheral blood lymphocytes from patients with adult and juvenile periodontis. *Clin. Immunol. Immunopathol.* 11: 77–82

190. Goodwin, J. S., Bromberg, S., Staszak, C., Kaszubowski, P. A., Messner, R. P., Neal, J. F. 1981. Effect of physical stress on sensitivity of lymphocytes to inhibition by prostaglandin E_2. *J. Immunol.* 127: 518–22

191. Wolinsky, S., Goodwin, J. S., Messner, R. P., Williams, R. C. Jr. 1980. Role of prostaglandin in the depressed cell-mediated immune response in rheuma-

toid arthritis. *Clin. Immunol. Immunopathol.* 17: 31–36

192. Goodwin, J. S., Messner, R. P. 1979. Sensitivity of lymphocytes to prostaglandin E2 increases in subjects over age 70. *J. Clin. Invest.* 64: 434–40

193. Goodwin, J. S. 1982. Changes in lymphocyte sensitivity to prostaglandin E, histamine, hydrocortisone and x-irradiation with age: Studies in a healthy elderly population. *Clin. Immunol. Immunopathol.* 25: 243–51

194. Shibata, Y., Volkman, A. 1985. The effect of bone marrow depletion on prostaglandin E-producing suppressor macrophages in mouse spleen. *J. Immunol.* 135: 3897–3904

195. Humes, J. L., Bonney, R. J., Pelus, L., Dahlgren, M. E., Sadowski, S. J., Kuehl, F. A. Jr., Davies, P. 1977. Macrophages synthesise and release prostaglandins in response to inflammatory stimuli. *Nature* 269: 149–51

196. Goodwin, J. S., Messner, R. P., Bankhurst, A. D., Peake, G. T., Saiki, J. H., Williams, R. C. Jr. 1977. Prostaglandin-producing suppressor cells in Hodgkin's disease. *N. Engl. J. Med.* 197: 263–69

197. Kunkel, S. L., Chensue, S. W., Phan, S. H. 1986. Prostaglandins as endogenous mediators of interleukin 1 production. *J. Immunol.* 136: 186–92

198. Staszak, C., Goodwin, J. S. 1980. Is prostaglandin a mediator for the inhibitory action of histamine, hydrocortisone and isoproterenol? *Cell. Immunol.* 54: 351–61

199. Marone, G., Kagey-Sobotka, A., Lichtenstein, L. M. 1979. Effect of arachidonic acid and its metabolites on antigen-induced histamine release from human basophils in vitro. *J. Immunol.* 123: 1669–77

200. Kantor, H. S., Hampton, M. 1978. Indomethacin in submicromolar concentrations inhibits cyclic AMP-dependent protein kinase. *Nature* 276: 841–44

201. Ciosek, C. P., Ortel, R. W., Thanassi, N. M., Newcombe, D. S. 1974. Inhibition of phosphodiesterase by nonsteroidal anti-inflammatory drugs. *Nature* 251: 148–50

201a. Peters, S. P., MacGlashan, D. W. Jr., Schleimer, R. P., Hayes, E. C., Adkinson, N. F. Jr., Lichtenstein, L. M. 1985. The pharmacologic modulation of the release of arachidonic acid metabolites from purified human lung mast cells. *Am. Rev. Respir. Dis.* 132: 367–73

202. Strom, T. B., Carpenter, C. B., Cragoe, E. J. Jr., Norris, S., Devlin, R., Perper, R. J. 1977. Suppression of *in vivo* and *in vitro* alloimmunity by prostaglandins. *Transplant. Proc.* 9: 1075–79

203. Strom, T. B., Carpenter, C. B. 1983. Prostaglandin as an effective antirejection therapy in rat renal allograft recipients. *Transplantation* 35: 279–81

204. Whittum, J., Goldschneider, I., Greiner, J., Zurier, R. 1985. Developmental abnormalities of terminal deoxynucleotidyl transferase positive bone marrow cells and thymocytes in New Zealand mice: Effects of prostaglandin E_1. *J. Immunol.* 135: 272–80

205. Anderson, C. B., Jaffee, B. M., Graff, R. J. 1977. Prolongation of murine skin allografts by prostaglandin E_1. *Transplantation* 23: 444–49

206. Tracey, D. E., Adkinson, N. F. 1980. Prostaglandin synthesis inhibitors potentiate the BCG-induced augmentation of natural killer cell activity. *J. Immunol.* 125: 136–41

207. Galasko, C. S. B., Bennett, A. 1976. A relationship of bone distinction in skeletal metastases to osteoclast activation and prostaglandins. *Nature* 263: 508–13

208. Santoro, M. G., Philpott, G. W., Jaffe, B. M. 1976. Inhibition of tumor growth *in vivo* and *in vitro* by prostaglandin E. *Nature* 263: 777–79

209. Goodwin, J. S., Bankhurst, A. D., Murphy, S., Selinger, D. S., Messner, R. P., Williams, R. C. Jr. 1978. Partial reversal of the cellular immune defect in common variable immunodeficiency with indomethacin. *J. Clin. Lab. Immunol.* 1: 197–204

210. Beer, D. J., Rocklin, R. E. 1984. Histamine-induced suppressor-cell activity. *J. Allergy Clin. Immunol.* 73: 439–52

211. Osband, M., McCaffrey, R. 1979. Solubilization, separation and partial characterization of histamine H1- and H2-receptors from calf thymocyte membranes. *J. Biol. Chem.* 254: 9970–72

212. Wang, Y., Kristensen, F., Joncourt, F., Slauson, D. O., DeWeck, A. L. 1983. Analysis of ^{3}H-histamine interaction with lymphocytes: Receptor binding or uptake? *Clin. Exp. Immunol.* 54: 501–08

213. Gajtkowski, G. A., Norris, D. B., Rising, T. J., Wood, T. P. 1983. Specific binding of [^{3}H]tiotidine to histamine H_2-receptors in guinea-pig cerebral cortex. *Nature* 304: 65–67

214. Foreman, J. C., Norris, D. B., Rising, T. J., Webber, S. E. 1985. The binding of [^{3}H]-tiotidine to homogenates of guinea-pig lung parenchyma. *Br. J. Pharmacol.* 86: 475–82

215. Cameron, W., Doyle, K., Rocklin, R. E. 1986. Histamine type I (H1) receptor radioligand binding studies on normal T cell subsets, B cells and monocytes. *J. Immunol.* 136: 2116–20

216. Khan, M. M., Melmon, K. L., Fathman, C. F., Hertel-Wulff, B., Strober, S. 1985. The effects of autacoids on cloned murine lymphoid cells: Modulation of IL-2 secretion and the activity of natural suppressor cells. *J. Immunol.* 134: 4100–6

216a. Lima, M., Rocklin, R. E. 1981. Regulation of human IgG synthesis *in vitro* by histamine receptor-bearing lymphocytes. *Cell. Immunol.* 64: 324–36

217. Berman, J. S., McFadden, R. G., Cruikshank, W. W., Center, D. M., Beer, D. J. 1984. Functional characteristics of histamine receptor-bearing mononuclear cells. II. Identification and characterization of two histamine-induced human lymphokines that inhibit lymphocyte migration. *J. Immunol.* 133: 1495–1504

218. Siegel, J. N., Schwartz, A., Askenase, P. W., Gershon, R. K. 1982. Suppression and contrasuppression induced respectively by histamine-2 and histamine-1 receptor agonists. *Proc. Natl. Acad. Sci. USA* 79: 5052–56

219. Khan, M. M., Marr-Leisy, D., Verlander, M. S., Bristow, M. R., Strober, S., Goodman, M., Melmon, K. L. 1986. The effects of derivatives of histamine on natural suppressor cells. *J. Immunol.* 137: 308–14

220. Burchiel, S. W., Hanson, K., Warner, N. L. 1984. Clonal heterogeneity of cyclic AMP responsiveness: A comparison of malignant murine lymphoid cell lines. *Int. J. Immunopharmacol.* 6: 35–42

221. Plaut, M., Lichtenstein, L. M., Henney, C. S. 1975. Properties of a subpopulation of T cells bearing histamine receptors. *J. Clin. Invest.* 55: 856–74

222. Hall, T. J., Chen, S-H., Brostoff, J., Lydyard, P. M. 1983. Modulation of human natural killer cell activity by pharmacologic mediators. *Clin. Exp. Immunol.* 54: 493–500

223. Lang, I., Torok, K., Gergely, P., Nekam, K., Petranyi, G. Y. 1981. Effects of histamine receptor blocking on human antibody-dependent cell-mediated cytotoxicity. *Scan. J. Immunol.* 13: 361–66

224. Hellstrand, K., Hermodsson, S. 1986. Histamine H2-receptor-mediated regulation of human natural killer cell activity. *J. Immunol.* 137: 656–60

225. Badger, A. M., Young, J., Poste, G. 1983. Inhibition of phytohaemagglutinin-induced proliferation of human peripheral blood lymphocytes by histamine and histamine H1 and H2 agonists. *Clin. Exp. Immunol.* 51: 178–84

226. Badger, A. M., Young, J., Poste, G. 1984. Reversal of histamine-mediated immunosuppression by structurally diverse histamine type II (H2) receptor antagonists. *Int. J. Immunopharmacol.* 6: 467–73

227. Bonnet, M., Lespinats, G., Burtin, C. 1984. Histamine and serotonin suppression of lymphocyte response to phytohemagglutinin and allogeneic cells. *Cell. Immunol.* 83: 280–91

228. Plaut, M., Kagey-Sobotka, A., Jacques, A. R. 1985. Modulation of cytotoxic T lymphocyte responses by histamine. In: *Frontiers in Histamine Research*, ed. R. Ganellin, J.-C. Schwartz, pp. 379–88. Oxford: Pergamon

229. Carlsson, R., Dohlstein, M., Sjogren, H. O. 1985. Histamine modulates the production of interferon-γ and interleukin-2 by mitogen-activated human mononuclear blood cells. *Cell. Immunol.* 96: 104–12

230. Center, D. M., Cruikshank, W. W., Berman, J. S., Beer, D. J. 1983. Functional characteristics of histamine receptor-bearing mononuclear cells. I. Selective production of lymphocyte chemoattractant lymphokines with histamine used as a ligand. *J. Immunol.* 131: 1854–59

231. Berman, J. S., Cruikshank, W. W., Center, D. M., Theodore, A. C., Beer, D. J. 1985. Chemoattractant lymphokines specific for the helper/inducer T-lymphocyte subset. *Cell. Immunol.* 95: 105–12

232. Mozes, E., Shearer, G. M., Melmon, K. L., Bourne, H. R. 1973. In vitro correction of antigen-induced immune suppression: Effects of Poly (A) Poly (U) and prostaglandin E1. *Cell. Immunol.* 9: 226–33

233. Rocklin, R., Habarek-Davidson, A. 1984. Pharmacologic modulation in vitro of human histamine-induced suppressor cell activity. *Int. J. Immunopharmacol.* 6: 179–86

234. Nair, M. P. N., Cilik, J. M., Schwartz, S. A. 1986. Histamine-induced suppressor factor inhibition of NK cells: Reversal with interferon and interleukin 2. *J. Immunol.* 136: 2456–62

235. Rocklin, R. E., Blidy, A., Kamal, M. 1983. Physicochemical characterization of human histamine-induced

suppressor factor. *Cell. Immunol.* 76: 243–52

236. Garovoy, M. R., Reddish, M. A., Rocklin, R. E. 1983. Histamine-induced suppressor factor (HSF): Inhibition of helper T cell generation and function. *J. Immunol.* 130: 357–61

237. Khan, M. M., Sansoni, P., Engleman, E. G., Melmon, K. L. 1985. Pharmacologic effects of autacoids on subsets of T cells. Regulation of expression/function of histamine-2 receptors by a subset of suppressor cells. *J. Clin. Invest.* 75: 1578–83

238. Khan, M. M., Silverman, E., Engleman, E. G., Melmon, K. L. 1985. Responses to autacoids in subsets of human helper T cells. *Proc. West. Pharmacol. Soc.* 28: 225–28

239. Cohen, M. G., Munro, A. J., Dracott, B. N., Ife, R. J., Vickers, M. R. 1985. Histamine receptors on leukocytes: The binding of histamine serum albumin conjugates is nonspecific. *Int. Arch. Allergy. Appl. Immunol.* 76: 9–15

240. Muirhead, K., Bender, P., Hanna, N., Poste, G. 1985. Binding of histamine and histamine analogs to lymphocyte subsets analyzed by flow cytometry. *J. Immunol.* 135: 4120–28

241. Osband, M. E., Lipton, J. M., Lavin, P., Levey, R., Vawter, G., Greenberger, J. S., McCaffrey, R. P., Parkman, R. 1981. Histiocytosis-X. Demonstration of abnormal immunity, T-cell histamine H2-receptor deficiency, and successful treatment with thymic extract. *N. Engl. J. Med.* 304: 146–53

242. Ahmed, T., Krainson, J. P., Yerger, L. D. 1983. Functional depression of H2 histamine receptors in sheep with experimental allergic asthma. *J. Allergy Clin. Immunol.* 72: 310–20

243. Beer, D., Osband, M. E., McCaffrey, R. P., Soter, N. A., Rocklin, R. E. 1982. Abnormal histamine-induced suppressor cell function in atopic subjects. *N. Engl. J. Med.* 306: 454–58

244. Meurs, H., Koeter, G. H., Kauffman, H. F., Timmermans, A., Folkers, B., de Vries, K. 1985. Reduced adenylate cyclase responsiveness to histamine in lymphocyte membranes of allergic asthmatic patients after allergen challenge. *Int. Arch. Allergy Appl. Immunol.* 76: 256–60

245. Hall, T. J., Hudspith, B. N., Brostoff, J. 1983. The role of prostaglandins in the modulation of histamine suppression of mitogen responses in atopic and normal subjects. *Int. J. Immunopharmacol.* 5: 107–14

246. Wang, S. R., Zweiman, B. 1978. Hista-mine suppression of human lymphocyte responses to mitogens. *Cell. Immunol.* 36: 28–36

247. Strannegard, I-L., Strannegard, O. 1977. Increased sensitivity of lymphocytes from atopic individuals to histamine-induced suppression. *Scand. J. Immunol.* 6: 1225–31

248. Brostoff, J., Pack, S., Lydyard, P. M. 1980. Histamine suppression of lymphocyte activation. *Clin. Exp. Immunol.* 39: 739–45

249. Deleted in proof.

250. Goldberg, E. H., Goodwin, J. S., Arritt, S. E. 1979. Prolongation of male skin graft survival by female mice treated with cimetidine. *Transplantation* 28: 432–33

251. Festen, H. P. M., Berden, J. H. M., Koene, R. A. P. 1980. Cimetidine does not accelerate skin graft rejection in mice. *Clin. Exp. Immunol.* 40: 193–96

252. Avella, J., Madsen, J. E., Binder, H. J., Askenase, P. W. 1978. Effect of histamine H2-receptor antagonists on delayed hypersensitivity. *Lancet* 1: 624–26

253. Wolfe, J. D., Plaut, M., Norman, P. S., Lichtenstein, L. M. 1979. The effect of an H2 receptor antagonist on immediate and delayed skin test reactivity in man. *J. Allergy Clin. Immunol.* 63: 208 (Abst.)

254. Nordlung, J. J., Askenase, P. W. 1983. The effect of histamine, anti-histamines and a mast cell stabilizer in the growth of Cloudman melanoma cells in DBA/2 mice. *J. Invest. Dermatol.* 81: 28–31

255. Weinstock, J. V., Chensue, S. W., Boros, D. L. 1983. Modulation of granulomatous hypersensitivity. V. Participation of histamine receptor positive and negative lymphocytes in the granulomatous response of *Schistosoma mansoni*-infected mice. *J. Immunol.* 130: 423–27

256. Ershler, W. B., Hacker, M. P., Burroughs, B. J., Moore, A. L., Meyers, C. F. 1983. Cimetidine and the immune response. I. *In vivo* augmentation of nonspecific and specific immune response. *Clin. Immunol. Immunopathol.* 26: 10–17

257. Friedman, H., Lee, I., Walz, D. T. 1982. Effects of histamine receptor antagonists metiamide and cimetidine on antibody formation in vitro by murine cells. *Proc. Soc. Exp. Biol. Med.* 169: 222–25

258. Descotes, J., Tedone, R., Evreux, J. C. 1983. Effects of cimetidine and ranitidine on delayed type hypersensitivity. *Immunopharmacology* 6: 31–35

259. Griswold, D., E., Alessi, S., Badger, A. M., Poste, G., Hanna, N. 1984. Inhibition of T suppressor cell expression by histamine type 2 (H2) receptor antagonists. *J. Immunol.* 132: 3054–57

260. Jorizzo, J. L., Sams, W. M. Jr., Jegasothy, B. V., Olansky, A. J. 1980. Cimetidine as an immunomodulator: Chronic mucocutaneous candidiasis as a model. *Ann. Int. Med.* 92: 192–95

261. White, W. B., Ballow, M., Desbonnet, C. R. 1984. Cimetidine modulates suppressor activity in common variable hypogammaglobulinemia. *Clin. Rest.* 32: 362A (Abst.)

262. Dy, M., Lebel, B., Kamoun, P., Hamburger, J. 1981. Histamine production during the anti-allograft response. Demonstration of a new lymphokine enhancing histamine synthesis. *J. Exp. Med.* 153: 293–309

263. Dy, M., Lebel, B. 1983. Skin allografts generate an enhanced production of histamine and histamine-producing cell-stimulating factor (HCSF) by spleen cells in response to T cell mitogens. *J. Immunol.* 130: 2343–47

264. Dy, M., Lebel, B., Schneider, E. 1986. Histamine-producing cell stimulating factor (HCSF) and interleukin 3 (IL-3): Evidence for two distinct molecular entities. *J. Immunol.* 136: 208–12

265. Moore, T. C. 1967. Histidine decarboxylase inhibitors and the survival of homografts. *Nature* 215: 871–72

266. Moore, T. C. 1969. Histamine decarboxylase inhibitors and second-set allograft survival. *Arch. Surg.* 99: 470–73

267. Moore, T. C., Lawrence, W. Jr. 1969. Suppression of antibody formation by histidine decarboxylase inhibitors. *Transplantation* 8: 224–34

268. Maeyama, K., Watanabe, T., Yamatodani, A., Taguchi, Y., Kambe, H., Wada, H. 1983. Effect of α-fluoromethylhistidine on the histamine content of the brain of W/Wᵛ mice devoid of mast cells: Turnover of brain histamine. *J. Neurochem.* 41: 128–34

269. Kamal, M., Rocklin, R. E. 1986. Augmented cellular and humoral immune responses in mice depleted *in vivo* of de novo synthesized histamine. *Int. J. Immunotherapy* 2: 127–35

SUBJECT INDEX

CONTRIBUTING AUTHORS

CONTRIBUTING AUTHORS, VOLUMES 1–5

A

Acuto, O., 2:23–50
Adams, D. O., 2:283–318
Adkins, B., 5:325–65
Alcaraz, G., 4:419–70
Allison, A. C., 1:361–92
Allison, J. P., 5:503–40
Alt, F. W., 4:339–68
Andersson, J., 4:13–36
Asherson, G. L., 4:37–68

B

Bailey, P. J., 2:335–57
Bartholow, T. L., 4:147–65
Benacerraf, B., 2:127–58
Benjamin, D. C., 2:67–101
Berzofsky, J. A., 2:67–101
Broder, S., 3:321–38
Brown, E. J., 2:461–91

C

Calame, K. L., 3:159–95
Cambier, J. C., 5:175–99
Capron, A., 3:455–76
Carson, D., 5:109–26
Cebra, J. J., 2:493–548
Chase, M. W., 3:1–30
Chedid, L. A., 4:369–88
Chen, P. P., 5:109–26
Chin, Y. H., 5:201–22
Clarke, L. M., 5:201–22
Clevinger, B. L., 4:147–65
Clift, R., 5:43–64
Colizzi, V., 4:37–68
Colten, H. R., 4:231–51
Croce, C. M., 5:253–77

D

Davidson, A., 5:85–108
Davie, J. M., 2:183–98, 4:147–65
Davies, D. R., 1:87–117
Davies, P., 2:335–58
Davis, M. M., 3:537–60
Denis, K., 3:213–35
Dessaint, J. P., 3:455–76
Diamond, B., 5:85–108
Dorf, M. E., 2:127–58

Durum, S. K., 3:263–87
Dustin, M. L., 5:223–52

E

East, I. J., 2:67–101
Eisen, H. N., 3:337–65
Eugui, E. M., 1:361–92

F

Fathman, C. G., 1:633–55, 5:477–501
Fauci, A. S., 3:477–500
Fearon, D. T., 1:243–71
Figueroa, F., 1:119–42
Flexner, C., 5:305–24
Flood, P. M., 1:439–63
Fong, S., 5:109–26
Ford-Hutchinson, A., 2:335–58
Fox, R. I., 5:109–26
Frank, M. M., 2:461–91
Frelinger, J. G., 1:633–55

G

Gallin, J. I., 5:127–50
Gallo, R. C., 3:321–38
Gaulton, G. N., 4:253–80
Geliebter, J., 4:471–502
Germain, R. N., 4:281–315
Gershon, R. K., 1:439–64
Getzoff, E. D., 3:501–35
Gleich, G. J., 2:429–59
Gluckman, J. C., 4:97–117
Goldenberg, M. M., 2:335–57
Goldfien, R. D., 5:109–26
Goodman, J. W., 1:465–98
Green, D. R., 1:439–64
Greene, M. I., 4:253–80
Greene, W. C., 4:69–95
Greenspan, N. S., 4:147–65
Gurd, F. R. N., 2:67–101

H

Hakomori, S., 2:103–26
Hamaoka, T., 4:167–204
Hamilton, T., 2:283–318
Hanley-Hyde, J. M., 4:621–49
Hannum, C., 2:67–101
Hardy, K., 4:593–619

Hashimoto, K., 3:87–108
Haskins, K., 2:51–66
Hayakawa, K., 1:609–32
Henkart, P. A., 3:31–58
Henson, P. M., 1:335–59
Herberman, R. B., 2:359–94, 4:651–80
Hercend, T., 2:23–50
Herzenberg, L. A., 1:609–32
Hodes, R. J., 1:211–41
Hohman, R., 4:419–70
Honjo, T., 1:499–528
Hood, L. E., 1:529–68, 4:529–91
Howard, M., 1:307–33
Humphrey, J. H., 2:1–22

I

Imboden, J., 4:593–619
Ishizaka, K., 2:159–82

J

Jensen, P. J., 3:87–108
Jirik, F., 5:109–26
Joiner, K. A., 2:461–91

K

Kabat, E. A., 1:1–32
Kang, A. H., 2:199–218
Kapp, J. A., 1:423–38
Kappler, J., 2:51–66
Kelly, K., 4:317–38
Kinet, J., 4:419–70
Kipps, T. J., 5:109–26
Kishimoto, T. K., 5:223–52
Kishimoto, T., 3:133–57
Klatzmann, D., 4:97–117
Klein, J., 1:119–42
Klinman, N. R., 1:63–86
Komisar, J. L., 2:493–548
Koshland, M. E., 3:425–53
Kronenberg, M., 4:529–91
Kuehl, M., 1:393–422

L

Lacy, P. E., 2:183–98
Lafferty, K. J., 1:143–73
Lane, H. C., 3:477–500

683

CHAPTER TITLES, VOLUMES 1–5

685

Annual Reviews Inc.

A NONPROFIT SCIENTIFIC PUBLISHER

4139 El Camino Way
P.O. Box 10139
Palo Alto, CA 94303-0897 • USA

Annual Reviews Inc. publications may be ordered directly from our office by mail or use our Toll Free Telephone line (for orders paid by credit card or purchase order, and customer service calls only); through booksellers and subscription agents, worldwide; and through participating professional societies. Prices subject to change without notice.   ARI Federal I.D. #94-1156476

- **Individuals:** Prepayment required on new accounts by check or money order (in U.S. dollars, check drawn on U.S. bank) or charge to credit card — American Express, VISA, MasterCard.
- **Institutional buyers:** Please include purchase order number.
- **Students:** $10.00 discount from retail price, per volume. Prepayment required. Proof of student status must be provided (photocopy of student I.D. or signature of department secretary is acceptable). Students must send orders direct to Annual Reviews. Orders received through bookstores and institutions requesting student rates will be returned. You may order at the Student Rate for a maximum of 3 years.
- **Professional Society Members:** Members of professional societies that have a contractual arrangement with Annual Reviews may order books through their society at a reduced rate. Check with your society for information.
- **Toll Free Telephone orders:** Call 1-800-523-8635 (except from California) for orders paid by credit card or purchase order and customer service calls only. California customers and all other business calls use 415-493-4400 (not toll free). Hours: 8:00 AM to 4:00 PM, Monday-Friday, Pacific Time.

Regular orders: Please list the volumes you wish to order by volume number.
Standing orders: New volume in the series will be sent to you automatically each year upon publication. Cancellation may be made at any time. Please indicate volume number to begin standing order.
Prepublication orders: Volumes not yet published will be shipped in month and year indicated.
California orders: Add applicable sales tax.
Postage paid (4th class bookrate/surface mail) by **Annual Reviews Inc.** Airmail postage or UPS, extra.

ANNUAL REVIEWS SERIES		Prices Postpaid per volume USA & Canada/elsewhere	Regular Order Please send:	Standing Order Begin with:
			Vol. number	Vol. number
Annual Review of ANTHROPOLOGY				
Vols. 1-14	(1972-1985)	$27.00/$30.00		
Vols. 15-16	(1986-1987)	$31.00/$34.00		
Vol. 17	(avail. Oct. 1988)	$35.00/$39.00	Vol(s). _____	Vol. _____
Annual Review of ASTRONOMY AND ASTROPHYSICS				
Vols. 1-2, 4-20	(1963-1964; 1966-1982)	$27.00/$30.00		
Vols. 21-25	(1983-1987)	$44.00/$47.00		
Vol. 26	(avail. Sept. 1988)	$47.00/$51.00	Vol(s). _____	Vol. _____
Annual Review of BIOCHEMISTRY				
Vols. 30-34, 36-54	(1961-1965; 1967-1985)	$29.00/$32.00		
Vols. 55-56	(1986-1987)	$33.00/$36.00		
Vol. 57	(avail. July 1988)	$35.00/$39.00	Vol(s). _____	Vol. _____
Annual Review of BIOPHYSICS AND BIOPHYSICAL CHEMISTRY				
Vols. 1-11	(1972-1982)	$27.00/$30.00		
Vols. 12-16	(1983-1987)	$47.00/$50.00		
Vol. 17	(avail. June 1988)	$49.00/$53.00	Vol(s). _____	Vol. _____
Annual Review of CELL BIOLOGY				
Vol. 1	(1985)	$27.00/$30.00		
Vols. 2-3	(1986-1987)	$31.00/$34.00		
Vol. 4	(avail. Nov. 1988)	$35.00/$39.00	Vol(s). _____	Vol. _____

ANNUAL REVIEWS SERIES	Prices Postpaid per volume USA & Canada/elsewhere	Regular Order Please send:	Standing Order Begin with:
		Vol. number	Vol. number

Annual Review of COMPUTER SCIENCE

Vols. 1-2	(1986-1987)................$39.00/$42.00		
Vol. 3	(avail. Nov. 1988)..............$45.00/$49.00	Vol(s). _____	Vol. _____

Annual Review of EARTH AND PLANETARY SCIENCES

Vols. 1-10	(1973-1982)................$27.00/$30.00		
Vols. 11-15	(1983-1987)................$44.00/$47.00		
Vol. 16	(avail. May 1988)..............$49.00/$53.00	Vol(s). _____	Vol. _____

Annual Review of ECOLOGY AND SYSTEMATICS

Vols. 2-16	(1971-1985)................$27.00/$30.00		
Vols. 17-18	(1986-1987)................$31.00/$34.00		
Vol. 19	(avail. Nov. 1988)..............$34.00/$38.00	Vol(s). _____	Vol. _____

Annual Review of ENERGY

Vols. 1-7	(1976-1982)................$27.00/$30.00		
Vols. 8-12	(1983-1987)................$56.00/$59.00		
Vol. 13	(avail. Oct. 1988)..............$58.00/$62.00	Vol(s). _____	Vol. _____

Annual Review of ENTOMOLOGY

Vols. 10-16, 18-30	(1965-1971; 1973-1985)........$27.00/$30.00		
Vols. 31-32	(1986-1987)................$31.00/$34.00		
Vol. 33	(avail. Jan. 1988)..............$34.00/$38.00	Vol(s). _____	Vol. _____

Annual Review of FLUID MECHANICS

Vols. 1-4, 7-17	(1969-1972, 1975-1985)........$28.00/$31.00		
Vols. 18-19	(1986-1987)................$32.00/$35.00		
Vol. 20	(avail. Jan. 1988)..............$34.00/$38.00	Vol(s). _____	Vol. _____

Annual Review of GENETICS

Vols. 1-19	(1967-1985)................$27.00/$30.00		
Vols. 20-21	(1986-1987)................$31.00/$34.00		
Vol. 22	(avail. Dec. 1988)..............$34.00/$38.00	Vol(s). _____	Vol. _____

Annual Review of IMMUNOLOGY

Vols. 1-3	(1983-1985)................$27.00/$30.00		
Vols. 4-5	(1986-1987)................$31.00/$34.00		
Vol. 6	(avail. April 1988)..............$34.00/$38.00	Vol(s). _____	Vol. _____

Annual Review of MATERIALS SCIENCE

Vols. 1, 3-12	(1971, 1973-1982)............$27.00/$30.00		
Vols. 13-17	(1983-1987)................$64.00/$67.00		
Vol. 18	(avail. August 1988)............$66.00/$70.00	Vol(s). _____	Vol. _____

Annual Review of MEDICINE

Vols. 1-3, 6, 8-9	(1950-1952, 1955, 1957-1958)		
11-15, 17-36	(1960-1964, 1966-1985)........$27.00/$30.00		
Vols. 37-38	(1986-1987)................$31.00/$34.00		
Vol. 39	(avail. April 1988)..............$34.00/$38.00	Vol(s). _____	Vol. _____

Annual Review of MICROBIOLOGY

Vols. 18-39	(1964-1985)................$27.00/$30.00		
Vols. 40-41	(1986-1987)................$31.00/$34.00		
Vol. 42	(avail. Oct. 1988)..............$34.00/$38.00	Vol(s). _____	Vol. _____